INTERNATIONAL SOCIETY FOR ROCK MECHANICS

SOCIETE INTERNATIONALE DE MECANIQUE DES ROCHES

INTERNATIONALE GESELLSCHAFT FÜR FELSMECHANIK

6 International Congress on Rock Mechanics

Congrès International de Mécanique des Roches

Internationaler Kongress der Felsmechanik

PROCEEDINGS / COMPTES-RENDUS / BERICHTE

VOLUME / TOME / BAND 3

Editors / Editeurs / Herausgeber
G.HERGET & S.VONGPAISAL

Montréal / Canada / 1987

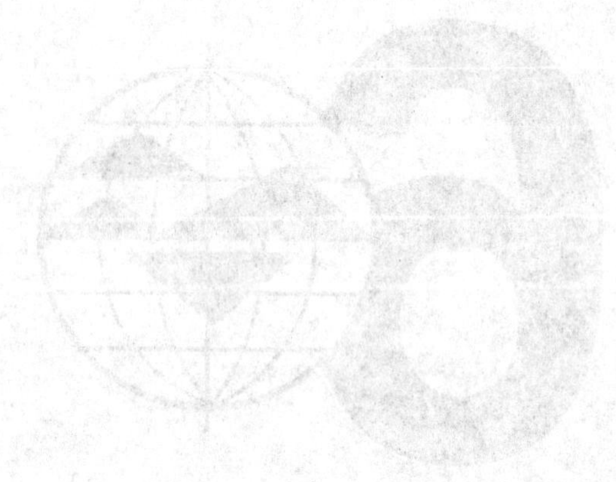

Proceedings
Comptes-rendus
Berichte

Sixth International Congress on Rock Mechanics

Sixième Congrès International de Mécanique des Roches

Sechster Internationaler Kongress der Felsmechanik

Montréal / Canada / 1987

VOLUME 3 Speeches
Theme lectures and discussions
Additional contributions
Complete subject and author indexes

TOME 3 Discours
Conférences thématiques et discussions
Contributions additionelles
Index des matières et index des auteurs complets

BAND 3 Ansprachen
Themavorträge und Diskussionen
Ergänzende Beiträge
Komplettes Themen und Autoren Verzeichnis

A.A.BALKEMA / ROTTERDAM / BROOKFIELD / 1989

and Canadian Rock Mechanics Association / CIM / CGS
et L'Association canadienne ae mécanique des roches
und Kanadische Vereinigung für Felsmechanik

For the complete set of three volumes, ISBN 90 6191 711 5
For volume 1, ISBN 90 6191 712 3
For volume 2, ISBN 90 6191 713 1
For volume 3, ISBN 90 6191 714 X
Published by A.A.Balkema, P.O.Box 1675, 3000 BR Rotterdam, Netherlands
A.A.Balkema Publishers, Old Post Road, Brookfield, VT 05036, USA

Printed in the Netherlands

Foreword

At the International Symposium on Weak Rock which took place in Tokyo, Japan, in 1981, the Council of the International Society for Rock Mechanics accepted the invitation of the Canadian Rock Mechanics Group – with support from the Canadian Institute of Mining and Metallurgy and the Canadian Geotechnical Society – to host the 6th International Congress for Rock Mechanics in Montreal, Canada. The final date for the Congress was set for August 30th – September 3rd, 1987.

Based on a number of inputs from national and international organizations, especially the ISRM Advisory Board, four general themes were chosen for the Congress, which would enable a comprehensive exchange of knowledge and experience related to recent advances in rock mechanics. The four themes are:

Fluid flow and waste isolation in rock masses
Rock foundations and slopes
Rock blasting and excavation
Underground openings in overstressed rock

The first two volumes of the proceedings contain all publications concerning the four themes which reached the office of the Organizing Committee in time. All papers which were submitted have been examined and selected by the respective national groups which have, in accordance with the statutes of ISRM, the responsibility for the content of the papers from their countries. Some of the papers were deficient in regard to the translation requirements. This was remedied wherever possible by the Proceedings Committee.

For each theme an internationally recognized speaker was asked to present a lecture of broad significance. These introductory lectures, including the lecture on rock engineering in Canada, are reproduced in volume three of the Congress proceedings. This volume also contains the discussion contributions, the official presentations for the Congress (opening speeches), one page summaries of poster sessions, and individual papers which unfortunately arrived too late to be included in volumes one and two. All discussion contributions are reproduced in the English section. They are reproduced in the French or German section only if they were submitted originally in one of these languages. All official addresses, theme lectures and moderator summaries are available in the three congress languages.

The proceedings of the Sixth International Congress for Rock Mechanics are of particular significance because they appear on the twenty-fifth anniversary since formation of the International Society for Rock Mechanics. These proceedings represent, therefore, a very important milestone in the development of the science of rock mechanics.

The compilation presented here would not have been possible without the generous help of many individuals and organizations, especially the Canada Centre for Mineral and Energy Technology (CANMET), that provided assistance in typing, translation and editing.

Branko Ladanyi
Honorary Chairman

Gerhard Herget
Congress Chairman

Norbert Morgenstern
Technical Program

Préface

Lors du symposium international sur les roches tendres qui avait lieu à Tokyo au Japon, en 1981, le comité de la Société internationale de mécanique des roches acceptait l'invitation du Groupe national canadien – appuyé par l'Institut canadien des mines et de la métallurgie et la Société canadienne de géotechnique – de tenir le 6e Congrès international de mécanique des roches à Montréal au Canada. Il fut décidé que le congrès aurait lieu du 30 août au 3 septembre 1987.

A partir de suggestions d'organisations nationales et internationales ainsi que du comité consultatif de la SIMR, on a choisi pour ce congrès quatre thèmes généraux qui devaient favoriser un vaste échange de connaissances professionnelles et de progrès récents dans le domaine de la mécanique des roches. Ces quatres thèmes sont les suivants:

Ecoulement de fluides et enfouissement de déchets dans les massifs rocheux
Fondations et talus rocheux
Sautage et excavation
Souterrains en massifs rocheux sous grandes contraintes

Les deux premiers tomes des comptes rendus contiennent toutes les publications se rapportant aux quatre thèmes du congrès qui sont parvenues à temps au comité organisateur. Toutes ces publications ont été examinées et sélectionnées par les groupes nationaux respectifs lesquels, selon les status de la SIMR, sont responsables du contenu de tous les travaux provenant de leur pays. Les articles qui ne rencontraient pas les exigences de traduction furent corrigés par le sous-comité des comptes rendus.

Un spécialiste de réputation internationale fut choisi pour présenter une conférence d'intérêt général pertinant à chacun des thèmes. Ces conférences ainsi que celle portant sur la mécanique des roches au Canada sont publiées dans le troisième tome des comptes rendus du congrès. Ce tome comprend aussi les notes des discussions soumises au comité organisateur du congrès, les présentations officielles du congrès (discours d'ouverture, etc), des sommaires d'une page des séances d'affichage ainsi que toutes les communications qui sont arrivées trop tard pour être publiées dans les tomes 1 et 2. Le comité organisateur a été dans l'impossibilité de traduire toutes les présentations écrites aux séances de discussion dans les trois langues du congrès. La politique mise en application a été de publier toutes ces présentations dans la section anglaise de même que dans leur section respective lorsque ces dernières ont été soumises en français ou en allemand. Les présentations des conférenciers thématiques, les sommaires des modérateurs et les discours officiels ont été traduits dans les trois langues.

Les comptes rendus du Sixième Congrès international de Montréal revêtent un intérêt tout particulier à cause de leur coïncidence avec le vingt-cinquième anniversaire de fondation de la Société internationale de mécanique des roches. Ainsi, ils devraient marquer une étape importante dans le développement de la science de la mécanique des roches.

La compilation présentée ici n'aurait pas été possible sans l'aide généreuse de nombreux individus et organisations, spécialement le Centre canadien de la technologie des minéraux et de l'énergie (CANMET), qui a assisté à la dactylographie, la traduction et l'édition.

Branko Ladanyi
Président honoraire

Gerhard Herget
Président du congrès

Norbert Morgenstern
Programme technique

Vorwort

Während des Internationalen Symposiums über Weichgesteine in Tokyo, Japan, 1981, entschied die Versammlung der Internationalen Gesellschaft für Felsmechanik (IGFM), die Einladung der kanadischen Felsmechanik Gruppe – unterstützt vom kanadischen Institut für Bergbau und Metallurgie und der kanadischen Gesellschaft für Geotechnik – anzunehmen, und den Sechsten Internationalen Kongreß der Felsmechanik in Montreal, Kanada zu organisieren. Als Datum für den Kongreß wurde der 30. August – 3. September, 1987 gewählt.

Auf Grund von Besprechungen mit nationalen und internationalen Organisationen, insbesondere dem Beirat der IGFM, wurden vier Themen ausgewählt, die einen umfassenden Austausch von Fachwissen und neuen Errungenschaften in der Felsmechanik ermöglichen. Diese vier Themen waren:

Flüssigkeitsbewegung und Abfallisolierung im Fels
Felsgründungen und Böschungen
Sprengen und Ausbruch
Untertägige Hohlräume im überbeanspruchten Gebirge

Die ersten zwei Bände enthalten alle Veröffentlichungen der vier Themen, die beim Organisationskomitee rechtzeitig eintrafen. Alle Beiträge sind von den entsprechenden Nationalgruppen überprüft und ausgewählt worden, da sie die Verantwortung für die Veröffentlichungen ihrer Länder haben. Einige der eingereichten Arbeiten hatten Mängel in Bezug auf Übersetzungen der Titel und der Zusammenfassungen, die soweit wie möglich behoben wurden.

Für jedes Thema wurde ein international anerkannter Sprecher gebeten, einen Vortrag von allgemeiner Bedeutung zu halten. Diese Einführungsvorträge, einschließlich des Vortrages über Felsbau in Kanada, erscheinen in diesem dritten Band der Kongreßveröffentlichungen. Dieser Band enthält auch die Diskussionsbeiträge, die Beiträge zur Eröffnungssitzung, Zusammenfassungen für die Postersitzungen und Veröffentlichungen, die zu spät eintrafen, um im ersten oder zweiten Band zu erscheinen. Alle Diskussionsbeiträge wurden nicht in die drei offiziellen Sprachen übersetzt. Im englischen Teil erscheinen alle Beiträge. Im französischen und deutchen Teil jedoch nur, falls der eingereichte Beitrag ursprünglich in einer dieser Sprachen vorgelegt wurde. Alle offiziellen Ansprachen, z.B. Beiträge der Sprecher zu den Themen und die Zusammenfassungen der Diskussionsleiter wurden in alle drei Kongreßsprachen übersetzt.

Die Sitzungsberichte des sechsten internationalen Kongresses für Felsmechanik sind von besonderer Bedeutung, weil sie zum 25. Jahrestag des Bestehens der IGFM erscheinen. Diese Sitzungsberichte stellen daher einen wesentlichen Markstein in der Entwicklung der Felsmechanik dar.

Die Zusammenstellung der Sitzungsberichte wäre nicht möglich gewesen ohne die großzügige Hilfe von vielen Personen und Organisationen, besonders dem kanadischen Zentrum für Mineral und Energie Technologie (CANMET), das Schreibkräfte und Übersetzer zur Verfügung stellte.

Branko Ladanyi
Ehrenvorsitzender

Gerhard Herget
General-Vorsitzender

Norbert Morgenstern
Technisches Program

Organization of the Sixth ISRM Congress

ORGANIZING COMMITTEE

Dr G.Herget, Congress Chairman
Dr B.Ladanyi, Honorary Chairman/Fundraising
Dr D.E.Gill, Congress Co-Chairman/Local Functions
Dr N.R.Morgenstern, Technical Program
Dr W.F.Bawden, Exhibits
Dr J.Bourbonnais, Hospitality
Mr T.Carmichael, Finance
Dr J.A.Franklin, Publicity
Mr L.Geller, Translations
Dr A.T.Jakubick, Technical Tours
Mrs J.Robertson, Accompanying Persons Program
Mrs D.Grégoire, Registration, CIM

ADVISORY COMMITTEE

Prof. E.T.Brown, President ISRM
Mr N.F.Grossman, Secretary General ISRM
Vice-Presidents:
Dr H.Wagner (Africa)
Prof. Tan Tjong Kie (Asia)
Dr W.M.Bamford (Australasia)
Dr S.A.G.Bjurström (Europe)
Mr A.A.Bello Maldonado (North America)
Dr F.H.Tinoco (South America)

CARMA EXECUTIVE

Dr J.E.Udd, Chairman
Dr J.Curran, Vice-Chairman
Mr T.Carmichael, Secretary Treasurer

I S R M C O N G R E S S 1 9 8 7

Time	Sunday August 30	Monday August 31	Tuesday September 1	Wednesday September 2	Thursday September 3
08:30	Pre-Congress Workshop: Young (Canada): Monitoring and interpretation techniques for mining induced seismicity (8:30 - 18:00)	Official Opening Hoek (Canada): Rock Engineering in Canada	Theme II: Rock Foundations and Slopes Chairman: Tinoco (Venezuela) Speaker: Panet (France) Reinforcement of rock foundations and slopes by active or passive anchors - 4 Presentations -	Theme III: Rock Blasting and Excavation Chairman: Tan Tjong-Kie (China) Speaker: McKenzie (Australia) Blasting in hard rock: Techniques for diagnosis and modelling for damage and fragmentation - 4 Presentations -	Theme IV: U/G Excavations in Overstressed Rock Chairman: Bamford (Australia) Speaker: Wagner (S. Africa) Design and support of underground excavations in highly stressed rock - 4 Presentations -
10:00		C O F F E E	C O F F E E	C O F F E E	C O F F E E
10:30		Theme I: Fluid Flow and Waste Isolation Chairman: Bjurstrom (Sweden) Speaker: Doe (USA) Design of borehole testing programs for waste disposal sites in crystalline rock - 5 Presentations -	- 9 Presentations -	- 9 Presentations -	- 9 Presentations -
12:00		L U N C H B R E A K and Poster Sessions	L U N C H B R E A K and Poster Sessions	L U N C H B R E A K and Poster Sessions	L U N C H B R E A K and Poster Sessions
13:30		- 9 Presentations -	Panel & Floor Discussions Moderator: Kovari (Switzerland) Panelists: MacMahon (Australia) Ribacchi (Italy) Yoshinaka (Japan)	Panel & Floor Discussions Moderator: Lindqvist (Sweden) Panelists: Favreau (Canada) Blindheim (Norway) da Gama (Brazil)	- 6 Presentations -
14:30		Coffee	Moderator Summary	Moderator Summary	Coffee
15:00		Panel & Floor Discussions Moderator: Barton (Norway) Panelists: Langer (FRG) Hudson (UK) Rissler (FRG)	Workshops Swelling Rock Constitutive Laws for Salt Rock Numerical Methods as a Practical Tool	Workshops Failure Mechanisms Around Underground Workings Rock Cuttability and Drillability Rock Testing and Testing Standards	Panel & Floor Discussions Moderator: Whittaker (UK) Panelists: Sharma (India) Kaiser (Canada) Maury (France)
16:00					Moderator Summary
16:30		Moderator Summary			Closing Address: Wittke (FRG)
17:00		Poster Sessions	Poster Sessions	Poster Sessions	Poster Sessions
18:00					
19:00	Reception by Congress Chairman	Reception ISRM President			Reception
19:30		Ballet Jazz			Dinner
20:30					Silver-Jubilee Addresses
21:30					Les Sortileges
22:00					

Organisation du Sixième Congrès de la SIMR

COMITE D'ORGANISATION

Dr G.Herget, Président du Congrès
Dr B.Ladanyi, Président honoraire
Dr D.E.Gill, Coprésident du Congrès
Dr N.R.Morgenstern, Programme technique
Dr W.F.Bawden, Exposition
Dr J.Bourbonnais, Acceuil
M. T.Carmichael, Finances
Dr J.A.Franklin, Publicité
M. L.Geller, Traduction
Dr A.T.Jakubick, Visites techniques
Mme J.Robertson, Programme pour les personnes accompagnant les participants
Mme D.Grégoire, Inscription, CIM

COMITE CONSEIL

Prof. E.T.Brown, Président SIMR
M. N.F.Grossman, Secrétaire général SIMR
Vice-présidents:
Dr H.Wagner (Afrique)
Prof. Tan Tjong Kie (Asie)
Dr W.M.Bamford (Australasie)
Dr S.A.G.Bjurström (Europe)
M. A.A.Bello Maldonado (Amérique du Nord)
Dr F.N.Tinoco (Amérique du Sud)

EXECUTIF DU CARMA

Dr J.E.Udd, Président
Dr J.Curran, Vice-président
M. T.Carmichael, Secrétaire général

C O N G R È S S I M R 1 9 8 7

Heure	Dimanche 30 août	Lundi 31 août	Mardi 1 septembre	Mercredi 2 septembre	Jeudi 3 septembre
08:30	Atelier pré-congrès Young (Canada): Séismicité induite par les mines: techniques de surveillance et d'interprétation (8:30 - 18:00)	Ouverture officielle Hoek (Canada): Ingénierie des roches au Canada	Thème II: Fondations et talus rocheux Président: Tinoco (Vénézuela) Conférencier thématique: Panet (France) Renforcement des fondations et des talus à l'aide d'ancrages actifs et passifs - 4 présentations -	Thème III: Sautage et excavation Président: Tan Tjong-Kie (Chine) Conférencier thématique: McKenzie (Australie) Sautage en roches dures: Techniques diagnostiques et de modélisation des nuisances et de la fragmentation - 4 présentations -	Thème IV: Souterrains en massifs rocheux sous grandes contraintes Président: Bamford (Australie) Conférencier thématique: Wagner (Afrique du sud) Conception et soutènement des souterrains dans les massifs rocheux sous grandes contraintes - 4 présentations -
10:00		P A U S E C A F E			
10:30		Thème I: Ecoulement de fluides et enfouissement de déchets dan les massifs rocheux Président: Bjurström (Suède) Conférencier thématique Doe (EU) Conception de programmes d'essais en forage pour les sites d'enfouissement des déchets dans les roches cristallines - 5 presentations -	- 9 présentations -	- 9 présentations -	- 9 présentations -
12:00		P a u s e d é j e u n e r e t s e s s i o n s d ' a f f i c h a g e			
13:30		- 9 présentations -	Discussions/ table ronde/plénière Modérateur: Kovari (Suisse) table ronde: MacMahon (Australie) Ribacchi (Italie) Yoshinaka (Japon)	Discussions/ table ronde/plénière Modérateur: Lindqvist (Suède) table ronde Favreau (Canada) Blindheim (Norvège) da Gama (Brésil)	- 6 présentations -
14:30			Résumé du modérateur	Résumé du modérateur	P a u s e c a f é
15:00		P a u s e c a f é Discussions table ronde/plénière Modérateur: Barton (Norvège) table ronde: Langer (RFA) Hudson (GB) Rissler (RFA)	A t e l i e r s Essais, analyse et conception en matière de roche gonflante Lois de comportement pour la modèlisation de sel gemme Les méthodes numériques, un outil commode	 Les mécanismes de rupture autour des ouvrages souterrains Essai de la taillabilité e de la forabilité des roches Essai des roches et normes d'essai	Discussions table ronde/plénière Modérateur: Whittaker (GB) table ronde: Sharma (Indes) Kaiser (Canada) Maury (France)
16:00		Résumé du modérateur			Résumé du modérateur
16:30					Cérémonie de clôture: Wittke (RFA)
17:00		S e s s i o n s d ' a f f i c h a g e			
18:00					
19:00	Réception du président du congrès	Réception du président de la SIMR			Réception
19:30		Ballet Jazz			Dîner
20:30					Présentations du jubilé d'argent
21:30					Les Sortilèges
22:00					

Organisation des Sechsten Kongresses der IGFM

ORGANISATIONSKOMITEE

Dr G.Herget, Kongreßvorsitzender
Dr B.Ladanyi, Ehrenvorsitzender
Dr D.E.Gill, 2. Kongreßvorsitzender
Dr N.R.Morgenstern, Wissenschaftliches
Dr W.F.Bawden, Ausstellungen
Dr J.Bourbonnais, Gastgeberfunktionen
T.Carmichael, Finanzen
Dr J.A.Franklin, Öffentlichkeitsarbeit
L.Geller, Übersetzungen
Dr A.T.Jakubick, Fachexkursionen
Frau J.Robertson, Programm für Begleitpersonen
Frau D.Grégoire, Unterkunft und Anmeldung, CIM

BERATUNGSKOMITEE

Prof. E.T.Brown, IGFM-Präsident
N.F.Grossman, IGFM-Generalsekretär
Vizepräsidenten:
Dr H.Wagner (Afrika)
Prof. Tan Tjong Kie (Asien)
Dr W.M.Bamford (Australien/Ozeanien)
Dr S.A.G.Bjurström (Europa)
A.A.Bello Maldonado (Nordamerika)
Dr F.H.Tinoco (Südamerika)

CARMA GESCHÄFTSFÜHRUNG

Dr J.E.Udd, Präsident
Dr J.Curran, Vizepräsident
T.Carmichael, Generalsekretär

6. INTERNATIONALER FELSMECHANIK KONGRESS 1987

Zeit	Sonntag 30. August	Montag 31. August	Dienstag 1. September	Mittwoch 2. September	Donnerstag 3. September
08:30	Pre-Kongress Arbeitskreis Young (Kanada): Technik der Messungen und Analyse für die Bestimmung der durch Bergbau verursachten Seismizität (8:30 - 18:00)	Begrüssung und Eröffnung Hoek (Kanada): Felsbau in Kanada	Thema II: Felsgründungen und Böschungen Vorsitzender: Tinoco (Venezuela) Sprecher: Panet (Frankreich) Felsverbesserungen für Gründungen und Böschungen mit vorgespannten und nicht vorgespannten Ankern - 4 Vorträge -	Thema III: Gesteinsprengen und mechanischer Vortrieb Vorsitzender: Tan Tjong-Kie (China) Sprecher: McKenzie (Australien) Sprengen im Hartgestein: Diagnose und Modelle für Sprengschäden und Haufwerk - 4 Vorträge -	Thema IV: Untertage Hohlräume im überbelasteten Fels Vorsitzender: Bamford (Australien) Sprecher: Wagner (S. Afrika) Planung und Ausbau von Untertagehohlräumen im Fels mit hohen Druckspannungen - 4 Vorträge -
10:00		P A U S E			
10:30		Thema I: Flüssigkeitsbewegung und Abfallisolierung Vorsitzender: Bjurstrom (Schweden) Sprecher: Doe (USA) Entwurf von Bohrlochuntersuchungsprogrammen für Abfallagerung im krystallinen Gestein - 5 Vorträge -	- 9 Vorträge -	- 9 Vorträge -	- 9 Vorträge -
12:00		M I T T A G S P A U S E und Poster Sitzung			
13:30		- 9 Vorträge -	Allgemeine Diskussion Diskussionsleiter: Kovari (Schweiz) Podiumsdiskussion: MacMahon (Australien) Ribacchi (Italien) Yoshinaka (Japan)	Allgemeine Diskussion Diskussionsleiter: Lindqvist (Schweden) Podiumsdiskussion: Favreau (Kanada) Blindheim (Norwegen) da Gama (Brasilen)	- 6 Vorträge -
14:30			Zusammenfassung vom Diskussionsleiter	Zusammenfassung vom Diskussionsleiter	P a u s e
15:00		P a u s e Allgemeine Diskussion Diskussionsleiter: Barton (Norwegen) Podiumsdiskussion: Langer (West-Deutschland) Hudson (England) Rissler (West-Deutschland)	Arbeitskreise Quellendes Gestein Materialgesetze für Salzgesteine Numerische Methoden als praktisches Hilfsmittel	Arbeitskreise Bruchvorgänge um untertägige Hohlräume Gesteinschneid- und Bohrfähigkeit Gesteinsprüfung und Prüfnormen	Allgemeine Diskussion Diskussionsleiter: Whittaker (England) Podiumsdiskussion: Sharma (Indien) Kaiser (Kanada) Maury (Frankreich)
16:00					Zusammenfassung vom Diskussionsleiter
16:30		Zusammenfassung vom Diskussionsleiter			Abschlussvortrag: Wittke (W-Deutschland)
17:00		P o s t e r S i t z u n g			
18:00					
19:00	Empfang des Kongress Vorsitzenden	Empfang des Präsidenten der IGFM			Empfang
19:30					Festessen
20:30		Jazz Ballett			Ansprachen zum 25. Jahrestag
21:30					Les Sortilèges
22:00					

Contents / Contenu / Inhalt
Volume 3 / Tome 3 / Band 3

1 English section

2 La partie française

3 Deutscher Teil

4 Additional contributions / Contributions additionelles / Ergänzende Beiträge

5 Annexes / Anhang

1

English section

Opening ceremony

Figure 1.

Figure 2.

Figure 3.

On Monday August 31, 1987 at 8:30 the delegates assembled in the Grand Salon of the Queen Elizabeth Hotel, Montreal. At 8:35 the head table guests marched into the auditorium guided by a group of soldiers dressed in 18th century French uniforms and consisting of a flute player, a drummer, two soldiers wearing muskets and a lance corporal (Figure 1).

The Chairman of the Congress, Dr. G. Herget, welcomed the delegates in the three official languages and introduced the head table. Those seated at the head table (Figures 2 and 3) were from right to left:

Mr. N. Grossmann, Secretary-General ISRM
Dr. Chapuis, Canadian Geotechnical Society
Prof. Dr. Gill, Congress Co-Chairman
Prof. Dr. N.R. Morgenstern, Technical Program
Prof. B. Maldonada, Vice President ISRM, North America
Prof. Dr. B. Ladanyi,, Honorary Chairman, 6th ISRM Congress
The Honourable R. Savoie, Minister for Mines and Native Affairs, Province of Quebec
The Honourable G. Merrithew, Minister of State for Forestry and Mines, Government of Canada
Dr. G. Herget, 6th ISRM Congress Chairman
Prof. Dr. E.T. Brown, President ISRM
Prof. Dr. E. Hoek, Introductory Lecture
Dr. J.E. Udd, Director Mining Research Laboratories
Mr. L. Milne, President Canadian Institute of Mining and Metallurgy
Dr. G. Miller, President Mining Association of Canada

Professor Ladanyi welcomed the delegates and asked Prof. Dr. E.T. Brown, President ISRM, to address the delegates on behalf of the International Society for Rock Mechanics.

Mr. Chairman, Distinguished Guests, Ladies and Gentlemen!

On behalf of the International Society for Rock Mechanics and the many visitors to Canada for this Congress, I should like to thank our Canadian colleagues for the warm personal welcomes that they have given us in the last few days, for their generous words of welcome this morning, and for the arrangements that they have made for what we confidently expect will be a truly memorable Congress. Those of us who are visitors to Canada are also greatly looking forward to having the opportunity of seeing something of this fascinating city of Montreal, and of the many and varied attractions of this large and exciting country.

Prof. Dr. E.T. Brown, President

The International Society for Rock Mechanics was founded in Salzburg, Austria, in 1962 by a group of far-sighted enthusiasts led by Prof. L. Müller. So during 1987, we are celebrating the 25th anniversary, or Silver Jubilee, of the founding of the Society. During those 25 years, rock mechanics has become a well established and vital component of professional practice in civil engineering, mining and energy resource exploitation as well as having application to the understanding of geological processes. Rock mechanics has developed a large literature, is a well established special subject in universities, and in consulting and research organizations, and is now sufficiently old that many of the current leaders in the subject have spent their entire careers working in it. In these and other respects, rock mechanics is beginning to assume some of the attributes of what is often referred to as a mature science.

It is fitting, therefore, that the year of the Society's Silver Jubilee is also the year of one of our quadrennial congresses at which we traditionally share our experiences and review the state of our subject as a whole. It is also fitting, I think, that we are holding this Congress in Canada. This country is the location of a number of major mining and rock engineering projects, and Canadians have always been counted among those making advances in the science and application of rock mechanics. Despite these facts, the Society has not previously held one of its international meetings in Canada. At long last we are correcting this omission in 1987.

I am pleased to say that a number of the founding members of the Society and members of its past Boards have made special efforts to be here for this Silver Jubilee Congress. We are delighted to have them with us, and hope that they take home with them pleasant memories of these few days spent together in Montreal.

I had hoped that the Society's Founding Father, Prof. Leopold Müller, would be able to join us for the Congress and to address us this morning. Regrettably, the pressures of a still active professional life have meant that he cannot be here. However, he has sent me a message which I should like to read on his behalf:

"Dear Mr. President,
Dear friends and colleagues!

It was my honest intention to participate in the Silver Jubilee of our Society, on the occasion of which we may look back upon its years of intensive activity: but unfortunately I am not able to attend this Congress.

Twenty-five years is actually a short period of time in the cultural life of the world: but in our age everything is developing very much faster than it did in earlier times and so also in rock mechanics. Within this quarter of a century, much has been mastered and our libraries are inflated with the results. Whoever takes it upon himself to reflect on how many of the terms and ideas that are used as a matter of course today were not known in 1962, can only be astonished.

It is regrettable to state that it was mostly catastrophes that gave the young science of rock mechanics this strong impulse for development especially in the Sixties, whereas in the foregoing decades only a small amount of interest existed in this field.

I am sorry to say that within these 25 years not only has a lot of research work been done, but also an astonishing amount has fallen into oblivion. I suppose that only a few percent of what our scientists produce will enter the practice of rock engineering. Only a small amount of that groundwork is really used in large projects throughout the world: seen from a geomechanical point of view, there are cases in which design is worse than it was twenty years ago.

In our discipline, an understanding of the fundamentals is more essential than sophisticated computational methods: but especially interest in the characteristics of the material rock, in the peculiarities of this partner in our work - that reacts to our interferences favourably or unfavourably according to our treatment of it - especially this interest has been lost in favour of the more attractive computations.

Let us hope that no new catastrophes are needed to revive this interest.

It is my sincerest wish to all of you that this Congress shall be conducive to closing the gap between theory and practice in rock mechanics.

Glückauf!

Leopold Müller"

Thank you Mr. Chaiman.

Dr. Herget thanked Drs. Brown and Müller for their address and asked Dr. C.G. Miller, President of the Mining Association of Canada to address the delegates on behalf of the Association.

"Mr. Chairman, Ladies and Gentlemen!

I am pleased to have this opportunity to add my greetings to those of the other speakers this morning. It is a great pleasure for me, on behalf of Canada's mining industry, to welcome delegates from many countries.

Dr. G. Miller, President Mining Association of Canada

The Canadian venue for this Sixth International Rock Mechanics Congress is highly appropriate. Canada is one of the world's great mining nations, and our mining industry owes a great deal to the science and practice of rock mechanics.

Canada produces about 60 minerals from over 300 mines, ranking first among producing countries on a per capita basis. Our country also ranked first in 1986 in the world as producer of uranium and zinc, second for potash, nickel, elemental sulphur, asbestos and gypsum, third for titanium concentrates, cadmium, aluminum, gold and platinum group metals, fourth for copper, molybdenum, lead and cobalt, fifth for silver and sixth for iron ore.

Canada is the world's largest exporter of minerals with more than 80 per cent of production shipped to markets around the world. The mining industry directly employs over 100,000 Canadians and related industries employ a further 300,000 persons.

The science of rock mechanics makes a great contribution, and a growing contribution, to the welfare of this important industry and its workers. The efficient and safe extraction of mineral deposits depends to a great extent on an adequate understanding of rock mechanics. And in the search for higher productivity, new mining methods are being developed. Ever more sophisticated rock mechanics models are required to support research into new methods which will lead to innovative design for new mines. Moreover, the practical application of rock mechanics in existing mines is making use of more extensive and accurate measurements together with predictive analyses, to permit more efficient and secure operations.

Thus, our industry commends the Canadian Rock Mechanics Association and the Canadian Institute of Mining and Metallurgy for organizing this important Congress. May your deliberations lead to more rapid advances in your science and more effective applications of your art!"

Dr. Herget thanked the speaker for his welcome address and asked the Honourable R. Savoie, Minister of Mines and Native Affairs of the Province of Quebec to address the delegates on behalf of the Province of Quebec (no official text submitted).

Following the address by the Honourable R. Savoie, the Honourable G. Merrithew, Minister of State for Mines, addressed the delegates on behalf of the Government of Canada. The text is reproduced below:

Good morning, Ladies and Gentlemen.

On behalf of the Government of Canada, and as Canada's minister responsible for mining, I welcome all of you, and especially those who may be visiting our country for the first time. I am sure that your plenary sessions, workshops and other discussions will be highly productive, and I know that you will find Montreal an incomparable setting for you social activities.

I realize that many of you have already seen parts of central and eastern Canada through pre-Congress technical tours, and that others will soon be visiting Niagara Falls. But for those who can take the opportunity, I invite you to spend some additional time exploring our country. Canada is an immense and spectacular land, and offers a great variety of scenic and cultural rewards. You will be warmly welcomed.

Like all disciplines that deal with the landmass and its use, rock mechanics is of vital importance to Canada. We are a resource-oriented nation. Every advance in rock mechanics has important applications in this country, as we strive to utilize our resources as efficiently and as safely as possible, and in an environmentally acceptable manner.

Canada prides itself, with good reason, on its work in rock mechanics. This country has always been a leader in high-productivity hard-rock mining, and our recent developments in vertical crater retreat mining are considered a major breakthrough in combining high productivity with safety. Rock mechanics plays a central role in this technique.

Also, Canada's hard-rock miners are now working at greater depths. It is not uncommon today to find them mining below 2000 metres. This could not be done without the contribution of rock mechanics.

I know that this Sixth International Congress on Rock Mechanics is a special one, as you are celebrating the silver anniversary of the International Society for Rock Mechanics. Canada is therefore doubly honored to have been chosen as the host country for this important milestone in the history of your organization.

I congratulate the ISRM for its outstanding record of accomplishments over the past quarter-century. Rock mechanics is a relatively young discipline; the progress that has been achieved in this field throughout the world over the past 25 years owes much to the ISRM and its activities.

One example that I would cite is the ISRM's contribution to the standardization of laboratory and field testing, site investigations and analytical methods. I am told that, to date, 80 suggested testing methods have been published, representing contributions from about 500 scientists.

I am also proud of the fact that Canadian specialists in rock mechanics have been active in ISRM since its inception. Through their participation in ISRM, our people have kept abreast of developments as they have occured throughout the world. And by the same token, they have been able to make their own contributions to the international fund of knowledge on rock mechanics.

The Honourable G. Merrithew, Minister of State for Mines.

One Canadian contribution that merits a few words is the pit slope manual developed by one of our outstanding pioneers in rock mechanics, the late Donald Coates. His comprehensive scientific and engineering guide has had a major impact on open pit mine design throughout Canada and elsewhere.

The four themes of this Congress - Fluid Flow and Waste Isolation in Rock Masses; Rock Foundations and Slopes; Rock Blasting and Excavation; and Underground Openings in Overstressed Rock - all have great relevance to current Canadian mining and engineering activities.

One serious problem in Canadian mining, for example, are rockbursts. From 1984 to 1985, no less than 217 rockbursts occurred in this country, with magnitudes ranging from 1.5 to 4 on the Richter scale. CANMET - Canada's federal mining research agency - is working in close collaboration with the mining industry, provincial mines departments and universities, in a five-year research program to develop technologies for detecting danger zones in mines where rockbursts are most likely to occur.

Details of this work will be presented during the course of the Congress, and those of you who had the opportunity to tour the mines at Sudbury last Tuesday and Wednesday will have had a firsthand look at this program.

Apart from their important role in contributing to safer and more efficient mining operations, Canadian rock mechanics specialists are involved in many other challenging studies and activities, ranging from permanent nuclear waste disposal to rock foundations, slopes and tunneling. No matter where their particular interests lie, I am sure that, like all of you, they will leave here at the end of the week with new insights and increased knowledge.

The planning and preparation that have been put into this Congress - from the technical sessions to the field tours, from the workshops to the exhibition - have obviously been superbly carried out, and for this I congratulate the Canadian Institute of Mining and Metallurgy, the Canadian Geotechnical Society, and the Canadian Rock Mechanics Association.

I now declare the Sixth International Congress on Rock Mechanics officially open!

Dr. Herget thanked the speaker for his welcome address and for the time he took out from a very busy schedule. Following the opening of the 6th International Congress on Rock Mechanics, Prof. Dr. N.R. Morgenstern, Chairman of the Technical Program Committee, introduced Prof. Dr. E. Hoek to deliver the lecture on Rock Engineering in Canada. The text of the lecture is reproduced below:

Introductory Lecture

Rock Engineering in Canada - E. Hoek (Canada)

Department of Civil Engineering, University of Toronto, Canada

Abstract: A number of projects in which rock engineering has played a significant role are described in order to illustrate the historical development and the present status of the subject in Canada

INTRODUCTION

Rock engineering, the application of the science of rock mechanics to practical engineering problems, has a long history in Canada. In preparing this review I have concentrated upon practical applications and will attempt to demonstrate the wide diversity of problems which were and continue to be encountered in the development of this huge country.

I make no claim that this is a comprehensive historical review of the development of rock engineering in Canada but, rather, I have arbitrarily selected topics which I consider to be significant and have tried to cover them in sufficient detail that interested readers may be able to follow up on them.

Prof. Dr. E. Hoek

BRIEF HISTORICAL REVIEW

The first significant Canadian tunnel to be constructed was a railway tunnel in Brockville, Ontario which was started in 1854 and completed in 1860. According to Legget (1985), Canadian tunnelling on a major scale may be said to have started in 1889 with the commencement of work on an underwater tunnel beneath the St. Clair River between Sarnia in Ontario and Port Huron in Michigan. This underwater shield-driven railroad tunnel was completed in 1890. Rock tunnelling did not get under way until the early 1900s with the completion of the Canadian Pacific Railway's Spiral Tunnels in 1909 and the Connaught Tunnel in 1916. These tunnels link the railroad through the Canadian Rockies.

The Register of Dams in Canada, published by the Canadian National Committee of the International Commission on Large Dams (1984), lists 613 dams of which 43 are in excess of 60m high. The first dam constructed in Canada to qualify for entry in this register is the 19m high masonry arch Jones Falls dam on the Rideau Canal in Ontario which was constructed in 1832. The highest dam in Canada is the 243m high Mica dam in British Columbia which was completed in 1972. Of the 613 dams completed by 1983, 441 are associated with hydroelectric projects.

The mining industry has played a vital role in the economic development of the country and mining technology has evolved to meet the special needs of the industry. The first mining rock mechanics research program was initiated at McGill University in the early 1950s. This program was concerned with stress induced failure and rockbursts in underground mines and the first paper reporting the results of this research was presented by Morrison and Coates at the 1955 Annual Meeting of the Canadian Institute of Mining and Metallurgy.

Rock mechanics research in the Mines Branch of the Department of Energy, Mines and Resources (later renamed the Canada Centre for Mineral and Energy Technology (CANMET)) started in 1950. In 1964, a laboratory devoted to hard rock underground mining was opened at Elliot Lake in Ontario and in 1967 the Mining Research Centre was established in Ottawa under Dr D.F. Coates. A useful summary of the rock mechanics research carried out in the Mining Research Laboratories of CANMET between 1964 and 1984 has been compiled by Udd and Hedley (1985).

One of the most significant events in the early development of rock mechanics in Canada was the publication by Coates, in 1965, of a textbook entitled "Rock Mechanics Principles". This was the first English language text in the subject and it became an important reference work for students and research workers during the formative years of the subject.

It is interesting to note that some important work in rock mechanics related to civil engineering and geology was being done in the early 1960s. This work, which was published in journals not often read by the rock mechanics community, is less well known than the mining related research but the contributions were no less important. Typical of this work was shear testing of sedimentary rocks by Ripley and Lee (1961) and a very detailed study of three-dimensional fabric diagrams (spherical projections) by Stauffer of the University of Saskatchewan (1966).

In order to provide a forum for rock mechanics research workers to present and to discuss their findings, the first Canadian Rock Mechanics Symposium was held in Toronto in 1962. A succession of such symposia followed and the next symposium in this series will be held in Toronto in September, 1988. From these early beginnings, rock engineering has developed into a mature subject which is recognized as an engineering discipline in its own right. Undergraduate and graduate courses are offered at a number of Canadian universities and rock mechanics techniques are being applied to practical engineering problems by consulting, industrial and government organizations across the country.

ATOMIC ENERGY OF CANADA LTD. UNDERGROUND RESEARCH LABORATORY

Atomic Energy of Canada Limited (AECL) has the responsibility for research and development of technologies for the safe and permanent disposal of Canada's nuclear fuel wastes. As part of this comprehensive program, AECL is constructing an underground research laboratory (URL) near Lac du Bonnet, Manitoba, to evaluate aspects of the concept of waste disposal deep in stable geological formations. No nuclear waste will be used in the URL program.

The URL is being excavated in the Lac du Bonnet granite batholith, an excellent quality rock mass, which is considered re-presentative of many of the granite intrusives in the Canadian Precambrian Shield. Before commencement of excavation, the rock mass was thoroughly characterized by detailed surface geological mapping, logging of diamond and percussion drilled boreholes, surface and borehole geophysics and hydraulic fracture stress measurements (Lang 1985). An extensive hydrogeological program was also carried out in order to characterize the site of the URL before the excavation commenced (Davison 1984).

During 1984 and 1985 a 2.8 x 4.9m rectangular cross section shaft was sunk by drill and blast methods to a depth of 255m. Shaft stations have been excavated at 130m and 240m depths and excavations for experimental laboratories have been excavated at these shaft station locations.

During sinking of the shaft, a number of experimental programs were carried out. These included shaft convergence measurements, in situ stress determination by means of both over-coring and hydraulic fracturing techniques and investigations of coupled geomechanical and hydrogeological behaviour of the rock surrounding the shaft (Lang and Thompson 1985), (Kuzyk, Lang and Peters 1986).

During the mining of some of the rooms in the lower shaft station, carefully controlled drill and blast excavation techniques were used in order to determine the amount of rock damage associated with high quality blasting in this type of rock. Rock damage of this type can be important in nuclear waste disposal because of the potential for creating a leakage path parallel to the excavation boundary. Kuzyk, Lang and Le Bel (1986) and Lang, Humphreys and Ayotte (1986) have described the blast design and the excellent results obtained in this excavation program.

A variety of experimental programs are continuing at the URL and some important results have already been obtained. Davison (1986) has compared observations and predictions based on a hydrogeological model developed by Guvanasen (1984). The model was found to overpredict the volume of seepage which occurred in the shaft but it gave good estimates of the spatial and time-variant piezometric drawdown in 171 piezometers surrounding the shaft excavation. In general it was found that the model was least accurate close to the shaft and that its accuracy improved with distance from the shaft. Further work is being carried out in an effort to resolve the differences between the observations and predictions and to test other hydrogeological models.

An interesting aspect of the hydrogeological work at the URL is the development, in cooperation with Westbay Instruments Ltd. of Vancouver, of specialized instrumentation for piezometric measurements and water sampling (Rehtlane and Patton 1982). This instrumentation is based on a very successful series of similar instruments developed by Westbay over the past decade. Piezometric pressures and water samples can be taken at close intervals at precisely controlled depths and, by keeping the sampling volume very small, the perturbation of the groundwater regime is minimized.

Lang et al (1986) have given details of the results obtained in an initial in situ stress measurement program which included 42 tests using the US Bureau of Mines technique. Subsequent tests have utilized the Australian CSIRO and modified South African CSIR triaxial stress measuring cells. Preliminary results indicate that both the maximum and minimum horizontal stresses undergo a sharp change at 185m below surface. This anomaly is associated with a fracture zone at this depth and it is hoped that further analysis of the triaxial stress measurements will provide an understanding of the correlation between measured stresses and the geological conditions encountered at different depths. The ratio of average horizontal stress to vertical stress was found to agree well with a relationship established by Herget (1980) for stress measurements carried out in the Canadian Shield.

In addition to its importance to Canada's nuclear waste disposal program, the work being carried out at AECL's underground research laboratory is of great interest to the general rock mechanics community in Canada. The information obtained from the very detailed experimental work in the URL will add considerably to our understanding of the behaviour of rock masses in situ.

SYNCRUDE'S OILSAND OPERATION IN NORTHERN ALBERTA

The Syncrude Canada Ltd. mine and plant site is located approximately 40km north of Fort McMurray in northern Alberta. Mining of the oilsands, utilizing four 65 cubic meter draglines, accompanied by four bucketwheel reclaimers and conveying systems, commenced in June 1977. At a production capacity of 129,000 barrels of synthetic crude oil per day (the present permit level), the daily mine production over the planned 25 year life of the mine is approximately 250,000 tonnes per day (Fair and Lord 1984).

The Athabasca oil sands in northern Alberta contain one of the largest accumulations of viscous bitumen in the world. The McMurray formation, which contains the oilsand deposit, is the sediment fill of a basin developed on eroded Devonian carbonates during the Pre-Cretaceous transgressive sequence. The various sedimentary facies reflect successive deposition in fluvial, esturine and marine environments resulting in the development of tidal flats dipping gently from the high water to the low water levels which existed during deposition (List, Lord and Fair 1985).

Drainage of the intertidal flats was accomplished by a network of interconnected meandering tidal channels which continuously reworked the tidal flat areas and which deposited inclined layers of sand, silt and clay. The inclined clay layers, which have dips of 20 degrees or more, form failure surfaces for block slides which have occurred in the 60 meter high slopes created by the draglines (Brooker and Khan 1980).

Development of a block slide on a part of the slope on which a dragline is positioned could have very serious consequences for the mining operation. Typically, these slides develop within a minute or two and, since it takes about 15 minutes to walk a dragline away from the highwall crest, it is essential that potential slides be identified and dealt with. Fair and Lord (1984) have described the techniques used to identify potential block slides and to monitor these slides during the mining operation.

One of the innovative techniques which has been used to stabilize slopes is the use of blasting to disrupt the continuity of the inclined clay layers along which the failure surfaces develop. This blasting is carried out in advance of the dragline mining operation in areas where potential block slides have been identified. In addition to increasing the friction angle of the materials along the potential failure surfaces, blasting also increases the permeability of these materials and allows pore pressures within the oil sands to dissipate. Both of these changes have a positive effect on the stability of potential block slides.

The desired effects are achieved by vertically displacing the oil sand utilizing spherical or squat charges in a crater blasting mode. Design of the blasts is based upon work carried out in association with Professor Alan Bauer of Queen's University and details of the blast design are described by List, Lord and Fair (1985). It has been found that the most cost efficient method of remedial stabilization blasting involves the use of Ammonium Nitrate and fuel oil (ANFO) in large diameter blast holes. Current blast designs utilize 0.76m diameter blastholes which are approximately 15m deep. The weight of ANFO per blasthole is approximately 1180kg and the peak particle velocities generated by the blasts are limited by controlling the charge per delay and the delay between successive blastholes.

In addition to the block slides discussed above, several other types of slope failure mechanisms have been identified (Brooker and Khan 1980). Block slides are the only ones which can be stabilized by the pre-blasting technique described. Since all highwall failures can have a serious impact upon the mining operation, efforts have to be made to identify and to deal with all potential failure modes. Fair and Lord (1984) and Fair and Isaac (1985) have described the extensive monitoring programs which are used to measure piezometric pressures and slope displacements during the mining operation. Interpretation of the measurements obtained from this monitoring program has assisted in making short term decisions on mining strategy and has contributed to an understanding of potential slope failure mechanisms in the Syncrude operation.

DEEP LEVEL UNDERGROUND MINING IN HARD ROCK

Canada is rich in mineral resources which are exploited by a large number of surface and underground mines. Underground gold and base metal mines in the Precambrian Shield operate at depths of up to 2300m and, in many of these mines, rockbursts can pose a threat to mine personnel and production.

The first record of a rockburst in the Sudbury area was in 1929. Rockbursts occurred regularly in the Kirkland Lake mines during the 1930s and, in 1939, Professor Morrison was retained by the Ontario Mining Association to advise them on measures which could be taken to reduce the rockburst hazard (Morrison 1942). Rockbursts have continued to present a challenge to the mining industry and to rock mechanics research groups which are attempting to gain a better understanding of this phenomenon.

The Ontario Regulations for Mines and Mine Plants defines a rockburst as "an instantaneous failure of rock causing an explosion of material at the surface of an opening or a seismic disturbance to a surface or underground mine". During 1984, a total of 105 rockbursts, with intensities in excess of 1.5 on the Richter scale, were recorded in Ontario mines (Scoble 1986). This figure can be regarded as typical of rockburst activity in recent years. While many of these bursts are too small to have significant consequences, a number of cases involving serious damage have occurred (Ontario Government 1986).

In 1985 CANMET initiated a major five year research project on rockbursts (CANMET 1986). This project will involve basic microseismic monitoring at several small mines and detailed studies, using state-of-the-art microseismic monitoring systems, on two Sudbury mines. In addition, studies with an advanced prototype microseismic system permitting detailed seismic event analysis will be carried out on one large Ontario mine. Associated with this project, a number of parallel activities, such as the development of a sophisticated three-dimensional numerical model (McDonald, Wiles & Vielleneuve 1988), are being undertaken by research groups which are partly funded by CANMET.

Two Commissions of Enquiry have prepared reports, one in 1981 and another in 1986, on the safety of underground mining operations in northern Ontario. Rock engineering, as it relates to rockbursts, destressing, backfilling, pillar recovery, blasting and ground support, received careful consideration by these commissions and their conclusions and recommendations make worthwhile reading for anyone interested in underground mining. As a direct result of the 1986 report prepared for the Government of Ontario by a committee chaired by Mr Trevor Stevenson, a mining research directorate has been established in Ontario with the aim of coordinating research in deep level hard rock mining. The initial thrust of this work will be in applied rock mechanics but it is intended that the activities of this organization will gradually expand to include other mining related research areas. The research directorate will operate in parallel with existing research organizations such as the universities, CANMET and the research groups which operate within technical services departments of the large mining companies.

TUNNELLING IN WEAK AND SQUEEZING GROUND

In contrast to the rockburst problems encountered in mining in the hard rocks of the Precambrian Shield, gradual closure or squeezing is a problem faced by tunnellers in other regions of the country. In Southern Ontario, where high horizontal stresses occur (Lo 1978), the relatively weak shales which are common throughout the region are particularly prone to squeezing (Lo, Cooke & Dunbar 1986), (Lo 1986).

Soft rock tunnelling problems also occur in other parts of the country and some of these problems have been described by Kaiser and McKay (1983), Kaiser, Guenot and Morgenstern (1985) and Boyd, Yuen and Marsh (1986). The last reference is particularly interesting in that it deals with a 4km long machine-bored under-sea tunnel in relatively weak sedimentary rocks in the coal mining area off the coast of Nova Scotia.

Theoretical studies on the behaviour of tunnels in squeezing rock have been undertaken by Ladanyi (1974, 1976), Ladanyi and Gill (1984), Lo and Yuen (1981), Emery, Hanafy and Franklin (1979), and Brown et al (1983). In research carried out under the direction of Professor K.Y. Lo at the University of Western Ontario, Ogawa (1986) has developed a three-dimensional solution for the stresses and displacements in the rock surrounding a tunnel face and the application of this solution to the interpretation of tunnel behaviour has been described by Lo, Lukajic and Ogawa (1984) and by Lo (1986). Another three-dimensional analysis of stresses and displacements surrounding an advancing tunnel face has been discussed by Eisenstein, Heinz and Negro (1984).

CANADIAN HYDROELECTRIC PROJECTS

The mountainous terrain and abundant supplies of water which occur in many parts of Canada provide the potential for the development of hydroelectric power generation facilities. A number of hydropower projects have been constructed but relatively few of these are known outside the country because details of the schemes have either not been published or they have been published in journals not normally read by the rock mechanics community.

Among these lesser known projects is the Spray hydroelectric project in Alberta (Eckenfelder 1952). Of particular interest is the design of the pressure tunnel which has a 45 degree inclined section which was excavated through massive limestone, dolomitic limestone, argillite and a very few bands of shale. By today's standards, the ratio of rock cover to static head used to select the length of steel lining at the Spray power plant would not be considered adequate unless pressure acceptance tests had been carried out as part of the design process (Humphreys and Hoek 1987). The fact that this tunnel has operated successfully for many years suggests that, under certain circumstances, current cover depth criteria may be conservative and that there is still room for research into pressure tunnel design.

Huber (1952, 1953) and Wise (1952) published construction details for the Aluminium Company of Canada's Kitimat-Kemano project in northern British Columbia. At the time of its construction, this was considered to be the world's largest underground power station.

The engineering geology and rock engineering design considerations for the W.A.C. Bennett (Portage Mountain) and Mica hydroelectric projects in British Columbia have been discussed by Imrie and Jory (1969) and Imrie and Campbell (1976). Underground stability problems associated with foliation shears at the Bersimis and Chute-des-Passes hydroelectric projects in Quebec are described in a paper by Benson et al (1974). Humphreys and Lang (1986) and Humphreys and Hoek (1987) have given details of the rock engineering aspects of the Cat Arm hydroelectric project in Newfoundland. This scheme is unusual in that it uses a carefully blasted unlined high pressure tunnel in very good quality granite.

Rock engineering played a major role in the Churchill Falls hydroelectric scheme in Labrador which, at the time, was a very large and ambitious underground project. Geotechnical investigations and design of the underground powerhouse have been described in a series of papers by Benson et al (1970a, 1970b, 1971, 1974) and some of the construction details have been published by Mamen (1969) and Gagne (1972).

Because of its remote location and large size, the La Grande project in the James Bay area of northern Quebec has attracted a great deal of attention. The project is designed to develop the hydroelectric potential of the La Grande River together with portions of the adjacent Caniapiscau and Eastmain-Opinaca drainage basins. The overall development includes nine powerhouses with a planned capacity of 13,753 MW and an annual energy generating capacity of 81.6 terrawatt-hours. The entire complex lies within the Canadian Shield and most of the project sites are underlain by granitic rocks ranging in texture from massive to gneissic. Because of the extensive body of mining and construction experience available from other areas of the Canadian Shield, the design and construction philosophy adopted for the La Grande development may be summarized as follows (Murphy & Levay 1982) :

1. Unless it can be demonstrated otherwise, rock quality is taken as good to excellent.
2. Drilling and blasting techniques are generally more significant than geological conditions in terms of minimizing overbreak and obtaining stable excavation profiles.
3. Experienced judgement based on field observations is usually more valid in the assessment of rock treatment requirements (reinforcement and surface protection) than any strictly theoretical approach.

In spite of the suggestion that rock mechanics was of little value in the design of the La Grande project, Murphy and Levay go on, in their paper, to describe the use of air photo interpretation, joint surveys, diamond core drilling with the use of borehole cameras and water pressure testing. In the case of the LG-2 underground powerhouse, a test shaft and drifts along the powerhouse arch were excavated to allow in situ modulus and stress determinations and the results of these investigations were used in finite element studies of pillar dimensions and cavern geometry. I find it fascinating that this amount of investigation is classed as "experienced judgement and field observation" as opposed to a "strictly theoretical approach". This suggests that, in the minds of the La Grande project's management, rock engineering has achieved the status of a practical design tool and I take this as a compliment to the many rock mechanics research workers who have struggled with their "strictly theoretical approach" for many years in order to achieve this goal.

The LG-2 underground powerhouse is claimed to be the world's largest. Excavated in massive granite at a depth of 160m below surface, the machine hall has a span of 28.4m while the parallel surge chamber has a span of 23.6m. Both of these excavations are approximately 47m high and the machine hall houses sixteen 333 MW turbine-generator units. In the powerhouse, two 405 tonne cranes are supported on rock ledges which were formed in high quality rock using carefully controlled excavation techniques.

The excellent quality of the rock also made possible the use of unlined diversion and tailrace tunnels which are designed to operate at water velocities of up to 14 m/s. Over 8000m of tunnels with dimensions up to 14 x 20m were excavated and concrete lining was only used in portals and for gate or stoplog structures.

The LG-2 spillway is another interesting piece of engineering. It is a cascade spillway which is designed to accommodate a peak flow of 16280 cms. The total length of the spillway channel is 2200m and it includes ten steps or benches ranging from 9.1 to 12.2m in height and from 127 to 200m in length. The nominal width of the channel is 122m but, because all the excavated rock was used for rockfill in the main dam, the width was increased to 183m for the last 1100m in order to provide more rock (Levay and Aziz 1978). The steps in the cascade channel were excavated in rock and only a few of the edges, where the rock was closely jointed, were reinforced with rockbolts and shotcrete. The spillway has been in operation since 1979 and has shown no significant deterioration in spite of the fact that water velocities have approached the design limit. This limit is a velocity of 11 m/s at the end of the first bench and 26 m/s on the last bench.

TUNNELLING THROUGH WESTERN CANADA'S MOUNTAINS

Roger's Pass is located in a beautiful wilderness area in the Selkirk mountains of British Columbia. It climbs to a maximum elevation of 1200m above sea level amidst mountains towering more than 3300m . In an average winter, more than 9m of snow falls here.

The first railroad through Rogers Pass was laid by Canadian Pacific in 1885. In spite of severe problems, such as an avalanche which killed 62 men working on the line in 1910, this rail line remained in operation for many years. A program to increase the capacity of the CP line across the country led to the construction of a tunnel through the base of Mount Macdonald in the Selkirk range. This tunnel, originally called the Selkirk tunnel and later renamed the Connaught tunnel, was started in 1913 and completed in 1916. The tunnel is approximately 8km long and was excavated from a small pilot tunnel parallel to the main bore which permitted access to a number of working faces from short connecting tunnels which were excavated at periodic intervals along the tunnel. The tunnel, still in operation today, not only shortened the line and reduced the maximum grades but also bypassed some of the most dangerous avalanche areas.

In order to eliminate bottlenecks in the main line route between Calgary and Vancouver, CP Rail has been working on a new upgrading scheme. One of the main components of this scheme is a new tunnel, known as the Mount Macdonald tunnel, which has been driven through Mount Macdonald and Cheops Mountain. This 14.7km long tunnel is the longest railway tunnel in North America. Construction started in 1984 and the two faces, driven from opposite ends of the tunnel, met earlier this year. This horseshoe shaped concrete lined tunnel will have a span of 5.18m and a height, from top of rail to crown, of 7.87m.

Tunnel excavation from the western portal was by full-face drill and blast methods with cast in place concrete lining following behind the advancing face (Steward 1987). A tunnel boring machine was used to drive the top heading from the eastern portal and this was followed by bench blasting to form the tunnel to full height (Knight 1987). The support used during tunnel excavation consisted of spot dowels and some mesh and shotcrete. It is interesting to note that the support requirements derived from rock mass classification studies called for significant lengths of the tunnel to be supported by means of pattern bolts, shotcrete and mesh. Based upon the experience gained from the parallel Connaught tunnel, which has stood unsupported for 70 years, the support requirements were downgraded to the minimal amount actually used. A full concrete lining is being placed in the tunnel and this has sufficient load bearing capacity to provide long-term stability for the rock mass surrounding the tunnel.

A brief description of the Rogers Pass tunnel project has been compiled by Gadsby (1985) and a good overview of the general engineering features of the project has been published by Tanaka (1987). A detailed discussion on the geological mapping and computer assisted data processing used on the project is given in a paper by Klassen et al (1987).

In another part of British Columbia, a number of tunnels are being evaluated as part of track upgrading of the Canadian National Railways line through the mountains. Benson, Hostland and Charlwood (1985) have described the geotechnical and ventilation requirements for these tunnels.

Kaiser and Gale (1985) have published the results of a very thorough study of the use of rock mass classification systems for estimating tunnel support requirements for BC Rail's Tumbler Ridge tunnels in British Columbia. A number of popular classifications schemes are compared and the advantages and shortcomings of each is discussed. I consider this paper to be a significant contribution to the growing body of literature on the use of rock mass classifications for the evaluation of rock support requirements in underground excavations.

MINING IN CANADA'S FAR NORTH

Located on Little Cornwallis Island, 77 degrees north of the equator, Cominco's Polaris mine is the world's most northerly mine. The lead-zinc orebody is contained within the permafrost zone which, at this latitude, extends to a depth of about 450m below surface. Temperatures on surface have ranged from a winter low of -55°C to a summer high of 15°C. Even in the warmest summer, only a few centimeters of the ground surface thaws (Leggatt 1982).

Because of the high ice content in the ore it is important to keep the rock mass frozen in order to maintain stability of the underground openings. This is achieved by ventilating the mine with cold air which keeps the underground temperature constant at approximately -7°C. A mixture of snow and gravel is dropped down raises to form fill in the stopes and this freezes into a very strong and effective fill.

Remarkably few geotechnical problems have been encountered since this mine began operation in 1982. Apart from the fact that personel have to dress a little more warmly than in a normal mine, the operation of the mine, the design of the stopes and pillars and the choice of excavation support systems are remarkably similar to those which would be found in a mine in a more temperate location.

ROCK SLOPE ENGINEERING

Transportation routes through the mountains and large open pit mines have in common the requirement for steep and yet stable rock slopes. Given the mountainous nature of parts of Canada and the size of its mining industry, it is not surprising that rock slope engineering has played an important role in the development of the country. The Canadian literature on this subject is too extensive to cover in this review and I will only mention a few of the highlights on this subject.

One of the first practitioners in the field of rock slope engineering was C.O. Brawner, now Professor of Geomechanics at the University of British Columbia. A practical engineer rather than a theoretician, Brawner has worked on rock slope problems associated with highways, railways and open pit mines around the world and has made some significant contributions to this field (Brawner 1986).

Coates, whose work on underground mining has already been mentioned, was also keenly interested in rock slope stability. This interest culminated in his initiation, in 1972, of a CANMET project which resulted in the publication of a multi-volume "Pit Slope Manual" (CANMET 1977). This massive 27 volume work covers all of the details which would have to be considered in designing rock slopes, waste dumps and tailings dam embankments associated with an open pit mine and it is an important reference work in rock slope engineering.

Professor Peter Calder and his colleagues at Queen's University in Kingston have made contributions in slope stability analyses, open pit blasting design and in slope monitoring (Calder 1969). Dr Douglas Piteau and his colleagues in Vancouver have use probabilistic techniques in the optimization of rock slope designs (Piteau et al 1985).

The analysis, monitoring and stabilization of toppling failures in slopes has been dealt with by a number of authors including Choquet and Tanon (1985), Wyllie (1980) and Brawner, Stacey and Stark (1975). The use of reinforcement in the form of tensioned cables and grouted dowels has been described by Martin (1987) and these techniques are expected to play an increasing role in rock slope engineering as more efficient and economical reinforcing techniques are developed.

SHOTCRETE DEVELOPMENTS

In response to the need to protect slopes against erosion in high rainfall environments and to carry out rapid repairs in busy railway tunnels, a considerable effort has been devoted to improving the strength and workability of shotcrete.

Improvements have included the use of large aggregates (MacRae 1973), the addition of steel fibre (Little 1985) and (Morgan 1985) and the addition of microsilica (Anon 1987), (Morgan et al 1987). Today, the use of steel fibre reinforced microsilica shotcretes is common in western Canada and is quickly becoming accepted across the country. In particular, the addition of microsilica results in significant improvements in both the workability and the strength of the shotcrete and makes it practical to use shotcrete is situations which, until a few years ago, required the use of steel sets or concrete lining.

CONCLUSION

I have attempted to illustrate, by means of a few practical examples, the current status of rock engineering in Canada. The development of this discipline from its theoretical base in rock mechanics has been slow and sometimes frustrating. While one or two individuals played a major role in getting the science started, the general evolution of rock engineering in Canada is characterized by the long-term involvement of a large number of people working in a number of different universities, government and private research groups and consulting organizations spread across the country. There is no recognized centre of Canadian rock engineering but, as I have tried to demonstrate in this review, there are a number of groups of engineers and geologists who are concerned with the application of the subject to the solution of practical problems. In my view, this parallel development is healthy in that it produces a diversity of approaches and opinions and results in the adoption of solutions which are judged on their practical merits instead of the reputation of the individual or organization which proposed them.

ACKNOWLEDGEMENTS

In preparing this review I sought and received the assistance of many individuals across Canada. It is not possible to name all of these persons but I wish to express my sincere gratitude to all of them for the notes, papers and photographs (not all of which I have been able to use) which were sent to me.

REFERENCES

GENERAL

The Register of Dams in Canada 1984. Montreal: Canadian National Committee of the International Commission on Large Dams.

Coates, D.F. 1965. Rock mechanics principles. Mines Branch Monograph 874. Ottawa: Mines Branch, Dept. Energy, Mines and Resources.

Legget, R.F. 1985. Tunnels and geology in Canada. In Z.E.Eisenstein (ed.), Canadian Tunnelling, p. 3-12. Tunelling Assoc. Canada. Vancouver: BiTech Publishers.

Morrison, R. & D.Coates 1955. Soil mechanics applied to rock failure in mines. Trans. CIM. 58: 401-11.

Ripley, C.F. & K.L.Lee 1961. Sliding friction tests on sedimentary rock specimens. 7th int. cong. on large dams, Rome, vol. 4, p. 657-671. Paris: Int. Commission on Large Dams.

Stauffer, M. R. 1966. An empirical-statistical study of three-dimensional fabric diagrams as used in structural analysis. Can. J. Earth Sci. 3: 473-498.

Udd, J.E. & D.G.Hedley 1985. Rock mechanics research in the Mining Research Laboratories of CANMET, 1964-1984. Mining Research Laboratories, Division Report MRP/MRL 85-19(TR). Ottawa: Mines Branch, Dept. Energy, Mines & Resources.

ATOMIC ENERGY OF CANADA LTD. UNDERGROUND RESEARCH LABORATORY

Davison, C.C. 1984. Hydrogeological characterization at the site of Canada's underground research laboratory. Proc. IAH int. symp. on groundwater resource utilization and contaminant hydrogeology, p. 310-335. Montreal: Can. Nat. Chapter, Int. Assoc. Hydrogeol.

Davison, C.C. 1986. URL drawdown experiment and comparison with model predictions. In AECL Tech. Rep. Tr-375, p. 103-124. Pinawa, Manitoba: Whiteshell Nuclear Res. Estab.

Guvanasen, V. 1984. Flow simulation in a fractured rock mass. Proc. IAH int. symp. on groundwater resource utilization and contaminant hydrogeology, p. 403-412. Montreal: Can. Nat. Chapter, Int. Assoc. Hydrogeol.

Herget, G. 1980. Regional stresses in the Canadian Shield. In Underground rock engineering, p. 9-16. Montreal: CIM.

Kuzyk, G.W., P.A.Lang & D.A.Peters 1986. Integration of experimental & construction activities at the underground research laboratory. In K.Y.Lo, J.H.L.Palmer & C.M.K.Yuen (eds), Recent advances in Canadian underground technology, p. 239-264. Toronto: Tunnelling Assoc. Canada.

Kuzyk, G.W., P.A. Lang & G.Le Bel 1986. Blast design and quality control at the second level of Atomic Energy of Canada's underground research laboratory. In K.H.O.Saari (ed.), Proc. int. symp. on large rock caverns, Helsinki, 6, p. 147-158. Oxford: Pergamon.

Lang, P.A. 1985. AECL's underground research laboratory under construction. TACNews, August, p. 1-3,12. Toronto: Tunnelling Assoc. of Canada.
Lang, P.A. & P.M.Thompson 1985. Geomechanics experiments during excavation of the URL shaft. In AECL Tech. Record TR-375. Chalk River, Ont.: S.D.D.O., Atomic Energy of Canada Ltd. Res. Co.
Lang, P.A., R.A.Everitt, L.K.W.Ng & P.M.Thompson 1986. Horizontal in situ stresses versus depth in the Canadian Shield at the underground research laboratory. In O.Stephansson (ed.), Proc. int. symp. on rock stress & rock stress measurement, Stockholm, p.449-456. Lulea: Centek.
Lang, P.A., R.W.Humphreys & J.G.Ayotte 1986. Blasting in the lower shaft station of Atomic Energy of Canada Ltd's underground research laboratory. Proc. 12th annual conf. on explosives & blasting techniques, Atlanta. Atlanta, GA.: Soc. Explosives Engrs.
Rehtlane, E. & F.D.Patton 1982. Multiple port piezometers vs standpipe piezometers: an economic comparison. In D.M. Nielsen (ed.), Proc. 2nd national symp. on aquifer restoration and groundwater monitoring, Columbus, p. 287-295. Worthington, OH: National Water Well Assoc.

SYNCRUDE'S OIL SAND MINING OPERATION

Brooker, E. & F.Khan 1980. Design and performance of oilsand surface mining slopes. Proc. oil sands geoscience conf., Edmonton, Alberta.
Fair, A.E. & E.R.F.Lord 1984. Methods used to monitor and control block slides in oilsands at Syncrude's dragline operation in Northern Alberta, Canada. Proc. 4th int. symp. on landslides, p. 103-112. Toronto: Int. Soc. Soils Foundn Engng.
Fair, A.E. & B.A.Isaac 1986. Highwall stability monitoring at Syncrude Canada Ltd. open pit mining operation. ISA Trans. 25(2):p. 45-54.
List, B.R., E.R.F.Lord & A.E.Fair 1985. Investigation of potential detrimental vibrational effects resulting from blasting in oilsand. In G.Gazetas & E.T.Selig (eds), Vibration problems in geotechnical engineering, p. 266-285. New York: Am. Soc. Civ. Engrs.

ROCKBURSTS IN HARD ROCK UNDERGROUND MINES

Canada and Ontario, Ministries of Labour. 1981. Towards Safe Production, The Report of the Joint Federal-Provincial Inquiry Commission into Safety in Mines and Mining Plants in Ontario. Two Vols.
CANMET. 1986. 1985-1986 Annual report of the Canada-Ontario-industry rockburst project. CANMET Spec. Publ. SP86-3E. Ottawa: Dept. Energy, Mines & Resources.
McDonald, P., T.Wiles & T.Vielleneuve 1988. Rock mechanics aspects of vertical retreat mining at 2000m depth at Creighton mine. Proc. conf. applied rock engineering, Newcastle, UK. in press. London: Instn. Min. Metall.
Morrison, R.G.K. 1942. Report on the rock burst situation in Ontario Mines. Trans. CIM. 45:225-272.
Ontario Government. 1986. Improving Ground Stability and Mine Rescue, The Report of the Provincial Enquiry into Ground Control and Emergency Preparedness in Ontario Mines. Toronto: Ontario Government, Publ. Services Sect.
Scoble, M.J. 1986. Strategic and tactical measures to alleviate rockbursting in Canadian underground mining. Mining Dept. Magazine, University of Nottingham. 38: 47-53.

TUNNEL DESIGN IN SQUEEZING ROCK

Boyd, J.M., C.M.K.Yuen & J.C.Marsh 1986. Design and performance of the Donkin-Morien Tunnels. In K.Y.Lo, J.H.L.Palmer, & C.M.K.Yuen (eds), Recent advances in Canadian underground technology, p. 199-219. Toronto: Tunnelling Assoc. Canada.
Brown, E.T., J.W.Bray, B.Ladanyi & E.Hoek 1983. Ground response curves for rock tunnels. J.Geotech. Engng Div., Am. Soc. Civ. Engrs 109:15-39.
Eisenstein, Z., H.Heinz & A.Negro 1984. On three-dimensional ground response to tunnelling. In K.Y.Lo (ed.), Tunnelling in soil and rock, p. 107-127. New York: Am. Soc. Civ. Engrs.
Emery, J.J., E.A.Hanafy & J.A.Franklin 1977. Finite element simulation of tunnels in squeezing ground. Proc. int. symp. on the geotechnics of structurally complex formations, Capri, 1, p. 219-228. Rome: Associazione Geotecnica Italiana.
Kaiser, P.K. & C.McKay 1983. Development of rock mass and liner stresses during sinking of a shaft in clay shale. In C.O.Brawner (ed.), Stability in underground mining, p. 790-809. New York: Soc. Min. Engrs, Am. Inst. Min. Metall. Petrolm Engrs.
Kaiser, P.K., A.Guenot & N.R.Morgenstern 1985. Deformation of small tunnels-IV. Behaviour during failure. Int. J. Rock Mech. Min. Sci. & Geomech. 22:141-152.
Ladanyi, B. 1974. Use of the long-term strength concept in the determination of ground pressure on tunnel lining. In Advances in rock mechanics, proc. 3rd congr. Int. Soc. Rock Mech., Denver 2B, p. 1150-1156. Washington: Nat. Acad. Sci.
Ladanyi, B. 1976. Quasi-static expansion of a cylindrical cavity in rock. In Engineering applications of solid mechanics, proc. 3rd symp., 2, p. 219-240. Toronto: Can. Soc. Civ. Engrs & Univ. of Toronto.
Ladanyi, B. & D.E.Gill 1984. Tunnel lining design in a creeping rock. In E.T.Brown & J.A.Hudson (eds), Design and performance of underground excavations, p. 19-26. London: Brit. Geotech. Soc.
Lo, K.Y. 1978. Regional distribution of in situ horizontal stresses in rocks of southern Ontario. Can. Geotech. J. 15:371-381.
Lo, K.Y. & C.M.K.Yuen 1981. Design of tunnel lining in rock for long term time effects. Can. Geotech. J. 18:24-39.
Lo, K.Y., B.Lukajic & T.Ogawa 1984. Interpretation of stress-displacement measurements. In K.Y.Lo (ed.), Tunnelling in soil & rock, p. 128-155. New York: Am. Soc. Civ. Engrs.
Lo, K.Y. 1986. Advances in design and performance evaluation. In K.Y.Lo, J.H.L.Palmer, & C.M.K.Yuen (eds), Recent advances in Canadian underground technology, p. 5-46. Toronto: Tunnelling Assoc. Canada.
Lo, K.Y., B.H.Cooke & D.D.Dunbar 1986. Design of buried structures in squeezing rock in Toronto, Canada. In K.Y.Lo, J.H.L.Palmer, & C.M.K.Yuen (eds), Recent advances in Canadian underground technology, p. 159-198. Toronto: Tunnelling Assoc. Canada.
Ogawa, T. (1986). Elasto-plastic, thermo-mechanical and three-dimensional problems in tunnelling. Ph.D. Thesis, University of Western Ontario.

HYDROELECTRIC PROJECTS

Benson, R.P., R.J.Conlon, A.H.Merritt, P.Joli-Coeur & D.U.Deere 1971. Rock mechanics at Churchill Falls. Proc. symp. on underground rock chambers, p. 407-486. New York: Am. Soc. Civ. Engrs.
Benson, R.P., T.W.Kierans & O.T.Sigvaldson 1970. In situ & induced stresses at the Churchill Falls underground power house, Labrador. Proc. 2nd congr. Int. Soc. Rock Mech., Belgrade, 2: 821-832.
Benson, R.P., D.H.MacDonald & D.R.McCreath 1974. The effect of foliation shear zones on underground construction in the Canadian Shield. In H.C.Patterson & E.D'Appolonia (eds), Proc. 2nd N. Am. rapid excavation and tunnelling conf., 1, p. 615-641. New: York: Soc. Min. Engrs, Am. Inst. Min. Metall. Petrolm Engrs.

Benson, R.P., K.K.Murphy & D.R.McCreath 1970. Modulus testing of rock at the Churchill Falls underground powerhouse, Labrador. Am. Soc. for Testing & Materials, Spec. Tech. Publ. No. 477, June, 89-116.

Eckenfelder, G.V. 1952. Spray hydro-electric power development. Engng J. 35:288-304.

Gagne, L.L. 1972. Controlled blasting techniques for the Churchill Falls underground complex. In K.S. Lane and L.A. Garfield (eds), Proc. 1st N. Am. rapid excavation and tunnelling conf., 1, p. 739-764. New York: Soc. Min. Engrs, Am. Inst. Min. Metall. Petrolm Engrs.

Huber, W.G. 1952. Alcan - British Columbia power project under construction. Civ. Engng., 22, p. 938-943. New York: Am. Soc. Civ. Engrs.

Huber, W.G. 1953. Complex excavation pattern cuts out underground powerhouse. Civ. Engng., 23, p. 396-401. New York: Am. Soc. Civ. Engrs.

Huber, W.G. 1953. Tunnels and underground penstocks require a million cubic yards of excavation (Alcan - British Columbia hydro project). Civ. Engng., 23, pp. 102-107. New York: Am. Soc. Civ. Engrs.

Humphreys, R.W. & P.A.Lang 1986. Rock support design in good quality rock; three case histories. In K.H.O.Saari (ed.), Proc. int. symp. on large rock caverns, Helsinki, vol.6, p. 97-107. Oxford: Pergamon.

Humphreys, R.W. & E.Hoek 1987. Rock engineering at Cat Arm hydroelectric development. Proc. waterpower '87 conf., Portland, Oregon. in press.

Imrie, A.S. & L.T.Jory 1969. Behaviour of the underground power house arch at the W.A.C. Bennett dam during excavation. Proc. 5th Canadian rock mech. symp. Toronto, p. 19-39. Ottawa: Mines Branch, Dept. Energy, Mines & Resources.

Imrie, A.A. & D.D.Campbell 1976. Engineering geology of the Mica underground power plant. In R.J.Robbins & R.J.Conlon (eds), Proc. '76 rapid excavation and tunnelling conf., Las Vegas, p. 534-569. New York: Soc. Min. Engrs, Am. Inst. Min. Metall. Petrolm Engrs.

Levay, J. & Z.Aziz 1978. Siting and excavation of the La Grande 2 spillway channel. Proc. Annual General Meeting of CANCOLD. Montreal.

Mamen, C. 1969. Rock work at Churchill Falls powerplant. Can. Min. J. 90(3):41-48.

Murphy, D.K. & J.Levay 1982. Rock Engineering: La Grande Complex, Quebec. J. Geotech. Engng. Div., Am. Soc. Civ. Engrs 108(GT8):1040-1055.

Wise, L.L. 1952. World's largest underground power station. Engng News-Record 149:31-36.

TUNNELLING THROUGH WESTERN CANADA'S MOUNTAINS

Gadsby, J. 1985. The Rogers Pass Tunnels. In Z.E.Eisenstein (ed.), Canadian Tunnelling, p. 13-22. Tunnelling Assoc. Canada. Vancouver: BiTech Publishers.

Benson, R.P., L.O.Hostland & R.G.Charlwood 1985. The Canadian National Railway tunnels in British Columbia. In Z.E.Eisenstein (ed.), Canadian Tunnelling, p. 23-36. Tunnelling Assoc. Canada. Vancouver: BiTech Publishers.

Kaiser, P.K. & A.D.Gale 1985. Evaluation of cost and empirical support design at B.C. Rail Tumbler Ridge tunnels. In Z.E.Eisenstein (ed.), Canadian Tunnelling, p. 77-106. Tunnelling Assoc. Canada. Vancouver: BiTech Publishers.

Klassen, M.J., C.H.MacKay, T.J.Morris & D.G.Wusyluk 1987. Engineering geological mapping and computer assisted data processing for tunnels at the Rogers Pass project, British Columbia. Proc. rapid excavation & tunnelling conf., New Orleans, p. 1309-1323. Littleton: Soc. Min. Engrs, Am. Inst. Min. Metall. Petrolm Engrs.

Knight, G.B. 1987. TBM excavation of the Rogers Pass railroad tunnel. Proc. rapid excavation & tunnelling conf., New Orleans, p. 673-683. Littleton: Soc. Min. Engrs, Am. Inst. Min. Metall. Petrolm Engrs.

Steward, A.P. 1987. Mount Macdonald tunnel. Proc. rapid excavation & tunnelling conf., New Orleans, p. 635-652. Littleton: Soc. Min. Engrs, Am. Inst. Min. Metall. Petrolm Engrs.

Tanaka, R.S. 1987. CP Rail tunnels in the Rogers Pass, British Columbia, Canada. Proc. rapid excavation & tunnelling conf., New Orleans, p. 1174-1184. Littleton: Soc. Min. Engrs, Am. Inst. Min. Metall. Petrolm Engrs.

MINING IN CANADA'S FAR NORTH

Leggatt, C.H. 1982. Polaris - world's most northerly mine. World Mining 35:946-951.

ROCK SLOPES

Brawner, C.O. 1986. Rock mechanics in surface mining - an update. In E.Ashworth (ed.), Research and engineering applications in rock masses, p. 49-66. Rotterdam: Balkema.

Brawner, C.O., P.F.Stacey & R.Stark 1975. Monitoring of the Hogarth Pit highwall, Steep Rock Mine, Atikokan, Ontario. Proc. 10th Canadian rock mech. symp., Queen's Univ. Kingston, p. 103-138. Kingston: Dept Min. Engng, Queen's Univ.

Calder, P.N. 1970. Slope stability in jointed rock. Bull. Can. Inst. Min. Metall. 63(697):586-590.

CANMET. 1977. Pit Slope Manual. Ten chapters with supplements, 27 volumes. CANMET: Dept. Energy, Mines and Resources Canada, Ottawa.

Choquet, P. & D.D.B.Tanon 1986. Nomograms for the assessment of toppling failure in rock slopes. In E.Ashworth (ed.), Research and engineering applications in rock masses. Rotterdam: Balkema.

Martin, D.C. 1987. Application of rock reinforcement and artificial support in surface mines. Proc. 6th cong. Int. Soc. Rock Mech., Montreal. in press.

Piteau, D.R., A.F.Stewart, D.C.Martin, & B.S.Trenholme 1985. A combined limit equilibrium and statistical analysis of wedges for design of high rock slopes. Proc. symp. behaviour of rock in engineering applications. ASCE Spring meeting, Denver.

Wyllie, D.C. 1980. Toppling rock slope failures. Examples of analysis and stabilization. Rock Mech. 13:89-98.

SHOTCRETE DEVELOPMENTS

Anon. 1987. Western Canada discovers the benefits of silica fume shotcrete. Concrete Construction. 37(6):525-528.

Little, T.E. 1985. An evaluation of steel fibre reinforced shotcrete for underground support. Can. Geotech. J. 22:501-507.

MacRae, A.M. 1972. Large aggregate shotcrete as tunnel support. Proc. 8th Can. rock mech. symp., Toronto, p. 251-264. Ottawa: Mines Branch, Dept. Energy, Mines & Resources.

Morgan, D.R. 1985. Steel fibre reinforced shotcrete in Western Canada. Geotech. News 3(1):24-25.

Morgan, D.R., N.McAskill, J.Neill & N.Duke 1987. Evaluation of silica fume shotcrete. Proc. CANMET/CSCE intnl. workshop on condensed silica fume in concrete, Montreal. Ottawa: Mines Branch, Dept. Energy, Mines & Resources.

Dr. J.E. Udd thanked Prof. E. Hoek for his well researched and well delivered presentation and as a token of appreciation presented him with an Eskimo carving from the organizing committee. Dr. Herget then closed the opening ceremony and asked the chairman of Session I, Dr. S.A.G. Bjurström (Sweden) to reconvene the delegates for the technical sessions after the coffee break.

Theme 1
Fluid flow and waste isolation in rock masses

Site investigation and rock mass characterization; subsurface water flow; contaminant transport; waste isolation; seepage control; thermal stress and excavation damage effects; monitoring and back-analysis

MONDAY, AUGUST 31, 1987

Chairman: S.A.G. Bjurström (Sweden)

Session Secretary: R. Chapuis (Canada)

Speaker: T.W. Doe (U.S.A.)

Design of Borehole Testing Programs for Waste Disposal Sites in Crystalline Rock.

(co-authored by
J. Osner (USA)
M. Kenrick (USA)
J. Geier (USA)
S. Warner (USA))

Abstract: The design of well testing programs for low permeability, fractured rock has two elements -- the selection of the well tests methods and the design of the sampling program.

Well testing methods have been developed somewhat independently from two sources, one being civil and mining engineering, the other petroleum engineering and hydrogeology. Civil-engineering well testing has focused on the constant-pressure injection test, or Lugeon test, which assumes steady flow. Petroleum and ground water testing has used mostly constant-rate withdrawal methods for higher conductivity rocks and pulse tests for lower conductivities. These tests use transient flow methods, which include storage terms, storage reflecting the compressibility and deformability of the water and rock.

The use of steady rather than transient flow methods does not result in major errors in the determination of hydraulic conductivity, however, transient methods potentially provide more information including compressibility effects in the well (wellbore storage), non-linear flow effects fracture deformation effects, influence of boundaries (such as intersections of fractures away from the hole, or termination at dead-ends), and the influence of drilling on conductivities near the borehole. Solutions for these factors have been well developed for constant-rate tests and can be adapted for constant-pressure tests, where transient flow is analyzed. For lower conductivity to a wide range of conductivities, elimination of wellbore storage, and relative speed, especially where very low-range flow meters can be used.

There are two distinct approaches to designing the sampling programs. The first, discrete zone testing, selects specific intervals of importance. This approach is used when a specific zone, such as a fracture zone or aquifer, is the main concern in the testing. The second, fixed-interval-length testing (FIL), samples the hole in contiguous zones having a set length. FIL testing is used when the testing is meant to provide a statistical sample of the rock's hydrologic properties. The distribution of conductivity values from a FIL testing program reflects both the frequency of fractures and the distributions of conductivity. FIL test data can be analyzed to provide information on both.

1 INTRODUCTION

This paper develops an approach for hydrologic testing in crystalline rock. The term crystalline rock covers high grade metamorphic and plutonic rocks, however, the methods discussed in this paper are generally applicable to any rock having a very low-permeability matrix where the flow is primarily in a system of fractures.

In designing a well testing program for crystalline rock one must consider both the selection of the test method and the approach to sampling. Rocks where flow is primarily in fractures may be extremely heterogeneous in their hydrologic properties. Thus the test methods chosen must be capable of providing data over a broad range of hydraulic conductivities. The heterogeneity also means that the sampling program must be designed to directly test or infer the properties of the conduits in the rock mass whether or not they are intersected by the boreholes.

This paper is divided into four sections. The first is a discussion of basic terms and concepts. The major factors influencing well tests are defined both in hydrologic and mechanical terms. This section also considers the distinctions between porous rock and fractures in basic definitions of hydrologic properties.

The second section describes the four major methods of testing crystalline rock (constant-rate, constant-pressure, pressure-recovery, and pulse/slug) and compares them in terms of reliability over a range of permeabilities, rapidity, and the information that can be obtained from each.

The third section looks in detail at the constant-pressure injection test, and the various factors that influence it such as boundary effects, fracture deformation, non-linear flow, and damage near the wellbore.

The final section of this paper discusses some sampling strategies for crystalline rock, in particular looking at the inference of fracture spacing and transmissivity distributions from a series of well tests. The problem of channelized flow is also considered.

An emphasis on single-hole testing reflects the likelihood that the early screening of rock masses will be conducted with limited numbers of boreholes. The best use of drilling resources will likely require distributing boreholes over large areas, rather than concentrating them in the small areas appropriate for multiple-hole testing.

2 BASIC ISSUES IN LOW-PERMEABILITY WELL TESTING

This section discusses basic terms and concepts, with definitions for the major factors influencing well tests given in both hydrologic and mechanical terms. In defining hydrologic properties, the important distinctions between porous rock and fractures are emphasized.

2.1 The Diffusion Equation Applied to Well Tests

The diffusion equation is the fundamental basis for all well testing methods for saturated, laminar flow. Simply stated, most well test methods, including those discussed in this paper, are based on solutions of:

$$\nabla^2 h = \frac{S}{T}\frac{\partial h}{\partial t} \qquad (1)$$

where: h = hydraulic head [L]
t = time [T]
T = transmissivity $[L^2/T]$
S = storativity [-]
∇ = divergence operator.

This equation is the linear form; it can be readily transformed to either a radial or spherical form for other flow geometries. The two basic material parameters obtained from well testing are the transmissivity, T, which describes the flow capacity of the aquifer or conduits intersected by the test zone, and the storativity, S, which describes the storage capacity of the conduits.

2.2 Transmissivity, Hydraulic Conductivity and Permeability

The fundamental basis of laminar, saturated flow theory is Darcy's law, which states that flow velocity and flux are a linear function of hydraulic gradient:

$$v = K\,\partial h/\partial l \qquad (2)$$

where: v = velocity [L/T]
K = hydraulic conductivity [L/T]
l = length [L]

The hydraulic conductivity is a material property of both the porous rock and the fluid. K can be separated into its rock and fluid terms by:

$$K = \frac{k\rho g}{\mu} \qquad (3)$$

where: k = permeability $[L^2]$
μ = dynamic viscosity [M/LT]
g = gravitational acceleration $[L/T^2]$
ρ = fluid density $[M/L^3]$

The fluid terms, density and viscosity, are temperature dependent. Particular care must be taken in well tests to know the temperature conditions and keep them constant to avoid mis-estimation of rock properties.

In dealing with the flow capacity of a porous aquifer, the term transmissivity, T, is commonly introduced as:

$$T = Kb \qquad (4)$$

where: b = aquifer thickness [L]

Transmissivity describes the capacity of a conduit or aquifer to transmit water, while hydraulic conductivity describes the conductive properties of the material comprising in the conduit or aquifer.

Well tests measure transmissivity. Hydraulic conductivity is calculated from the transmissivity by taking b as the thickness of the aquifer or aquifers intersected by the well. For stratified aquifers, the value of b is simply the thickness of the water bearing stratum. For well tests in massive crystalline rock, b is not so clearly defined. Commonly, the thickness used for calculating hydraulic conductivity is simply the test zone length, despite the fact that only a small portion of the test zone is conducting fluid. Unless the test zone length is very long relative to the fracture spacing, such conductivity values are probably not representative of the rock mass as a whole. An accepted method of calculating rock mass conductivity does not exist. Carlsson et al. (1983) suggested using the geometric or harmonic mean of the conductivity measurements in a well as a measure of rock mass conductivity. More complicated estimations can be prepared using fracture network models (e.g., Long et al., 1982, Dershowitz, 1985).

The distinction between transmissivity and hydraulic conductivity is important in considering fracture flow. A single fracture may be viewed conceptually as a single aquifer or conduit. The transmissivity of the single fracture can be described similarly to a porous aquifer as:

$$T = K_f e \qquad (5)$$

where: K_f = fracture conductivity [L/T]
e = fracture aperture [L]

If the fracture is filled, K_f is the hydraulic conductivity of the filling material. If the fracture is not filled, then the fracture conductivity is a measure of the frictional drag of the flow along the fracture walls and the tortuosity of the flow paths within the fracture. The frictional aspect of fracture conductivity may be idealized by the so-called cubic law, which describes flow between parallel plates:

$$Q \propto T = \frac{\rho g e^3}{12\mu} \qquad (6)$$

where: Q = flowrate $[L^3/T]$

This cubic equation gives the largest possible transmissivity a fracture with a given aperture can have. Single open fractures are commonly not ideal parallel plates, and the measured transmissivities usually give apparent apertures that are much less than the actual wall separations of the fractures. The distinctions between the definitions of K, S, and T for fractured and porous media are shown schematically in Figure 1.

2.3 Storativity

In a porous aquifer, the storativity describes the volume of fluid per unit area of aquifer that is added or removed from storage for a unit change in hydraulic head. Storativity is a dimensionless lumped parameter originating from the analagy between grondwater flow and thermal diffusion (Theis 1935). In saturated, confined flow, the storage has its source in the compressibility of the rock and fluid. For porous media the storativity is related to the rock and water compressibilities by:

$$S = \rho g b(C_r + nC_f) \qquad (7)$$

where: C_r = rock compressibility $[LT^2/M]$
C_f = fluid compressibility $[LT^2/M]$
n = porosity [-]

For a single fracture, this equation may be rewritten as (Doe and Osnes, 1985):

$$S = \rho g(\frac{1}{k_n} + eC_f) \qquad (8)$$

where: k_n = fracture normal stiffness $[M/L^2T^2]$

As with a porous medium, the storativity has two components — one due to fluid compressibility, and one due to aperture changes. In porous media, the rock compressibility component is usually much smaller than the fluid component and is generally neglected. Single fractures, on the other hand, often have very small fluid volumes and the stiffness component may then be dominant.

Deformation presents a problem in these definitions Specifically, in the derivation of the diffusion equation, pressure-dependent permeabilty terms are neglected and permeability is assumed constant (Snow, 1968). Thus there is an inherent inconsistency in the definition of fracture storativity: storage must be due primarily to aperture changes, but permeability varies with the square of the aperture.

Storativity is a property of an aquifer or conduit. The specific storage, obtained by dividing the storativity by the aquifer thickness, is a property of the material contained in the conduit. For unfilled single fractures, specific storage, which is a material property, has no physical significance.

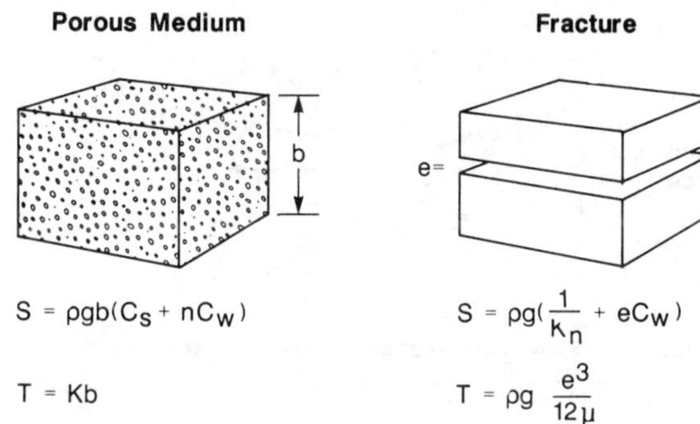

Figure 1. Illustration of transmissivity, permeability (hydraulic conductivity) and storativity in porous and fractured rock.

2.4 Wellbore Storage

The primary purpose of well testing is to determine the properties of the reservoir of fluid in the rock surrounding the borehole. The transient well test response is caused by the removal or addition of fluid to the storage of this reservoir. However, the wellbore and the test equipment also serve as fluid reservoirs. These additional reservoirs have storage properties which under some circumstances can dominate the transient behavior of the well test. As the wellbore storage has nothing to do with the rock reservoir properties, it is important to recognize its effect so that wellbore behavior is not misinterpreted as rock reservoir behavior. It is also important to design tests to limit or eliminate the wellbore storage effect.

Wellbore storage was examined in the hydrogeologic literature by Papadopolus and Cooper (1967) and may be defined as the volume of fluid added to (or released from) storage in the wellbore for a unit change in pressure (Earlougher, 1977):

$$C = \frac{\Delta V}{\Delta P} \tag{9}$$

The terms used to mathematically describe wellbore storage are analogous to the terms that make up the storage coefficient for a porous aquifer. These terms include:

- C_1, the change in storage for a change in water level
- C_2, the change in storage due to fluid compressibility, and
- C_3, the change in storage due to deformation of the wellbore, tubing, and other well test equipment.

The wellbore storage is the sum of the components, or:

$$C = C_1 + C_2 + C_3 \tag{10}$$

The water-level term, C_1 occurs only in open hole tests, such as constant-rate withdrawal tests (or pump tests). The open-hole case is analogous to unconfined aquifers where storage is primarily the result of water level fluctuations. For an open-hole test in an unconfined aquifer, C_1 will be considerably greater than the other terms, hence C_2 and C_3 are generally neglected. The wellbore storage term, C_1 is simply the volume of water stored in a unit length of hole divided by the pressure per unit length of head, or:

$$C_1 = A/\rho g \tag{11}$$

where: A = cross-sectional area over which the fluid level fluctuates [L^2]

For tests where there is no water-level fluctuation, the wellbore storage consists only of the compressibility terms, C_2 and C_3. Most packer tests, especially pressurized injection tests using packers, fall into this category. In these tests the wellbore storage is analogous to the storativity of a confined aquifer. The fluid compressibility term, C_2, may be defined as:

$$C_2 = V_f C_f \tag{12}$$

where: V_f = the volume of water pressurized (or depressurized) in the wellbore [L^3]

The last wellbore-storage term C_3, reflects the deformability of the borehole wall and the test equipment. For this discussion, we will use the terms equipment compliance and wellbore compliance, where compliance is simply a change in volume for a given change in pressure. The equipment compliance includes the deformation of the packers, of the tubing conducting fluid downhole, and of any other component of the test system. The wellbore compliance is the deformation of the open borehole in the test zone. The term C_3 is the sum of the wellbore and the equipment compliances. Compliance values for test tubing and the wellbore may be calculated if the elastic properties of the materials are well known. The compliance of non-rigid tubing and packers should be measured directly in a controlled laboratory test.

2.5 Skin Effects

A wellbore skin is a thin zone near the wellbore where the permeability of the rock has been affected by drilling or by treatments to enhance productivity as shown in Figure 2 (Earlougher, 1977; Agarwal et al., 1970; Almen et al., 1986). Invasion of drilling muds or cuttings into the formation results in a positive skin effect. An increase in permeability near the borehole due to drilling damage or washing out of pore fillings appears as a negative skin. Natural effects associated with aquifer or fracture heterogeneities may also appear as skin effects. In a highly channelized fracture the location of the borehole in an island of low permeability may appear as a positive skin in the well test, while intersection of a short fracture or a channel within a fracture may appear as a negative skin.

The skin may be viewed as a zone of altered conductivity, K_s, extending a radius of r_s from the hole, for which the skin factor, ξ is given as (Earlougher, 1977):

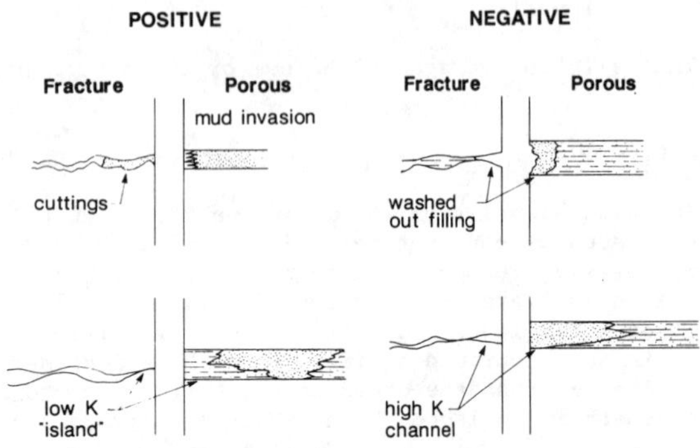

Figure 2. Illustration of skin effects in porous rock and fractures.

$$\xi = (K/K_s - 1) \ln (r_s/r_w) \tag{13}$$

where: K = formation hydraulic conductivity
K_s = skin hydraulic conductivity
r_w = wellbore radius

Estimation of skin factors in injection and withdrawal tests is discussed in Section 3.

2.6 Thermal Effects in Well Tests

Temperature variations may have a significant effect on test results, especially in low-permeability rocks. Thermal effects arise due to the temperature dependence of fluid viscosity and density and the effects of thermal expansion.

Temperature changes affect the viscosity and density of the fluid, which in turn change the hydraulic conductivity. Variable temperature in the rock surrounding the well can introduce variations in hydraulic conductivity, even where the rock is completely homogeneous. Unless recognized, temperature variations can be misinterpreted as permeability changes with distance from the well.

Thermal expansion may be especially significant in low-permeability well testing. For an injection test or a pressure-pulse test, where the test system is shut in, any temperature change can result in increased pressure of the fluid. The pressure change is given for a closed system by:

$$P = \frac{\Delta\theta\alpha}{C_f} \tag{14}$$

where: $\Delta\theta$ = change in temperature
α = thermal expansion coefficient of the fluid

Minor changes in temperature can cause large changes in pressure due to the relatively low compressibility and high thermal expansion coefficient of water as noted by Grisak et al. (1985). In a constant-pressure injection test at low flowrates, the volume increase due to thermal expansion can cause an underestimation of the flowrate, or even result in a flow from the hole. Temperature control is clearly an important factor in low permeability well test design.

2.7 Flow Geometry

Classical interpretation methods for well tests generally consider homogeneous aquifers (or fractures) and very simple geometries. The three major flow geometries are radial flow, linear flow, and spherical flow (Figure 3). In this paper we will primarily consider radial flow, as this is the flow that will occur for fractures or aquifers that are normal to the borehole. The hydraulic response of a given test may exhibit one or all of these forms

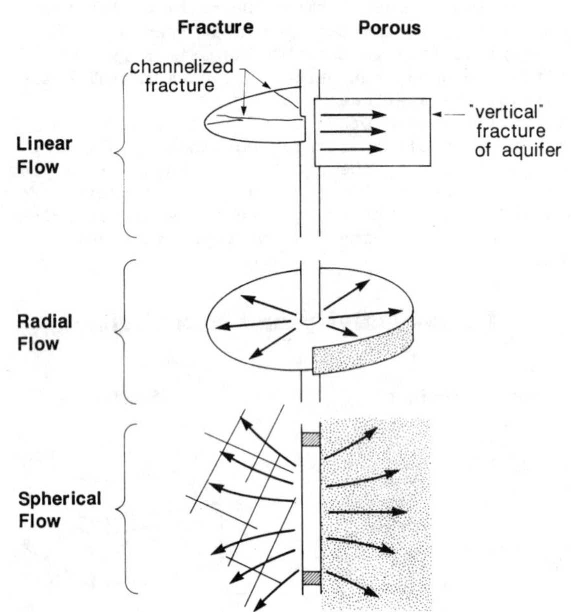

Figure 3. Flow geometries in well tests.

during a test, the usual progression being from lower dimension (linear) to higher dimension (spherical).

Linear flow is some interest as this case may occur when fractures in the well are parallel to the borehole (i.e. vertical fractures in vertical holes) or where the flow is resticted to a few channels. Karasaki has developed an approach that considers linear flow near the well and radial flow in a porous-equivalent network away from the well. This approach may a good analog for channel flow. Almen et al (1986) report that most tests in a research program on hydrologic methods showed evidence of radial or linear, and not spherical flow. A good discussion of flow-geometry effects on constant-rate tests is contained in Ershagi and Woodbury (1985).

2.8 Outer Boundary Effects

In addition to the inner boundary effects (such as skin factor and wellbore storage), a well test may be influenced by outer boundaries. Two basic types of outer boundaries exist: a closed, or no-flow boundary, and an open or constant-pressure boundary (Figure 4). Closed boundaries occur where the fracture terminates in impermeable material. Open or constant-pressure boundaries indicate a connection to a reservoir that is sufficiently extensive and conductive that the well test has no effect on its pressure. A constant-pressure boundary in crystalline rock would include any fracture that is significantly more conductive and extensive than the fracture intersecting the well, including major fracture zones. Man-made features can act as constant-pressure boundaries. Flow to an underground opening or back to the wellbore itself may appear as a constant-pressure boundary.

Leakage into surrounding rock may appear as a boundary effect if the rock matrix permeability is sufficiently high. Hantush (1956, 1959) developed solutions for leakage in both constant-rate and constant-pressure tests. The Hantush solutions did not consider the storativity of the rock surrounding the aquifer; this factor was added by Neuman and Witherspoon (1969). In crystalline rock, leakage to or from the intact rock may occur despite the low rock conductivity if the fracture surface area is sufficiently large.

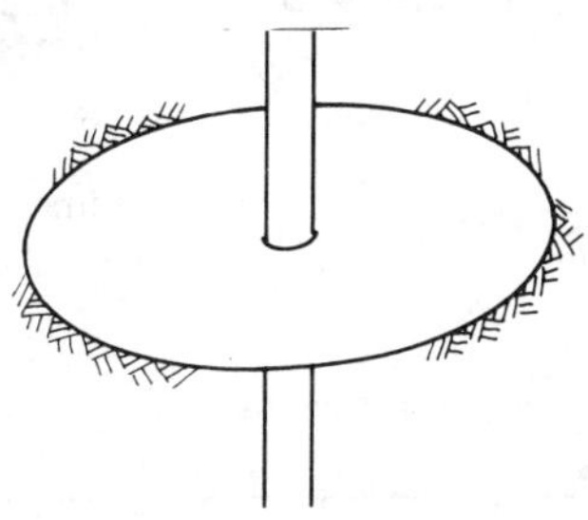

No Flow or Closed Boundary

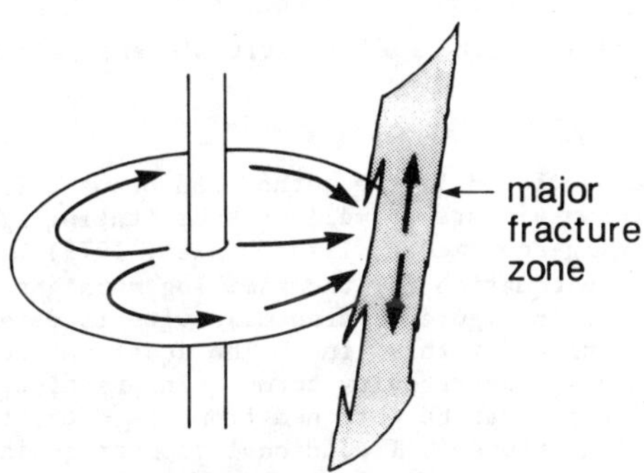

Open or Constant Pressure Boundary

Figure 4. Illustration of outer boundary types.

3 SINGLE-HOLE WELL TESTING METHODS

This section describes the basic types of single-hole well tests that are commonly used in crystalline rock, and discusses their relative utilities. The main features one desires in a well test method for crystalline rock are (1) the ability to test a wide range of conductivities and (2) relative speed.

The typical range of conductivities encountered in crystalline rock may range from high values of about 10^{-5} m/s to low values near 10^{-12} m/s (Brace, 1980). One does not necessarily need to be able test the full range of conductivity values — the higher conductivity zones are clearly more important than the lower for transport problems. However, if one wishes to formally determine the distribution of conductivity values for statistical analyses, the method should be able to measure below the median and preferably as low as the 25th percentile of conductivities.

As to time considerations, clearly the more rapidly a test can be completed the better. If one is testing a large number of zones to assess variability in a rock mass, the preferred test should be capable of testing the desired low conductivities within one day.

3.1 Definitions of Test Types

A single-hole test measures the hydraulic properties of the rock by observing the well's response to a pressure perturbation. In this paper we will consider four methods of obtaining hydraulic data:

- constant-rate tests
- constant-pressure tests
- pulse tests
- recovery analyses.

Typical variations of pressure and flow versus time for each of these tests are shown in Figure 5.

Pressure and Flow Versus Time for Various Tests

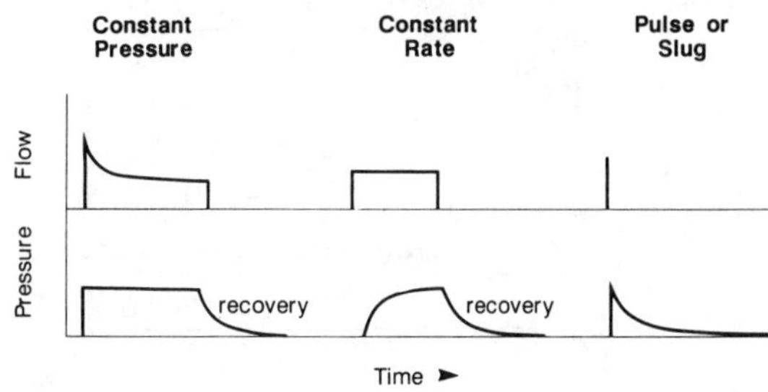

Figure 5. Generalized pressure and flow behavior of constant-rate, constant-pressure, and pulse/slug tests.

Constant-rate tests perturb a well by injecting or withdrawing fluid for a period of time at a constant flowrate. The hydraulic characteristics of the rock are evaluated from the pressure changes in the well.

Constant-pressure tests hold the pressure in the well constant and use the flowrate into or out of the well to determine the hydraulic characteristics.

Recovery methods use the pressure changes in the well after a constant-pressure or constant-rate perturbation ceases and the well returns to the static formation pressures. Recovery methods employ essentially the same test theory as constant-rate tests.

Pulse tests involve an instantaneous pressure change in the well, and the hydraulic characteristics of the rock are determined from the pressure recovery.

The test system is referred to as closed if the test zone is isolated from free fluid surfaces in the well. All injection tests, either constant pressure or constant rate, are closed system tests provided the injection pump is connected directly to the test zone. Withdrawal tests are not commonly run in a closed-system configuration unless the well is naturally flowing, that is, the total head in the aquifer is at a higher elevation than the point of fluid withdrawal.

Pulse tests may be run in either an open-system or a closed-system configuration. In the former case the test is known as a "slug" test; the latter is a "pressure-pulse" or simply "pulse" test.

The major difference between open- and closed-system tests is the nature of the wellbore storage. In open-system tests, the wellbore storage is dominated by the open-hole water-level change, and water compressibility effects are secondary. In closed-system tests, the wellbore storage has only water and equipment compressibility terms, which are much smaller than the open-hole water-level effect.

Constant-rate tests may be run in either closed or open configuration. Traditional pump tests (constant-rate withdrawal tests) are nearly always run in an open system. Constant-pressure tests are commonly run in a closed-system configuration as injection tests or as withdrawal tests in flowing wells.

3.2 Constant-Pressure Injection Test

The constant-pressure injection test, also known as the pressure test, the packer test, and the Lugeon test, is extensively used in engineering applications for evaluating the transmissivity of poorly conductive crystalline rock. The analysis of test data may utilize one of several analytical methods including the traditional steady-state flowrate approach, the transient pressure-flowrate type curve approach (Jacob and Lohman, 1952; Hantush, 1959), or the transient flowrate-pressure recovery approach (Uraiet and Raghavan, 1980; Ehlig-Economides and Ramey, 1981).

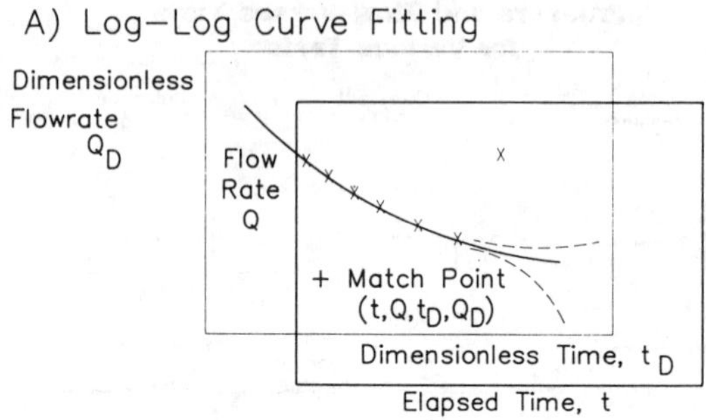

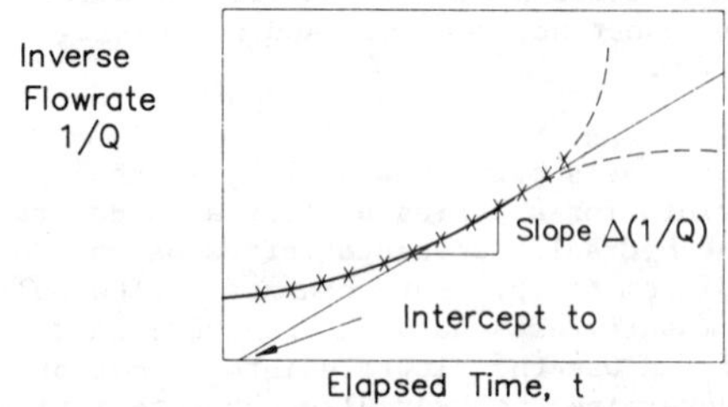

Figure 6. Illustration of logarithmic type curve matching and semi-logarthmic analysis.

Constant-pressure tests are analyzed in engineering practice using the observed steady flowrate to give transmissivity:

$$T = C_s Q/\Delta h \tag{15}$$

where: C_s = shape factor which varies depending on the assumed flow geometry.

Ziegler (1976) and Hoek and Bray (1974) have compiled lists of common shape factors. One commonly used shape factor is based on the steady-flow equation for radial, laminar flow with a constant-pressure boundary at a radius, R, from the hole, or:

$$C_s = \frac{1}{2\pi} \ln(R/r_w) \tag{16}$$

The radius to the boundary (or radius of influence) is never known very precisely. However, since the term appears as a logarithmic variable, large variations in R result in only small changes in T. For example, assuming a wellbore radius, r_w, of 38 mm, varying the value of R from 20 to 500 m results in calculated transmissivities which differ by only about 1/3. This error is small compared to the large range of variation for transmissivity measurements in fractured rock.

The transient flow solution for constant-pressure tests presented by Jacob and Lohman (1952) for infinite aquifers and later expanded by Hantush (1959), can be described by:

$$Q = 2\pi \Delta h T Q_D(t_D) \tag{17}$$

and

$$t_D = \frac{Tt}{S r_{w^2}} \tag{18}$$

where: $Q_D(t_D)$ = dimensionless flowrate
t_D = dimensionless time.

The logarithmic plot of $Q_D(t_D)$ is known as a type curve.

The analysis of transient data may be performed using type curves or a semi-log straight-line approximation (Figure 6). The application of the semi-log method for constant-rate tests is discussed

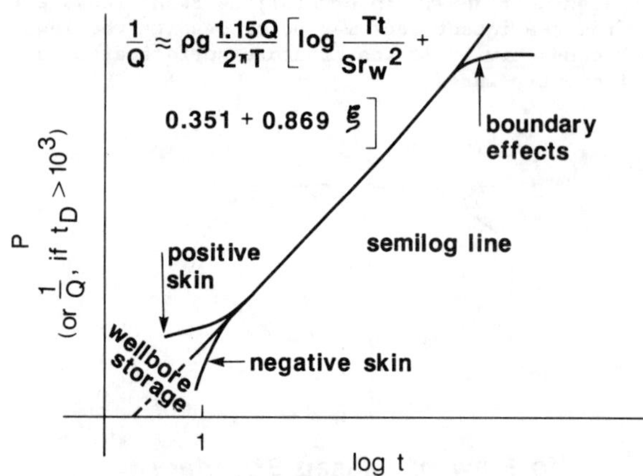

Figure 7. The semi-logarithmic analysis method.

in Section 3.3; the method can be used for constant-pressure tests as well by substituting 1/Q for Δh (Ehlig-Economides, 1979; Uraiet, 1979).

The equation for the semi-log straight line is given in Figure 7; transmissivity is determined from the slope of this line. The equation includes both skin and storativity terms. In practice, only one of these can be obtained from the data, the other being assumed. Traditional geohydrologic practise is to assume no skin factor ($\xi = 0$) and thus obtain storativity. Conversely, petroleum reservoir engineers calculate the skin factor assuming a storativity value based on porosity and fluid compressibility.

In comparing steady flow and transient flow calculations, Doe and Remer (1981) have concluded that the errors associated with using steady flow assumptions to calculate T are relatively small.

The lower limit of transmissivity which can be evaluated by the constant-pressure injection test is dependent upon the lowest flowrate which can be measured (see Section 5.3). The lowest-range flowmeters can resolve about 0.1 mL/min (Almen et al., 1986). For a head of 100 m and an assumed radius of influence of 30 m, this translates into a lower limit for transmissivity measurement of about 10^{-10} m²/s. The upper limit is usually governed by frictional losses due to the limited capacity of equipment, however, these can usually be reduced by applying a smaller head difference.

3.3 Constant-Flowrate Tests (Including Recovery Analysis)

Constant-rate testing is the classic pump test of hydrogeology and the basic test used in petroleum reservoir engineering. The constant-rate test involves injecting or producing the well at a constant flowrate. The hydraulic characteristics of the formation are determined from the transient pressure response. The standard methods of analyzing constant-rate tests and pressure recovery after such tests are well documented in Earlougher (1977) and Kruseman and De Ridder (1979).

The basic solution to radial flow due to constant-pressure production is the exponential integral solution (Ei). If we define dimensionless time as in Equation 18, the dimensionless pressure in the well is given by:

$$p_D = -Ei(-1/4t_D)/2 \quad (19)$$

and

$$p_D = 2\pi T\Delta h/Q \quad (20)$$

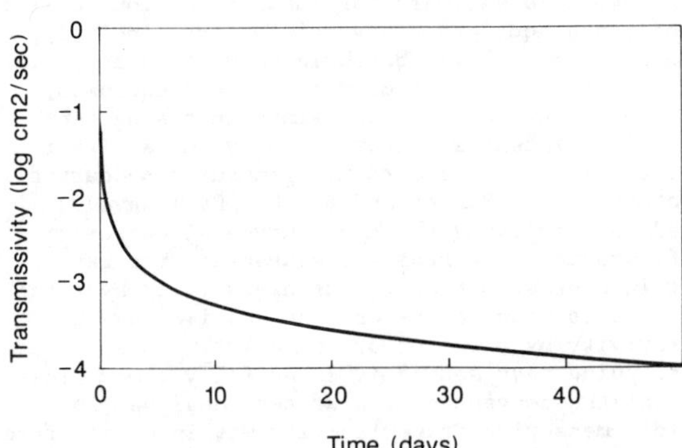

Figure 8. Duration of wellbore storage effects in an open system constant-rate test.

By taking advantage of the logarithmic approximation to the exponential integral for large values of t_D, this reduces to:

$$p_D = (0.809 + \ln t_D)/2 \quad (21)$$

Equation (19) is the familiar Theis curve to which log-log plots of pressure-change versus time may be matched (Theis, 1935). The exponential form gives the well-known semi-log straight line in a plot of pressure versus log-time (Figure 7). Strictly speaking, the Theis curve is for an infinitesimal-diameter well; modifications for finite-diameter wells are given in Mueller and Witherspoon (1965).

The pressure recovery after pumping is modeled by superimposing and equal source of opposite sign (Earlougher, 1977) and modifying the logarithmic approximation, Equation 21, by using "Horner" time, t_H:

$$t_H = \frac{t_p + \Delta t}{\Delta t} \quad (22)$$

where: t_p = pumping time
Δt = recovery time

The major limitation on using constant-flowrate tests and pressure recovery tests in low-permeability rock is the wellbore storage effect, which masks the early-time rock-mass response. Wellbore storage effects are progressively more acute for lower-conductivity rock.

Wellbore storage effects may be recognized in early-time data from a unit-slope straight line on a log-log plot of pressure and time. The time required to overcome wellbore storage effects is the major factor controlling the lower limit of transmissivity resolution from constant-rate tests. The duration of wellbore storage depends strongly on whether a closed-system or an open-system test is performed. In dimensionless time, the duration of wellbore storage is approximately given by Earlougher (1977) as:

$$t_{wD} = (60 + 3.5\xi)C_D \quad (23)$$

where the dimensionless wellbore storage, C_D, is defined as:

$$C_D = \frac{C\pi g}{2\pi S r_w^2} \quad (24)$$

In terms of the wellbore-storage coefficient described in Section 2.4 and neglecting skin effects, this reduces to give the duration of wellbore storage:

$$t_w = 30C\rho g/\pi T \quad (25)$$

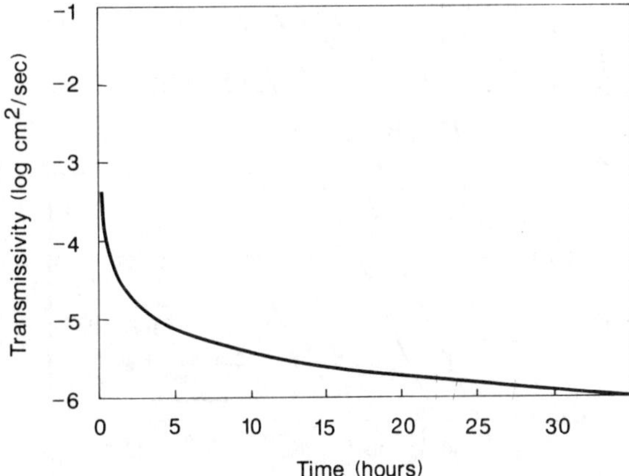

Figure 9. Duration of wellbore storage effects in a closed system constant-rate test.

As discussed in Section 2.4, the wellbore storage, C, for an open-system well test, is given by the factor C_1 which, when substituted into Equation 23 gives:

$$t_w = 30A/T \quad (26)$$

where: A = the cross-section area of the well over which the fluid level is fluctuating.

Figure 8 shows the duration of wellbore storage as a function of T for a well with an effective cross-sectional area of 45 cm^2. Nearly five days would be required to overcome wellbore storage for a test zone with a T of 10^{-7} m^2/s. Open-hole pump tests are clearly not viable for most low permeability well testing conditions.

The duration of wellbore storage for a closed system test (such as a pressurized constant-rate injection test) can be calculated using Equation 23 and wellbore-storage coefficients C_2 and C_3. Neglecting equipment compliance effects, the duration of wellbore-storage is then:

$$t = \frac{30V_w C_f \rho g}{\pi T} \quad (27)$$

The wellbore-storage term for the closed-system case is considerably smaller than that of the open-system test, thus leading to shorter wellbore-storage duration for the closed-system test. Inclusion of an equipment-compliance term would increase the wellbore-storage duration. As an example, the wellbore-storage duration for a test with a fluid volume of 0.3 m^3 (which corresponds to a 20-mm diameter tube string in a 1000-m hole), is shown in Figure 9. Closed-system tests can measure approximately two orders of magnitude lower transmissivity for comparable durations of wellbore storage. The lower limit of transmissivity resolution, based on wellbore-storage effects can be improved further by reducing the volume of fluid in the test system.

3.4 Pulse and Slug Tests

The slug test was first introduced to evaluate the transmissivity of an aquifer by observing the water level response in an open well resulting from the instantaneous change in head in the well. Modifications to account for the finite diameter of the well, as well as the development of type curves for estimating transmissivity from head-response data, were proposed subsequently (Cooper et al., 1967). More recent modifications to the analysis of

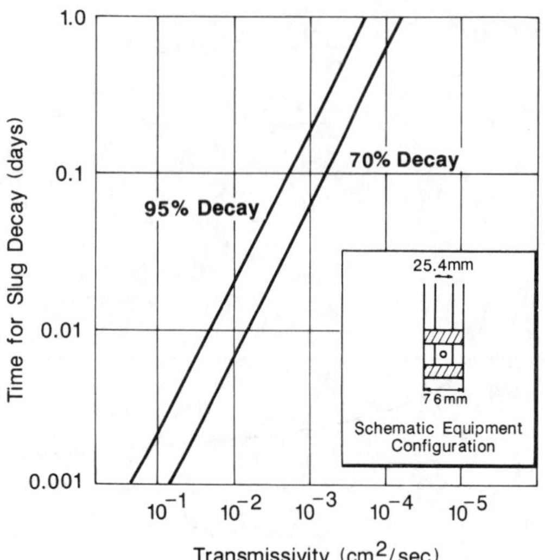

Figure 10. Time required for 90% and 70% decay for a slug test for the conditions shown in inset.

slug-test data have been introduced to account for the potential effect of finite-thickness skins on well response (Faust and Mercer, 1984), and to apply slug-test analysis to fissured aquifers (Black, 1985; Barker and Black, 1983). Karasaki (1986) has developed slug-test approaches for systems that are fracture-dominated near the well but behave as porous media at a distance. Moench and Hsieh (1985) have considered the effect of skin on slug tests. Pulse and slug tests have gained broad acceptance in recent years and have been extensively used in the Canadian (Davison and Simmons, 1983) and Swiss (Grisak et al., 1985) radioactive-waste programs.

The estimate of the storativity is often unrealistic because the perturbed volume of aquifer can be relatively small. Caution must also be taken when applying the standard homogeneous radial flow analysis to a fractured medium or to an aquifer where the ratio of horizontal to vertical permeability approaches unity (Papadopulos et al., 1973). In the case of a fractured or fissured aquifer, the quantitative formulation of the head response must take into account the head changes not only within the borehole, but also within both the rock matrix and the fissures (Black, 1985; Barker and Black, 1983).

Slug tests are usually of short duration (minutes to days) and are more frequently applied to formations in which the range of transmissivity is between about 10^{-5} to 10^{-9} m^2/s. For aquifers characterized by a much higher transmissivity, the slug tends to decay so rapidly that meaningful data are unobtainable. If the transmissivity is too small (e.g., low-permeability formations such as granite), the instantaneous head change may take an exceptionally long time to decay.

The percentage of slug decay considered necessary to obtain an accurate fit of the data to the standard type curves of Cooper et al. (1967) (Figure 10) is 70 percent (Wilson et al., 1979). No indication is made as to the percentage of decay required to obtain a fit to the modified "fissure" type curves of Black (1984), although the 70-percent mark may be considered as adequate. Figure 10 shows the time span necessary for a 70-percent slug decay in a homogeneous aquifer as a function of aquifer transmissivity. For comparison, the 95-percent time curve is also shown. The curves have been drawn for a well configuration consisting of a 25-mm diameter casing in a 76-mm diameter borehole. A reasonable time span of 1 minute to 1 day is also considered. Under these conditions, an aquifer zone with a transmissivity of 10^{-8} m^2/s and a storativity of 10^{-5} could be tested in just under 24 hours if the 70 percent decay curve is considered. For a test zone of 10 m, the resultant hydraulic conductivity is 10^{-9} m/s.

The above results are for ideal conditions only in homogeneous aquifers. Several factors, including nonhomogeneous media, borehole conditions and skin effects, temperature fluctuations, and equipment compliance may induce uncertainty into slug test analysis. Potential error in analyzing a fissured aquifer with the standard homogeneous type curves is also possible. Barker and Black (1983) have surmised that applying the type curves of Cooper et al. (1967) to the data from a fissured aquifer may result in overestimating the transmissivity by a factor of 2 to 3 and over- or under-estimating the storativity by a factor of up to 10^5.

The pulse test method is essentially a modification of the conventional slug test designed for rapidly measuring transmissivity within single fractures of small aperture (Wang et al., 1978). Instead of monitoring the slow decline of head in an open standpipe resulting from an instantaneous change in water level, a small volume of water is pressurized in a shut-in interval, and the pressure decay (or recovery if the interval is underpressurized) is observed with time (Bredehoeft and Papadopulos, 1980). The mathematical formulation of the pulse-test analysis using a porous media assumption is similar to that of the slug test and only differs in that the rate at which water flows from the well is equivalent to the rate at which the initially pressurized volume of water expands as the pressure dissipates (Grisak et al., 1985). The same type curves used to estimate transmissivity and storativity for slug-test data are used for the analysis of pulse-test data with the exception that these hydrologic parameters must now incorporate terms for the compressibility and the density of water at the formation temperature and pressure at which the test is conducted. Under the assumption of transient flow in fractured media, the continuity of flow velocities in the fracture and at the wellbore fracture/contact is used as a boundary condition at the wellbore radius (Wang et al., 1978).

If both the rock and the test equipment are assumed to be perfectly rigid, the pressure decay in the pulse test is a function of the hydraulic properties of the rock and the compressibility of the pressurized water stored in the wellbore (Wilson et al., 1979). Under ideal conditions, transmissivity can be determined after a 10 percent pressure decay in the initial wellbore pressure pulse. Figure 11 shows the range of formation transmissivity which can be determined over a test time of up to 1 day, given pressure decays of 10 percent to 90 percent at a formation temperature and pressure of 40°C and 100 kPa, respectively. The wellbore configuration for the curves shown consists of a borehole diameter of 76 mm over a pressurized section of 100 m (the test interval is not considered for transmissivity determination). For the 10 percent pressure decay, a transmissivity of less than 10^{-12} m^2/s can be tested (hydraulic conductivity of 10^{-13} m/s for a porous test interval of 10 meters), a value much lower than that which can be determined using the conventional slug test over the same time frame. Assuming the fracture media approach of Wang et al. (1978), a single fracture aperture of 1 μm can, in theory, be tested in under two hours.

An additional advantage to the pulse test is that later time response data may be used to evaluate fracture geometry beyond the wellbore. Theoretical decay curves constructed by Wang et al. (1978) show that for both impermeable and infinitely conductive boundaries, pressure decay curves are a function of fracture aperture, packer separation, fracture volume, and wellbore volume.

Several difficulties encountered in pulse test analysis need to be discussed. One difficulty concerns the assumption that the dissipation of a pressure pulse applied instantaneously occurs only due to the flow of fluid from the isolated test interval (Wilson et al., 1979). Since the pulse is realistically applied over a finite, though short, time interval, some energy transmitted during the pulse is spent on the compression of rock and downhole equipment. Neuzil (1982) has investigated the problem of accounting for equipment compressibility in the shut-in well, and has concluded that the compressibility of test equipment may in fact be significantly larger (by a factor of six) than the compressibility of water. The effect of using values for equipment compressibility rather than water compressibility in the calculation of transmissivity would result in proportionally larger values of transmissivity and would cause the decay curve of Figure 12 to shift to the right.

Further investigation into the effects of equipment compliance has shown that a large pressure build-up within a borehole over a relatively short time span could have a large packer-compliance effect, resulting in a pronounced effect in pressure response (Thackston et al., 1984) and ultimately in transmissivity and storativity determination. To some extent, however, the energy expended in compressing downhole equipment (and the rock matrix) is recoverable. This results in transmitting energy back into the fluid, thus sustaining high pressure for longer periods of time (decay curve shifts to left). Left uncorrected, such an effect would introduce errors to fracture aperture determination. Both Wilson et al. (1979) and Neuzil (1982) suggest pre-test analysis of potential equipment compliance effects to generate equipment-specific type curves.

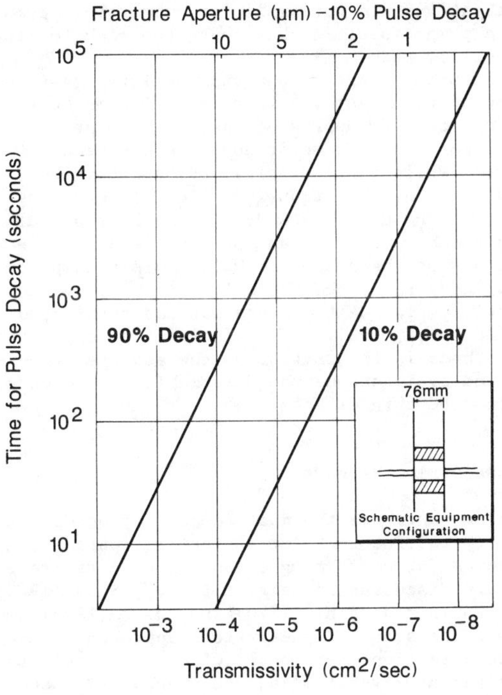

Figure 11. Time required for 90% and 10% decay for a pulse test for the conditions shown in inset.

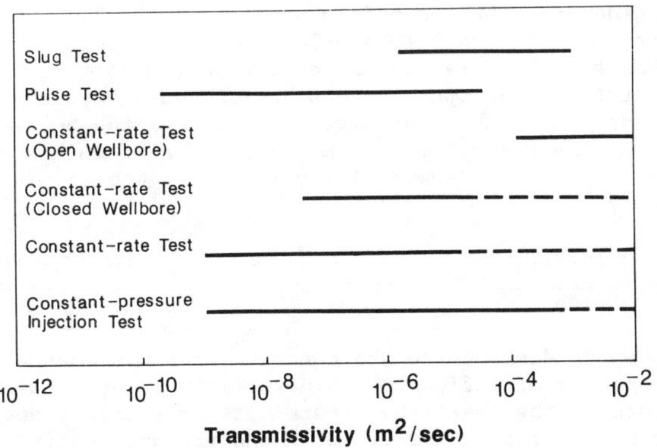

Figure 12. Practical range of transmissivities that can be measured by well test methods in crystalline rock.

3.5 Comparison of Test Methods

The general range of transmissivities which can be determined by the various test methods described previously is illustrated in Figure 12. The ranges of measurable transmissivities shown in the figure are based on an equivalent porous medium with a conservative storativity (10^{-5}). The effects of wellbore storage have been included for the constant-rate tests to more realistically portray the values of transmissivity which can be measured under specific time limitations.

The results shown are consistent with the ranges of hydraulic conductivities considered achievable for the various test methods as described by Wilson et al. (1979) and as determined by the Swedish site characterization study of Almen et al. (1986). The constant-pressure injection test and the pulse test are shown to be applicable over a wide range of transmissivities including those values below 10^{-8} m^2/s (typical for crystalline rock) while the constant-rate test and the slug test should not be used to test formations with transmissivities below about 10^{-5} to 10^{-8} m^2/s, respectively. The dashed lines of Figure 12 represent those ranges of transmissivity in which a higher storativity would allow the specific test to be applicable over larger ranges of transmissivity.

Based upon the above discussion, and in consideration of the hydraulic information which can be obtained from each test method, the constant-pressure injection test is an appropriate single hydrologic test method for low-permeability crystalline rocks. Advantages of this method include the wide range of transmissivity which may be measured, the ability to deduce flow geometry and boundary conditions, the lack of wellbore-storage effects, and the relatively large volume of rock which may be tested in a short time. The Swedish program (Almen et al., 1986) has also recommended the use of the pressure-recovery test after the injection phase to confirm the data obtained during the injection testing, although the recovery test may be limited in its application due to wellbore storage.

The constant-rate withdrawal test is inappropriate for quickly testing low-permeability rock because of the persistence of wellbore storage. The constant-rate injection test, though fundamentally similar to the constant-pressure injection test, is also affected significantly by wellbore storage. Additionally, unless pre-test transmissivity values have been accurately estimated for the formation of interest, an incorrect injection rate over an extended period of time may lead to fracture deformation or even hydraulic fracturing.

Pulse tests also meet the general criteria for use in crystalline rock, particularly if complemented by slug tests to accommodate the higher transmissivities that may be encountered. The disadvantages to pulse tests relative to constant-pressure tests is that they generally affect a smaller volume of rock and they may require more time in the lower-conductivity zones.

One advantage of pulse tests is relative ease of interpretation when the background hydraulic gradient is changing during a test. The assumed

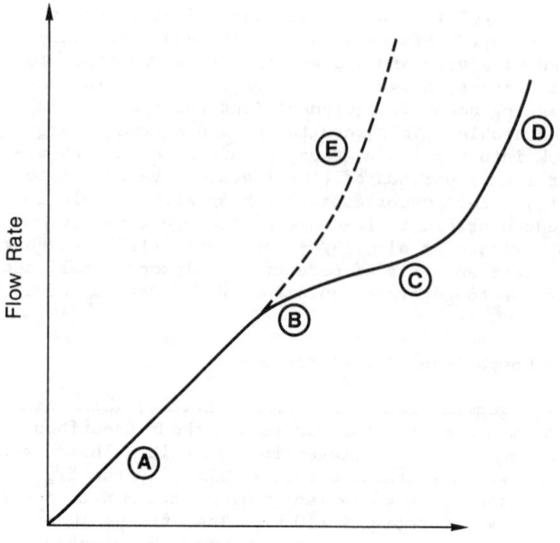

A. Laminar Flow
B. Turbulent Flow
C. Turbulent Flow Compensated by Fracture Deformation
D. Fracture Deformation Predominant
E. Very Deformable Rock

Figure 13. Pressure-flowrate relationship for an idealised seried of steady flow injection steps (after Louis and Maini, 1970).

initial condition of static pressure conditions is frequently violated in low-permeability well tests. The problem is especially noticed in testing boreholes from underground test facilities which undergo large pressure changes whenever the hole is open and allowed to flow. Another advantage of pulse tests is the small volume of water used, which minimizes the impacts on other borehole testing activities, such as geochmical sampling.

In comparing tests, it should be remembered that all of the methods described here are based on solutions of the diffusion equation and the differences are simply ones of assumed boundary conditions and the type of perturbation. Not surprisingly, comparisions generally show that various methods yield comparable transmissivity results provided the instrumentation used and the test durations chosen are appropriate for the transmissivity of the zone being tested (Almen, et. al., 1986). The main factors that discriminate between various methods are largely practical, that is the simplicity of the equipment, ease of test control and time required to perform the measurement.

4 CONSTANT PRESSURE INJECTION TESTS

This section discusses the application of the constant-pressure injection test in crystalline rock. Particular emphasis is placed on the independent development of test methods in civil engineeering, on one hand, and in hydrogeology and reservoir engineering, on the other.

4.1 Brief History

The theory of constant pressure injection testing has been developed somewhat independently in the civil engineering, hydrogeologic, and petroleum engineering literatures. The civil engineering approach can be traced to Lugeon (Serafim, 1968) and has been based on injection testing.

The hydrologic uses of constant-pressure tests have been for wells where the formation head is above the wellhead elevation, that is flowing wells (Jacob and Lohman, 1951; Hantush, 1959). In this case the constant pressure is atmospheric pressure.

Constant pressure methods have been studied in petroleum engineering partly for the cases where the formation head has been drawn down to the elevation of the pump — and thus the well is maintained at a constant head (Ehlig-Economides, 1979; Uraiet, 1979) — and partly because reservoir engineers have run out of other problems to solve.

The more extensive use of constant-pressure tests in civil engineering reflects the equipment commonly used and the types of rocks encountered. Traditional constant-rate testing developed largely because producers of underground fluids were dealing with permeable rocks which they produced by installing pumps that worked at a more-or-less constant rate. On the other hand, engineering site investigations, particularly in hard rocks, generally resorted to injection tests using the drill pumps which work at more-or-less constant pressure. Furthermore, standard exploration coreholes for site investigations are too small for submersible pumps and the pump capacities have been too high for low permeability rock.

The constant-pressure approaches differ most in the use of transient versus steady-flow methods. Civil engineering approaches have relied primarily on analyses that assume steady flow has been achieved, while other disciplines have developed methods that use the transient flowrate or pressure recovery. The use of steady-flow methods presumes the existence of a constant-pressure boundary, which may or may not be present. While this assumption may be inaccurate, the transmissivity and hydraulic conductivity calculations are not badly affected (Doe and Remer, 1980). What is lost by using steady flow methods is information on the storage properties (and by inference the deformability), boundary effects, and skin effects.

4.2 Steady-Flow Methods

Injection tests are the most common method of evaluating crystalline rock for engineering purposes. Procedures for performing steady-flow tests are extensively discussed in Zeigler (1976), Snow (1966, 1968), Louis and Maini (1970) and Banks (1972) among others. The standard packer test approach is to isolate a section of borehole with packers and inject water at several injection pressures. Each pressure step lasts several minutes until steady flow is presumed to have been achieved. For laminar flow, the steady flow rates should be linearly proportional to the injection pressure. Deviations from linearity can be interpreted either as turbulent flow or fracture deformation, depending on the nature of the deviation, as shown in Figure 13.

For the steady-flow analysis, the test zone is considered to be a confined aquifer with full penetration by the well (a single fracture in impermeable rock is a reasonable approximation of this condition). An open or constant-pressure boundary is assumed to exist at a radius, R, from the borehole. The steady flow is a consequence of this boundary. The transmissivity for this geometry is simply:

$$T = \frac{Q \ln(R/r_w)}{2\pi\Delta h} \tag{28}$$

The exact distance to the constant pressure boundary (or radius of influence) is never known, but since R occurs in the logarithmic term, its uncertainty does not have a strong influence on the transmissivity value.

Correction factors are readily available for other flow geometries, including spherical or elliptical rather than radial flow among others. These are summarized in Zeigler (1976) and Hoek and Bray (1981). Employing factors for alternate flow geometries generally does not affect the transmissivity calculation by more than about 30% (Zeigler, 1976) which is small considering the several orders of magnitude over which transmissivity values vary from fracture to fracture in a rock mass.

4.3 Analysis of Transient Tests

The transient-flow analysis of constant-pressure test data is analogous to the transient-pressure analysis used in conventional pump tests, the only difference being which parameter (flow rate or wellbore pressure) is held constant.

4.3.1 Background

Transient flow solutions have been developed independently in the hydrologic and reservoir-engineering literature.

In the hydrologic literature, Jacob and Lohman (1952) presented the transient flowrate curve for infinite aquifers as a method of analyzing data from flowing wells. They also noted (1) that the transient flowrate curve was approximately the inverse of the transient pressure (Theis) curve if flow is substituted for pressure, and (2) that pressure recovery after constant-pressure production could be analyzed the same way as for constant-rate production. Hantush (1959) presented the equations for constant-pressure tests with open (constant- pressure) or closed (no-flow) boundaries, as well as for leakage with an aquitard having no storage. More recently, Jaiswal and Chuhan (1978) have calculated curves for spherical flow from partially penetrating wells.

Transient-flow analyses can cover a wide range of effects including fracture deformation (Noorishad and Doe, 1982), non-linear flow (Elsworth and Doe, 1986), skin effects (Uraiet, 1979), and boundary effects.

What is notably absent in the list of considerations of effects is wellbore storage. Constant-pressure tests effectively have no wellbore storage because once the test is started there are no pressure changes in the wellbore or the equipment, hence the only storage effects are those in the aquifer or fracture. The mass of water in the wellbore and equipment is constant, barring temperature changes.

This absence of wellbore storage is the constant-pressure test's major advantage in testing low permeability rock. Unlike constant-rate tests, there is no period of time at the beginning of the test where the data are unusable.

In reservoir engineering the same problem of transient flow was attacked by Moore et al (1933) and Hurst (1934). Fetkovich (1980) used the closed-boundary case to describe the depletion of reservoirs in their later stages of production. Ehlig-Economides (1979) and Uraiet (1979) developed the theory of transient-rate analysis as well as analysis of pressure recovery after constant-pressure production. The former author used analytical solutions, the latter numerical. Both considered boundary effects and both concluded that pressure-recovery analysis was the same for constant-pressure tests as for constant-rate tests.

4.3.2 Boundary Effects

Boundary effects are potentially a major feature of transient rate curves. If one views single fractures as confined aquifers, then the terminations of the fractures should appear as hydraulic boundaries.

The termination of a fracture within solid rock effectively creates a no-flow boundary. No-flow or closed boundaries have distinctive forms in constant-pressure tests. As shown in Figure 14, the transient rate response follows the curve for an infinite aquifer until the pressure front from the injection reaches the edge of the fracture. At this point the flow should fall off and eventually decline to a relatively low rate reflecting leakage into the surrounding low-permeability rock. Type curves for closed conditions have been prepared using a numerical inversion of the Laplace-space solutions. The data are given in Appendix 1.

Open or constant-pressure boundaries are present when the fracture being injected at the well intersects another, more permeable fracture. As shown in Figure 15 this condition ultimately creates steady flow. As with the closed boundary case, the transient flowrate should decline along the curve for an infinite aquifer until the pressure front from the injection reaches the intersection. This condition is the steady flow often observed in constant-pressure tests.

The distance to the boundary is the elusive radius of influence used in the steady-flow calculation. This length can be estimated from matching the observed transient-rate data to the type curve. The time required to observe boundary effects can be estimated by (Uraiet, 1979)

$$t_D = 0.809\ R_D^{1.99} \tag{29}$$

where: $t_D = Tt/Sr_w^2$
$R_D = R/r_w$

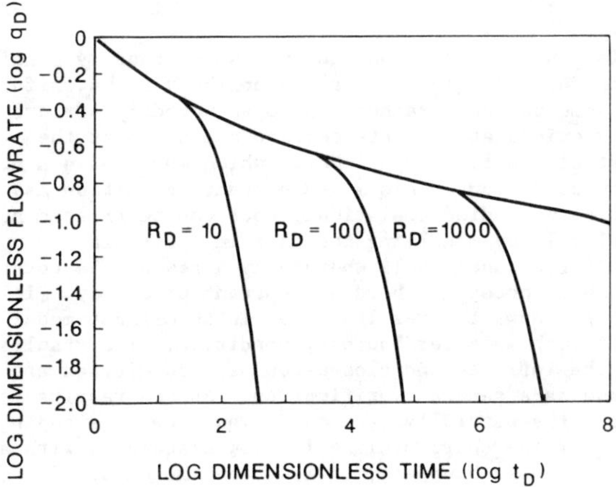

Figure 14. Transient flowrate type curves no-flow boundaries.

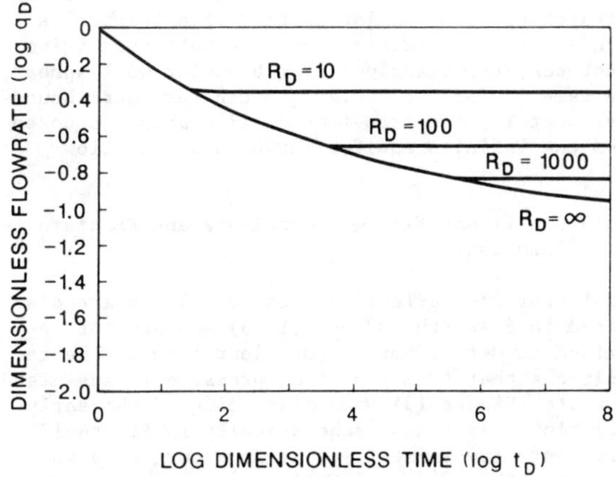

Figure 15. Transient flowrate type curves for constant-pressure (open) boundaries.

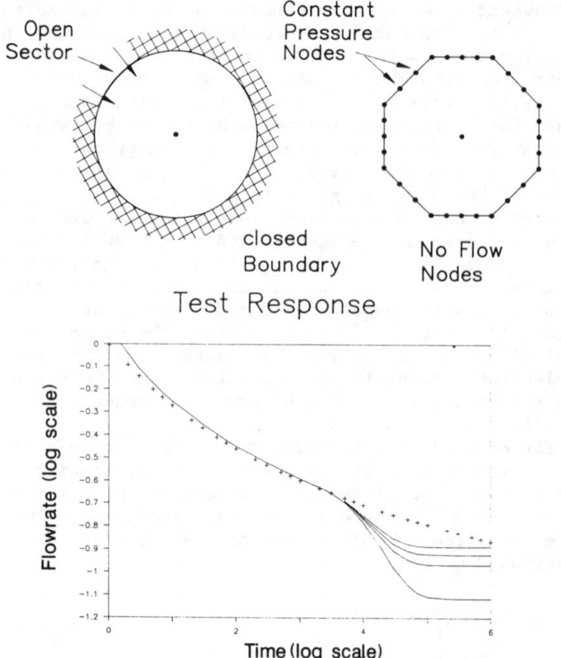

Figure 16. Transient flowrate curve for a single fracture with a mixed open-closed outer boundary.

In practice, the constant-pressure boundary condition for a single fracture is unlikely to be uniform at some radius. Rather, the open boundary strictly only exists at the intersection and not over the rest of fracture's periphery, which would have a no-flow condition. This mixed boundary condition has not been studied analytically but can be treated numerically by modelling the flow in a circular aquifer around a well where only a few of the outer boundary nodes are held at constant pressure. Figure 16 shows the results of a finite-element model with such an outer boundary condition. The results of the infinite and closed-boundary conditions are shown as a general verification. As the results show, the partially open condition shows the beginning of the sharp decline in flow associated with a closed outer boundary, followed by the onset of steady flow due to the constant pressure boundary.

In these examples, the number of boundary nodes held at constant pressure was varied between one and four, representing at most only 15 percent of the periphery. Each case reaches steady flow at progressively higher rates, with the effect of the closed-boundary being relatively subtle. This suggests that, if any major portion of a fracture's boundaries are intersections with more conductive fractures, the transient-flow behavior will appear the same as for a circular constant-pressure boundary; that is, the transient decline should proceed from the infinite aquifer curve to steady flow.

4.3.3 Non-linear Flow — Turbulence and Fracture Deformation

Non-linear flow effects due to turbulence are discussed in Elsworth and Doe (1986) and are not described in detail here. Turbulent flow may be recognized either from a plot of pressure versus steady flowrate (Figure 13) or a flattening of the early-time flowrates (shown schematically in Figure 17). Turbulent flow is most likely to occur early in a test when the pressure gradients near the borehole are greatest; flow may become laminar later in the test as the gradients lessen. Elsworth and Doe (1986) also note that the distance to a constant-pressure boundary may be inferred from the results of a stepped-pressure test where both laminar and turbulent stages can be recognized.

Fracture-deformation effects also may be recognized in a pressure-flowrate plot from a series of steady flow steps (Figures 13 and 18). Alternately, fracture deformation may have recognizable effects in the transient flowrate itself. As to the latter, Noorishad and Doe (1982) used a coupled stress-flow model to show that fracture deformation causes an increase in flow with time rather than a decrease. This effect is shown schematically in Figure 17. Noorishad and Doe noted that fracture-deformation effects would occur only after the pressure had propagated some distance along a fracture, recognizing that the fracture deforms not in response to the pressure in the borehole but due to the pressure integrated over the fracture face, i.e. the total load. Alternatively, Sabarly (1968) suggested the pressure-flowrate curve may have a 4th-power relationship if the fracture deformation is linear over the injection range.

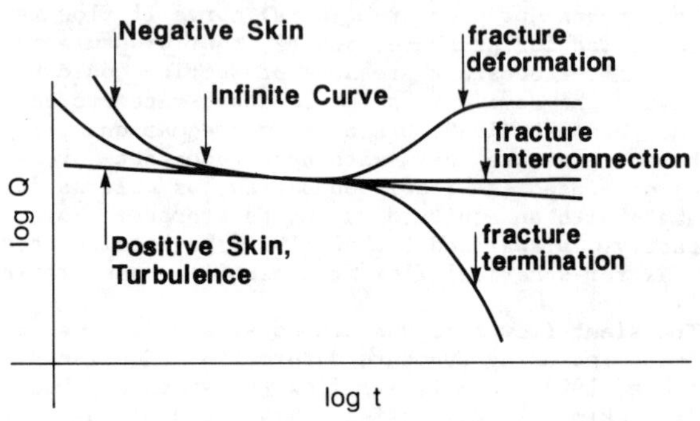

Figure 17. Idealized transient flowrate curves for constant-pressure tests.

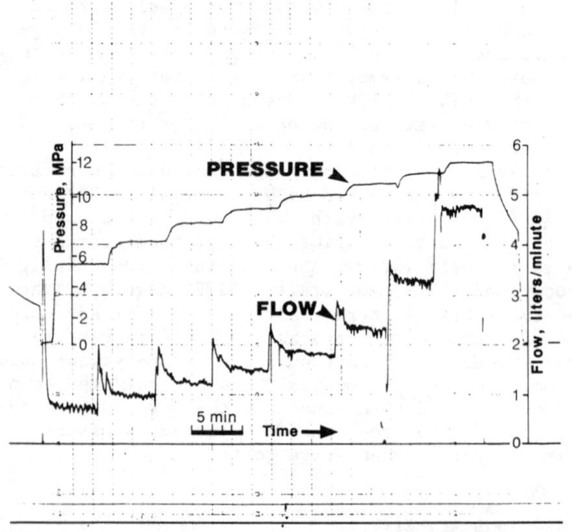

Figure 18. Pressure and flow versus time for a stepped injection test. Note the transient increase in flow in the highest pressure step.

4.3.4 Skin Effects

Skin effects (changes in rock or fracture properties at or near the borehole) may be incorporated into the constant-pressure type curves. Skin effects may be interpreted either from the logarithmic type-curve or from semi-log plots. Logarithmic type-curves showing the influence of skin effects are shown in Figures 19 to 21 for the infinite aquifer case and two boundary conditions. Unfortunately, the skin-effect type curves do not have markedly different shapes than the curves without skin effects. Data from skin-affected tests can easily be matched to non-skin curves.

The alternate method of interpreting skin uses the semi-logarithmic analysis as is commonly done in petroleum engineering. The Swedish radioactive waste program (Almen et al, 1987) has used this approach to interpret skin values from constant pressure tests in crystalline rocks. Unfortunately, as discussed in Section 3.1, storativity and skin cannot be separated in this approach, thus an interpretation of skin must assume a storativity value.

4.4 Constant-Pressure Test Summary

A summary of the various effects that may be observed in the transient-flowrate curve for a constant-pressure test is shown in Figure 16. The early-time flowrate may be increased or decreased by either skin effects or turbulent flow. The later

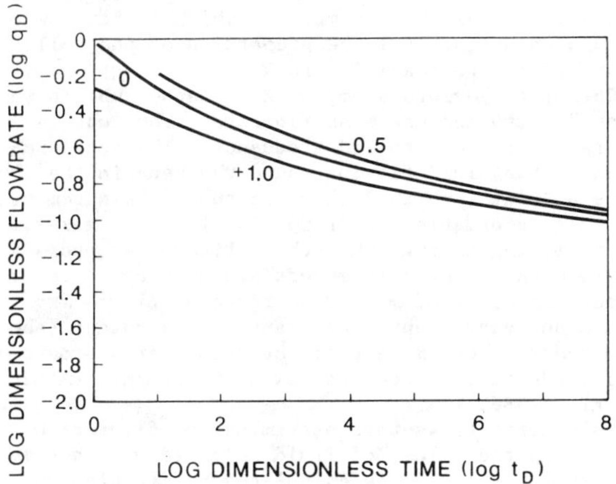

Figure 19. Transient flowrate type curve for an infinite aquifer with skin effects.

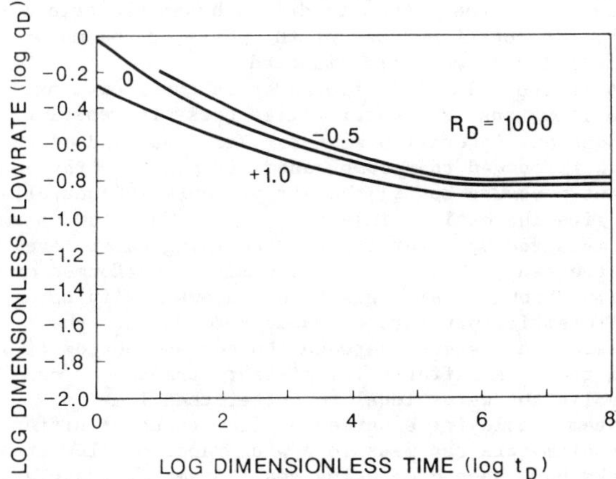

Figure 20. Transient flowrate type curve for an aquifer with a constant pressure outer boundary and skin effects.

flowrate may reflect either boundary effects or fracture deformation. A fracture that does not intersect other fractures may show a sharp decrease in flow as the closed boundary is affected. Intersection with a more permeable fracture should result in steady flow. Fracture deformation should exhibit a characteristic increase in flowrate with time.

5 EQUIPMENT FOR BOREHOLE TESTING

The appropriate design and choice of equipment is an essential to ensure that constant-pressure tests produce useable data. This section presents a practical review of equipment considerations for borehole testing.

5.1 Zone Isolation — Packers

Most tests in low-permeability rock are performed in open holes using inflatable rubber packers to provide temporary seals isolating the test zone. The packers typically provide a seal between 1 and 2 m long.

Packers may be inflated using either air or water. Air provides a very convenient method for inflating packers, especially when compressed air bottles are used. However, air-inflated packers are more compressible than water-inflated packers and can contribute to wellbore-storage effects. Air inflation can be used only to the depth where the air-bottle pressure can overcome the pressure of the fluid column in the borehole. As a practical matter, air inflation is not feasible below depths of about 500 m.

Water inflation provides a much stiffer packer system. The only real disadvantage of water inflation occurs at sites where the water table is more than 20 to 30 m below the surface. In such cases the static head in the packer-inflation line may be sufficient to keep the packers set after the packer pressure has been relieved at the surface. This problem may be overcome by employing a downhole pressure-relief valve or by filling the borehole with water while the packers are being deflated and moved in the hole.

In tight formations, the expansion of the packers may squeeze the water in the test zone, causing a pressure buildup that can approach the packer inflation pressure. This problem can be overcome either by opening the injection line to atmospheric pressure during packer inflation, or by employing a downhole valve to relieve the inflation-induced build-up in the test zone.

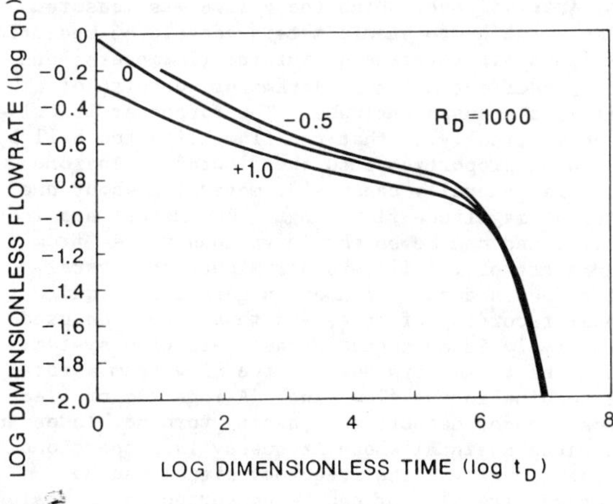

Figure 21. Transient flowrate type curve for an aquifer with a no flow outer boundary and skin effects.

Several options are available for routing the inflation pressure to the packers. The most common and simple one is to use an inflation line separate from the injection line. For test systems using rigid pipe, commercially available drill-stem test tools have a mechanical valve that routes the flow from the tubing to either the packers or the test zone.

Most packer systems employ only two packers; additional packers may be added to isolate guard zones. These may be used to isolate separate injection zones above and below the main injection to "radialize" the flow (Maini et al., 1971). Guard zones may also be used to monitor for packer leakage. Pressure changes in guard zones may reflect factors other than packer leakage. Such conditions may include flow through the rock back to the well or pressure transients associated with sealing off zones that may have been communicating with one another through the open hole.

There are considerable complications in designing multiple-zone packer systems. Each zone requires its own pressure monitoring, as well as the sections above and below the packer string. Furthermore, downhole pressure relief valves may be required to relieve pressure build-up due to packer inflation.

5.2 Fluid-Injection Lines

Traditionally, well testing has been performed using a rigid steel injection tubing or drill rod assembled in sections. Unless high-quality rods are used, the joints in such a system provide potential leaks which are almost impossible to locate.

In recent years, wireline systems have become popular, such as the Swedish umbilical test system (Almen et al., 1986). The wireline systems use hydraulic hoses or flexible steel tubes, bundled with electrical cables and a strain element to take the weight of the system. Despite early concerns about losing wireline tools or having them stuck, the application of these systems has been quite successful. One advantage of the wireline system is the speed of testing. Additionally, costs are lowered by eliminating workover drill rigs and crews.

5.3 Measurement and Control of Flow

Flowmeters are of two basic types which can be termed hydrodynamic and volumetric (Figure 22). Hydrodynamic flowmeters measure the flowrate directly using some measurement of the energy of the flowing fluid. Volumetric flowmeters simply record the volume or mass of fluid injected or withdrawn; flowrate is obtained by dividing the volume of flow by the time interval over which the volume was measured.

Among the hydrodynamic meters capable of measuring low flows are rotameters, turbine flowmeters, and coriolis-effect meters. Rotameters consist of a ball in a transparent tube. The rotameter is oriented vertically so that the flow lifts the ball by an amount proportional to the flowrate. Any one meter is generally capable of measuring about one order of magnitude flow range. Rotameters are inexpensive and can be bought in various ranges from hundredths of milliliters per minute to several liters per second. Rotameters generally require manual recording of data, and thus cannot be used effectively in automated data-acquisition systems.

Turbine flowmeters measure the flow from a rotating turbine in the flow line. A magnetic or electrical sensor detects the passing turbine blades and generates a signal whose frequency is proportional to the flowrate. The frequency signal can be recorded directly, or can be converted to an analog current or voltage signal for recording. Like rotameters, a single turbine flowmeter is effective for flows ranging within only one to two orders of magnitude. Furthermore, at the high end of the range, the pressure losses due to the flowmeter can become significant. Large pressure losses require that pressure sensors or controllers be located downstream of the meter. Turbine meters can detect flows as low as about 4 mL/min, which is not low enough to satisfy a large proportion of the well tests in low-permeability rock.

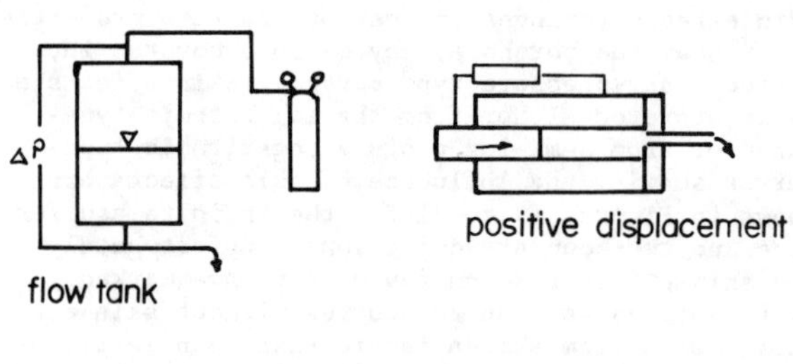

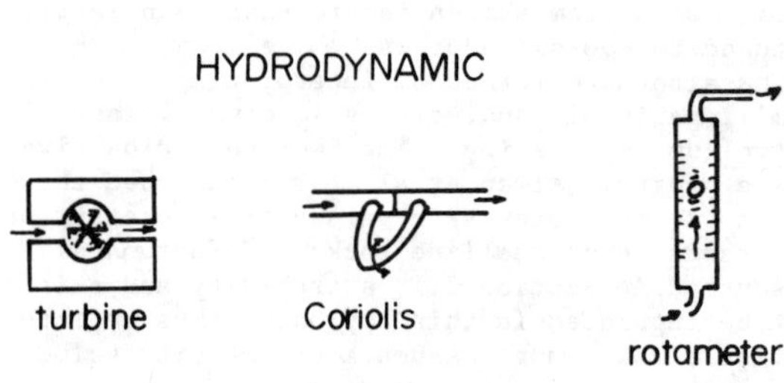

Figure 22. Illustration of various types of flowmeter.

Coriolis flowmeters employ a U-shaped tube in the flow line to measure mass flowrate. The tube is vibrated at a controlled frequency. The acceleration of the fluid flowing round the bend in the tube puts a twist into the vibrating tube. Measurement of this deformation is proportional to the mass of fluid moving through the tube. Like other hydrodynamic meters, coriolis meters are accurate over about one order of magnitude range of flow. Provided one can accept lower ranges of accuracy (about 5 percent) these meters can be used over 4 orders of magnitude to flowrates as low as 0.1 mL/min (Almen at al., 1986).

Volumetric flowmeters determine the flowrate by measuring the volume of fluid injected over a time interval. Flowrate is calculated by dividing the volume by the duration of measurement. Volumetric flowmeters are generally an integral part of the pumping system. In simple volumetric systems, the displacement of a single piston can be monitored using an LVDT (linear variable displacement transducer). If the piston is driven by an electric motor, the rotation speed of the gears or the motor itself can be monitored instead.

A common volumetric flowmeter is the flow tank. The flow tank is a water-filled pressure vessel with a constant internal diameter. The upper end of the tank is opened to a compressed-air source. For constant-pressure tests, the air pressure is controlled to give the desired injection head. The flow volume is measured by observing the declining water level in the tank. This observation may be performed manually through a sightglass, or automatically using a differential pressure transducer to monitor the pressure difference between the top and bottom of the tank. A differential pressure transducer can resolve the water level to better than 1 mm. Systems employing a series of flow tanks of different diameters can measure a wide range of flowrates. Tanks have been made using small diameter stainless steel tubing (Haimson and Doe, 1983).

Volumetric flowmeters have virtually infinitesimal resolution provided the flowrate can be measured over a long period of time. The accuracy of the method decreases with increased frequency of measurement points, hence volumetric meters are best for measuring low, steady flowrates, and are less desirable for measuring rates that may be changing rapidly. They are generally inappropriate for measuring early-time flows in transient injection tests.

5.4 Measurement and Control of Pressure

Pressure measurement and control is vital in any well testing. As well-test solutions require the test-zone pressure to be known accurately, pressure measurement at the wellhead is less than ideal: care must be taken to calibrate the test system for pressure losses from the point of measurement down to the test zone. Downhole pressure measurement has therefore become an desireable feature of low-permeability test systems.

Numerous downhole pressure-transducer systems have been built in recent years; all require separate electrical lines. With several instruments downhole, the electrical cable can become cumbersome unless the signals are multiplexed. For protection, pressure transducers may be housed in a water-tight container downhole. Alternatively, sealed pressure transducers are available that can withstand high water pressures. For the fullest control of pressure measurement, all transducers should be temperature-compensated and paired with temperature transducers such as thermistors or RTDs (resistance temperature devices).

5.5 Downhole Valving — Methods of Supplying Pulses

A key requirement for low-permeability well testing is a downhole shut-in valve for the flow line. This valve is used to isolate the test zone from the injection tubing prior to testing. By closing the valve and pressurizing the injection tubing before the test, one can eliminate most of the equipment-compressibility effects. The downhole valve also aids in giving a sharp initiation to each test and speeds recovery by shutting in the test zone directly.

The downhole valve is also essential for pulse tests. The simplest way of running a pulse test is to pressurize the injection tubing with the test zone shut in, and then instantaneously create a pressure pulse in the test zone by quickly opening and closing the valve.

6 SAMPLING STRATEGIES

The success of a well testing program depends both on the accuracy of the individual well tests and on the efficiency of the sampling program. This section describes two major approaches to sampling in boreholes: fixed-interval-length (FIL) sampling and discrete-zone sampling (DZ). The section concludes with a discussion of the influence of channelling on the accuracy of well-test inferences.

6.1 FIL AND DZ Testing

FIL sampling is testing a borehole in contiguous intervals having the same packer separation. In FIL testing, the selected interval length is intended to be short enough to provide some detail on the variability of the rock, but long enough that the hole can be tested in a reasonable period of time. If the packer spacing is too short, many intervals may contain no conduits, hence the testing time may be wasted. Yet, as Snow (1970) showed, the number of no-flow tests is a useful parameter for inferring

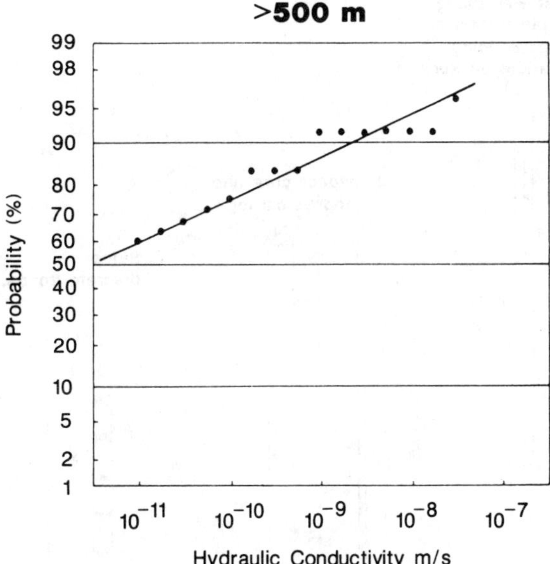

Figure 23. Log-probability plot of the 25-m interval data from crystalline rock well in Sweden (from Carlsson and Winberg, 1983)

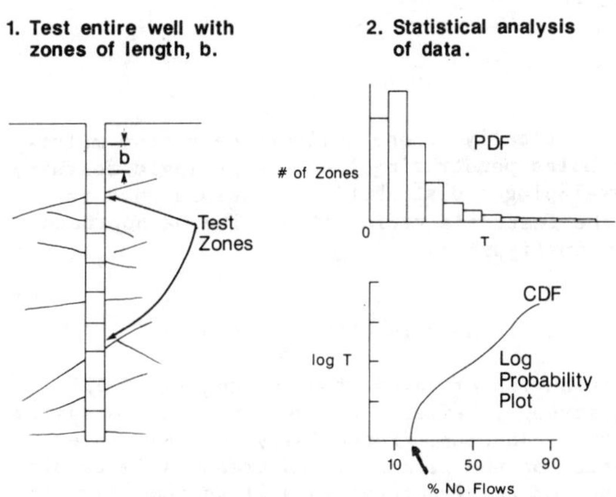

Figure 24. Illustration of the Fixed-Interval-Length (FIL) testing approach.

conduit frequency. Ideally, a testing program should yield a small number of no-flow zones, perhaps about 20%, for the program to be efficient. Further approaches to analyzing FIL test data are discussed in the following section.

The main use of FIL testing is to provide data for a statistical treatment of the hydraulic properties of the rock. Each borehole essentially becomes a sampling line for hydraulic testing. The FIL test program yields a distribution of conductivity values which is commonly lognormally distributed. An example of a lognormal distribution is shown in Figure 23. It is from this distribution that one may infer information about variability of conductivity and some information on the geometry of the conducting fractures. The FIL approach is shown in Figure 24.

The other major testing approach is discrete-zone sampling (DZ). The DZ approach is used when one is concerned about specific geologic features. For example, DZ testing might be used to probe the hydraulic properties of a specific shear zone along a tunnel alignment. In this case, only specific zones are tested; the overall properties of the rock mass are unimportant. The discrete-zone approach can be

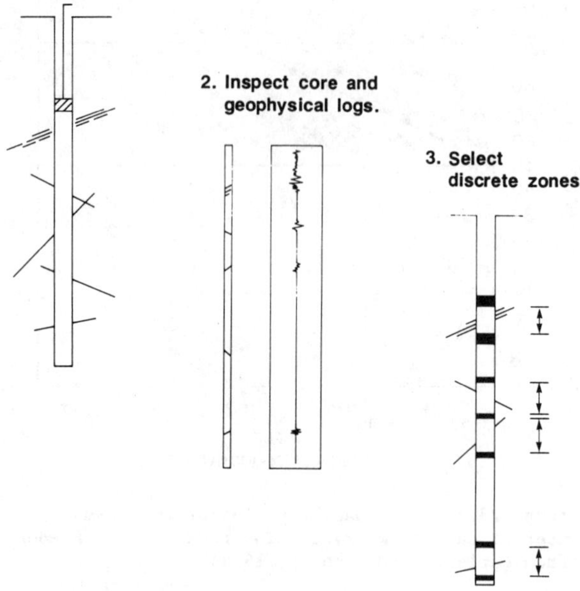
1. Test well using single packer or large spacing straddle packer.
2. Inspect core and geophysical logs.
3. Select discrete zones.

Figure 25. Illustration of the Discrete Zone (DZ) testing approach.

used statistically if one collects data from multiple boreholes penetrating the same geologic feature, thus developing a distribution of values that reflects the feature's variability. The DZ approach is shown in Figure 25.

6.2 Statistical Analysis of FIL Test Data

The distribution of values from a program of FIL testing strongly reflects the geometry of the system of fracture conduits. Potentially, one can infer information on the frequency and transmissivity distributions of the fractures as well as some data on channelling. This section discusses the status of theoretical development for these approaches.

The use of FIL data for inferring fracture geometry was suggested by Snow (1970).
The basis for Snow's derivation was the observation that the discharge from a well test is equal to the sum of the flows into each of the conductive fractures intercepted by the test interval.
Consequently, estimators of the mean and the variance of the transmissivities of individual conductive fractures intercepted by a series of fixed-interval well tests can be derived on a basis similar to Snow's:

$$T = \sum_{i=1}^{N_f} T_i \tag{30}$$

where: T = transmissivity measured,
T_i = transmissivity of the ith conductive fracture intercepted
N_f = number of conductive fractures intercepted

Note that the number of conductive fractures intercepted, as well as their transmissivities and the resultant transmissivity measurement, are inherently random.

Snow (1970) made the following assumptions in his derivation:

- the rock mass characterized by a series of well tests is statistically homogeneous;
- the transmissivities of and spacings between conductive fractures are statistically and mutually independent;
- the spacing between conductive fractures is exponentially distributed, implying that the number of conductive fractures intercepted is Poisson distributed.

Based on the relationship defined in Equation 30 and on the assumptions of statistical homogeneity and independence, the moments (the mean and the variance) of the transmissivity measurements can be expressed in terms of the moments of the number of fractures intercepted and their transmissivities. Under the Poisson-distribution assumption, the mean and the variance of the number of conductive fractures intercepted per test, and the probability of a "no-flow" test in which no conductive fractures are intercepted, can be stated explicitly in terms of the mean conductive-fracture frequency and the packer spacing.

By intuition, the relative frequency of "no-flow" tests in a series of well tests should provide an estimate of the probability of "no-flow" tests. Furthermore, the arithmetic mean and the variance of the transmissivities measured in a series of well tests can be used to estimate the mean and the variance of the transmissivity measurements, respectively. By substituting these estimators into the expressions for the moments of the transmissivity measurements and for the number of conductive fractures intercepted per test, estimators for the mean conductive-fracture frequency and for the mean and the variance of the transmissivities of conductive fractures can be derived. The estimators are:

$$\hat{\lambda}_c = \frac{\ln(N/n_0)}{b} \tag{31}$$

$$\hat{\mu}_T = \frac{\hat{\mu}_T}{\ln(N/n_0)} \tag{32}$$

$$\hat{\sigma}_T^2 = \frac{1}{\ln(N/n_0)}\left[\hat{\sigma}_T^2 - \frac{\hat{\mu}_T}{\ln(N/n_0)}\right] \tag{33}$$

where: $\hat{\lambda}_c$ = estimator of mean conductive-fracture frequency
N = number of well tests in series
n_0 = number of "no-flow" tests in series
b = packer spacing
$\hat{\mu}_T$ = estimator of mean transmissivity of conductive fractures
$\hat{\mu}_T$ = arithmetic mean of transmissivities measured in a series of well tests
$\hat{\sigma}_T^2$ = estimator of variance of conductive-fracture transmissivities,
$\hat{\sigma}_T^2$ = variance of transmissivities measured in a series of well tests

The estimators defined by Equations 31, 32, and 33 are essentially equivalent to the estimators derived by Snow (1970), although Snow worked in terms of discharge instead of transmissivity.

Carlsson et al. (1984) proposed an alternative method that can be used to estimate the mean conductive-fracture frequency. Like Snow (1970), these investigators made the statistically homogeneous and independent assumptions and used the frequency of "no-flow" tests in their estimation technique. However, they made no assumption as to the distribution of the number of conductive fractures intercepted. Instead, the information from well tests is supplemented by the actual number of fractures intercepted by each well test based on core recovered from the test zones.

The corelog information is used to separate the series of well tests into *m* groups, where each well test in the ith group intercepted *i* fractures. For each group of tests, a maximum-likelihood estimator of the probability that a fracture is nonconductive can be defined in the following form:

$$\hat{p}_i = \left(\frac{n_{0i}}{N_i} \right)^{1/i}, \quad i = 1, \ldots, m \tag{34}$$

where: $\hat{p}_i$ = estimator of the probability that a fracture is nonconductive based on tests that intercepted *i* fractures
n_{0i} = number of "no-flow" tests that intercepted *i* fractures,
N_i = total number of tests that intercepted *i* fractures.

Equation 34 defines *m* estimators of the probability that a fracture is nonconductive. The number of estimators actually evaluated is chosen based on the overall fracture frequency in the area tested. An aggregate estimator of the probability that a fracture is nonconductive is obtained through a weighted summation of the *m* estimators, where the weighting factors are chosen to minimize the variance of the aggregate estimator. This summation and the weighting factors are defined by Equations 35 and 36, respectively.

$$\hat{p} = \sum_{i=1}^{m} a_i \hat{p}_i \tag{35}$$

$$a_i = \left[\operatorname{Var}(\hat{p}_i) \sum_{j=1}^{m} \frac{1}{\operatorname{Var}(\hat{p}_j)} \right]^{-1}, \quad i = 1, \ldots, m \tag{36}$$

where: $\hat{p}$ = estimator of the probability that a fracture is nonconductive
$\operatorname{Var}(\hat{p}_i)$ = variance of estimator $\hat{p}_i$

The variance of estimator $\hat{p}_i$ can be approximated as:

$$\operatorname{Var}(\hat{p}_i) \approx \frac{p^{2-i}(1-p^i)}{n_i i^2}, \quad i = 1, \ldots, m \tag{37}$$

where: p = probability that a fracture is nonconductive.

Estimating the probability that a fracture is nonconductive according to Equations 35, 36, and 37 is an iterative process since the probability being estimated in Equation 35 is the same probability needed to calculate the variances in Equation 37.

The mean conductive-fracture frequency is the product of the mean fracture frequency and the probability that a fracture is *conductive*. Hence, an estimate of the mean conductive-fracture frequency can be obtained from an estimate of the mean fracture frequency based on corelog information and the estimated probability that a fracture is *nonconductive*, as follows:

$$\hat{\lambda}_c = (1 - \hat{p}) \left(\sum_{i=1}^{m} n_{0i} i \right) \Big/ \left(\sum_{i=1}^{m} n_{0i} b \right) \tag{38}$$

The practical difficulty with estimators that are based on the number of "no-flow" tests is that the lower measurement limit in well tests, albeit very low in terms of appreciable values of transmissivity (e.g., 10^{-10} m²/s for 25-m test sections), is nonetheless nonzero. Tests in sections in which the transmissivity is less than the measurement limit must be treated as "no-flow" tests, even though the sections may still contain a few conductive fractures. This ambiguity is particularly acute in Snow's estimator of the mean conductive-fracture frequency because of the logarithmic relationship between the relative frequency of "no-flow" tests and the estimator.

6.3 FIL Probabilistic Model

The mean values and variances whose estimators are discussed in the preceding section characterize central tendencies and dispersions of the hydraulic properties of fractures intercepted in a series of fixed-interval well tests. A probabilistic model provides complete distributional information; that is, not only can the statistics (such as mean and variance) be derived from a probabilistic model, but also the probability of a property exceeding any given value can be evaluated. Furthermore, the statistical characteristics (such as bias and mean-squared error) of estimators can be derived from a probabilistic model that describes the underlying distribution of the estimators. These characteristics may be useful in the design of optimal well-test programs that reduce or minimize the mean-square errors in the estimates, based on the tests performed in the program.

In addition to the assumptions made by Snow (1970) in his derivation of estimators of the hydraulic properties of fractures intercepted by well tests, an assumption regarding the distribution of the transmissivities of conductive fractures must be made to develop a probabilistic model that describes the distribution of transmissivities measured in well tests. Several investigators have noted that the lognormal distribution yields a good fit to transmissivity data from well tests (Snow, 1970). However, an analytical function for the sum of lognormally distributed variables, such as is required in the summation over conductive-fracture transmissivities shown in Equation 30, does not exist.

The gamma distribution has a similar shape to the lognormal distribution. This similarity has been noted in fitting fracture-length data, although the lognormal distribution usually fits that data better (Baecher et al, 1977). Unlike the lognormal distribution, the gamma distribution is regenerative under addition. This additive property is precisely what is needed to derive the distribution of the sum of conductive-fracture transmissivities. Consequently, it is assumed that the transmissivities of conductive fractures can be adequately described by a gamma distribution because the gamma distribution's shape is similar to the shape of the lognormal distribution but its form is more tractable mathematically.

Under the assumptions stated previously, the following probabilistic model can be derived by applying the total probability theorem to Equation 30:

$$\Pr[T \le \tau] = \exp(-\lambda_c b) \left[1 + \sum_{n=1}^{\infty} \frac{(\lambda_c b)^n \gamma(n\upsilon_T, \lambda_T t)}{n! \Gamma(n\upsilon_T)} \right] \tag{39}$$

where: υ_T = shape parameter for gamma distribution describing conductive-fracture transmissivities
$\Pr[T \le \tau]$ = probability that the transmissivity measured is less than or equal to τ
$\Gamma(n\upsilon_T)$ = gamma function.
λ_T = scale parameter for gamma distribution describing conductive-fracture transmissivities
$\gamma(n\upsilon_T, \lambda_T t)$ = incomplete gamma function,

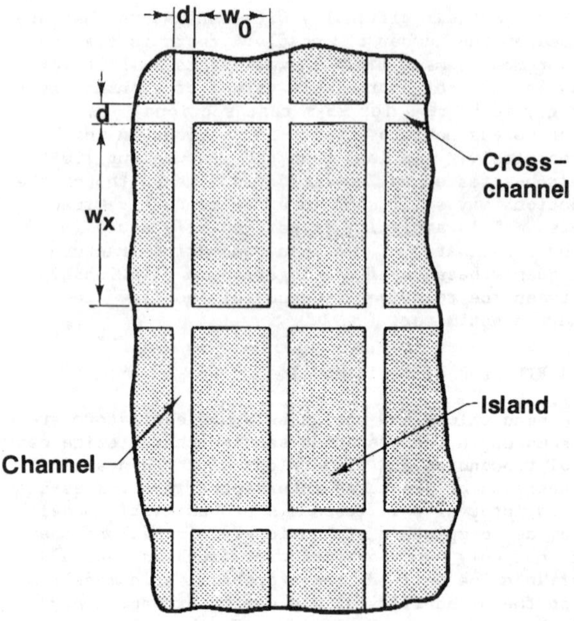

Figure 26. Conceptual model of fracture channels.

6.4 Effects of Channeling

Channelling is a geological factor which could have a significant effect on transmissivity measurements made using single-hole injection tests (Tsang and Tsang, 1987). Specifically, it was concluded that, if the open areas in fractures can be described as narrow, widely separated channels, then single-hole injection tests would be a very inefficient means of sampling the hydraulic properties of the fracture system. In this section, a simple conceptual model for channels is defined. Based on the model, the effects of channeling on single-hole injection tests are quantified.

By definition, a model is an idealized representation of a natural entity or phenomenon. Hence, a model of channels should represent the areal distribution of the open and contact areas within fractures. However, no field-scale observations of the geometry of channels in fractures have been made directly because such observations would require measuring the areal distribution of a fracture's aperture without first disturbing or separating the fracture, a virtually impossible task. Therefore, any model of channels is highly speculative since it must be based on indirect evidence of the geometry of channels.

Rasmusson and Neretnieks (1986) have proposed the idealization of channels illustrated in Figure 25. As stated in the reference, their "idea is that the water flows in channels which are quite widely separated and which may not for quite considerable distances intersect other channels." This concept is supported by casual observations and careful measurements of the distribution of natural flow into an excavation (Neretnieks, 1985), by the uneven distribution of sorbed tracers on fracture surfaces observed in a field experiment (Abelin et al, 1985), and by analyses of tracer experiments performed on lab-size samples of natural fractures (Moreno et al, 1985). The idealization means that channels are roughly parallel to each other with occasional cross-channels connecting the primary set of channels.

As shown in Figure 25, the areas between channels will be referred to as "islands". An island is simply an area where the flowrate is insignificant compared to the flowrate through the channels. Therefore, islands include not only contact areas but also areas where the apertures are significantly smaller than the apertures of the channels.

The impact of channels on transmissivity measurements made using single-hole injection tests is a question of scale: how narrow and how widely separated are the channels with respect to the borehole's diameter? For example, Figure 26 could be an illustration of a photomicrograph of the tortuous flowpaths between asperities on a fracture's surfaces. If this were the case, generally a borehole would intercept many channels and an injection test would accurately indicate the fluid transmission characteristics of the fracture even though the channels are extremely narrow. This example illustrates that channeling is tolerable when the probability of intercepting a channel is relatively high. If this probability can be estimated, then the effect of channeling can be assessed.

Let us represent channels as a regular grid of orthogonal flowpaths with uniform widths equal to d The islands formed by the channels are rectangular in plan view with dimensions w_0 by w_x, where w_x (the distance between cross-channels) is greater than or equal to w_0. The fraction of the fracture area which is composed of islands and which essentially is closed to flow can be easily shown to be:

$$\theta = \frac{w_0 w_x}{(w_0+d)(w_x+d)} \tag{40}$$

where: θ = fraction of fracture area composed of islands.

Determining the probability of intercepting a channel with a borehole of radius r_w is a straightforward problem. Assuming that any point on the fracture is equally likely to be the borehole's center, the probability Pr[I] that some portion of the borehole will intercept a channel is:

$$\Pr[I] = \begin{cases} 1 - \theta\left(1 - \frac{2r_w}{w_0}\right)\left(1 - \frac{2r_w}{w_x}\right) , & \frac{2r_w}{w_0} < 1 \\ 1 , & \frac{2r_w}{w_0} \geq 1 \end{cases} \tag{41}$$

Equation 41 quantifies the qualitative statements made previously: the probability that a borehole will intercept at least one channel is a function of the width of the channels and the distance between channels relative to the borehole's diameter. Equation 41 states that a borehole will certainly intercept a channel (Pr[I]=1) if the borehole's diameter is larger than the distance between channels ($2r_w/w_0>1$) regardless of how narrow the channels are.

Based on Equation 41, the probability of intercepting at least one channel is graphed in Figure 27 as a function of the borehole's diameter relative to the distance between channels. As shown in this figure, an envelope enclosing all of the possible geometries of rectangular islands can be defined for a given value of θ by the lower-bound curve given by $w_x = \infty$ (no cross-channels) and the upper-bound curve given by $w_x = w_0$ (square islands). The maximum difference in probability of interception between these two extremes is $\theta/4$ at $2r_w/w_0 = 0.5$. Hence, island geometry has a secondary effect on the probability of intercepting at least one channel.

Two values of θ are considered in Figure 26. Iwai (1976) performed laboratory experiments on flow through fractures and reported that 10% to 20% of a fracture surface was in contact with its mate.

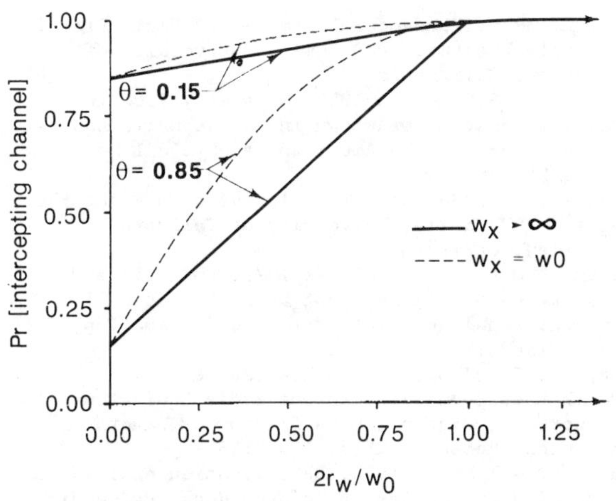

Figure 27. Probability of a borehole intercepting channels in a fracture as a function of the relative contact area, θ.

Furthermore, the upper value was approached only when moderately high stresses (20 MPa) were applied across the fracture. Hence, $\theta = 0.15$ may be representative of the island area within fresh, planar, well-mated fractures such as the tension-induced fractures investigated by Iwai were. Much higher values of θ are suggested by Rasmusson and Neretnieks (1986). Their observations of flow into excavations in Swedish bedrock indicate that only 5% to 20% of a fracture's area carries most of the water. In a transport model with parallel channels, Rasmusson and Neretnieks assumed that each channel was 0.2 m wide separated by impermeable zones 1 m wide, which is equivalent to $\theta = 0.83$. Hence, $\theta = 0.85$ may be representative of ancient fractures which have undergone significant mechanical and geochemical alteration, such as the fractures in the Precambrian bedrock of Sweden undoubtedly have.

Equation 41 indicates that the minimum probability of intercepting a channel is $1 - \theta$, a function of island area alone. Therefore, if the island area is a small fraction of the fracture area, then the probability of intercepting at least one channel with a borehole is high regardless of how small the borehole's diameter is relative to the distance between channels. The case of $\theta = 0.15$ in Figure 27 illustrates this point. However, if the island area is a large fraction of the fracture area, such as the case of $\theta = 0.85$ in Figure 27, then the probability of intercepting a channel depends strongly on the borehole's diameter relative to the distance between channels.

In summary, the model of channel geometry shown in Figure 26 indicates that:

- the shape of the islands between channels has a secondary effect on the probability of intercepting a channel;
- if the islands compose a small fraction of the fracture area, then the probability of intercepting a channel is high regardless of the distance between channels;
- if the islands compose a large fraction of the fracture area, then the probability of intercepting a channel depends strongly on the borehole's diameter relative to the distance between channels.

It must be emphasized that the model from which the preceding inferences are drawn is a simplified idealization based on indirect evidence of the geometry of channels. Although the approach to quantifying the effect of channels on transmissivity measurements in terms of the probability of interception may be useful in general, the specific model presented here and the inferences based on it must be regarded as speculative.

This section has dealt with only one of the potential effects of channeling on transmissivity measurements. Specifically, only the efficiency of sampling the hydraulic properties of channeled fractures with single-hole injection tests has been addressed. Even if the probability of intercepting channels is relatively high so that an injection test generally will indicate a conductive fracture is indeed conductive (good sampling efficiency), the conditions of the test must still be interpreted according to some theoretical model of the flow regime to determine the transmissivity.

ACKNOWLEDGEMENTS

This paper reflects several years' effort pursuing truth in packer testing. Many persons have provided considerable aid through discussion, including past and present colleagues, Derek Elsworth, Jahan Noorishad, Gen-Hua Shi, Janet Remer, Charles Wilson, and William Dershowitz. In particular we should mention our Swedish colleagues, Leif Carlsson, Karl-Erik Almen, Jan-Erik Andersson, and Bjorn Rossander with whom we share a number of prejudices.

REFERENCES

Abelen, H., I.Neretnieks, S.Turbrant & L.Moreno 1985. Final Report on the migration in a single fracture: Experimental results and calculations. SKB Technical Report 85-83, Swedish Nuclear Fuel and Waste Management Co. Stockholm, Sweden

Agarwal, R.G., R.Al-Hussainy & H.J.Ramey Jr. 1970. An Investigation of Wellbore Storage and Skin Effect in Unsteady Liquid Flow: I. Analytical Transport. Society of Petroleum Engineers Journal 10:279-290.

Almen, K.E., J.E.Andersson, L.Carlsson, K.Hansson & N.A. Larsson 1986. Hydraulic Testing in Crystalline Rock: A Comparative Study of Single Hole Test Methods. SKB Technical Report 86-27, Swedish Nuclear Fuel and Waste Management Co. Stockholm, Sweden.

Barker, J.A. & J.H.Black 1983. Slug Tests in Fissured Aquifers. Water Resources Research 19(6):1558-1564.

Black, J.H. 1985. The Interpretation of Slug Tests in Fissured Rocks. The Quarterly Journal of Engineering Geology 18(2):161-171.

Brace, W.F. 1980. Permeability of Crystalline and Argilleceous Rocks. International Journal of Rock Mechanics, Mineral Science, and Geomechanics Abstracts 17:241-251.

Bredehoeft, J.D. & S.S.Papadopulos 1980. A Method for Determining the Hydraulic Properties of Tight Formations. Water Resources Research 16(1):233-238.

Carlsson, L. & T.Olsson 1985. Hydrogeological and Hydrogeochemical Investigations in Boreholes--Final Report. Technical Report 85-10, Swedish Geological Company, Uppsala, Sweden.

Carlsson, L., A.Winberg & B.Grundfelt 1983. Model Calculations of the Groundwater Flow at Finnsjon, Fjalleuden, Glider and Kamlunge. KBS Technical Report TR 83-45.

Cooper, H.H., J.D.Bredehoeft & I.S.Papadopulos 1967. Response of a Finite Diameter Well to an Instantaneous Charge of Water. Water Resources Research 3(1):263-269.

Davison, C.C. & G.R.Simmons 1983. The Research Program at the Canadian Underground Research Laboratory. Proceedings of the NEA Workshop on Geological Disposal of Radioactive Waste: In Situ Experiments in Granite. (October 25-27 1982, Stockholm, Sweden), Nuclear Energy Agency, Organization for Economic Cooperation & Development, Paris, France. p.197-219.

Dershowitz, W.S. 1985. Rock Joint Systems. PhD Thesis, Massachusetts Institute of Technology, Cambridge, Mass.

Doe, T.W. & J.D.Osnes 1985. Interpretation of Fracture Geometry from Well Tests. International Symposium on Fundamentals of Rock Joints, September 15-20 1985, Bjorkliden, Sweden.

Doe, T. & J.Remer 1981. Analysis of Constant-Head Well Tests in Nonporous Fractured Rock. Proceedings of the Third Invitational Well-Testing Symposium — Well Testing in Low Permeability Environments (March 26-28 1980), University of California, Lawrence Berkeley Laboratory, Berkeley, California.

Earlougher, R.C. Jr. 1977. Advances in Well Test Analysis. Monograph Volume 5, Society of Petroleum Engineers, American Institute of Mining, Metalurgical and Petroleum Engineers, Dallas, Texas.

Ehlig-Economides, C.A. 1979. Well Test Analysis for Wells Produced at Constant Pressure. PhD Thesis, Stanford University, Stanford, California.

Ehlig-Economides, C.A. & H.J.Ramey Jr. 1981. Transient Rate Decline Analysis for Well Produced at Constant Pressure. Transactions of the Society of Petroleum Engineers 271:98-104.

Elsworth, D. & T.W.Doe 1986. Application of Non-Linear Flow Laws in Determining Rock Fissure Geometry From Single Borehole Pumping Tests. International Journal of Rock Mechanics, Mineral Science, and Geomechanics Abstracts 23(3):245-254.

Ershagi, I. & J.J.Woodbury 1985. Examples of Pitfalls in Well Test Analysis. Journal of Petroleum Technology 37:335-341.

Faust, C.R. & J.W.Mercer 1984. Evaluation of Slug Tests in wells containing a finite-thickness skin. Water Resources Research 20(4):504-506.

Forster, C.B. & J.E.Gale, J.E. 1981. A Field Assessment of the Use of Borehole Pressure Transients to Measure the Permeability of Fractured Rock Masses, Swedish-American Cooperative Program on Radioactive Waste Stored in Mined Caverns in Crystalline Rock. Technical Information Report SAC-34, Lawrence Berkeley Laboratory, Berkeley, California.

Gale, J.E. 1982. Assessing The Permeability Characteristics of Fractured Rock. In:- T.Narasimhan (ed.), Recent Trends in Hydrogeology. Geological Society of America, Special Paper 189, p.163-182.

Grisak, G.E., J.F.Pickens, D.W.Belanger & J.D.Avis 1985. Hydrogeological Testing of Crystalline Rocks During the NAGRA Deep Drilling Program. Technisher Bericht 85-08, GTC Geologic Testing Consultants, Ottawa, Ontario and Intera Technologies, Houston, Texas.

Haimson, B.C. & T.W.Doe 1983. State of Stress, Permeability and Fractures in the Precambrian Granite of Northern Illinois. Journal of Geophysical Research 88:7355-7371.

Hantush, M.S. 1956. Analysis of Data from Pumping in Leaky Aquifers. Transactions of the American Geophysical Union 37:702-714.

Hantush, M.S. 1959. Nonsteady Flow to Flowing Wells in Leaky Aquifers. Journal of Geophysical Research, 64(8):1043-1052.

Hoek, E. & J.Bray 1974. Rock Slope Engineering. London: Institute of Mining and Metalurgy.

Iwai, K. 1976. Fundamental Studies of Fluid Flow in a Single Fracture. PhD Thesis, University of Berkeley, California.

Jacob, C.E. & S.Lohman 1952. Nonsteady Flow to a Well of Constant Drawdown in an Extensive Aquifer. Transactions of the American Geophysical Union 33(4):559-569.

Karasaki, K. 1986. Well Test Analysis in Fractured Media. PhD Thesis, University of California, Berkeley, California.

Kruseman, G.P. & N.De Ridder 1979. Analysis and Evaluation of Pumping Test Data. Bulletin 11, International Institute for Land Reclamation, The Netherlands.

Long, J.C.S., J.S.Remer, C.R.Wilson & P.A.Witherspoon 1982. Porous Media Equivalents for Networks of Discontinuous Fractures. Water Resources Research 18(3):1253-1265.

Louis, C. & T.Maini 1970. Determination of In-Situ Hydraulic Properties in Jointed Rock. Proceedings of the 2nd Congress International Society of Rock Mechanics (Belgrade, Yugoslavia) 1:235-245.

Maini, Y.N.T., J.Noorishad & J.Sharp 1971. Theoretical and Field Considerations on the Determination of In Situ Hydraulic Parameters in Fractured Rock. Symposium on Percolation Through Fissured Rocks, International Society of Rock Mechanics, Stuggart, West Germany.

Moench, A.F. & P.A.Hsieh 1985. Analysis of Slug Test Data in a Well with a Finite Thickness Skin. In:- Hydrogeology of Rocks of Low Permeability, Proceedings of the 17th International Congress (January 7-12 1985, Tucson, Arizona). Memoirs of the International Association of Hydrogeologists 18(1):52-71.

Moreno, L., I.Neretrieks & T.Erikson 1985. Analysis of some laboratory tracer runs in natural fissures. Water Resources Research 21:951-958.

Mueller, T.D. & P.A.Witherspoon 1965. Pressure interference effects within reservoirs and aquifers. Journal of Petroleum Technology 27:471-474

Neuman, S.P. & P.A.Witherspoon 1969. Theory of Flow in a Confined Two-Aquifer System. Water Resources Research 5(4):803-816.

Neuzil, C.E. 1982. On Conducting the Modified 'Slug' Test in Tight Formations. Water Resources Research 18(2):439-441.

Noorishad, J. & T.Doe 1982. Numerical Simulation of Fluid Injection into Deformable Fractures. Proceedings of the 23rd US Rock Mechanics Symposium: Issues in Rock Mechanics. Society of Mining Engineers, p.645-663.

Papadopulos, I.S. & H.H.Cooper 1967. Drawdown in a Well of Large Diameter. Water Resources Research 3(1):241-244.

Rassmuson A. & I.Neretnieks 1986. Radionuclide Transport in Fast Channels in Crystalline Rock. Water Resources Research 22:1247-1256.

Sabarly, F. 1968. Les injections et les drainages de fondation de barrages. Geotechnique 18:229-249

Serafim, J.L. 1968. Influence of interstitial water on the behavior of rock masses. In:- K.Stagg & O.Zienkiewicz (ed.), Rock Mechanics in Engineering Practice, p.55-97. New York: Wiley.

Snow, D.T. 1968. Fracture deformation and changes in permeability and storage upon changes in fluid pressure. Colorado School of Mines Quarterly 63:201-244.

Snow, D.T. 1970. The Frequency and Apertures of Fractures in Rock. Journal of Rock Mechanics, Mineral Science, and Geomechanics Abstracts 7:23-40.

Thackston, J.W., L.M.Preslo, D.E.Hoexter & N.Donnelly 1984. Results of Hydraulic Tests at Gibson Dome No.1, Elk Ridge No.2, and E.J.Kubat Boreholes, Paradox Basin, Utah. Report ONWI-491 prepared by Woodward-Clyde consultants for Office of Nuclear Waste Isolation, Battelle Memorial Institute, Columbus, Ohio.

Theis, C.V. 1935. The Relation Between the Lowering of the Piezometric Surface and the Rate and Duration of Discharge of a Well Using Ground-Water Storage. Transactions of the American Geophysical Union 16:519-524.

Tsang, Y. & C.Tsang 1987. Channel Model of Flow through Fractured Media. Water Resources Research 23:467-479.

Uraiet, A.A. 1979. Transient Pressure Behavior in a Cylindrical Reservoir Produced by a Well at Constant Bottom Hole Pressure. PhD Thesis, University of Tulsa, Tulsa, Oklahoma.

Uraiet, A.A. & R.Raghaven 1980. Unsteady Flow to Well Producing at Constant Pressure. Journal of Petroleum Technology 32:1803-1812.

Wang, J.S.Y., T.N.Narasimhan, C.F.Tsang & P.A.Witherspoon 1978. Transient Flow in Tight Formations. Proceedings of the Invitational Well-Testing Symposium (October 19-21 1977), Report LBL-7027, Lawrence Berkeley Laboratory, Berkeley, California.

Wilson, C.R., T.W.Doe, J.C.S.Long & P.A.Witherspoon 1979. Permeability Characterization of Nearly Impermeable Rock Masses for Nuclear Waste Repository Siting. Proceedings of the OECD Workshop on Low Flow, Low Permeability Measurements in Largely Impermeable Rocks, (March 19-21), Nuclear Energy Agency, Organization for Economic Cooperation and Development, Paris, France.

Ziegler, T. 1976. Determination of Rock Mass Permeability. US Army Engineers Waterways Experiments Stations, Technical Report S-76-2, Vicksburg, Mississippi.

APPENDIX 1

The data for dimensionless type curves for constant-pressure tests are given below for three boundary conditions (infinite, no-flow, and constant-pressure) and for a slight positive and negative skin effect. The data were generated by a Golder Associates' program for PC, CHQP, which performs a numerical inversion of the Laplace transform space solutions for the constant-pressure differential equations. The Laplace transform space solutions may be found in Ehlig-Economides (1979).

Table 1. Effects of open and closed boundaries. For boundaries at the given radii, data are given from the time when the type curves begin to deviate from the infinite case, until the dimensionless flowrates for the closed case reach 10^{-3}.

Log tD	Dimensionless Flowrate		
	Infinite reD -	Open reD 10	Closed reD 10
1.0	0.5339	0.5339	0.5338
1.1	0.5077	0.5080	0.5071
1.2	0.4836	0.4850	0.4811
1.3	0.4614	0.4658	0.4545
1.4	0.4409	0.4510	0.4254
1.5	0.4219	0.4410	0.3927
1.6	0.4043	0.4351	0.3554
1.7	0.3880	0.4325	0.3136
1.8	0.3729	0.4318	0.2678
1.9	0.3587	0.4319	0.2195
2.0	0.3456	0.4324	0.1710
2.1	0.3332	0.4328	0.1248
2.2	0.3217	0.4331	0.0841
2.3	0.3109	0.4333	0.0512
2.4	0.3008	0.4335	0.0275
2.5	0.2912	0.4336	0.0124
2.6	0.2822	0.4338	0.0044
2.7	0.2737	0.4339	0.0010
2.8	0.2657		
2.9	0.2581	reD	reD
3.0	0.2510	100	100
3.1	0.2442		
3.2	0.2377	0.2378	0.2376
3.3	0.2316	0.2321	0.2310
3.4	0.2257	0.2271	0.2240
3.5	0.2202	0.2231	0.2163
3.6	0.2149	0.2203	0.2072
3.7	0.2098	0.2185	0.1965
3.8	0.2050	0.2176	0.1839
3.9	0.2004	0.2172	0.1691
4.0	0.1959	0.2171	0.1521
4.1	0.1917	0.2171	0.1332
4.2	0.1876	0.2171	0.1126
4.3	0.1837	0.2172	0.0912
4.4	0.1800	0.2172	0.0700
4.5	0.1764	0.2171	0.0501
4.6	0.1730	0.2171	0.0330
4.7	0.1696	0.2171	0.0195
4.8	0.1664		
4.9	0.1633	reD	reD
5.0	0.1604	1000	1000
5.1	0.1575		
5.2	0.1547	0.1548	0.1547
5.3	0.1521	0.1522	0.1519
5.4	0.1495	0.1499	0.1489
5.5	0.1470	0.1480	0.1456
5.6	0.1445	0.1466	0.1417
5.7	0.1422	0.1457	0.1370
5.8	0.1399	0.1451	0.1314
5.9	0.1377	0.1449	0.1246
6.0	0.1356	0.1448	0.1166
6.1	0.1335	0.1448	0.1072
6.2	0.1315	0.1448	0.0965
6.3	0.1296	0.1448	0.0845
6.4	0.1277	0.1448	0.0715
6.5	0.1259	0.1448	0.0579
6.6	0.1241	0.1448	0.0444
6.7	0.1223	0.1448	0.0319
6.8	0.1207	0.1448	0.0210
6.9	0.1190	0.1448	0.0124
7.0	0.1174	0.1448	0.0064

Table 2. Data for type curves showing skin effects for the case of a boundary at infinity.

Log tD	Dimensionless Flowrate		
	Skin 0	1	- 0.5
2.0	0.3456	0.2603	0.4099
2.1	0.3332	0.2531	0.3933
2.2	0.3217	0.2462	0.3778
2.3	0.3109	0.2396	0.3633
2.4	0.3008	0.2334	0.3499
2.5	0.2912	0.2275	0.3373
2.6	0.2822	0.2219	0.3255
2.7	0.2737	0.2165	0.3145
2.8	0.2657	0.2114	0.3041
2.9	0.2581	0.2065	0.2944
3.0	0.2510	0.2018	0.2852
3.1	0.2442	0.1973	0.2766
3.2	0.2377	0.1930	0.2684
3.3	0.2316	0.1889	0.2607
3.4	0.2257	0.1849	0.2534
3.5	0.2202	0.1812	0.2465
3.6	0.2149	0.1775	0.2399
3.7	0.2098	0.1740	0.2336
3.8	0.2050	0.1707	0.2277
3.9	0.2004	0.1674	0.2220
4.0	0.1959	0.1643	0.2167
4.1	0.1917	0.1613	0.2115
4.2	0.1876	0.1584	0.2066
4.3	0.1837	0.1556	0.2019
4.4	0.1800	0.1529	0.1974
4.5	0.1764	0.1503	0.1931
4.6	0.1730	0.1477	0.1890
4.7	0.1696	0.1453	0.1851
4.8	0.1664	0.1429	0.1813
4.9	0.1633	0.1406	0.1776
5.0	0.1604	0.1384	0.1741
5.1	0.1575	0.1363	0.1708
5.2	0.1547	0.1342	0.1675
5.3	0.1521	0.1322	0.1644
5.4	0.1495	0.1302	0.1614
5.5	0.1470	0.1283	0.1585

(Table 2 continued)

5.6	0.1445	0.1264	0.1557
5.7	0.1422	0.1246	0.1530
5.8	0.1399	0.1229	0.1503
5.9	0.1377	0.1212	0.1478
6.0	0.1356	0.1195	0.1454
6.1	0.1335	0.1179	0.1430
6.2	0.1315	0.1163	0.1407
6.3	0.1296	0.1148	0.1385
6.4	0.1277	0.1133	0.1363
6.5	0.1259	0.1119	0.1342
6.6	0.1241	0.1105	0.1322
6.7	0.1223	0.1091	0.1303
6.8	0.1207	0.1077	0.1283
6.9	0.1190	0.1064	0.1265
7.0	0.1174	0.1051	0.1247
7.1	0.1159	0.1039	0.1229
7.2	0.1143	0.1027	0.1212
7.3	0.1129	0.1015	0.1196
7.4	0.1114	0.1003	0.1180
7.5	0.1100	0.0992	0.1164
7.6	0.1087	0.0981	0.1149
7.7	0.1073	0.0970	0.1134
7.8	0.1060	0.0959	0.1119
7.9	0.1048	0.0949	0.1105
8.0	0.1035	0.0938	0.1091

Table 3. Data for type curves showing skin effects for the case of an open boundary at a dimensionless radius of 1000.

Log tD	Dimensionless Flowrate		
		Skin	
	0	1	- 0.5
5.0	0.1604	0.1384	0.1741
5.1	0.1575	0.1363	0.1708
5.2	0.1548	0.1342	0.1676
5.3	0.1522	0.1323	0.1646
5.4	0.1499	0.1305	0.1619
5.5	0.1480	0.1291	0.1598
5.6	0.1466	0.1280	0.1581
5.7	0.1457	0.1272	0.1571
5.8	0.1451	0.1268	0.1564
5.9	0.1449	0.1265	0.1562
6.0	0.1448	0.1265	0.1561
6.1	0.1448	0.1265	0.1561
6.2	0.1448	0.1265	0.1561
6.3	0.1448	0.1265	0.1561
6.4	0.1448	0.1265	0.1561
6.5	0.1448	0.1265	0.1561
6.6	0.1448	0.1265	0.1561
6.7	0.1448	0.1265	0.1561
6.8	0.1448	0.1265	0.1561
6.9	0.1448	0.1265	0.1561
7.0	0.1448	0.1265	0.1561

Table 4. Data for type curves showing skin effects for the case of a closed boundary at a dimensionless radius of 1000.

Log tD	Dimensionless Flowrate		
		Skin	
	0	1	- 0.5
5.0	0.1604	0.1384	0.1741
5.1	0.1575	0.1363	0.1707
5.2	0.1547	0.1342	0.1675
5.3	0.1519	0.1320	0.1642
5.4	0.1489	0.1298	0.1607
5.5	0.1456	0.1273	0.1568
5.6	0.1417	0.1243	0.1522
5.7	0.1370	0.1208	0.1468
5.8	0.1314	0.1165	0.1402
5.9	0.1246	0.1113	0.1324
6.0	0.1166	0.1051	0.1232
6.1	0.1072	0.0978	0.1124
6.2	0.0965	0.0893	0.1003
6.3	0.0845	0.0797	0.0868
6.4	0.0715	0.0690	0.0723
6.5	0.0579	0.0575	0.0575

(Table 4 continued)

6.6	0.0444	0.0458	0.0431
6.7	0.0319	0.0344	0.0300
6.8	0.0210	0.0240	0.0191
6.9	0.0124	0.0152	0.0108
7.0	0.0064	0.0086	0.0052

11:15 Short presentations

Dr. E.T. Brown, president of the International Society of Rock Mechanics introduced Dr. D. Elsworth, this year's Manual Rocha medal winner. The conference paper "Physical and Numerical Analogues to Fractured Media Flow", was co-authored by A.R. Piggott, U.S.A.

A.A. AZEVEDO, D. CORËA FO, E.F. QUADROS and P.T. da CRUZ, Brazil "Hydraulic Conductivity of Basaltic Discontinuities in the Foundation of Taquaruçu Dam, Brazil".

T.L. BY, Norway "Geotomography for Rock Mass Characterization and Prediction of Seepage Problems for Two Main Road Tunnels Under the City of Oslo.

L. CHARO, A. ZÉLIKSON and P. BÉREST, France "Saltdoming Hazard Related to HLW Disposal in Salt Layer".

12:00 **FOYER, CONVENTION FLOOR,** Opening of Trade Exhibition.

12:00-13:30 Lunch Break.

13:30 DULUTH-MACKENZIE - Short presentations

CHEN GANG and LIU TIAN-QUAN, China "A Comprehensive Analysis of the Influential Factors and Mechanism of a Water-Inrush From the Floor of a Coal Seam".

F. CROTOGINO, K.-H. LUX, R. ROKAHR and H.-J. SCHNEIDER, FRG. "Geotechnical Aspects of Hazardous Waste Storage in Salt Cavities".

D.C. HELM, Australia "Prediction of Subsidence due to Groundwater Withdrawal in the Latrobe Valley, Australia".

S. JOHANSSON, Finland "Excavation of Large Rock Caverns for Oil Storage at Neste Oy Porvoo Works in Finland".

K. KOJIMA and H. OHNO, Japan "Connection between Farfield and Nearfield in Relation to the Fractures and/or Flowpaths for the Geological Isolation of Radioactive Wastes".

V. MAURY, J.-M. SAUZAY and D. FOURMAINTRAUX, France "Geomechanical Approach for Oil and Gas Production. Main Problems, First Results".

K. NAKAGAWA, H. KOMADA, K. MIYASHITA and M. MURATA, Japan "Study on Compressed Air Storage in Unlined Rock Caverns".

A. PAHL, L. LIEDTKE, B. KILGER, S. HEUSERMANN and V. BRÄUER, FRG "Investigation of Permeability and Stress Field in Crystalline Rock to Assess the Barrier Efficiency".

L.J. PYRAK-NOLTE, L.R. MYER and N.G.W COOK, P.A. WITHERSPOON, U.S.A. "Hydraulic and Mechanical Properties of Natural Fractures in Low-Permeability Rock".

K. WATANABE, K. ISHIYAMA and T. ASAEDA, Japan "Instability of the Interface Between Gas and Liquid in an Open Fracture Model".

15:00-15:30 FOYER, Coffee Break

15:30 DULUTH-MACKENZIE, Panel and Floor Discussions

Moderator: N. Barton (Norway)

Panelists: M. Langer (FRG), J.A. Hudson (UK); P. Rissler (FRG).

Panel contributions:

M. Langer (FRG) - Evaluation of barrier efficiency of rock masses for waste isolation

Abstract: According to the multi barrier principle, in waste repositories the geological setting must be able to contribute significantly to the waste isolation over long periods. The assessment of the integrity of the geological barrier can only be performed by making calculations with validated geomechanical and hydrogeological models. The proper idealization of the host rock in a computational model is the basis of a realistic calculation of thermal stress distribution and excavation damage effects. The determination of water permeability along discontinuities is necessary in order to evaluate the barrier efficiency of crystalline host rock.

1. PRINCIPLES OF MODELLING

Safety considerations regarding underground waste disposal are first of all based on the efficiency of the natural geological barrier. This barrier is expected to be able to isolate the wastes for very long periods of time. Therefore, a qualitative evaluation of the efficiency of the geological barrier is necessary [1].

Due to the multi barrier principle, the geological setting must be able to contribute significantly to the isolation over long periods of time. Although qualitative understanding, description models and expert judgement are important factors for a reasonable assurance that the integrity of the geological barrier - within the system 'waste product disposal facility, geological situation' will be valid, the assessment of the integrity of the geological barrier should be performed by calculations with validated geomechanical and hydrogeological models. The proper idealization of the host rock and the surrounding geological formations into a computation model is the basis of a realistic calculation. The natural system has to be considered with given properties as permeability, thermo-mechanical behaviour, tectonic fractures as well as in situ stress. Initial conditions given by the operation of the repository have to be taken into account, whereby the repository as a geotechnical system can be optimized to the existing geological situation. This optimization must regard systemimmanent conditions, e.g., geological and geochemical long-term processes, limiting values for temperature, rock failure modes. The adequacy of the computation model with respect to predict the barrier efficiency has to be proved by a validation procedure.

It is obvious that modelling can only reach a certain level of accuracy, since the actual behaviour of a complex geological structure will always remain unknown up to a certain extend.

The geoscientific approach to overcome this general difficulty is a continuous improvement of the model, appropriate to the improved knowledge of the input data. The main features of this approach are the establishment of a consistent constitutive relationship for the mechanical and hydrological behaviour, validation of this model, and quantification of site-relevant input data.

Herein, model validation has to follow a strict scientific procedure:

- prior to validation, the numerical code used for computation has to be verified. This means it has to prove that the code gives mathematically correct answers,
- model validation is achieved through successful predictions for laboratory tests or in situ tests, taking into account a consistent constitutive model, proper boundary conditions, and initial conditions. Model validation in this sense is not curve fitting by back-analysis but a demonstration to what extent a particular consistent model is able to describe the reponse of the host rock, although the constitutive model perhaps does not take into account the entire mechanical and hydrological behaviour, and
- a validated constitutive relationship for the rock mass is then the proper basis for modelling of the site-specific geological situation. However, site-specific units of equal behaviour and related parameters have to be determined and are to be confirmed by field tests.

The numerical models used in the assessment of the efficiency of the geological barriers have to represent the physical system of the geological medium. The degree of representation (simulation) is significant because the misapplication of mathematical models may produce erroneous results. Anyway, the use of mathematical models must be justified by the site data. Even if these data were based on a comprehensive site investigation, there is always a certain amount of uncertainties modelling a geological system. Therefore, boundary approaches must be used in a conservative manner. The extent to which boundary approaches can be justified and demonstrated as being conservative will affect the reliance of the estimation of the barrier efficiency.

Nevertheless, numerical calculations based on models are of particular significance in the case of the final storage of wates since licensing procedures require the prior reliable and convincing demonstration of safety.

2 GEOMECHANICAL MODELS AND CALCULATIONS

The main aim of geomechanical modelling is directed to stability calculations in order to demonstrate that stress redistribution due to mining operations and possible thermally induced stresses do not endanger the failure-free state of equilibrium in the rock, do not cause any inadmissable convergence or support damage during the operative period, and maintain the long-term integrity of the rock formations. It is necessary to calculate the distribution of stress and deformation in the formations surrounding the mine, under consideration of the temperature-dependent rheological properties of the rock and to compare them with the limiting load-bearing capacity of the rock mass. Above all, this requires the formulation of the geomechanical model and the associated calculation models, the study of parameters, and the definition of failure criteria [2].

To build up a calculating model, certain idealizations are unavoidable. In the terms of this model one can distinguish between:

- rock-mechanical models, and
- static models.

The rock-mechanical model encompasses:

- the geological structure of the rock mass,
- material laws to describe the time, temperature and stress dependent deformation and strength behaviour of the rock mass, and
- the primary state of the rock mass (stresses, temperature).

In the static model the following conditions have to be determined:
- geometry (establishing the load-bearing capacity of the rock mass by spatial, planar or rotation-symmetric substitute systems), and
- stress states and their changes over time:
- thermal effects (e.g. storage and heat-generating wastes, ventilation), and
- mechanical effects (e.g. leaching, backfilling, static loads, etc.).

A calculating method which is particularly suitable for the solution of such geomechanical problems is the method of the finite elements (FEM). This numerical calculation method which is specifically tailored to the requirements of automatic data processing even allows consideration of many important factors, e.g. tectonic features, operating conditions, mine geometry, construction methods etc. (Fig. 1) to an accuracy closely approaching the real situtation. The FEM method is also suitable for the calculation of non-linear limiting values. For the specific purpose of the analysis of waste disposal facilities, the Federal Institute of Geosciences and Natural Resources (BGR), F.R.G., has developed the ANSALT computer program (analysis of non-linear thermo-mechanical analysis of rock salt) [3].

Another capable computer code used for rock mechanics purposes is the ADINA code.

The long-term assessment on the geological barrier integrity cannot be evaluated from experiments alone but is only possible by computations. In addition geomechanical computations on the rock behaviour of a waste mine have the following objectives:
- analyzing thermo-mechanical processes by calculations shall lead to a proper assessment of consequences,
- experience-based conclusions can be extended by computations,
- rock mechanical criteria for a stable mine design can be developed from computational parametric studies,
- such criteria are necessary to adapt a preliminary mine design to the real geological situation.

SAFETY CRITERIA	NATURAL INFLUENCES	TECHNICAL INFLUENCES	MEASUREMENTS
DEFORMATIONS	GEOLOGICAL CIRCUMSTANCES	GEOMETRY OF CAVITY	GEOLOGICAL INVESTIGATION
STRESSES	TECTONIC	CONSTRUCTION PROCEDURES	GEOTECHNICAL TESTS
FAILURE MODE	PRIMARY STRESS	TYPE OF USAGE	STATIC CALCULATION
BEARING CAPACITY	MECHANICAL PARAMETERS	OPERATING CONDITIONS	CONTROL MEASUREMENTS
BRINE AND GAS LEAKS	BRINE AND GAS PRESENCE	TEMPERATURE	MINING MEASUREMENTS

Fig. 1: Geotechnical stability concept

3 HYDROGEOLOGICAL MODELS AND CALCULATIONS

The main aim of hydrogeological modelling is directed to calculation in order to [4]:
- better understand the processes by which toxic materials can be transported by ground-water,
- demonstrate that no harmful concentrations of toxic materials is transported to the biosphere, and
- proof that a long-term isolation of the waste is guaranteed.

Hydrogeological modelling must therefore especially take in to account factors that affect transport by ground-water. These include:
- convective transport by the flow,
- dispersive and diffusive transport,
- chemical interactions with the rocks along the flow path, and
- rates at which the wastes come into solution.

The concept of hydrogeological modelling must be based on the main types and areas of recharge, where water is added to the system and points of discharge, where ground-water leaves the system, including the qualitative amounts of water. It also has to consider the general flow paths of water passing through the system, quantitative data of water transport velocity and chemcial and physical characteristics of the ground-water in various parts of the whole system.

Adsorption and chemical reactions, including adsorption, ion exchange, and precipitation, retard or restrain the movement of some contaminants. Adsorption and ion exchange are strongly dependent on the chemical form of the contaminants, including their oxidation state and extent of complexing. The sorptive properties of various chemical substances vary through many orders of magnitude and have yet to be determined for many of the elements for various endorsing media and under repository conditions. However, the basic principles of these reactions are well-known.

In order to construct a complete model of contaminant flow, the availability of the waste to percolating ground-water must be known. This availability will depend on the form of the wastes. In the simplest one-dimensional case in which ground water flows through a porous medium (soils), the velocity of flow is given by Darcy's Law.

Regarding the modelling of the barrier of crystallline host rock, the consideration of water conductivity and water permeability along discontinuities (fissures, joints, faults) is necessary because the inflow of water capable of leaching and/or the outflow of contaminated water from the final deposit facility into the biosphere is a critical feature. In fissured or jointed rock the discontinuities are the major factors influencing the permeability of the rock mass, whereas the pore-volume of the intact rock is of secondary, or even negligible importance. The hydraulic conductivity depends for the most part on the discontinuity surfaces, their width and filling material, as well as on their origins and intersections.

In this case, hydrogeological modelling and calculation (as well as the testing of the rock) serve to quantify the direction-dependent permeability of the rock with respect to rock pressure and water pressure. They also strengthen the understanding of the applicability of equivalent porous media conceptualizations of ground-water flow in fractured rock. The controlling factors and the appropriate scale of application of Darcy's Law - if any - can be determined.

Computer codes for hydrogeological modelling on the basis of DARCY's LAW are world-wide available (e.g. computer program SWIFT, PHOENICS). Unfortunately, regarding the modelling of water flow in fissured media, several large gaps exist in knowledge of the details of transport systems. Indeed - in spite of recent big international research efforts in this area - the physical basis (that is the discontinuum mechanics) is missing which would allow reliable and validated calculations of water flow and transport mechanism in fractured media.

A promising approach to fracture flow may come from statistical analysis of fracture patterns. Statistical information is already used in discrete fracture modelling in an attempt to statistically reproduce the hydraulic characteristics of a field site by numerical modelling. This approach is only possible in cases, where a lack of information on fracture patterns within the body of the rock mass is stated. It is considered in this approach that the range of results obtained from numerical calculations will approximate the range of results that would be obtained from random sample volumes on the field.

The finite element program ADINA or the newly developed FE program DURST (BGR, Hannover) can be used to simulate flow processes in fissured rock. The fissures are represented three-dimensionally, the finite elements can be one-, two-, or three-dimensional in these codes [5].

4 ROCK TESTING

The proper idealization of the repository mine in the geological formation into a computation model is the basis for a realistic evaluation of the barrier efficiency. The geological environment has to be considered with complex properties, as internal structure, thermo-mechanical behaviour, hydrogeological features, and initial conditions. Thereby, the correct description of the behaviour of rock within a constitutive model can only be as good as the in-put data. Therefore, rock testing is an indispensable part of any modelling.

The rock properties of principal interest for waste disposal are those related to strength, deformation, and hydrological characteristics.

These properties are:

- strength characteristics: compressive strength, tensile strength, shear strength (cohesion, angle of friction);
- deformation characteristics: elasticity (E-modulus, G-modulus, Poisson-ratio), rheology (V-modulus, plasticity, retardation time, creep parameters);
- hydrologic characteristics: potential head of aquifers, permeability of aquifers, water table elevation, groundwater temperature, groundwater viscosity and density, transmissivity (storage coefficient, specific capacity), ground water recharge and discharge points.

Substantial strength is necessary for engineering design of subsurface repository facilities, especially in maintaining the integrity of underground openings. Strength properties provide the durability or resistence of material to processes such as erosion and weathering.

In general, the greater the strength, the greater the ability to resist weathering. Parameter representative of strength include cohesion or friction angle, unixial compressive strength, tensile strength, and yield limit.

Stress-strain properties indicate the deformation characteristics that a material will exhibit under stress. Parameters that describe the nature of the deformation of a disposal medium include elastic parameters (Young's modulus, Poisson's ratio, bulk modulus, and shear modulus) and rheological parameters (creep parameters, plasticity). They are significant in the analysis of earth material's properties for mineability and isolation. The ability of an earth material to deform and seal discontinuities to fluid flow is desirable. Conversely, a rigid earth material is important to the stability of the repository tunnel opening.

Hydrologic properties are essential to assessing the potential for fuid flow. They are evaluated by the parameters of permeability, hydraulic gradient, and porosity. Restriction to transport of radionuclides requires a permeability as low as possible.

A special testing procedure for the measurement of the hydrologic properties of fissured rock has recently been developed [6]. The principle is given in figure 2.

The equipment consists of the following items:

- probes and packers in the borehole,
- pumps
- computers.

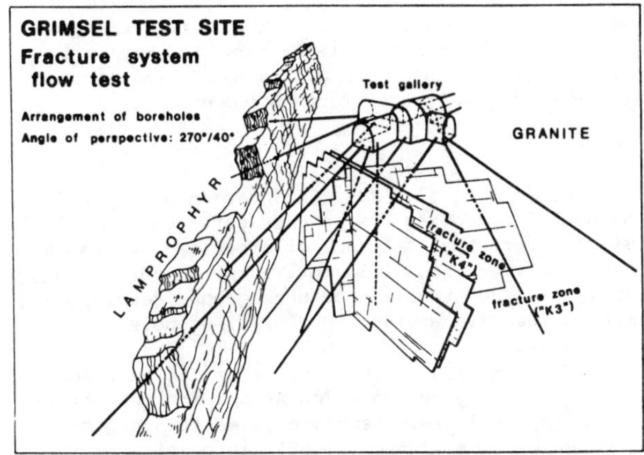

Fig. 2: Generalized diagram of the BGR-fracture system flow test

Pressure sensors are used to measure water pressure (5 - 100 bar). Temperature and electrical conductivity are also recorded. A probe consists, therefore, of three sensors. It is placed in an observation borehole (diameter 86mm) between two pneumatic packers (length ca. 1.0m). In general, up to 4 probes, each separated by a packer, are introduced one after the other into the borehole. In exceptional cases, however, 10 probes may be placed one after the other in the borehole. The data transmission cable passes through a water-proof, stainless steel pipe (length 1 - 2 m) between the probes and the computer.

For the data processing two computers, a plotter and a multiprogrammer are used.

The petrography, open joints, estimates of the aperture width, orientation of the joints, material filling the joints, faults, and foliation are determined along each borehole. The data are processed using special programs. These plot the joint systems and extrapolate from one borehole to the next and to the test gallery. The injections are done preferably in the highly jointed sections of the borehole. Changes in the pressure are measured in the surrounding boreholes in sections between packers, especially in the direction of the joint system that is being injected into.

The in situ stress state of the rock is particularly important in its influence on the calculation results and hence on the engineering safety data. The determination of the natural stress state in the rock mass is one of the most difficult tasks in rock mechanics.

It is for this reason that throughout the world a large research effort is currently underway to develop effective stress measurement methods. Several methods are available and used world-wide, one of them being the overcoring method.

The overcoring method could previously only be used in boreholes of less than 100 m. The further development of measurement methods and specific device development by the BGR [7] now makes it possible to carry out deformation and stress measurements in boreholes to depths of 300 m and more. This has been achieved in particular by running a mini-computer into the borehole, together with the measurement probe, which records the measurement values without needing a cable link through the drill string.

Stress measurements are made at a single point and extrapolated to larger sections of rock. The joint evaluation of the stress measurements and the tectonic maps of the joints and fissures yields information that will show whether such data can be extrapolated to other areas. This is appropriate only if, as demonstrated in [7], enough data is available for geologically uniform areas. The planning of stress measurements should, therefore, be done only in connection with a detailed tectonic mapping of the volume of rock in question.

References

[1] Langer, M. et al. 1986. Engineering-geological methods for proving the barrier efficiency and stability of the host rock of a radioactive waste repository. - Proc. IAEA Int. Symp. on the siting, design and construction of underground repositories for radioactive wastes, SM 289/23, Hannover, S. 463-475

[2] Wallner, M. 1986. Stability verification concept and preliminary design calculations for the Gorleben high level waste repository. Waste Management ´86 (Proc. Symp. Tucson, 1986) (in print)

[3] Bundesanstalt für Geowissenschaften und Rohstoffe 1983. Entwicklung eines optimalen FE-Programmes ANSALT zur Berechnung thermomechanischer Vorgänge bei der Endlagerung hochradioaktiver Abfälle, Final Report to BMFT-Forschungsvorhaben, KWA 2070/8, BGR, Hannover

[4] Langer, M. 1986. The impact of waste disposal mines of the geoenvironment. roc. UNESCO-workshop "Impact of Mining on the Geoenvironment", Tallin (USSR) (in print)

[5] Pahl, A. et al. 1986. Results of Engineering geological research in granite. Bull. 34 of IAEG, S. 60-65, Paris

[6] Liedtke, L. & A. Pahl. Water Injection Test and Finite Element Calculations of Water Percolation Through Fissured Granite. Proc. on a Joint CEC/NEA Workshop on "Design and Instrumentation of in-situ Experiments in Underground Laboratories for Radio active Waste Disposal". Brüssel 15.05.-17.05.1984

[7] Pahl, A. & ST. Heusermann. Stress Measure ments at the Grimsel Rock Laboratory, Geol. Jb., Hannover (1986) (in print)

R. Rissler (FRG) - Rock Hydraulic Models and Reality

Conceptual models have been developed in rock hydraulics for the past two decades. They reflect the engineer's approach who is always forced to simplify complex processes or conditions as the only means of arriving at solutions within an acceptable period of time. Any simplification is at the expense of attention to detail. Occasionally, this may therefore give rise to arguments with those who believe that essential elements are lost through simplification and that it is impossible to reduce nature in all its complexity to models. Let us take up this controversy as it lends itself to site determination.

Models are not used for their own sake but as problem solving tools. The value of a model in general as well as in rock hydraulics depends on:

- whether, despite simplification, the essential characteristics present in nature are represented in the model, and
- the requirements of the terms of reference.

DEGREE OF SIMILITUDE BETWEEN MODEL AND REALITY

The question as to the degree of similitude between the model and reality can only be answered for a given situation and for the respective problem to be solved. Various characteristics are essential:

- We distinguish between the continuous and the discontinuous model. The former is based on the assumption that the relative distances d between the flow paths are very small in relation to the dimensions a of the structure, thus assuming a homogeneous flow. The continuous model becomes discontinuous when d/a = 1/10 to 1/20.
- Another factor is whether seepage flow occurs in open pores in the rock mass, in the parting plane system or in both. In that case the permeability is regionally iosotropic from the start and the flow is more laminar. If the parting plane system is involved, the flow may also be anisotropic and/or even turbulent in certain regions.
- If the parting planes are the main flow paths, it must be determined in each case whether they are more areal or more linear.
- For ease of treatment and in the absence of corresponding data it is often assumed that the spatial orientation of the parting planes is strictly in the form of sets, that the fissure apertures are constant, the surface roughness is the same, and the flow paths are uniform over large areas. This apparently is the main objection of the critics. They fail to understand that simplified modelling is required as a first step to make the problem amenable to calculation. I am pleased to note that we have meanwhile advanced one step further. The recent paper by Koppelberg (RWTH Aachen - Rhine-Westphalian Technical University, Aachen) provides a very detailed study that allows us to estimate the error inherent in this simplification. I consider it extremely remarkable that this study was done by a geologist. As far as I am concerned, the ideal approach is to start with a model which is of necessity very simplified at first and to adapt it to more realistic conditions in the next stage or to determine the maximum allowable error in the model in the next stage.

TERMS OF REFERENCE

Rock hydraulics deals with the following problems:

1. Determination of permeability.
2. Prognosis of potential, pressure and gradient distribution.
3. Prognosis of the dynamic effects of seepage water.
4. Prognosis of the amounts to seepage water.
5. Prognosis of groutability.

Problems 2 to 4 are based on the permeability of the rock mass. The auxiliary parameters of fissure width and surface roughness are not required for determining the permeability. Inadequacies in the description or estimation of fissure width and surface roughness are therefore not critical for determining water permeability and, consequently, some of the objections by the critics do not apply.

For a prognosis of groutability, however, it would be desirable to have better data on fissure widths, surface roughness, and fissurization.

In general it is felt that rough hydraulics has meanwhile reached a stage where it provides realistic models for most phenomena of fluid flow in rock masses. It appears that the usual problems in dam construction which I am primarily addressing here, do not require any urgent research at this time. However, the prognosis of groutability calls for reliable data on the structure of fluid conductors. In addition, the old problem regarding the relationship between the capability of accepting water and accepting the grouting material has not yet been resolved.

The essential difficulty in assessing permeability lies usually not so much in poor rock hydraulic modelling but in our lack of knowledge of the rock properties within the large areas that have not been investigated. The following example is a case in point: Sixteen test boreholes up to 50 m in length were sunk at the cutoff level of a 470 m long dam. The continuously conducted water pressure tests yielded data on the permeability behaviour in an area of only 1000 to 2000 m^2 of the total sealing area of 30,000 m^2, i.e. 306% of the total area. Thus, more than 90% of the rock mass was not explored but merely estimated by means of analogue techniques. The presence of inhomogeneities in the unexplored area has a much greater adverse effect on the conclusions than shortcomings in the models. In my view, more research must be focused on this aspect.

Another area of difficulty which, in my view, has remained largely unresolved is the long-term permeability behaviour. It is important for dam safety and for long-term storage of wastes.

Limestones in the foundation dissolve in the presence of soft dam water. Fissure fillings may leach out at high gradients in the course decades. Chemical substance dissolve sealing materials. Professor Langer will no doubt report on the role of water in connection with rock salt. The effects of thermal loading on the permeability behaviour are also not clear. It seems to me that a great deal of research is required in this field.

REFERENCE (See German section)

FLOOR CONTRIBUTIONS:

Paul R. La Pointe (USA)

My contribution/question is directed to:

T.W. Doe, "Design of well testing programs for waste disposal sites in crystalline rock"

You have described three types of wellbore tests to detemine rock mass hydrologic properties, and recommend the constant pressure test based upon theoretical and pragmatic grounds. Nothing was said as to which test or interpretation was more accurate or useful for design. I am concerned that our theory is, perhaps, out too far in advance of large-scale field testing. Hydrofracturing stress measurement was given an immeasurable boost in credibility through the Rangely experiment. Where are our Rangely to sort out the useful models and theory from the useless ones? Shouldn't we be giving first priority to undertaking very large-scale tests, rather than developing more theory and models?

Response by T.W. Doe (USA)

Dr. La Pointe has made two comments, one regarding the relative accuracy of well test methods and the other regarding large-scale versus small scale testing approaches. In answer to the first question, all of the methods commonly in use today are simply variations on solutions of the diffusion equation with different boundary conditions and perturbations. Provided the tests are properly designed with appropriate instrumentation all should provide reasonably accurate determinations of near-hole rock transmissivity. The methods do vary in the time requirements and in their sensitivity to boundary conditions.

Next is the question of large-scale versus small-scale testing appraoches. Single-borehole well tests provide essentially a line sample of the hydraulic properties of a rock mass. Because the rock is not a homogeneous continuum, the result of a single test cannot be considered a measurement of the large scale hydraulic properties of the rock, however, a series of single hole well tests can provide data reflecting the properties of a rather large mass of rock. The thrust of the discussions in statistics portion of the paper is that we can make inferences of the large scale hydraulic properties of the rock from our hydrologic "line samples". This approach is similar to that used for geomechanical problems where one uses fracture survey data to infer large-scale mecahnical properties of rock masses.

Large-scale testing can be very difficult in highly heterogeneous rock - indeed it is not clear that one can describe the hydraulic behaviour of a crystalline rock mass using a simple set of property values. Large scale tests using borehole arrays are difficult to perform due to (1) ambiguity of results when some or most of the boreholes fail to intersect conducing portions of the rock mass. This was the point I tried to make in my presentation by posing the following problem: how would you characterize the hydraulic system of your home by drilling holes through it? Large-scale tracer tests can be used to confirm transport properties, but these tests require information on the details of the flow paths in order to properly locate injection and withdrawal points.

In conclusion, I believe both large-and small-scale testing approaches will be required. Small-tests provide details of flow paths that are smeared in large-scale exercises. Large scale tests, on the other hand, will ultimately be required to verify the conclusions of the more detailed well-testing programs.

Frank Patton (Canada)

My contribution/question is directed to: T.W. Doe, "Design of well testing programs for waste disposal sites in crystalline rock"
While Dr. Doe described the advantages of constant head packer tests for hydro-geologic characterization of low permeability rock masses, I felt that it would be appropriate to describe the disadvantages of such tests at the same time. Namely:

(1) Packer tests in an open hole permit flow in those parts of the hole not being tested thereby allowing the natural formation (or fracture) fluids to be altered or displaced. This has an adverse effect upon water quality tests (on samples) which generally are at least equal in importance to the hydraulic testing.
(2) Packer tests (with or without guard packers) always have the real connections through fractures from the test zone to the adjacent parts of the open drill hole not being tested. Thus the tests would measure an unnatural condition that is an artifact of the drill hole and not of the rock mass.
(3) Equilibrium in the test pressure or test flow may only represent equilibrium with adjacent unnatural conditions in the open drill hole due to conditions noted in 2 above.
(4) Holes left open for packer tests can affect results in nearby holes with respect to samples, pressure heads and other hydraulic parameters. There are other serious disadvantages of packer testing in open holes, but perhaps these are the main ones. I would appreciate your comments on these points.

Response by T.W. Doe (USA)

Dr. Patton has raised some key concerns regarding testing in open holes versus testing in completed piezometers. As Dr. Patton points out there are clear technical, if not always practical, advanatages to performing tests through piezometers rather than in an open hole.

I should point out that the methods described in the presentation could be carried out equally well in a multiple-ported piezometer system, such as that produced by Westbay Instruments. The usual objection to testing in piezometers is that conventional configurations do not allow testing a sufficient number of zones to characterize the rock's heterogeneity, however, the Westbay system with its capacity for a large number of packers and ports is clearly a promising approach to overcoming this problem.

As to the question of flow back to the borehole (point 2 and 3) the existence of a flow path from the test zone back to the hole should effect mainly the estimation of boundary distance and not the near-hole transmissivity value.

Akira Kobayshe Hazama-Gumi, Japan
My contribution/question is directed to theme I, "Fluid flow and waste isolation in rock masses". (Proceedings pp. 207-210)
The coupled thermal, hydraulic and mechanical analysis of the macro-permeability experiment and the buffer mass test conducted in the Stripa project.

We used the newly developed code of which the governing equations are derived with the fully coupled thermal hydraulic and mechanical relationships. As conclusions we got:

(1) Reduction of permeability near an underground opening seems to indicate the closure of fracture due to circumferential stress increase.
(2) As the temperature of rock mass heats up, rock blocks expand and fractures close. As a result the permeability of rock mass reduces.
(3) The change of inflow rate to the opening is strongly influenced by the temperature dependency of the kinematic viscosity.
(4) The heat conductivity in buffer materials is influenced by the porosity more effectively than water content.
(5) The water content in buffer material may strongly be influenced by water movement due to evaporation.

We believe that the qualitative interpretation of the behaviour of the rock mass subjected to complex loading can be done by the powerful numerical tool.

Dave Goodall, SINTEF, Norway
My contribution/question is directed to:
Watanabe et al. "Instability of Interface ..."
I am concerned about the quality of the apparatus and believe that the results may be in error.
Q(1) How thick were the acrylic plates?

I am concerned that the deformation of the acrylic plates may have been sufficient to influence the results of experiments 2, 3 and possibly others.

My testing with one inch thick glass plates with a tube steel supporting frame suggests that experimental points should fall on the theoretical line of Figure 6. I should like to have equation (2) clarified before making a more positive conclusion.

Q(2) What are the intervals D? Presumably they are the widths of the air fingers.

Response by: Kunio Watanabe, Keji Ishiyama, and Takashi Asaeda, Japan
The authors present a laboratory study of a highly idealized rock fracture geometry. It is somewhat unlikely and, in fact, undesirable that a gas storage cavern should be constructed intersecting, and parallel to, an extensive vertical fracture. Furthermore a perfectly horizontal interface is highly improbable. Nevertheless studies such as the one under discussion are considered necessary to appreciate the fundamentals of gas containment in underground rock caverns. This discussor has conducted similar tests using smaller apertures (50 and 200 microns, see Goodall, 1986).

The model which the authors have tested is expected to represent a vertical one dimensional flow condition although there are no details in the paper which indicate attention to uniformity of aperture. Acrylic plate is quite deformable and deflections of the order of millimeters are possible for the pressures existing in the tests. Details of aperture control and an estimate of aperture variability from the authors could relive this concern which arises from the apparent lack of support of the plates as indicated by Figures 3 and 5 in the paper.

The procedure which the authors have followed presumes that fractures are initially dry adjacent to the air pocket. Their observation of "comb like fingers" on Figure 3 is directly a consequence of this presumption. The laboratory experience of this discussor (Goodall, 1986) indicates that, had the model been initially saturated with glycerol, there would have been no observation of fingering or any air entry between the plates during the first stage of any of the experiments (air pocket maintained at atmospheric pressure). It is suggested that for initially saturated apertures no fingering is observed for $H_s>0$; for the authors' procedure fingering will always be observed in the lower part of the model unless $H_s>7L_i$ (the symbols used here are those of the authors). In the first case the limiting fluid potential gradient is zero; in the second it is one (downward flow) in the lower part of the model. Clearly one must carefully reflect upon which case is appropriate in the field.

The second part of the experiments, namely the determination of the air pressure at which gas escape occurs, may or may not be influenced by fingering in the lower part of the model. This discussor has found (Goodall, 1986) that, in the absence of fingering, the fluid potential gradient for this geometry will be zero at gas escape initiation. This corresponds to the line $H_a = H$ on Figure 6 (neglecting capillary pressure). The data obtained for experiment Exp-4 disagrees directly with this experience because no fingering could exist in experiment Exp-4 prior to gas escape.

The data reported from experiments Exp-1, Exp-2 and Exp-3 do not consistently show an effect of fingering. Data for Exp-1 agrees well with this discussor's experience although one must hasten to add that the authors report the disappearance of the fingering in the lower part of the model just prior to the point of gas escape. This is entirely consistent with the criterion $H_a = H$ on Figure 6 since it corresponds to hydrostatic conditions and no flow exists to support the fingers. On the other hand results from Exp-2 and Exp-3 do suggest a possible influence of fingers but their inconsistency with experiment Exp-1 is bothersome. The only real difference between experiment Exp-1 and experiments Exp-2 and Exp-3 is the fluid pressures if it is assumed that the interface perturbations due to fingering are identical for all experiments. This leads this discussor to conclude that plate deformation may have influenced the results for experiments Exp-2, Exp-3 and Exp-4.

In conclusion one should note that the maximum critical hydraulic gradient observed in this test is about 0.42. Real rock fractures should be expected to have roughness and irregularities which result in higher critical gradients for this one dimensional flow geometry.

Equation (2) in the original paper is assumed to be in error. This discussion assumes it to be as follows:

$$H = (H_s+L_i) / (\gamma_w/\gamma_g)$$

Reference:

Goodall, D.C., 1986. Containment of Gas in Rock Caverns. Ph.D. thesis, Dept. of Civil Eng., University of California, Berkely. 371 pp. Paper copies available from University Microfilms International, 300 N. Zeeb Road, Ann Arbor, MI 48106, USA.

D.C. Goodall (Norway)

My contribution/discussion is directed to:

K. Nakagava, H. Komada, K. Miyashita and M. Murata (Japan), Study on compressed air storage in unlined rock caverns.

The broad scope of this paper and consequently the lack of details make it difficult to evaluate the results which are given. Few papers have been published in referred forums on this topic, however, and this opportunity is taken to review this paper in comparison with this discussor's findings.

The criterion which the authors have used to present their results for the parallel plate tests is scale dependent. To appreciate this statement consider the sketch of their model as shown on the Figure below (not to scale).

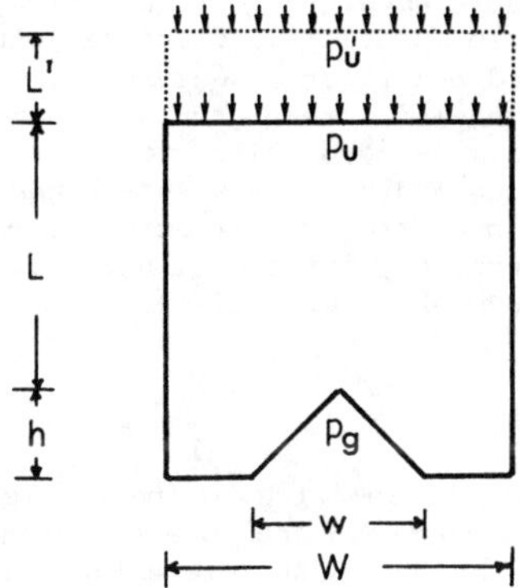

Representation of Parallel Plate Model

The results are reported in terms of a ratio P_{ws}/pa where $P_{ws} = Pu + \rho_w g\ L$ (the symbols used here are those of the authors). Suppose that a test has been conducted and that the conditions at gas escape have been defined by the symbols. Then suppose that the experiment is repeated except with an additional length L' of plates as indicated on the above Figure. Gas escape in the new model must occur when P_u and P_g assume the same value as were obtained for the original model. In general one should expect that some downward water flow is occurring at gas escape and that this flow, for the dimensions under discussion here, is characterized by a one dimensional flow gradient, I. It can thus be demonstrated that

$$P'u = Pu + \rho_w gL' \ (I-1) \qquad (1)$$

and that

$$(P_{ws}/P_a)' = [P_u + \rho_w g(L'I + L)]/P_a \qquad (2)$$

Clearly this is not the same as that computed directly from the authors' criterion for the original test namely

$$P_{ws}/P_a = (P_u + \rho_w gL)/P_a \qquad (3)$$

unless I = 0, which is not generally true unless significant capillary pressures exist. More specifically, I ≠ 0 unless the capillary pressure at the base of the model exceeds a water pressure equivalent to the height of the notch (h on Figure 1) (see Goodall, 1986). Thus the criterion which the authors have used is not generally applicable.

Unfortunately the authors do not present adequate information to evaluate the capillary pressure of the model or the impact of capillary pressure on the results. The capillary pressure for a 1 mm aperture between acrylic plates is expected to be of the order of 10 mm for an air/water interface. Consequently one should expect that significant hydraulic gradients will exist as long as h > 10 mm. According to Figure 2 of the paper (h = 30 mm) this is true for at least some of the models. But for the larger α angles listed on Table 1 the width of the model becomes restrictive and h must be reduced. For α = 180°, for example, h = 0 which confirms that capillary pressure is overwhelming for such a test and that one can expect hydrostatic conditions for such geometry. This is presumably the reason that the lowest P_{ws}/P_a ratio was observed for α = 180° (see Table 1). The results of the α = 180° test suggest a capillary pressure equivalent to 33 mm of water which seems unreasonably high and is possibly a consequence of pressure measurement inaccuracy (difference of large numbers).

Irrespective of the capillary pressure, the data on Table 1 do confirm the existence of hydraulic gradients as high as 0.09 at gas escape initiation. This gradient may not seem particularly high in practical terms given the uncertainty of field conditions and normal factors of safety. The problem is that dramatically different results can be obtained for different fracture mouth geometries and capillary pressures (Goodall, 1986; Aberg, 1977). In the field the fracture mouth irregularities may be substantially larger than 30 mm, at a variety of angles and with a large variety of capillary pressures (apertures). If the authors had tested a geometry with w/W = 1 (see definition of symbols on the Figure included here) it can be shown that the hydraulic gradient at gas escape for a model otherwise similar to theirs would be about 0.65 (for a capillary pressure of 10 mm, α = 30°, h = 30 mm) (see Goodall, 1986). This is substantially different from hydrostatic conditions and, if such fracture geometries exist in the field, design strategies must be modified accordingly. If this example is carried one step further to the case of negligible capillary pressure, the critical gradient will be one which is the design criterion recommended by Aberg (1977).

In view of the rather demanding conditions for gas containment imposed by the examples discussed above one must ask what role such geometry will play in a full scale cavern in fractured rock. This discussor considers that such nearly one dimensional geometry is appropriate only on the scale of individual fractures intersecting the gas cavern; it is not a scaled model of a gas cavern storage. The second model used by the authors (see their Figures 3 and 4) is more typical of the overall field condition except in its presumption of homogeneity, isotropy and continuity.

Experience of this discussor indicates that for geometry similar to that given on Figures 3 and 4, the critical gas pressure (water head equivalent) is very close to the depth of the cavern beneath the water table. The allowable gas pressure is not sensitive to capillary pressure and it appears to be slightly less than the depth in the case of negligible capillarity (Goodall, 1986). These remarks are considered appropriate to conditions of equal water levels in the left and right reservoirs and where the depth of the cavern is at least 10 times the cavern diameter. Otherwise different results are anticipated. When the left and right reservoirs are at differing levels the perpendicular distance from the cavern to the phreatic line is considered more appropriate. In the most extreme case examined by the authors (I-4) the operative water head is estimated to be 81 percent of that used by the authors. Unfortunately the authors do not present the results of their tests so that an evaluation of such effects can be made. Considering that over 60 percent of their tests are conducted with unequal elevations in the left and right reservoirs, this discussor is moved to question the strength of the authors' conclusion that "I_{cr} values gathered around zero" and that the allowable gas pressure is that corresponding to the depth of water above the crown of the cavern. It must be remembered that capillary effects are much more dominant in laboratory models and that the significant capillarity of the porous slabs used in the models may result in rather optimistic observations.

With respect to the two phase flow analysis which the authors have undertaken there appears to be inconsistency in the approach. When treating two phase flow it is common to assign the material a permeability and compute the conductivities on the basis of fluid properties (air and water) as is done in the authors' equation (1). The relationship is as follows.

conductivity = (fluid unit weight/ viscosity) x permeability

From this it should be clear that the conductivities for various fluids in a given material are fixed by the fluid properties and that the ratio between various fluid conductivities is similarly fixed.

In the case of water and air the following laboratory (19°C) parameters are thought appropriate:

	Air	Water
units weight kN/m³	0.0123	9.8
viscosity N-s/m²	1.84×10^{-5}	1.006×10^{-3}

From this information one may conclude that the ratio of the conductivities water : air should be about 15. The corresponding ratio of the reported conductivities (which the authors refer to as permeabilities) ranges from 0.1 to 0.6. The reason for this large discrepancy appears to be turbulent effects rather than some subtle effect of Knudsen diffusion. Additional information regarding these tests (Komada and Nakagawa, 1985) clearly shows that a plot of water flow versus head is strongly concave down, thus indicating that water conductivity results are significantly reduced.

The impact of the above observation on the authors' results in the paper is uncertain. The hydraulic gradients in their model of gas storage are somewhat less than those used in the water permeability tests. It is not clear what permeability was assumed for their model material although this should not impact final steady state conditions. More details of assumed relative permeabilities (these are different for air and water saturation) are needed as well as the remaining parameters assumed in the model. This discussor does not agree that the computed results agree well with measured values (Figures 5). Furthermore it seems that disproportionate air and water conductivities have been assumed in the authors' compressed air storage system simulation.

This discussor believes that the two conclusions arrived at in this paper are not well supported by the research which has been reported and that there appears to be a lack of appreciation of the fundamentals. These fundamentals should be appreciated prior to undertaking the even more difficult task of extending the results of such studies to a real complex geologic environment.

References:

Goodall, D.C., 1986. Containment of Gas in Rock Caverns. Ph.D. thesis, Dept. of Civil Engineering, University of California, Berkeley, 371 pp.

Komada, H. and Nakagawa, K., 1985. Study of Compressed Air Storage in Model Experiments. Research Rept. 384044, Central Power Research Laboratory, Japan.

Aberg, B., 1977. Prevention of Gas Leakage from Unlined Reservoirs in Rock. Proc. Ist. Int. Symposium on Storage in Excavated Rock Caverns, Stockholm, Sweden.

Fluid Flow and Waste Isolation - Theme I

Moderator Summary: N. Barton (Norway)

Abstract: Papers submitted to Theme I are grouped under six main sub-themes: Network modelling (permeability tensor problems), discrete joint modelling (coupled behaviour), sealing of rock masses (water inflow problems), creep problems, and measurement techniques (hydraulic and geophysical). Major issues raised under the six sub-themes are listed. A short moderator contribution is made on the themes of discrete joint modelling, hydro-thermo-mechanical coupled behaviour, and model validation problems.

1 INTRODUCTION

The fifty two papers submitted to Theme I were represented by only fourteen oral presentations. Inevitably, the choice of these interesting papers left several themes untouched. A particular void was felt due to the exclusion of papers on coupled numerical models, an area of great concern to those involved in nuclear waste isolation problems. In order to redress the balance slightly, this moder ator's comments will dwell specifically on some of the problems identified in this area.

2 LISTING OF SUB-THEMES AND AUTHORS

In order to aid readers of the proceedings (Vol-(umes 1 and 2) the following listing of authors and sub-themes is submitted. The listing is not complete but will provide initial guidance to those seeking papers on specific topics.

2.1 Network modelling

* Elsworth and Piggot
* Dershowitz and Einstein
* Long, Billaux, Hestir, Chiles

Permeability tensors

* Brown and Boodt
* Barbraux, Cacas, Durand, Fenga
* Oda, Hatsuyama, Kamemura

2.2 Discrete joint modelling/testing

* Pyrak-Nolte, Myer, Cook, Witherspoon
* Gentier
* Nakagawa Komada, Miyashita, Murata
* Watanabe, Ishiyama, Asaeda

2.3 Sealing of rock masses (water inflow problems)

A. Jointed rock

* Azevedo, Correa, Quadros, Cruz
* Matsumoto, Yamaguchi
* Semprich, Speidel, Schneider
* Gang, Tian, Quan
* Mineo, Hori, Ebara
* Johansson

B. Salt/potash

* Charo, Zelikson, Berest * Langer
* Crotogino, Lux, Rokahar, Schneider
* Schwab, Fischle * Mraz

2.4 Coupled numerical models

* Chan, Scheir
* Curran, Carvalho
* Izquierdo, Romana
* Ohnishi, Kobayashi, Nishigaki
* Sato, Ito, Shimiza
* Shaffer, Heuze, Thorpe, Ingraffea, Nilson
* Thiercelin, Roegiers, Boone, Ingraffea
* Kuriyagawa, Matsunaga, Yameguchi, Zyvdloshi, Kelkar
* Charo, Zelikson, Berest
* Côme
* Crotogino, Lux, Rokahr, Schneider
* Langer

2.5 Creep problems

(i) Salt creep

* Balthasar, Haupt, Lempp, Natau
* Langer
* Preece
* Charo et al.
* Crotogino et al.
* Mraz

(ii) Well stability problems

* Maury, Sauzay, Fourmaintraux
* Rongzun, Zuhui, Jingen
* Preece
* Geunot

2.6 Investigation Methods (hydraulic, geophysical etc.)

* By
* Pahl, Liedtke, Kilger, Heusermann, Bräur
* Barbreau, Cacas, Durand, Fenga
* Brown, Boodt
* Cornet, Jolivet, Mosnier
* Andrade
* Haimson
* Heystee, Raven, Belanger, Semec
* Azevedo, Corréa, Quadros, Cruz
* Matsumoto, Yamaguchi
* Mineo, Hori, Ebara
* Agrawal, Jain

3 PANELIST CONTRIBUTIONS

1. Dr. Peter Rissler (Ruhr Reservoirs Association) "Rock hydraulic models and reality, basis for further research" (Felshydraulische Modelle und Wirklichkeit, Ansätze für die weitre Forschung)
2. Prof. Michael Langer (BGR, Hannover) "Evaluation of the barrier efficiency of rock masses for waste isolation" (Berechnung der Barrierenwirksamkeit von Fels zur Abfallisolierung)
3. Dr. John Hudson (Imperial College, London) "Coupled mechanisms - the matrix approach to the overall engineering objective"

4 MAJOR ISSUES AND KEY QUESTIONS

A short list of key questions is given below to focus attention on some of the major issues. This list is inevitably incomplete and subjective, but does reflect some of the questions raised when reading the proceedings.

4.1 Joint network modelling (2D and 3D)

a) How much error do we introduce by not modelling deformable joints?
b) Can we model the channels at block edges (joint intersections)?
c) Can super-conductors be detected and modelled?
d) Relevance of equivalent media permeability tensors?
e) What representative test volume?

4.2 Discrete joint modelling

a) Are linear joint stiffness approximations acceptable?
b) How important is it to model shear?
c) How many definitions of joint aperture do we need?
d) Contact area → channelling → scale effects → cubic law?

4.3 Sealing of rock masses

a) How well do permeability measurements indicate groutability in jointed foundations?
b) How well can we seal bulkheads in salt rocks?
c) How well can we seal tunnels in jointed rocks?
d) Brine migration problems?

4.4 Coupled numerical models

a) It is worth performing hydro-thermomechanical studies if our simplifications of one process introduce errors larger than the total effect of ignoring one of the processes?
b) What are the key input parameters and how will we obtain them?

4.5 Creep problems

a) Closure of boreholes and stress redistribution effects?
b) What is a "plastic" zone in reality?
c) Salt creep predictions - why are they not more accurate?

4.6 Measurement techniques

a) Hydraulic testing - how representative of the rock mass as a whole is the result?
b) Geophysical testing - how much geological detail is detectable?

Discussion to Theme I: from left to right: P. Rissler, J.A. Hudson, N. Barton, M. Langer, G. Herget, W. Bawden.

5 MODELLING DIFFICULTIES

The modelling of fluid flow through rock masses using regular spaced, rigid networks of intersecting parallel plates is unfortunately, a long way from reality. Variable joint aperture, orientation and continuity, chanelling, pipe flow (along intersections out of the 2D plane) and effective stress - aperture coupling are some of the factors that make simulation difficult.

For obvious reasons there appears to have been a division of labour in the modelling of rock joints. Hydrologists and radionuclide migration specialists on the one hand have developed fascinating models some of which are now capable of simulating several of the real features that are deemed important, i.e. chanelling, and super conductors to name but two. Rock mechanics specialists on the other hand have developed discrete element or discontinuum models which account for the effect of stress change on rock joint aperture, whether this be caused by compression, tension or shear.

Since the equivalent parallel plate smooth wall aperture (e) is directly related to joint conductivity (k) by the well known relation

$$k = e^2/12$$

it would appear in principle to be straightforward to relate (k) to the joint closure, opening and shear caused by changing compressive, tensile and shear stresses.

It does not require too much imagination to envisage the computer limitations presently encountered when the best 3D rigid network hydrology models are coupled to the best deformable rock mechanics models. Fortunately it may sometimes be in order to neglect one of the processes in a coupled system if it can be shown that only small changes (i.e. heating) are likely to occur in the time span involved.

6 DEFORMABLE JOINT MODELLING

An illustration of a sophisticated single process model presently operated at NGI is illustrated in Figure 1. The simulation is performed using the discrete element code UDEC (Cundall, 1980) in its micro-machine version (µDEC). This code has a new sub-routine BB to describe the non-linear details of joint behaviour based on the coupling of joint shear strength, deformation and conductivity (Barton, Bandis and Bakhtar, 1985). The code can be operated with coupled joint flow when using a main frame computer.

The conducting aperture of the joints shown for the right-hand tunnel in the lower figure, is based on the continuous tracking of mechanicl apertures (E). These joint apertures are a function of cloure, opening, shear and dilation caused by the excavation process. Mechanical apertures (E) are converted to conducting apertures (e) based on the data sets shown in Figure 2. These inequalities between (e) and (E) are caused by the tortuous fluid flow path and channelling around contact areas in the joint plane; factors which increase the inequality at higher stresses.

The code contains a sub-routine for fluid flow on the joints. This has not been activated in this tunnelling study, due to the provision of pregrouting and the application of fiber reinforced shotcrete; initial measures designed to limit water in flow (and subsequent fluid pressure changes) to very small values.

The ultimate goal will be to incorporate a realistic non-linear joint code in a 3D discrete element code with joint water flow and thermal gradient modelling. This goal lies some years in the future due firstly to the very limited amount of generic (or specific) input data.

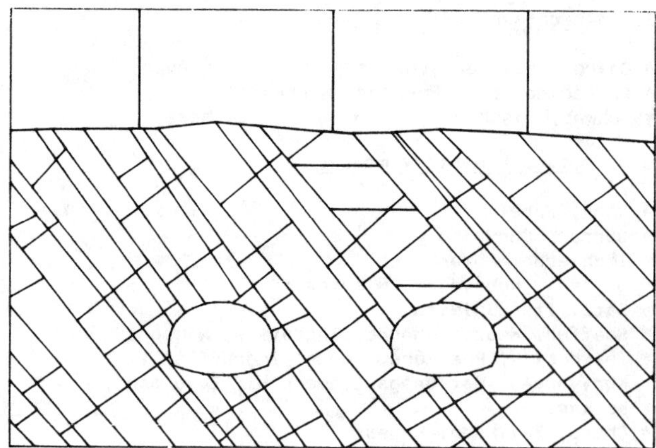

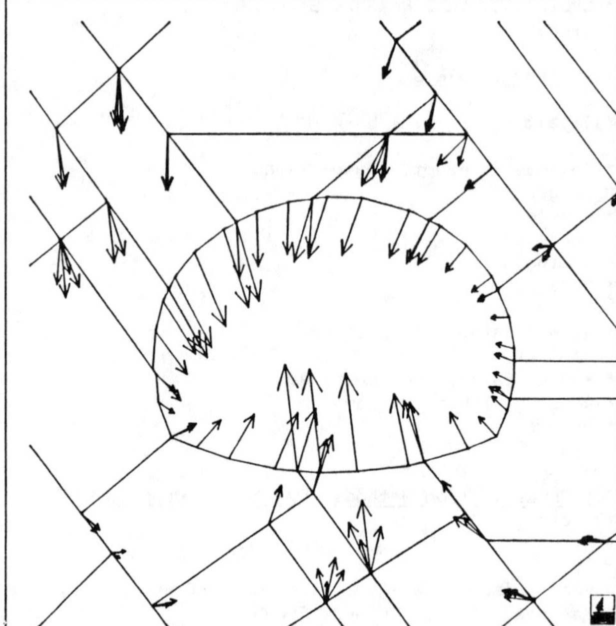

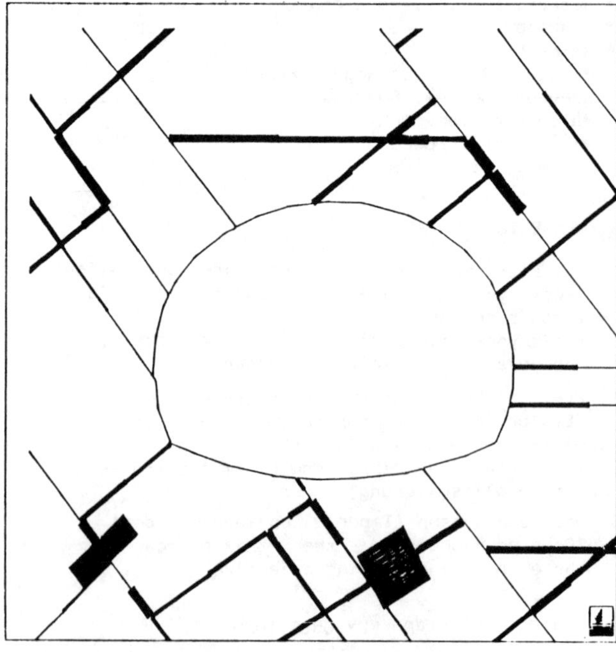

Figure 1. µDEC-BB discrete element model of tunneling effects on block deformations and conducting apertures of joints. (Makurat et al, 1987).

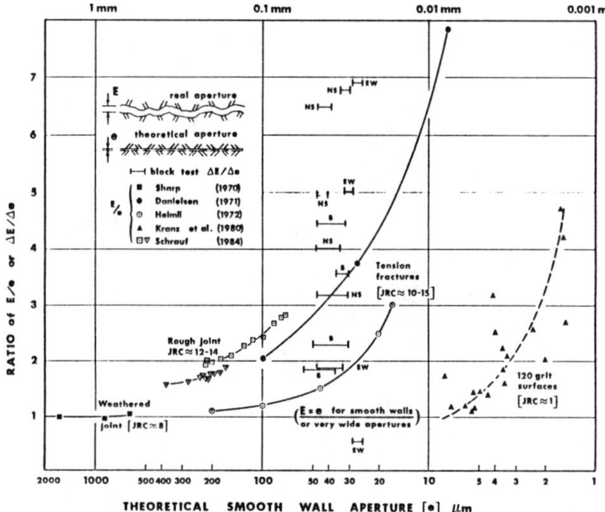

Figure 2. Inequality of real joint aperture (E) and theoretical conducting aperture (e), due to flow path tortuosity, contact areas and channelling. (Barton et al, 1985).

7 PROBLEMS WITH FULL COUPLING

Joint behaviour under full hydro-thermo-mechanical coupling is little understood at present, due to the limited number of studies performed. The CSM heated block test performed at the Colorado School of Mines experimental mine by Terra Tek (Hardin et al. 1982) showed us that joints may have the potential for fitting together more tightly when the temperature is raised (Fig. 3).

Whether this effect also increases the shear resistance is unknown, but appears likely. The behaviour may be such that shear resistance increases with temperature under pure normal loading, together with the reduced conductivity. If differential stress changes caused by the heating are large enough to cause shearing, the post-peak shear resistance may drop back to ambient values (or lower) due to the loss of interlock and due to

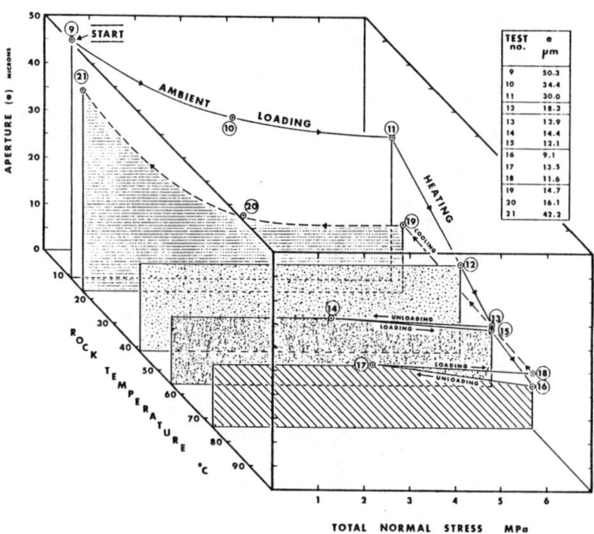

Figure 3. Hydro-thermo-mechanical coupling in a heated block test. Joint conducting aperture (e) shows improved fit of joint walls at 75°C under pure normal loading. (Hardin et al, 1982.)

the reduced joint wall strength at high temperature. In situ test programmes and constitutive modelling schemes are a long way from the ultimate goal, but significant steps have been taken in the performance of such block tests.

8 PROBLEMS WITH CODE VALIDATION

Recent experiences with code validation at NGI have focussed attention on a dilemma. If a test object such as an in situ test is to be modelled with a new code, where the objective is code validation against a set of measured data, how much simplification and pre-knowledge of the test object is allowable?

Figure 4 illustrates a simplified μDEC-BB model of the CSM block test, using fluid pressure to load the jointed block in a similar manner to the real flatjack loading. The simplification adopted was to model the block with only three joints (four blocks). This is only a fraction of the total

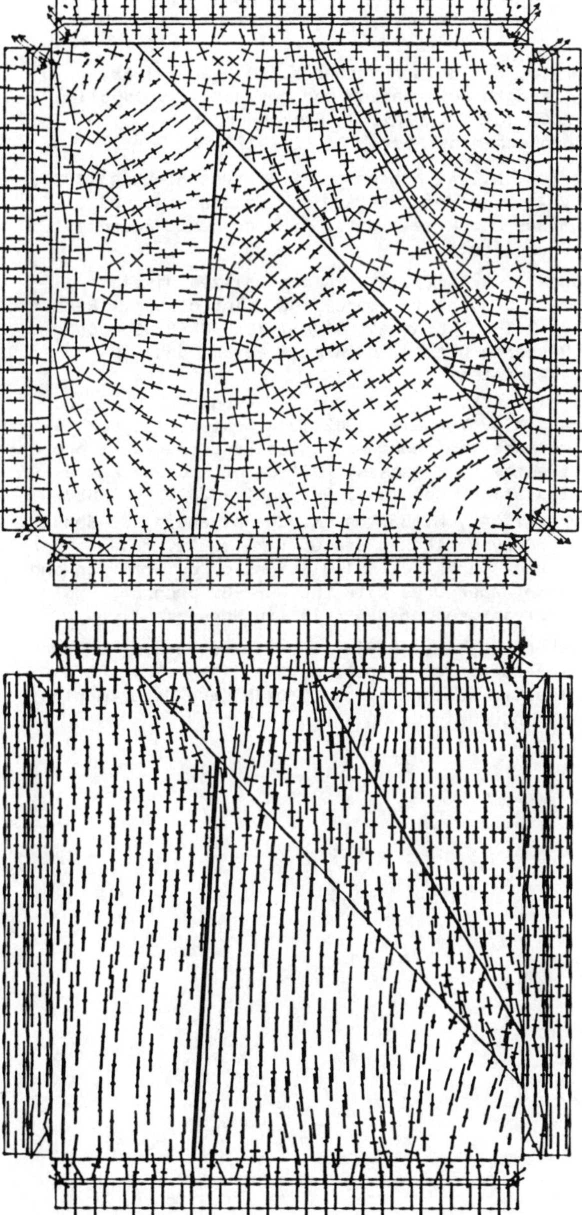

Figure 4. Principle stresses for biaxial and uniaxial loading of an in situ block of jointed rock. Barton et al, 1987.

number of joints observed on the block surface. However the four blocks have been identified by Richardson (1986) as those in which significant differential displacements occur. They are identified as the active blocks.

In forward predictions one will not have the advantage of knowing which blocks are "active" in a given rock mass. The problem therefore focuses on the need for improved remote methods of distinguishing continuous from discontinuous joints, in parallel with the need to remotely distinguish the smooth joints from the rough joints. The combination of characteristics "continuous" and "smooth", is obviously more likely to define the active (shearable) joints than the combination "discontinuous" and "rough".

The pressing need for in situ investigation methods that will distinguish the important jointing from the unimportant jointing must be clear, and represents a real challenge to both geophisicists and hydrologists.

REFERENCES

Barton, N., S. Bandis and K. Bakhtar, (1985). Strength deformation and conductivity coupling of rock joints. Int. J. Rock Mech. Min. Sci. Geomechanics Abst., 22 (2): 121-140.

Barton, N., K. Monsen and P. Chryssanthakis (1987). Validation of μDEC against Colorado School of Mines block test data. SKB project "Bergets Stabilitet".

Cundall, P.A. (1980). A generalized distinct element program for modelling jointed rock. Report PCAR-1-80, Contract DAJA37-79-C-0548, European Research Office, US Army. Peter Cundall Associates.

Hardin, E.L., N.R. Barton, R. Lingle, M.P. Board and M.D. Voegele (1982). A heated flatjack test to measure the thermomechanical and transport properties of rock masses. Office of Nuclear Waste Isolation, Columbus, OH, ONWI - 260, p. 1983.

Makurat, A., N. Barton, M. Christianson, S. Bandis, P. Chryssanthakis and G. Vik (1987). Fjellinjen - rock mechanics calculations of deformation and joint apertures with the μDEC-BB program. NGI contract report 85614-13 (in Norwegian).

Richardson, A. (1986). In situ mechanical characterization of jointed crystalline rock. Ph.D Thesis. Colorado School of Mines, Department of Mining Engineering.

Theme 2
Rock foundations and slopes

Site investigation and rock mass characterization; slope stability and open-pit mining; foundations for bridges, buildings and dams; probabilistic approaches to design; prediction, monitoring and back-analysis

8:30 LE GRAND SALON - September 1, 1987

Chairman: F.H. Tinoco (Venezuela)

Session Secretary: P. Rosenberg (Canada)

Speaker: M. Panet (France)

Reinforcement of Rock Foundations and Slopes by Active or Passive Anchors.

ABSTRACT

The reinforcing of rock masses with prestressed anchors is a well-known, proven technique. Preliminary testing and monitoring of provisional and final tensioning ensure that each anchor is working properly. At the same time, reinforcement with untensioned bars, a widely used technique because of its lower cost, has demonstrated its effectiveness on many building sites. However, the conditions of interaction between the reinforcement and the rock mass are more difficult to analyse, and current calculation procedures are not entirely satisfactory. Furthermore, there is no control method that fully guarantees the proper installation, correct functioning and durability of the reinforcement. Research efforts should be continued to establish the characteristics of a reinforced rock mass.

In many cases, the stability of rock masses cannot be ensured with an appropriate factor of safety unless corrective measures are taken, such as modifications to the geometry of excavations, drainage of the rock mass, construction of retaining works or reinforcement of the rock mass.

The best solution is chosen on the basis of a stability analysis in which the behaviour parameters of the discontinuity surfaces play an essential role. In rock masses, failure starts with detachment and slippage along discontinuity surfaces. Thus any method of reinforcement tending to restore continuity to a rock mass improves its stability very significantly. This objective may be reached by using steel bars or cables, and this kind of reinforcement is becoming common. There are two competing techniques:

- reinforcement with active (prestressed) steel bars;
- reinforcement with passive (untensioned) steel bars.

An examination of the practices followed in various countries indicates that designs vary and favour one or other of these two techniques. Each obviously has its advantages and disadvantages, which I will examine with the benefit of a detailed analysis of their performance in a discontinuous rock mass.

1 ACTIVE REINFORCEMENT

Reinforcement of rock masses by prestressed anchors or bars is an old technique. High-strength anchors (up to 10,000 kN) were used by André Coyne and Eugène Freyssinet in the 1930s.

A prestressed anchor consists of the following (Fig 1):

- the anchoring zone, where the anchor is embedded in the rock;
- a plug to isolate the embedment zone from the unsupported section to prevent the return of grouting material along the drill hole during injection to embed the anchor;
- the unsupported length of the anchor bar;

  the anchor head and the bearing mass, with the locking device.

In France, the recommendations distinguish:

- maximum tension T_1, corresponding to the breaking point of either the anchor bar or the embedment;
- tension T_G, corresponding to the elastic limit of the anchor bar;
- T_a, the maximum allowable tension in service.

The following must be complied with:

$T_a < 0.6\ T_G$ for permanent anchors;

$T_a < 0.75\ T_G$ for temporary anchors in use for less than 18 months.

There is still no satisfactory theory that will determine what tension will lead to failure of the embedded portion. However, compliance testing and tensioning procedures allow us to check that anchors are performing properly.

According to the French recommendations, tensioning is done in successive stages up to the test tension T_e, such that:

$T_e = 1.2\ T_a$ for temporary anchors;

$T_e = 1.3\ T_a$ for permanent anchors.

In these tests, the critical creep tension can also be checked, as can the equivalent unsupported length.

The force applied by prestressing counteracts opening of the discontinuities or increases their resistance to shear; in the former, it acts through its perpendicular component and must therefore be

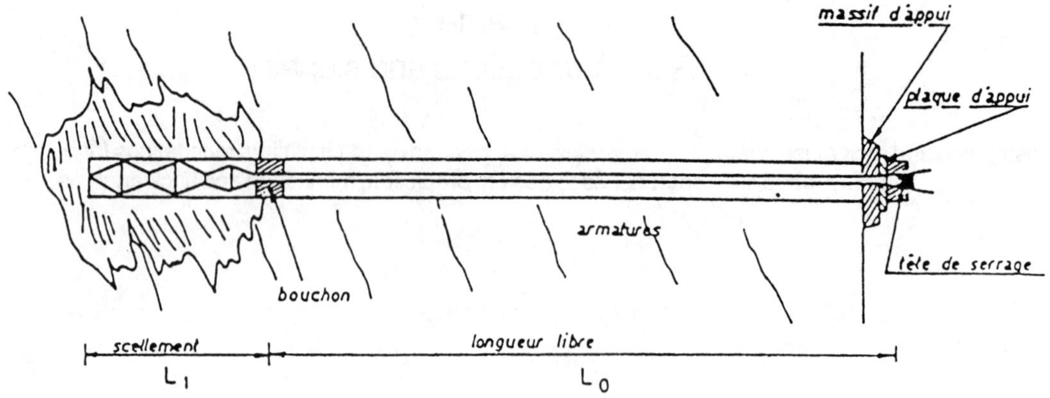

Figure 1 - Prestressed anchor

oriented at right angles to the corresponding discontinuities; in the latter, the perpendicular component and the tangential component are involved, so the optimum orientation must be determined, naturally taking into account what can actually be done in practice (W. Wittke).
Stability calculations for rock masses reinforced with prestressed anchors are simple in principle, since they consist of an analysis, in the stability range, involving the introduction of prestressing forces as outside forces of known constant direction and intensity.

However, these hypotheses must be checked to determine whether they actually hold true in practice. The assertion that the forces are outside the mass to be stabilized means that the embedment zones are beyond the potential failure surfaces and, in particular, that the values for the unsupported lengths of anchors hold true. The retensioning of anchors has all too often demonstrated that the unsupported length is much shorter than believed, because of undesirable local embedding effects caused by inadequate control of grout movement.

Similarly, prestressing forces can usually be considered as constants, since deformations in the rock mass after tensioning result in only very minor relative displacement between the anchoring zone and the bearing head, by comparison with the unsupported length of the anchor. This is not always the case, however; a displacement of 1 cm for an unsupported length of 20 m causes a variation in stress of 100 MPa in the anchor. Moreover, when the anchors are installed before the earthwork is completely finished, the deformations which may be produced must be evaluated to decide whether the anchors should be given a temporary tension that is less than the final service tension (Fig 2). Examination with extensometers in the drill hole is thus very useful. Retensioning allows us to recheck the unsupported length of the anchor and to evaluate the change in anchor tension between the two tensioning phases. Engineers have two major concerns about the behaviour with time of prestressed anchors: reduction or increase in tension by relaxation creep, and corrosion. Enough time has passed now for us to assert that carefully installed prestressed anchors are safe and reliable. It is interesting in this regard to note the observations made in Hong Kong and reported by Brien-Boys and Howells. They concerned 214 anchors in 17 projects and provided the following principal results:

- the average loss in tension over 4 years was 9%;

- 16% of the anchors had a tension equal to or greater than 1.2 times the service tension.

Compliance with existing standards and recommendations of several countries provides effective protection against corrosion.

It is apparent that both the design and installation aspects of the technique of reinforcing with prestressed anchors can be completely mastered by qualified engineers.

However, where the rock masses to be stabilized are large, this technique is expensive. In fact, to significantly improve the conventional factor of safety obtained by traditional stability analyses - for example, to increase it from 1.1 to 1.5 - the prestressing force is often in the neighbourhood of 10% to 25% of the weight of the rock mass to be stabilized, which is considerable.

2 PASSIVE REINFORCEMENT

Passive reinforcement refers to a reinforcing device that is embedded in the rock mass over its entire length and is not put under tension during installation. If there is no deformation in the reinforced rock, passive steel reinforcing is not stressed.

To my knowledge, one of the earliest examples of an application of this technique for reinforcing a rock mass is the support for the Chaudanne dam in France.

The Chaudanne dam is a thin arch 70 m high built between 1949 and 1952 on the Verdon River, in southern France. The arch is supported on a limestone rock mass containing discontinuity planes dipping 45° upstream and toward the river. On the right bank, strengthening on the downstream side of the support was accomplished with horizontal 50-mm bars, 25 to 30 m long, and concrete buttresses. On the left bank, the rock mass was consolidated with cables, composed of 115 to 185 5-mm strands, sloping at 15° from the horizontal. These cables were not tensioned, but had a cement grout injected over their entire length (Fig 3).

The action of passive reinforcements is much more difficult to analyse than that of prestressed reinforcements, because this type creates a composite environment with local interactions between the rock and the reinforcements, particularly at the intersections with the discontinuity surfaces. Let us examine the torque of the stresses on a passive bar at 0, the point of intersection with a discontinuity plane, where

No is the component of the force perpendicular to the discontinuity plane;

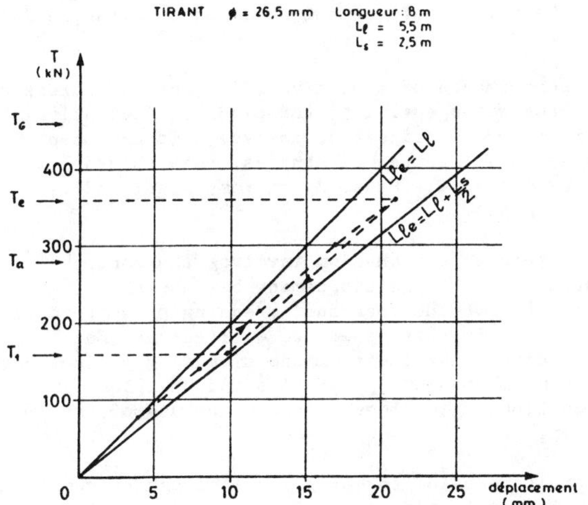

a) Courbe d'épreuve et de mise en tension provisoire

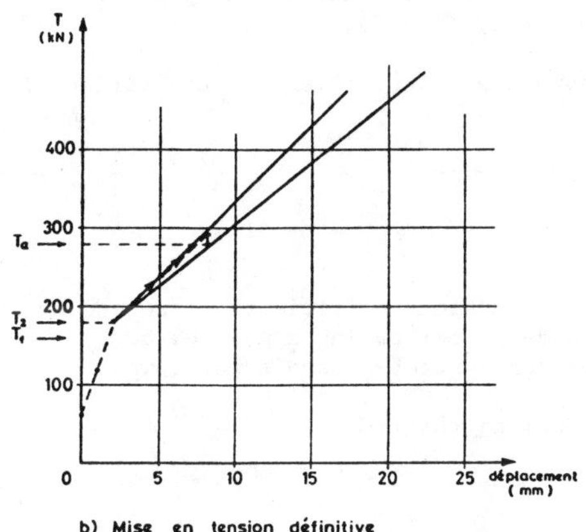

b) Mise en tension définitive

Figure 2 - Tensioning of an anchor in two phases

So is the component of the force tangential to the discontinuity plane;

Mo is the bending moment.

It may be hypothesized that Mo is zero, since, if there is tangential movement, the bending point of the bar is at 0 (Fig 4). I will therefore examine in succession the behaviour of a fixed bar under the influence of the two elementary stresses No and So, assuming that the bar is perpendicular to the plane of discontinuity.

2.1 Passive reinforcement stressed by a perpendicular force

Let us examine the behaviour of a bar of diameter d embedded over its entire length and affected by a perpendicular force N.

As long as we remain within the elastic range and the bar embedment remains intact, the shear stress at the contact of the steel and embedment material decreases with distance in the plane of discontinuity, in accordance with an exponential law (Farmer). However, when the perpendicular force increases, gradual loosening of the bar occurs and spreads in depth until there is either complete loosening or the bar breaks (Fig 5).

Studies done in Canada by Ballivry and Ben Mokrane on short bars 36.8 mm in diameter, fixed in cement mortar in drill holes 76 mm in diameter, indicate that local loosening occurs when the perpendicular force is only 25% to 30% of the maximum force (Fig 6).

For bars that have a diameter between 20 mm and 40 mm and are properly embedded in cement or resin, an embedded length of only 2 m is generally enough for the bar to break before complete loosening occurs.

The elastic limit of the bar is reached with very little displacement; for example, within a rock mass, only a small opening in a discontinuity surface is needed to stress the bar passing through it beyond its elastic limit.

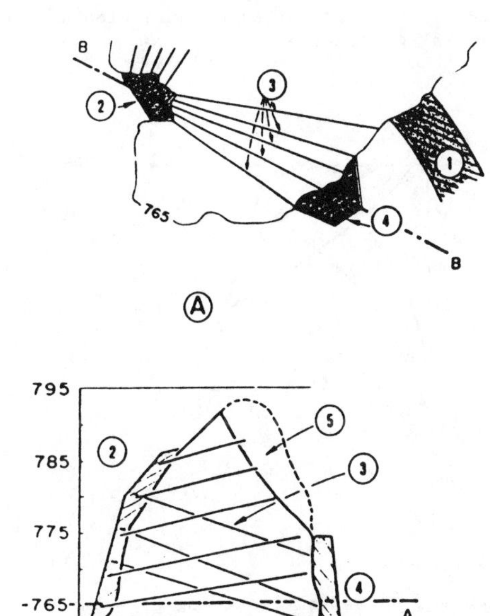

a
A - Section horizontale
B - Section verticale
1 - Barrage voûte
2-4 - Contrefort en béton
3 - Ancrages
5 - Glissement

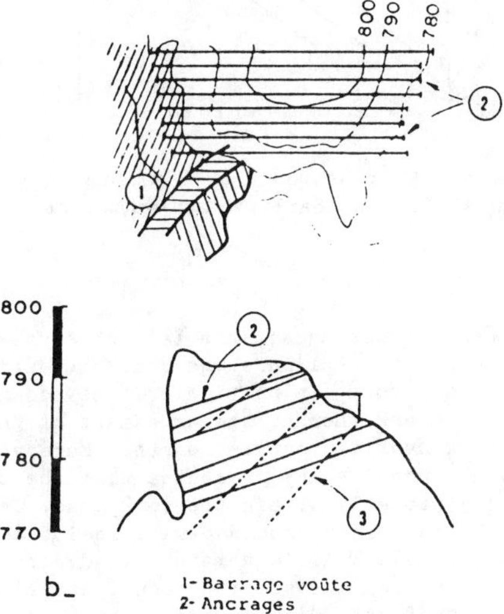

b_
1- Barrage voûte
2- Ancrages
3- Stratification

Figure 3 - Reinforcement of the Chaudanne dam with passive anchors

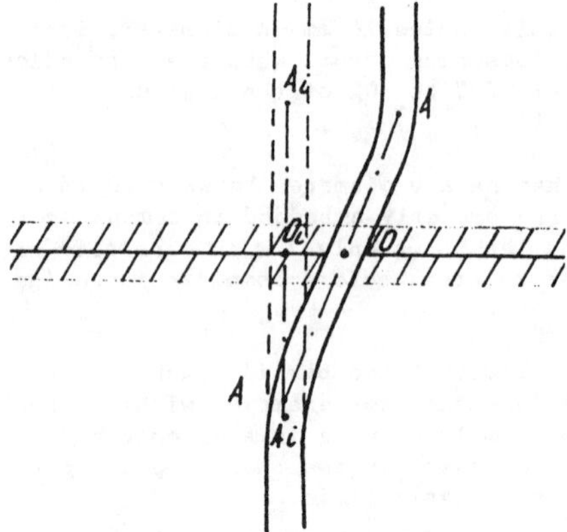

Figure 4 - Location of zero inflection at 0 with tangential displacement

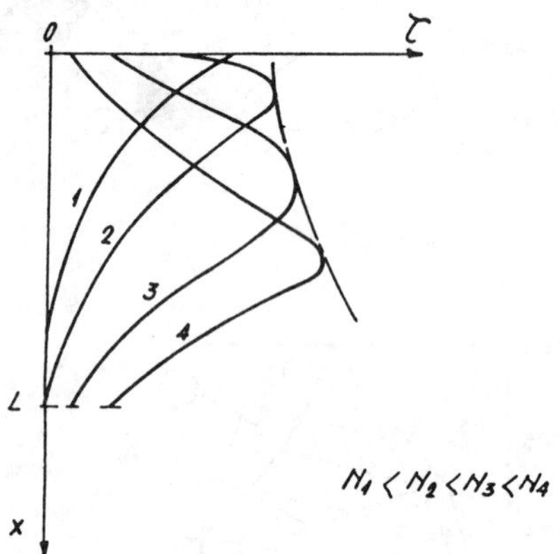

Figure 5 - Gradual loosening of a bar under increasing pull

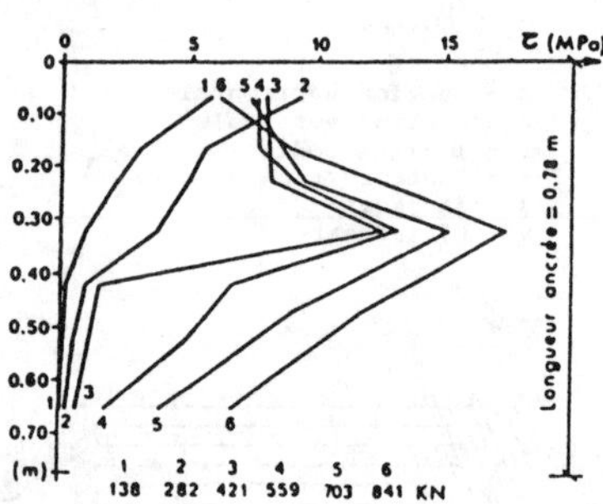

Figure 6 - Pulltest on a 0.78 m long bolt which failed at 950 kN (Ballivy and Benmokrane)

Under these circumstances, the tension tests commonly used in compliance and control testing are hardly satisfactory. In particular, they do not allow us to check whether the embedment of the bar is effective over its entire length. Moreover, the tests do not provide any guarantee that the bar is functioning properly within the rock mass, nor that bars which are merely embedded will resist corrosion for a long time. This is a major disadvantage, especially in rock masses with open fracturing, in which losses of embedding material are hard to control. It is then often necessary to make a preliminary injection of grout to block the open discontinuities.

2.2 Passive reinforcement stressed by a tangential force

The performance of a passive reinforcement stressed by a force tangential to the plane of discontinuity is much more difficult to analyse. It has been examined by a number of studies, both theoretical and experimental (Haas, Bjurstrom, Azuar and Dight-Egger).

The interaction mechanism involves the bending behaviour of the reinforcement bar and the resistance of the rock and embedding product to its movement. In this paper, we will assume, for simplicity, that the rock and embedding product have the same mechanical properties and that the discontinuity is closed without the use of any filling material.

If the tangential force is small enough to remain within the elastic range of the medium in which the bar is embedded, a standard analysis will determine, as a function of the tangential force S, the deformation y and the bending moment of the bar as a function of the distance in the plane of discontinuity (Fig 7).

The tangential displacement at the surface is given by:

$$yo = \frac{2\ S}{kd\ \lambda}$$

where

k is the reaction coefficient of the medium in which is embedded a bar having a modulus of deformation E_m and Poisson's ratio ν_m.

Assuming that:

$kd = mE_m$

λ is called the "transfer length".

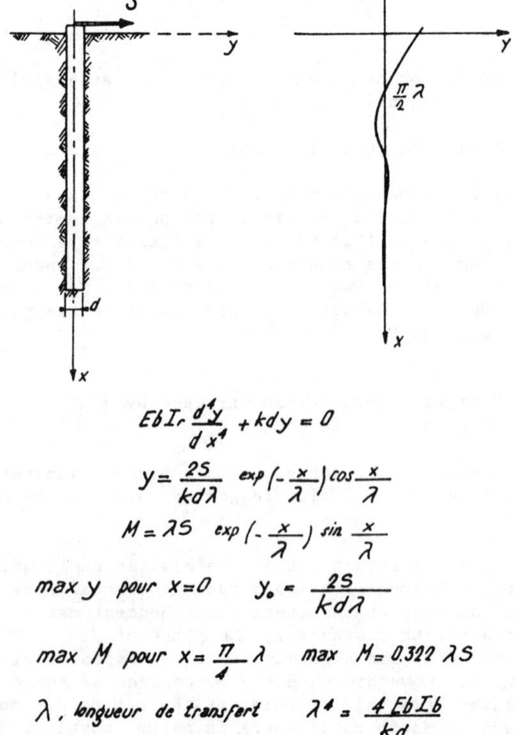

S
y
d
x
y
$\frac{\pi}{2}\lambda$
x
$E_bI_r \frac{d^4y}{dx^4} + kdy = 0$
$y = \frac{2S}{kd\lambda} \exp(-\frac{x}{\lambda}) \cos\frac{x}{\lambda}$
$M = \lambda S \exp(-\frac{x}{\lambda}) \sin\frac{x}{\lambda}$
max y pour x=0 $y_o = \frac{2S}{kd\lambda}$
max M pour $x = \frac{\pi}{4}\lambda$ max $M = 0.322\ \lambda S$
λ, longueur de transfert $\lambda^4 = \frac{4 E_bI_b}{kd}$

Figure 7 - Behaviour of a passive bar subjected to a tangential force S under elastic conditions

$$\lambda^4 = \frac{4\, E_b\, I_b}{kd}$$

E_b is the Young's modulus of the reinforcement steel.

I_b is the moment of inertia of the reinforcement bar.

thus

$$\lambda^4 = \frac{\pi}{16} \; \frac{1}{m} \; \frac{E_b}{E_m} \; d^4$$

Assuming that:

$$\alpha^4 = \frac{\pi}{16} \; \frac{1}{m} \; \frac{E_b}{E_m}$$

we obtain

$$\frac{yo}{d} = \frac{2\, S}{mE_m\, \alpha\, d^2}$$

and

$$\lambda = \alpha\, d$$

The maximum bending moment M_{max} is equal to:

$$M_{max} = 0.322\, \alpha\, ds$$

and the point of maximum bending moment is at a distance 1 from the discontinuity surface, such that:

$$1 = \frac{\pi}{4}\, \lambda$$

or

$$\frac{1}{d} = \frac{\pi}{4}\, \alpha$$

To establish the orders of size, let us examine the case of steel bars of the prestressed type with Young's modulus $E_b = 2 \times 10^5$ MPa and elastic limit $\sigma_e \equiv 840$ MPa. The bars have diameters between 20 mm and 50 mm.

Figure 8 gives the variation of the ratio 1/d as a function of E_m in semilogarithmic co-ordinates. It shows that the point of maximum moment is at a distance from the surface of discontinuity equal to less than two diameters and even less than one diameter when E_m is greater than 38,000 MPa.

Figure 9 shows, as a function of modulus E_m, the variation in tangential displacement yo for a value of tangential force Se corresponding to the elastic limit, assuming, in accordance with the Tresca criterion that:

$$Se = \frac{1}{2}\; Ne = \frac{\pi}{8}\, d^2\, \sigma_e$$

hence

$$\frac{yo}{d} = \frac{\pi}{4} \; \frac{\sigma_e}{\alpha\, mE_m}$$

For this value of the tangential force, the displacement yo remains less than d/4, and even less than d/10 when E_m is greater than 10,000 MPa.

Figure 10 illustrates the variation in the maximum bending moment corresponding to the value of the limit shear force Se.

The maximum bending moment M_{max} is then related to the plastic moment Mp of the steel bar.

For the plastic moment Mp, I have adopted the

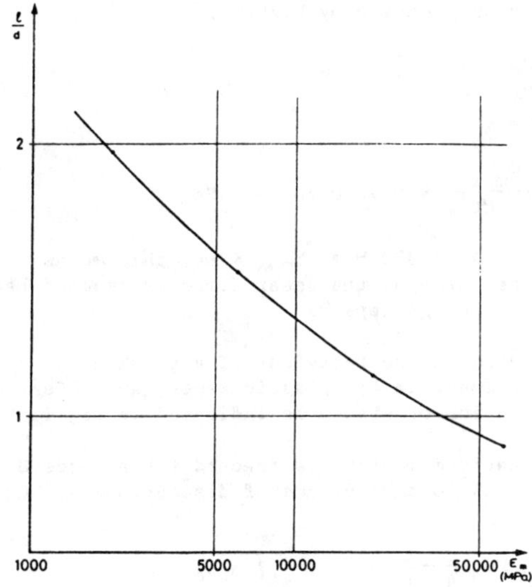

Figure 8 - Change in the 1/d ratio with variation in deformation modulus, Em

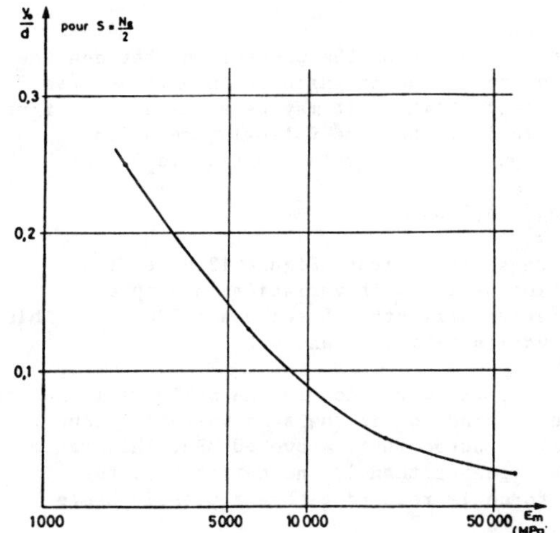

Figure 9 - Tangential displacement Yo versus Em corresponding to tangential Se force

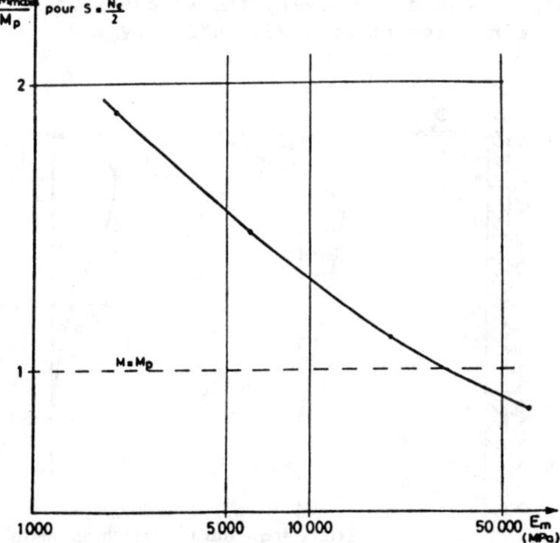

Figure 10 - Variation in maximum bending moment corresponding to peak Se

expression proposed by Dight:

$$Mp = 1,7\,\frac{d^3}{32}\,\sigma_e$$

hence

$$\frac{M_{max}}{Mp} = 0,76\ \alpha \text{ for } S = Se$$

where $E_m > 32{,}000$ MPa, $M_{max} < Mp$; this means that the value of the shear force is reached before a plastic hinge appears.

Dight studied the behaviour of a passive reinforcement in the plastic range, and inferred a distribution of stress as indicated in Fig 11.

The plastic moment Mp is reached for a force S_1 tangential to the surface of discontinuity, such that

$$S_1 = \frac{d^2}{4}\sqrt{1,7\,\pi\,\sigma_e\,P_u}$$

the distance l from the point of maximum bending moment at the discontinuity surface is given by:

$$\frac{l}{d} = \frac{1}{4}\sqrt{\frac{1,7\,\pi\,\sigma_e}{P_u}}$$

The determination of the pressure p_u between the bar and the medium in which it is embedded was discussed by Dight. It may be assumed, in a first approximation, that the following relationship holds between p_u and the simple compressive strength:

$$p_u = 6,5\,\sigma_c$$

Under these conditions, Figure 12 gives the variation of l/d with variation in simple compressive strength between 6 and 200 MPa. This ratio varies between 3 and 0.5.

In Fig 13, the variation in the Sℓ/Sa ratio has been plotted against σ_c in the same variation range. It will be noted that, above 60 MPa, this ratio becomes greater than 1, indicating that the limit shear force is reached before a plastic hinge appears.

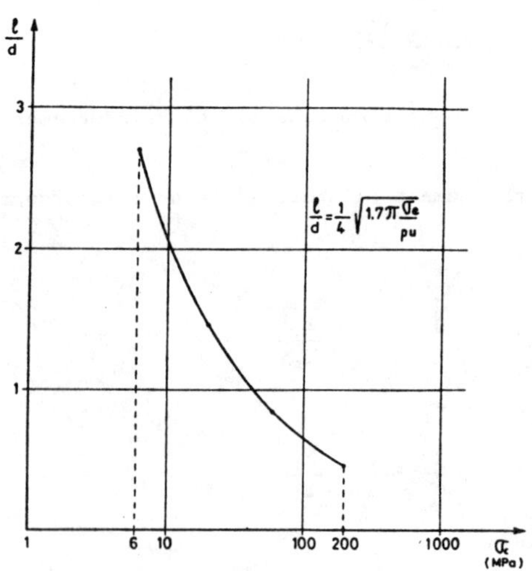

Figure 12 - Change in the 1/d ratio versus σ_c in the 6 to 200 MPa range

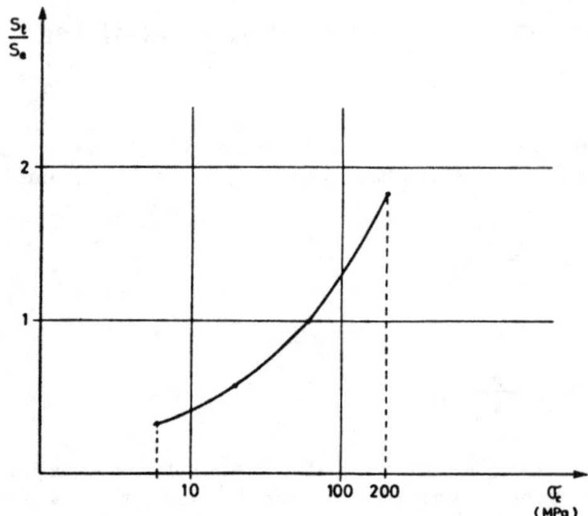

Figure 13 - Change in the Sl/Sa ratio versus σ_c

2.3 Contribution of passive reinforcement to the shear strength of a discontinuity

In the foregoing paragraphs, we analysed the behaviour of a bar embedded in a rock mass subjected to simple stresses. However, the stresses on a passive reinforcement at a discontinuity are complex. During shear that follows a discontinuity surface, the displacement has, in addition to a tangential component Ut, a perpendicular component Un at the discontinuity because of its expansion characteristics.

Consider a passive reinforcement bar that intersects a discontinuity within a rock mass, forming an angle θ with the perpendicular to the discontinuity surface; the resultant R of the forces acting on the cross-section of the bar at the discontinuity makes an angle β with the direction of the bar. We can resolve R into a perpendicular force N parallel to the bar and a shear force S perpendicular to the bar direction (Fig 14a).

It is generally accepted that the contribution of the bar to shear strength is represented by the expression

$$C_b = R\cos(\theta + \beta)\ \text{tg}\ \psi + R\sin(\theta + \beta)$$

In practice, the most commonly used methods involve making very simple hypotheses.

In one method, we assume that:

$$\beta = 0$$

$$R = Ne$$

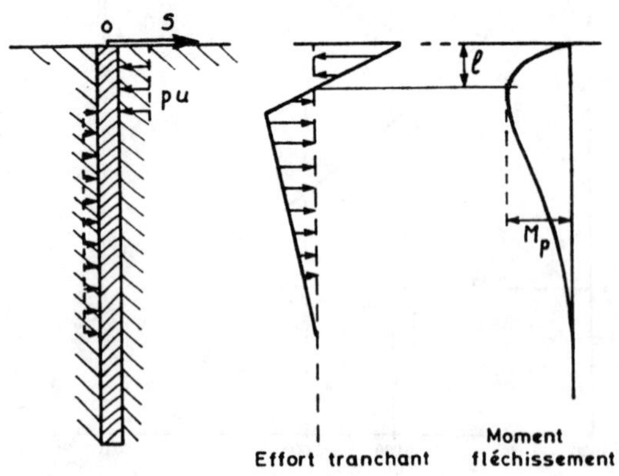

Figure 11 - Force and behaviour of reinforcement in the plastic domain

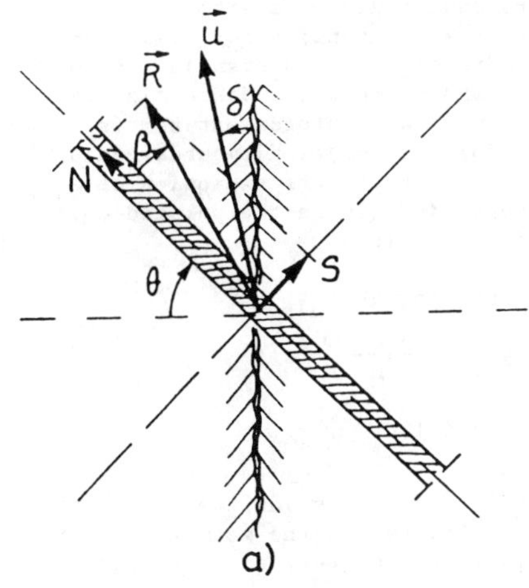

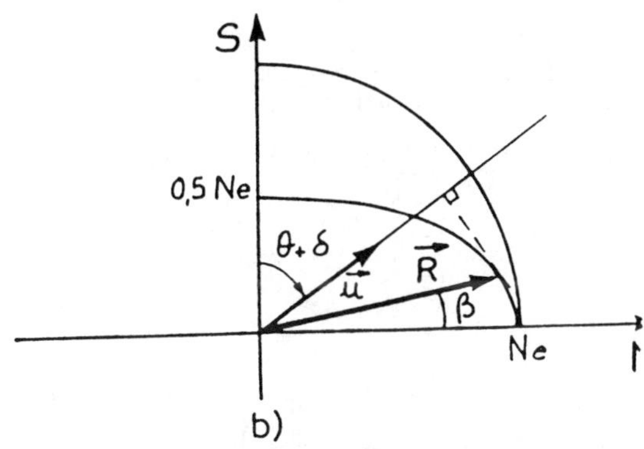

Figure 14 - Separation of resultant R into forces parallel to and perpendicular to the reinforcing bar direction

The bar is stressed to its elastic limit under tension parallel to its direction (Gasiev). A factor of safety for the steel bar is often arrived at by taking 2/3 of the elastic limit as the allowable stress (Colombet - Bonazzi).

The value of C_b is maximum if

$$\theta = \frac{\pi}{2} - \delta$$

In another method, the hypotheses are as follows:

$$\beta = \frac{\pi}{2} - \theta$$

$$R = \frac{N_e}{2}$$

This method is used mostly when θ is very low; the bar acts only through its resistance to the shear force, and the expansion effect is not taken into account. Although this method is a priori pessimistic, approximately the same or very slightly higher results are obtained when direct shear tests are done on discontinuities with very little expansion containing bars perpendicular to the shear plane (Azuar-Groupe Français). The bar is actually sheared (Fig 15). This also assumes that the rock walls are in contact without an intervening filling material. In this regard, Haas has suggested the comparison with scissors that have too much play and cut poorly.

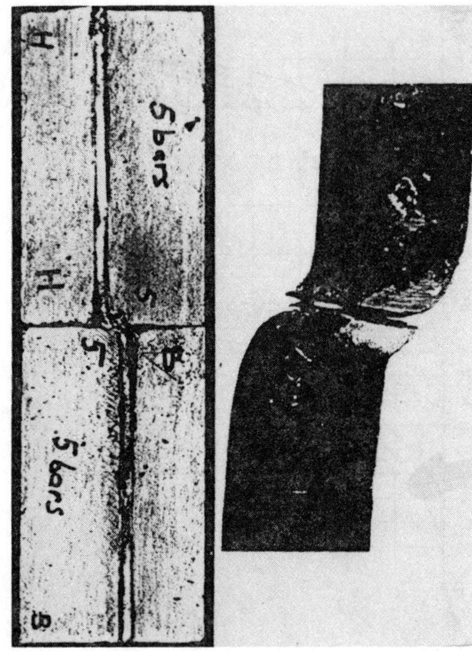

Figure 15 - Failure of a passive bar indirect shear

A third method is provided by the simple application of the principle of maximum work (Fig 14b). It was introduced in certain nailing programs in soil mechanics (Schlosser).

The limit stress R of the reinforcement bar at its end on the ellipse:

$$\left\{\frac{N}{Ne}\right\}^2 + \left\{\frac{S}{0,5\ Ne}\right\}^2 = 1$$

For a displacement having direction u forming angle σwith the surface of discontinuity, the principle of δaximum work tells us that, for any permissible value R, the limit force R must confirm the relationship:

$$(R - R)\ u > 0$$

Consequently, the end of vector R is the point on the ellipse whose tangent is perpendicular to u .

It may be deduced from this that:

$$\frac{R}{Ne} = \left[\frac{1 + \frac{m^2}{16}}{1 + \frac{m^2}{4}}\right]^{1/2}$$

$$tg\beta = \frac{m}{4}$$

$$m = cotg\ (\theta + \delta)$$

Fig 16 provides graphs giving R/Ne and β for values of angle δ of 0°, 5°, 10° and 20° and for θ varying between 0° and 60°. They clearly show that, if θ is very low, depending on the value of δ (expansion), R/Ne is between 0.5 and 0.8, which agrees with the experimental results.

On the other hand, if θ > 45°, angle β is less than 15°, and R/Ne is greater than 0.9.

It is helpful to determine the bar orientation θ which will give the maximum value of C_b.

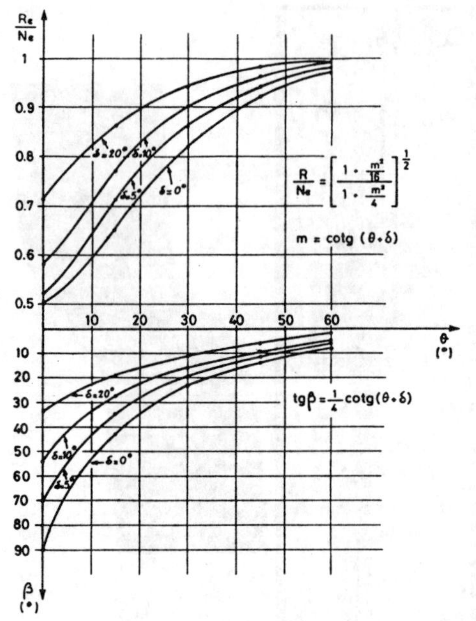

Figure 16 - Re/Ne and β values for δ = 0°, 5°, 10°, 20° and θ = 0° à 60°

Figure 17 shows how C_b varies with θ for three cases.

Case A ψ = 30° δ = 0

Case B ψ = 40° δ = 5°

Case C ψ = 50° δ = 10°

In these three cases, the maximum value of C_b is found, respectively, when θ > 60°, θ = 60° and θ = 45°.

The lower the values are for δ and ψ, the greater angle θ is for the maximum value of C_b.

However, it should be pointed out that this optimization does not necessarily correspond to an economic optimization that also considers bar length and must take into account the practical possibilities for implementation. It is a matter of specific cases. Under many circumstances, however, a value of 45° for angle O is close to the economic optimum.

A number of comments should be made concerning the use of the above formulae to determine the contribution of a passive reinforcement bar to the shear resistance of a discontinuity.

(a) It is assumed that we can simultaneously apply the limit stress of the steel bar and the angle of friction considered for the discontinuity.

In actual fact, we have seen that, regardless of the type of stress on an embedded bar, a very small displacement is required to reach the elastic limit of the steel. These minor displacements may be satisfactory for highly imbricated discontinuities without any fill material (Fig 18a); however, this is certainly not the case for poorly imbricated or loosened discontinuities (Fig 18b) or those with rock walls separated by a filling material with poor deformability.

(b) The choice of angle of expansion is also a delicate one. On the one hand, depending on the nature and condition of the discontinuities, expansion for small tangential displacements can vary substantially. On the other, the expansion for a discontinuity reinforced with a reinforcement bar is locally less than for a discontinuity under natural conditions. In fact, the expansion of a discontinuity depends on the perpendicular forces acting on it.

Barton assumes the law that

$$d_n = 0,5 \ K \log \frac{\sigma c}{\sigma n}$$

$$\text{if } 1 < \frac{\sigma c}{\sigma n} < 100$$

K - coefficient expressing the nature of the discontinuity and the geometry of the walls (K varies between 0 and 20)

σ c - simple compressive strength of the rock forming the walls

σ n - perpendicular stress

For low values of σ_n $\frac{\sigma c}{\sigma n} > 100$

dn = dn° = K

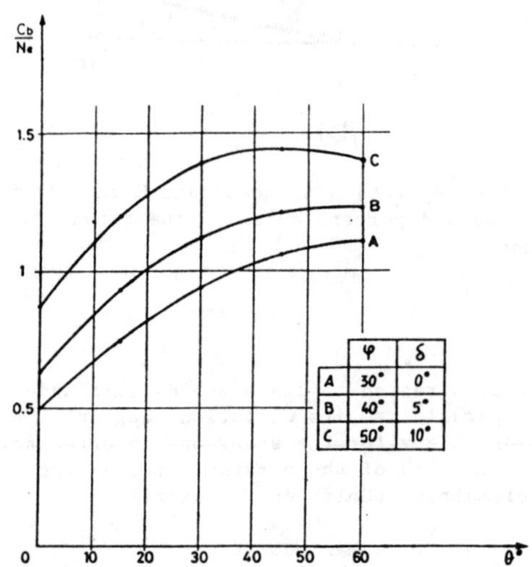

Figure 17 - Change of Cb with variation of θ for three cases

The form of the law proposed by Gaziev is different:

$$d_n = d^\circ_n \left[\ell - \frac{\sigma n}{\sigma c} \right]^{10}$$

Considering the effect of the δ parameter in the above analyses, we should be very cautious; it is recommended that a δ value be chosen that is less than the value for d_{n-} which may be determined by the usual methods. (c) Finally, the above analyses assume that the bar embedment is effective. If the embedding is inadequate, a gradual loosening occurs, causing progressive deformation of the bar, with two plastic hinges that move apart (Fig 19). In this case, the contribution of the bar to shear resistance is greatly reduced (Azuar et al).

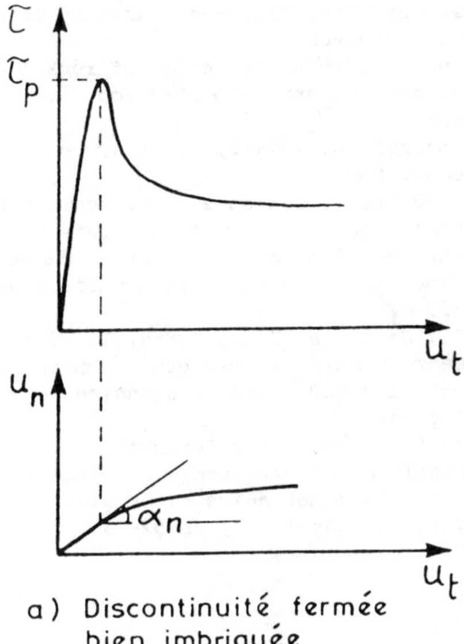

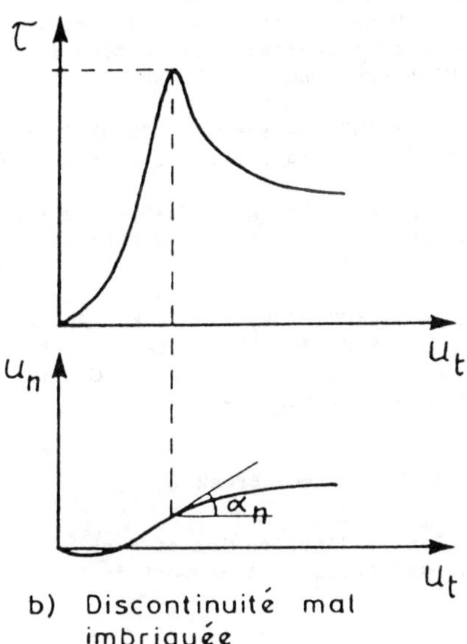

Figure 18 - Shear values for discontinuities subjected to constant vertical pressure

2.4 General design of reinforcement for a rock mass using passive reinforcement bars

From the analyses outlined above, the principles for reinforcing a rock mass are readily apparent.

(a) Minor displacements at discontinuities are enough to stress passive anchors locally to their elastic limit. These displacements are even smaller if the rock mass is more rigid or if the discontinuities are tighter and hence more subject to expansion. In actual practice, therefore, it is possible to prevent deformations at the discontinuities by reinforcing a rock mass as excavation progresses.

(b) An analysis of the deformation kinetics of potential rupture processes and an assessment of the properties of discontinuities allow proper orientation of reinforcement bars and optimization of reinforcement projects.

(c) It is important to pay special attention to proper embedding of bars on either side of a discontinuity; this becomes increasingly difficult with more open discontinuities and more pronounced pressure release effects. While tension testing allows determination of the quality of the grout and of the embedding method, it does not guarantee the entire embedment.

(d) In large excavations, where deformations cannot be ignored, the stresses on passive bars may exceed the elastic limit and cause local loosening of the embedding. Although this approach is entirely acceptable for temporary installations, we must be far more cautious in permanent structures because of the danger of corrosion. An examination of a reinforced rock mass using extensometers in drill holes is the best way of analysing its behaviour during the excavation and service start-up and throughout the life of an installation. When work is finished, it may prove necessary to complete the reinforcing to secure the structure permanently. However, in large excavations, sometimes reaching a hundred metres in depth, the highest parts

are practically inaccessible at reasonable cost, unless access has been maintained as the excavation progressed.

(e) For reasons having to do with the performance of the installation (greater homogeneity of stress distribution) and with its security (less probability of deterioration), preference should be given, for a given weight of steel, to a greater number of bars over bars with a thicker cross-section, even if this arrangement results in higher cost because of increased drilling expenses.

The successful use of distributed anchor bolts in underground works has reached into the field of surface installations, in which reinforcement with passive bars is increasingly common. In 1974, at the Denver convention, P. Londe and D. Bonazzi introduced the concept of "reinforced rock". Since

Figure 19 - Progressive detachment of a passive bar during a direct shear test

then, considerable progress has been made in the study of the performance of passive bars in a rock mass. However, the concept is not entirely clear. In fact, contractors would like to have a method for determining overall mechanical properties related to the deformability and breaking conditions of a reinforced rock mass - information which they could, for example, enter into a digital model.

There is very active research being done on composite materials in many fields. As far as I know, not enough research is being done on reinforced rock masses, although the few attempts that have been made using block models are very interesting.

3 CONCLUSIONS

Prestressed anchors and passive bars used to reinforce rock masses are very different in their design, function and approach to problems of safety and durability.

The opinions expressed in this paper are, of course, my own. However, I would like to thank all my colleagues at French research centres and departments with whom I have worked for many years on this theme; they have offered many valuable ideas that I have incorporated into my own thinking.

BIBLIOGRAPHY

Azuar, J.J. 1977 Stabilisation des massifs rocheux fissurés par barres d'acier scellées - Rapport de Recherche n° 73 - Laboratoire Central des Ponts et Chaussées Paris

Azuar, J.J., Panet, M. 1978 Le comportement au cisaillement des aciers passifs dans les massifs rocheux. Industrie Minérale. Document SIM B4 pp 93-98

Azuar & al 1979 Le renforcement des massifs rocheux par armatures passives. Proc. 4th ISRM Cong. Montreux. Vol. 1 pp. 23-29

Ballivy, G., Benmokrane, B. 1984 Les ancrages injectés et la consolidation des parois rocheuses Int. Conf. Renforcement en place des sols et des roches. ENPC. Louisiana State University Paris pp. 213-218

Bjurstrom, S. 1974 Shear strength of hard rock joints reinforced by grouted untensioned bolts Proc. 3rd Cong. ISRM. Denver, Vol. 2B, pp 1194-1199

Bonazzi, D., Colombet, G. 1984 Réajustement et entretien des ancrages de talus. Int. Conf. Renforcement en place des sols et des roches ENPC. Louisiana State University. Paris pp. 225-230

Descoeudres, F., Gencer, M. 1979 Rupture progressive en versant rocheux stratifié et fissuré. Proc. 4th ISRM Cong. Montreux, Vol. 1, pp 613-619

Dight, P.M. 1983 Improvements to the stability of rock slopes in open pit mines. Ph.D Thesis. Monash University Australia, 348 p.

Dight, P.M. 1985 The theoretical behaviour of full contact bolts subject to shear and tension. Int. Symp. on the Role of Rocks Mechanics. Zacatecas Mexico, pp 290-297

Egger, P., Fernandes, H. 1983 Nouvelle presse triaxiale. Etude de modèles discontinus boulonnés. Proc. 5th ISRM Cong. Melbourne, pp A171-A175

Gaziev, E.G., Lapin L.V. 1983 Passive anchor to shearing stress in a rock joint. Proc. Int. Symp. on Rock. Abisko, pp 101-108

Haas, C.J., & al 1978 An investigation of the interaction of rock and types of rock bolts for selected loading conditions. University of Missouri Rolla. US Bureau of Mines PB 293988

Heuzé, F.E. 1981 Analysis of bolt reinforcement in rock slopes. 3rd Int. Conf. on Stability in Surface Mining. Vancouver

Hoek, E., Londe, P. 1974 The design of rock slopes and foundations. General Report. Proc. 3rd ISRM Cong. Denver

Londe, P., Bonazzi, D. 1974 La roche armée. Proc. 3rd ISRM Cong. Denver

Panet, M., 1978 Stabilisation et renforcement des massifs rocheux par aciers précontraints et aciers passifs. Atti del Seminario sur consolidamento di terreni e rocé in posto nell' ingegneria civile. Stresa pp 33-62

Schlosser, F., Jacobsen, H.M., Juran, I. 1983 Soil reinforcement. General Report 8th European Conf. on Soil Mechanics and Foundation Engineering. Helsinki, pp 83-104

Wittke, W. 1965 Verfahren zur Berechnung der Standsicherheit belasteter und unbelasteter Felsböshungen. Rock Mechanics and Engineering Geology. Vol. 30, Suppl. II, pp 52-79

9:20 LE GRAND SALON - Short presentations

A. CARRERE, C. NURY and P. POUYET, France "The Contribution of a Non-Linear, Three-Dimensional Finite Element, Model to the Evaluation of the Appropriateness of a Geologically Complex Foundation for a 220 m High Arch Dam in Longtan, China".

P. EGGER and K. SPANG, Switzerland "Stability Investigations for Ground Improvement by Rock Bolts at a Large Dam".

T.Y. IRFAN, N. KOIRALA and K.Y. TANG, Hong Kong "A Complex Slope Failure in a Highly Weathered Rock Mass".

A.N. MARCHUK, A.A. KHRAPKOV, Y.N. ZUKERMAN and M.A. MARCHUK, U.S.S.R. "Contact Effects at the Interface between Rock Foundations and Concrete Dams with Power Plants at Their Toes".

10:00-10:30 FOYER, Coffee Break.

D.C. MARTIN, Canada "Application of Rock Reinforcement and Artificial Support in Surface Mines".

B. NILSEN, Norway "Flexural Buckling of Hard Rock - a Potential Failure Mode in High Rock Slopes?".

K.J. ROSENGREN, R.G. FRIDAY and R.J. PARKER, Australia "Preplaced Cable Bolts for Slope Reinforcement in Open Cut Mines".

R.K. ROWE and H.H. ARMITAGE, Canada "A New Design Method for Drilled Piers in Soft Rock: Implications Relating to Three Published Case Histories".

J.P. SAVELY, U.S.A. "Probabilistic Analysis of Intensely Fractured Rock Masses".

T. SERRA de RENOBALES, Spain "Strata Buckling in Foot Wall Slopes in Coal Mining".

J.R. SHARP, M. BERGERON and R. ETHIER, U.K., Canada "Excavation, Reinforcement and Monitoring of a 300 m High Rock Face in Slate".

T. SUEOKA, M. MURAMATSU, Y. TORII, M. SHINODA, M. TANIDA and S. NARITA, Japan "Economical Design and Construction of a Large Scale Mudstone Cut Slope".

R. WIDMANN and R. PROMPER, Austria "Influence of the Foundation on the Stresses Acting Near the Base of an Arch Dam".

12:00 MARQUETTE, Lunch Break and Poster Sessions.

13:30 LE GRAND SALON, Panel and Floor Discussions.

Moderator: K. Kovari (Switzerland)

Panelists: B.K. MacMahon (Australia); R. Ribacchi (Italy); R. Yoshinaka (Japan).

Panel and floor discussions:

B. McMahon (Australia)

Evidence from building foundation settlements suggests that rock mass modulus in the fully confined condition may be similar to core modulus. Reduction factors have been obtained by correlation with plate bearing tests on tunnel walls. These may represent only a relatively thin zone of rock around the tunnel. If reduction factors are applied to all the materials in a finite element analysis, it could lead to overestimatation of stresses around the openings. Has anyone made similar observations?

N. Barton (Norway)

My remark is directed to Rosengren et al., Preplaced cable bolts for slope reinforcement in open cut mines.

Concerning optimum bolt or anchor installation angles, it should be pointed out that dilation is not mobilized at zero shear displacement. It requires shear, more shear as the scale of the problem increases. Correct scaling of dilation (which is a part of shear strength) and consideration of the mobilized (displacement dependent) shear strength gives an improved approach to optimum anchor installation angles.

A. Bolle (Belgium)

My question is directed to J.P. Savely, Probabilistic analysis of intensely fractured rock masses.

The Author formulates the probability calculation as a capacity-demand problem, and uses the standard normal PDF. This process implicates two assumptions :

1. The capacity C and the demand D are statistically independant,

2. The computing procedure using normally distributed input variables don't distort the shape of the PDF of the output variables.

The first assumption is seldom allowable, because the same variables are used to compute C and D, for example the gravity forces and some geometrical data. A better way is to use directly the safety margin SM=C-D, in which case that implied correlation is automatically taken into consideration. Disregarding the functional correlation could modify widely the results, in various directions.

The second assumption is correct for linear combinations of normally distributed random variables. It is also admissible with first order-second moment methods, but the Monte-Carlo simulation as well as the Rosenblueth's method give results beyond the second moment. The functional skewness, due to the computing procedure, can thus be taken in consideration. The choice of the normal PDF influences directly the estimated Pf (probability of failure).

Based on the "maximum entropy of information" principle, the best choice for the PDF knowing the first three statistical moments (mean, variance and skewness) is the gamma distribution.

To estimate the composite probability of failure, the Author calculates the reliability of a series-connected system as the product of the reliabilities of each separate element.

This implicates also a strong assumption on the independance of the various modes of failure, because the product formula concerns only independant events. That assumption is difficult to justify in general, and particularly in slope stability problems. Indeed, the same variables or correlated variables are used for the calculation of several modes of failure, giving again an implicit correlation between these modes.

In the meaning of the Writer, a better way is, for each set of data of the Rosenblueth's method, or for each random sample in a Monte-Carlo simulation, to consider simultaneously all the possible failure modes, with a view to obtain an unique "safety margin".

REFERENCES

Bolle, A., 1987. "A probabilistic approach of slope stability in fractured rock", Proc. 6th ICRM, Montreal, pp. 301-303.

B. Voight (USA)

My contribution relates to accurate time forecasts of slope failure with the aid of the phenomenological law.

The simple law $\dot{\Omega}^{-\alpha}\ddot{\Omega}-A=0$ describes the terminal stages of failure under constant stress, where Ω is an arbitrary fluent such as strain or displacement, and A and α are constants of experience. The law seems broadly applicable under diverse loading conditions and proportional stress histories to a variety of materials, such as metals and alloys, ice, plastic, concrete, rock and soil. Numerous independently-discovered relationships including the Rabotnov-Kachanov equations emerge as special cases, and insight is cast on limitations and extensions of these relationships. The law provides for finite values at failure for strain, strain rate, and strain acceleration, and seems to fit experimental and field data as well or better than any phenomenological theory of comparable complexity yet proposed. The law may be extended to provide multiaxial constitutive equations, and may be wielded to enable forecasts of time to failure. Its simplicity aids mathematical manipulation and keeps minimal the requirements for parameter determination.

Applied to slope failure, one may express Ω = displacement. For $\alpha=1$, $\Omega=\Omega_o$ at $t=t_o$:

$$\dot{\Omega} = \dot{\Omega}_o e^{A(t-t_o)}$$

(1a)

for $\alpha\neq1$,

$$\dot{\Omega} = [A(1-\alpha)(t-t_o) + \dot{\Omega}_o^{1-\alpha}]^{1/1-\alpha}$$

(1b)

or

$$\dot{\Omega} = [A(\alpha-1)(t_f-t) + \dot{\Omega}_f^{1-\alpha}]^{1/1-\alpha} ,$$

(1c)

for $\alpha>1$, where t_f is time of failure, $\dot{\Omega}_o$ is arbitrary initial rate and $\dot{\Omega}_f$ is rate at failure. For simplification by neglect of Ω_f in (1b), the velocity equation of Fukuzono (1985) is obtained. Commonly $\alpha=2 \pm 0.3$. The constants A and α may be determined by regression of rate against time (1a), or by plotting logarithm of acceleration against logarithm of velocity of slope movement.

Expressions for Ω arise from integration. For $\alpha>1$, $\alpha\neq 2$,

$$\Omega-\Omega_o = \frac{1}{A(\alpha-2)}\left\{\dot{\Omega}_o^{\,2-\alpha} - [A(1-\alpha)(t-t_o) + \dot{\Omega}_o^{\,1-\alpha}]^{\frac{2-\alpha}{1-\alpha}}\right\} \tag{2a}$$

or

$$\Omega-\Omega_o = \frac{1}{A(\alpha-2)}\left\{[A(\alpha-1)(t_f-t_o) + \dot{\Omega}_f^{\,1-\alpha}]^{\frac{2-\alpha}{1-\alpha}} - [A(\alpha-1)(t_f-t)+\dot{\Omega}_f^{\,1-\alpha}]^{\frac{2-\alpha}{1-\alpha}}\right\} \tag{2b}$$

For $\alpha>1$, $\alpha\neq 2$, $t_o=\Omega_o=0$, and neglecting $\dot{\Omega}_f$, normalized displacements are given by

$$\lambda^* = \frac{\Omega}{\dot{\Omega}_o t_f} = \frac{\alpha-1}{\alpha-2}\left[1-(1-t/t_f)^{\frac{\alpha-2}{\alpha-1}}\right] = \lambda[1-(1-{}_t/t_f)^{1/\lambda}] \tag{3}$$

Displacements normalized to $\dot{\Omega}_o t_f$ may also be determined from (2). Divergence between simple and general curves occurs for $t/t_f>0.5$, showing the influence of finite Ω_f. The effect of $\dot{\Omega}_f$ is particularly significant for $\alpha<2$.

At any arbitrary $t=t_*$, where $\dot{\Omega}=\dot{\Omega}_*$, remaining time to failure is:

$$t_f-t_* = \frac{\dot{\Omega}_*^{\,1-\alpha}-\dot{\Omega}_f^{\,1-\alpha}}{A(\alpha-1)} \tag{4}$$

Practical estimates of $\dot{\Omega}_f$ are feasible from consideration of slope geometry, materials, and groundwater conditions. Neglect of $\dot{\Omega}_f$ in time prediction leads to non-conservative forecasts (predicted remaining time to failure too large). This effect may or may not be of importance to practical decision-making, depending on circumstances. Plots of inverse rate against time, nearly linear for α nearly 2, may be usefully employed for forecasts. Concave-up curvatures occur for $\alpha<2$, and are more difficult to accurately extrapolate by graphical means. Equation (4) may however be used to advantage. Failure occurs when the inverse rate value of $\dot{\Omega}_f$ is obtained, and is not precisely the point of intersection of the inverse rate curve with the time abscissa.

Predictions may be updated with acquisition of new data. Accuracy of the method is conditioned by the precision and frequency of observations and the regularity of slope loading conditions. Variable stress histories will produce rate changes and variation in predicted time to failure, but this variation is encompassed by a life fraction rule. Because slope movements generally include complex factors requiring consideration, at the experience of the interpretor is an important factor. The method applies well to diverse categories of landslides in rock and soil, including first-time and reactivated slides, pit slope failures, toppling, and waste pile failures.

References

Fukuzono, T., Proc. IV Int. Conf. and Field Workshop on Landslides, Tokyo, 145-150, 1985

Rock Foundations and Slopes - Theme II

Moderator Summary: K. Kovari (Switzerland)

The rock structures slopes and foundations exhibit a great variation both in size and the risk involved. Cuts for roads and railroads generally do not exceed several tens of metres because otherwise a tunnel would provide the economical solution. Large dams may reach 300 m in height but it is the capacity of the reservoirs which is decisive for the risk involved. Also open pit mining is proceeding more and more to large scales leading to slopes with several hundreds of metres in height. The nature of the risk involved here is, however, completely different from that of dam sites. Finally, landslides - such as the Vajont slide - can also be of main concern, particularly if they border the reservoir of a dam. The variation of size and risk involved largely determines the approach of the engineer to the problem and also the criteria upon which one accepts the existence of such structures. Obviously the means of influencing the behaviour of such rock structures decreases with their increasing size and in some cases the only way out is to abandon the site or to refrain from any impact on the rock mass such as reservoir level variation. The performance criteria of surface structures can be expressed in quantitative terms such as: Factor of Safety, Probability of Failure, Stress level in the Concrete and Loss of Water. Loss of water may also affect the stability of the dam .

Constructional Measures for Rock Slopes

Consider first the slope angle which may largely affect the economy and the safety. During construction the sequence of excavation must be such as to avoid local failure but still allow sufficient advance of the work. In cases of large slopes, controlling the water on the ground surface and in the rock mass is the only corrective measure one can take. Lowering the phreatic surface or decreasing water pressure in potential sliding surfaces can be achieved by drillholes and drainage galleries. Piezometric head monotoring will show the effectiveness of those measures. Considering the reinforcement there is a tendency to favour fully grouted anchors due to economy and the expectation that they will be more resistant against corrosion than the active ones. The statical effect of active anchors is simple to understand, one just has a pair of forces the size of which can be controlled any time. Fully grouted anchors are only effective after deformations in the rock mass have taken place. If deformations are confined to only a few joints then stress concentrations will occur at these locations in the grouted anchors, whereas the stress level in the active anchor would not be largely increased. Further research into the behaviour of passive anchors is definitely needed. In many instances slope construction can and should be rejected in favour of a tunnel.

Constructional Measures for Dam Foundations

They consist of removing weak rock, frequently leading to slope stability problems in steep valleys. The consolidation grouting is expected to increase the stiffness and strength of the rock mass and the curtain grouting is designed to control seepage in the rock. One of the largest grouting operations in history was recently carried out at the El Cajon dam site in Honduras involving 11 km of grouting galleries and approx. 500 km total length of drill holes. It is extremely difficult to predict the grouting requirements in carstic areas resulting in some cases in

60% more work than according to the design. Apart from slope stability problems in many instances it is the concrete body itself which must have a sufficient safety against sliding. In some instances fully grouted anchors may provide a solution. In other cases a system of galleries is established and filled with concrete in order to establish shear keys. Shear keys are sometimes applied to increase stiffness of the rock mass.

Decision Making

There are so many decisions of far-reaching consequence to be made in surface workings in rock that it is worth looking closer at the decision making process. In many instances the situation is considered so "obvious" that decisions are made merely referring to experience rejecting even reasoning. At the other end of the spectrum one encounters the "complex" situation where experience alone does not seem to be sufficient and one has to resort to what we call the rock mechanics "tool box". Before looking closer at this box it must be noted that frequently a situation is only considered to be "obvious" because people do not have an adequate knowledge of rock mechanics. Anyway the notion of "ovious" and complex is rather subjective and to which extent advantage is taken from this "tool box" is also largely determined by the experience and background of responsible engineers involved in the job. The "tools" may help to assess

- the relevant rock mass properties
- the ground water conditions
- the factor of safety and
- the deformation of the structures.

It has already been pointed out that, despite the availability of powerful tools, so-called sound judgement is required at all levels of the approach and also in the acceptance of the results. Ideally, all decisions should be made prior to construction because then a correct cost estimate, correct contract documents, selection of adequate equipment and correct time schedules can be made. This approach is in many cases not feasible, so some people advocate the philosophy of "design as you go" which is certainly applicable in many open pit mine situations and whenever the owner and the contractor belong to the same organisation. Turning to the rock mechanics "tool box" consider just one aspect of material behaviour. For joint controlled slope or dam stability problems the deformation capacity of discontinuities may be crucial. In case of **little** deformation capacity, brittle mode of failure will occur and already small displacements may entail a significant decrease in strength and safety. In this case pretensioned anchors are really more favourable than the passive ones and it will be advisible to introduce residual strength values rather than peak ones into the calculations. On the other hand, large displacement capacity with yielding mode of failure inducing progressive failure may call for fully grouted anchors and will allow peak strength parameters to be considered. One little known aspect of this issue is of course the determination of material properties on the scale of the structures under consideration! Another persisting problem of rock mechanics concerns the effect of discontinuities on the deformability of a rock mass. One way of dealing with this problem experimentally is to carry out large scale load tests (a cable jacking test for example) and to observe the strain profile along a borehole in the direction of the acting load. The sharp peaks in the strain distribution curve clearly reveal the presence of active joints and weak layers intersecting the borehole.

Let us have a look at the computational methods currently being used in the analysis of slopes and dam foundations. Any statical consideration relies on models, there are models for material behaviour for the statical system and for the load quantities. Modelling means reduction of reality to the essentials of a problem. The rigid body concept prevails in the study of ultimate behaviour and safety. Linear elastic assumptions are made for stress and deformation analysis. Non-linear rock mass behaviour may be assumed for selected computational runs but such efforts do not aim at determining the ultimate (collapse) conditions.

Real structures are seldom 2-dimensional but in some cases 2-dimensional models are fully justified such as the plane failure of a slope or the stress analysis of a gravity dam. The load quantities are static or dynamic and the question in which cases should a dynamic response analysis be carried out and in which instances it suffices to consider just an equivalent static load to simulate earthquake is still a matter of dispute.

Analytical techniques rely heavily on a thorough understanding of the mechanisms which govern the structural behaviour of slopes and foundations. There are some practical rules to observe before starting with computations and before accepting a set of results: Often many different behaviour patterns must be studied to detect the worst possible case as well.

The next item to be discussed is monitoring which is intended

- to check equilibrium
- to detect unexpected behaviour at an early stage
- to explain unexpected behaviour in order to take appropriate corrective measures
- to help assess safety of existing structures and
- to predict failure.

Often these purposes are taken in combination. The most important physical quantities are deformations (such as absolute or relative displacements, strains and change in inclination) together with the piezometric head and anchor forces.

In most cases deformations are measured pointwise. The movement of selected points on a slope surface can indicate if stable or unstable conditions prevail. A considerable increase of information is obtained if so-called "linewise observations" are carried out. They involve the measurement of the distribution of a quantity along a straight or curved line. For example, from the distribution of the horizontal displacements along a borehole in an unstable slope the slip surface my be determined which is the cause of the undesirable behaviour. Similarly, from the measured distribution of strain along a borehole in the rock foundation of a dam we can find out where existing joints become active i.e. they open and close as the reservoir water level fluctuates. The sharp peaks in the strain profile reveal such active joints intersecting the borehole. Knowledge of the presence of active joints is important for seepage and for safety assessment.

Another aspect of slope monitoring refers to failure prediction. In slopes showing high deformation capacity and creep, progressive failure is enhanced enabling in many cases prediction of slope failure but also the effectiveness of corrective measures. On the other hand, if heavy rainfall or earthquake cause the triggering, deformation monitoring will be of limited use. In the analysis there is an inherent uncertainty in the assumptions regarding the nature of discontinuities, regarding rock mass properties and regarding ground water conditions. Additionally, there is the unavailability of data. Computational results are therefore, as a rule, hypothetical. But they are explanatory showing causes and effects. On the other hand, monitoring suffers from the limited number of observations, from time periods without any observations at all but also from the lack of suitable measuring techniques. Results of observations reflect reality but do not necessarily give an understanding of it. Without special effort monitoring may remain non-explanatory. However, if the combination of analysis and monotoring is feasible it yields both understanding and reality.

In order to obtain the greatest possible advantage from analysis the computational models should be adequate to the problem involving a minimum number of parameters, which should all, at least conceptually, be measurable. Increasing model sophistication leads to sophisticated interpretations, which do not necessarily facilitate engineering decisions. Regarding monotoring it should be mentioned that it is only useful if it is based on clear concepts. Or, in other words, "No observation without theory".

It was the intention of the moderator to close this brief survey pointing to some of the existing shortcomings in our approach to the problems. Our rock engineering community will continue to make progress both in the theoretical and the practical fields. But is it not the case in rock engineering that the engineer has to make decisions and assume heavy responsibility even in the absence of facts and adequate knowledge ? This Congress has definitely contributed to our endeavour to reduce uncertainties and find more facts when decisions have to be made.

Theme 3
Rock blasting and excavation

Blast vibration analysis, response and control; drilling technology and control; fragmentation; performance of full face boring and roadheader machines; large diameter shaft drilling

08:30 LE GRAND SALON - September 2, 1987

Chairman: Tan Tjong-Kie (China)

Session Secretary: W. Comeau (Canada)

Speaker: C.K. Mackenzie (Australia)

Blasting in Hard Rock: Techniques for Diagnosis and Modelling for Damage and Fragmentation.

ABSTRACT: Models for prediction of rock responses to blasting have been developed to evaluate alternative blast designs for specific applications and rock types. Design optimisation, however, further requires monitoring and measurement to ensure control over detonation sequencing, and to identify the more important influences controlling rock response.

1. INTRODUCTION

A model of rock fragmentation by blasting has been developed by the Julius Kruttschnitt Mineral Research Centre in Australia, based on the results of measurements and monitoring at more than 25 mine sites, totalling more than 200 weeks of site work over the past three years. The model has permitted the definition of rock damage from blasting, and is being used to optimise blast design for the achievement of specific blasting objectives.

The model has the primary feature of directly utilising fracture data from fracture mapping and stereoplots, and identifies the existing fracturing as the primary influence on both fragmentation and damage. Perhaps of equal importance is the development of monitoring and measurement techniques used for the collection of data required by the model. The use of these monitoring techniques has highlighted some of the basic faults in blasting, including lack of control over charge initiation, variability in explosive and initiator performance, lack of detailed design assessment procedures, and an inadequate understanding of the mechanisms of damage and the factors controlling damage. In most applications to current operations, modelling is considered to be of secondary importance to basic control - over charging procedures, design, and product specifications.

The results of some basic blast monitoring are presented to emphasise the application of monitoring to the optimisation of blast performance. The improvements to productivity resulting from the monitoring have been substantial, and beyond the current scope of modelling. Mines have reported savings in excess of $2 million per year in explosive and drilling costs alone, with powder factors being reduced by up to 50%, while improving fragmentation, reducing damage, and controlling ground vibration.

2. FRAGMENTATION MODELLING

The JKMRC fragmentation and damage model is based on the following observations :
1. fracturing has the largest single influence on fragmentation and damage ;
2. the degree of breakage of a rock is dependent on the size of the particle and the energy imparted to it ;
3. during blasting, breakage or fracturing is primarily produced by shock energy or brisance, and rock displacement is produced by gas energy.

The basic assumption of this mechanistic model of breakage can therefore be stated : if the size of particle and the input energy to it are known, then the degree of breakage of the particle can be predicted. This follows from observations of the breakage response of thousands of rock particles in laboratory breakage tests. Perhaps the primary feature of the model is that it is based on measurements rather than theoretical rock or fracture properties, and that all predictions of the model can be verified through measurement.

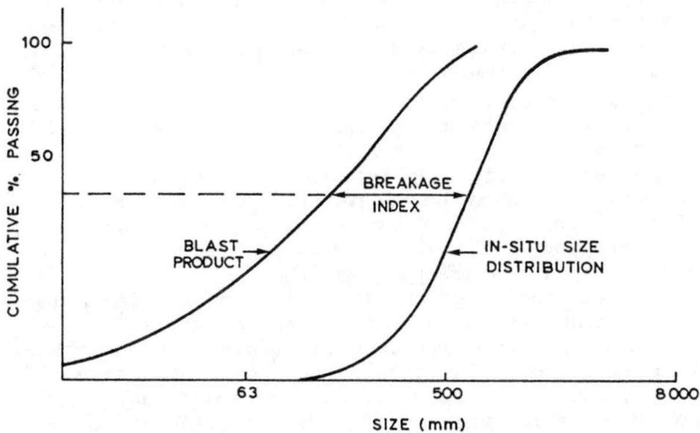

Figure 1. Size distribution before and after blasting, showing extent of breakage.

2.1 Fracture Measurement

The intensity of fracturing is measured by line scanning techniques, whereby the points of intersections of all natural fractures with a scan line are measured, together with the dip and dip directions of the fractures. These fractures are then represented on an equal-area stereo plot and a Monte Carlo simulation used to calculate insitu block size distributions for three different joint spacings distributions. The rock is therefore described as a system of irregularly shaped blocks, with a size distribution of blocks being defined by the natural macro-fracturing.

The fracture measurement technique immediately indicates the maximum size of particle which could be observed in the muckpile, and the percentage of particles greater than any particular size. If the desired muckpile fragment size distribution can be defined, a comparison of the two distributions can be made as shown in Figure 1. This comparison provides an immediate indication of the degree of breakage required by the explosive, from which a "breakage index" could be defined.

2.2 Distribution of Shock Energy

Shock energy is measured as ground vibration using velocity gauges (geophones). An array of gauges is located around the blast, and the information recorded either digitally or on analogue tape recorders. Full vibration waveforms are used to calculate energy at a point from knowledge of the rock density, the p-wave velocity of the rock, and the particle velocity time history at the point of interest. The measurements of particle velocity are made close to the blast (from 5m to 25m) to avoid excessive errors in the calculation of vibration energy due to extrapolation.

From multiple recordings of particle velocities, the vibration propagation equations can be defined for the particular rock/explosive combination in the blast. The equations used at the JKMRC take both frictional and geometric attenuation mechanisms into account for spherical waves, and permit detailed extrapolations concerning the shape and amplitude of ground vibrations at any location. Through measurement and calculation it is therefore possible to determine the shock or vibration energy at any location in or around a blast, and to establish contours of ground vibration around a blasthole pattern. It is this vibration energy which is considered responsible for the introduction of new fractures or breakage of the rock mass.

2.3 Fragmentation Measurement

Any program to predict blast fragmentation must be verified by a measurement of fragmentation. One technique being used at the JKMRC is a photographic process where many photographs of a muckpile are taken at various stages of excavation, and all discernable particles in the photograph digitised manually to obtain basic dimensions such as area, perimeter, maximum and minimum projected chord lengths. Extensive calibrations have developed a relationship between the digitised measurements and the particle size, permitting a size distribution to be calculated. The technique has been shown to accurately describe observed changes in size distribution, but is extremely labour intensive, requiring up to 1 hour to digitise several hundred particles in a single photograph.

The measurement of fragmentation is becoming more and more necessary in modern blasting where specific fragmentation objectives are being sought to permit the application of more cost effective mining technology such as continuous mucking, mobile crushing units, and conveyor transport of materials after blasting. Although it is possible to define the required fragmentation for these applications, it appears that it is not practically feasible to measure fragmentation so that mining engineers can further optimise the process of blasting and primary size reduction. This is clearly one area where computer aided analysis could be used, if the problems of boundary discrimination can be overcome.

2.4 Model Predictions

Using all of the above techniques, it therefore becomes possible to model the process of explosive breakage, using field observations and measurement to determine the model parameters. Several normal blasts were monitored, in terms of vibration response, muckpile fragmentation, and pre-blast fracture distribution and the results used to determine values for critical parameters. Blast design was then modified and the model used to predict the changes in muckpile fragmentation from two different design configurations. In the first, blasthole diameter was decreased from 165mm to 114mm, with powder factor being held constant by reducing both burden and spacing proportionately. In the second experiment, powder factor was reduced by increasing both burden and spacing, reducing the powder factor by 20% using the standard hole diameter of 165mm. Figure 2 presents the agreement between the observed and predicted fragmentation using the parameters determined from previous blasting measurements.

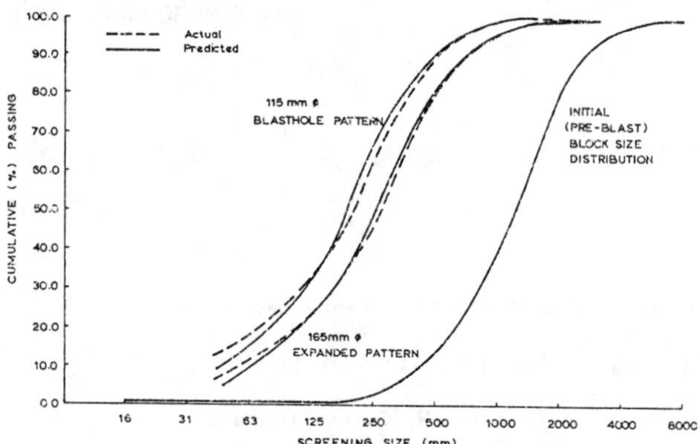

Figure 2. Predicted and observed size distributions from fragmentation model applied to actual blast designs.

2.5 Damage Measurement and Definition

The mechanistic modelling approach can clearly give accurate predictions of fragmentation, and can easily be extended to predict damage. Damage can now be defined as the change in the insitu block size distribution due to adjacent blasting, and can be related to the "breakage index" concept presented earlier, ie. the shift in the pre- and post-blast block size distributions. Depending on the application of the measurements of damage, this definition may or may not be adequate, highlighting the need for a more explicit definition of damage. The definition used above is adequate to compare the effects of different blast designs on back-cracking, or overbreak, but gives little indication of the resulting loss of stability which may ultimately lead to failure, ore dilution and lost recovery.

3. THE LIMITATIONS OF MODELLING

Predictive modelling, in contrast to back analysis, must make assumptions concerning the behaviour of explosive and the accuracy of blast design specifications such as blast-hole diameter, hole accuracy, pattern size, etc. Without some form of monitoring programme, the real performance of a blast can not be known, and the real causes of the observed performance can not be determined.

Current modelling of explosive rock fragmentation assumes conformability between actual and designed charge performance. This will be shown to be a major limitation of modelling, and a major impediment to blast design optimisation programmes. The advances made in modelling and computing are lost if blast initiation control can not be strictly maintained. Simple monitoring of charge performance, and the ability to detect initiation malfunctions, can lead to far greater gains in cost effective blasting than predictive modelling. Whereas mathematically it is possible to vary blasthole diameters, burdens, spacings, explosive strengths, and stemming lengths independently over wide ranges, in practice this can easily lead to sympathetic charge initiation, charge desensitisation, or charge dislocation, which can not be detected without careful monitoring. Delay intervals can also be theoretically modified over a wide range, but modelling does not consider the effects of an inappropriate interval on sequence reversal, vibration enhancement, or dislocation of adjacent charges.

4. BLAST MONITORING TECHNIQUES

Instruments located near to a blast can provide clear records of the detonation of explosive charges. Figure 3 shows the response to a single charge, recorded from a geophone located approximately 20 metres from the charge and permanently grouted to the rock in a borehole. The precise initiation time, initiation error, and vibration amplitude can easily be measured from this record. Initiation timing provides information concerning sequencing, misfires and delay scatter, while vibration amplitude provides information concerning performance of explosive and energy available for breakage.

Figure 4 presents the response to a multi-delayed blast, showing also the blast design. In this example, the charges are decked, ie. several charges within the one blasthole, and the nominal and actual initiation times for each deck are shown. Clearly shown is the scatter in the initiation time for the two 300 ms charges. Despite the delay errors, these charges are performing as well as can be expected - observed delay scatter is within expected limits and all charges have initiated with strong vibration response.

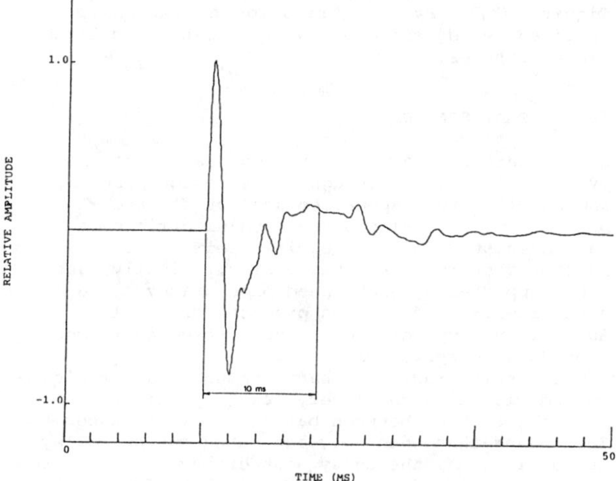

Figure 3. Vibration response to a single explosive charge, recorded within 20 m of a blasthole.

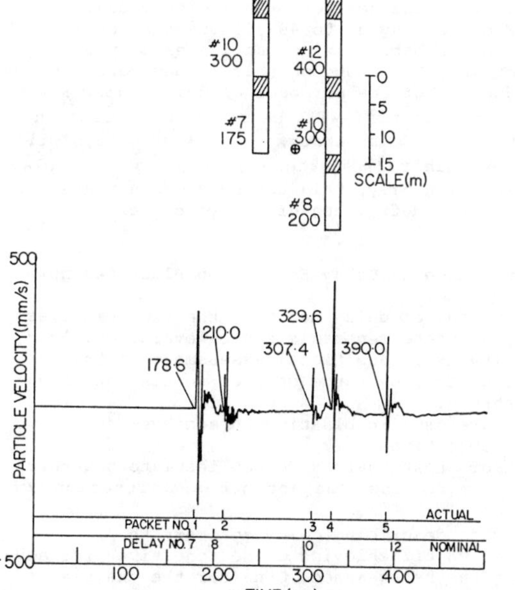

Figure 4. Vibration response to a multi-delayed blast.

4.1 Initiation Malfunctions

Figure 5 presents three of the most common forms of charge malfunction which have been observed to occur commonly in blasts throughout Australia, Canada, USA and Chile. The extent of charge misfires is strongly influenced by the complexity of blast design, number of delayed charges, duration of blast, and geological structure. Overcharging is the most common cause of initiation problems, poor fragmentation, and excessive overbreak. A common response to poor fragmentation is to increase powder factor - but this may aggravate the situation by promoting misfires and increasing overbreak which tends to be blocky as defined by pre-existing fractures.

The extent to which misfires and other initiation problems can be present is shown in Figure 6, where only 9 out of a total of 20 charge have initiated as designed. Of the total number of charges, 20% failed to initiate, and 35% initiated sympathetically with delays 5 and 8, producing 2 tight bunchings of vibration pulses.

Once identified, charge malfunctions can be eliminated, restoring initiation control and invariably using less explosive. Figure 7 presents a well controlled blast which resulted from modification of the blast design used in the record of Figure 6. Misfires have been eliminated, the number of small deck charges has been reduced, powder factor has been reduced by approximately 50%, and vibration levels, damage and overbreak have been reduced.

A fourth common charge malfunction is the reduced performance of charges, and relates particularly to the application of emulsion based explosives. These explosives have frequently been observed to produce variable vibration levels, variable fragmentation, and variable muckpile displacement, and have also been commonly observed to misfire. These features suggest that the explosives are not always detonating at maximum or designed rates, and that the energy available for breakage is therefore significantly reduced. For the emulsion-based explosives to maintain their cost effectiveness, they must consistently achieve rated performance levels.

In the light of the evidence concerning the extent to which initiation problems are occurring in production, and development or tunnel blasting, instrumentation to quickly identify the problems would appear to have greater application to blast design optimisation than modelling. Blast performance monitoring has frequently reduced explosive consumption by around 50%, and averages approximately 25% with improved control over fragmentation, vibration and damage.

5. BLAST DESIGN ASSESSMENT

Blast performance monitoring has amply demonstrated the importance of maintaining tight control over the distribution and concentration of explosive. However, even a low average powder factor can produce localised initiation faults indicative of overblasting. Simple computer algorithms can be developed to give blast design engineers a better assessment of a blasthole and charging pattern than is currently obtained.

Figure 8(a), for example, shows how the concentration of explosive in one section of a blast has been increased to be more than double the average. In this case, the concentration was caused by an increase in blasthole diameter from 70mm to 115mm in order to achieve the increased hole length without loss of drilling accuracy. In this section of the blast, sympathetic detonation of charges is more likely to occur, resulting in overbreak and increased damage. Figure 8(b) shows how this charge concentration can be reduced by using a lower strength explosive in the larger diameter holes.

Similar applications to fan patterns in underground blast designs have resulted in powder factor reductions in excess of 10%, by adjusting the lengths of explosive columns in blastholes.

Simple design assessment techniques such as these

also offer considerable scope for blast design optimisation beyond the scope of current models. Future modelling should be able to determine the point at which sympat-hetic detonation of charges and charge disruption are likely to occur.

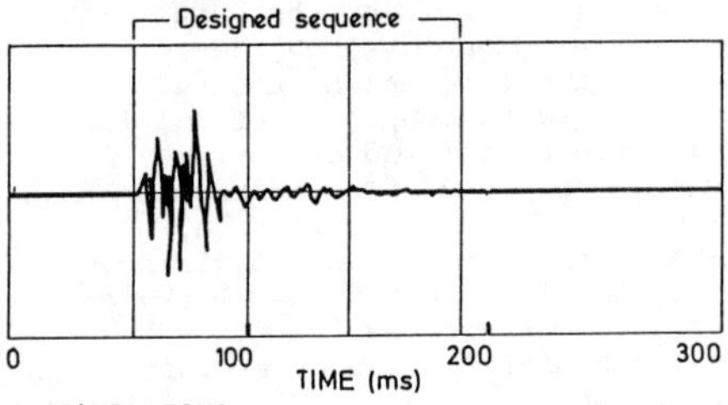

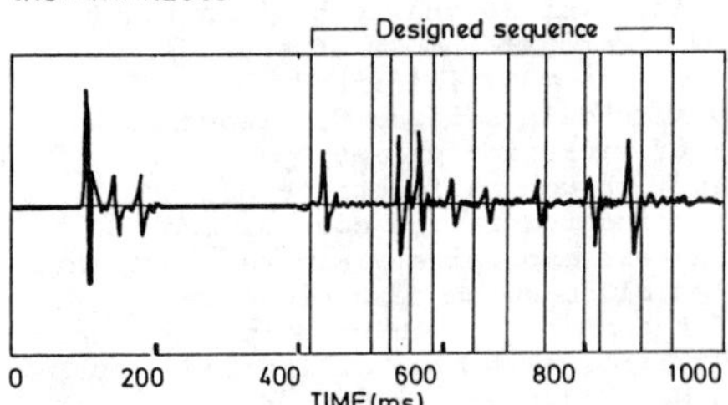

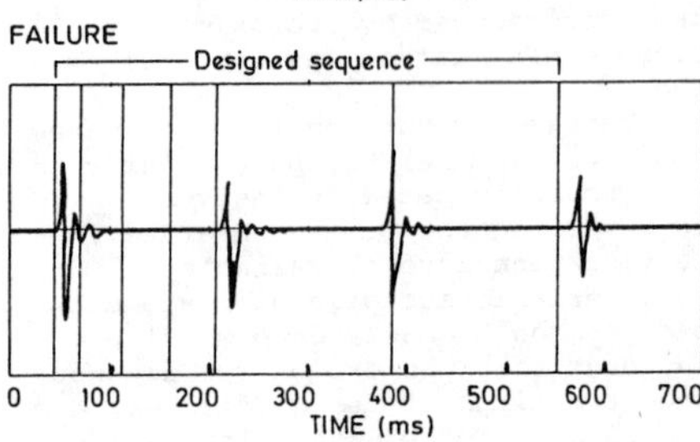

Figure 5. Common forms of explosive initiation malfunctions - actual records.

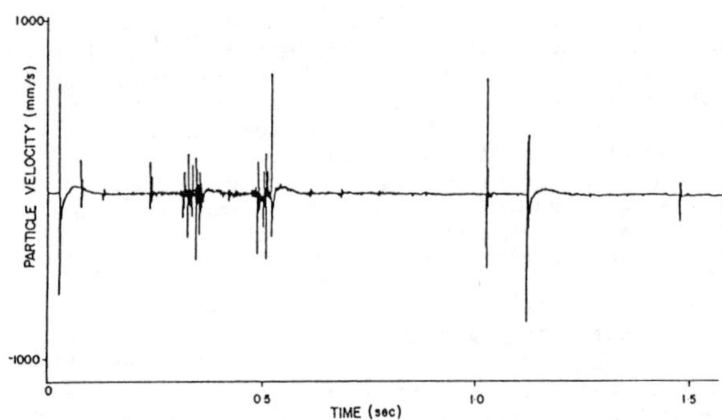

Figure 6. Extent to which misfires can be present - actual underground blast record.

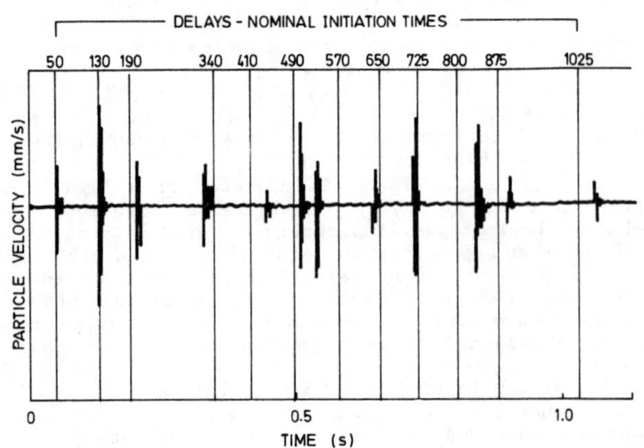

Figure 7. Controlled initiation of charges, free from misfires - actual underground blast record.

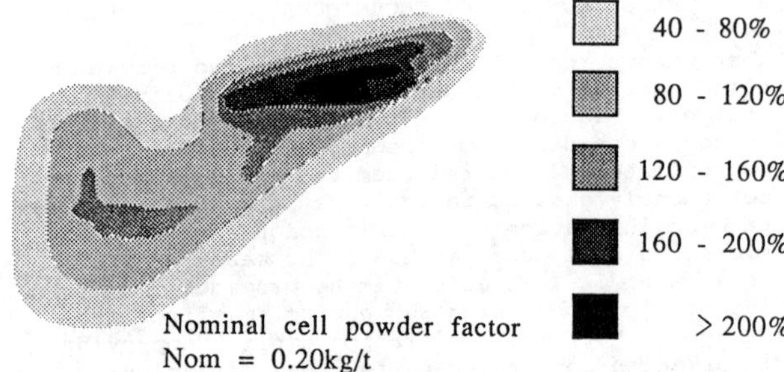

Figure 8(a). Concentration of explosive during loading caused by increased hole diameter.

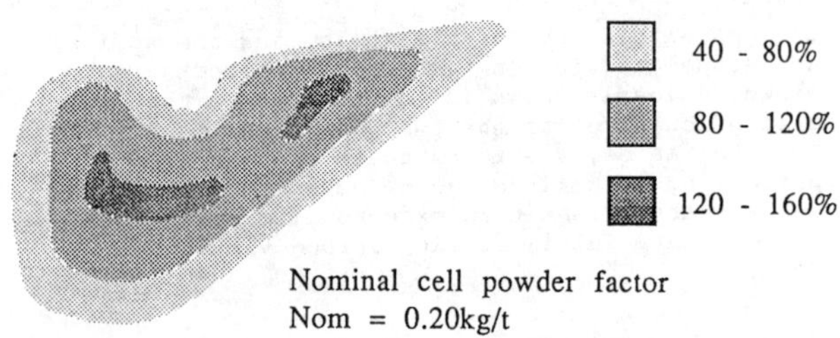

Figure 8(b). Even distribution of explosive obtained by adjusting explosive strength in large diameter holes.

6. DELAY SCATTER

Delay scatter is a well accepted feature of pyrotechnic delays, though blast design programs and models rarely appear to account for the phenomenon. Several thousand non-electric delays have been studied to determine their precise scatter patterns, each timed electronically, and many supported by high speed photography. The studies were conducted on products from all Australian suppliers, covering most delay ranges from 25ms to approximately 10s.

The results indicate that although scatter within an individual batch is very low (approximately 3 to 4%), the scatter between batches is much higher. This means that when only one batch and one delay number is used, the delay scatter may be considered to be minimal and around 3 or 4%, but if more than one delay number or batch is used, then the delay scatter will increase to a much higher level, up to 13%, depending on the product. Figure 9 shows the delay scatter for two batches of 500ms delays. The total scatter when these two batches are mixed is at least twice the scatter of each individual batch. From testing up to 40 different batches, scatter due to batch mixing can increase by a factor of up to 3, although this is dependent on the product. For the purposes of this paper, a delay scatter of ±13% will be used to describe the distribution of delay error, ie, 95% of delays will detonate within ±13% of the nominal or quoted delay time. Experimentally, this has been demonstrated to be a realistic definition of delay scatter.

6.1 Influence of Delay Scatter on Blast Design

From knowledge of delay scatter, the probabilities of delay sequence reversals can be evaluated. This probability is an additional assessment of the level of control over the correct initiation of a blast pattern.

Delays are used in blasting to achieve one of two specific objectives :

1 to ensure that charges do not initiate together;
2. to promote the instantaneous initiation of charges.

Using data describing delay scatter, the probabilities of achieving these objectives for any situation can be assessed. Consider the example of a tunnel or development round shown in Figure 10 which presents the hole pattern and delay sequencing used in a typical underground round. In particular, the analysis will consider the

following probabilities :
1. that the perimeter holes sharing a common delay will interact to promote smooth blasting and smooth walls to the drive ;
2. that the perimeter holes will initiate after the inner holes have fired and cleared, minimising damage to the walls and back.

For adjacent blastholes to interact to promote smooth blasting, they must initiate within a critical interval, determined by the crack propagational velocity of the rock. Assuming a p-wave velocity of approximately 6000m/s, a crack propagational velocity of one-third of the p-wave velocity, and a blast-hole separation of no more than 1 metre, then adjacent holes must initiate within 0.5 ms of each other to achieve any interaction between charges and to achieve a smooth blasting effect. Figure 11 shows the probability of interaction as a function of delay time, for a delay scatter of approximately 13%. For the high order detonators around the tunnel perimeter, the probability that adjacent holes will detonate within 0.5ms is effectively zero, and the probability that they will detonate within 5 ms is less than 2%. There will therefore be no smooth blasting effect in tunnel blasting using high order, long period delays.

A similar analysis of smooth blasting along stope walls using millisecond delays is shown in Figure 12, where the easer hole represents an intermediate hole between rows, on the stope periphery. Ideally this hole should detonate at the same time as the wall hole in the row behind to promote a smooth stope wall. The probability of achieving this effect is shown to be approximately 3% for 600ms delays, and increases to only 10% for 100ms delays.

The probability of sequence reversals and perimeter holes initiating before the adjacent inner holes has also been investigated. Figure 13 shows the overlap probability function for the delays used in the blast, with a delay scatter of approximately 13%. The probability refers to the overlap potential between successive delay numbers. A maximum overlap probability of approximately 14% is shown for delay periods around 6 seconds. There is therefore a reasonably high probability that a perimeter hole will initiate before an inner hole, promoting wall damage, irregular drives, and increased ground support costs. By ensuring that perimeter holes are separated from inner holes by at least two intervals, the probability of overlap decreases to effectively zero.

Placing charges on separate delays does not necessarily ensure that they will initiate separately or in the desired sequence, particularly when standard intervals have been reduced by combining delays. Models of blasting must first be able to account for the influence of delay intervals on fragmentation, and secondly, account for the influence of out-of-sequence initiations on fragmentation and damage.

6.2 Effect of Delay Scatter on Vibration Control

One of the main reasons for using delays in modern blasting is to control and minimise vibrations and vibration-induced damage. For this reason, many modern long hole stoping operations are using multiple charges within each blasthole, each charge separated by inert stemming and separately delayed. This leads to large numbers of delays required to achieve desired levels of production tonnage per blast. Since standard delay series are limited in size to no more than approximately 40 delays, it is common practice to extend the available series by combining delays, using a collar delay at the top of the hole to initiate the downline to a second delay at the bottom of the blasthole. This increases the number of available delays but decreases the interval between delays and therefore increases the probability of overlap between successive delays.

Currently available vibration prediction equations relate to the maximum charge weight per delay, and place little emphasis on defining the minimum period of the delay. In practice, this period is controlled totally by the rock type, being relatively short for high modulus, elastic rock types, and being much greater for low modulus, plastic rock types. The period can be related to the induced vibration response of the rock. The minimum vibration interval is the time after detonation before the rock returns to a near-zero vibration state as shown in Figure 3. Vibrations produced by individual charge detonations are not influenced by any other charges, and vibration levels are well controlled.

Vibration level prediction using the above concepts can then consider the probability of enhancement of vibrations from different charges by evaluating the probability that delay scatter will cause charges to initiate within any particular interval, determined from vibration monitoring using instrumentation capable of recording the full time history of ground vibration at any location. This approach, for example, has been used to simplify blast design where large numbers of delayed deck charges are involved. Probabilistic analysis has been used to compare designs containing large numbers of deck charges with small delay intervals, against a reduced number of decks (larger charge weights) with large delay intervals. In many cases, the larger charge weights and longer delay intervals can be demonstrated to produce lower vibrations, less variability in ground vibration (improved vibration control), and improved ease and speed of charging.

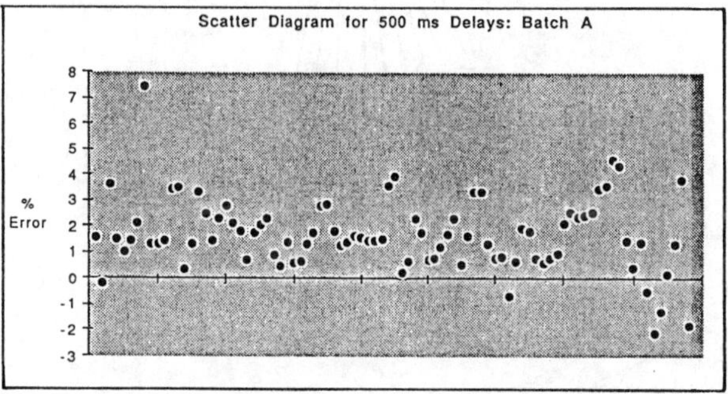

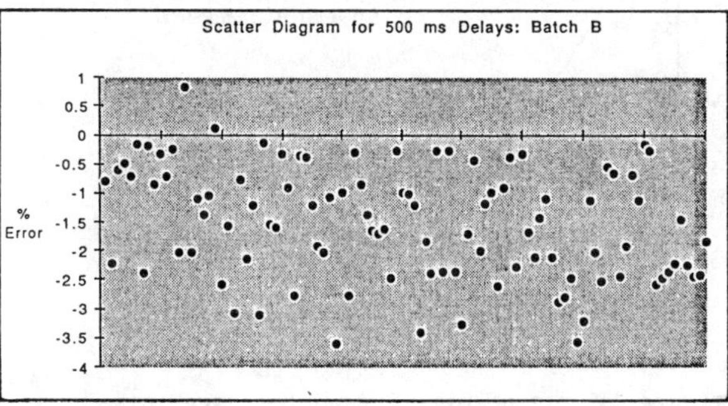

Figure 9. Delay scatter for different batches of detonators.

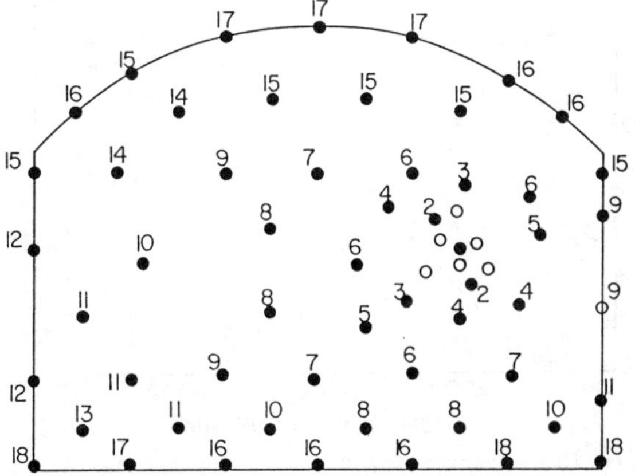

Figure 10. Development blast pattern and sequencing design for 52 mm hole diameter.

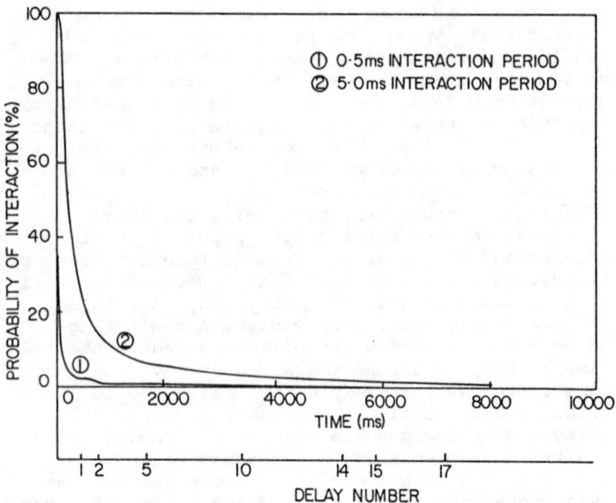

Figure 11. Probability function describing potential for constructive interaction between adjacent peripheral holes in a development blast.

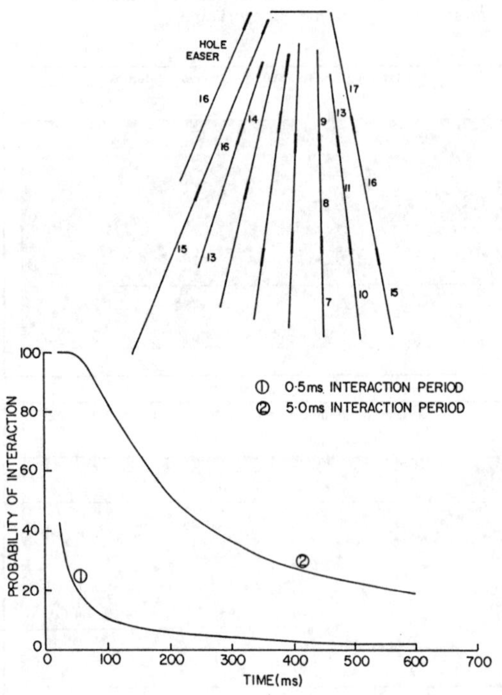

Figure 12. Probability function describing potential for constructive interaction between adjacent peripheral holes in an open stope blast.

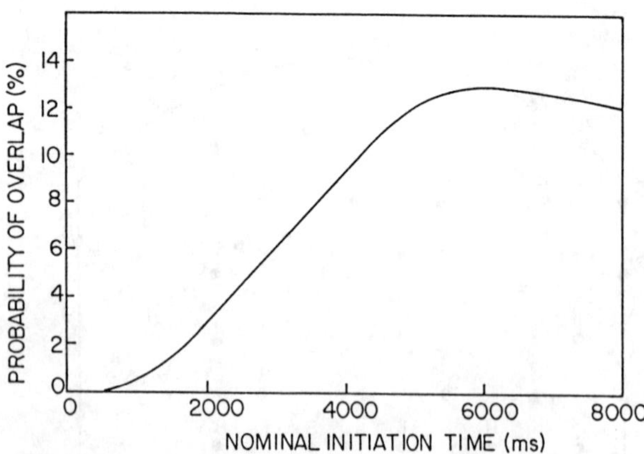

Figure 13. Probability function describing potential for overlap between successive delay numbers, likely to produce out-of-sequence initiation of wall charges in tunnel blasting.

7. DAMAGE THROUGH GAS PENETRATION

Observations of blasting have indicated that high pressure explosive gas products penetrate along fractures in all directions around a blasthole. The forward penetrating gases promote movement and swell of the muckpile, but the backward penetrating gases cause swelling behind the blast and the subsequent loss of friction along fracture surfaces, decreasing stability. The influence of the gas is strongly dependent on the state of fracture of the rock mass.

Simple experiments have been conducted to demonstrate the flow of gas behind blasts, yielding information concerning the velocity of propagation, the timing of burden movement, and its dependence on blast design variables. In these experiments, boreholes are drilled behind the blast pattern, pressure gauges are inserted, and the top of the borehole is sealed with a concrete plug. Adjacent to, or even in the borehole, is a vibration gauge which is used to determine the precise initiation time of the blasthole near the pressure gauge.

Figure 14 shows the pressure response in a borehole located 6 metres from the end of a pre-split line of blastholes. The gauge, which has been vibration compensated, still shows a small vibration response as the nearest blasthole detonates, and then shows a sudden increase in pressure followed by a gradual release. In this experiment, the high pressure gas path has clearly been along the pre-split line which has extended beyond the last blasthole, rather than through natural fractures.

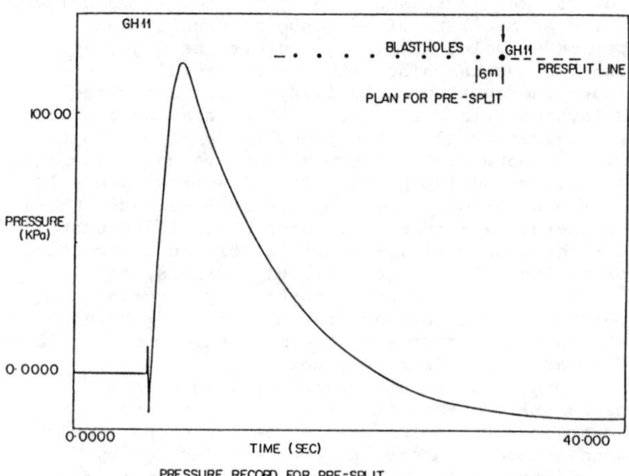

Figure 14. Gas pressure response in uncharged borehole located 6 m from the end of a pre-split line of blastholes.

In a second experiment, gas pressure and vibration monitoring were conducted behind a normal overburden blast in a strip coal mine, producing the responses shown in Figure 15. The blast was fired in rows parallel to the free face, in five rows, with the vibration response from each row clearly shown on the vibration record from the radial guage in the BM 11 hole. These vibration peaks can also be seen on the gas pressure records of the other two signals, indicating tht the piezo-electric gauges are not fully vibration compensated. Also clearly visible, however, is the gas pressure response in the GM 7 hole, showing first an increase in pressure followed by a decrease resulting in a negative borehole pressure. Final return to ambient pressure takes approximately 1 second after gas first started flowing into the borehole. Gas propagation velocities have been calculated from many experiments to be approximately 8 to 12 metres/second.

The negative pressure indicates that the blasthole volume is increasing, due probably to a release-of-load relaxation as the burden in front of the last blasthole moves away from the bench face. Measurements of burden movement times based on this assumption indicate that the rock takes

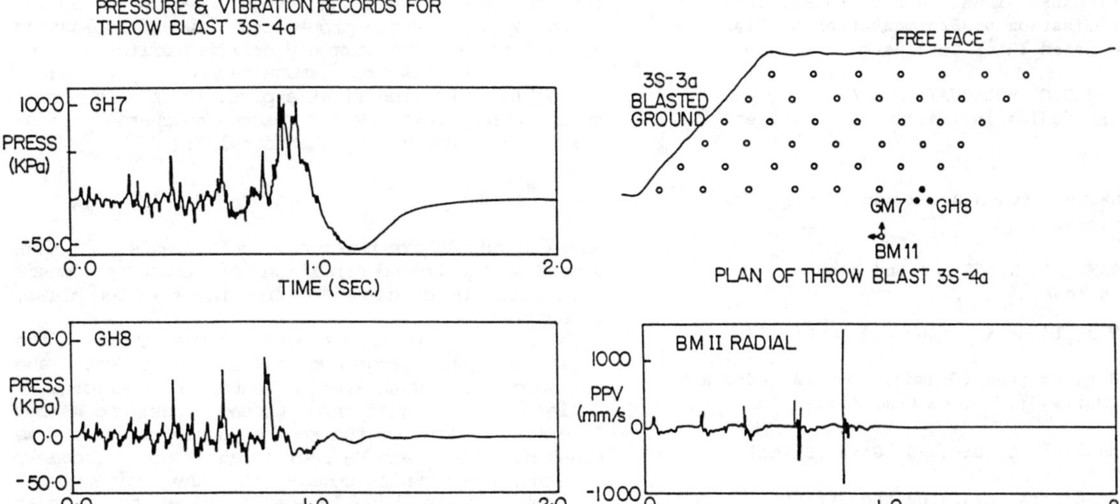

Figure 15. Gas pressure response in uncharged borehole located behind a production blast in heavily fractured rock.

approximately 90 ms before any burden movement occurs after detonation of the explosive charges.

Gas penetration has been measured at distances of 8 metres behind a blast, and is proposed as a major cause of damage and lost stability. The penetration of gas can be minimised through :

1. minimisation of the time over which the pressure is acting (maximise burden movement velocity) :
2. minimisation of peak borehole pressure ;
3. providing an alternative gas vent such as a presplit line.

Of these techniques, reduction in the time over which the gas pressure is acting is expected to be the most effective means of controlling gas penetration. Burden movement velocities can be maximised, without increasing peak borehole pressure, by minimising burden dimension. Blast designs to minimise blast induced damage should therefore have the following features :

1. few blast rows, promoting unrestricted burden movement;
2. small burden dimension;
3. pre-split along critical boundaries.

Modelling of blasting must ultimately be able to predict damage as well as fragmentation. Both factors are critically dependent on the existing state of fracturing. Gas pressure plays a critical role in determining the final muckpile shape, and also in the extent of gas related damage. Modelling must therefore be able to predict the flow of gas as well as the volume and pressure of gas produced in blasting.

8. CONCLUSIONS

Mathematical models have contributed significantly to our understanding of some of the mechanisms involved during blasting. However, unlike most other applications of mathematical models, blasting models have developed without monitoring and measurement procedures required to validate predictions or even confirm the major mechanisms. Explosives and initiators are assumed to function exactly as designed or specified, and little importance is placed on evaluating factors which may detract from their optimum performance. Fragmentation and damage, primary factors in modelling exercises, can not be accurately measured on a large scale. The lack of basic measurement techniques and instrumentation, which can be used by the mining industry, is a major limitation to the application and further development of blasting models. Instrumentation to monitor the performance of explosive charges is available and is the bare minimum of monitoring instrumentation required to ensure that any blast design has the best chance of producing the desired, or predicted, blast product.

9:20 LE GRAND SALON - Short Presentations

B. BONAPACE, Austria "Advance and Limits of TBM-Driving during the Excavation of a 22 km Long Pressure Tunnel".

J. BRYCH, NGOI NSENGA and XIAO SHAN, Belgium, Zaire, China "Destructibility of Rocks with Rotary Drill Bits".

I.W. FARMER and P. GARRITY, U.K. "Prediction of Roadheader Cutting Performance from Fracture Toughness Considerations".

GINN HUH, KYUNG WON LEE and HAN UK LIM, Korea "Determination of Vibration Equation by Empirical Methods".

10:00 FOYER, Coffee Break.

10:30

Y. GOTO, A. KIKUCHI and T. NISHIOKA, Japan "Dynamic Behaviour of Tunnel Lining due to Adjacent Blasting".

J.R. GRANT, A.T. SPATHIS and D. BLAIR, Australia "An Investigation of the Influence of Charge Length upon Blast Vibrations".

H. HERAUD and A. REBEYROTTE, France "Presplitting Tests and Measurements of the Induced Back-Breaks in Granitic Rocks of the French Massif Central".

T. INAZAKI and Y. TAKAHASHI, Japan "Evaluation of Rock Mass Quality Utilizing Seismic Tomography".

N.H. MAERZ, J.A. FRANKLIN, L. ROTHENBURG and D.L. COURSEN, Canada, U.S.A. "Measurement of Rock Fragmentation by Digital Photoanalysis".

A.M. MIRANDA and F.M. MENDES, Portugal "Rock Weathering as a Drillability Parameter".

S. MITANI, T. IWAI and H. ISAHAI, Japan "Relations between Conditions of Rock Mass and TBM's Feasibility".

R.D. SINGH, VIRENDRA SINGH and B.P. KHARE, India "A Study for Optimization of Fragmentation by Blasting in a Highly Cleated Thick Coal Seam".

W.H. WILSON and D.C. HOLLOWAY, U.S.A. "Fragmentation Studies in Instrumented Concrete Models".

12:00 MARQUETTE, Lunch Break and Poster Sesssions.

13:30 LE GRAND SALON, Panel and Floor Discussions.

Moderator: P.A. Lindqvist (Sweden)

Panelists: R.F. Favreau (Canada); O.T. Blindheim (Norway); C.D. da Gama (Brazil).

Panel contribution: C. Dinis da Gama (Brazil)

ABSTRACT: Six papers regarding the effects of rock blasting vibrations and four papers on the subject of fragmentation constitute the kernel of the author's contribution for the development of his task as panelist for Theme III of this Congress.

After a brief description of each paper's main aspects, specific comments about their merits are pointed out.

General ideas on both topics are then presented to complement the author's view points.

1 INTRODUCTION

There is no doubt that the subject of Rock Blasting and Excavation deserves an important share of modern Rock Mechanics, due to the many topics that need further research, both theoretical and applied, in order to better control field operations and to lessen incorrect or excessive explosive consumptions, thus reducing excavation costs.

The complex mechanisms of rock disruption under explosive action are still a challenge to many Rock Mechanics experts, who feel the necessity to better understand the time sequence of events that contribute to reach a certain fragmentation, which should be previously set, according to economic optimization criteria. But not only engineering design is affected by uncomplete or erroneous knowledge on this subject; the so-called secondary effects from rock blasting (vibrations, flyrock, airblast, dust and gases) also require scientific contributions aimed to reduce their detrimental consequences.

In the panelist's task to tackle this subject, a couple of blasting topics were distributed by the General Reporter: rock vibrations and fragmentation. Both are full of physical and mechanical interrelated phenomena, with very short time intervals of occurrence and with many things in common: in effect, when blasting design is adequate (perfect balance of explosive quantities and rock volumes) the two aspects are the best measures of blasting efficiency.

But when design is mistaken or field operations go wrong, either the excessive amount of explosive causes vibration problems and/or overfragmentation , or subcritical explosive weights produce many rock boulders, thus revealing the need for corrections of several types.

Therefore, rock vibrations and fragmentation are key factors for the success of any blasting operation, either in mining or civil engineering works, as confirmed by the review of ten excellent papers on these topics.

2 PAPERS ON BLASTING VIBRATIONS

2.1 Seismic control of mine and quarry blasting in the USSR, by A.B. Fadeev, L.M. Glosman, M.I. Kartuzov, and L.V. Safonov

The article describes the methods used in the Soviet Union to quantify the propagation laws of blasting vibrations, and the commonly utilized criteria to protect structures from damage caused by nearby detonations. For the first aspect, it is reported that particle velocity (V) relates to charge mass (Q) and distance (r) by the formulae:

$$V = K_1 K_2 \frac{Q^{1/3}}{r}$$

where K_1 and K_2 are numerical coefficients that depend on the ground conditions of structure foundations, and on the number of free faces in the blast, respectively.

Values of K_1 and K_2 are given in the paper, the first one with average and maximum magnitudes, and the second only with average values. Furthermore, K_1 is indicated for just three ground types, and K_2 for 1 to 4 free faces of the rock mass subjected to blasting. The K_1 and K_2 coefficients are probably assuming values which comply with the dimensional coherence of the above indicated expression, in order to provide V in cm/s, after using r in meters and Q in kg.

The most curious characteristic of that formulae is the indication that the particle velocity is supposable decreasing with the first power of the distance, in all cases.

As the explosive charge Q is considered per delay, the authors proposed another empirical expression to determine the time interval T between consecutive charges, as given by:

$$T = 2 W \left(\frac{\gamma}{q}\right)^{1/2}$$

where W stands as the shortest distance (in meters) from a charge to a free surface, γ is the specific weight of the blasted rock (in t/m^3) and q is the powder factor, expressed in kg of explosive per m^3 of blasted rock. Upon the utilization of this criterion, examples are given in the paper for blasts ranging from 320 kg to 11,000 kg of an undisclosed type of explosive.

As far as damage criteria are concerned, the authors suggest (in Table 4 of their paper) a series of values for limiting vibration velocities in several kinds of structures, from hospitals to mine openings. It is interesting to note that those limits are reduced to half or 1/3 of their single blast values, when they result from repeated blasts. The article refers only 5 russian publications on the subject, the most recent one being of 1976.

2.2 A determination of vibration equation by empirical methods, by G. Huh, K.W. Lee and H.U. Lim (Korea)

The need to protect different types of structures located in the vicinity of two subway lines in Seoul, whose excavation was accomplished by means of explosives, led the authors of this paper to determine the propagation laws of blasting vibrations.

Using field data of more than 100 blasts, they considered the mutual interference of the following variables:

- distance between the detonation and registration points;
- weight of detonated explosive per delay;
- type of explosive (3 types);
- blasting pattern (4 kinds);
- rock types and corresponding weathering degrees.

The general expression of vibration propagation is proposed as:

$$V = K \left(\frac{W^{\frac{1}{3}}}{D}\right)^n$$

where V (cm/s) is the particle velocity, W (kg) is the weight of explosive detonated per delay, D (m) the distance, and K and n two parameters which depend both on the explosive and the rock under consideration. Table 6 of the article indicates the values of K for a range of explosives and blasting patterns, and for its calculation it is proposed the expression:

$$K = E_i \, (R_i \, S_c + Q_i)$$

where E_i is a correction ratio dependent on the explosives, R_i a constant of each rock type, S_c the compressive strength of rock and Q_i a correction factor according the kind of blasting pattern. In the field work K varied between 48 and 138, while n (the exponent of distance) ranged from 1.5 to 1.7.

All the empirical expressions resulted from experimental field data, but no correlation coefficients among the variables are reported. Regarding damage criteria, the authors have used the one proposed by the German Standard DIN 4150, on the basis of peak particle velocities of vibration. They conclude the paper by suggesting the execution of more blasts in order to better understand the essential mechanisms of blasting vibrations under the prescribed conditions.

2.3 Dynamic behavior of tunnel lining due to adjacent blasting, by Y. Goto, A. Kikuchi, T. Nishioka (Japan)

The excavation and lining of two parallel tunnels oppened in weathered diorite rock in Ogitsu (Japan) was performed with continuous monitoring of blasting effects, using accelerometers and strain meters. Reciprocal effects were measured in each tunnel, as the other one was excavated, including the performance of concrete linings that were used to support rock walls.

Due to the short distance between both tunnels (from 1.5 to 1.8 meters) special attention was given to the maximum amount of explosive charge that could be detonated safely, without damage to the nearby cavities and their linings.

As for damage criteria, the authors utilized three methods:

a) correlation of particle velocities and distances;
b) deformations of tunnel linings;
c) strains induced by the stress waves on the rock walls and additionally caused by the tunnel linning deformations.

Several graphs of the distributions of strain, deformation and particle velocity are presented in the paper, for different explosive weights detonated in the adjacent tunnel, and a comparison between measured and calculated strains is accomplished.

Unfortunately, no correlations between those variables and damages are proposed, neither a measure of explosive charge weight limits is suggested. Lack of quantification is noted as far as relating distances, charges and vibration variables (either deformation, strain or particle velocity). However, Fig. 4 of the paper gives a general idea of that correlation, which could be improved by means of multiple variable regression analysis. In addition to this, it is suggested that the authors proceed their research in order to develop reliable criteria for the stability of nearby excavations.

2.4 An investigation of the influence of charge length upon blast vibrations, by J.R. Grant, A. T. Spathis and D.P. Blair (Austrália)

The paper describes a series of full scale experiments for the measurement of stress waves in a central hole, surrounded by 8 boreholes located at equidistant points of a 20 m radius circle. In each of these points, explosive charges of a PETN-TNT mixtures were detonated, having lengths from 0.09 m to 8.0 m.

The purpose of this research was to detect the effect of charge length on the characteristics of vibrations recorded 20 m away, especially the wave frequencies and amplitudes.

Results of the experiments have shown that as the charge length increases, with a constant charge diameter, the dominant energy transmited by the stress waves shifts to lower frequencies. The dominant frequency decreased from 2 kHz (for the 0.09 m long charge) to 0.2 kHz (for the 8 m long charge), but the corresponding peak amplitudes did not increase in proportion to the length, or the weight, of the explosive. This last aspect is probably the result of the fixed distance for monitoring detonation effects (20 meters) which involves only near-field consequences of the blasts. The fact is that detonation pressure is independent of charge weight, while the total released energy increases with charge weights, showing its effects at larger distances. The authors emphasize the aspect that to reduce vibrations, in order to protect nearby structures, decreasing charge weights to be detonated by delay results favorable to their stability, mostly as a consequence of frequency increases (away from structures ressonant frequencies).

However, it is not clear if this behavior is maintained for other distance ranges, namely in the far-field effects of blast vibrations. Another interesting conclusion of this research regards the identification of the P-wave peaks as the most important ones in the near-field range of blasting.

On the other hand, modelling techniques have proved adequate to simulate explosive behavior, and further efforts will be done by the authors in order to apply dynamic finite element codes to determine the differences between near and far-field effects, and also to establish the mutual influences of site parameters and blast patterns on the transmitted vibrations to the surrounding media.

2.5 Essais de prédécoupage et measures des arrière-bris dans les roches granitiques du Massif Central Français, by H. Heraud and A. Rebeyrotte (France)

The paper concerns a comparison of two explosive substances as presplitting agents in a granitic rock excavation. The explosives were: a detonating cord (with a density of 70 g/m) and a water gel sausage type (cisalite), with different physical and chemical properties. In terms of detonation velocities, while the first one is 6,500 m/s, the second is just 2,850 m/s, according to Table I of the article.

As for loading densities, the detonating cord is 70 g/m and cisalite is almost six times greater (400 g/m). Another important aspect of presplitting, the decoupling ratio, is also very different: 0.12 for the detonanting cord and 0.24 for cisalite, as taken from the data given in Table I. Corresponding hole spacings were 0.8 m and 1 m, respectively.

The excavated rock is described as a fractured granite and the presplitting tests were conducted in conjunction with a 3 m x 3 m grid of 0.115 m diameter blast holes, with powder factors of the order of 400 to 410 g/m³.

To measure the quality and efficiency of presplitting in the field experiments, the authors used three methods:

a) definition of quality indices;
b) pseudo-frequency measurements;
c) microseismic monitoring.

Two quality indices were proposed to monitor the state of the rock face after the blasts: P (profile index) as the ratio of the real to the ideal profile lengths, and C (resulting borehole index) as the ratio of visible borehole lengths to the theoretical length of all boreholes. The value of 100% is the optimum for both indices, although their variation is opposite: bad presplitting leads to higher than 1 values of P, and smaller than 1 values of C.

Comparison of these indices for the two fore-mentioned explosives indicate advantages for cisalite.

With respect to black-break determinations, techniques of measuring pseudofrequences and microseismic events were utilized. The first one is based on the correlation between pseudofrequence and duration of wave propagation to the fracturing status of the rock mass, so that it is as sound (or harder) when the frequency is higher and sign duration smaller. By plotting these two variables in a (x, y) graph, distinct point clusters identify the results of each blast and a comparison of explosives is accomplished as Fig. 4 in the paper shows.

Microseismic monitoring, by means of continuously logging the remaining rock, in terms of propagation velocities, also provides a method to measure the damage caused in the remaining rock. This logging was performed in water filled boreholes, drilled perpendicularly to the presplitted rock face, and also concluded for the superiority of cisalite.

The only unfavorable comparison for cisalite was the one regarding costs, revealing a slight increase with respect to the 70 g/m detonating cord. However, other types of charging densities (for example 40 g/m) were not subjected to the same type of analysis.

Results of this research, which is obviously valid for the specific conditions of the rock under study, as well as the particular explosives and blasting patterns, made possible a valid comparison of their actions.

The interesting combination of presplitting performance indices and geophysical techniques for the assessment of black-break, are the outstanding aspects of the paper.

2.6 Low frequency vibrations from surface mine blasting over abandoned underground mines, by D.E. Siskind (U.S.A.)

The author describes field work and interpretation of results of blasting vibrations in a particular situation where low frequency waves were a reason for concern.

The detonations occurred in a surface coal mine, and their effects were measured at residential structures located at distances ranging from 300 to 3,000 meters. In between the mine and the houses there is an abandoned underground coal mine, whose cavities are partially filled with sandy gravely drift. Because of the abnormal low frequency waves reaching the houses and causing damage, an investigation was conducted by the U.S. Bureau of Mines on the participation of the following factors: a) blast characteristics, namely charge weight per delay and multideck fired, as well as two single-charge detonations that were specially conducted in order to identify the generation and wave propagation mechanisms; b) the influence of ground type (a thick zone of glacial till) on the reduced frequency of transmitted stress waves; c) subsurface composition and extent of abandoned mine workings over the propagation distances; d) residential structures performance and strength under dynamic stresses.

As much as 235 detonations were subjected to the analyses, by means of recording their amplitudes, frequencies and durations, at 7 houses in a period of 11 months. In addition, seven blasts were specially prepared for research purposes, and their effects monitored 17 m away, so that wave characteristics were measured before modifications imposed by the medium occurred.

Main conclusions of the study, as pointed out by the author, are: a) single charge blasts showed normal vibration amplitudes (with respect to previous USBM experience) while vibrations from production blasts, for identical charge weights per delay, averaged three times greater, indicating a superposition of effects from adjacent charges; b) with respect to frequencies, it was demonstrated that in general as the distances increased there was a reduction of dominant frequency, although that tendency changed from site to site; c) Rayleigh waves, generated along the ground-air interface, have been identified as the main component of the low frequency wave-trains, and a combination of Rayleigh and Love waves could be traced in several cases; d) results of interpretation of single charge blasts indicated a propagation pattern of systematic attenuation with distance, accompanied by a progressive reduction of dominant frequencies, and an increase in pulse duration. In conclusion, it is suggested that a combination of geological, local and technical factors have contributed to the peculiar situation under analysis, and although additional studies are proposed by the author, particularly on the subsurface conditions of the region, it is suggested to the mine operator that additional care must by taken in order to avoid the superposition effects of multiple hole delayed blasts, that contribute to the generation of low frequency vibration waves.

3 PAPERS ON FRAGMENTATION

3.1 Measurement of rock fragmentation by digital photoanalysis, by N.H. Maerz, J.A. Franklin, L. Rothenburg (Canadá) and D.L. Coursen (U.S.A.)

The measurement of fragmentation, or the size distribution of blocks resulting from rock blasting, is one of the most challenging questions in contemporary Dynamic Rock Mechanics. As the authors pointed out, there have been attempts to solve this problems by several techniques, ranging from sieving in small scale blasts, or by means of empirical formulae and computer simulations, as well as photographic methods. Due to difficulties of these techniques, particularly the last ones, the authors propose a new concept of interpreting photos taken on the fragments from blasting. The technique is based on digital photoanalysis and aims to measure sizes of all fragments (both overlapping and non-overlapping ones) in order to build their size distribution. In steads of taking photos vertically over the muck pile (with a camera located within a balloon) or through a vertical cut of the pile, they selected to photograph the fragments inside haulage trucks, at regular intervals. The blocks were loaded along the centerline of the muck pile and were taken perpendicularly to the bench subjected to detonation.

The subsequent phases of this process are:

a) manual measurement of oversize blocks, not included in the photos;
b) digitalization, through manual methods, of the photos, in order to transfer a two dimensional image to the computer memory;
c) calculation of block areas from the digitized image, and evaluation of their "sizes", by means of the equivalent circles (equal area ones). Ten-class histograms are then constructed for the presentation of frequency distributions, in terms of number of blocks a each size, or equivalent diameter, per square meter of the muck pile surface;
d) determination of real block size distribution, using two alternative methods: analytical and empirical. The first one uses unfolding functions to convert from two to three dimensions, which are based on geometric probabilities and stereology theory, while the second method utilized small scale particles of the same rock that were sieved, as well as photographed and digitized, so that a correlation between 2-D and 3-D diameters was established. Relevant problems such as fragment overlapping,

missing of fines, and anisotropic stacking were delt with, in order to develop a reallistic approach to the size distribution of fragments from blasting.

Although the authors state that this is an early state of research on the topic, further developments on the photoanalysis technique (namely, automatic digitalization and better unfolding expressions) will make it a more reliable methodology.

3.2 An investigation into the effect of blast geometry on rock fragmentation, by D.P. Singh and V.R. Sastry (India)

Small scale blasts in sandstone blocks were conducted by the authors to determine the mutual consequences for fragmentation that can be attributed to three geometric dimensions of rock blasting: burden, spacing and bench height. Two types of experiments were performed: single hole and multiple hole simultaneous detonations, the first being dedicated to study the relative effects of burden and height, while the second ones were aimed to analyse the mutual interferences of the three dimensions. A total of 83 tests were executed, and in each one the resulting fragments were sieved in order to determine their size grading, using detonating fuse as explosive in all cases.

In terms of fragmentation, the authors have defined several interesting concepts:

a) optimum fragmentation burden, as the distance that maximizes mass surface area after the single hole blasts, and for each bench height. Mass surface area is defined as the product of total fragment mass and the new surface area created by the blast;

b) suboptimum and over optimum burdens resulted in comparison with optimum values, when multiple hole simultaneous detonations were accomplished, and after varying spacing to burden ratios;

c) optimum breakage burden, as the distance corresponding to maximum mass of fragments, for single hole tests;

d) In addition, measurements of some comminution parameters were also established: average fragment size, new surface area, mass surface area, fine and coarse fragmentation indices, and apertures or sizes at 25%, 50% and 75% through which the total mass of debris had passed.

Upon optaining size distribution histograms after all tests, the authors discussed the results and presented several interesting conclusions, as follows:

a) while optimum fragmentation burden was around 51% of bench height, optimum breakage burden was about 72%;

b) increases of bench height produced increases of optimum fragmentation burdens, all other factors constant;

c) increases of burden caused a reduction on fragmentation, particularly for short benches;

d) ratios of 3 to 4 in spacing to burden gave optimum fragmentation results;

e) when these ratios were greater than 5, each hole behaved independently, with the formation of humps between them;

f) for suboptimum as well as for optimum burdens, the mass surface area increased proportionally to bench height dimensions. These general conclusions are of qualitative nature, and can be utilized to improve or control fragmentation on real scale rock blasting.

3.3 A study for optimization of fragmentation by blasting in a highly cleated thick coal seam, by R.D. Singh, V. Singh and B.P. Khare (India)

An Indian coal surface mine uses bench blasting techniques to excavate and produce the ore out of a 18 m thick seam, that is mined in two benches. The coal is banded, with layers ranging from 0.30 m to 2.74 m thick, and a system of discontinuities is present, with two principal sets of cleats intersecting each other at an angle of 112°. Compressive strength of the coal was found to be (in laboratory tests) of the order of 10-15 MPa along bedding planes and 8-20 MPa in the direction perpendicular to these planes.

In terms of breakage the authors indicated that the coal used to fragment along cleats and bedding planes, and this fact led to the formation of many lumps that did not pass a 4.5 m x 3 m grizzly. Furthermore, the pre-existing blast patterns were not adequate in terms of fragmentation, because they produced many boulders.

The patterns were formed by 6" diameter holes, with 4 m burden and 2.5 m spacing, and 4 m distance between successive rows. ANFO explosive was used with Dynak, and deck loading was practiced, with simultaneous firing of each row of holes, and a delay between 15 and 51 ms between rows. Explosive consumption ranged for 0.4 kg/ton in the top bench and 0.2 kg/ton in the bottom one.

In order to improve breakage the authors changed the drilling mesh to 3 m burden and 4 m spacing and taking into account the presence of cleats and bedding planes, thay obtained an oversize fraction (greater than 0.3 m blocks) of 10 to 15% of total blasted coal, in stead of 25-30% with the previous patterns.

The changes were based in studies conducted with cratering tests, trial blasts and comparison of results. The cratering experiments gave indications about theoretical burdens and spacings, but modifications of these were ajusted to account for the existence of discontinuities in the coal.

They observed also that coarser fragmentation and overbreak occurred when the direction of blasting was out of the lesser angle of cleat intersection. Consequently, better breakage was achieved when blasting direction was out of the open angle between cleat sets.

In terms of microdelays between adjacent holes, the authors recommend a delay interval of 9 ms per meter of burden, while between rows they suggest a delay of 2 to 3 times greater than the one existing between holes of the same row.

A composition of the above mentioned solutions, provided the way to obtain better fragmentation degrees in the coal mine under analysis.

3.4 Fragmentation studies in instrumented concrete models, by W.H. Wilson and D.C. Holloway (USA)

The paper describes a series of ten laboratory tests on concrete blocks subjected to explosive action, in order to study the mechanisms of fracture and fragmentation. These phenomena were monitored by means of various instruments, such as strain gages, crack detection gages, accelerometers, fiber optic probes and high speed filming.

In the tests, eight were conducted in 0.5 m^3 concrete blocks for single-hole blasts, and two others had two time-delayed holes, all of them loaded with PETN charges from 1.5 to 5 grams in 6.4 mm diameter boreholes.

Charge burdens varied between 7.6 mm to 130 mm, and the ratio burden/diameter ranged from 11.5 to 20.5, so that several situations were covered, although corresponding fragmentation results are not reported.

The experimental setting included additional cement blocks to constraint the block under study, working as momentum traps.

Fracture development in both the top and bench faces of the cement model are the main aspects of this research. Collected data showed that the most important cracks in terms of breakage are the ones formed early at the bench face in front of the charged borehole, as a result of the tensile stresses induced by the reflexion of P-waves at that surface. Also the top surface above the drillhole is subjected initially to the same mechanism, and later on the gas expansion action develop these fractures, rather than creating new ones.

This important conclusion, which was confirmed by measurements of dynamic strains at the bench face, indicated the important role of the stress waves in the fragmentation process. Another evidence was that lateral areas of the bench face, away from the borehole position, sustain high compressions, starting immediately after the passage of stress waves, and caused by the gas pressure which bends plate-like fragments created by the radial fracturing process.

Duration times of gas expansion were measured at 1 to 1.5 ms after detonation, and extrapolating this information to field blasting the authors concluded that the whole phenomenon is finished after 13 to 16 ms per meter of burden. In terms of fragmentation, high speed films showed that while the largest fragments were formed early in the process as radial fractures, most of the smallest ones come from the area around the boreholes. However, the final crater of the bench face is explained by the radial fractures.

The general conclusion of this research indicates that laboratory blasting tests, using monitoring techniques can be of great help to model real scale blasting with the aim of optimizing this important operation.

4 GROUND VIBRATIONS FROM BLASTING

Although a very small portion of the total explosive energy released in normal rock blasting operations is converted into seismic motion (typically about 1%), their damage potential to nearby structures and a cause for complaints from neighbours, makes this effect of blasting a very important one.

The theoretical aspects of the problem are very complex if the real 3- dimensional situation is analysed, because of the interference of many variables (explosive properties, charge loading parameters and mechanical characteristics of rock). The experimental approach to the problem is based upon instrumentation to record vibrations away from the shot point and to correlate their magnitudes with observed effects on points of interest, such as houses, structural foundations, slopes, tunnels,etc. In general, this approach is considered sufficiently accurate to develop relationships between the amounts of detonated explosives, the distances of propagation and the vibration intensities. Empirical formulae are thus utilized to predict effects of future detonations in the same rock mass, for a certain range of distances and taking into consideration the directions of wave propagation, including their orientations, because of geometric variations along the rock mass.

As Fig. 1 depicts, for two geophones located at the same distance d from the blast, one can expect higher amplitudes of vibration in the record taken behind the bench than in front of it.

To characterize vibration levels as measured by engineering seismographs it is common to record particle velocities, though accelerations, displacements and frequencies are also of interest.

The reason to preferably utilize vibration velocity is mainly due to the fact that it correlates best with observed damage, as a result of extensive research conducted at the U.S. Bureau of Mines since 1942, as well as Nitro Nobel of Sweden, and other institutions.

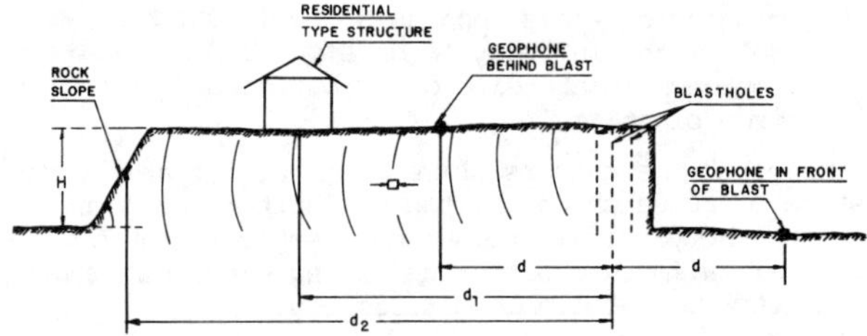

Fig. 1 - Geometric constraints of vibration monitoring in bench blasting.

It must be pointed out that the proposition of safety levels of vibration (for example, 5 cm/s as safe peak particle velocity) is independent of the seismic waves frequency, at least in the interval of 2 to 500 hertz. Many questions arise from these empirical rules:

a) are the safe particle velocities valid for all types of structures?

b) at which point of the structure should these values be considered?

c) how do the propagation laws change with the direction (anisotropy), rock type (heterogeneities), stress-strain behavior (anelasticity) and the presence of faults or fractures (discontinuities)?

d) what is the effect of explosive charge geometry (spherical or cylindrical), decoupling ratio, and depth of burden, on those expressions relating charge weight, distance and vibration velocity?

Many other doubts exist in the engineer's mind, especially when he is asked to establish the maximum charge per delay, in order to protect a certain structure located near the blast. A careful study has to include the dynamic strength properties of that structure, which is a complex interaction of its geometry, mechanical characteristics of their constituent materials and distance to the shot point. Therefore, the common methods of safety prediction are not adequate for application, mainly because structural reliability is not expressed in terms of particle velocity of vibration. An alternative method described by the author elsewhere (Gama, 1978) proposes the dynamic tensile strength of the weakest members of the structure to be protected, as the safety criterion under blasting vibrations.

In order to develop this approach it is necessary to obtain the charge - weight relationships in terms of dynamic stresses, and this done by the use of the well known equation:

$$\sigma = \rho c v$$

where σ is the stress associated with the wave travelling in a medium of mass-density ρ, with a propagation velocity $\underline{c}$ and for a vibration velocity $\underline{v}$. Although this expression is originally valid for uni-directional propagation, it can be approximately extended for field operations if velocity $\underline{v}$ is given in terms of experimental data, namely:

$$v = a Q^b D^{-c}$$

where Q is the detonated charge weight, a distance D from the point of interest, and a, b and c are numerical constants, depending on field conditions and blasting pattern.

Under this approach, the stress wave pulse in given by:

$$\sigma = \rho c a Q^b D^{-c}$$

thus making possible to establish safety criteria in

function of stresses. This concept allows additional advantages:

a) it accounts for the type of foundation rock and material properties forming the structure to be protected;
b) it agrees with experimental evidence that a magnification factor of the order of 2 to 5 (and sometimes more) for stresses transmited in rock in opposition to soil. The ratio of characteristic impedances (the products ρc) between these two media explains properly that difference of effects;
c) it confirms failure limits of rocks under blasting actions, as described by several authors. For example, Brawner (1974) indicates, that rock blasting design range is to be accomplished for particle velocities of 50 to 100 in/s, and using typical characteristic impedances of rocks we obtain dynamic tensile strengths of the order 5 to 50 MPa, which is confirmed by the values given by Attewell and Farmer (1976). Another evidence comes from Langefors (1963) who specified particle velocity limits for granitic rocks, showing the initiation of cracks for peak velocities of 60 cm/s. Assuming a density of 2.65 and P-wave velocity of 6,000 m/s for those rocks, we obtain a dynamic tensile strength of 10 MPa, which is compatible with "in situ" determinations.

Consequently, the criterion of structural safety can be implemented through the evaluation of safety factors, defined as the ratio:

$$F_s = \frac{\sigma_t}{\sigma}$$

where σ_t is the dynamic tensile strength of the structure to be protected (or its weakest member) and σ is the stress associated with the seismic wave originated by the blast. Two additional aspects must be kept in mind: if particle velocity is considered as peak values, so σ is the resulting peak stress, and the change from compressive to tensile stress is due to wave reflection mechanisms because of the well-known smaller resistance of solids in tension than in compression.

Applications of dynamic stress criterion have been described by Gama, 1978, and later on utilized in the prediction of vibration effects on residential buildings of different types, located in the vicinity of quarry blasts.

In the case of an earth dam (see Fig. 2) that had to be protected from the demolition by explosives of a large concrete wall, the propagation law in terms of peak particle velocity was determined as:

$$v = 0.058\ Q^{1.98}\ D^{-0.97} \quad \text{(m/s)}$$

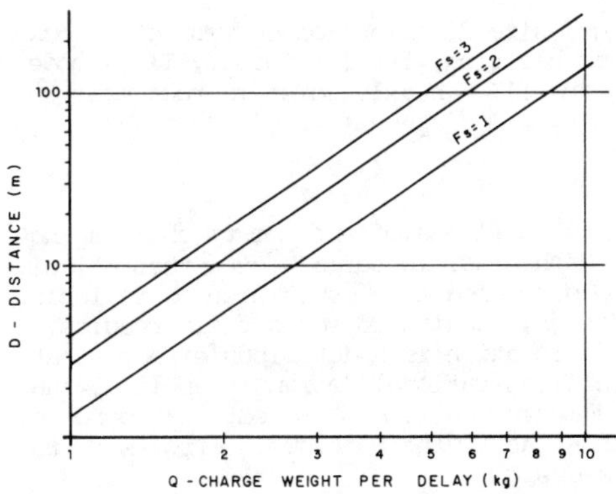

Fig. 2 - Safety criterion for the Sobradinho cofferdam, submited to vibrations from nearby demolition.

The peak stress equation was obtained, after substituting for a cofferdam core material density of 2.3 and a seismic velocity of 600 m/s, thus resulting:

$$\sigma = 0.08\ Q^{1.98}\ D^{-0.97} \quad \text{(MPa)}$$

and estimating a dynamic tensile strength of the clay core as 62 KPa, the various safety factors were represented as Fig. 2 shows.

Another case study regards the evaluation of minimum width for a basaltic wall constituting the intermediate roadway linking the two margins to the deviation canal in the Paraná river, during the construction of Itaipu Dam. That basaltic wall was 62 m high, with vertical sides, and it was designed to resist ground vibrations from adjacent explosions performed to excavate the canal.

"In situ" measurements revealed a propagation law of the type:

$$v = 0.53\ Q^{1.46}\ D^{-2.19} \quad \text{(m/s)}$$

and using the properties of basalt (ρ = 2.8 and c = 4,800 m/s) the expression for dynamic peak stresses is:

$$\sigma = 7.26\ Q^{1.46}\ D^{-2.19} \quad \text{(MPa)}$$

Using a dynamic tensile strength for basalt of the order of 35 MPa, it was possible to establish a safety criterion for that structure, which is represented in the graph of Fig. 3.

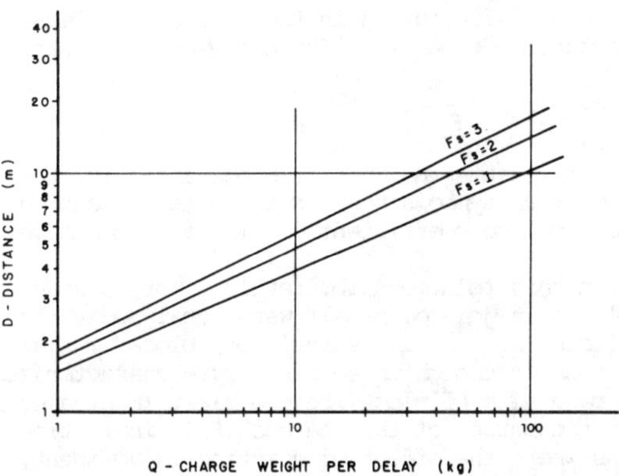

Fig. 3 - Safety criterion for Itaipu dam intermediate roadway of deviation canal, submited to vibrations from nearby excavation.

These examples demonstrated the possibility of taking appropriate decisions, with respect to the dimensions of a struture to be designed, or to the amount of explosive to be detonated per delay, within a new framework of analysis for ground vibrations resulting from rock blasting. Those decisions can take into account the choice of an adequate safety factor and its consequences in terms of costs, allowing the possibility of protecting in different ways, the various structures subjected to the phenomenon.

Furthermore, the use of stresses in stead of particle velocities seems to be a more familiar concept to engineers, even though this a simplified approach, which order of magnitude is not greater than the experimental errors commonly recorded in vibration monitoring.

Additional efforts have to be developed in order to better quantify reliable criteria for damage/failure/stability of rock structures submited to ground vibrations, in order to protect them adequately.

5 ROCK FRAGMENTATION BY BLASTING

An increasing awareness of the importance of fragmentation in rock blasting can be confirmed by the organization of international symposia on the subject, every four years (the first was held in Lulea, Sweden in 1983 and the second in Keystone, Co in 1987). Many research groups are now involved in studies for better understanding the phenomenon, so that field operations can be adequately designed to reach a certain fragmentation degree that will minimize total costs of rock excavation (drilling and blasting), transportation (loading and hauling) and mechanical breakage (crushing and milling).

Early in 1971 the author published a paper proposing the utilization of comminution theory to quantify the size distribution of fragments as a result of a blast. The governing law, which was shown to be consistent with Bond's theory of comminution, indicated that the percent cumulative undersize weight of the fragment fraction size S, is given by:

$$P_S = a\,W^b\,\left(\frac{S}{B}\right)^c$$

where W is the total explosive energy espended per unit weight of blasted rock, B is the burden of the charges and a, b, c and numerical factors dependent on explosive type, rock properties and blasting pattern. This empirical relationship was established on the basis of laboratory scale tests, and it has been confirmed in real scale blasts (for example, Tognon, 1976 and Borquez, 1981).

Due to the fact that ordinary blasting operations are conducted in jointed rock, ajustments of that law have been proposed (Gama, 1983) to include the natural state of fracturing in the rock mass subjected to the detonation. The new expression was:

$$P_S = a\,W^b\,\left(\frac{S}{B}\right)^c\,\left(\frac{1}{F_{50}}\right)^d$$

where F_{50} represents the average block size in the bench, the other symbols keeping the same meaning as before, and the coefficient d is a positive number.

To obtain this relationship, results of seven quarry blasts in jointed basalt were analysed in detail, including a manual survey of block size distributions. Although these cases were characterized by an average size of blocks before blast (F_{50}) always less than the burden of the charges (B), indicating situations where the effect of jointing is dominant, it seems probable that the expression contains the general trend of variation for the phenomenon.

Many authors stress the importance of discontinuities in the consequences of rock blasting, namely fragmentation and flyrock, leading the research in this area to concentrate on sophisticated instrumentation, high speed cinematography,etc. For example, Anderson, Winzer and Ritter (1985) state that "the fracture density, measured on the face prior to the tests, can be related in a general way to gross changes in fragment size distribution". However, the mechanics of breakage are so complex, with wave reflexions at external and internal free faces, microdelay influences, time sequence of gas expansion and their mutual interferences, that more research in needed to fully explain this phenomenon.

In a qualitative way, it can be confirmed that it exists a close relation between the size distributions of blocks before and after detonation, as Fig. 4 depicts. The hatched area is proportional to the fraction of explosive energy effectively used in fragmentation, or in pratical terms, to the powder factor (specific energy consumption), which is the variable W in the above formulae.

The technique to establish the natural block size grading of a rock bench was described elsewhere (Gama, 1977) and it is based on volumetric distribution of blocks, upon a survey of discontinuity attitudes and spacings. Microcomputer programs are available to establish the size distributions of blocks in a rock bench that is subjected to such a survey.

As far as breakage is concerned, one can develop a model of the influence of natural fracturing, as Fig. 5 represents.

For increasing degrees of natural joint density it can be observed the corresponding size distribution curves, indicating the use of less explosive consumption to reach a desired fragmentation. It is envisaged that for each real blasting situation it is possible to develop the two curves depicted in Fig. 4, thus leading the design of blasting rounds to high-level engineering. The process can be even more sophisticated if adequate consideration is given to the spacial location of bigger blocks within the bench to be excavated, so that the placement of explosive charges is determined on the basis of their individual effects in terms of reducing the size of those blocks, in order to build a desired size distribution of fragments for the total blast.

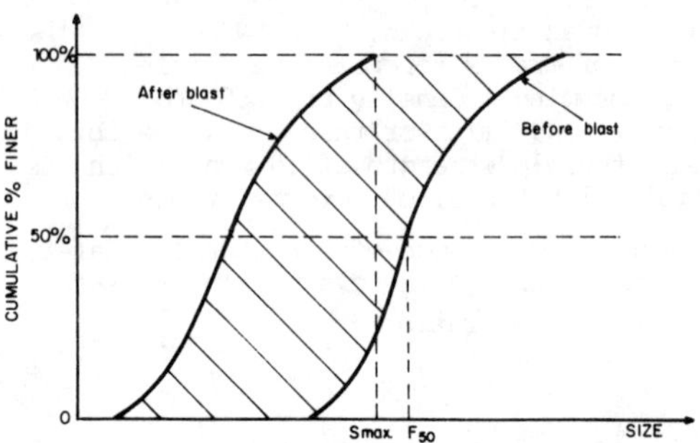

Fig. 4 - Block size distribution curves prior and after blast, and hatched area proportional to explosive energy used in fragmentation.

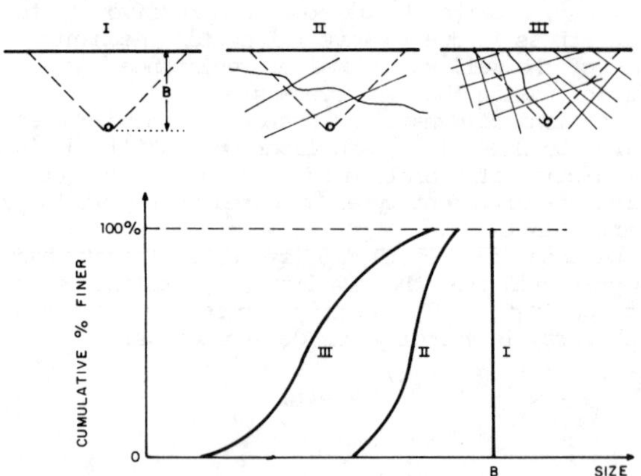

Fig. 5 - Typical size distribution curves of rock blocks prior to blasting, for I - intact, II - moderate jointing and III - highly jointed rock mass.

6 CONCLUSIONS

Rock blasting is still one of the most challenging areas of Rock Mechanics, in terms of theoretical research to be developed to subsidise reliable engineering design, leading to sound field results.

While ground vibration is a deterious effect that reduces naturally as rational design is applied, the subject of fragmentation is of great economic importance, thus justifying continuous efforts to improve such a design.

Creativity and innovation have a great role to participate in this task to make better rock blasting methods, for the benefit of civil and mining engineering.

7 REFERENCES

Anderson, D.A.; WINZER, S.R. and Ritter, A.P. 1985 - The relationships between rock structure and delay timing, and fragment size and distribution in explosively - loaded rock. Fragmentation by Blasting (1st Edition). Society of Experimental Mechanics, p.41-62.

Ash, R.L. 1985 - Flexural rupture as a rock breakage mechanism in blasting. Fragmentation by Blasting (1st Edition). Society of Experimental Mechanics, p.24-29.

Attewell, P.B. and Farmer, I.W. 1976 - Principles of engineering geology. Chapman and Hall.

Borquez, G.V. 1981 - Estimating drilling and blasting costs - An analysis and prediction model. Engineering and Mining Journal, Jan. p.83-89.

Brawner, C.O. 1974 - Rock mechanics in open pit mining. Proc. 3rd. Congress of Int. Soc. Rock Mechanics. Vol. I-A. p.755-815, Denver.

Coates, D.F. 1981 - Rock mechanics principles (3rd. ed.). Dep. Energy, Mines and Resources, Ottawa.

Dick, R.A.; Fletcher, L.R. and D'Andrea, D.V. 1983 - Explosives and blasting procedures manual. U.S. Bureau of Mines, Information Circular 8925. Washington.

Gama, C.D. 1971 - Size distribution general law of fragments resulting from rock blasting. Transactions Society of Mining Engineers of AIME. Vol. 250. p.314-316.

Gama, C.D. 1977 - Computer model for block size analysis of jointed rock masses. Proc. 15th APCOM Symposium. AIMM. Brisbane, p.305-315.

Gama, C.D. 1978 - Dynamic stresses from blasting vibrations. A better damage criterion. Proc. Int. Symp. Rock Mechanics Related to Dam Foundations. Rio de Janeiro, CBMR/ABMS, p.III.1-III.14.

Gama, C.D. 1983 - Use of comminution theory to predict fragmentation of jointed rock masses subjected to blasting. Proc. 1st Int. Symp. Rock Fragmentation by Blasting. Vol. 2, Lulea, p.565-580.

Gama, C.D. 1984 - Microcomputer simulation of rock blasting to predict fragmentation. Proc. 25th U.S. Symp. on Rock Mechanics, Ed. AIME, Urbana.

Hagan, T.N. 1983 - The influence of controllable blast parameters on fragmentation and mining costs. Proc. 1st Int. Symp. Rock Fragmentation by Blasting, Vol. 1, Lulea, p.31-52.

Kolsky, H. 1963 - Stress waves in solids. Dover Publications, New York.

Langefors, U. and Kihlstrom, B. 1963 - The modern technique of rock blasting. John Wiley & Sons, New York.

Tognon, A.A. 1976 - Analytical study of blasts in basalt from several dam quarries (in portuguese). Proc. 1st Brazilian Congress of Engineering Geology. Rio de Janeiro, ABGE, Vol. 2, p.283-296.

Floor Discussions:

Tore L. By (Norway)
My contribution/question is directed to:

Y. Goto, Dynamic behaviour of tunnel lining, --

Maximum acceptable particle velocity on the fresh concrete lining was discussed. The critical importance of the blast vibration frequency content was not treated. At the very short distance in this actual project, the frequencies probably were very high, that is many hundred hertz. At this high frequency level, damage will easily occur. The measured level of 900 mm/s would definitely be far more dangerous if the distance from blasting to lining was 50 m instead of 2-5 m. What is the experience of Mr. Goto concerning the frequency content of blast vibration?

Tan Tjong Kie (China)
My contribution/question is directed to:

Y. Goto, A. Kikuchi and T. Nishioka (Japan), Dynamic behaviour of tunnel lining due to adjacent blasting

The authors present an interesting paper on the excavation by blasting of twin parallel tunnels strengthened by the rock-bolt-shotcrete method. What makes the problem attractive is that it seems daring to keep the mutual distance between the linings only as small 1.5 to 1.8 m, whereas a Young's modulus of the rock mass of 1-3 GPa and its low propagation velocity of 0.9-1.2 km/s and further V_p = 1.5-3.0 km/s and V_s = 0.8-1.5 km/s for rock samples are clear indications that the rock mass must be fissured and not of good quality. In order to study the possible risk of damage to the separating rock wall the authors have performed many dynamic measurements with the help of piezoelectric accelerometers and strain gauges; using these results the displacements and particle velocities have been computed.

On the basis of table 1 and figures 5 to 12, I will now present the following analysis: The dynamic stress σ_d due to a propagation stress wave in a solid can be given by the following formula:

$$\sigma_d = \rho . C_p . V_p$$

whereby the density $\rho = \gamma/g$ = volume weight/981 cm/s²;

C_p = velocity of wave propagation (1.5-3.0 km/s)
V_p = particle velocity within the rock mass.

The authors have computed the particle velocities at the free surface, which we denote by $V_p$surf which are in the order of 10-90 cm/s. The internal particle velocity V_p is known to be:

$$V_p = \tfrac{1}{2} V_p\text{surf}$$

With the above data I get $\sigma_d \sim 2 - 35$ kg/cm²; these stresses may be consequently compressive and tensile due to multiple reflections. As the rock wall concerned does not seem very sound, dynamic tensile stresses of this order of magnitude may generate vertical fissures in the separating rock wall and the fresh concrete lining and further spalling at the unlined cavity surface. Under the influence of the vertical stresses due to the overburden, long term cracking leading to possible instabilities is not unlikely. Hence, in addition to the rockbolting-shotcrete technique, I recommend cement grouting to prevent further disintegration; furthermore rock bolting prior to and not after the excavation will be preferable.

Rock Blasting and Excavation - Theme III

Moderator Summary: P.A. Lindqvist (Sweden)

1 INTRODUCTION

Theme III of the Sixth International Congress of Rock Mechanics is rock blasting and excavation. In the proceedings, a total of 30 papers on drilling, mechanical rock excavation and blasting were presented. A summary of technical areas covered in the papers is shown in table 1.

Table 1. Technical areas of papers in Theme III

Drilling	No of papers
Drilling - general	3
Percussive drilling	2
Rotary drilling	1
Oil well drilling	2
	8
Mechanical excavation	
TBM	3
Roadheader	4
Blind raise boring	1
Static demolisher	1
	9
Blasting	
Rock mass characterization	1
Blasting - general	4
Blasting - Open pit	3
Presplitting	1
Tunneling - various aspects	4
	13

As can be seen there is a good balance between drilling, mechanical excavation and blasting.

It is more interesting perhaps to look at the problems treated by the different authors. In table 2 papers on drilling and mechanical excavation have been grouped under four main headings.

Table 2. Selected problems - drilling and mechanical excavation

	No of papers
Prediction of penetration rate/performance/feasability	
Bernaola & Oyanguren	
Howarth & Rowlands	
Miranda & Mendes	
Farmer & Garrity	
Bonapace	
Mitani, Iwai & Isahai	6
Design of bits/drill steel/equipment	
Brysch, Nsenga & Shan	
Brighenti & Mesini	
Nishimatsu, Okubo & Jinno	
Xu, Tang & Zou	4
Borehole stability	
Kaiser & Maloney	1
Cutting mechanisms/water jet	
Hood, Geier & Xu	
Iihoshi, Nakao, Torii & Ishii	
Ip & Fowell	
Kutter & Jütte	
Sekula, Krupa, Koci, Krepelka & Olos	
Fukuda, Kumasaka, Ohara & Ishijima	6
	17

In Table 3, finally, blasting problems taken up by the authors have been classified under six headings. It is notable that no papers on computer modelling of blasting and empirical modelling of fragmentation were presented.

Table 3. Selected problems - blasting

Blasting theory	
Singh & Sastry	
Wilson & Holloway	2
Blasting design	
R D Singh, V Singh & Khare	1
Vibration criteria	
Fadeev, Glosman, Kartuzov &Safonov	
Huh, Lee & Lim	
Goto, Kikuchi & Nishioka	
Grant, Spathis & Blair	
Siskind	5
Presplitting	
Heraud & Rebeyrotte	1
Measurement of fragmentation	
Mærz, Franklin, Rothenburg & Coursen	1
Rock mass characterization/cost model	
Inazaki & Takahashi	
Pöttler & John	
Mikura	3
	13

In summary the papers of Theme III show that active and successful research and technical development is being carried out throughout the world in this area. Since the last Rock Mechanics Congress a few major developments must be mentioned. Among these are the increased use of medium high pressure water jets to assist roadheader cutting and the wide establishment of blast vibration criteria based on local assumptions but often influenced by international experience. Papers of Theme III also include a few important case studies that contribute to the long term development of rock blasting and excavation technology.

2 HIGHLIGHTS

Considering the high quality, it is a difficult task to select and comment on specific papers. A few are however worth highlighting.

The paper "Fragmentation studies in instrumented concrete models" by W H Wilson and D C Holloway uses a combination of different observation techniques which makes it possible to draw new and firm conclusions. This is a key paper and gives, in my opinion, a major contribution to the understanding of mechanics leading to fragmentation.

The paper by J R Grant and co-workers entitled "An investigation of the influence of charge length upon blasting vibrations" indicates a possibility to detect detontation cut off, at present a serious problem in underground mining.

The paper "Measurement of rock fragmentation" by N H Mærz and co-workers is an instructive example of how the use of computers opens new ways to solve difficult engineering problems.

The paper by J Brych and co-workers shows the impressive result of several MSC and PhD theses under the title "Destructability of rocks with rotation drilling bits".

3 FUTURE DEVELOPMENTS

When looking into the future there are several aspects to consider. One of them is the length of time. In the discussion of future research needs I have taken a fairly short term perspective. It is arguable whether scientists should consider the near future. It is my opinion, however, there are coming or soon will come changes that will have a major impact on the selection of research problems.

Befor starting your research you first have to choose the technical area to work in. Then you have to select the scientific problem and finally you have to consider the kind of method you are going to use, such as laboratory testing, field testing and computer modelling. Ideally all should be used on the same problem, but this is certainly not always the case.

In the following presentation a few important technical developments will be identified and the consequences specified. Future research needs will then be given.

Future developments	Consequences
New materials Medium high pressure water jets High capacity equipment	High speed drilling and high efficiency equipment will reduce the cost of holes TBM and Roadheader mine development and mining in hard rock will be introduced High availability of equipment will be very important
Increased scale in underground mining and excavation	Larger holes (diameter 50-100 mm, length 10-50 m) will be introduced even in small scale mining
Automation and robotization will increase	Increased accuracy in drilling and blasting
Monitoring and hole logging will increase Data handling	Dramatic increase in aquisition, analysis storage and presentation of geodata
Precision caps and pumpable, variable strength explosives	Dramatic increase in blasting design possibilites

4 RESEARCH NEEDS

With identified developments and consequences in mind the following research needs are proposed.

4.1 Drilling and mechanical excavation

Case studies with emphasis on new experience.

Development of cutting removal models.

Development of Roadheader prediction models in hard rock.

Development of fast methods to measure hole length and hole deviation, models and techniques to reduce hole deviation.

Development of new methods by which geodata can be obtained:

Measurement while drilling (MWD) for hammar drilling including model development.

Sampling of cuttings while drilling and chemical and physical analysis including model development.

Hammar drilling followed by geophysical hole logging including model development.

Development of drilling and mechanical fragmentation models.

Development of integrated systems for efficient aquisition, analysis, storage and presentation of geodata.

4.2 Blasting

Data collection and further development of empirical models and blast design practice with emphasis on actual problems and new explosives.

Development of fast methods to monitor blast performance with respect to delay time, total charge detonation, vibrations and fragmentation.

Further development of presplitting and smooth blasting methods with new explosives.

Further development on theoretical (first principal) blasting models, including influence of rock type, rock mass and explosives behaviour.

Theme 4
Underground openings in overstressed rock

Site investigations and rock mass characterization; design methods; rock support and reinforcement; rockbursts and seismicity; rupture mechanisms of underground openings; gas outbursts and other catastrophic hazards; prediction, monitoring and back analysis

8:30 LE GRAND SALON - September 3, 1987

Chairman: B. Bamford (Australia)

Session Secretary: J. Nantel (Canada)

Speaker: H. Wagner (South Africa)

Design and Support of Underground Excavations in Highly Stressed Rock.

ABSTRACT: In mining and to a lesser extent civil engineering many situations arise where rock fracturing around underground excavations is unavoidable. In these circumstances the emphasis of the excavation design is on the engineering of the fracture zone and the support of the fractured rock. A general lack of factual information on the strength and in particular the post-failure properties of rock masses indicates the need for semi-empirical design procedures. In hard brittle rock elastic analysis coupled with an understanding of basic rock fracture phenomena provide a useful basis for design. The control of the post-failure deformation of fractured rock is the basis for support design in highly stressed rock masses as it not only preserves the inherent strength of the rock but also limits the size of the fracture zone. These requirements are most readily achieved with integrated support systems comprising of support tendons, wire mesh and shotcrete. Seismic loading of highly stressed excavations does not greatly influence the design and support aspects except in those cases where excavations are situated in close proximity to the epicentre of tremors.

INTRODUCTION

Man, in his endeavour to extract minerals, to provide transport infra-structures and to meet the energy needs of modern society, is penetrating the earth crust to ever increasing depths. Exploratory boreholes have been successfully drilled to depths in excess of 12 000 m, tunnels developed through mountains with a cover of more than 3 000 m, and mining of gold bearing reefs reached depths below surface in excess of 3 800 m. Rock excavation at these depths creates unique technological problems and hazards.

Possibly the greatest wealth of information on rock excavation at depth exists in South Africa where more than 100 million tons of gold bearing reef are being mined annually at an average rock breaking depth of 1 600 m. The deepest working levels are 3 800 m below surface, and shaft systems to extract the reefs at depths well in excess of 4 000 m are presently being developed. Annually in excess of 800 km of shafts, ore passes, tunnels and haulageways are being developed in South African gold mines to facilitate the extraction of the narrow gold bearing reefs at these depths. While the rock mechanics problems related to the narrow tabular reef bodies are thus unique, there are many other rock pressure problems encountered in these deep mines which are of more general interest and importance to the rock engineering fraternity.

The purpose of this overview paper is to report on some of the more important findings concerning the behaviour of excavations in highly stressed ground, and to derive some general criteria for the design and support of such excavations. Much of the information on the behaviour of highly stressed excavations relates to excavations situated in relatively strong quartzitic rock. However, information on excavation behaviour in other highly stressed geological formations has been included for purposes of comparison and to allow for more general conclusions.

Since most experience on the behaviour of excavations in highly stressed rock exists in deep mines, the emphasis of the paper will be on this aspect of rock engineering. In mining many aspects of the behaviour of highly stressed excavations can be studied, which, for a variety of reasons, could not be observed in civil engineering underground excavations.

ENGINEERING ASPECTS OF HIGHLY STRESSED EXCAVATION

In the context of this paper rock engineering in highly stressed rock masses refers to those situations where the stresses which result from the construction of an underground opening exceed the strength of the rock mass. Consequently, fracturing will take place in the rock mass surrounding the excavation. In such a situation the objective of rock engineering is two-fold: to design the shape and size of the excavation so as to minimize the extent of rock fracturing, and, to select a functional and cost-effective support system for the control of the fracture zone surrounding the excavations.

From this follows that rock fracturing is an inevitable consequence of the underground excavation process in a highly stressed rock mass. Therefore, detailed knowledge of both the strength properties of the rock mass and the rock fracturing processes are essential aspects of the design process. Equally important is the knowledge of the stress situation within the rock mass in which the opening is created. In mining there is a further aspect which requires attention, that is, the stress changes that take place after excavation due to additional mining activities.

An important aspect of the design and support of highly stressed excavations is that distinct fracture zones form around the opening. The objective of the rock engineering process is to design the excavation sequence and geometry so as to obtain the most favourable fracture zone from a support point of view. In contrast to this, the design of excavations in a low to moderate rock stress environment is strongly controlled by existing geological discontinuities along which movement can occur owing to gravity.

Any discussion on the behaviour of highly stressed excavations must consider the formation of new rock fractures and also the factors which control these fractures, and cannot merely confine itself to the analysis of already existing geological discontinuities. Whilst the latter are of importance in any underground excavation design, they tend to be less significant in a high rock stress environment. Nevertheless, these discontinuities, and in particular major differences in the mechanical properties of different rock layers, can have an overriding influence on the mechanism of fracture formation in a high rock stress environment. Examples of this can be found in deep coal mines where considerable differences exist in the mechanical properties of the coal seam and the surrounding strata. However, even relatively minor differences can have a marked effect on fracture formation around excavations in deep hard rock mines.

ROCK STRENGTH AND ROCK FRACTURING

Strength properties

It is fair to say that more research has been devoted to the clarification of the strength properties than any other aspect of the rock behaviour. Nevertheless, it is still doubtful whether reliable predictions can be made with regard to the failure of rock surrounding underground excavations.

There are a number of reasons for this difficulty. Firstly, the strength properties of rock are usually determined from small samples of intact pieces of rock which are tested under artificial loading conditions. For design purposes, extrapolation of the results obtained in this manner is made difficult, if not impossible, by the presence of joints, fissures, and other planes of weakness which may have a significant influence on the strength of the rock mass. There have been noteworthy attempts in recent years to account for these influences. One of the more promising approaches is the empirical failure criterion given by Hoek and Brown (1982, p.137).

$$\sigma_1 \leqq \sigma_3 + \sqrt{(m\sigma_c \sigma_3 + s\sigma_c^2)} \qquad (1)$$

where σ_1 and σ_3 are the major and minor principal stresses at failure, σ_c is the uniaxial compressive strength of the intact rock material, and 'm' and 's' are constants which depend upon the properties of the rock and upon the extent to which it has been broken before being subjected to the stresses σ_1 and σ_3. For intact rock material the value of s = 1 and the value of m ranges from 7 to 25 depending on the rock type.

The second source of difficulty is the definition of strength itself, as was pointed out by Salamon in 1974. The most commonly used definition of strength involves the inequality

$$\sigma_1 \leq a\sigma_3^b + \sigma_c \qquad (2)$$

where the parameter 'a' describes the increase in rock strength with confining stress σ_3; the exponent is a measure of the linearity of the relationship and is generally close to one. In many instances it is therefore appropriate to simplify Equation (2) to

$$\sigma_1 \leq a\sigma_3 + \sigma_c \qquad (2a)$$

The important implication of this inequality, and incidentally also that of the Hoek and Brown failure criterion, Equation (1), is that they describe the upper limit of the strength of rock for a given confining stress, but do not allow any conclusions to be drawn with regard to the resistance or state of rock once this inequality has been satisfied.

A third source of difficulty is that the rock properties undergo significant changes once these inequalities have been satisfied. To account for these changes, and in particular to allow for the reduced confinement provided by the broken rock mass surrounding a highly stressed excavation, Hoek and Brown (1982), revised their original criterion, Equation (1), by introducing a failed or broken rock mass rating value. In terms of this concept the values for m and s of the broken rock mass are reduced depending on the extent of fracturing and jointing that has taken place.

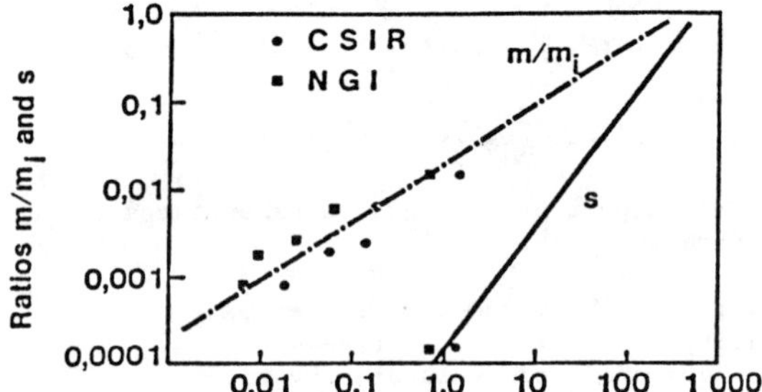

Fig. 1. m and s as a function of rock mass rating

They suggest that the values of m_r and s_r for a fractured and jointed rock mass can be estimated using either the NGI or the CSIR rock mass classifications systems. Figure 1 gives the relationship between the CSIR rock mass rating and the factors by which the values of m and s have to be reduced to take the extent of jointing and weathering of the rock mass into account. The Hoek-Brown criterion for fractured and jointed rock masses has been employed with a considerable degree of success to analyse excavations in deep mines. However, much care has to be taken when applying such corrections, since they are strongly dependent on confining geometries and the effects of support. For example, studies by Wagner and Schümann, (1971), have shown that the actual extent of the fracture zone above and below highly stressed pillars tends to be narrower than that calculated using either Equation (1) or (2). The reason is the fact that these Equations do not take into consideration the build-up of localized confining stresses resulting from the dilatation of the fractured rock.

Fourthly, the spacing of joints, and other structural weaknesses such as bedding and parting planes, in relation to the critical linear dimension of the excavation, are important factors which control the extent of the fracture zone. This aspect is of particular significance when designing excavations in sedimentary rock formations such as roadways in deep collieries, or large underground openings in quartzites or laminated shale and sandstone formations, Jacobi (1976), Piper (1984).

A basic shortcoming of most of the rock mass classification systems that are in use is that they do not account for the influence of excavation dimensions, Piper, (1985).

In massive rock formations good use can be made of the relationship between the uniaxial compressive strength and specimen diameter which was devised by Hoek and Brown, (1982).

$$\sigma_c = \sigma_{c50}(^{50}/d)^{0,18} \qquad (3)$$

where σ_{c50} is the uniaxial compressive strength for a 50 mm diameter rock specimen and 'd' is the diameter of the specimen.

The relationship given in Equation (3) has been applied successfully to estimate the fracturing of tunnels with diameters of two to three metres which were developed in massive quartzites, Figure 2.

Fifthly, the presence of soft rock layers in the walls of highly stressed excavations can exercise a major influence on the behaviour of the excavation and on the mode of fracturing, Jacobi, (1976); Kersten et al, (1983); Reuther, (1987). Soft layers influence excavation behaviour in several ways. Firstly, if the soft layer is relatively thick most of the rock deformation will take place in this layer. Secondly, because of the low modulus of deformation and the generally high Poisson's ratio of soft rock materials, high shear stresses develop at the interface between hard and soft layers if the loading conditions change as is the case during the excavation process, Brummer (1984), Figure 3.

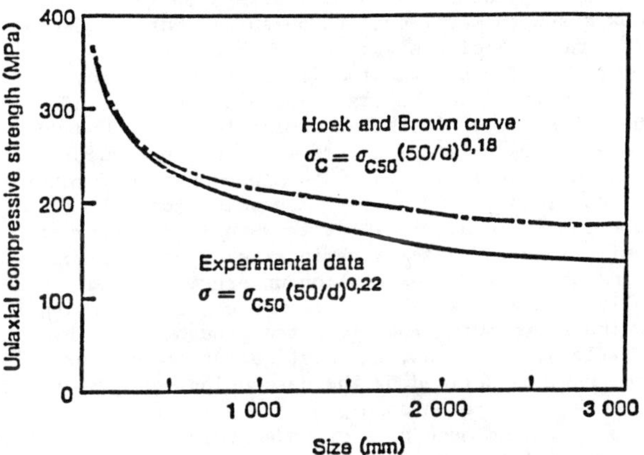

Fig. 2. Effect of excavation rise on rock strength (after Johnson, 1986)

Since rock is very much weaker in tension than in compression it is found that tensile fractures develop in hard layers that are sandwiched between soft layers, and that the rockwalls fail at much lower stress levels than would be expected from the results of compression tests made on material from the hard layers. Laubscher, (1984), published a nomogram from which can be estimated the strength of rockwalls which comprise of weak and strong rock layers, Figure 4.

Rock fracturing

Figure 5 shows the various modes of rock fracturing that are observed in excavations which are subjected to high vertical stresses. In massive brittle rocks, Figure 5(a), extension fractures form in the sidewalls of the excavation. The extent and spacing of these fractures depends on the magnitude of the stresses and the brittleness of the rock. Rock deformation around excavations situated in jointed rock masses is controlled by the shear stresses and gravitational forces, Figure 5 (b). The presence of soft layers in the sidewalls of excavations facilitates the formation of extension fractures in brittle rock layers which are sandwiched between the soft layers. These extension fractures tend to occur at very much lower stress-levels than is found in the absence of the soft layers. Furthermore, these fractures extend much further into the sidewalls than is the case in massive brittle rocks, Figure 5(c). Bedding planes with a low coefficient of friction, Figure 5(d), promote the formation of wedge fractures which can extend deep into the sidewalls of the excavation. In addition, high lateral stresses are induced in the surrounding strata. Depending on local circumstances, extension fractures or buckling and folding of thinly laminated strata are observed, Jacobi (1976), Reuther (1987).

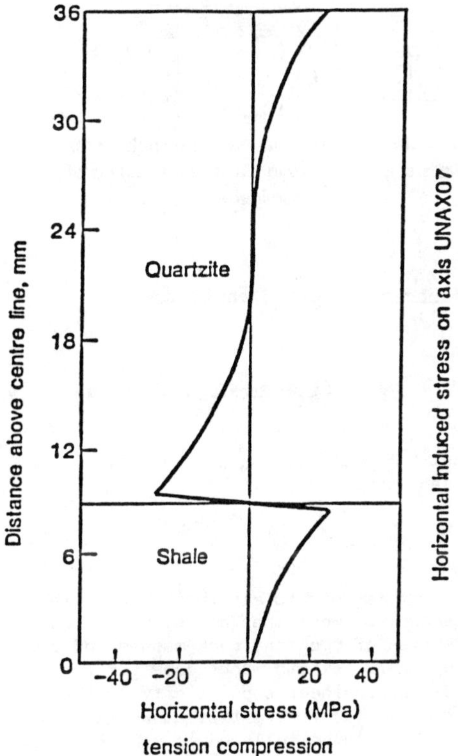

Fig. 3. Effect of soft rock layer on the horizontal stresses in the centre of a cylindrical rock specimen.

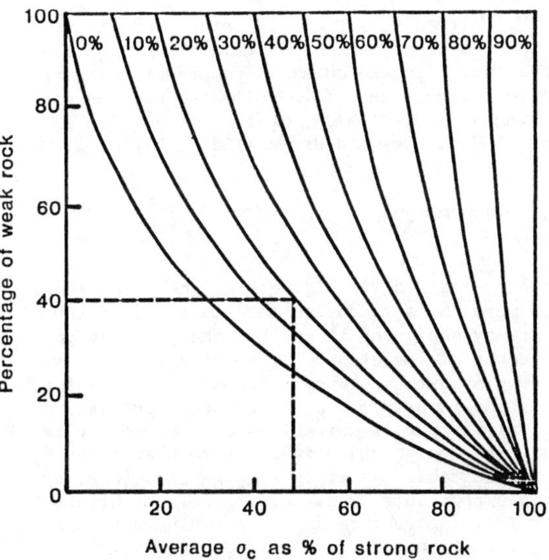

Fig. 4. Reduction of rock strength by weak bands. (After Laubscher, 1984)

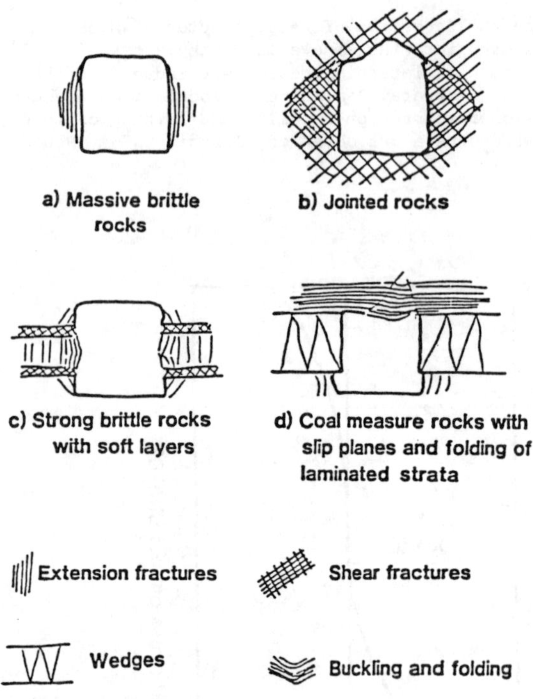

Fig. 5. Typical rock failures in highly stressed tunnels

In conclusion it is fair to say that despite the considerable amount of work that has been done on the strength properties of rock and rock masses and on rock fracturing there exists much scope for further work. From a design engineer's point of view the present situation is far from satisfactory and new approaches will have to be followed to make real progress. The development of sophisticated numerical models and the availability of powerful computers suggest that back-analysis of actual cases is the route to pursue. The inherent power of this approach was demonstrated by Salamon (1967) when he developed a procedure for the design of bord and pillar workings in South African collieries based on the analysis of 125 cases of intact and failed bord and pillar workings. Mention must be made in this connection of attempts by rock mechanics practitioners in South African gold mines to relate the behaviour of tunnel walls to the uniaxial compressive strength of the rock formation and the stresses which act on these tunnels, Cook (1972), Ortlepp et al. (1972), Wiseman (1979), Hoek and Brown (1982), Piper (1984).

DESIGN OF UNDERGROUND EXCAVATIONS

General

In mechanical and civil engineering the traditional design approach is to ensure the stability of the structure by designing it so that the stresses in each element of the structure are always less than the strength of that element. Depending on the importance of the structure and the consequences of a structural failure, a suitable factor of safety is defined. The design objectives are then achieved by selecting suitable materials for the structure, by choosing appropriate cross-sections of the stressed members in the structure and by avoiding unfavourable loading conditions. In rock engineering, however, this freedom of choice is severely restricted by the specific purpose of the underground excavation which often determines the site of the excavation and consequently the geological formation in which it is situated. Furthermore, the stress field in which the opening has to be excavated is often pre-determined by the site constraints, and the only freedom left to the design engineers is the orientation of the excavation relative to geological structures and the stress field. However, there are many situation, where even this freedom does not exist. In the extreme case the only design freedom left to the engineer is the choice of the shape of excavation, the selection of the support system and the sequence of excavation.

Stable and unstable rock fracturing

Regarding the design and support of highly stressed excavations it becomes immediately apparent that the traditional engineering approach cannot apply since in terms of the earlier definition the stresses acting on the rock surrounding the excavation exceed the strength of the rock material, and fracturing is unavoidable. However, this does not necessarily imply that the stability of the excavation is endangered. Indeed, there are many situations where rock has failed in parts surrounding underground excavations and where these excavations have remained stable often without even introducing supports. The criterion that the stresses must always be less than the rock strength may therefore be sufficient to ensure the stability of a structure, although this is not necessary. The explanation for this is that most rocks in their failed state are still capable of supporting some stress. According to Jaeger and Cook (1979, p.80) a material is said to be in a ductile state as long as it can sustain permanent deformation without loosing its ability to resist load. A material is in a brittle state when its ability to resist load decreases with increasing deformation.

In the course of excavation underground, and especially in highly stressed rock masses, some portion of the surrounding rock is often brought into the brittle state. Hence, the post-failure behaviour of the rock, as defined by the descending portion of the stress-strain relationship, must be taken into account if the mechanics of underground excavations are to be fully understood.

Figure 6(a) shows an idealized load-compression diagram for a brittle rock. Also shown in this diagram is the force displacement characteristic of the loading system as defined by the slope, k, of the loading line.

Figure 6(b) shows the variation in slope, λ , of the force-displacement curve of the rock and the value of the slope ,k, of the loading system. According to Salamon, (1974) the system is in a stable equilibrium if, and only if,

$$k + \lambda > 0 \tag{4}$$

Since, by definition, k is positive the inequality in (4) indicates that the equilibrium will be stable, regardless of the stiffness of the loading system, as long as the resistance of the rock to deformation increases with deformation, OB region. In the post-failure region, BC, the equilibrium will be stable up to the point at which the loading line of the system becomes tangential to the force-displacement curve, $P = P_s$. At this point violent failure of the rock will occur, however, violent failure will not take place as long as the minimum slope of the force-displacement curve of the rock, λ_{min}, satisfies the inequality $k + \lambda_{min} > 0$.

From this discussion follows that the design of excavations in highly stressed rock must not be confined to the analysis of the fracture zones around the excavations. There also needs to be considered the stiffness of the surrounding rock mass which together with the post-failure behaviour of the rock determine whether or not violent rock failure is likely to occur.

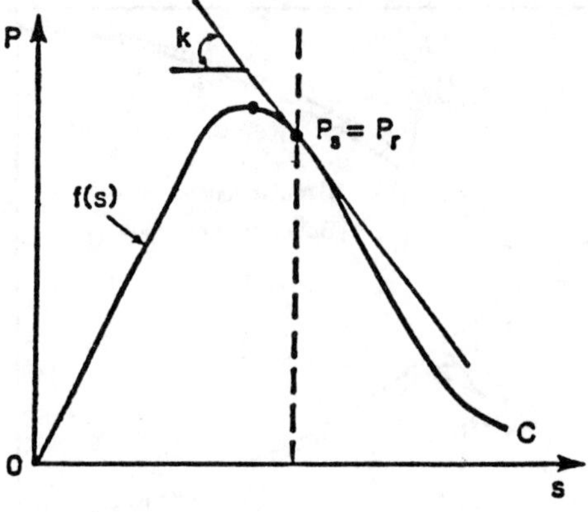

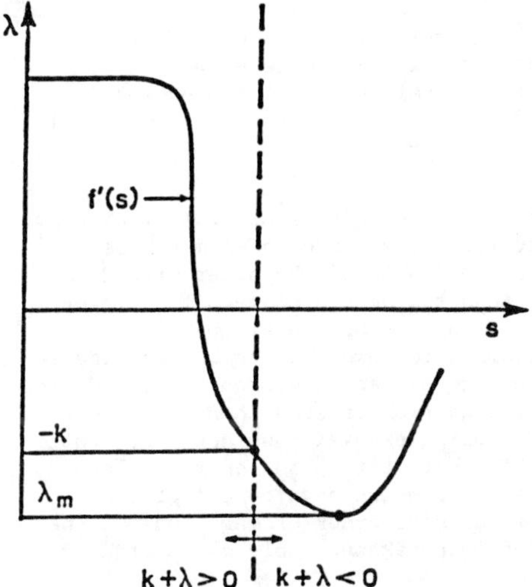

Fig. 6. Idealised load compression diagram for a brittle rock. (After Salamon, 1974)

In addressing the latter Jaeger and Cook (1979, p. 472-474) examined the idealized case of a circular opening, subject to a hydrostatic field stress, p, such that the stresses in a thin annulus of rock of thickness, t, around the interior of the opening of radius R have just exceeded the stength of the rock. Assuming that for r > R the rock around the tunnel behaves linearly and elastically, the stiffness of the radial stress applied to the annulus by the surrounding rock is given by

$$k_R = \sigma_R/\varepsilon_R = 2G \qquad (5)$$

where G is the modulus of rigidity of the rock. This yields the inequality

$$\left|\frac{t}{R}\frac{d\sigma}{d\varepsilon}\right| < \left|2G\right| \qquad (6)$$

as the criterion for the stability of a failed annulus of rock with respect to the stresses applied to it by the surrounding rock. Since in most instances $t/_R \ll 1$ it follows that a thin annulus of fractured rock is 'soft' in relation to the stiffness of the surrounding rock, and is therefore stable.

Although this is a very simplified example it nevertheless illustrates that rock fracturing around tunnels in a geologically undisturbed rock mass is likely to be stable. This finding is supported by many underground observations which show that fracturing around tunnels and caverns tends to be a stable process.

Attention is drawn to the work by Deist (1965) and by Ulgudur (1973) who studied in considerable detail the problem of stable or unstable fracture formation around highly stressed tunnels.

It is important to note that these observations do not apply to underground excavations which have geometries which differ significantly from those of tunnels. In particular, failure of support pillars in tabular mining situations can result in unstable catastrophic regional collapses with the release of large amounts of seismic energy. Similarly, rock fracturing around highly stressed tabular excavations can be unstable, especially in the presence of geological disturbances, Heunis (1980)

Energy considerations

In the early 1960's Cook (1963) called attention to the significant energy changes that take place during mining. In particular he stated that 'the excess potential energy causes the damage noticed as rockburst'. Since then various energy quantities, and especially the energy release rate, have become widely used design tools in deep level mining.

In 1974 and 1984 Salamon examined some of the fundamental aspects of the energy changes that take place as a result of the enlargement of an excavation. Salamon points out that as a result of the increased size or number of mining excavations, displacements are induced in the surrounding rock. Acting through these displacements, the external and body forces do some work, W. This work is commonly referred to as 'gravitational' or 'potential' energy change. Also, a certain amount of strain energy is stored in the rock, which is removed during the process of mining, U_m. The sum $(W+U_m)$ is the energy input that must be accounted for or has to be expended in some manner.

This energy is partially accounted for by an increase in the strain energy content of the surrounding rock mass, U_c. If the excavations are backfilled or supported, then some work is performed in deforming the supports, W_s. Assuming that the rock mass is an elastic continuum then it follows that no energy is consumed in fracturing or in non-elastic deformation. Thus, the energy that has been accountably expended during a mining step is (U_c+W_s).

Salamon concluded that these simple processes do not account for the full energy input, and that some energy,

$$W_r = (W+U_m) - (U_c+W_s) > 0 \qquad (7)$$

must be released or dissipated by other means. The lower limit of the released energy is thus given by

$$W_r \geq U_m > 0 \qquad (8)$$

In the case of tabular excavations the energy quantity U_m and hence the lower limit of W_r can be estimated from the components average stress, $\sigma_{k3}^{(p)}$, and convergence, $S_k^{(i)}$, before and after the removal of the mineral from the small area ΔA

where $\sigma_{k3}^{(p)} (k = 1,2,3,)$ are the components of the stress vector in A prior to mining this area and $s_k^{-(1)}$ are the components of the relative displacement vector after mining the same region.

$$\Delta U_m = -\int_{\Delta A} \tfrac{1}{2} S_k^{(i)} \sigma_{k3}^{(p)} \Delta A \qquad (9)$$

these components can be calaculated readily using either the electrical resistance network analog, or one of the many digital tabular mining simulators that were developed in recent years, Ryder (1986).

Practical experience gained in deep gold mines shows that the $\Delta U_m/\Delta A$ rate is a significant parameter for predicting the problems of maintaining satisfactory strata control in deep level stoping, Hodgson and Joughin (1966), Heunis (1980), Salamon and Wagner (1979). An example of the relationship between

$\Delta U_m/\Delta A$ and the number of rockbursts per 1 000 m^2 of stoped out area is given in Figure 7.

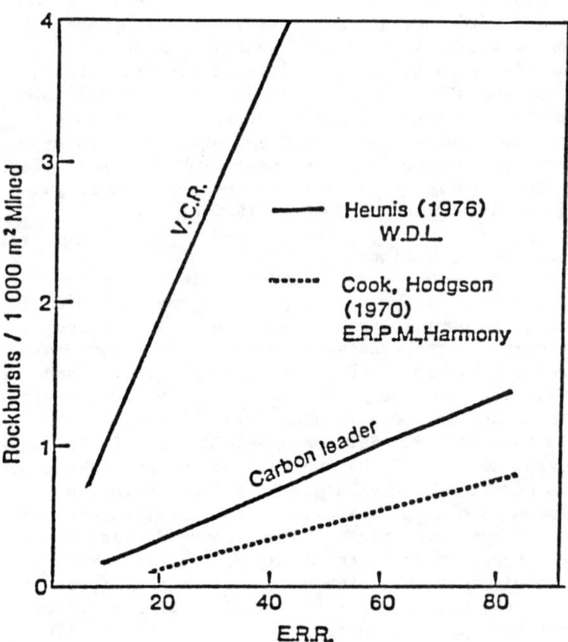

Fig. 7 Relationship between released energy ERR and a number of rockbursts per 1 000 m^2 mined.

However, little work has been done as far as energy changes in non-tabular underground excavations are concerned. Salamon (1984) studied the energy changes that result when the radius of a circular tunnel is suddenly increased from a value, a, to a value, c,. The following expression for the energy release rate, W_r, has been obtained for an unsupported tunnel subjected to a hydrostatic pressure of magnitude p

$$W_r = \frac{2(1-\nu^2)p^2}{E} V_m \qquad (10)$$

whereby V_m is the volume of rock to be mined per unit length of tunnel

$$V_m = \pi(c^2 - a^2) \qquad (11)$$

The potential kinetic energy that is released as a result of a sudden increase in the radius of the tunnel is given by

$$W_k = \frac{(1+\nu)p^2}{E}\left(1 - \frac{a^2}{c^2}\right) V_m \qquad (12)$$

This result is of considerable importance since it indicates that a significant amount of kinetic energy can be released if the size of a tunnel is increased suddenly, Figure 8.

Finally, a comparison is given below of the energies released during the extraction of a tabular deposit situated at a depth of 3 000 m with that encountered during tunnelling at the same depth.

The upper limit of the energy release rate that is likely to be encountered in extensive stoping operations is given by

$$\Delta U_m/\Delta A = S_m \sigma_v \qquad (13)$$

whereby S_m is the extracted mining height and σ_v is the vertical component of the virgin stress tensor. Assuming a mining height of $S_m = 1$ m and a vertical stress σ_v = 80 MPa the value of $\Delta U_m/\Delta A$ = 80 MJ/m^2.

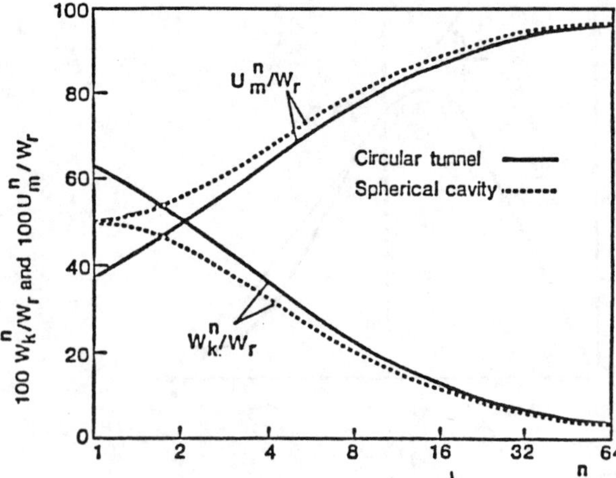

Fig. 8. Split of total released energy, W_r, into total (potential) kinetic energy, W_k^n, and total strain energy (removed with the excavated rock) (After Salamon, 1984)

In comparison, the energy release rate that results from the excavation of a tunnel of 2 m radius in a stress field of 80 MPa is of the order of 0,2 MJ/m^3, that is, more than two orders of magnitude lower than that experienced in stoping operations.

From this example follows that energy considerations are likely to be less important in the design of deep tunnels and rock caverns than they are for the design of stoping excavations. However, mention must be made of the fact that the energy release rates that result from the development of deep tunnels are of the same order of magnitude as the energy required to break one cubic metre of hard rock in compression.

Excavation shape and stresses

The choice of excavation shape is one of the most important decisions in rock engineering, as mentioned earlier. This is especially so in the case of excavations that are subjected to high rock stresses. The shape of the excavation and its orientation to the field stresses determines not only the boundary stresses but, in addition, governs the formation of the fracture zone and influences support decisions.

In a low stress environment it is normally the geological and structural discontinuities that control the excavation behaviour and determine the support requirements. Under these circumstances the shape of the excavation is usually selected so as to achieve the most favourable geometry from the point of view of controlling movements along rock joints and of preserving the natural strength of the rock formations. For example, in stratified rock considerable benefits can be gained by utilizing well developed bedding or parting planes to form the roof of the excavation or by selecting one of the stronger and competent beds to form a natural roof, even if this does not result in the most favourable geometry from a stress point of view.

As the field stresses increase relative to the strength of the rock mass in which the excavation is situated, more attention needs to be given to the selection of the most favourable geometry from a stress point of view. Experiences gained from tunnels in deep South African mines show that in the case of square tunnels of three to four metre widths, minor sidewall spalling is observed when the vertical component, σ_v, of the field stress tensor exceeds a value of about 0,2 of the uniaxial compressive strength, σ_c, of the rock as determined in the laboratory. Severe sidewall spalling is observed when $\sigma_v/\sigma_c > 0,3$, and heavy support is required when this ratio exceeds a value of about 0,4; Ortlepp et al (1972).

While the simple field stress criterion based on the vertical component of the field stress tensor served the gold mines well it suffers the disadvantage that it does not take into account the effect of the other stress components upon the behaviour of underground excavations. Wiseman (1979) accounted for these by formulating a sidewall stress concentration factor, SF, which he defined as

$$SK = \frac{3\sigma_1 - \sigma_3}{\sigma_c} \tag{14}$$

where σ_1 and σ_3 are the maximum and minimum principal stresses acting on the excavation, and σ_c is the uniaxial compressive strength of the rock as determined in the laboratory. In a detailed survey of more than 20 km of typical gold mine tunnels Wiseman observed that the condition of unsupported tunnels deteriorated markedly when the sidewall stress concentration factor reached a value of about 0,8, Figure 9.

These examples show that in massive rock formation the boundary stresses, when normalized with respect to the uniaxial strength of the rock formation, can serve as good indicators of potential excavation problems. Attention must be drawn to the difficulties in obtaining a reasonable estimate of the uniaxial compressive strength of the rock mass. These difficulties were discussed previously but need to be re-emphasized.

A first step in the design process is, therefore, to determine the stresses that act on the boundary of the proposed excavation. A necessary requirement is a knowledge of the field stresses that exist at the proposed excavation site. In the case of mining it is appropriate to model the effects of the generally very extensive stoping excavations on the stresses that exist at the proposed site of the opening. Numerous boundary element and finite element programs have been developed for this purpose. In the case of civil engineering excavations it may become necesary to conduct stress measurements on site. Where this is not feasible regional trends in the stress field should be taken into account when estimating the stress situation at the planned excavation site. As a general rule it has been found that the vertical component of the field stress tensor is in good agreement with the overburden stress as calculated from the depth of cover and the density of the rock formations. However, much uncertainty often exists as far as the direction and magnitude of the horizontal stresses is concerned, Hoek and Brown (1982).

A first, and often sufficient, approximation of the boundary stresses around an underground opening can be made on the basis of the following relationships, Brady and Brown (1985, p. 195).

$$\sigma_A = p\left(1 - K + \sqrt{\frac{2W}{\rho_A}}\right) \tag{15}$$

$$\sigma_B = p\left(K - 1 + K\sqrt{\frac{2H}{\rho_B}}\right) \tag{16}$$

where σ_A and σ_B are boundary circumferential stresses in the sidewall (A) and crown (B) of the excavation, ρ_A and ρ_B are the radii of curvature at points A and B, and K is the horizontal to vertical stress ratio, Figure 10.

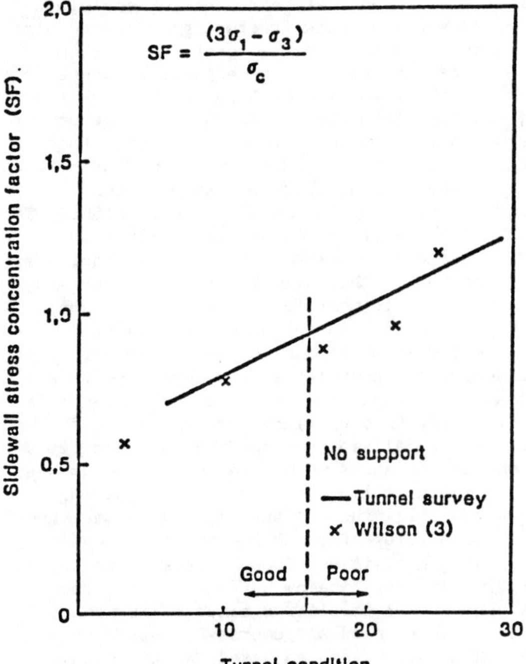

Fig. 9. Conditions in unsupported gold mine tunnels as a function of sidewall stress concentration factor. (After Wiseman, 1979)

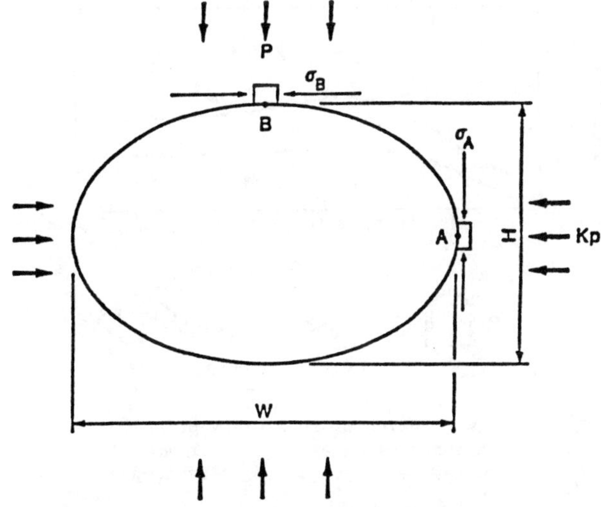

Fig. 10. Definition of geometry and stresses around elliptical openings.

Massive rock formations

In massive rock formations no stress related excavation problems are likely to be encountered when the values of the boundary stresses which have been normalized with respect to the compressive strength of the rock mass are well below one. In those cases where the normalized boundary stresses approach a value of one, changes to the shape of the excavation should be considered so as to reduce the boundary stresses. However, in many instances the field stresses are so high that even the most favourable excavation geometry is insufficient to prevent rock failure at the boundary of the excavation. These situations arise when the maximum value of the subsidiary principal field stress, $\bar{\sigma}_1$ approaches a value of about half of the uniaxial compressive strength of the rock $\bar{\sigma}_1 \cong 0{,}5\sigma_c$. In these instances the excavation shape has to be designed so as to minimize the extent of the fracture zone and to facilitate the support of the excavation.

Experience in highly stressed tunnels situated in massive quartzites shows that in these situations the most favourable excavation geometry departs greatly from that indicated by an elastic stress analysis.

Typically it is found that in tunnels of equidimensional cross-section, rock fracturing commences with the formation of extension fractures in the tunnel walls that are subjected to the highest compressive stresses. As the formation of these rock slabs migrates further into the tunnel walls the length of the slabs tends to reduce, and at a certain stage an equilibrium situation is reached where no further fracturing takes place. This process has been described in detail by Fairhurst and Cook (1966). The resultant geometry of the tunnel is an elongated ellipse with the major axis orientated perpendicular to the direction of the maximum compressive stress, that is, opposite to most favourable elastic shape.

Possibly the most comprehensive analysis of a tunnel situated in a ultra high stress field is that of an experimental tunnel that was developed in solid rock ahead of a stope face in a deep gold mine to study seismicity in deep mines, Ortlepp and Gay (1984). This tunnel was developed in a stress field which ranged from about 50 MPa in the mined-out area to 140 MPa in the solid rock ahead of the mining area. Subsequently, the stresses acting on the tunnel increased to a value of about 230 MPa as a result of further mining activities. These stresses correspond to an equivalent depth below surface of more than 8 000 m. The tunnel itself was situated in massive quartzites with a uniaxial compressive strength of about 350 MPa which, when adjusted for the size of excavation using Equation (3) reduced to about 200 MPa. Figure 11 shows the variation of the field stresses along the axis of the tunnel at the time of tunnel development, and the blasted and resultant tunnel cross-sections at several points along the tunnel.

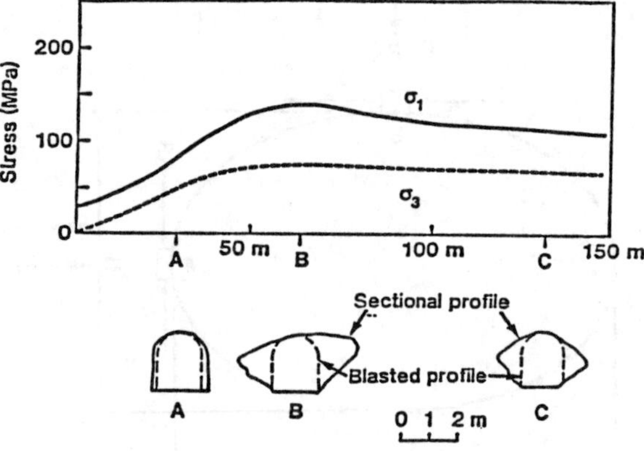

Fig. 11. Distribution of subsidiary principal stresses along axis of experimental tunnels and typical cross-section

In the high stress zones a localized band of intense fracturing was observed at about mid-height at the tunnel face. This crushed zone formed itself into a stress-raising 'notch' which rapidly penetrated more deeply into the sidewall as the face advanced, by a process of buckling and ejection of 'micro-slabs' at the notch tip. This essentially non-violent process of fracturing continued until, after some time, a characteristic sharp-ended, elliptical profile developed which stabilized at a width of some 3 m to 4,5 m. An important observation was that, ultimately, a final stable tunnel geometry was reached with a minimum amount of support. In a section of the tunnel where support was installed at an early stage its main function was to contain the fractured rock in the sidewalls of the tunnel. Another important finding was the absence of extensive fracturing in the roof and floor of the tunnel and the remarkably good roof condition.

These observations are typical for tunnels situated in massive brittle rocks which are subjected to high vertical stresses. It should be noted that a similar mode of rock fracturing is also observed in tunnels that are situated in areas of high horizontal rock stresses. In these cases, however, the fracturing of tunnel walls occurs in the roof and the floor of the tunnel.

In general it can be stated, therefore, that rock fracturing in highly stressed tunnels situated in massive brittle rocks is largely confined to the tunnel walls which are parallel to the direction of the maximum subsidiary compressive field stress. As a result of rock fracturing in these walls the preferred tunnel geometry changes to that of an ellipse whose long axis is orientated perpendicular to the direction of the maximum compressive field stress. The aspect ratio of the ellipse tends to increase as the value of the maximum compressive stress increases relative to the strength of the rock. From the point of view of controlling the extent of rock fracturing, the most favourable geometry of excavations situated in highly stressed brittle rocks is therefore, opposite to that based on the analysis of the stresses acting on the boundary of an opening in an elastic medium.

A first estimate of the extent of the fracture zone in the sidewalls of highly stressed tunnels can be obtained by performing an elastic stress analysis, and by determining calculating lines where the stresses around the excavation satisfy one of the well-known failure criteria for rock, Equation (1) or (2). The shortcoming of this approach is, however, that the changes in rock properties in the 'failed' zones are not taken into account in the analysis. Nevertheless, experience has shown that this idealized approach is well suited for comparing the relative merits of different excavation shapes in a high rock stress environment. Figure 12 shows the extent of the fracture zone, ΔW_r in the sidewalls of four tunnels of different shape as a function of the ratio of the vertical component of the field stress tensor, σ_v, and the uniaxial compressive strength of the rock. The results shown have been obtained for a horizontal to vertical field stress ratio $\sigma_h/\sigma_v = 0,5$. It is seen that at low stress levels, $\sigma_v/\sigma_c < 0,5$ the vertical elliptical and the square tunnel shapes are more favourable. However, once fracturing has taken place in the tunnel walls, $\Delta W_R/w > 0$, the horizontal elliptical and the circular tunnel geometries are more favourable, especially if $\sigma_v/\sigma_c > 1$.

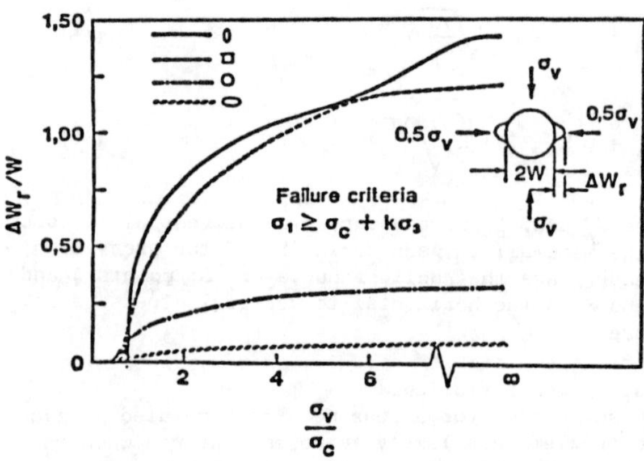

Fig. 12. Width of fracture zone in sidewall of tunnels of different shape as a function of σ_v/σ_c

An explanation for the more favourable excavation behaviour of tunnels with sharp curved sidewalls can be found in Figure 13 which shows the distribution of the principal stresses σ_1 and σ_3 along the horizontal axis of the tunnel. The most notable feature of these stress distributions is the flat gradient $\sigma 1/\Delta x$ for the tunnels which have a very large radius of curvature of the sidewall, namely, the vertical elliptical and square tunnels. The values of the minimum principal stress, σ_3, close to the sidewalls of the four tunnels are very different. The tunnels with the smallest radius of curvature have high minimum principal stresses close to the tunnel walls. Since the strength of rock increases rapidly with confining stress it follows that sidewall fracturing around excavations with small radii of curvature will be confined to a very narrow zone.

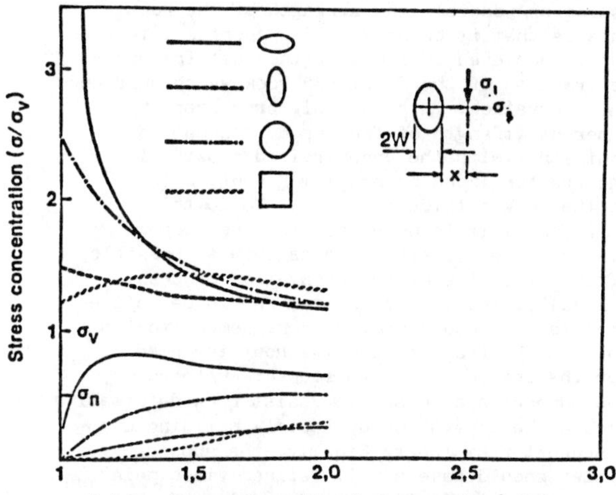

Fig. 13. Variation in principal stresses σ_1 and σ_3 with distance from the sidewall of the excavation.

Another factor which is in favour of tunnels with sidewalls that have small radii of curvature is the effect of stress-gradients on the strength of rocks. This effect is well known from attempts to measure the tensile strength of rocks indirectly, Cook and Jaeger (1979, p. 197-198) where it is found that the tensile strength of rock increases with stress-gradient. A similar effect was observed by Wagner and Schümann (1979) who studied the strength of rock under high contact stresses. These authors found a marked stress-gradient effect on the strength of brittle rocks. From these observations and those reported by Johnson (1985) it follows that the effects of stress-gradients cannot be ignored when designing excavations at depth.

Laminated strata

The presence of hard and soft rock layers and the existence of well developed planes of weakness are features which are frequently found in sedimentary geological formations. As these can have a marked effect on the behaviour of excavations it is necessary to consider them in the design and support process. By far the greatest knowledge on excavation behaviour in laminated strata exists with respect to coal mines and much benefit can be gained from a study of experiences in deep European coal mines, Jacobi (1976), Reuther (1987).

As a result of the excavation process the stress distribution in the rock mass is changed and the rock surrounding the opening deforms. In the case of excavations in massive brittle rock formations the main emphasis was placed on the formation of fracture zones due to overstressing of the rock. In laminated strata on the other hand, it is found that the presence of well defined bedding planes along which rock movement can take place has a strong effect on excavation behaviour. This problem has been studied by many authors on laboratory-scale models using equivalent materials, Everling (1962), Buschmann (1964). Recently, Purrer (1984) developed numerical models to analyze the effects of soft coal seams and planes of low frictional resistance on excavation behaviour.

Using simple stress analysis Reuther (1987) determined zones around a typical coal mine roadway where frictional sliding can take place along horizontal bedding planes, Figure 14. According to this example, which assumes a hydrostatic stress field, sliding along bedding planes takes place in the upper portion of the roadway and can extend well into the roof and haunches of the excavation if the coefficient of friction, μ, is low. In the case of thick beds, any slip and cracking over the haunches could lead to the possibility of failure by shear displacement of the bed. For thinly laminated strata, the implied problem is one of stability of the roof beds under gravity loading, and more importantly the lateral thrust acting on the beds which could lead to extensive buckling or folding of the beds which is a commonly observed phenomenon in deep collieries.

The axial stress σ_a at which a rock slab of thickness, t, and length, ℓ, will buckle is given by

$$\sigma_a = \frac{\pi^2 E}{12\zeta^2 (\ell/t)^2} \tag{17}$$

where E is the modulus of elasticity of the rock,

ℓ/t is the slenderness ratio of the rock slab, and

ζ is a constant which depends upon the end condition of the slab and varies between 0,5 for a slab which is clamped on both ends and 1 for a slab with both ends pin-jointed.

Equation 17 shows that the axial stress that can be supported by a rock slab before it buckles is inversely proportional to the square of the slenderness ratio. Consequently, thin rock slabs will buckle more easily than thick slabs.

From a design point of view it is desirable to carry a thick bed in the roof of the excavation to minimise the possibility of buckling of the roof beds. Where this is not possible consideration should be given to artificially create a competent roof by extensive roof bolting and rock reinforcement.

One possible means of reducing the potential for buckling and folding of thinly laminated strata is to drill de-stressing slots into the haunches of the tunnel so as to permit a certain degree of bedding plane slip without thrusting the roof beam. The partially de-stressed beds are then bolted to the upper strata and form a competent roof. The potential advantages of this approach have been demonstrated in model tests and deserve further attention. Figure 15 shows an example of de-stressing boreholes in a coal mine roadway, Roest and Gramberg (1981).

Although much of the damage observed in excavations which are situated in horizontally laminated strata occurs in the roof and floor of the excavation the height of excavation walls is a critical design parameter as it controls

Zones of potential bedding plane slip

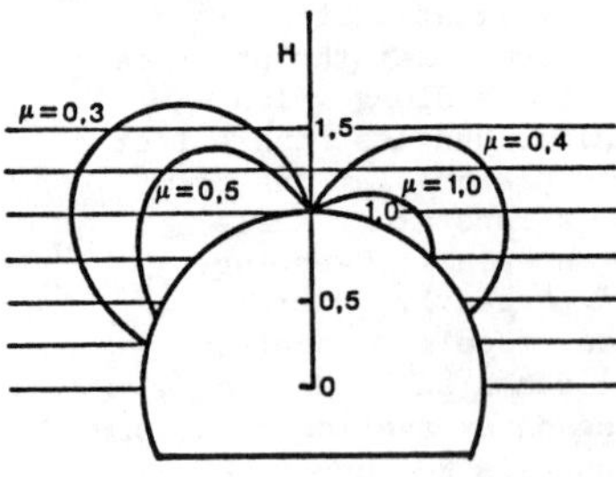

Tunnel failure as a result of bedding plane slip

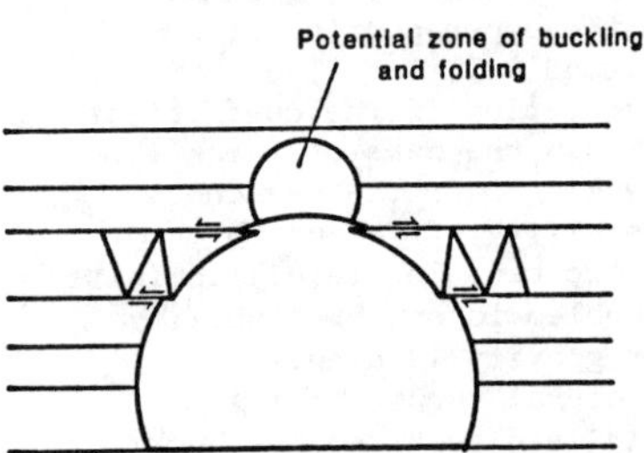

Fig. 14 Potential slippage zones around tunnel in horizontally laminated strata.

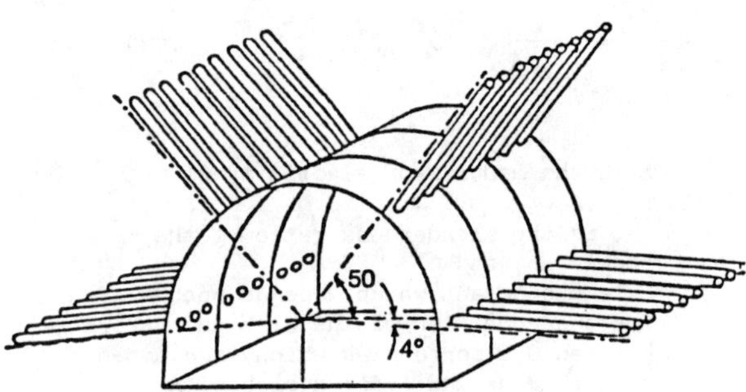

Fig. 15. Destressing boreholes in a coal mine roadway

to a large extent the depth to which sidewall fracturing can take place. This in turn influences the effective roof spans and roof support design requirements. From this point of view the height of excavations in horizontally laminated strata should be kept as low as possible. This requirement is more important in cases where slip can take place along low friction bedding planes. Where no such plances exist the height requirement of the excavation walls is less critical.

So far the discussion has concerned the design of excavations in horizontally laminated strata subjected to high vertical stresses. In the case of high horizontal stresses the thickness of the strata in the immediate roof and floor of the excavation is a critical factor as it determines the spans at which buckling occurs. In horizontally layered, thinly laminated, carbonaceous shales subjected to relatively high horizontal stresses roof falls extending up to ten metres or more have frequently been observed in six metre wide roadways in South African collieries. Usually these falls terminated at the first competent sandstone layer.

SUPPORT OF EXCAVATIONS IN HIGHLY STRESSED ROCK

In the discussion of the general design considerations for excavations in highly stressed rock masses the importance of the selection of the correct geometry of the underground opening was pointed out. In Figure 12 it was shown that the choice of cross-section is particularly critical in all of those cases where the value of the maximum compressive field stress is equal to or greater than the compressive strength of the rock mass. The magnitude of the support problem in highly stressed rock masses is, therefore, greatly dependent on the design of the excavation geometry as this not only determines the shape but also the depth of the fracture zone which surrounds the opening.

The primary function of support in highly stressed rock masses is to mobilize and preserve as far as possible the strength of the rock mass so that it becomes self-supporting. This can be achieved by limiting the post-failure deformation of the fractured rock which surrounds the excavation. This not only preserves the inherent strength of fractured rock but in addition limits the uncontrolled growth of the fracture zone by providing confinement to the rock outside this zone. To do this efficiently it is necessary for the support system to become effective as soon as possible after completion of the excavation process.

Ideally, the support system should be active, that is, it should provide confinement to the rock in the fracture zone without the need for the fracture zone to loosen. Furthermore, the support system should resist rock deformation around the excavation by rapidly building up a support resistance. Finally, the support system should have a well defined yield point which prevents it from being overloaded either as a result of a growth in the size of the fracture zone or by dynamic loading as is sometimes observed in deep underground mines.

Since the fracture zone around highly stressed underground openings does not form instantaneously but requires some time to develop, the initial or primary support must be capable of accommodating some rock deformation without loss in support capability. This aspect is of particular importance in the case of the support of mining excavations as the number, size and relative position of excavations in a mine change continuously. Consequently, the field stresses which act upon a mining excavation will change throughout the useful life of the excavation.

In civil engineering underground structures, where changes in field stresses after completion of the excavation process are unlikely to occur, it has become common practice to delay the installation of the final support or lining until the displacements of the excavation walls have come to rest, Rabcewicz et al (1972).

Rock support interaction

The design of the support of excavations is made difficult by the fact that the structure, composed of the support and the surrounding rock, is in general statistically indeterminate, that is, to say, the load acting on the support is determined by the deformation of both components of the structure. However, there are only a few idealized situations where this complex interaction has been solved numerically, Salamon (1974); Brady and Brown (1985).

Practical considerations

In practice, many decisions concerning the support of underground excavations are thus made on the basis of experience and deformation measurements. However, serious errors of judgement of the support requirements for highly stressed underground excavations can be avoided by following a few simple rules.

Firstly, it is recommended to start the assessment with an analysis of the stress distribution around the excavation. In many instances, and especially in the case of excavations in massive brittle rocks, this analysis can be made on the basis of the theory of elasticity. A comparison of the resultant stresses and the strength properties of the rock gives an indication as to whether stress failure can be expected around the excavation and what the likely extent of this failure is. In performing this analysis the sensitivity of the extent of the fracture zone to uncertainties in the strength properties of the rock mass should be examined, Figure 16.

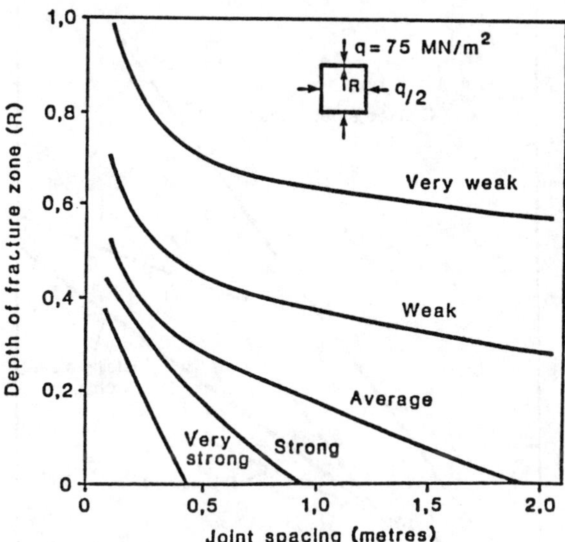

Fig. 16. Effect of rock strength on fracture zone around a square tunnel (After Piper, 1984)

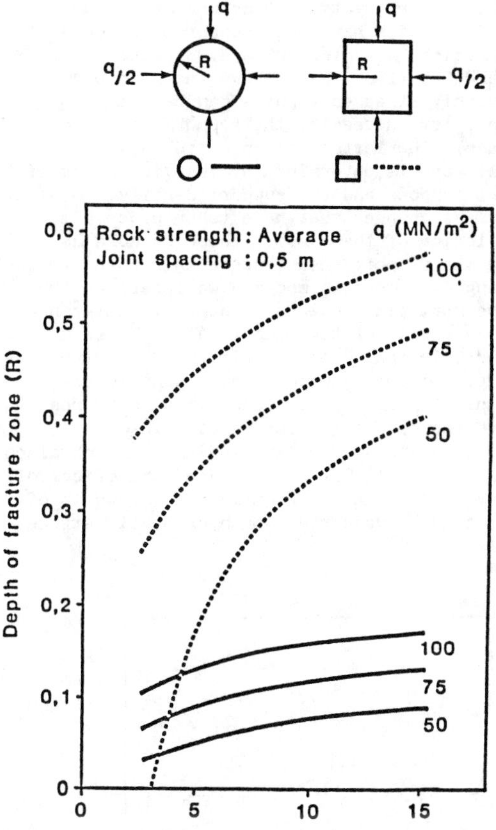

Fig. 17. Effect of excavation dimensions and field stress on depth of fracturing around excavations.

Secondly, a comparison should be made with other excavations whose behaviour in similar rock types is already known. In order to be meaningful this comparison should be done on the basis of a stress and failure zone analysis. When comparing the support requirements of excavations which are situated in the same rock formation but have different sizes, it is necessary to make allowance for the decrease in rock mass strength with increasing excavation size, Piper (1984). For example this can be done by adjusting the joint-spacing in the CSIR geomechanics classification system according to the height of excavation, Figure 17.

Thirdly, the effect of known geological discontinuities on the behaviour of the excavation walls should be assessed using either stress analysis techniques or block or wedge models. Lastly, conceptual models of possible excavation wall failures should be drawn and the support design evaluated against these models.

Proven support methods

Practical experiences on Witwatersrand gold mines and in deep tunnels have shown that an integrated support system comprising of rockbolts, wire mesh and rope lacing meets most support requirements for highly stressed excavations situated in massive brittle rocks. Sometimes this support system is supplemented by a thin layer of shotcrete.

In designing support systems for these conditions the following general guidelines have proved useful:

(i) The length of the rock bolts or rope anchors should not be less than one half the width of the excavation in the case of roof support or about one half its height in the case of sidewalls. This general guideline can be amended in the light of the results of the analysis of failure zones around the excavation, which could indicate that a bolt length of one half the excavation height is inadequate in the case of very highly stressed **tunnels**.

(ii) The maximum distance between bolts should preferably not exceed one half the bolt length. In the case of very long bolts or rope anchors it is recommended that shorter bolts be incorporated into the support pattern so as to reduce the bolt spacing at the excavation surface.

(iii) In the case of roof support, the strength of the bolts and the spacing between bolts should ensure a total support load equal to about twice the dead weight of the strata traversed by the bolts.

(iv) In laminated or bedded rock the bolts should preferably intersect the laminations at an angle of not less than 45 degrees.

(v) The rock tendon support employed in highly stressed excavations should be supplemented by wire mesh or welded mesh and/or wire lacing to prevent rockfalls between the rock tendon supports. In the case of more permanent support a thin layer of shotcrete should be applied to reduce weathering and to provide a keystone effect.

Using the support systems described above tunnels measuring 4 m by 4 m have been supported successfully in stress fields exceeding 120 MPa. Studies by Hepworth and Gay (1986) have shown that the convergence in highly stressed tunnels can be reduced significantly by increasing the support resistance in the tunnel. The latter is expressed in kN/m^2 of tunnel wall, and is determined by the yield load of the support tendons and the support density. To allow for the stress changes that have taken place after the installation of the support these authors have normalized the support resistance with respect to the stress changes. Figure 18 shows a well defined trend between the normalized values of support resistance and the convergence of the tunnel walls. These results indicate that below a critical support resistance value, the convergence in mine tunnels increases rapidly, while an increase in support resistance above the critical value results in only marginal improvements. Similar findings were obtained by Brady and Brown (1985) who analysed the effect of support pressure on the deformation of the walls of a circular tunnel subjected to a hydrostatic stress field.

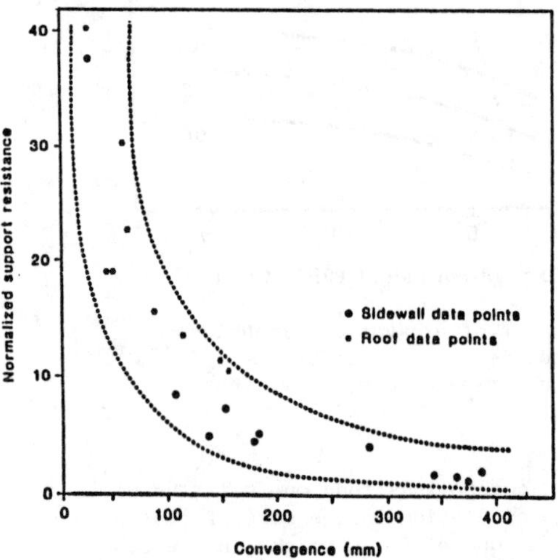

Fig. 18. Effect of support resistance on tunnel convergence (After Hepworth and Gay, 1986)

The effectiveness of different tunnel support systems employed in deep South African mines have been assessed by Wiseman (1979). The results of his study are shown in Figure 19. Of interest is the rather poor performance of steel sets and yielding steel arches compared to the performance of the various types of rock tendon support. In the light of these findings it is not surprising that the use of steel sets and yielding steel arches has been discontinued for all practical purposes in deep hard rock mines.

Shotcrete when combined with full column grouted rockbolts and wire mesh has proved itself as a most effective support under high stress conditions, Hepworth and Gay (1986).

In weak strata such as found in most European collieries yielding arches are widely used, Jacobi (1976). This support system has proved to be effective in situations where the roadways of longwalls come under severe stresses as a result of the extraction of the coal seams by the longwall mining operation. Roadway convergences of more than 50 per cent are frequently observed under these conditions. However, even in situations where the roadways are outside the influence of the abutment stresses of longwall faces the convergence often exceeds 30 per cent. This can be attributed to the practical difficulties in achieving good contact of the arches with the surrounding rock mass and the low radial stiffness of the arches. These factors facilitate a fairly unrestricted growth of the fracture zone surrounding the roadways, Figure 20.

Initial trials with resin-bonded rockbolts in deep German collieries have been successful from a technical, safety and economical point of view, Reuther (1987). This type of support will, therefore, find wider application in collieries and replace steel arches on a significant scale.

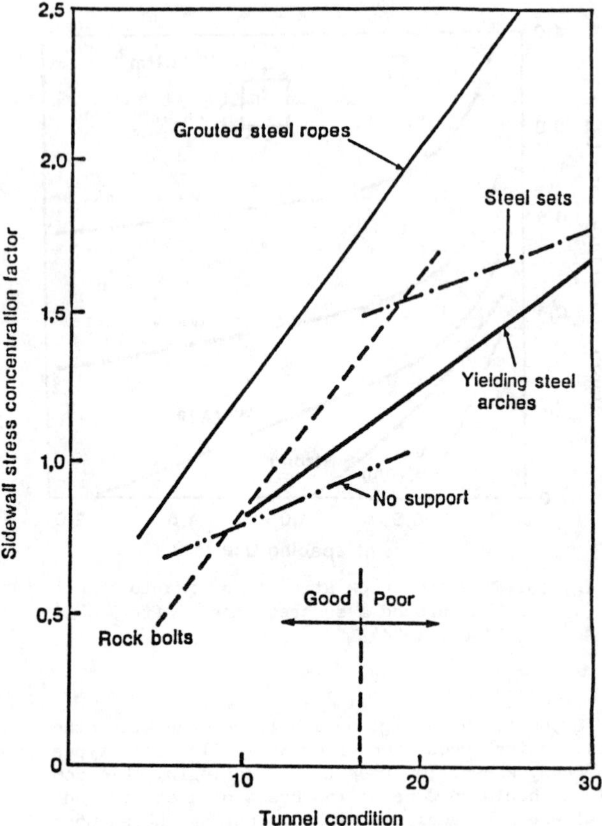

Fig. 19. Effectiveness of various support systems under high stress conditions. (After Wiseman, 1979)

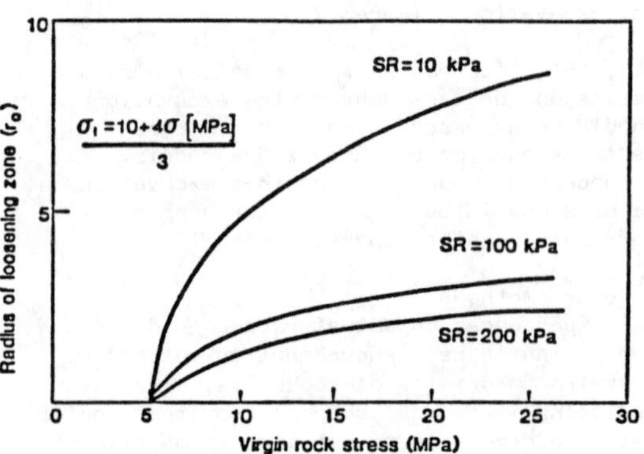

Fig. 20. Effect of rock stress and support resistance on radius of loosening zone around a coal mine tunnel. (After Jacobi 1976)

EXCAVATION BEHAVIOUR UNDER SEISMIC LOADING

Seismic loading of highly stressed underground excavations is a phenomenon which is frequently observed in deep mines but is also important in many other situations. Seismicity can affect underground excavations in several ways. Firstly, as seismic waves propagate through the rock they interact with the excavation and can give rise to dynamic stress increases. Secondly, ground motions can distort the excavation and stress the lining of the excavation. Thirdly, the ground motions can accelerate loose slabs of rock and exert additional forces on the supports.

Dynamic stress changes

Mow and Pao (1971) investigated the interaction of steady-state P-waves with cylindrical cavities in cases where the wave front travels parallel to the axis of the tunnel. These authors concluded that the peak stress concentration due to harmonic loading may exceed the static values by as much as 10 - 15 per cent for P-waves and 5 per cent for S-waves. These peak values occur at relatively long wave-lengths - on the order of 10 to 20 tunnel diameters. The maximum normal, σ_p, and shear, τ_{xy}, stresses due to a P-wave are given by

$$\sigma_p = \frac{(1 - \nu)\ E}{(1 + \nu)\ (1 - 2\nu)} \cdot \frac{V_p}{c_p} \qquad (18)$$

and

$$\tau_{xy} = \frac{G\ V_p}{2\ c_p}$$

where E and G are the elastic modulus and the shear modulus of the material, ν is the Poisson's ratio, V_p is the peak particle velocity and c_p is the P-wave velocity.

The stresses σ_p and τ_{xy} give an indication of the variation in the field stresses due to a passing P-wave. Assuming values of $E = 50$ GPa, $\nu = 0,2$, $V_p = 1\ ms^{-1}$ and $c_p = 5.10^3\ ms^{-1}$ the change in the value of the compressional field stress is approximately 5 MPa which is small when compared to the stresses that act at depth. It should be noted that the dynamic field stress change is proportional to the peak particle velocity, V_p. According to McGarr et al (1981) the peak particle velocity V_p of mine tremors is given by

$$\log R\ V_p = 3,95 + 0,57\ M_L \qquad (19)$$

where R is the distance from the focus of the tremor and M_L is the magnitude of the event.

According to Equation 19 the peak particle velocity is inversely proportional to the distance of excavation from the source of the tremor, and for most practical situations is less than 1 m/s. It follows, therefore, that dynamic stress changes due to remote seismic loading are negligible for all practical purposes. However, dynamic stress changes can be significant if the excavation is situated in close proximity to the source area of a large tremor. The maximum compressive strain, ε_p, due to a P-wave approaching an excavation is given by

$$\varepsilon_p = \frac{V_p}{c_p} \qquad (20)$$

Since Vp is of the order of ms^{-1} and c_p of the order of km^{-1}, it follows that the compressive strains due to seismic loading $\varepsilon_p < 10^{-3}$ except for excavations situated close to the source area of large tremors.

Support requirements

Wagner (1984) examined the support requirements in tunnels subjected to seismic loading. He based his model of the support requirements on the assumption that rock slabs which surround the tunnels and which have a mass, $\bar{m}$, are accelerated to the value of the peak particle velocity, Vp, and that the support must be capable of absorbing the kinetic energy, W_k, of these slabs without fear of failure.

$$W_k = \frac{1}{2}\ \bar{m}\ V_p^2$$

Assuming a yield load of the support tendons, Fy, of 100 kN and an elongation, $\Delta \ell$ of 20 mm, which are typical values for support tendons used in gold mine tunnels, the thickness of rock slabs that can be supported under dynamic loading conditions can be estimated from Figure 21. This diagram shows that the ability to support thick rock slabs with conventional rockbolts is rather limited. Wagner (1984) concluded that the only practical means of improving the dynamic performance of support systems is to increase the yield capabilities. This can be achieved either by using grouted smooth tendons or by incorporating yielding elements into the tendon design.

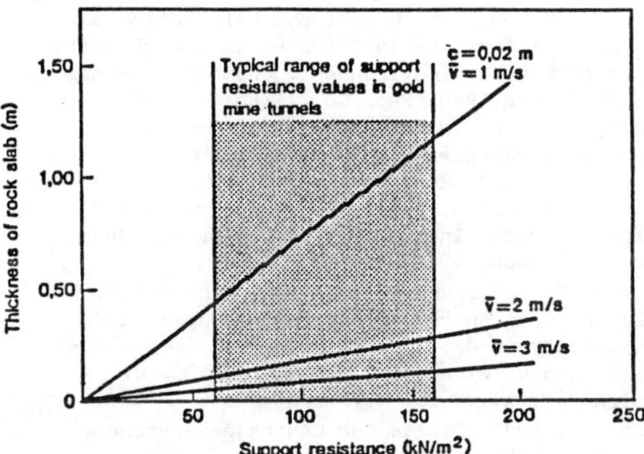

Fig. 21. Performance of gold mine tunnel supports under seismic loading conditions.

CONCLUSIONS

Practical experiences in deep mines have shown that it is feasible to design and support excavations at field stress levels which approach the compressive strength of the rock mass. From a design engineer's point of view the most significant yet unknown factor is the assessment of the strength of the rock mass and its dependence of excavation size. The only practical means of making real progress in this respect is to embark on an international programme of back-analysis of highly stressed excavations. Much work has been done in deep hard rock mines but this work will have to be extended to other rock types.

In the design of excavations in highly stressed rock masses the emphasis has to be on the engineering of the fracture zone surrounding these excavations so as to facilitate the support of the fracture zones.

Active early acting and relatively stiff support systems have the greatest potential as they preserve the inherent strength properties of fractured rock by minimising the post-failure deformation, and in addition allow the build-up of a triaxial state of stress in the vicinity of the excavation walls.

Seismic loading of underground excavations is in general not a serious problem. However, in mining where tremors can occur in close proximity to excavations, high dynamic stresses can be generated and supports can be overloaded. In this situation the peak particle velocity, Vp, is the most critical parameter as it controls both the dynamic stress changes and the work performance of the supports.

BIBLIOGRAPHY

Brady, B.H.P. and Brown, E.T. 1985. Rock mechanics for underground mining. George Allen and Unwin (Publ.) Ltd., London.

Brummer, R.K. 1984. The significance of the layered nature of the rock surrounding mining excavations from the point of view of numerical modelling. Chamber of Mines of South Africa, Res. Rep. 39/84, Johannesburg.

Buschmann, N. 1970. Modellversuche über den Einfluss von Gleitlösen im Streckenstoss. Bergbau-Archiv Vol. 25, Nr. 5, p. 35/39.

Cook, N.G.W. 1963. The basic mechanics of rockbursts. J.S.Afr. Inst. Min. Metall., Vol. 64, p. 71-81.

Cook, N.G.W. 1972. The siting of mine tunnels and other factors affecting their layout and design. Assoc. Mine Managers of South Africa, Papers and Discussions, 1973-73, p. 199-215, Johannesburg.

Deist, F.H. 1965. A non-linear continuum approach to the problem of fracture zones and rockbursts. J.S. Afr. Inst. Min. Metall., Vol. 65, p. 502-522.

Everling, G. 1962. Modellversuche über das Zusammenwirken von Gebirge und Ausbau in Abbaustrecken. Glückauf 98 p. 1471/79.

Hepworth, N. and Gay, N.C. 1986. Deformation and support performance in a tunnel in a high stress, hard rock mining environment. Chamber of Mines of South Africa, Res. Rep. No. 11/86.

Heunis, R. 1980. The development of rockburst control strategies for South African gold mines. J.S. Afr. Inst. Min. Metal. Vol. 80, April 1980.

Hoek, E. and Brown, E.T. 1982. Underground excava tions in rock. Institution of Mining and Metallurgy, London, 1982.

Hodgson, K. and Joughin, N.C. 1966. The relation ship between energy release rate, damage and seismicity in deep mines. Proc. 8th Symp. Rock Mech., Univ. Minnesota, p. 194-209, Ed. C. Fairhurst, A.I.M.E., New York, 1967.

Jacobi, I. 1976. Praxis der Gebirgsbeherrschung. Verlag Glückauf GmbH, Essen.

Jaeger, J.C. and Cook, N.G.W. 1979Fundamentals of Rock Mechanics, Chapman and Hall London.

Kersten, R. Piper, P. and Greef, H. 1983. Assessment of support requirements for a large excavation at depth. Proc. Symp. on Rock Mechanics in the design of tunnels. S. Afr. Nat. Group on Rock Mechanics, Johannesburg.

Laubscher, D.H. 1984. Design aspects and effectiveness of support systems in different mining conditions, Trans. Inst. Min. Metall, Sect A, Vol 93, p. A70-A81.

McGarr, A., Green, R.W.E. and Spottiswoode, S.M. 1981. Strong ground motion of mine tremors : some implications for near source ground motion parameters. Bull. Seism. Soc. Am, Vol. 71, No. 1, 1981, p. 295-319.

Mow, C.C. and Pao, Y.H. 1971. The diffraction of elastic waves and dynamic stress concentrations, R-482-R. The Rand Corp. Santa Monica, California.

Ortlepp, W.D. and Gay, N.C. 1984. Performance of an experimental tunnel subjected to stresses ranging from 50 MPa to 230 MPa. I.S.R.M. Symposium : Design and performance of underground excavation, Edited by Brown and Hudson, British Geotechnical Society, London, 1984, p. 337-346.

Ortlepp, W.D., More O'Ferrall, R.C., and Wilson, J.W. 1972. Support methods in tunnels. Assoc. Mine Managers of South Africa. Papers and Discussions, 1972-73, p. 167-195, Johannesburg.

Piper, P.S. 1984. The effect of rock mass charac teristics on fracturing around large excavations at depth. Proc. Seminar on Design and Construction of large underground openings. S. Afr. National Committee on Tunnelling, Pretoria, Nov. 1984, p. 7-16.

Piper, P.S. 1985. The application and evaluation of rock mass classification methods for the design and support of underground excavations at depth. Proc. Symp. Rock Mass Characterization, S. Afr. Nat. Group on Rock Mechanics, Johannesburg.

Purrer, W. 1984. Rechenmodell für das Verhalten einer tiefliegenden Flözstrecke bei festem von Lösen durchsetztem Begleitgebirge unter Berücksichtigung der In-Situ und an Modellen festgestellten Bruchmechanismen der Weichschichten. Dissertation, Montan-Universität Leoben/Austria.

Rabcewicz, L., Golser, J. and Hackl, E. 1972. Die Bedeutung der Messung im Hohlraumbau. Part 1. Bauingenieur. Vol. 47. p. 225/234.

Reuther, E.U. and Hermühlheim, W. 1987. Untersuchung zur Anwendbarkeit von Anker-Verbundausbau im Steinkohlenbergbau. Report published by the Department of Mining, Rheinisch-Westfälische Technische Hochschule, Aachen, Germany.

Roest, J.P.A. and Gramberg, J. 1981. Deep mining project. Rock Mechanics Laboratory, Technical University Delft, Netherlands.

Ryder, J.A. 1986. Application of numerical stress analysis to the design of deep mines. Gold 100, Proc. of the Int. Conf. on Gold, Vol. 1 : Gold Mining Technology, p. 245-253, S.Afr. Inst. Min. Metall. Johannesburg 1986.

Salamon, M.D.G. 1967. A method of designing bord and pillar workings. J.S. Afr. Inst. Min. Metall., Vol. 68, p. 55-67.

Salamon, M.D.G. 1974. Rock mechanics of underground excavations. Proc. 3rd Congress, International Society for Rock Mechanics. Vol. 1, Part B, p. 951-1099. Washington, D.C. Nat. Acad. Sci.

Salamon, M.D.G. 1984. Rockburst hazard and the fight for its alleviation in South African gold mines. Proc. Symp. Rockbursts : prediction and control. Inst. Min. Metall. London. p. 11-36.

Salamon, M.D.G. 1984. Energy considerations in rock mechanics : fundamental results. J.S. Afr. Inst. Min. Metall., Vol. 84, p. 233-246.

Salamon, M.D.G. and Wagner, H. 1979. Role of stabilizing pillars in the alleviation of rockburst hazard in deep mines. Proc. 4th Congr. Int. Soc. Rock Mech., Sept. 1979, Montreaux, Vol. 2, p. 561-566. Balkema, Rotterdam.

Ulgudur, S. 1973. A non-linear continuum theory of fracture for rocks. Thesis, University of Newcastle upon Tyne.

Wagner, H. 1984. Support requirements for rockburst conditions. Proc. 1st Int. Congress on Rockbursts and Seismicity in Mines. S.Afr. Inst. Min. Metall. Johannesburg, 1987, p. 209-221.

Wagner, H. and Schümann, E.H.R. 1971. The stamp-load bearing strength of rock : An experimental and theoretical investigation. Rock Mechanics, Vol. 3, 1971, p. 185/207.

Wiseman, N. 1979. Factors affecting the design and condition of mine tunnels. Chamber of Mines of South Africa, Res. Rep. No. 45/79, Johannesburg.

9:20 LE GRAND SALON - Short Presentations

G. BARLA, C. SCAVIA, M. ANTONELLIS and M. GUARASCIO, Italy "Characterization of Rock Mass by Geostatistical Analysis at the Masua Mine".

T. BORG and A. HOLMSTEDT, Sweden "A Rock Mechanical Investigation of a Longitudinal Opening at the Kiirunavaara Mine".

S. BUDAVARI and R.W. CROSSER, South Africa "Effects of the Extraction of Tubular Deposits Around Vertical Shafts in Deep-Level Mines".

R.T. EWY, J.M. KEMENY, ZHENG ZIQUONG and N.G.W. COOK, U.S.A. "Generation and Analysis of Stable Excavation Shapes Under High Rock Stresses".

10:00 FOYER, Coffee Break.

10:30 LE GRAND SALON

W.J. GALE, J.A. NEMCIK and R.W. UPFOLD, Australia "Application of Stress Control Methods to Underground Coal Mine Design in High Lateral Stressfields".

A. GUENOT, France "Stress and Rupture Conditions Around Oil Wellbores".

Y. ISHIJIMA, Y. FUJII and K. SATO, Japan "Microseismicity Induced by Deep Coal Mining Activity".

A.J. JAGER, P.S. PIPER and N.C. GAY, South Africa "Rock Mechanics Aspects of Backfill in Deep South African Gold Mines".

R.W.O. KERSTEN and H.J. GREEFF, South Africa "The Influence of Geology and Stresses on Shaft Excavations: A Case Study".

M. KUSABUKA, T. KAWAMOTO, T. OHASHI and H. YOICHI, Japan "Integrated Stability Assessment of an Underground Cavern".

KYUNG WON LEE, HO YEONG KIM and HI KEUN LEE, Korea "Measurement of Rock Displacements and its Interpretation for the Determination of a Relaxed Zone Around a Tunnel in a Coal-Mine".

K. MATSUI, M. ICHINOSE, K. and K. UCHINO, Japan "Deformational Behaviour and its Prediction in Longwall Gate Roadways under Pillars and Ribs".

I.M. PETUKHOV, U.S.S.R. "Forecasting and Combating Rockbursts - Recent Developments".

12:00 MARQUETTE, Lunch Break and Poster Sessions.

13:30 LE GRAND SALON

C. TANIMOTO, S. HATA, T. FUJIWARA, H. YOSHIOKA and K. MICHIHIRO, Japan "Relationship between Deformation and Support Pressure in Tunnelling through Overstressed Rock".

J.E. UDD and D.G.F. HEDLEY, Canada "Rockburst Research in Canada - 1987".

M.L. VAN SINT JAN, L. VALENZUELA and R. MORALES, Chile "Flexible Lining for Underground Openings in a Block Caving Mine".

A. VERVOORT, J.-F. THIMUS, J. BRYCH, O. de CROMBRUGGHE and E. LOUSBERG, Belgium "Verification by the Finite Element Method of the Influences on the Roof Conditions in Longwall Faces".

C.M.K. YUEN, J.M. BOYD and T.R.C. ASTON, Canada "Rock-Support Interaction Study of a TBM-Driven Tunnel at Donkin Mine, Nova Scotia".

ZHU WEISHEN, LIN SHISHENG, ZHU JIAQUIAO, DAI GUANYI and ZHAN CAIZHAO, China "Some Practical Cases of Back Analysis of Underground Opening Deformability Concerning Time and Space Effects".

14:30 FOYER, Coffee Break

15:00 LE GRAND SALON, Panel and Floor Discussions.

Moderator: B.N. Whittaker (UK)
Panelists: V.M. Sharma (India); P.K. Kaiser (Canada); V. Maury (France).

Panel contribution: P.K. Kaiser (Canada);

Detection of rock mass rupture modes.

1. INTRODUCTION

At the time, when the theme" Underground Openings in Overstressed Rock" was selected, it was clear that the topic would be of interest to many. However, we did not suspect that the topic of rock failure would create so much excitement as we have witnessed here during:

- the ISRM-Commission Meeting on Rock Failure Mechanisms in Underground Openings;
- the related work shop held yesterday;
- todays keynote address; and
- the many excellent papers (Volume II of the Congress proceedings).

A majority of papers deals with some aspects of detection, prediction, understanding or control of underground opening behaviour.

Seldom do engineers dare to talk about failures and scientists make an attempt to meet practitioners in an effort to advance our knowledge so that we may become more successful in preventing failures. I believe this Conference will be remembered by many for exactly this reason.

Yesterday, after showing us the most fascinating manifestations of rock mass failures, Mr. Ortlepp, stated (I hope I quote correctly) " ... to learn more about the behaviour of underground openings in highly stressed rock, we must understand rock mass failure modes much better." I strongly agree with this view point and wish to add a few comments to the problem of "Detection of Rock Mass Rupture Modes", with particular emphasis on field monitoring.

Persistent pressures on the market for natural resources have forced the mining and underground construction industry to develop innovative procedures for rock excavation, rock fragmentation, rock removal, support of underground openings and construction sequencing. In this manner, productivity has been increased significantly over the last decade and the economic limits in terms of opening size, shape and overburden depth as well as the rate of extraction have been extended. Furthermore, the application of these new and improved technologies has permitted the engineers to advance into higher stress fields and into poorer ground. Unfortunately, as a consequence, more rock mass failures have occurred over the last few decades. For example, in Canada, rock bursts in hard rock mining have recently become an area of major concern (Udd and Hedley; Theme IV, same conference).

At the same time, a tremendous need for safer construction has emerged because of more respect for human life and because of higher financial risks resulting from large up-front investment costs for new, more complex technologies. The consequences of a major failure in underground construction are unbearable and methods must therefore be found to reduce the risk by a better prediction and control of the actual ground behaviour. Only then can Brunel's advice be followed and "...... *the best technology made to work because it fits the ground*".

It will never be possible to eliminate all risks in underground construction, but much more can be done to reduce and minimize the risk of failure by improving our understanding of rock mass behaviour and rupture modes. We will only be able to develop improved methods for controlling or preventing rock failures after we can clearly detect and interpret signs of impending rock mass rupture.

Of the various alternatives for risk reduction, I will concentrate in the following on the use of in-situ stress change, strain and displacement monitoring techniques for the detection and comprehension of rock mass rupture modes. However, it should be understood that an acceptable risk can only be achieved when such field observations are combined with careful site investigation, providing detailed geological data about the rock mass quality and structure, and with other techniques that permit reduced exposure to hazardous conditions (e.g., microseismic monitoring in rockburst prone environments).

2. OBJECTIVES OF STRESS AND DEFORMATION MONITORING (Why, What and Where?)

Why?

Monitoring has become an integral part of any well managed

underground construction project but the objectives of conventional monitoring techniques are often not well defined. Measurements should:

- provide circumstantial evidence or proof of proper understanding of rock mass behaviour; or
- assist in unveiling the actual character of the rock mass; and help
- to establish warning criteria; and
- to assist in modifying and improving design and construction techniques.

What and Where?

We install instruments to deliver conclusive evidence and to provide the links that permit arrival at a unique and correct conclusion. A proper monitoring program for the purpose of stability assessment provides evidence:

- to permit identification and classification of rupture modes, and
- to allow localization and quantification of a kinematically possible mode of instability,

such that means for prevention or control of failure or rupture initiation and propagation can be implemented.

The underground design and construction process is complex and a monitoring program (during construction) may have many purposes. However, immediate feedback and clearly defined early warning indicators are essential for a successful stability assessment.

Because of geological complexities and the uncertainties inherent in underground construction, it is impossible to install instruments for the detection of all possible mechanisms. Consequently it becomes a first priority of a monitoring program and its data interpretation to verify and confirm one or several assumed modes of rock mass behaviour (e.g., elastic or yielding, failing in tension, compression or by localized shear, etc.).

"Safety rests . . .", as de Mello in his 1977-Rankine Lecture stated, *". . . . not upon the accuracy of our calculation, but upon the adequacy of our hypothesis"*. Hence, monitoring for stability assessment must first of all identify the mode of behaviour (or rupture). All other objectives are of subordinate importance. A monitoring program must be designed for a very specific purpose.

Furthermore, because rock seldom behaves as an elastic continuum or as a homogenous yielding material, monitoring should attempt to capture the most frequently encountered, but least understood, phenomena of brittle failure, discontinuum behaviour and localized shearing (Fig.1).

3. RUPTURE MODE IDENTIFICATION

Many modes of behaviour can be identified and may be divided into one of two categories:

- stable modes, where deformations are limited and acceptable, and
- unstable modes, where deformations are large and excessive for the purpose of the underground structure under consideration (Fig.2).

Instabilities may be further subdivided into brittle, sudden collapse, and ductile yielding modes. Whether failure occurs or not depends on the properties and structure of the rock mass.

3.1 Brittle Rupture Mode Identification:

The nature of these sudden collapse mechanisms (gravity falls, violent slabbing, rock bursting, etc.; C in Fig.2) requires that they are detected by careful mapping (i.e. keyblock identification) or by 'energy-build-up' monitoring methods (e.g., for rock bursting, stress change measurements, micro-seismic monitoring, etc.). Alternatively, support systems capable of withstanding the effects of brittle failure modes (e.g., cable lacing, etc.) may be designed. This topic has been well covered by the pre-congress workshop on "Mining-Induced Seismicity, Monitoring and Interpretation Techniques", the theme lecture by Dr. Wagner and several papers (e.g., Udd and Hedley) and no further comments are offered here.

3.2 Ductile Rupture Mode Identification:

Ductile rupture can be a result of either global yielding or localized shearing (A and B in Fig. 2). Both modes may be detected by conventional methods but it is necessary to differentiate between them.

• *A* • *Global Yielding*

Several authors have dealt with the problem of detecting and handling highly squeezing or yielding ground. One aspect which has not yet received sufficient attention during the oral presentations is the problem of early detection of excessively yielding ground and early prediction of the ultimate extent of the yield zone, e.g., for the selection of the proper anchor or bolt length. In our contribution to this congress (Barlow and Kaiser), we have expanded on the earlier work by Sulem *et al.* (1987) and have shown that the normalized convergence rate is an excellent indicator allowing prediction of the ultimate extent of a yield zone very close to the advancing excavation face. The normalized convergence rate is the ratio of the current rate of convergence, at a certain distance from the tunnel face, to the maximum rate measured at or close to the face. As shown in Figs 6 and 7 (Barlow and Kaiser, Vol. 2), the normalized convergence rate is a much more sensitive indicator than the convergence itself. Unfortunately, this technique requires that very frequent convergence measurements are made close to the advancing face. The economic benefits resulting from the early detection of very difficult ground conditions may, however, often balance the inconvenience of interruptions during construction. Practitioners are encouraged to evaluate this concept in their projects.

• *B* • *Localized Shearing or Yielding (resulting in a ductile response of the opening)*

A rock mass is normally heterogenous and its strength and failure modes depend often, as for a chain, on the location, orientation and strength of the weakest link. However, yielding of a single weakness near an underground opening does not necessarily lead to collapse because of geometric restraints (see schematic diagrams in Figs 3 and 4). Only for Class I - type problems such as plane slope or keyblock failures can movement along a single weakness occur or propagation from one weakness (including induced local failure due to stress concentration) lead to a kinematically possible collapse mechanism. For Class II - type problems at least two weaknesses are required to establish a kinematically possible failure mechanism. Some examples showing the propagation of localized yielding and the formation of a rupture mechanism (creation of a non-yielded rock wedge) are illustrated in Fig.5 (Shading indicates the shape of the propagating yield zone). Completely different modes of failure propagation develop, for otherwise identical conditions, if two weaknesses provoke the propagation of a kinematically possible rupture mechanism. Furthermore, these weaknesses must not necessarily be preexisting but may be created by local stress concentrations.

4. DETECTION OF CLASS II - TYPE (WEAKNESS DOMINATED, DUCTILE) RUPTURE MODES

• *By Convergence Measurements ?*

Ground convergence curves for the roof and springline in a circular tunnel calculated by finite element simulations (Kwong, 1988) are shown in Fig.6. The computer code SAFE, developed by Chan (1986) at the University of Alberta, was used for this analysis. Five cases, including elastic, homogeneous plastic and three weakness-dominated deformation patterns, are presented (Fig.5). It is most interesting to observe that all roof convergence curves tend toward the same point (about 1.5 u_e at 50% support pressure) indicating that the convergence at this stage does not permit differentiation amongst the various modes of failure. At the springline, where the non-elastic deformations are much larger, the same observation can be made (at about 10 u_e). Only the global yielding case (II) deviates and indicates that much higher support pressures would be needed (at more than 6.5 u_e) to prevent collapse if this mode prevailed. At small displacements, Case V with a weak joint needs the most support to maintain good ground control.

In general, it can be concluded that convergence measurements are good witnesses of instability but poor indicators for the identification of the type and location of the rupture mechanism. Laboratory measurements (Kaiser *et al.*, 1985) confirm this conclusion, but it has also been shown that it is often difficult or impossible to detect yield initiation from convergence data (Kaiser, 1981).

Localized measurements are needed to establish the actual mode of behaviour and the location of instabilities. This is illustrated in Fig.7 by the simulated ground strain data detected by the four Extensometers A to D (Case V). Extensometers A (in the roof) and B and D (at the springline) are in non-yielding rock and, hence, give little or no indication that a major instability is approaching. Even at collapse ($p_s < 50\%$), only 0 to 0.24% average strain is recorded. The strategically well placed Extensometer C, however, provides early evidence of instability (> 2% strain at $p_s = 75\%$). It follows that localized measurements at appropriate locations are needed for stability assessments. Pattern monitoring will likely miss critical conditions and only instruments placed strategically, based on a clear vision of the potential failure mode, can be helpful. Furthermore, localized measurements of relative displacement are needed to properly locate and identify failure modes.

5. WASHUUZAN TUNNEL (Japan) ; CASE HISTORY

An excellent example illustrating the rationale presented above is given by the data shown in Figs 8 and 9 (Sakurai, 1983). Some rupture clearly occurred at about 70 days as indicated by the sudden jump in some of the convergence records (Fig.8). However, the rupture mode cannot be localized accurately and hence, proper remedial measures cannot be designed based on these convergence measurements only. A comparison of local bolt strain measurements, presented in Fig.9 shows that sudden extensional straining, probably due to extension fracturing or initiation of slabbing, occurred near the right springline. Stability could be achieved easily by some supplemental bolting. No further support was required in the roof and left springline.

Another case history, reflecting a similar type of localized failure mechanism was recently described by Morrison (1987). In this case, large scale brittle failure processes were induced during mining by slippage along a pre-existing fault and by the rupture of propagating off-shoots.

6. CONCLUSIONS

- Because rock masses are seldom homogeneous, weaknesses cannot be ignored for the assessment of underground opening stability. However, sufficient yielding must normally occur, often along more than one weakness, to create a kinematically acceptable failure mode.
- A monitoring program and the instrumentation layout must be designed for the specific purpose of behaviour mode identification; e.g., energy-storage-type monitoring for brittle failure detection and local strain monitoring for ductile failure modes. Instruments must be located at strategic positions based on a hypothesis of several likely behaviour and rupture modes. Pattern monitoring for the purpose of stability assessment is to be discouraged.
- The first priority of any data interpretation is to confirm the assumed mode of behaviour. For this purpose, measurements from carefully selected instrument locations, providing unique data, are to be compared with predictions from numerical or analytical models.
- Back-analyses based on continuum models, neglecting weaknesses when rupture modes develop, are often misleading.
- For ductile failure modes, convergence records are generally poor indicators for the detection of yield initiation but they are good indicators of yield zone propagation after large non-recoverable strains accumulate.
- For the detection of brittle rupture mode initiation, convergence measurements may be a good indicator.
- Localized measurements of strain are needed to identify and locate a ductile failure mechanism. They are essential for a conclusive interpretation of field measurements.

These brief comments on a few specific aspects of rupture mode detection do not completely cover this topic but they are intended as a stimulus for further discussions on this rather important component of risk reduction in underground construction.

References: (other than in proceedings of this congress)

Chan, D., 1986. Finite Element Analysis of Strain-Softening Materials. Ph.D. thesis, Department of Civil Engineering, University of Alberta, Edmonton, Alberta, 355 p.

Kaiser, P.K., A. Guenot and N. R. Morgenstern, 1985. Deformation of Small Tunnels - Behaviour during Failure. *International Journal of Rock Mechanics and Mining Sciences & Geomechanics Abstracts*, Volume 22, pp.141-152.

Kaiser, P.K., 1981. Monitoring for the Evaluation of the Stability of Underground Openings. *Conference on Ground Control in Mining, Morgantown, West Virginia*, pp.90-97.

Kwong, A., 1988. Assessment of Borehole Stability. Ph.D. thesis, Department of Civil Engineering, University of Alberta, Edmonton, Alberta, (in preparation).

Morrison, D.M., 1987. Rockburst Research at Falconbridge Limited. *89th Annual General Meeting of the CIM, Toronto*, 22 p. and 13 figs.

Sakurai, S., 1983. Personal Communication.

Sulem, J., M. Panet and A. Guenot, 1987. Closure analysis in deep tunnels. *International Journal of Rock Mechanics and Mining Sciences & Geomechanics Abstracts*, Volume 24, pp.145-154.

Symbols:

p ... Fictitious support pressure or force (required to maintain stability at a given displacement)

u ... Displacement of a given point (often at the tunnel wall)

u_e ... Elastic wall displacement at zero support pressure.

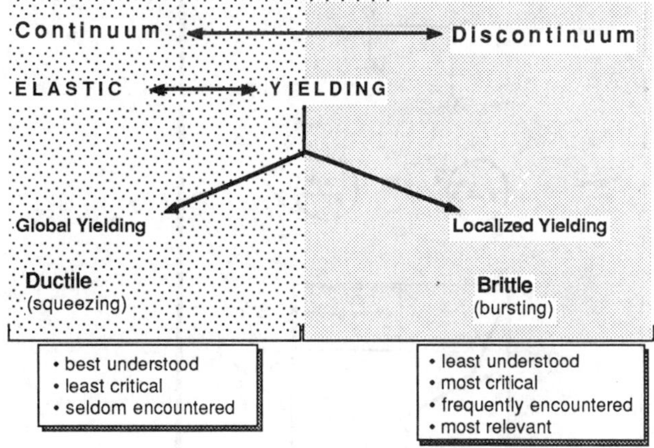

Figure 1 Behavioral Modes of Rock Masses

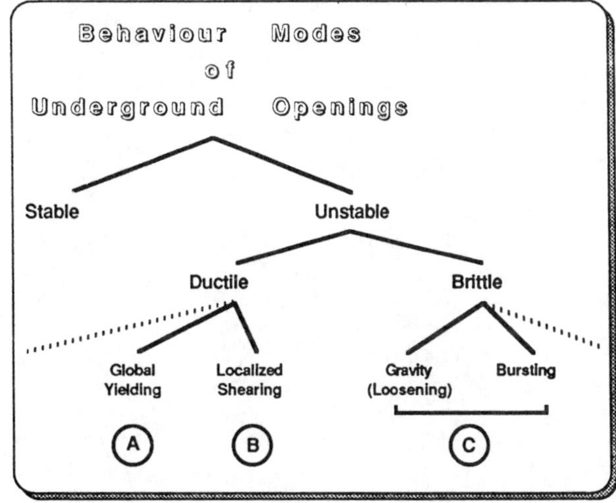

Figure 2 Behavioral Modes of Underground Openings

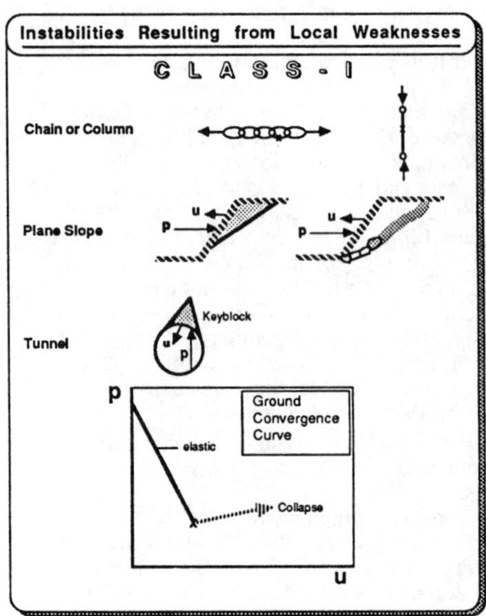

Figure 3 Class I - Instabilities Resulting from Local Weaknesses

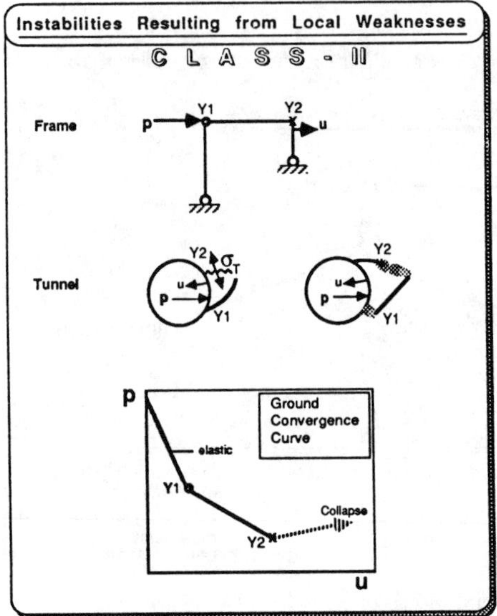

Figure 4 Class II - Instabilities Resulting from Local Weaknesses

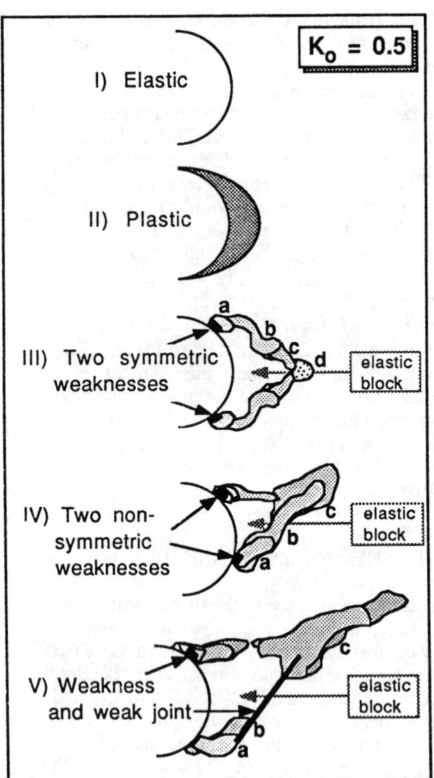

Figure 5 Yield Patterns for Five Rock Mass Configurations

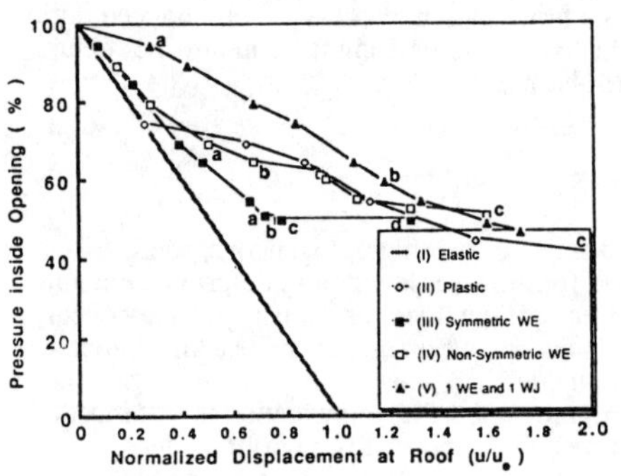

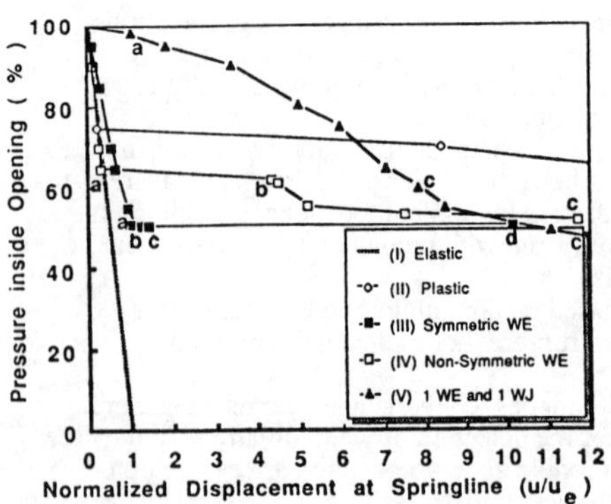

Figure 6 Ground Convergence Curves (Top: Roof; Bottom: Springline)

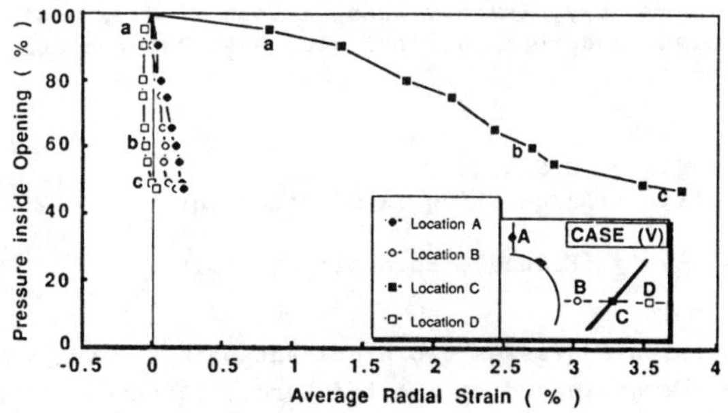

Figure 7 Ground Strain Curves for Four Locations in Case V

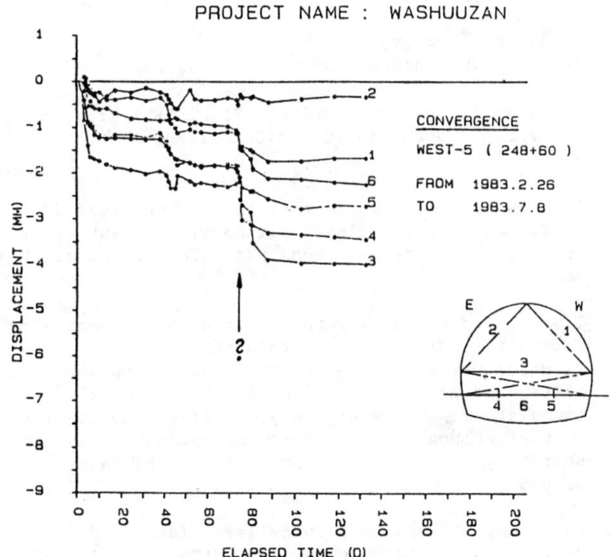

Figure 8 Convergence Development at Washuuzan Tunnel (Japan) (Sakurai, 1983)

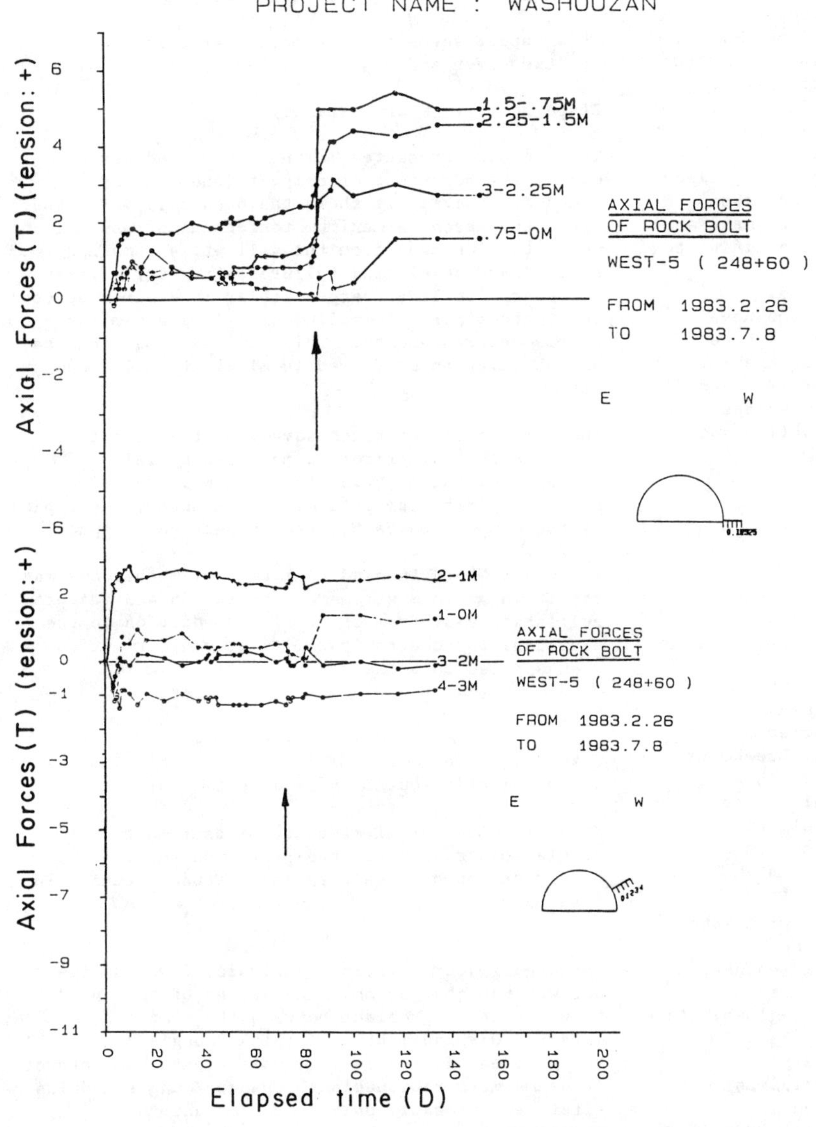

Figure 9 Axial Bolt Force Development at Washuuzan Tunnel (Japan) (Sakurai, 1983)

Floor discussions:

N. Barton (Norway)
My contribution/question is directed to:

Van Sint Jan et al., (Chile), Flexible lining for underground openings in a block caving mine.

I think that I detected that the author resisted admitting that his flexible support (consisting of bolting shotcrete and mesh) was NATM. I support his stand.

I think it is time for the international tunnelling community to resist being told that they have used NATM just because they have used shotcrete and rock bolts, and have not applied the whole support immediately. There are many countries who use these support components. It does not automatically imply that they believe in or have applied the NATM method.

For example, will NATM tunnellers claim that Norwegians use NATM, when they (NATM) also finally give up labour intensive mesh reinforcement, and go over to using wet process fiber reinforced shotcrete?

Response by Van Sint Jan et al. (Chile):

In our paper, we do declare that we followed the design philosophy which has generally been associated with the New Austrian Tunnelling Method. However, I agree with Dr. Barton's position that the simple fact of using rock bolts and shotcrete to support and underground excavation does not mean that one is following the NATM procedure. Such is the case, for example, of loosening ground conditions, where the whole support must be applied near the face shortly after the excavation takes place. In such case it is not necessary to make any special accomodations in the design of the lining to allow movements in the loosening rock.

I believe that there may be many people who have built tunnels in highly stressed ground and have used the design procedure associated with the NATM without even knowing of its publication. However, we think that it is reasonable to acknowledge credit to those who were the first to widely publish such design philosophy.

Ziqiong Zheng (U.S.A.)
My contribution/question is directed to:

V. Maury (France), Failure around u/g openings.

Figure 2 shows different failure modes caused by different stress conditions. In case B, when σ_a (axial stress) becomes maximum and the maximum tangential stress $3\sigma_H-\sigma_h-p$ becomes intermediate, as internal pressure p is zero or small, breakouts as shown in Case A could take place; this is at least theoretically possible, as a final stable cross section shape. Spalling on the wall of a circular opening could take place as shown in fig. 2B initially when failure criterion is satisfied. This initial spalling causes the formation of an elliptical cross section with long axis parallel to σ_h. The elongation in σ_h direction increases the stress concentration near the ends of the long axis and raises up the tangential on the boundary around those regions. As σ_A stays relatively constant, these will be a transition point when the maximum boundary tangential stress becomes the maximum principal stress locally. As a consequence breakouts generated from an elliptical opening, (as shown in fig. 10, R. Ewy, Kemeny, Zheng & Cook, 1987), could take place and reach a final stable shape of cross section.

J.C. Rogiers (U.S.A.)
My contribution/question is directed to:

Dr. V. Maury (France), Borehole stability.

My question addresses two excellent points made by Dr. V. Maury regarding the fracture initiation location and the fracturing periodicity. By developing the fully-coupled poro-elastic solution of a borehole, two phenomena battle each other. On one side, at the beginning of the loading the rock behaves as an undrained material. But as diffusion sets in, the behaviour moves towards drained conditions. The consequence of this is that although the radial stress profile is hardly affected the circumferential stress distribution varies in a non-monotonic fashion as a function of time.

Such a phenomena can explain:
(i) failure initiation inside the rock mass instead of the borehole wall

(ii) thickness of the slabs are an immediate consequence of the classic failure criterion and shows an extreme regular periodicity.

This aspect seems to have been overlooked in the reviews given today.

Response by V. Maury (France):

The approach presented in the review addresses mainly the behaviour of compact (and most of the time) dry rocks. It shows the possibility for these rocks to present a maximum tangential stress concentration away from the wall while remaining elastic and developing failures along this stress concentration line (perpendicular to minimum stress) and further periodic failures following similar stress rearrangements. This could address this case and observation performed in mines, tunnels, and lab.

The poroelastic approach adversing the effect of pore pressure distribution may also explain a failure initiation, away from the wall and give birth to periodical phenomenon. However, the topics is too complex to be treated in such short space.

It must be born in mind that several mechanisms may result in maximum stress concentration and failure initiation away from the wall, in addition to the classical but controversial plastic distribution, when applied to rocks.

Sakurai (Japan)
My contribution/question is directed to:

Kyung-Wan Lee, et al. (Korea), Measurement of rock displacements and its interpretation for the determination of a relaxed zone around a tunnel in a coal mine.

Concerning your numerical analysis, I would like to ask why you changed only the values of "m" and "s" to get better agreement between the computed and measured displacements. I think Young's modulus has the greatest influence on displacement. Therefore, it seems more reasonable to change Young's modulus first, and probably only then, "m" and "s", if necessary.

Response by Kyung-Wan Lee et al. (Korea):

In this paper, the stress state around a tunnel was defined by the boundary conditions, such as virgin stresses, geometry of the opening and support pressure on the walls. Therefore, the stress state is independent of the deformation of the surrounding rockmass. The deformation of the rocks was calculated at the defined stress state by referring to the elastic properties, such as Young's modulus and Poisson's ratio. The formula for rock failure criteria depends only on the state of stresses in the analysis, so the Young's modulus has no effect on the calculations determining the relaxed zone around the tunnel (which is one of the weak points in elastic analysis). In case of elastoplastic analysis, the deformation properties can be considered for the determination of the relaxed zone.

A. Guenot and F.J. Santarelli (France)
Our contribution/question is directed to:

R.T. Ewy, J.M. Kemeny, Z. Zheng and N.G.W. Cook (U.S.A.), Generation and analysis of stable excavation shapes under high rock stresses.

In the oil industry, the designer of a wellbore is faced with two major challenges which are first, to determine whether or not the borehole wall will undergo failure, and second, what must be done if the well fails. In so far as this latter point is concerned, the common practice consists in increasing the mud density and therefore the support pressure of the wall. However, the economical consequences of such an increase are always negative. This is where the contribution made by EWY et al. (1987) becomes extremely important. If the failure process around a borehole were able to stabilize itself in all cases by the creation of a stable shape as these authors have shown, the practice of drilling for oil wells could be dramatically changed. However the interesting approach described in the above mentioned paper calls a few questions which will be addressed in the next four paragraphs.

How is failure initiation determined ?

Is the final shape of the borehole stable with respect to another failure mode ?

What happens when the entire wall of the borehole is initially subjected to failure ?

What happens in the case of the formation of a shear zone ?

FAILURE INITIATION

The analyses made by EWY et al. consists in comparing the linear elastic stresses around the hole with a peak strength criterion and to assume that the rock has failed when the stress becomes larger than the corresponding strength. This method has been tested during laboratory tests on small scaled models of underground openings and it has been shown (e.g. GUENOT, 1987) that the wall of such model excavations fails when the maximum stress is several times larger than the corresponding strength. Possible explanations for this phenomenon are presented elsewhere (SANTARELLI and BROWN, 1987 and VARDOULAKIS, SULEM and GUENOT, 1987) and are based on the use of more appropriate stress strain laws. Under these circumstances, the designer is faced with the choice of an adequate value for the uniaxial compressive strength of the rock Q if he wants to use the model proposed by EWY et al. But this choice is made very delicate by the large scatter of the data gathered by GUENOT (1987). It is therefore suggested to determine an apparent strength of the wall by means of a more appropriate model. However, the question which remains is how much the use of linear elastic stress fields as described by the authors will also influence the propagation process of failure ?

SHAPE STABILITY WITH RESPECT TO ANOTHER FAILURE MODE

As mentioned by EWY et al., the final shape obtained during their stability analysis has been observed in the laboratory and in the field. However, the stability of such a shape, for all conditions, is not sufficiently assessed. As these authors rightly recognize, the tip of the breakout is submitted to high stress concentrations which in turn limit the stress concentration elsewhere. In fact, at such a singular point, the stress concentration should even become infinite and in the case of non circular openings, the authors recognize that some level of shear or splitting could occur. Therefore, even though the final shape obtained by the authors in the case of a circular opening is stable with respect to their criterion, one can wonder whether or not a small shear crack will not develop in this singular point. Then, as shown by an experiment by POISEL (1984), the tip of this crack will itself become the siege of very high stress concentrations which may in term induce shear failure. There is in fact no available evidence to determine if such a crack growth will stabilize itself. Such a pattern with breakouts and a shear crack developping from its tip can hardly be observed in the field by wellbore calipers or even televiewers. However, HOECK and BROWN, 1980, p. 214, present a study of small scale model where the formation of a single crack from the tip of the breakout leads to the complete collapse of the specimen ; the propagation was unstable. In the oil industry, the existence of such shear cracks can be of prime importance in terms of mud circulation. The stability of the excavation in terms of such cracks must therefore be studied.

FAILURE OF THE ENTIRE WALL

Another problem which is not covered by the analysis by EWY et al. is the case where the entire wall of the excavation is initially broken. This situation is of great importance in the case of deep boreholes for which the in situ stresses are not too anisotropic and far in excess of the strength of the material. The analysis as presented in the paper predicts that the hole would be entirely broken and the excavation entirely enlarged. However, experimental studies on hollow cylinders (e.g. SANTARELLI, 1987) show that this is not necessarily the case and that for certain rocks, failure can start to develop from a very localized cavity and then give birth to oriented breakouts. Furthermore, it has been also observed in the laboratory, but also in deep mines, that not only the depth but also the width of the breakouts increase during rupture propagation (e.g. SANTARELLI and BROWN, 1988).

DEVELOPMENT OF ANOTHER RUPTURE PROPAGATION PROCESS

The numerical scheme used by the authors to simulate the breakout propagation seems appropriate to describe the rupture process observed in sandstone or in quartzite. However, in fine-grained materials, experimental evidence show that another mode of propagation is more prone to happen. This has been shown for instance for coal (KAISER et al., 1987) or for limestone (GUENOT, 1987). Cracks propagate within the material, but the failed material still participates to the overall stability. In that case, the final shape is different, with no pointed tip, and the approach proposed by the authors does not seem to be appropriate.

CONCLUSION

The model presented by EWY and al. (1987) brings an interesting light in the area of wellbore stability which could be of great consequences for the day to day drilling practice of the oil industry. However, four improvements should be considered before industrial application :

- First, the value of the uniaxial compressive strength Q to be used by the model should be clarified. The use of classical linear elasticity is, in that respect, of primary importance and the choice of Q made consequently.

- Second, the stability of the final shape obtained should be tested for the development of a shear fracture from the tip of the breakout.

- Third, the analysis in the case of a deep well where the entire wall is affected by failure should be improved to be able to account for failure localized along a diameter of the well or for failure developping all around the well.

- Last, a similar analysis should be conducted in simulating a softened-broken material.

BIBLIOGRAPHY

EWY R.T., J.M. KEMENY, Z. ZHENG and N.G.W. COCK (1987) - Generation and analysis of stable excavation shapes under high rock stresses. Proc. 6th Congr. Int. Soc. Rock Mech., 2 : 875-881. ROTTERDAM : Balkema.

GUENOT A. (1987) - Contraintes et ruptures autour des forages pétroliers. Proc. 6th Congr. Int. Soc. Rock Mech., 1 : 109-118. ROTTERDAM : Balkema.

HOEK E. and E.T. BROWN (1980) - Underground excavations in rocks, 527 pp. LONDON : Inst. Min. Met.

KAISER P.K., GUENOT A. and N.R. MARGENSTERN (1985) - Deformation of small tunnels : IV Behaviour during failure. Int. J. Rock Mech. Min. Sci. & Geomech. Abstr., 22 : 141-152.

POISEL R. (1984) - Design philosophies of the NATM system bolting In E.T. BROWN and J.A. HUDSON eds, Proc. Symp. Int. Soc. Rock. Mech. on Design and performance of underground excavations : 131-138. LONDON : Brit. Geotech. Soc.

SANTARELLI F.J. (1987) - Theoretical and experimental investigation of the stability of the axisymmetric wellbore. PhD thesis, University of LONDON, 472 pp.

SANTARELLI F.J. and E.T. BROWN (1987) - Performance of deep wellbores in rock with a confining pressure-dependent elastic modulus. Proc. 6th Congr. Int. Soc. Rock Mech., 2 : 1217-1222. ROTTERDAM : Balkema.

SANTARELLI F.J. and E.T. BROWN (1988) - A macroscopic study of the failure of rocks in triaxial and hollow cylinder tests. To be submitted to the Int. J. Rock Mech. Min. Sci. and Geomech. Abstr.

VARDOULAKIS I., J. SULEM and A. GUENOT (1987) - Stability of deep borehole as a bifurcation phenomenon. To be submitted to the Int. J. Rock Mech. Min. Sci. and Geomech. Abstr.

Response by: R.T. Ewy et al. (USA):

I would like to thank Mr. Guenot for his valuable comments. With regard to the stability of the ultimate shapes presented in the paper and the possibility of a switch to another mechanism, such as a shear fracture emanating from near the pointed tip into the surrounding rock, it is true that the regions above and below the tip are subject to high shear stress. As shown in the paper, however, these regions are also subject to high mean stress, which has a stabilizing effect. The stresses in the rock surrounding all of the ultimate shapes were compared to shear failure criteria. For some of the ultimate shapes resulting from non-circular initial shapes, the shear failure criterion was met in a small area above and below the pointed tip, as mentioned in the paper. Usually this was not the case, however. With regard to observations of shear fractures of this type, I would like to caution that results of laboratory tests may be influenced by the proximity of the loaded outside boundary and by the stiffness of the loading system.

With regard to the validity of the ultimate shapes for other possible modes of failure initiation and progression, the ultimate shapes, if achieved, are stable regardless of how they are achieved. Other failure modes, such as shear-dominated failure, may be possible depending on rock type, stress conditions, and possibly other factors. A confining stress, from a fluid pressure in a borehole, for example, would tend to inhibit extensile splitting, as shown by the fracture mechanics model presented in the paper. If the tangential stress were higher than the confined strength of the rock, shear failure might then occur. If future research shows that shear failure will not progress to a stable ultimate shape, then our results indicate that, if failure must occur, it would be beneficial to promote extensile splitting to increase the chances of ultimate stability.

The observation that there is no unique relationship between the shape of a failed borehole and the values of the in situ stresses is an important one. Zheng, Myer & Cook (1987) have analyzed this further and find that the depth of any breakout depends mainly on the initial breakout angle (circumferential extent of hole wall failed). The same initial breakout angle can result from many different combinations of stresses, and this angle also is influenced by the sequence in which holes are drilled and the stresses applied. They conclude that breakout shapes cannot be used to infer the magnitudes of the stresses orthogonal to a borehole. It should also be pointed that when the two stresses are nearly equal, the direction of failure may be determined by any strength anisotropy of the rock mass.

An important implication of the above findings is that perhaps stable design shapes for mining excavations are not highly dependent on the magnitudes of the in situ stresses but mostly dependent on the direction of maximum stress orthogonal to the axis of the opening, as long as it is substantially greater than the minimum orthogonal stress.

Reference:

Zheng, Z., L.R. Myer & N.G.W. Cook 1987. Borehole breakout and stress measurements. Submitted to 29th

U.S. Symposium on Rock Mechanics, University of Minnesota, June, 1988.

J.-C. Roegiers (U.S.A.)
My contribution/question is directed to:

R.T. Ewy, J.M. Kemeny, Z. Zheng and N.G.W. Cook (U.S.A.) General and analysis of stable excavation shapes under high rock stresses.

Since your final 'stable' shape is highly stress path-dependent, how critical is the choice of the volume of rock removed at each step; i.e. does the thickness of the slabs influence the final geometry?

Response by: R.T. Ewy et al. (USA):

As stated in the paper, the thickness of rock removed at each step was varied by factors of 0.5 to 2, and this was found to produce no change in the final shape. Further analysis (Zheng, Myer & Cook 1987) has revealed that the depth of failure is determined mainly by the circumferential extent of the initial failure (see response to question by A. Guenot).

U/G Excavations in Overstressed Rock - Theme IV

Moderator Summary: B.N. Whittaker (U.K.)

ABSTRACT: The major issues and questions which are regarded as the common ground for consideration and discussion on underground openings in overstressed rock are reviewed. Particular contributions to the congress are referred to in the summary that focusses on where future developments are recommended to be directed.

INTRODUCTION

The term 'overstressing' is generally understood to mean levels of rock stress which under normal conditions exceeds their natural strength and results in some form of rock failure. Particular examples are used to contrast the consequences of overstressing: firstly roadways in deep coal mines can be located in overstressed rock by virtue of the weakness of the surrounding strata, with the result that rocks fail around the opening and usually cause a degree of gradual closure. Conversely, an overstressed pillar of coal can exhibit sudden failure behaviour if its slenderness ratio favours such instability although most coal pillars are designed to ensure that if they are overstressed then the consequences will be that of controlled deformation. Deep hard rock mine excavations experience overstressing effects of differing degrees of magnitude, from localised damage to the roof or sidewalls through to sudden closure as can occur with rockbursts. The nature of the rock properties, particularly their strain energy capacity, the geological setting and stress field are important factors.

A common feature of all excavations in overstressed rock is the desire to predict the resulting closure and degree of stability, and the need for any special measures to combat problems associated with the overstressed conditions. A range of predictive tools is available to examine the anticipated stress field around excavations, and these have been extended to consider the influence of different support forms in promoting stability. The formation of excavations in deep overstressed rock conditions still presents considerable uncertainty as to the nature of particular problems especially that of sudden rock failure or even rockburst occurrence. Special efforts have been made to refine the use of models as design tools. Fracture modelling is playing an important role and no doubt will be used more widely in the future in view of the need for designers to have greater insight of the potential zone of fracture. Fracture modelling allows the consequences of overstressing to be better appreciated in terms of excavation design and support.

BASIC PROBLEMS AND ANSWERS SOUGHT

Table 1 presents a summary of the key basic problems and the answers sought. Rock engineering requires as a prerequisite to the provision of satisfactory solutions to such problems, knowledge of the stresses produced within the rock mass. Equally important is the consequence of such a state of stress on the response of the rocks to the presence of the opening.

Table 1. Outstanding basic problems and answers sought in relation to underground openings in overstressed rock

Some basic problems:

a) to predict the stresses induced within the rock mass due to the presence of the opening

b) from the stress analysis to predict accurately the movements experienced around the opening

Answers required to:

1. How does the original stress field effect opening behaviour?
2. How does stratification effect the stress redistribution and opening behaviour?
3. To what extent does the rock mass fail around the opening under given stress fields?
4. What are the plastic-deformational properties of the broken rock and their significance to opening design and performance?
5. What role does time play in influencing stress response of rock, in respect of deformation, failure and rate of movement?

Table 1 shows five areas where answers are required, and each is commented upon below.

1. Thorough identification of the geological factors, such as major defects and discontinuities needs to be fulfilled, otherwise misleading assessments of the stress field can arise. There is a pressing need to properly evaluate the significance of various forms of geological defect and discontinuity on the general stress field.

2. Stratification plays a major role in influencing the pattern of fracture development and accompanying deformation associated with underground excavations. There is a need to examine how stratification can arrest propagation of rock fractures and induce asymmetric deformation effects.

3. Rock failure around underground openings generally develops with time and more needs to be understood regarding its significance.

4. Overstressing implies failure so that in many situations problems arise due to the behaviour of material in various post-failure states.

5. Weathering and stress-strain aging effects are limited with time in the case of rocks. This leads to the question of how do the rock properties change with time? Designing excavations in rock must be done within acceptable limits which are governed by such factors as shape, size and support. An important question that arises here therefore is of what significance are these factors in overstressed rock conditions?

MODERATOR'S SUMMARY OF PRINCIPAL AREAS ADDRESSED BY VARIOUS CONTRIBUTORS

A wide range of problems associated with the design and stability of underground excavations has been reported upon by the various contributors to the Congress, and the following summary brings to attention points regarded as of special importance to the major issues of Theme IV. The contributors referred to are all authors whose papers appear in Volume 2 of the Proceedings of the ISRM Sixth International Congress on Rock Mechanics. For detailed consideration of the contributions of the authors, readers are referred therefore to Volume 2 of the Proceedings. The principal areas addressed by various contributors to Theme IV can be subdivided into four main sections and each is commented as follows. Many authors have made outstanding contributions to the Congress, and only a few are selected here for the purpose of identifying areas of major current concern in rock mechanics as applied to excavations in overstressed rocks.

1 GROUND CONTROL AND DESIGN METHODS FOR IMPROVED UNDERSTANDING AND STABILITY BEHAVIOUR OF EXCAVATIONS

Bawden and Milne in Canada effectively demonstrate the application of modelling (numerical and empirical) to mine design for improved stability. Also of importance has been the contribution of Duncan Fama and Wardle of Australia in using numerical modelling to extend knowledge on coal pillar behaviour.

Del Greco and colleagues from Italy have shown how excavation roof stability can be assessed in respect of pressure arching.

Vervoort and colleagues in Belgium have successfully demonstrated the useful role of finite element modelling to predicting roof conditions in mining excavations. Cunha of Portugal also effectively shows how models can be used as an important design tool in tunnel stability assessments.

Ewy and colleagues in the USA have made a significant contribution in their successful simulation of formation of stable shapes through progressive failure.

An important contribution has been made by Herget and Mackintosh of Canada who demonstrate the role of instrumentation to assess convergence and stress for improved design purposes.

In this section due acknowledgement must be made of Maury's report on the stability of single galleries - especially in respect of selecting the appropriate theory to explain stability.

2 SUPPORT ASSOCIATED ASPECTS OF EXCAVATIONS IN OVER-STRESSED CONDITIONS

Shaft pillar design and subsequent extraction has occupied the attention of engineers world-wide for many years. It is timely to have the type of contribution made by Budavari and Croeser from South Africa on shaft pillar extraction, and their work demonstrates the feasibility of more efficient extraction by using the principles of rock mechanics. McKinnon's paper, also from South Africa, makes a complementary contribution on shaft pillar aspects.

Support still continues to be a major area of attention of engineers world-wide. Tan from China makes an interesting contribution of NATM applied in difficult conditions. Rock reinforcement continues to be an important support function and noteworthy papers have been given by Thompson from Australia, Aydan and colleagues from Japan and Ballivy from Canada.

Backfill has been reported by Jager and colleagues from South Africa as being highly effective in reducing the stresses acting on pillars thereby promoting increased safety.

Interesting contributions have also been made on ground support from Van Sint Jan of Chile, Barroso and Lamas of Portugal, Fine of France and Korini of Albania, and Yuen and colleagues from Canada in respect of rock-support interaction.

3 EXCAVATION BEHAVIOUR, ASSESSMENT AND CONTROL WITH RESPECT TO OVER-STRESSED CONDITIONS

The work of Barlow and Kaiser in Canada has clearly demonstrated the importance of tunnel convergence measurements interpretation for improved assessment of excavation and support aspects of tunnel construction operations. Changing stress field conditions can be assessed in respect of the effect on tunnel behaviour as shown by Esterhuizen of South Africa.

Tunnel behaviour is characterised by time dependent response curves, and Gill and Ladanyi of Canada make an important contribution in showing how the time element can be taken into account in predictions of tunnel behaviour. Time effects on tunnel behaviour is also effectively examined by Minh and Rousset of France. Tunnel stability continues to receive special attention as reported by Sharma and colleagues in India.

The development of fractures in relation to mine excavations has been comprehensively investigated by Ortlepp and Moore (South Africa and UK) who report important findings on the rate, nature and general character of associated movements. Yield zone interpretation is examined by Lee and Kim from Korea who make important comments on relaxation patterns around tunnels. Ishijima and colleagues from Japan draw attention to micro-seismic developments in assisting fracture interpretation phenomena around excavations.

4 ROCK CHARACTERISATION, BEHAVIOUR AND STRESS CONTROL IN RELATION TO UNDERGROUND EXCAVATIONS

Rock mass characterisation by Barla and Scavia from Italy draws our attention to the important role which can be played by geostatistical analysis in interpreting the significance of rock discontinuities to rock behaviour in mine excavations.

An important feature of the work of Bandis and colleagues from Norway, is their explanation of stress and fracture behaviour of cavities in poor quality ground. The valuable progress made by them no doubt justifies further in depth examination for other rock conditions.

Gale and colleagues in Australia have demonstrated the value of stress control for improved mine design especially in respect of optimisation of mining design and extraction orientation to mitigate the effects of stress.

Rock burst phenomena continue to be of major concern to engineers and significant progress has been made by Udd and Hedley of Canada with respect to understanding rock burst occurrence and control. Equally important has been the work of Sugawara and colleagues in Japan in explaining the mechanisms and control aspects of coal outbursts.

The interaction between geology and mining is emphasised by Kersten and Greef of South Africa and their work clearly shows how a better understanding becomes possible regarding stability in shaft excavation work in deep mining conditions. Their interpretation and principal observations on interaction between geology and mining deserve special considerations in application to allied problems of stability in other over-stressed underground excavation situations. Our principal speaker Horst Wagner of South Africa also emphasises importance here and draws our attention to the contribution which excavation shape can make in promoting improved stability in overstressed conditions.

In conclusion I return to the list of "answers required to" as shown in Table 1 of this report. These five points deserve your special attention in the future.

APPENDIX

REVIEW OF STABILITY PERTAINING TO TUNNELS IN SOFT ROCK CONDITIONS

This Appendix is included in the Moderator's Report to demonstrate how some of the problems and questions mentioned are being researched at a rock mechanics research centre in the UK.

Problem Definition

Rock Mechanics, like any other branch of mechanics, deals with motion, and as a result it can be broken down into two fundamental elements:

(i) the difference in a medium property between two points, causing an inequality in that medium, known as the 'driving potential', and

(ii) the movement of the medium from the higher potential to the lower potential, until a state of equilibrium returns within it.

In rock mechanics, the driving potential is the state of stress in the ground that is produced as a result of forming an excavation, with stress equilibrium being returned due to rock movement into that excavation, which may or may not be externally supported.

The problem of predicting tunnel behaviour can be classified into two distinct sections:

a) to predict the stresses induced within the rock mass due to the presence of the opening, and deciding where rock failure takes place,

b) to predict accurately the movements experienced around the opening as a result of the stress analysis.

There are several factors which effect tunnel stability within the soft rocks of the Coal Measures and they are:

(i) the initial state of stress in the ground prior to excavation,

(ii) the strength of the rock mass in its intact or pre-failure state resulting in the subsequent formation of a zone of broken rock (yield zone) as a result of its properties and the strength of the rock within the yield zone, referred to as its post-failure behaviour,

(iii) the strata sequence in which the tunnel is sited, and the relative strengths of the rock types within it,

(iv) the tunnel shape and size along with the type and strength characteristic of the support system that is used within the tunnel,

(v) the elastic (pre-failure) and plastic (post-failure) movements that are caused as a result of the stress change within the rock mass. Elastic movement is a consequence of Poisson's effect, and is relatively insignificant in comparison with the plastic movement of the broken rock, which is a result of its volume increase, caused by the change of state (Whittaker and Frith 1987).

Realistically, it is desirable to consider all of these factors in one design method. However, if this is to be achieved, much ground work needs to be done, in the establishing of the behavioural laws for each of the factors. Very often, theoretical laws have been found to be inadequate in describing such behaviour and empirical laws based on measurements need to be established. It is on this basis that the prediction model developed by Nottingham University is heavily biased.

A further aspect of the development of design models is in using actual tunnel data for both convergence and yield zone size to assess their accuracy. These data only become available from many hours of underground monitoring and to this end Nottingham University have been actively involved at several collieries around the UK (Whittaker et al 1981).

Present State-of-the-Art at Nottingham University

Early work was aimed at establishing procedures and standards for the post-failure testing of Coal Measures rocks from which the data to be used in prediction work is obtained. Using such standards, a whole variety of rock types were tested, with special emphasis on their pre- and post-failure strengths and their volumetric strain characteristics. From the results of these tests, behavioural laws were established for rock strength and the plastic movements associated wtih broken rock (Whittaker et al 1985). These two laws have been instrumented in the success of the design model that will now be referred to.

At present the design model, developed by Nottingham University is based on the following assumptions:

(i) it applies to a circular tunnel of any given size,

(ii) initially hydrostatic stress conditions are considered,

(iii) the support installed in the tunnel exerts a constant pressure to the rock mass, regardless of its deformation,

(iv) stratification is accounted for by using computerised storage of such information, splitting up the rock mass surrounding the tunnel into small, mutually independent elements, and assigning rock properties to such elements according to which rock type they are situated in,

(v) rock strength is characterised by the equation

$\sigma_1 = A\sigma_3^B + C$, where A, B and C are unique

to a rock type and failure state; this equation has been established from the rock testing program, previously mentioned,

(vi) the volumetric strain of the broken rock is given by the equation:

$V_e = ae^{-\sigma_3/b} + c$ where a, b, and c are

unique to a rock type. This equation represents the maximum volumetric strains which are likely to occur and therefore represents the worst possible case. This is desirable in terms of initial design, (Whittaker et al 1984).

Stress analysis is performed using the differential equation for cylindrical symmetry under equilibrium conditions. This equation is a simplistic stress equation and it requires a circular tunnel with the initial stress conditions being hydrostatic. Hence, assumptions (i) and (ii) are necessary. Assumption (iii) is a consequence of the need for a boundary condition to facilitate a unique solution to the resultant stress equation. Assumptions (iv) to (vi) all require the use of a micro-computer in that an analytical solution cannot be found with such equations. The technique is therefore numerical rather than analytical and has been termed as the Independent Element Method (IEM).

A common problem to all such prediction methods is in rock mass classification. It is commonly accepted that the intact rock strengths, found from laboratory testing are orders of magnitude higher than those found in situ, as a result of the selective sampling of a fractured rock mass. Various methods have been proposed to alter the strength values to that of the in situ rock mass. However, most of them require detailed geological information, which is not always available. Therefore some other approach is required, and a possible solution has been found, by assuming that the intact rock mass is also characterised by the residual or post-failure rock strength. This once again represents the worst possible case, but from an evaluation made between measured and predicted tunnel convergences the correlations were found to be very satisfactory (see Figure 1) (Kapusniak et al 1984). Indirectly, this justifies the assumption that has been made, and it is now an accepted practice in tunnel prediction work at Nottingham University.

What has been established thus far is a relatively simple design model based on sound rock behavioural laws established from rock testing measurements, that can take account of the stratified nature of the rock mass. The validity of this approach has been justified by comparing predicted and measured convergences for a circular tunnel. However, the model is limited to circular roadways in hydrostatic stress conditions with the effect of the support being represented as a single support pressure. It is in these three areas that future developments must take place, given that the fundamental basis has now been established.

Future Developments

The support that is placed in the tunnel is an integral part of the rock mechanics problem and deserves full consideration in such analysis. Essentially, the support of whatever type will have its own stress-deformation characteristics which should be included in the design model as there are certain finite limits of deformation which the support can safely accommodate. Deformation outside of this range are considered to be detrimental to tunnel stability and a different support type may be required in such situations. At present, research is underway to accurately obtain these support characteristics by scale testing and to incorporate them into the design model by storing them in computer memory and recalling them as necessary. This will allow the accurate prediction of tunnel deformation when the support loses some or all of its strength after failure, an occurrence which is fairly common in deep Coal Measures strata.

The most significant advancement that is required is in being able to consider different shaped tunnels under a variety of initial stress conditions. This implies departing from the present stress equations and using some other technique. The idea that is presently being formulated is to utilise the finite element method (FEM) of analysis, but adapt it specifically to the rock strength equation that has already been defined. This will allow acceptable stress analysis of different shaped tunnels under various initial stress conditions to be performed. However, the FEM is limited when considering only elastic deformation, which is known to be insignificant where a large yield zone is formed. Therefore, the analysis of movements would still be performed using the volumetric strain law, that has already been established as valid, but based on the improved method of stress analysis. This 'hybrid' method will represent a very powerful design tool indeed.

The inclusion of varying tunnel shapes and stress conditions is especially significant. The circular tunnel is still relatively rare in the UK coal mines and hydrostatic stress conditions are considered only as a generalised rule.

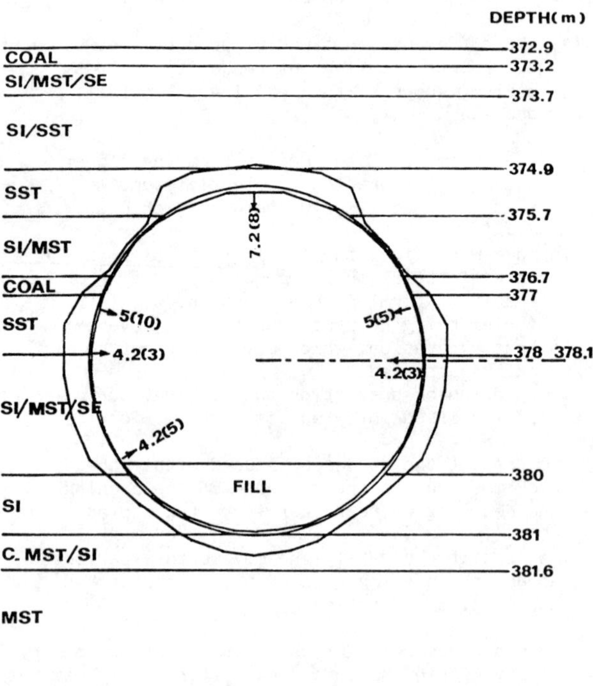

Figure 1. Radial deformation and yield zone calculation for a circular tunnel in stratified rocks

Recently, anisotropic stress conditions have been seen to be responsible for tunnel failure, these conditions being a consequence of the presence of geological faulting and disturbances. Therefore, it is essential that these advances are made as soon as is possible so that remedial support measures can be designed with confidence in the results.

Concluding Remarks

The research work at Nottingham University has established a realistic basis for rock mechanics problem solving in terms of rock behavioural laws and computing techniques. At present the design model is limited, due to the stress equations that it uses, but is has still given an excellent correlation between predicted and measured convergence data. The potential for future developments is enormous, which if realised, will result in a powerful tunnel design technique.

REFERENCES

Whittaker, B N & R C Frith 1987. Design of support systems for mining tunnels in Carboniferous rock conditions. 6th Int Conf on Ground Control, Morgantown, W Virginia, USA, pp 258-270.

Whittaker, B N; M J White & C J Bonsall 1981. Design and stability of tunnels in Carboniferous rocks. Proc Int Symp on Weak Rock, Tokyo, pp 921-926.

Whittaker, B N; M R Carter; S S Kapusniak & A J Townley 1985. Design and selection of support systems in mine roadways and tunnels with reference to UK coalfields. Int Bureau of Strata Mechanics, Sofia, Bulgaria.

Whittaker, B N; C J Bonsall; D G Branch & S S Kapusniak 1984. Stability aspects of major coal mining tunnel projects. Proc ISRM Symp, Cambridge.

Kapusniak, S S; N Riggott & S F Smith 1984. Studies of mine tunnel stability in UK Carboniferous rock conditions. 2nd Int Conf Stability in Underground Mining, AIMMPE, Lexington, Kentucky, USA.

CLOSING CEREMONY - SEPTEMBER 3, 1987

The technical sessions closed on Thursday, September 3, 1987. The chairman of the Congress, Dr. G. Herget, asked the Honorary Chairman, Prof. Dr. B. Ladanyi, to thank on behalf of the Canadian organizing committee, the chairmen, speakers, moderators, and panelists for their excellent contributions to the technical sessions. Professor Ladanyi thanked all speakers and presented all senior officials with one eskimo carving.

Following Prof. Ladanyi's words of thanks, Prof. Dr. Wittke, West Germany was asked to present his closing address with his personal views on the contributions made in rock mechanics since the founding of the International Society for Rock Mechanics 25 years ago and the needs which are to be addressed now.

Professor E.T. Brown, president of the International Society for Rock Mechanics, then presented the following closing remarks:

"Mr. Chairman, Ladies and Gentlemen!

At the opening ceremony, on behalf of the International Society and the delegates, I thanked our Canadian colleagues for the warm welcome that they had given us and for the preparations that they had made for the Congress. Now that we have come to the end of the technical sessions, I have the pleasure of thanking them, once again on your behalf, for their efforts in organizing what has proven to be an excellent Congress.

As our distinguished Past-President, Professor Wittke has told us, our subject has made considerable advances in the 25 years of the existence of our International Society altough, as ever, much still remains to be done. During the conference I have spoken to a number of "old hands" in the business - that is to people of my own generation and some who are just a little bit older - and we have marvelled at the capabilities and achievements of many of the younger people who have followed us into the subject, including a number of the high calibre candidates for the Society's Manuel Rocha Medal. Rock mechanics is becoming much more of a high technology subject than it has been hitherto. With this, and the talented young people in it - particularly here in North America - I am confident that we can look forward to learning of further exciting advances when we meet for our next Congress in Aachen in 1991. I would urge you all to make a special effort to join us there.

As I am sure we all realize, a Congress like this just doesn't happen - it has to be planned and then organized in very great detail by a dedicated group of hard-working enthusiasts, exemplified by the members of our Organizing Committee and their many helpers. We are particularly grateful to their Godfather - Professor Branko Ladanyi - who initiated the whole thing and has overseen their activities in a manner which might possibly be described as benign, but I'm not entirely sure about that. Several members of the Organizing Committee have made special contributions to the organization of the technical aspects of the Congress. Professor Norbert Morgenstern was the brains behind the Technical Program which has been of a high standard. Everyone to whom I have spoken on the subject has agreed that the specialty workshops were a particular success.

Dr. Will Bawden was responsible for the excellent Trade Exhibition which was one of the best that I have seen at a rock mechanics meeting. In this context Will bears particular personal responsibility for the bagpiper who accompanied the Minister around when he opened the exhibition on Monday.

Will was also given responsibilitiy for the detailed organization of the Technical Sessions which involved him in giving instruction to session chairmen and speakers and seeing that they were carried out. Someone made an inspired choice in allocating this task to Will - he was just the man for the job.

Professor Denis Gill has been the Congress Co-chairman with special responsibility for the many details of what were described as local functions but which, from my personal observations, included just about everything. Denis has been a tower of strength and we are all most grateful to him.

The whole thing was held together by the man with two right hands, the Congress Chairman, Dr. Gerhard Herget. Gerhard's labours, and those of his Committee, began some years ago and are far from over yet. Those of us, like the Secretary-General and myself, who have been in close contact with Gerhard over an extended period of time, know only too well the great personal effort that he has put into ensuring that this Congress was a great success.

Branko, Gerhard, and the other members of the

Organizing Committee, we salute and thank you.

Ours is a truly international society having three official languages and several unofficial ones, including my own idiosyncratic Australian English. Accordingly, we rely greatly on the quality of the simultaneous translation for effective communication during our technical sessions. The contribution made by the translators in so skilfully carrying out their demanding task is often overlooked, but we shall not do so on this occasion. Ladies and gentlemen, we thank you very much indeed.

This brings this Session and the technical proceedings of this Congress to a close. We meet next for ISRM sponsored symposia in Minneapolis, USA, in June 1988, and in Madrid, Spain in September, 1988. And as I have said, we meet for our next Congress in Aachen, West Germany, in 4 years time.

Ladies and gentlemen, this Session is closed. Thank you!"

The 7th International Congress on Rock Mechanics will be held in Aachen, West Germany, from September 16-20, 1991. Standing in front of the City of Aachen display are from left to right: Dr. W. Bawden, Prof. Dr. D. Natau, The Honourable R. Savoie, Province of Quebec, Mr. M. Dillo, Mr. J. Gaydos, Dr. G. Herget.

Even after a congress has ended, discussions continue. Here a delegate listens to a discourse by Prof. Kovari, Switzerland.

Silver Jubilee Banquet

The Silver Jubilee Banquet commemorated the 25th Anniversary of the founding of the International Society for Rock Mechanics. It was attended by 575 persons and required Le Grand Salon, the Marquette and Jolliet rooms of the Queen Elizabeth Hotel. The evening commenced with the serving of cocktails in a side room. Then the delegates proceeded to the dinner tables. After the head table had been seated, the Congress Chairman, Dr. G. Herget, welcomed the delegates in three languages and wished them an enjoyable dinner. The menu is reproduced below and the Queen Elizabeth Hotel served all meals at the table in quick succession under the able guidance of Mr. J. Druda, director of banquets for the Queen Elizabeth Hotel.

CROSS CANADA MENU

*

Arctic char and Pacific salmon bouquet

*

Ontario small game consommé

*

Filet of beef from the Prairies with bison sauce

Mixed vegetables from the Okanagan valley

Fiddle heads and small potatoes of the Maritimes

*

Caged apple temptation sweetened with Quebec maple syrup

*

Coffee

*

Wine

Cuvée Spéciale
Le Reine Elizabeth

Digestive * To your choice

While the coffee and liqueurs were being served the Chairman, Dr. G. Herget introduced the head table guests:

Mr. P. Michaud, Montreal, Canada, Executive Director Canadian Institute of Mining and Metallurgy
Mr. N. Grossmann, Lisbon, Portugal, Secretary-General ISRM
Prof. Dr. Langer, Hannover, West Germany, Past-President International Society of Engineering Geologists
Prof. Bello M., Mexico City, Mexico, VP-ISRM North America
Dr. M.D. Everell, Ottawa, Canada, Assist. Dep. Minister EMR
Mrs. J. Udd, Ottawa, Canada
Prof. Dr. Tan T.-K., Beijing, China, VP-ISRM for Asia
Mrs. L. Gill, Montreal, Canada
Prof. Dr. B. Ladanyi, Montreal, Canada, Honorary Congress Chairman, Past President Canadian Rock Mechanics Association
Mrs. U. Herget, Ottawa, Canada
Prof. Dr. E.T. Brown, London, United Kingdom, President International Society for Rock Mechanics
Dr. G. Herget, Ottawa, Canada, General Congress Chairman
Mrs. N. Ladanyi, Montreal, Canada
Prof. Dr. D.E. Gill, Montreal, Canada, Congress Co-Chairman
Mrs. Tan T.-K., Beijing, China
Dr. J. Udd, President Canadian Rock Mechanics Association
Mrs. A. Reid (Franklin), Orangeville, Canada
Prof. Dr. B. Bamford, Melbourne, Australia, VP-ISRM Oceania and Australia
Prof. Dr. Tinoco, Caracas, Venezuela, VP-ISRM, South America
Prof. Dr. Wagner, Johannesburg, South Africa, VP-ISRM, South Africa
Dr. J. Franklin, Orangeville, Canada, President elect ISRM
Dr. M. Bozozuk, Ottawa, Canada, President Canadian Geotechnical Society

Following this Dr. G. Herget thanked all his members of the organizing committee who had given so freely of their time for three years to prepare this Congress. All these members had taken on extra duties during these years while they still had to perform their normal duties. Every organizing committee member received a round of applause when asked to come to the head table to receive their awards:

Dr. B. Ladanyi, Honorary Chairman and Fund Raising (Eskimo Carving);
Dr. D.E. Gill, Congress Co-Chairman and local functions (Plaque);
Dr. N.R. Morgenstern, Technical Program (Plaque);
Dr. J.A. Franklin, Publicity (Plaque);
Mr. T. Carmichael, Finance (Plaque);
Dr. W.F. Bawden, Exhibits (Plaque);
Mr. J. Gaydos, CIM, Tradeshow Manager (Eskimo Carving);
Dr. A.T. Jakubick, Technical Tours (Plaque);
Mr. L. Geller, Translations (Plaque);
Mrs. J. Robertson, Spouses Program (Plaque);
Mrs. D. Grégoire, Registration, CIM (Eskimo Carving);

Dr. J. Bourbonnais, Hospitality (Plaque);
Mr. J. Nantel, Fund Raising (Plaque);
Dr. S. Vongpaisal, Assistant Editor (Plaque).

The Chairman stressed that these people all represent only the leaders of groups of people who had to work hard especially in the last few months, to prepare the Congress.

Having presented and thanked his organizing committee, the Chairman turned the meeting over to the outgoing president of the International Society for Rock Mechanics, Prof. Dr. E.T. Brown:

Thank you Gerhard. Good evening ladies and gentlemen!

Well, here we are approaching the end of this Congress and of our time together in Montreal and here I am on my feet yet again, but for the last time as President of the International Society. No doubt it is a time of great relief for us all - for our Canadian colleagues who organized the Congress and who will be relieved that it is all over, for me because I can now pass my responsibilities over to someone else, and for all of you who will all be relieved that in a few minutes time you won't have to listen to people talking any more, particularly me.

Before I finally shut up and sit down, I want to add the thanks of the visitors to Montreal to those which Gerhard has given to the members of the Organizing Committee. Especially this evening I wish to thank those people who organized the programme for social events and the accompanying person's programme - notably Mrs. Joan Robertson and Dr. Jacques Bourbonnais. We have enjoyed your hospitality, we have enjoyed the special events that you have arranged for us including the excursions, receptions and the performance by Les Ballets Jazz de Montreal on Monday evening - especially the pas de deux performed by Gerhard Herget and Jacques Bourdonnais - and we are certainly enjoying ourselves this evening. We have particularly enjoyed the opportunity to meet in Montreal. It is a most attractive city and was a wonderful choice for a Congress location.

At the closing ceremony this afternoon I thanked the Honorary Chairman, Professor Branko Ladanyi, and the Chairman, Dr. Gerhard Herget, for their central roles in the organisation of this Congress. Under these less formal circumstances, I wish to thank them once again on behalf of all of you. When I came into the hall this evening, Gerhard was here with his wife working away in his customary methodical and inconspicuous manner ensuring that everything was in readiness for the banquet. And I was pleased to note that he has now sorted out his left hand from his right.

As you know, in 1987 we are celebrating the Silver Jubilee of the Society and tonight's dinner is the Silver Jubilee Banquet. To mark that important event we have asked all members of the Past Boards of the Society who were able to do so, to join us for the Congress and to sit together, with their wives, at a special table this evening. Although they are a distinguished body of men who have made vitally important contributions to the life and well-being of our Society, I must admit that they do not represent an especially pretty sight. Nevertheless, I am going to ask them to stand for just a few seconds so that we can welcome them and acknowledge their special contributions to our subject and our Society. Gentlemen please.

For the past 20 years, the Society has been administered by a Secretariat located at the LNEC in Lisbon, Portugal. The Society has been particularly fortunate to have been served by a number of distinguished Secretaries-General, two of whom - Fernando Mello Mendes and Ricardo Oliviera - were among the crew whom you have just seen. For the past four years, the Secretary General has been Nuno Grossmann. Those of you who have attended our meetings during this time will be accustomed to seeing this stocky figure bustling about, organizing all manner of things and ensuring that all of us, especially me, act correctly and in the best interests of the Society. As your President, I think that I know better than almost any of you just what devoted, hard-working and loyal service he has given to the Society. Nuno, I thank you personally and on behalf of the Society.

Dr. J. Franklin, newly elected president of the International Society for Rock Mechanics, thanks the past vice-president of North America, Prof. Bello-Maldonado, for his distinguished service to the society. In the foreground Mrs. U. Herget and Prof. Dr. E.T. Brown.

For these twenty years the other important member of the Secretariat has been the Executive Secretary, Mrs. Maria de Lurdes Eusébio. If the Secretaries General have kept the Presidents and Boards in line, then in her quiet and charming way, Maria de Lurdes has kept the Secretaries General in line. One of the greatest pleasures in being an officer of this Society is having the opportunity to work with Maria de Lurdes. She is always tactful, charming, elegant and most efficient. We are all greatly attached to her and simply couldn't get along without her. To mark her 20 years of outstanding service to the Society and as a small token of our abounding esteem, I would like to make a presentation to her on the Society's behalf.

The work of the Society also depends heavily on the members of the Board - which includes the Vice-Presidents for each of our six regions. I am personally most grateful to the members of the outgoing Board for the support and advice that they have freely given me in the past four years. I am especially grateful to our First Vice-President, Dr. Sten Bjurström of Sweden. Sten has had to return to Stockholm to negotiate some highly important and valuable business and, regrettably, can't be with us tonight.

Some of you will have noticed that during our week here in Montreal there has been a certain amount of, shall we say, discussion, leading up to the election of the new Board which took place at the Council meeting held earlier this evening. As my final act as your President, it is my great pleasure to introduce to you the members of the new Board which will guide the affairs of the Society from now until the end of the 7th Congress in Aachen in September, 1991. They are:

V-P South America - Dr. Carlos Dinis da Gama (Brazil)
V-P North America - Dr. Jim Coulson (USA)
V-P Europe - Mr. Marc Panet (France)
V-P Australasia - Dr. Ian Johnston (Australia)
V-P Asia - Mr. S.L. Mokhashi (India) (who is unfortunately not present)
V-P Africa - Dr. Oscar Steffen (South Africa)

Finally, and most importantly, I should like to introduce, congratulate, and hand over the affairs of the Society to my successor as your President, Dr. John Franklin, of Canada.

Following this presentation the incoming president of the International Society for Rock Mechanics, Dr. J.A. Franklin, was asked to address the delegates.

Dr. J. Franklin thanks the past vice-president of Australasia, Prof. Dr. W.M. Bamford, for his distinguished service to the society. In the foreground Dr. G. Herget.

"Dear Colleagues, retiring members of the ISRM Board, incoming Board members, ladies and gentlemen!

We are all very much looking forward to being entertained by the Sortileges folk dancing ensemble now waiting in the wings, and to retiring after what has been a most stimulating yet, as always, tiring meeting. Therefore, I will take very little more of your time.

Rock mechanics is full of mysteries, not least of which tonight is the question "Who is Anne Reid?" - the lady introduced to you earlier as our head table guest. That question is easily answered if I introduce her to you as my dear wife.

The elections for the ISRM presidency, as you may know, were closely contested between myself and two well-known and most distinguished colleagues. It is particularly an honour to have been selected as your representative, and I pledge myself to serve you and the ISRM to the very best of my ability during my coming four-year-term of my office. We have already held our first Board meeting, and have various plans afoot to better serve our member nations during the second twenty-five years in our history. Details of these plans, however, will wait until another day.

In closing, it is my great pleasure to present to each of the outgoing ISRM Board a certificate, as a token of our most sincere appreciation of their services to the Society".

(Dr. John Franklin then read the text of the certificates to the assembled delegates).

With this presentation by Dr. Franklin, the Chairman, Dr. G. Herget, ended the official proceedings of the 6th ISRM Congress. He thanked the delegates for coming to Montreal and invited them to enjoy the performance of "Les Sortilèges", Canada's National Folk Ensemble. Their program is reproduced below:

Dr. J. Franklin thanks the past vice-president of Africa, Dr. H. Wagner, for his distinguished service to the society. In the foreground from left to right, Dr. Herget and Mrs. N. Ladanyi.

CANADA'S NATIONAL FOLK ENSEMBLE

LES SORTILÈGES

*

Program

* Castor P'tits Chars: Quebec
* Brandy Frotté: Quebec
* Danse de la jarretière: Quebec
* Gigue pour ti-Jean: Quebec
* Jarabe Tapatio: Mexico
* Lindjo: Yugoslavia
* Upton on severn: England
* U Balle du Criviello: Italy
* Rapper Sword dance: England
* Highland Sword dance: Scotland
* Hrechaniki: Ukraine
* Quagigue: England-Ireland-U.S.A.-Quebec
* 6e partie du quadrille: Quebec
* Danse du capitaine: Quebec
* Gigue des balais: Quebec

Canada's National Folk Ensemble, Les Sortilèges.

After the performance the members of the ensemble asked the delegates to join them and soon the dance floor in front of the stage was filled to capacity with a happily dancing crowd. This concluded the 6th International Congress for Rock Mechanics and most of the delegates departed at about 23:00.

Workshops

A pre-congress workshop was organized on mining induced seismicity; Six additional workshops were held during the congress. The topics covered: Swelling rock; Constitutive laws for salt rock; Numerical methods as a practical tool; Failure mechanisms around underground openings; Rock cuttability and drillability; Rock testing and testing standards

FRED LEIGHTON MEMORIAL WORKSHOP ON MINING INDUCED SEISMICITY

Chairman: R.P. Young (Canada)

This workshop was first suggested by the rockburst sub-committee of the CIM and the organizing committee of the 6th ISRM Congress. Stan Bharti of Falconbridge and Will Bawden of Noranda approached me with the request to organize a one day workshop as part of the 6th ISRM International Rock Mechanics Congress, held in Montreal August 30 - September 3, 1987. We thought the timing of the meeting was appropriate because it was five years since the 1st International Symposium on Rockbursts and Seismicity in Mines, held in Johannesburg, South Africa. In addition, we felt that this type of workshop would generate ideas which could be developed further at the 6th ISRM Congress and the 2nd International Symposium on Rockbursts and Seismicity in Mines, to be held at the University of Minnesota, June 1988.

I decided to dedicate the workshop to Fred Leighton, a scientist and former head of the US Bureau of Mines Microseismic Applications Group at the Denver Research Center, Colorado, who died in 1986. Fred was born and educated in Colorado and attended the Colorado School of Mines. For over 20 years, he made significant contributions in the application of AE/MS techniques to problems in mine health and safety. Fred is probably best known for the several volumes on geotechnical applications of AE/MS techniques, which he co-edited with Dr. Reg Hardy of the Pennsylvania State University. The last scientific visit made by Fred was to my laboratory at Queen's University in the summer of 1985, as part of a Natural Sciences and Engineering Research Council of Canada scientific review panel. We discussed my research programme to investigate mining induced seismic phenomena, which was then at the design stage and I remain indebted to him for his valuable comments and advice.

The Fred Leighton Memorial Workshop was divided into three consecutive sessions. The objective of the morning session was to focus in on selected topics within the field of mining induced seismicity, to probe the extent of knowledge in those areas and to highlight the direction for further work. Papers were presented by four authors: Dr. W. Blake on microseismic instrumentation; Dr. J. Niewiadomski on source location techniques; Dr. A. McGarr on seismic processing and interpretation techniques; and myself on applications of geotomographic imaging in the study of mining induced seismicity. The lunchtime poster session allowed mining companies, research organizations and universities who are carrying out work in the field, the opportunity to present their approach, observations and interpretations in the form of poster displays. It was very encouraging to see that 17 poster presentations were made. The objective of the afternoon session was to provide an overview of selected international/national research programmes in the field of mining induced seismicity. Papers were presented by four authors: Dr. H.R. Hardy gave an international review of AE/MS techniques, whilst Drs. D. Hedley, S. Spottiswoode and B. Brady provided national perspectives for Canada, South Africa and the USA respectively. The workshop concluded with a discussion by a panel of experts from mining companies, national research organizations and universities. The panel, chaired by Mr. D. Ortlepp, addressed key questions and problems in the field of mining induced seismicity.

I was both delighted and encouraged by the positive response the workshop received. This was highlighted by the fact that there were over 125 delegates from 5 continents, who attended the workshop and dinner. The proceedings (distributed at the workshop and now available from Queen's University), over 300 pages from 25 papers, reflect the current awareness of and necessity to understand phenomena associated with mining-induced seismicity. As the World's mines become deeper and mining induced seismicity increases, the greater potential for rockbursts will necessitate a better understanding of the complex interactions between mine design, rock mass physical properties, local and regional stresses, structural geology and seismicity. It was unfortunate that Professor S.J. Gibowicz of the Polish Academy of Sciences, a person who I and fellow scientists consider to be the guru in the field of mining induced seismicity, was not able to be with us at the meeting. Professor Gibowicz is editing a special issue of Pure and Applied Geophysics on "Seismicity in Mines". He has expressed a keen interest to include an edited version of the proceedings in his special issue and several edited papers from the proceedings will appear in the Journal of Pure and Applied Geophysics as part of this special issue in 1988/89.

Finally, I should like to thank the many individuals who helped me make the workshop a reality, including the authors, Dr. H. Brehaut for his talk at the workshop dinner, the members of the expert panel and the delegates.

PAPERS PRESENTED AT THE WORKSHOP:

TOPICS IN MINING INDUCED SEISMICITY

Microseismic instrumentation
W. Blake

Source location techniques for seismic activity in mines
J. Niewiadomski

Analysis of exceptionally large tremors in two gold mining districts of South Africa
A. McGarr, J. Bicknell, E. Sembera and R.W.E. Green

Geotomographic imaging in the study of mining induced seismicity
R.P. Young, D.A. Hutchins, J. McGaughey, J. Towers, D. Jansen and M. Bostock

INTERNATIONAL AND NATIONAL PERSPECTIVES ON MICROSEISMIC/ROCKBURST RESEARCH

A review of international research relative to the geotechnical field application of acoustic emission/microseismic techniques
H.R. Hardy, Jr.

The Canada-Ontario-industry rockburst project
D.G.F. Hedley and J.E. Udd

Perspectives on seismic and rockburst research in the South African gold mining industry: 1983 to 1987
S.M. Spottiswoode

Microseismic/rockburst research at the Galena mine, USA: a perspective
B.T. Brady

POSTER PAPERS: NATIONAL RESEARCH ORGANIZATIONS

Seismic monitoring systems being used in the Canada/Ontario/industry rockburst project
D. Hanson, P. Rochon and T. Semadeni

Mine tremor studies at a South African gold mine
J.M. Churcher, S.M. Spottiswoode and D. Brawn

Stress control engineering for rockburst control
F.M. Jenkins

POSTER PAPERS: MINING INDUSTRY

Rock mechanics at Campbell Red Lake Mine
T. Makuch

Rockburst research at Falconbridge Limited
G.R. Davidge

Research related to mining induced seismicity at INCO Limited
P. McDonald and L. Cochrane

Mining induced seismicity: monitoring and interpretation
W.J.F. Quesnel and R. Hong

The CANMET-Noranda seismic monitoring system using fibre optic signal transmission
V. Labuc and O.A. Momoh

The microseismic monitoring system at Brunswick Mining and Smelting Corporation Ltd #12 mine, Bathurst, New Brunswick
B. Kristof, T. MacDonald and G. Landry

MA/AE research at PCS Mining operations in Saskatchewan, Canada
P. Mottahed and J.B. Vance

Coal-mining-induced seismicity in central Utah, USA: Seismic events due to pure shear and shear-implosional (?) failure
I. Wong

POSTER PAPERS: UNIVERSITIES

Application of acoustic emission in geological materials
F.P. Hassani, M. Betourney, M. Kat, V. Mlarak and M. Momayezzadeh

Seismicity associated with a large scale gas outburst
K. Sato and Y. Fujii

Field oriented geotechnical acoustic emission/ microseismic research at the Pennsylvania State University
H.R. Hardy, Jr.

Laboratory and field investigations of rockburst phenomena using concurrent geotomographic imaging and acoustic emission/microseismic techniques
R.P. Young, D.A. Hutchins, S. Talebi, T. Chow, S. Falls, L. Farrell, D. Jansen, J. McGaughey, J. Towers and T. Urbancic

Development of microseismic monitoring software
T.K. Yeo

Seismic studies in Saskatchewan potash mines
D.J. Gendzwill and A.F. Prugger

PANEL DISCUSSION:

KEY QUESTIONS IN MINING INDUCED SEISMICITY

1. What should future microseismic monitoring instrumentation consist of and how can we best utilize the latest technologies?

2. What should be the critical path for the processing of microseismic data?

3. How should microseismic data be interpreted?

4. What are the major mechanisms involved in mining induced seismic events and how can these be recognized?

5. What role does geological structure play in mining induced seismicity?

Panel Chairman - Dr. D. Ortlepp, Anglo American Corp. SA
Panelists - Dr. W. Blake, Microseismic Consultant USA
- Dr. B. Brady, US Bureau of Mines
- Dr. C. Fairhurst, Univ. of Minnesota
- Dr. D. Hedley, CANMET
- Dr. D. Gendzwill, Univ. of Saskatchewan
- Dr. P. McDonald, INCO
- Mr. D. Morrison, Falconbridge/INCO
- Dr. M. Salamon, Colorado School of Mines
- Dr. S. Spottiswoode, Chamber of Mines, SA
- Dr. R.P. Young, Queen's University

TRANSCRIPTION OF TAPED DISCUSSION

David Ortlepp (Anglo American Corp. SA): Paul Young has referred to this workshop as a stepping stone between the first international symposium on mining induced seismicity and that which Charles Fairhurst is going to organize for us in June of next year. I think it is more than a stepping stone; I think it's one very substantial bridge and in identifying the leap between the first symposium and the second, Paul has identified the main issues that provide the important steps to be taken in crossing this bridge and continuing with the very necessary research which we have embarked upon. While the ultimate aims of workshops such as these must not be lost sight of, that is to finally provide guidelines to mine operators to allow them to improve mine planning or design mine support which will reduce the effects of rockbursts, Paul has correctly chosen issues which are more relevant to the fundamental or research aspects of mine seismicity problems. Because of time constraints I would like to open discussion with a brief opportunity for those who have major questions to be resolved, or if there is anything to add to the list of issues or changes in priority to be made, then you will have that opportunity. At the end of the bridge is Charles Fairhurst and the second international seismicity in mines symposium in Minneapolis in June next year and we will close with a few minutes from Charles to provide a foretaste of next year.

Terry Wiles (INCO, Canada): There is a new emphasis on waveform capture which disturbs me a little bit because I see the emphasis at the meeting two years ago in Sudbury was on source location and we are getting very good at source location, but we are still not making good use of the data and I don't want to see us go in another direction when we are at a critical point in making headway with the data we have. It seems we are going to branch out rather than focus.

Don Gendzwill (University of Saskatchewan): There's perhaps a similarity here between waveform capture and source location because unless you have very impulsive events, the timing of the event is a critical issue and waveform capture allows the seismologist to do more accurate picks on the wave. The phase identification of picking which we find with emergent events, is sometimes very critical. So for accurate and detailed work I think the two go hand in hand and having the waveforms is an essential part of this for any type of comprehensive and accurate analysis including source location.

Steve Spottiswoode (Chamber of Mines, SA): Last week I heard a presentation at a major geophysics symposium in Vancouver, where one of the authors from Japan showed that by looking at phase shifts of wave arrival time on the surface, it is possible to distinguish a time difference of down to a milli-second. This implies that there is the potential, by looking at waveforms recorded at surface, to obtain relative location accuracies of the order of 10 meters or so. That's with events that locate close to one another with similar mechanisms.

David Hedley (CANMET, Canada): I said earlier in my presentation that we have 14 microseismic systems at present in Ontario owned by companies and they are the ones responsible for looking at that data which is coming in daily. It can only be looked at on site and is difficult to send away to another place. It is the responsibility of the mining companies to do more with the existing data. On top of that they are putting in these seismic whole-waveform systems but these are being put in as research instruments. Essentially the companies are not paying for them, they are being paid for by the provincial and federal governments and the analysis will be done by CANMET and university researchers to see what use these systems can be to the mining companies. Hopefully by 1990 we will know which of the things we are trying to do with these systems can be done.

Brian Brady (Bureau of Mines, USA): Paralleling what Dave just said is that in the Northern Idaho area we are primarily looking at the analog data and hopefully will be able to develop better relocation algorithms, and get better velocity structures of the material of the rock mass where these events are occurring. From this we find that we get better and better accuracies of locations and that the microseismic events are falling on pre-existing fault planes or zones of weakness within the geologic mass around the mine opening. The digital data analysis is primarily being done as a research tool. Whether this will have imminent value is something we really don't know. However, the ability to go back and understand what the actual source function is, is very desirable. We are a long way from that and that is going to be a major area of research.

Miklos Salamon (Colorado School of Mines, USA): To try to present the other side, I think we all have a tendency to take advantage of developments in technology and since it has been improving very rapidly in recent times, both in computing and electronics, we have tended to use fancier equipment and hopefully we are getting better data, but not necessarily making better use of it always. Our record since the mid-60s is not particularly good. I would not be surprised if the Chamber of Mines in South Africa might have produced several tons of records which are totally useless and have never been looked at, so I think it's our responsibility to try to find the right compromise between these two pressures. One is to try to get better research which is absolutely necessary if you want to understand the problem better, but at the same time not to neglect and be guilty of a form of escapism where we try to use better electronics, better software or better computing techniques. It's a very difficult and delicate problem to find this compromise, but I think we do have to face this problem and deliberately make sure that we don't make the mistake of falling on one or the other side.

Paul Young (Queen's University, Canada): Very simply, in source location, which was the point of Terry's question, there is the algorithm and the input data and in many of the systems we use in Canada, the input data is somewhat questionable in that the way the arrival time is estimated from a simple trigger. I agree with you that a lot of effort is being put into developing better and better source location algorithms, but the same input data is used. I am somewhat concerned that we can't go back and look at the raw data, because it is thrown away by something like an Electrolab MP250 which picks a very simplistic amplitude detection algorithm to decide where the wave starts. Depending on the orientation of the transducer, it might pick the S wave or the P wave and this is automatically then assumed to be the P wave arrival time and put into sophisticated source location algorithms. My point is simply that if you put garbage in you get garbage out. Effort should be in some way directed to try to be assured that the data we are putting in to these algorithms which people have been developing for years is good. If we are not certain about the quality, we should have the opportunity to go back and check it. The only way you can do that is by having the waveforms recorded. I don't think it's a duplication of effort. I think that if you want to be assured that the advances made in algorithms are significant, you have to make sure the data going into them is good.

Paul McDonald (INCO, Canada): As a research specialist, perhaps I can just put industry's point of view, even if it's in contrast to most of what has been said before. INCO is a major consumer of microseismic equipment. We have 4 ElectroLab systems and we'll be putting in another 3 in the coming year and it comes down basically to one per mine. They are there almost strictly for operational purposes and we just get what else we can out of them. Of course it's been said before, we lose the waveforms, but we get the source locations and we've done our best with those. If we have to justify the purchase of a microseismic system, then it seems absolutely out of the question that we are going to get any more than one more system for research purposes which we will be doing at Creighton Mine as well, which will make 4 in the next year. When we justify the purchase of a system it is because the mine management know that they only have to turn to the output from that system even if it's only a printer attached to an Electrolab MP250 and they can have a pretty fair estimate of where any particular event of concern has taken place. If we say that we'll install a system for maybe $200,000 that can allay their concerns like that so they can turn immediately and know where their problems are happening or may be likely to occur, then that's just great because we are converting them to an acceptance of seismic systems as an important operational tool. Once they have accepted that, and don't forget that this is an industry that only accepted computers within the last three years, then we can make more and more strides in the research aspects. I think though that they are going to be asking us rather why don't we get more out of the equipment we have already got in in the various geological and mining contexts in which they are installed, rather than going and taking more and more equipment and putting it in the same place that we already have this under-utilized installation that is sitting there now.

Doug Morrison (INCO, Canada): I think that you can continue to improve the accuracy and precision of data ad infinitum and there comes a point when you have to say enough is enough. While it is certainly useful to go back and look at the first motion of the signal to see whether or not you have a good source location, really the only way to validate source location from a miscroseismic system is how well it correlates with actually what happens on the ground. The other way to check data is to actually go underground and look and see what is actually there. You can play with computers for ever and ever, but the only validation is what happens on the ground and the correlation with the geological features on the ground. I think that geological factors are beginning to be ignored. We really have not done enough with the data that is available to us already and perhaps we should spend more time working with it before we make another technological leap which will produce a plethora of more un-examined data.

David Ortlepp: The discussion has touched on a fundamental issue which is what do we need to do in order to make more use of the available data.

Terry Wiles: My main concern is that I see that our source algorithms now are, I believe, accurate to within 20 ft. or so and to put that into perspective, I don't think that my modelling is any more accurate. If I want to look at the whole picture I ask what is the objective, what are they working for. They are working for predictive capability and the way INCO plans to use that is by using microseismic results to correlate the modelling and ensure that you are seeing what the model is telling you so you are in a position to do some predicting. I have no doubt that with the new equipment and new approaches you'll get even more accurate data, but what you will be doing then is exceeding the present capability of another field which has to go along hand in hand and what I would like to see is a swing the other way. Let's not go shooting ahead and exceed the level of accuracy of microseismic monitoring.

David Ortlepp: A suggestion was made that it could cost a lot more to bring this extra understanding, which is an increase in quality rather than precision of data, inasmuch as it will tell us the nature of the event that is happening. It's important for the purpose of modelling as well to know whether it's a strike slip event or an implosion that is located with this alleged accuracy of 20 ft. or so. But I think Paul has something to say about the cost of the search for extra quality of the data.

Paul Young: From personal experience in Canada, a lot of systems are installed and it isn't necessary to say to the mine operators that the system they have is no good. You can hook on things which allow you to get extra data while the system is running in parallel and this we have done. We have a system installed at Falconbridge's Strathcona Mine which hooks on to an MP250 while the MP250 continues to run as usual. We have improved software which emulates an MP250 in terms of what the miners are used to and then in addition to that, we have the whole waveform data which can then be used for advanced processing. The key thing here is that we are then able to look at the very source mechanisms, the very things that are causing this activity and get a more fundamental understanding of what is going on. This system has been installed at Falconbridge and will be installed at two of Noranda's mines. They all have MP250s. In addition, the equipment can also be installed as a stand alone microseismic monitoring system. It's an extremely cheap and off the shelf system. It's based on a single board that's made in California by RC Electronics and this board will sample at 1 MHz with 12 bit resolution and you can basically put this in any standard IBM-type PC. The software has been developed in collaboration with RC Electronics and ourselves and this software is made available to the mining companies. For about $20,000 to $25,000 they are able to put this system on the back of their MP250. So this is an idea of enhancing what's already there which has been very appealing to the management of the mines I have been involved with and it doesn't throw away the existing equipment. As new developments come along, you can slowly move over to the other analysis and so the mine which is used to one particular form of analysis can then see how the other side develops before they get more involved in it. In terms of cost, the only expense is a standard IBM-type PC and this board which costs about $5,000, so it's very low cost and you can plug a ribbon cable straight into the back of the MP250. This is one particular piece of equipment and there is nothing complex about it. It will allow you to record the waveforms and then if you have the need to do the further work, you can do so.

Dave Hedley: We have on four occasions attached a Gould waveform recorder to the back of the Electrolab systems and once to other systems. What we found when we did this was that you certainly could get more accurate first arrival times. But the sensors that have been used by Electrolab for very small events, more often than not saturate and you cannot get the complete peaks. If you are going to add on to the Electrolab, you're going to have to start looking at what sensors are being used.

Paul Young: On the systems we've installed, we have changed the amplifier units on the accelerometers to give dual gain output. We use two of these RC boards to increase the dynamic range of the present system by recording on two levels, so we get effectively 100 dBs of dynamic range. A slight modification to the pre-amplifier system underground on the sensors allows you to take that opportunity, so Dave is absolutely right, but it's not something that can't be overcome. Electrolab now provide a modified pre-amplified unit which we developed in conjunction with them for that purpose and it can also be used for tri-axial recordings. However, first motion studies can still be done on an MP250 with standard accelerometers.

Terry Wiles: My point simply is that we have enough accuracy now and we don't need any more.

Jeff White (Spokane Research Centre, Bureau of Mines, USA): I would like to focus attention on the problem of rockbursting and how we approach it. It would seem to me that there are two approaches that we follow: one is to change the mine so that fewer rockbursts occur and they are of less severity; the other way is to try and predict where and when they will occur so we can evacuate personnel and adapt in that way. I would be interested to know what the panel and participants think of the progress in these areas and the relative potential for progress in these two areas.

Miklos Salamon: I would add a third approach. There is a lot of effort to install better ground support which if the others fail can still give you protection, so there are essentially three approaches.

Dave Hedley: In our rockburst project in Ontario, we have concentrated on reducing the severity and in trying to control damage. We have downplayed the prediction of rockbursts. That is not to say that research is not going ahead in this area, but it's tending to be done at the university level. Prediction includes two items: location and time. Most of the mines know rough locations of where they are going to run into rockburst problems. When we have pillar type rockbursts, more often than not there's an indication of a build-up of activity prior to a major rockburst so that there is advance warning. There has been a number of occasions especially in fault slip rockbursts, and I am thinking particularly of the Falconbridge mine where there was no warning whatsoever of that first major rockburst. My thoughts are that to promote a fault slip rockburst you either have to increase the shear stress along the fault or decrease the clamping forces across the fault. It appears at Falconbridge that it was more likely a decrease in the clamping forces, so if you have a decrease in stress initiating something, then you may not have any microseismic noises prior to that particular rockburst.

Comment from the floor: The papers printed have given a very informative outlook on the advancements made in this area. I have been associated with rockbursts for a long time and am reminded of Dr. Phillips's idea for accumulation of energy prior to a rockburst. I wonder whether any of our present techniques have made an approach as to the assessment of the re-accumulation of the energy so that it may be correlated to the phenomena and its timing.

Brian Brady: - The research we have seen up at the Galena Mine, for example, substantiates what Dave said. Most of our events are occurring in the hangingwall. There are some in the footwall, but most are in the hangingwall. If you do numerical modelling of this and put pre-existing fault zones in, you find out that the shearing stress gradually builds up and when the pillar is about 80 ft. or so from the sill, the shear stress reaches a maximum..... The other point that is very interesting is the idea of decreasing normal stress along the fault that can lead to a slip event with little or no precursors. We have seen that in the Idaho area, where we have seen a precursory build-up and then a drop-off and then the rockburst itself and then a few weeks later we saw very low level activity, then another slip event with no build-up so that would substantiate what Dave was saying. Also, Dave was talking about strain localization, where and when strained localization occurs; if we can see the effects of it then it might be possible to give a forecast of where and when such a slip event is going to occur. To do this we need local network arrays because they are often very low magnitude events which are not going to be picked up with a typical mine wide system, so down the line we have to address what we really want to do. If we wish to accurately forecast when a rockburst is going to occur, we'll have to have stope monitoring and that's expensive. At the Galena Mine now there are 17 active stopes. In the system we're deploying there, we'll be monitoring this activity. We'll take basically an 8 node network to do it and we plan to go in the next few years to 26 to 29 stopes and that's going to be a lot of data analysis. So we must find out what we want the system for and that's going to require a lot of data analysis if we're going to go to stoke monitoring. If we just want an idea of the mechanism of the events, then maybe a mine-wide system is valuable; if we wish to predict or control, then we may have to go to a much more expensive system. That's a question that each mine has to answer for itself.

Miklos Salamon: Most of you will remember Steve referring to one or two of the diagrams with the abbreviation ERR. For some years, the South Africans have discovered that there is a general relationship between the seismicity hazard and this energy release rate. This is an empirical relationship which is not obviously exact, but I think you can say generally that as the ERR number increases the hazards tend to increase and subsequently it was noted on a theoretical basis that this number is actually equivalent to the energy content of the rock which is to be mined out, provided you assume that no fracture takes place which is of course practically nonsense. But it does in other words give you an indication that this ERR is a measure of the stress concentration around the area you are trying to mine. Looking at it that way, it makes sense that you would expect a larger hazard generally where there are larger stress concentrations. Going into this a little deeper, clearly you are not likely to have a seismic event unless in the rock volume that suddenly changes, some energy is stored. It's almost a truism to say if the rock is not strained and does not have energy stored in it, then it's not going to be the cause of a seismic event. So there is obviously a fundamental need to have energy stored in a rock before you can have nasty events.

David Ortlepp: (addressing comment by Dr. Verma from the floor) Do I understand that what you are looking for is some kind of a physical indicator of where strain energy may be localized or concentrated which will then be the potential source of a rockburst.

Paul McDonald: I was going to suggest the reasons why we don't do it at INCO. Referring briefly to what Miklos was saying that in the past they used an empirical relationship between stored elastic energy around the mine areas. It's quite interesting they were able to develop that relationship over a long period of time from results in situations that were of fairly consistent mining geometry, a very brittle rock that stored loads and loads of strain energy before it would let go. It's interesting to contrast that with the situation in many of the steeply dipping wide and narrow veins in Ontario, for example, where we have obviously had a lot of seismicity, but have really different mining geometries and I guess we don't always understand at first why those events have taken place. It took a long time for people to realize that structurally controlled seismicity was the dominant factor in the events at least in Falconbridge No. 5 shaft and in Strathcona and we have had other circumstances where we've just had thousands and thousands of events both damaging and not bursting, just in the host rock around the ore body, due I guess to reasons that are not all that clear. We know that if we mine, then we throw the stresses out into the abutments and into the associated host rocks, then surely if those host rocks are at all jointed, which they are guaranteed to be, then it's far more likely that any displacement is going to occur on an existing discontinuity than it is to generate a new one. If we can possibly develop techniques where we can estimate the strength and other geo-mechanical parameters of the rock mass and simulate our mining and see where the over-stressing is likely to cause failure on the rocks in situ, then we can go back to our seismic data as we did in our poster presentation and say, yes, now this is what the model says, under these circumstances we get violent failure. Other circumstances which may at first appear to be very similar do not cause violent failure. It's all very well just to take an empirical relationship and say, there's the stress, there's the deformation, let's work out the stored strain energy, when it might be completely out of wack with what's going on there. We might have events that are structurally controlled which may have nothing to do with stored strain energy. We may have events that are caused by subsidence in the hanging walls, again due to a tensional effect rather than a high amount of stored strain energy. So I think there is far too wide a range of circumstances in our minds to use something so simple.

Miklos Salamon: Certain points have to be made. First of all, empirical research is legitimate. In other words, if you can establish empirically that something does work, just because it's empirical I don't think you should throw it out. You're right - this was established in a fairly special geometrical situation which in the South African mines is prevailing - many of the mines have similar geometry. Even there, no-one pretends that this is a predictive method. It really essentially compares the probability if you like between two geometries. In other words, it enables you to make comparisons between alternative mining lay-outs. It's never been pretended that it will tell you that at such and such a place and such and such a time there will be a burst. It's not a predictive method in that sense and also we do know now that many of the events that do occur don't come into this category associated with the edge of the mined out area that has very high stress concentration. So those you might say come from a different family and in certain fields those in fact dominate, even in South Africa, so there is still a long way to go.

David Ortlepp: I understood the original question to have been directed in a more general way to the necessary requirement that there should be a store of strain energy somewhere in the space and not necessarily as the energy release rate concept has it, close to the mining face in the plane of the excavation. Miklos has already answered that by saying the truism, that in order to generate a seismic event some of which cause rockburst damage, there must be stored strain energy which is released in some unstable equilibrium manner. In order to try and elucidate the manner of that instability, it seems to me necessary to understand the source mechanism which seems to require more than automatically selected arrival time, but to require seismic data which is collected in such a form, called seismograms, that enable the seismologists to really think about it and get back to the source. I see Wilson looking at me eagerly and it struck me that for someone who has had so much input into the problem in North America, he has remained singularly quiet, but he is waiting for his chance to bat!

Wilson Blake (Microseismic Consultant, USA): I just wanted to say something about prediction and maybe put the prediction thing into perspective. Everyone hopes to have prediction, but I think the number of mines that are instrumented now, we have something like 500 years of monitoring world-wide. There have been less than probably 10 predictions in 500 years of mine monitoring and that's by predicting that it's going to occur here and within a certain time period. So we have a long way to go before we can predict with any kind of reliability and obviously to do that we don't understand the mechanisms. This is why we need some of these supplementary whole waveform systems to look at what's going on so maybe we can figure out the mechanisms. But right now our record is extremely bad.

Steve Spottiswoode: While we are in this downbeat mood, one further complicating factor is that we know that the rock deformation doesn't all take place with seismicity. There's an awful lot that goes on where you might get a certain amount of microseismicity, but if one looks at the total amount of accumulated movement or energy release, it doesn't account for the full amount of deformation. This in fact can be argued actually happens within a seismic event itself. Art McGarr's talk this morning illustrated that one expects the high energy part of an individual seismic event, which in earthquake terms would be called an asperity and that's where much of the energy change takes place. It's very conceivable that the rest of the fault slip zone quietly creeps away on its own. Regarding this energy release rate story, we have estimated the total shear slip that has accumulated over the regular shear fractures ahead of an advancing face and found that the regular small microseismic events could account for only maybe a few % of the total shear deformation that we observe, so there is no doubt that deformation takes place without really significant seismicity. It's important to consider the aseismic deformation as well as rockbursts.

Don Gendzwill: I'd like to add to that some of the induced earthquakes in Saskatchewan potash mines have reached magnitudes of 3.6 which are larger than many of the events that have been put on the screen here earlier. We haven't had any fatalities or injuries from these events because the energy is dissipated in limestone rock well above the mine opening, that is the faults do not intersect the mine openings. In attempting to justify the relatively large magnitude of these events, I calculated some of the seismic moments based on measured closures in the mines and came up with seismic moments that have to be dissipated over quite a bit in excess of the equivalent moments for the Richter magnitudes of these events. We did studies on the rock properties of the limestone and did laboratory tests on the elastic and creep properties. The viscosity of the limestone that we measured, when taken into account the proper way, diminished the seismic moment considerably. So between creep and perhaps microfracturing, displacements and small fractures, we ended up with much smaller seismic moments than would at first thought be obtained from the closure and the volume of rock that had been extracted. The other side of the coin is the third energy accumulation. Here we have certainly identified a significant means of energy loss which is a good thing, but we still get earthquakes.

David Ortlepp: That seems to point almost paradoxically to the realization that something can go on deep in the rock mass which totally alters our appreciation of the rockburst problem which is in fact something that happens to a mine opening. That in turn makes one appreciate some of the suggestions that have been implicit earlier on; that it's necessary to understand the processes that are going on in the rock mass, not purely location-wise, but in terms of what the propagation path and the structure does to the source mechanism.

Ackbar Sheikh (Dennison Mines, Canada): I was a little upset by Dave's remark that data-handling is still the problem of the operating mines. I think it's like giving a little shovel to a guy and asking him to clean up that big muck-pile. My question is actually to Paul Young. When you described that the velocity is distorted because of the geological discontinuities and changes in the structure and other things, would it be possible that we should be paying more attention to the calibration of the unit itself, if time is available. What I am thinking is that is it possible to account for these effects in our algorithms.

Paul Young: Maybe in answer to that question, I can outline an experiment that Noranda are attempting to do at their Norita Mine. They are installing a similar system to the one I described earlier, but in each location they are putting in not just one hole for the sensor, but another hole fairly close at each of the various sensor locations. Once a week somebody at the mine goes around and sets off controlled blasts at each of these locations, big enough to be detected on the array. Now what this does, is that without the miners themselves having to carry out a tomographic image, you have all of these potential ray paths from every source to every sensor. It doesn't take very long to do and just by looking at the changes in velocity between these various sensors and various sources, you can get an idea of how things are changing and also help to calibrate your system. One of the problems here that I outlined earlier is that calibrating by blasting is one thing that should be done, but in addition some of the source mechanisms of mining induced events produce different components of P and S waves. This can result in the problem I mentioned earlier that if you use single component detectors, then you could be in a position where the arrival times that you detect might not necessarily be the ones you think they are. So in this particular mine, because the idea was to just cover an area with a series of sensors, we are installing hydrophones and this I think is an approach to overcome the directionality, especially when the objectives are to look at first motion and to do source location. We are actually putting in omni-directional hydrophones which have got uniform coupling to the ground through water filled boreholes. This also allows us to look at amplitude data which is something that couldn't have normally been done. So here is an example where something can be done by the mine and the miners are very keen to do that. This has been a suggestion from them so they are actually going to carry out this work on the basis of discussion with Noranda Research and ourselves. They're installing the system at the present time and are going to carry out the experiment in the near future.

Paul McDonald: I think that's great and it almost reinforces my disappointment in what Ackbar said before he asked his question. Ackbar is chief mine engineer for Dennison and if he doesn't think that the responsibility for the data handling from his monitoring system is the mine's responsibility, then I wonder whose he thinks it is. Because surely if he was interested enough in installing the system, he should be interested enough to get as much as he can out of it, even if it means hiring somebody extra.

Ackbar Sheikh: I didn't mean to say that it's not the responsibility of the operators. What I am saying is that there should be some intelligent system which one can use to scan the data.

Paul McDonald: What does that mean?

Ackbar Sheikh: That means that when there are reams and reams of data coming out, blasting, calibration and events - what I was thinking is is there any way for just taking out the blasting information.

Paul McDonald: Well, I think you have to accept the limitations of microseismic monitoring at the moment and you can't just look at a sheet of paper with a summary of events that have occurred over the last 24 hours and say that's a blast and that's a blast, without knowing how your mine operates. I just think that in my mind if I'm going to put a seismic system in and I'm going to tell somebody to make sure they have records of all the blasts, take those out of the records before they present them to the people who are actually interested in the summaries of where the events took place. At the moment there isn't a consistent and reliable method of taking blasts out except to get the operators to understand why they should record when and where they do blast, because of course if you record a very large event in a working place, it's not until you find out whether somebody has blasted that you know whether or not you should be ringing underground to tell them that a regular event has taken place. I know that there are some that still arise in the seismic output, especially from Electrolab systems that we work with and we're able to say that that was a blast and don't worry about it - certainly not all the time, maybe only 50% of the time.

Ackbar Sheikh: My point was that we are looking for improvement to all areas and at the same time we should look at data acquisition and intelligent data acquisition.

David Ortlepp: I promised to let Charles have a few words about the perhaps unresolved issues that we haven't finally put to bed and he can give us some foretaste as to how the next mining induced seismicity symposium will address some of these issues.

Charles Fairhurst (University of Minnesota, USA): I have been here today largely to listen and watch and see how we should construct the next international symposium. We are trying to emulate the superb work done by the South African group in the first symposium. We have deliberately associated the symposium with the annual US rock mechanics symposium which next year will also be held in Minneapolis and that symposium itself, although held in North America, is intended to be an international meeting which will focus on key questions in rock mechanics. I think we have heard today a number of people express disappointment or concern that progress is not being made more rapidly and having been associated with rock mechanics for a long time and actually having heard colleagues in other disciplines talk about the relatively slow rate of progress they have seen over a professional life-time, it is true that things can become a fad. We can see people go off in one direction, and the central problem may perhaps not progress as rapidly towards a solution as we would like. We have deliberately stated in this symposium what are the key issues in rock mechanics that need to be addressed, that have been with us for a long time and we are not making much progress on. So, when we deal in our case at the meeting before that, the International Symposium on Rockbursts and Seismicity in Mines, we hope to ask those questions that have been touched upon here today and others too dealing with the actual engineering problems of how to deal with rockbursts, because we know they are going to stay with us. Dr. Salamon said there are some fundamental ideas about strain energy that almost go without saying, and one thing that equally goes without saying is that the problems are not going to get any easier. There is going to be a constant search to do more and to stretch our engineering talent more. I have been very impressed with what has gone on here today and want to thank Paul Young and his colleagues who have organized this. Without a doubt it has been very stimulating and got us to the point where I know that there is a lot for us to do next year. Hope to see you and your friends in Minneapolis.

David Ortlepp: In closing, I would like to add my own personal thanks to the members of the panel and my thanks and congratulations to Paul whose superb organization made this all possible, so we should express our gratitude to him particularly and to the rest of the panel. Dr. Udd will now make some closing remarks.

John Udd (CANMET, Canada): There's absolutely no question that this workshop has been tremendously successful. There are 125 of you here and I think the first thing I'd like to say is that this is a tremendous tribute to the late Fred Leighton after whom this workshop was named and I am sure that he would have been very pleased at the enthusiastic turn-out. Rockbursting in Canada is a tremendously serious problem and has been for many years, but it re-exploded literally and figuratively back into prominence about three years ago just at about the time I joined CANMET. Consequently I have had the opportunity to be involved, not in doing rockburst research, but on the other side in steering proposals along and encouraging research and trying to push the money on the federal scene and trying to get things together. So there has been a tremendous satisfaction in seeing the evolution of rockburst research in Canada suddenly blossom. David Hedley presented an overview today on the Canada/Ontario industry rockburst research project which was started a mere 18 months ago and our best estimate now is that in the next 5 years, something of the order of $20 million is going to be expended in Canada on rockburst research and that's tremendous. I don't really want to say any more than that except that it has been a tremendously exciting workshop; it's an exciting time in this area of research and your enthusiasm has been absolutely evident. On behalf of all of you, I'd first of all like to present a small token of your appreciation to David Ortlepp for chairing this afternoon's session and to Paul Young for organizing this workshop. There are two other things I have to do and I was asked in my capacity of Chairman of the Canadian national group to find an appropriate occasion to honour a couple of individuals who seem to be tremendously elusive. They are both here at the moment. The Rock Mechanics and Strata Control Committee of the Canadian Institute of Mining and Metallurgy has asked me to confer upon two individuals, certificates in recognition of outstanding achievement and distinguished service in the field of rock mechanics and ground control in Canada and I can think of no finer occasion than this to do it. First of all, I'd like to ask Dave Hedley to come up and receive his award. We have in the federal government, a staff member who I have hardly seen in the last year - apart from David - but this other person has been involved in being the general chairman of this particular congress. That's a huge responsibility and so consequently it hasn't been possible to pin him down. I would ask Gerhard Herget to come up please.

LIST OF DELEGATES

J. Archibald, Department of Mining Engineering,
Queen's University, Kingston, Ontario, K7L 3N6

Jean Arteau, 505 W. de Maisonneuve,
Montreal, Quebec, H3A 3C2

Eileen Ashworth, 501 E. St. Joe,
Rapid City, SD. U.S.A. 57701

Michel Aubertin, C.P. 700,
Rouyn, Quebec, J9X 5E4

William Bawden, 240 Hymus Blvd.
Pointe Claire, Quebec, H9R 1G5

Stan Bharti, Falconbridge Limited,
Sudbury Operations, Falconbridge, Ont. P0M 1S0

Eugene P. Binnall, Lawrence Berkeley Laboratory,
Bldg. 50b, Rm. 4235, Berkeley,
California, U.S.A. 94720

Wilson Blake, P.O. Box 928,
Hayden Lake, ID., U.S.A. 83835

Torgny Borg, 58249 Linkoping, Sweden.

Michael Bostock, Department of Geological Sciences,
Queen's University, Kingston, Ontario, K7L 3N6

B. Brady, U.S. Bureau of Mines, P.O. Box 25086,
Building 20, Denver Federal Center,
Denver, CO. U.S.A. 80401

H. Brehaut, Dome Mines Ltd.

Jean Breniaux, 11 Avenue d'Altkirch,
F68100 Mulhouse, France.

K. Bullock, Department of Mining Engineering,
Queen's University, Kingston, Canada.

Mary Cajka, 1 Observation Cres. Ottawa, Ont.

J.P. Chauvin, Noranda Norita Mine, Quebec.

Dai Choi, 4000 Brownville Rd. Pennsylvania.

Lawrence B. Cochrane, 680 Camelot Drive,
Sudbury, Ontario, P3B 3N1

Pierre Colin, Tour Aurore Cedex 5,
Defense 2, Paris, France, 92080

Glen Davidge, Falconbridge Limited,
Sudbury Operations, Onaping, Ontario, P0M 2R0

Ruth Debicki, 100 Ramsey Lake Road, Sudbury, Ont.

Jozef Descour, Earth Mechanics Institute,
Colorado School of Mines,
Golden, CO, U.S.A. 80401

Denis Fabre, 38402 St. Martin d'Mere,
Cedex, France.

Charles Fairhurst, University of Minnesota,
500 Pillsbury Drive SE,
Minneapolis, MN, U.S.A. 55455

Steve Falls, Department of Geological Sciences,
Queen's University, Kingston, Ontario, K7L 3N6

Veronique Falmagne, Minnova Inc.,
Division Lac Dufault, Rouyn-Noranda,
Quebec, J9X 5B4

Loraine Farrell, Department of Geological Sciences,
Queen's University, Kingston, Ontario, K7L 3N6

P.M. Ferrigan, U.S. Dept. of Energy,
9800 South Cass Ave. Argonne, Ill. 60517, U.S.A.

Robin Friday, 11 Irving Ave., Box Hill,
Victoria 3128, Australia.

Yoshiaki Fujii, Mizumoto, Muroran, Japan 050.

Don Gendzwill, Dept. of Geological Sciences,
University of Saskatchewan, Saskatoon,
Saskatchewan, S7N 0W0

David Goodall, 386 Rang 40,
Ormstown, Quebec J0S 1K0

D. Hanson, P.O. Box 100, Elliot Lake, Ont.

H. Reginald Hardy, Jr., 117 Mineral Sciences Building,
University Park, Pennslyvania, U.S.A. 16802

Ferri P. Hassani, McGill University,
Mining and Metallurgical Engineering,
3480 University St., Montreal, Quebec, H3A 2A7

D. Hedley, Elliot Lake Laboratories, CANMET,
P.O. Box 100, Elliot Lake, Ontario, P5A 2J6

Thomas Herbst, 65 Bells Rd. Unit 8, Concorde, Ont.

G. Herget, Mining Research Labs. 555 Booth St.
Ottawa, Ontario K1A 0G1

Richard Hong, P.O. Box 550,
Kirkland Lake, Ontario, P2N 3J7

David Hughson, 217 Rosedale Heights Drive,
Toronto, Ontario, M4T 1C7

David Hutchins, Department of Physics,
Queen's University, Kingston, Ontario, K7L 3N6

Satoshi Imamura, Lawrence Berkeley Laboratory,
50E Building, Berkeley, CA. U.S.A. 94720

Takeshi Iwamoto, Institute of Weak Rock Engin.
2-13-12, Honkomagome, Bunkyoku, Tokyo, Japan 113.

Dion Jansen, Department of Physics,
Queen's University, Kingston, Ontario, K71 3N6

F.M. Jenkins, U.S. Bureau of Mines,
Spokane WA. 99207-2291 U.S.A.

Peter K. Kaiser, Geo. M.P. Research Centre Sudbury, Ont.

M. Kat, McGill University, Mining & Metallurgical Eng.,
3480 University Street, Montreal Quebec, H3A 2A7

Johannes Klokow, Private Bag X2011,
Carletonville, South Africa 2500

Sassa Koichi, Dept. of Min. Science & Tech.
Kyoto University, Kyoto, Japan

Brent A. Kristoff, Mining Division,
P.O. Box 3000, Bathurst, New Brunswick, E2A 4N5

Denis Labrie, 2700, Rue Einstein, E-RC-225,
Ste. Foy, Quebec.

Vladimir Labuc, 623 Main Rd., Box 20,
Hudson, Quebec, J0P 1H0

David A. Landriault, 84 St. Brendan Street,
Sudbury, Ontario, P3E 1K5

Peter Andrew Lang, URL, Atomic Energy Canada Ltd,
Pinawa, Manitoba, R0E 1L0

P. Lappalainen, Pyhasalmi Mine,
86900 Pyhakumpu, Finland.

Christopher Laughton, Groupe GC - Div.Lep,
Cern Meyrin, Ch 1211, Geneva 23, Switzerland

Gerald Lefrancois, C.P. 751-1216, 3e Ave.,
Val d'Or, Quebec, J9P 4P8

Gordon Ley, 3151 Wharton Way,
Missisauga, Ontario, L4X 2B6

T. Macdonald, Mining Division, P.O. Box 3000,
Bathurst, N.B. E2A 4N5

P. McDonald, Mines Research Department,
Inco Ltd., Coppercliff, Sudbury, Ontario.

A. McGarr, U.S. Geological Survey, 345 Middlefield Rd.,
Menlo Park CA. 94025 U.S.A.

J. McGaughey, Department of Geological Sciences,
Queen's University, Kingston, Ontario, K7L 3N6

D. Madsen, Department of Mining Engineering,
Queen's University, Kingston, Ontario, K7L 3N6

Ashrof Mahtab, 918 Mudd, Columbia University,
New York, NY. U.S.A. 10027

Tony Makuch, Campbell Red Lake Mines Ltd.,
P.O. Box 10, Balmertown, Ontario, P0V 1C0

Juri Martna, S-16287 Vallingby,
Stockholm, Sweden.

James Mathis, Div. of Rock Mechanics,
University of Lulea, Lulea, Sweden 951 87

Vincent Maury, Elf-Aquitaine, Cedex, France.

Aleksander Mendecki, Rock Mechanics Dept.,
P.O. Box 2083, Welkom, South Africa.

V. Mlarak, McGill University, Mining & Metall.Eng.
3480 University St. Montreal, Quebec, H3A 2A7

M. Momayezzadeh, McGill University, Mining & Metall. Eng.
3480 University St. Montreal, Quebec, H3A 2A7

O. A. Momoh, Noranda Research Center,
Pointe Claire, Quebec H9R 1G5

Brian Moore, P.O. Box 1167,
Johannesburg, South Africa

D. Morrison, Mines Research Department,
Inco Ltd., Coppercliff, Sudbury, Ontario.

Parvis Mottahed, PCS Tower, Suite 500,
122-1st Avenue South, Saskatoon, Saskatchewan, S7K 7G3

Dennis Mraz, 410 Jessop Avenue,
Saskatoon, Saskatchewan, S7N 2S5

S.N. Muppalaneni, P.O. Box 1500, Elliot Lake, Ont.

J. Niewiadomski, Mining Research Laboratories,
CANMET, P.O. Box 100, Elliot Lake, Ontario, P5A 2J6

R.C. More O'Ferrall, P.O. Box 641,
Stilfontain, South Africa 2550

Brian O'Hearn, Cleland Bldg., Falconbridge Ltd,
Onaping, Ontario, P0M 2R0

Hideo Ohtomo, 2-2-19, Daitakubo, Urawa,
Saitama, Japan 336

D. Ortlepp, Anglo American Corp.,
P.O. Box 61587, Marshaltown 2107, South Africa.

Jack Pierre Piguet, Lab.de Mecanique des Terrains
Ecole des Mines, 54042 Nancy Cedex, France.

T.F. Pugsley, Falconbridge, Commerce Court West,
Toronto, Ontario, M5L 1B4

W. Quesnel, Lac Minerals Ltd., Hemlo Division,
P.O. Box 580, Manitouwadge, Ontario, P0T 2C0

Neil Rauert, P.O. Box 444,
Broken Hill, N.S.W., Australia 2880.

Roger Revalor, Lab. de Mecanique des Terrains
Ecole des Mines, 54042 Nancy Cedex, France.

Kevin Riemer, West Driefontein Goldmine,
Private Bag X2011, Carletonville,
South Africa 2500.

P. Rochon, P.O. Box 100, Elliot Lake, Ont.

Miklos Salamon, Mining Engineering Dept.,
Colorado School of Mines, Golden, CO, U.S.A. 80401

Kazuhiko Sato, Mizumoto, Muroran, Japan 050.

Sean Seldon, West Driefontein Goldmine,
Private Bag X2011, Carletonville, South Africa.

T. Semadeni, P.O. Box 100,
Elliot Lake, Ont.

Robert Sharp, P.O. Box 503,
Los Alamos, NM, USA 87544

A. Sheikh, Denison Mines

David Spencer, Private Bag 82279,
Rustenburg, South Africa 0300.

S.M. Spottiswoode, P.O. Box 91230,
Auckland Park 2006, South Africa.

Irving Studebaker, Montana Tech.,
Butte, MT. U.S.A. 59701.

Z. Sun, China.

P. Swanson, U.S. Bureau of Mines, P.O. Box 25086,
Building 20, Denver Federal Center,
Denver, CO. U.S.A. 80401

Tim J. Swendseid, 209 Emerald,
Kellogg, ID., U.S.A. 83837

Shahriar Talebi, Department of Geological Sciences,
Queen's University, Kingston, Ontario, K7L 3N6

Soichi Tanaka, Tokyo Office, Oyo Corporation,
3-2-1, Ohtsuka, Bunkyoku, Tokyo, Japan 112.

Chikosa Tanimoto, Dept. of Civil Engineering.
Kyota University, Japan

Paul Thompson, Whiteshell Nuclear
Research Establishment, Pinawa, Manitoba, R0E 1L0

Jeff Towers, Department of Geological Sciences,
Queen's University, Kingston, Ontario, K7L 3N6

J. Udd, Mining Research Laboratories,
CANMET, 555 Booth Street, Ottawa, Ontario, K1A 0G1

Ted Urbanic, Department of Geological Sciences,
Queen's University, Kingston, Ontario, K7L 3N6.

Marcus van Bers, South Africa.

J.B. Vance, Potash Corp. PCS Tower, Suite 500,
122 1st Ave. Sth., Saskatoon, Sask.

Ed van Eeckhout, Los Alamos, NM, USA 87545

Bhagwat Sahai Verma, 43 Golf Links Road,
Bedford, Nova Scotia, B4A 2J1

T. Vladut, 300, 24-21, 7th Ave, Calgary.
Jeff Wyatt, Research Centre, Spokane, Washington.

T. Wiles, Mines Research Department,
Inco Ltd. Coppercliff, Sudbury, Ontario.

Ivan G. Wong, Woodward-Clyde Consultants,
250-1390 Market Street, San Francisco, CA U.S.A. 94102

T.K. Yeo, Department of Mining Engineering,
Queen's University, Kingston, Ontario.

R. Paul Young, Department of Geological Sciences,
Queen's University, Kingston, Ontario, K7L 3N6

Thian Yu, P.O. Box 2082, Timmins, Ontario, P4N 7K1

Submission received after workshop:

T. Vladut (Geo-Technology Ltd., Canada)

MINE INDUCED SEISMICITY RISK ASSESSMENT

The workshop on mine induced seismicity dealt with the extended mine monitoring programs to reduce the risk of seismic events related to deep mines. This program focuses on improvement of monitoring capabilities and by this to obtain and enhanced understanding of triggering conditons of mine induced seismicity. A significant component of the program is associated to enhancement of geotomographic contribution to knowledge of the areas prone of seismic activities.

Mine induced seismicity is on environmental concerns which affect mine developments mostly unexpectedly generated in complex field conditions controlled by the geological environment and the mine development often associated with significant depletion of the one bodies.

Recurrance of mine seismicity on Ontario mobilize significant attention on the monitoring side which will allow inplementation of active mitigation procedures.

As dealing with an indirect man-made associated activity along the monitoring several elements may be assembled using data related to other past historical data which affected other mines.

Mine represents significant local underground developments which modify stress condition in the earth crust. Similar modification of stress conditions are associated to reservoirs and dams which also sometimes are related to induced seismicity.

Assembling data on reservoir induced seismicity related to historical data allowed estimation of risk of man-made induced seismicity and a similar procedure of evaluation is suggested for mine related seismicity. Some of the elements of application of the geotechnical observational method applied to risk evaluation of reservoir induced seismicity (RIS) may be relevant in comparing mines subjected to seismic activity. Elements of the reservoir induced seismicity evaluation were related to a world survey of RIS made by the International Commission of Large Dams, initiated in 1980 (International Water Power and Dam Construction, May, August 1980, R & D Section) and summarized in a report for the 1988 Congress in San Francisco (Q60R40).

Application of the geotechnical observational method for mine induced seismicity require collection of historical data on mine affected by seismicity and a simplified analysis will deal with comparison of geomorphological conditions and estimation of field data of seismic activities.

1. Evaluation of morphological situation of mine development could be defined by the geometry of the extraction shape of the mines. Considering a uniform three dimensional shape (a cube) the geometrical elements would define an adimensional number of one, related to three equal dimensions (a = b = c). Practically any mine development could be expressed in a simplified manner using ratios associated to size of the mine development. The morphological shape of the mine development would be broadly related to the ore body shape.

   1a. Evaluation of the mine development could be associated to degrees of extraction, which would be zero before mining and may reach values close to one for fully extracted ore bodies, e.g., $V = \alpha \nu$ in which "V" is the extracted volume and "ν" the ore body volume, and would refer to the degree of extraction. In practical terms the extraction value would be always less than one, e.g. probably 0.4 to 0.5 for pillar extraction for coal and greater for longwall mining, etc..

2. Orientation of mine development related to principal stresses would provide reference if the mine development may enhance (or may not) fracture development. To exemplify mines developing perpendicular on principal stresses may enhance fracture developments.

3. To estimate seismic condition around mines simplified synthetic means should be utilized. A simple approach was suggested by McGarr (1984) in which the seismic activity (by seismic moment S) is related to mine volume or its variation (Δ V): Sm $<$K Δ V.

4. Several other elements associated to field condition, mainly geology should be structured in the survey on past seismic data which affected mine developments.

It is suggested that a world survey on past seismic activities associated to mine developments may provide an estimation of seismic risk associated to mining. Such development may be an associated alternative to provide passive mitigation procedures for mine seismic hazards which will complement the active means associated to enhancement of monitoring procedures. This proposal was developed by the author and Fred Leighton along the review of the USBM/CANMET joint undertaking for microseismic monitoring of slope instabilities in open pit operations.

WORKSHOP ON SWELLING ROCK

Chairman: H.H. Einstein (USA)

This workshop was held on September 1, 1987. It consisted of two short presentations on the work of the ISRM Commission on Swelling Rock, followed by formal talks on new developments and concluded by an extensive discussion on all these topics. Specifically the format and context of the workshop was as follows:

Introduction (Einstein)

<u>Presentation of Commission Work</u>

Laboratory Testing of Swelling Argillaceous Rock* (Madsen) (Presented by Einstein)

Review of Analysis Methods* (Gysel)

<u>New Developments</u>

Constitutive Models and Laboratory Testing:

Anisotropic Swelling of Claystone* (Fröhlich)
Elasto-plastic Constitutive Model* (Bellwald)

Design and Field Testing:

Laboratory and In-situ Swelling Tests for the Freudenstein Tunnel (Kirschke)

Special Problems

Investigation of Swell Pressure* (Bucher)
Highly Compressed Bentonite

Swelling Rock in Fills (Lovell)

Discussion

The preceding presentations and future work of the Swelling Rock Commission were discussed.

The workshop was a great success. An overflow crowd estimated at 70 people attended. The discussion lasted for an hour and would have continued if the late hour of the day had not forced us to stop.

In order to have a reasonably permanent record of the workshop, written versions of some presentations (marked with an asterisk), were prepared. They are reproduced below in the sequence of the above listed program. The reader is made aware of the fact that for the presentations of work by the ISRM Commission on Swelling Rock, only abstracts are given. This is so because the Suggested Methods for Laboratory Testing of Argillaceous Swelling Rock will be printed in the near future and because a review of Analysis and Design Procedures for Structures in or on Swelling Rocks has been published in Volume I, pp. 377-381 of these Proceedings.

SUGGESTED METHODS FOR LABORATORY TESTING OF ARGILLACEOUS SWELLING ROCKS

Abstract

Professor Dr. F. Madsen
Eidgenossische Technische Hochschule (ETH)
Zurich, Switzerland

The engineering problems caused by swelling rocks are well recognized and so is the necessity to test them in the laboratory and in the field to determine the appropriate engineering parameters.

The ISRM Commission on Swelling Rock in cooperation with the ISRM Commission on Testing Methods in working a new systematic treatment of the swelling methods for laboratory testing of argillaceous swelling rocks were developed. They consist of four parts:

Part I Sampling, storage and preparation of test specimens

Part II Laboratory testing for determining the maximum axial swelling stress

Part III Laboratory testing for determining the maximum axial and radial free swelling strain

Part IV Laboratory testing for determining axial swelling stress as a function of axial swelling strain

The tests described in Parts II and III are simple tests providing first estimates on the order of magnitude of swelling stress and swelling strain. For rigorous analysis and design only the complete stress-strain behavior as obtained with the test described in Part IV is acceptable. For the analysis-design methods described on the right only such tests are acceptable. Since such complete tests can last many months, first estimates obtained with the simpler tests will be useful, for example in preliminary design. Also, the simple tests allow one to quickly estimate where the most critical zones exist and to collect additional samples for use in the complete test. Finally, the simple tests can be used as field control tests during construction. It will thus be the decision of the engineer to select the appropriate tests in the context of design and construction of a particular project.

Future development in testing of swelling rock will have to address anhydrite containing rocks, true three-dimensional behavior and field testing.

ANALYSIS AND DESIGN PROCEDURES FOR STRUCTURES IN OR ON SWELLING ROCK

Abstract

Dr. M. Gysel, Chief Engineer
Motor Columbus Engineering, Inc.
Baden, Switzerland

Laboratory tests, from which swelling parameters and mathematical swelling laws (constitutive relations) are derived, are the basis of current analysis and design methods for structures on or more frequently in swelling rock.

The presentation demonstrates that the layout and the interpretation of the swelling tests mostly do not cover the full 3-dimensional behaviour. Furthermore, inelastic material response cannot be taken into account. Nevertheless, a swelling law has been established for reversible swelling of argillaceous rock both in 1-dimensional and 3-dimensional formulation.

According to the inherent inaccuracies of the measured swelling parameters and swelling laws, the design calculations have to be considered as one step of a more comprehensive design procedure for structures in swelling rock. The proposed design procedure comprises 5 steps:

1. Laboratory swelling tests.
2. Analysis of test structure (this step only for large structures).
3. Construction and monitoring of the test structure (adaptation of swell parameters, this step only for large structures).
4. Analysis of the actual structure.
5. Construction and monitoring of the structure: back analysis to obtain swell parameters.

The development of design methods (Step 4) mainly went on during the past 15 years. The work and the relevant design methods or procedures by H. Grob, G. Lombardi, H.H. Einstein, M. Gysel, W. Wittke and K. Kovari and others, mark some points along the line of development. The presentation and the paper in these proceedings (Vol. Ik pp. 377-381) briefly review these methods which range from purely analytical to numerical (finite elemnt) solutions. Some case histories for calculations are given.

While the ISRM Commission on Swelling Rocks concludes that some appropriate methods exist for analyzing structures in argillaceous swelling rock, almost any future improvement of the analysis methods has to be based on improved testing techniques in order to establish improved swelling laws (constitutive laws) and relevant swelling parameters. Furthermore, anisotropic swelling and non-reversible swelling (anhydrite) or partly non-reversible swelling (mixed swelling of clay minerals and anhydrite) still represent a large field for future development of swelling tests and analysis and design methods.

ANISOTROPIC SWELLING BEHAVIOUR OF DIAGENETIC CONSOLIDATED CLAYSTONE

B.O. Fröhlich (West Germany)

SUMMARY: On the basis of observations in tunnelling and mineralogical data one can expect that swelling of claystone is anisotropically. Unconfined orientated swelling tests and radial confined orientated swelling pressure tests show that the swelling behaviour is very anisotropically. On this basis a material law including the swelling anisotropy is implemented in a finite element program. The results of the calculations which are described correspond very well with the in situ observations.

Since the last century invert heave in tunnelling in swelling rock with nearly horizontal bedding is known. The swelling behaviour causes damage of the tunnel lining. v Rabcewicz reports about side wall deformations in swelling rock with a dipping of 45° resp. 90° . It seems that swelling normal to the bedding is larger than parallel to the bedding and that the curvature of the invert and the construction method are not the most important influence to explain the invert heave.

Swelling of claystone is caused by an annexation of water and exchange of ions in resp. on the clay minerals. Clay minerals are very flat; therefore the anisotropy of swelling depends on the degree of mineral adjustment. On the basis of the structure and the orientation of the clay minerals one can expect an anisotropic swelling behaviour.

The methods to test the potential swellability are to divide in the group with quantitative results and in the group with index results. Mineralogical tests like DTA, TG, specific inner surface, X-ray and the soil and rock mechanics standard tests like plasticity, water capability and free swelling belong to the index value group. During the most tests the internal structure and the water content will be disturbed. A correlation of the index values with a quantitative swelling of claystone is not sufficient. The determination of the swelling anisotropy is not possible. The test with orientated samples is necessary to find out the swelling behaviour. Therefore, unconfined swelling tests and oedometer tests, sometimes with the measurement of the confining pressure were carried out.

The unconfined swelling deformations are normal to the bedding planes up to 20 times or more larger than parallel to the bedding planes; the mean value is about 8,8 (Fig. 1). Different rock types from various locations had been tested.

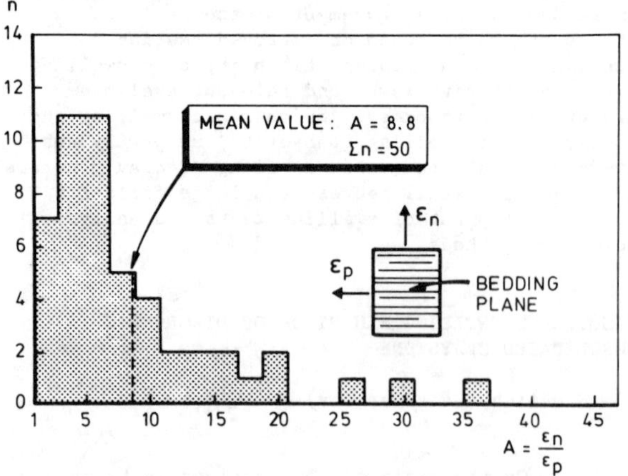

Fig. 1: Anisotropy of the unconfined swelling deformations $\varepsilon_{n,p}$ of diagenetic consolidated claystone

The tests to determine the swelling pressure depending on the orientation were carried out with cylindrical, lateral confined specimens. The testing unit is developped and constructed by the Lehrstuhl für Felsmechanik at the University of Karlsruhe. The testing procedure includes a cycle of loading, unloading and reloading of the specimen before the cell will be flooded with water on a high stress level. The specimen will be unloaded stepwisely while the time depending swelling deformations will be measured on each stress level. The maximum pressure with swelling deformations is called the maximum possible swelling pressure, a value different from the maximum swelling pressure under constant volume.

The total axial deformations (Fig. 2) are to divide into the terms of elastic deformations and of swell deformations. Between the swell deformations and the logarithm of the axial pressure exists a linear correlation. As a result of the orientated tests one can see that the maximum possible swelling pressure normal to the bedding planes is about hundred percents larger than the corresponding value parallel to the bedding planes. The ratio of the swelling pressure normal to the bedding to parallel to the bedding depends on the swelling deformations and

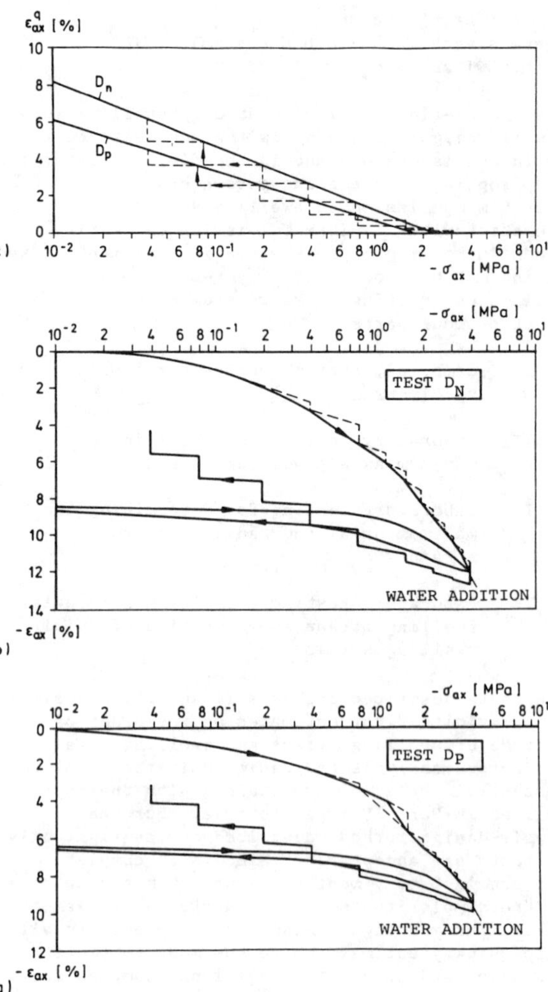

Fig. 2: Results of an orientated swelling pressure test

the stress level. The lower the stress level the larger is the anisotropy of the swelling pressure. The mean value of the anisotropy of the swelling pressure is about 30 under a low axial pressure of -10 kPa (compression: negative). A comparison of the test results has shown that the swelling pressure anisotropy on low stress level correlates with increasing of the unconfined swelling deformations.

The information about the unconfined swelling deformations and the maximum possible swelling pressure is not enough. You need the complete dependence of the swelling deformations on all stress levels. As the schematical diagram in Fig. 3 shows the swelling pressure of a stiff rock is higher than the value of a soft rock if the swelling pressure is developing under a constant volume.

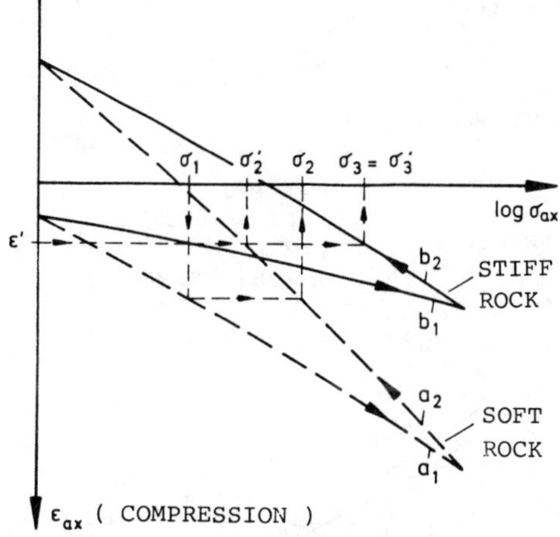

Fig. 3: Swelling pressure depending on the stiffness of the rock/rock mass

The higher the stress level at the beginning of the swelling the swelling pressure under constant volume will increase. A swelling process under constant volume means that the increase of volume caused by the swelling is equal to the elastic compression caused by the increase of the pressure. Therefore, you can say that the swelling behaviour depends on the deformation parameters of the rock. An increase of pressure parallel to the bedding planes is caused by a high swelling pressure normal to the bedding planes. Therefore, the anisotropy of the swelling pressure will decrease if the stress level increases and the Poisson ratio is high.

The test results of the anisotropic swelling and the theoretical deducations (e.g. by MÜHLHAUS and FECKER) lead to a material law as a reduced simplified material model which describes the orthotropic swelling behaviour superposed on an elastic material behaviour.

$$\varepsilon_{ij}^{ges} = \left(\frac{1}{2G} \cdot \sigma_{ij} + \frac{2G - 3K}{18\,K \cdot G}\,\sigma_{kk}\delta_{ij}\right) + \beta n_i n_j f\,(\sigma_n)$$

$$\sigma_n = \sigma_{ij} n_i n_j$$

$$f(\sigma_n) = :\ \ln \frac{\sigma_n}{\sigma_{n0}} \text{ für } \sigma_n \geq \sigma_{n0}$$

$$f(\sigma_n) = :\ 0 \text{ für } \sigma_n < \sigma_{n0}$$

σ_{n0} : maximum possible swelling pressure

ß : swelling coefficient

This model is implemented in a finite element program. The mathematical model describes the finished swelling process and has no information about the time dependence. While the material behaviour is hyperlinear the initial strain method is assumed.

Different models are calculated with this FE-program. In the following few examples are explained. Fig. 4 shows the displacements of the top and of the side wall depending on the swelling deformation capacity on low stress level.
The displacements of the top/invert are about ten times larger than the displacements of the side wall. If the overburden pressure increases the top/invert displacements decrease without change of the swelling parameters. The horizontal wall distance will be enlarged if the overburden surcharge increases. The top/invert displacements decrease and the horizontal convergence increases if the ratio of the horizontal to the vertical stress increases. This calculation describes a tunnel without lining and fixed boundary conditions. The complete range of the rock mass is swelling.

To estimate the influence of the swelling behaviour in tunnelling the following basical questions are to be answered:

- Which range of the rock mass is swellable?
- Which range of the rock mass will be influenced by tunnelling?
- How much has the rock mass swelled before tunnelling starts?

To study the influence of the extent of the swelling rock range around a tunnel, a tunnel model with a flat invert and an overburden of 110 m is calculated (Fig.5).

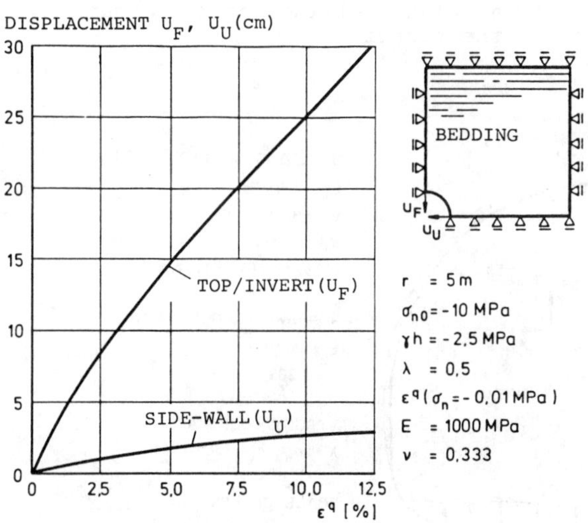

Fig. 4: Displacement of the top/invert and the side-wall depending on the swelling deformation capacity ε_q (σ_n = -0,01 MPa)

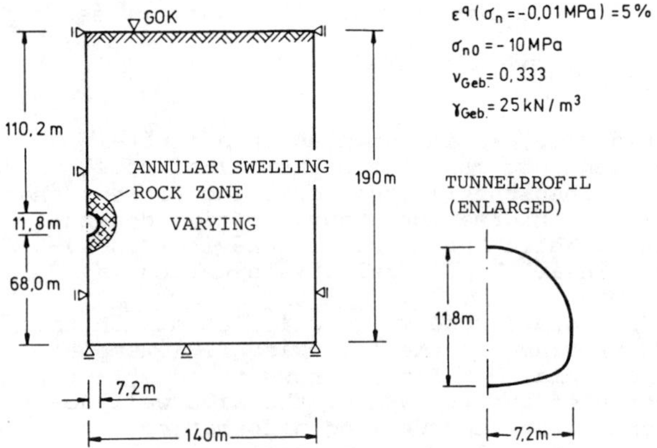

Fig. 5: Rock mass section with tunnel profil
(Swelling rock mass: E-Modul E_{ij}
non-swelling rock mass: E-Modul E_a)

If the extent of an annular range of swelling rock increases up to a thickness of about 1,5 diameter the vertical convergence increases (Fig. 6).

The complete rock mass over the tunnel will be heaved. The magnitude of the displacements of the tunnel and the surrounding rock depends on the stiffness of the non swelling rock. A heave of the ground surface can be observed in situ. Although the swelling rock range is annularly around the tunnel the vertical convergence is extremely higher than the horizontal convergence (Fig. 7). The distance of the side walls can enlarge.

To estimate the pressure on a tunnel lining the vertical load/deformation-line of a concrete tube is compared with the vertical tunnel convergence of the calculations of the FE-model (Fig. 8).

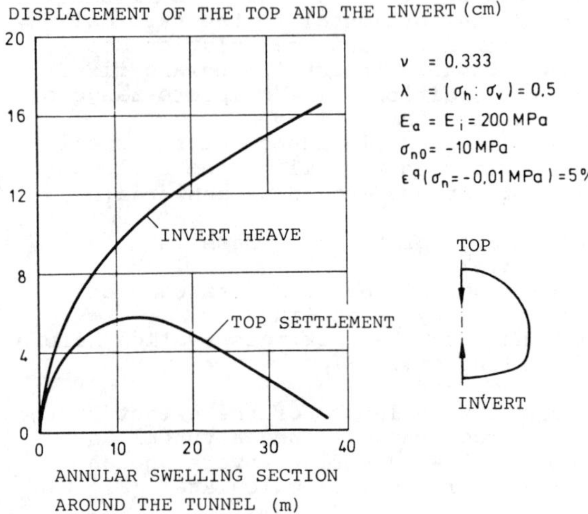

Fig. 6: Top settlement and invert heave depending on the extention of the annular swelling section around the tunnel

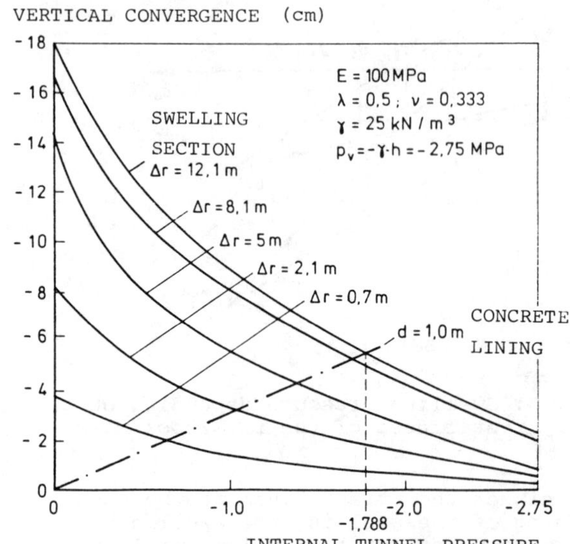

Fig. 8: Vertical tunnel convergence depending on the internal tunnel pressure (as a reference for the vertical pressure)

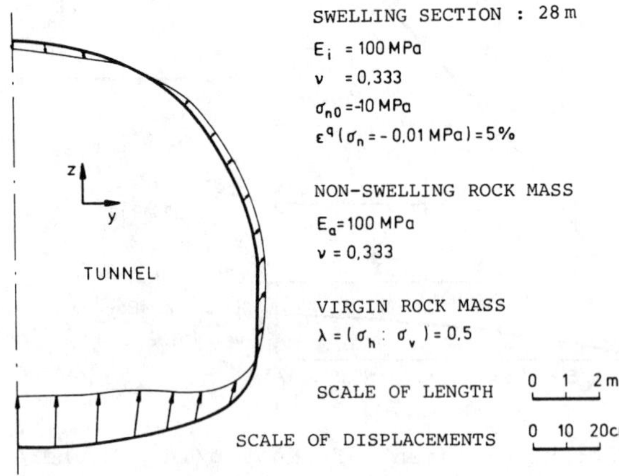

Fig. 7: Displacement of the tunnel cross-section caused by swelling

The vertical tunnel convergence is calculated for different extent of the annular swelling rock range and for different internal tunnel pressures. The vertical convergence of the concrete tube is calculated with a radial pressure p depending on the angle of the circumference (p = "1" $\sin\varphi$; top: $\varphi = 90°$) and considering the forces caused by the modulus of subgrade on the side wall. The effective pressure on the tunnel lining (about - 1,8 MPa) is significantly lower than the maximum possible swelling pressure (σ_{no} = - 10,0 MPa) (Fig. 8). If the thickness of the concrete lining increases the loading of the lining increases too.

The described tests and calculations show that the swelling behaviour of claystone is extensively anisotropic. The anisotropy of the swelling is the most important cause for the characteristic of the measured displacements in tunnelling. The finite element model gives the possibility for the first time to calculate the influence of the anisotropic swelling behaviour.

FRÖHLICH, B. (1986): Anisotropes Quellverhalten diagenetisch verfestigter Tonsteine. Veröffentlichung Nr. 99 des Instituts für Bodenmechanik und Felsmechanik Karlsruhe, Germany.
MÜHLHAUS, H.-B. and FECKER, E. (1979): Beitrag zum Quell- und Schwellverhalten horizontal gelagerter Sedimentgesteine. Ber. 2. Nat. Tag. Ing.-Geol. der DGG, Fellbach 1979, pp. 127 - 134.
v. RABCIEWICZ, L. (1961/62): Aus der Praxis des Tunnelbaues: Einige Erfahrungen über echten Gebirgsdruck. Geologie und Bauwesen, Jg. 27, H 3-4, pp. 153-167.

ELASTO-PLASTIC CONSTITUTIVE MODEL

P. Bellwald (USA)
H.H. Einstein (USA)

ABSTRACT: In this paper swelling in shales is treated analogously to a consolidation problem. Three main features have to be considered when designing underground structures in swelling shales: (1) the initial state of stress in the ground, (2) the behavior of the swelling shales during the undrained phase and, (3) the behavior of the swelling shales during the drained phase. Yielding can occur during either the undrained phase or the drained phase, leading to a plastic behavior of the swelling shales. In order to better understand this behavior it is necessary to perform undrained and drained laboratory testing under 3-d conditions.

1 INTRODUCTION

The swelling shale problem will be considered in the context of analyzing the behavior of an underground opening in such a medium. Specifically the behavior of swelling rock around an opening will be discussed first. This will be followed by a review of the current laboratory practice with emphasis on its drawbacks. Finally suggestions will be made on further investigations.

2 MECHANICAL BEHAVIOR OF SWELLING SHALES AROUND UNDERGROUND OPENINGS

Swelling in shales is treated analogously to a consolidation problem. Three main phases have to be considered when designing underground structures in swelling shales: (1) the original (natural) phase with the initial state of stress in the ground, (2) the behavior of the swelling shales during the undrained phase and, (3) the behavior of the swelling shales during the drained phase.

The stress path method of Lambe (Lambe and Whitman, 1969) will be used to describe what goes on in each of these phases.

2.1 Initial state of stress

The knowledge of the initial state of stress is one of the key issues in the predicting of the rock mass behavior around an excavation. For our discussion we consider a homogeneous isotropic plate of unit thickness (2-d) (Figure 1).

This plate is subjected to the initial state of stress which is assumed to be completely known at the level of the future excavation. The principal stress directions are defined in Figure 1. Three assumptions are made in this discussion:

1. the computations below will not include the principal stress σ_L acting perpendicular to the plate. This is done to simplify the discussion, and does not affect the validity of the conclusion,
2. the initial coefficient of lateral stress at rest K_o ($= \frac{\sigma_h'}{\sigma_v'}$) is assumed to be 1.5, a value which can often be observed in sedimentary rocks, and
3. the future opening lies at a depth of 100m below the surface and 40m below the ground water table (Figure 2).

These specific values have been chosen to provide meaningful examples, any other values could have been selected, however.

Figure 2 shows the initial state of effective stress in a p-q diagram where p is the mean total stress $\frac{\sigma_r + \sigma_h}{2}$, p' is the mean effective stress $\frac{\sigma_r' + \sigma_h'}{2}$ and q is the deviatoric stress $\frac{\sigma_r - \sigma_h}{2}$.

The Mohr-Coulomb envelope is defined by 2 parameters, a' and α' which are directly obtained from the mechanical parameters c' and ϕ'[1]. c', the drained cohesion is assumed to be 2.0 MN/m2 and ϕ', the drained internal angle of friction is 30°.

2.2 Undrained phase

The excavation of an underground opening changes the stress distribution inside the rock mass. The low permeability of the rock (10^{-10} to 10^{-12}m/s) and

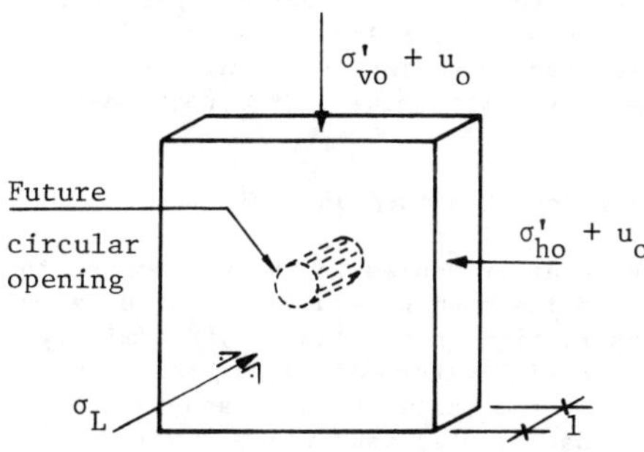

Figure 1. 2-dimensional model. Homogeneous isotropic plate of unit thickness.

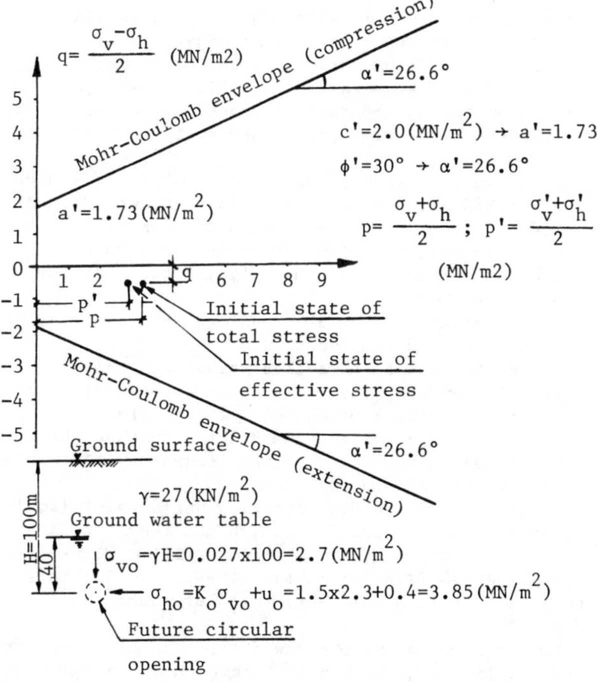

Figure 2. p-q diagram. Initial state of stress.

the relatively high rate of excavation (3 to 10 m/day) cause the rock mass to behave in an undrained manner just after excavation. Undrained behavior means that no change in the water content (w=constant), but a change Δu in pore pressure will occur.

The stress paths of two points located at the circumference of a circular opening; point I in the invert and point S in the springline, are shown in Figure 3 for purposes of providing an example. The total stress paths represent vertical extension loading in the invert and vertical compression unloading in the springline (Lambe and Whitman, 1969). The effective stress paths are assumed to follow the general trends observed in undrained tests on highly overconsolidated uncemented clays (Parry, 1960). The shear induced pore pressure changes Δu are negative at both points I and S because of the dilatant behavior of the material upon loading or unloading. The exact form of the stress paths depends on the particular case but two generally valid points can be made:

1. if the effective stress path for either point I or S or both touch the Mohr-Coulomb envelope, yielding will take place: yielding in extension in the invert and yielding in compression in the springline.
2. cavitation may develop when the pore pressure drops below zero absolute pressure. High negative pressures can occur as shown by Chenevert (Chenevert, 1957) , who reported negative pressures of several MN/m2 in shales of low porosity. Cavitation would lead to an instantaneous partial drainage at the particular point of the rock mass.

2.3 Drained phase/consolidation and swelling

Unbalanced pore water pressures in the rock mass, the drainage effect of the opening and inflow of external water (from construction operations or air humidity) will change the initial water content of the rock mass with time. If the excess pore pressures are positive, consolidation will take place, while swelling will occur where the excess pore pressures are initially negative. This drained phase will last until the entire rock mass reaches a new equilibrium. One knows, from practical experience, that this can go on for many years.

The following discussion will concentrate on the zone in the invert in which most of the swelling will take place because of the large negative pore pressures developed during the undrained phase (Figure 3). The release of negative excess pore pressures during the drained phase will decrease the effective stress drastically and swelling will occur.

In a p-q diagram the point I can only move inside the hatched area towards the origin (Figure 4). At present one does not know the exact form of this path. Laboratory testing is needed to obtain a better understanding of the swelling behavior during the drained phase.

Nevertheless, two conclusions can be drawn from the preceding considerations:

1. the horizontal effective stress σ_h' remains always larger than the vertical effective stress σ_v' because the point I moves below the p-axis,
2. once the stress path of point I touches the Mohr-Coulomb envelope, yielding occurs; swelling will thus lead to failure.

An example of yielding occuring during swelling is illustrated in Figure 5 in which the result of an oedometer test on an undisturbed swelling clay-shale are presented (Sun Jun et al., 1984). During this test, the vertical and horizontal stresses were recorded.

Figure 5a presents the raw data. The same data are plotted in a p-q diagram in Figure 5b which shows that yielding take place followed, in this particular case, by a progressive reduction of the mechanical parameters a' and α', thus of c' and ϕ'; c' goes to zero while ϕ' decreases towards its residual value ϕ'_{res}. Ultimately, the rock will reach this residual shearing resistance.

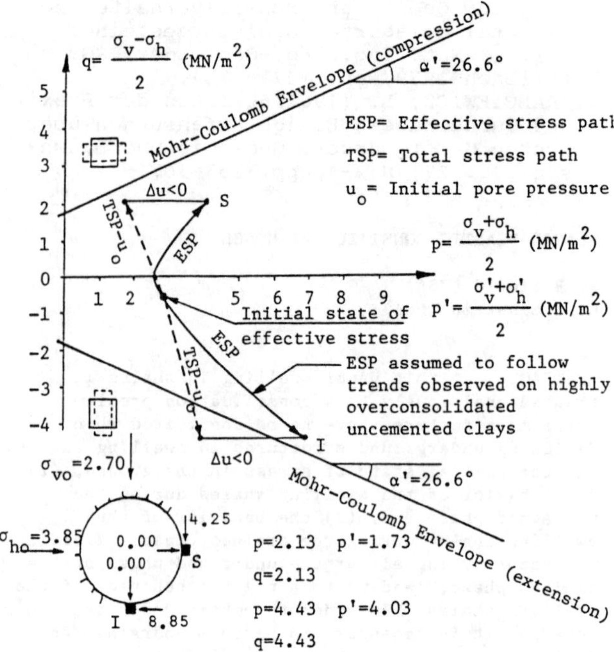

Figure 3. p-q diagram. Stress path of points I and S during undrained phase.

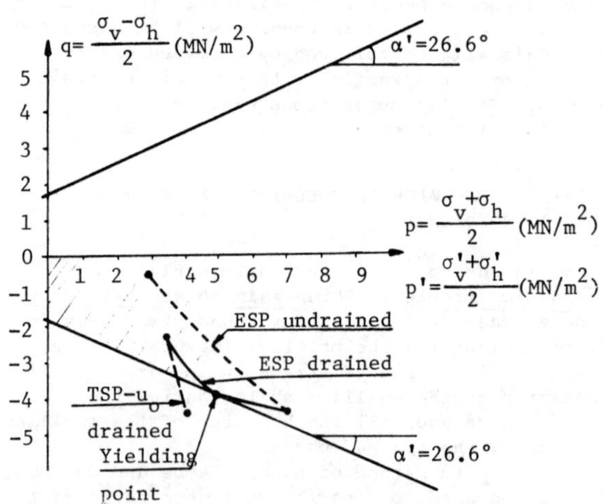

Figure 4. p-q diagram. Stress path of point I inside hatched area during drained phase.

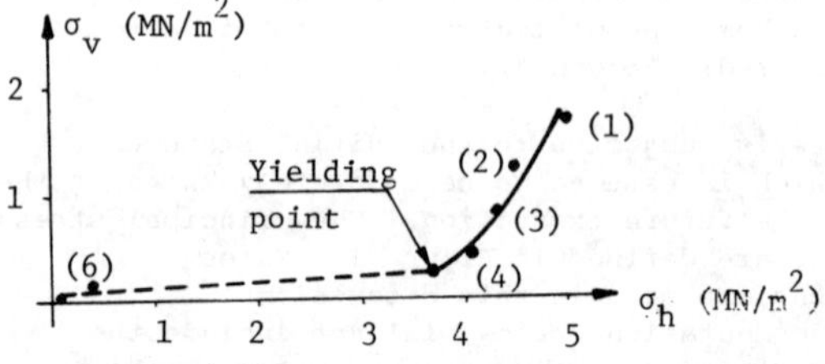

Figure 5a. 1-dimensional swelling test. Vertical and horizontal stresses as recorded during test (after Sun Jun et al., 1984).

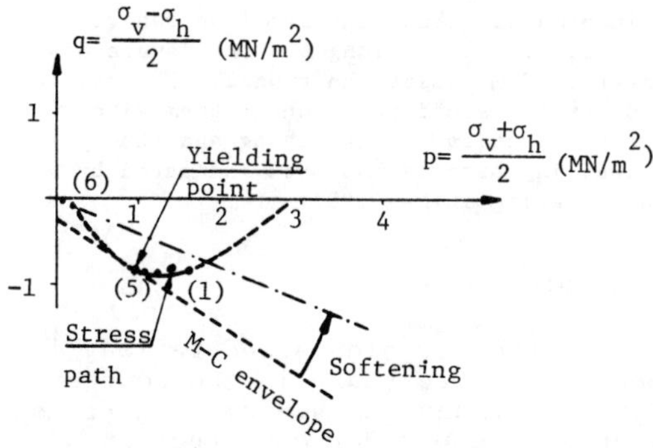

Figure 5b. p-q diagram. 1-dimensional swelling test (after Sun Jun et al., 1984).

3. PRESENT LABORATORY PRACTICE

As shown in the preceding discussion, undrained and drained tests should be performed to predict the complete swelling behavior.

3.1 Undrained tests

To the author's knowledge, no systematic research has been undertaken to understand better the undrained behavior of swelling shales. As one can see in Figure 3, important stress changes occur inside the rock mass when an opening is excavated. In particular we note a dramatic increase in the mean effective stress. Undrained tests should tell us how important this increase will be. The estimation of the state of effective stresses in the ground just after excavation is important because it is the starting point of the effective stress path of the drained phase and thus considerably affect the behavior of the rock during the drained phase.

3.2 Drained tests

Drained tests are mostly, if not exclusively, performed with the oedometer by following the Huder-Amberg procedure (Figure 6) (Huder and Amberg, 1970). As can be seen in this figure the resulting swelling curve relates the axial vertical strain ε_v to the axial vertical stress σ_v. The major drawback of this test is its inability to record the radial horizontal stress (except for the version developed by Sun Jun et al., 1984), an issue which will be addressed in section 4 below.

It would be desirable to have testing equipment and procedures allowing one to conduct both drained and undrained tests in a consistent manner and under 3-d conditions. At MIT, we have therefore developed a new computer controlled triaxial cell in which undrained and drained tests on swelling shales will be performed along predetermined stress-or strain paths.

4 THREE DIMENSIONAL BEHAVIOR

In order to extend the 1-d swelling law as obtained with the oedometer tests to three dimensions, one generally assumes that the horizontal stress is proportional to the vertical stress by the coefficient $\nu/(1-\nu)$ where ν is the Poisson's ratio. This expression is derived from the theory of elasticity (Einstein et al., 1972). To check this assumption, the recorded values of K_o obtained by Sun Jun et al. (1984) have been plotted versus the vertical strain ε_v, and they are compared with K_o as one would obtain based on Poisson's ratio as described above (Figure 7). By making the assumption of proportionality, one greatly underestimates the K_o-value both in the elastic and in the plastic ranges. 3-d behavior therefore needs to be considered not only in the analysis but by conducting appropriate laboratory tests. The new MIT approach is to perform pseudo-oedometer tests in the triaxial cell by controlling the strain path of the specimen during swelling.

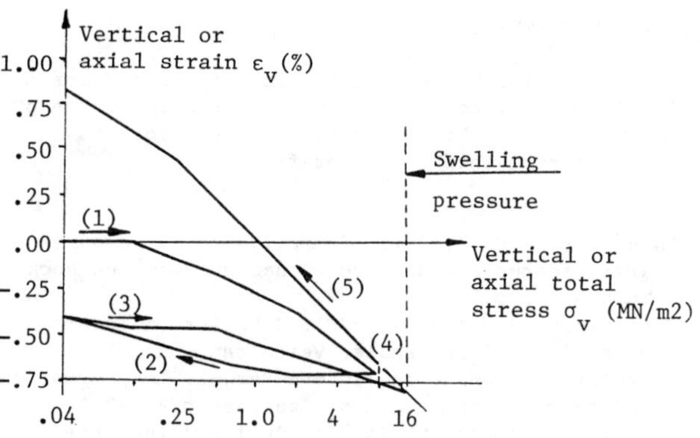

(1) First loading, dry
(2) Unloading, dry
(3) Second loading, dry
(4) Strain due to swelling after water is added at constant load
(5) Unloading, final strain due to swelling as function of the compressive axial stress

Figure 6. 1-dimensional swelling test (after Huder and Amberg, 1970).

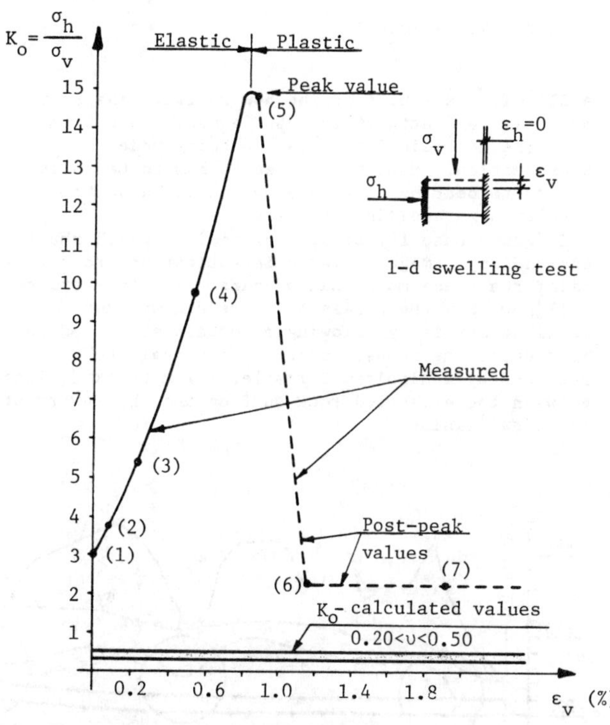

Figure 7. K_o-value as a function of the vertical strain. Comparison between measured and K_o-computed values (after Sun Jun et al., 1984 for measured data).

5 CONCLUSIONS

It was demonstrated that in addition to the initial state of stress, two basic phases should be considered and analyzed when excavation in swelling shales is performed: an undrained phase and a drained phase. In order to obtain a good understanding of the behavior of swelling shales around underground openings, laboratory tests need to be performed for these two phases and under 3-d conditions.

ACKNOWLEDGEMENTS

The author thanks Herbert H. Einstein, Professor of Civil Engineering at the Massachusetts Institute of Technology, for his encouragements and for his meticulous reviews of the drafts.

Chenevert, M.E. (1957). Adsorptive pore pressures of argillaceous rocks. Proc. Canadian Symp. on Rock Mechanics. p.599-627.

Einstein, H.H. et al. (1972). Behavior of invert slabs in swelling shale. Verhalten von stollen=sohlen in quellendem Mergel. Int. Symp. on Underground Construction, Lucerne. p.296-319.

Huder, J., Amberg, G. (1970). Quellung in Mergel, opalinuston und anhydrit. Schweiz. Bauzeitung (88)43:975-980.

Lambe, T.W, Whitman, R.V. (1969). Soil Mechanics. John Wiley & Sons. 553 p.

Parry, R.H.G. (1960). Triaxial compression and extension tests on remoulded saturated clays. Geotechnique 10: 166-180.

Sun Jun et al. (1984). The coupled-creep effect of pressure tunnels interacting with its water-osmotic swelling viscous elasto-plastic surrounding rocks. Proceedings, Peking.

LABORATORY AND IN SITU SWELLING TESTS FOR THE FREUDENSTEIN TUNNEL

D. Kirschke (Germany)

ABSTRACT: As a part of the new railway line from Mannheim to Stuttgart in West Germany the 6.8 km long Freudenstein-Tunnel is now being under construction. Most of its length has to be built in anhydrite bearing claystone, which is known to develop high swelling stresses.

To get a base for an optimal design, which shall exclude any damage, a large investigation program is being run since more than 6 years. Results obtained till now show the necessity to reduce maximum swelling stress by allowing a certain strain of the rock under the tunnel invert. This shall be realized by a highly deformable, 1.2 m thick filling between the excavated rock surface and the invert of the final lining.

To get information about the swelling behaviour of the rock in situ, a 100 m long testing cavern has been excavated 50 m beside the tunnel. The cavern is divided into 15 sections, each of them with different construction of the lining and the invert. Swelling process has been initiated by artificial watering in May 1987.

1 INTRODUCTION

As a part of the new high speed railway line from Mannheim to Stuttgart in the southern part of West Germany the 6.8 km long Freudenstein-Tunnel is now under construction. The tunnel has to be built in the triassic formation of Middle Keuper, which consists of claystone, marlstone, Gypsum and Anhydrite. Such a ground is wellknown for the danger of swelling. The central part with a length of 4.3 km is estimated to be extremly problematic, because of the high content of finely distributed Anhydrite in the claystone(Fig.1)

2 DESIGN PHILOSOPHY

In a first stage of design it has been assumed as possible to avoid swelling by absolutely dry working and by prevention of any water inflow. If this concept should be successful, the stable rock would need a very light support only. Meanwhile we know that water can never totally kept away. There are several places where ground water reaches the tunnel or its vicinity. Even a large program of cement grouting, which has been executed during the last year, has not made the ground watertight, but only reduced the permeability. The reduction is enough to drive the tunnel without water problems, however the long time danger of swelling is as big as before.

Another way to prevent damages by swelling can be to install a tunnel lining strong enough to withstand the maximum swelling stress. When we started design according to that principle,the maximum value ever found in laboratory tests was 4 MPa. That is about 10 times the usual loading of a tunnel in weak rock. Till now there is no example for a tunnel which has been designed for such a load. On base of parametric studies we decided for a circular lining 1 m thick in the upper part and 2 m thick in the invert. (Fig. 2a). The ultimate swelling pressure is about 3 MPa. We chose that construction, because it seemed to be the limit of that,what can practically be built. At the same time we hoped, that the swelling stress develloping below the actual tunnel would be considerably lower than in laboratory test. Of course this was only an assumption, and we still

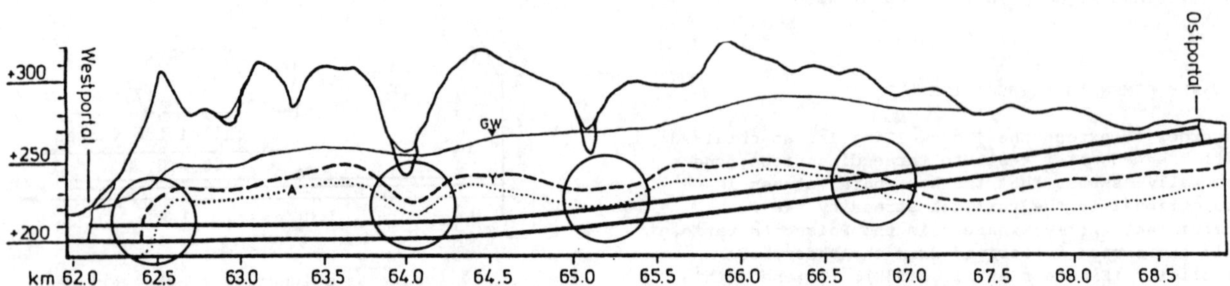

Figure 1. Freudenstein-Tunnel

GW: ground water surface, Y: gypsum surface, A: anhydrite surface, ◯: water problem zone

had to check it. If we should find, that the in-situ swelling pressure is more than 2 MPa, the principle of a rigid strong lining would have to be abandoned.

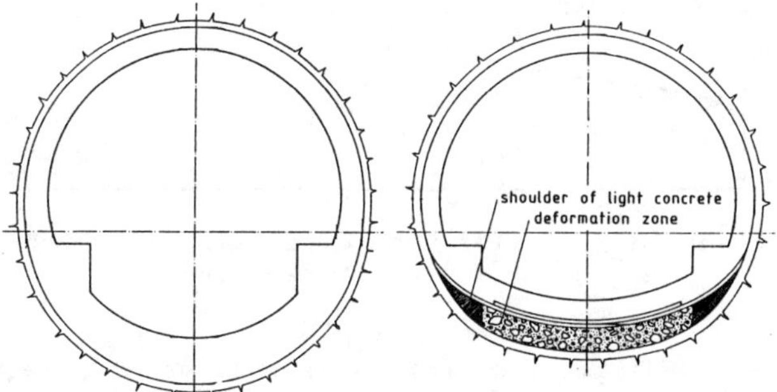

Figure 2. Possibilities of tunnel lining

a (left): Strong lining, ultimate bottom load 3 MPa
b (right): Invert with deformation zone, ultimate bottom load 1 MPa

As an alternative we started to investigate possibilities of stress reduction by allowing a certain strain of the swelling rock. Possibly the usual method of reconstruction of the invert after a time of swell deformations and provisional repairs is the most economic solution for tunnels in Anhydrite. However, it can be applied only, if occasional closures of the tunnel are acceptable. Just for this reason that method is not appropriate for a high speed railway tunnel, which must be absolutely stable for a century from the very beginning. If we want to reduce swelling stress by allowing swelling strain, this has to occure without any visible influence on the tunnel lining and the position of the rails. The necessary space for swelling deformations must therefore be kept free below a strong invert arch. This can be realized, for example, by filling the room between excavated bottom and final concrete lining with compressible material, according to a suggestion by Prof. Kovari (Fig. 2b).

The rigid tunnel lining as well as the stress reduction measures had to be further investigated by laboratory tests and also by large field experiments. Construction of the tunnel lining should not be started before a optimum solution has been found and checked for function and practicability. However, as the tunnel must be finished within 3 years, there is no time to wait for final results of all our tests and only then start with tunnelling. So we have to do all things together, now. While top headings are driven from several points of a pilot gallery,experiments to to find the best construction of the invert are being executed in a seperate test cavern beside the tunnel. Within a year we need a final decision about the shape of the lower part of the tunnel to start with benching and excavation of the bottom.

3 LABORATORY TESTS

Laboratory tests for the Freudenstein-Tunnel have been started 6 years ago.

In a first series we made a rough test whether the rock tends to swell or not by measuring the strain of cores in a water basin. The length of the specimens was about 10 cm. Within a year the length increased for 10 % to 36 %, whereas the core diameter extended for less than 3 %, only. This result showed a high swelling capacity of the rock, and at the same time an extremly anisotropic deformation behaviour. However, such tests are absolutely useless for a prediction of swelling stress, as there is no relation between strain, elasticity and stress. Besides, the measured strain was only partly due to an increased volume. A considerable part has been caused by bending of some layers, producing hollow spaces between them. To avoid such effects, swelling tests should be always done under a small axial load.

With the second series of tests we tried to find the maximum axial stress under total restraintof deformation.For this purpose special oedometer have been constructed, similar to the swell pressure device, which is shown in the suggestions of the commission on swelling rock. Our samples are 4 cm high and have a diameter of 8 cm. A 20 KN electric load measuring cell was regarded to be sufficiently dimensioned, because we expected swelling stresses of less than 4 MPa.

All the tests have been started with a loading-unloading cycle before watering to get the deformation behaviour of the dry sample. After that procedure tests have been executed in two different ways:

The first way is to block deformation after application of an axial stress of 0.4 MPa, what is 10 % of the capacity of the measuring cell. In this case the specimen has to develop swelling stress by compressing itself with increasing content of Gypsum. This process can take a long time.

To reduce this time, as a second way specimens have been watered under a prestress of 4 MPa. If maximum swelling stress was higher, this could be measured after a comparably short time. If maximum swelling stress was less than the applied prestress, load had to be reduced, until stress increased again by swelling.

The solid line in Fig. 3 shows a typical development of stress according to the first way. After half a year most of the specimens had reached a stress of 4 MPa.

The respective curve according to the second way is shown as dotted line. Already after some days, at the most after a month we knew,that maximum swelling stress would be more than the prestress of 4 MPa.

As we expected, that swelling potential of the specimens would be exhausted soon, we reduced the stress by allowing a small heaving and blocked the system again. However, stress climbed over the capacity of the measuring

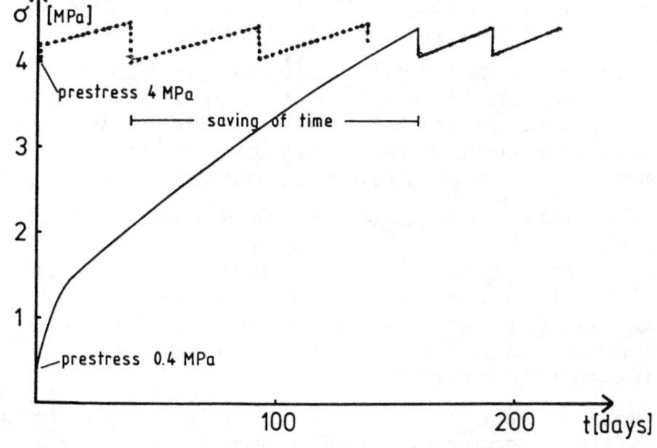

Figure 3. Development of swelling stress of specimens watered under different prestresses

cell again, and that procedure had to be repeated periodically to prevent destruction of the cell (Fig. 4). Finally, we chanced the cells to stronger ones, which could measure

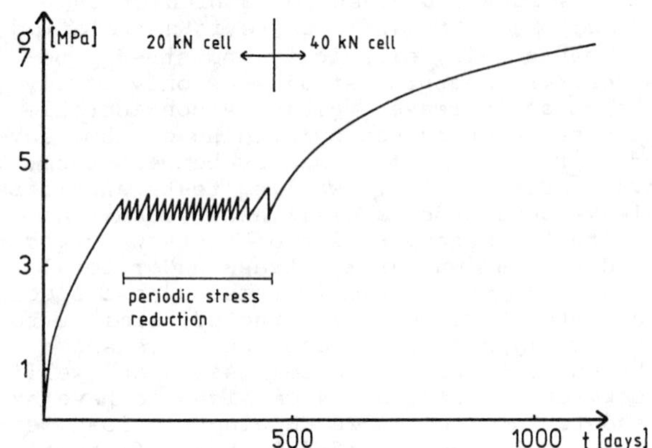

Figure 4. Typical development of swelling stress before and after the extension of measuring range

a stress of 8 MPa. It seems to be remarkable, that the further development of stress did not give any indication, whether prestress had been low or high. Apparently the swelling potential is so big, that the small loss in the beginning doesn't influence the final result at all.

After nearly 4 years of swelling most of the specimens have developed a stress of more than 7 MPa, and there is no end in sight, yet.

Because of this result the chances to get along without any stress reducing measures become smaller and smaller. So we are no longer especially interested in the final value of maximum swelling stress, but more in the strain necessary to limit the load to an acceptable size. We decided to dimension the order of magnitude of 1 MPa. Most of the old swelling tests have been changed according to the new question. Additionally a lot of new tests has been started, using samples from the pilot tunnel. The thickness of specimens for these tests is 20 mm, as against to 40 mm before.

There are three general possibilities to control stress or strain during a test, until swelling comes to an end:

a) The test starts with free strain of the specimen and changes to the development of stress under total restraint.

b) The test starts with the development of stress under total restraint and continues with strain under constant stress.

c) The specimen swells against an elastic system, which allows a certain strain during development of stress. This is the only way to run the test under realistic conditions, however, it is difficult to control.

Till now no swelling test has reached a final state. So we cannot say yet, whether the different possibilities of test performance will lead to the same result or not.

In the following some preliminary results and observations from the third series of tests are described:

- Most of the tests show a clear acceleration after 3 months. After 6 months usually a final velocity of stress or strain development has been reached (Fig. 5).

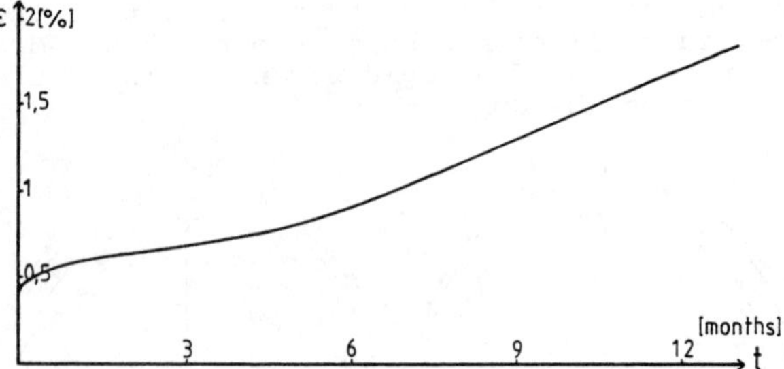

Figure 5. Acceleration of strain rate after some months of swelling

- Development of stress as well as of strain can be very different, even if specimens seem to be absolutely equal. Some specimens start rapidly, others don't show any reaction during the first time. However, this initial swell behaviour doesn't give any indication on the further development. Slowly starting specimens can still overtake specimens supposed to be the fastest ones (Fig. 6).

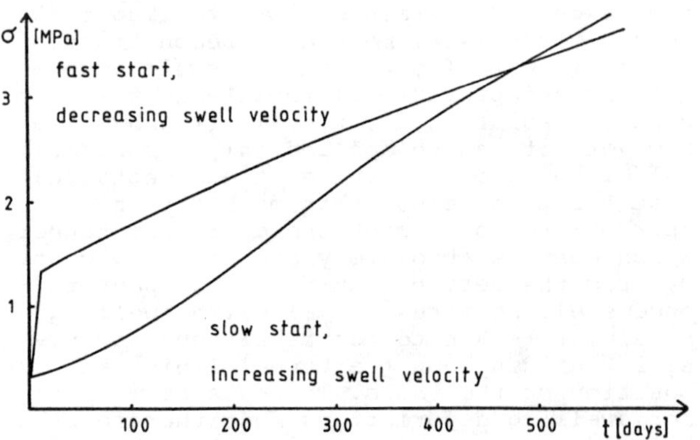

Figure 6. Different types of development of swell velocity

- Specimens with very low swelling capacity develop high stress during the first hours or days after watering, when they had lost their natural water content before. After the first increase stress stays constant or even decreases. Such specimens should be released and then blocked again. Mostly they will swell very little after that (Fig. 7).

On the other hand, specimens with actually high swelling capacity will reach and exceed the previous stress level after some time again.

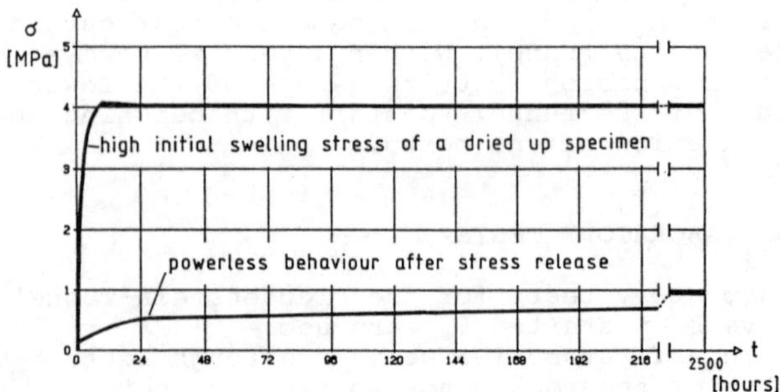

Figure 7. Test to eliminate initial effects due to watering

- Old specimens, which had been watered some years ago and tested for maximum swelling stress in the second series, expand much faster under constant stress than freshly watered specimens. Obviously they didn't loos their capability to swell during the previous test, but are just now in best condition (Fig. 8).

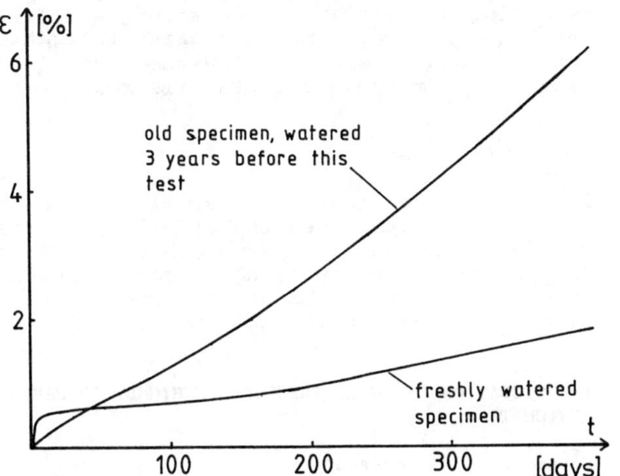

Figure 8. Swelling under constant axial load, σ = 0.5 MPa

Because of this effect we couldn't find any system in the results for a long time. Finally a distinct evaluation of old and new specimens lead to two different groups of stress-strain-curves, each of them for another duration of swelling (Fig. 9).

The continuous curves belong to the group of old specimens. Out of the group of new specimens the 12-months-curve is drawn as dotted line. It is placed between the 3 months and 6 months curves of the other group!

4 FIELD TESTS

800 m from the western end of the tunnel a testing cavern has been built during the last year. It is placed 50 m beside the main tunnel and can be reached through an already finished tunnel section and an access gallery (Fig. 10).

All the structures in the cavern have half the size of those in the original tunnel. The cavern is subdivided in 15 sections, each of them with a different shape and thickness of lining and constructed according to another philosophy.

There are sections with an extremly strong lining which possibly will resist to the maximum swelling stress (Fig. 11 a).

Some sections have the same shape, but a much weaker lining. We expect, that they will break after some time (Fig. 11 b).

Other sections have a lining with smaller curvature of the invert, however, there is a deformable filling of puffed up burned clay at the bottom as a means of stress reduction (Fig. 11 c).

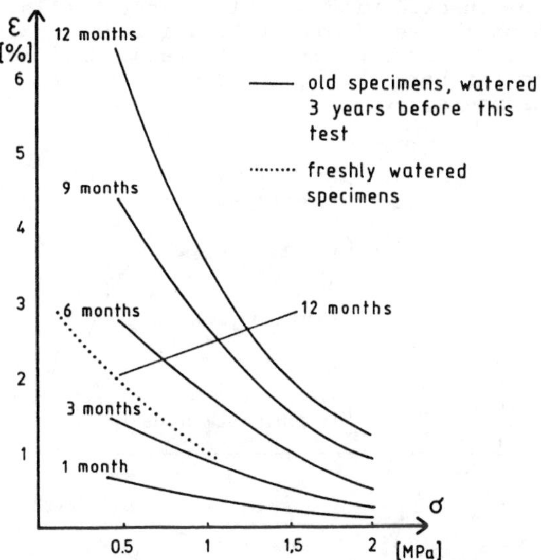

Figure 9. Strain as a function of axial stress after different duration of swelling

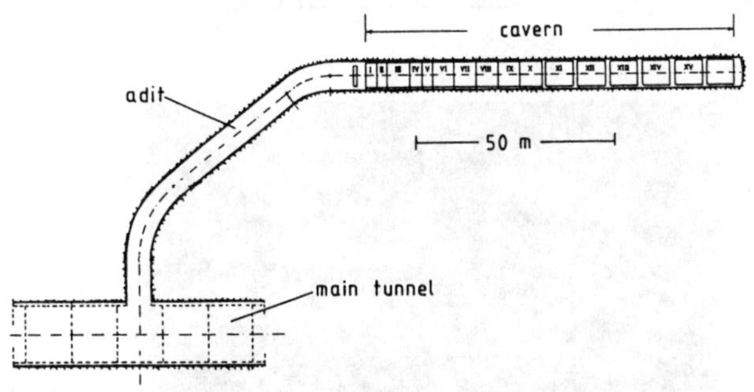

Figure 10. Position and layout of experimental cavern

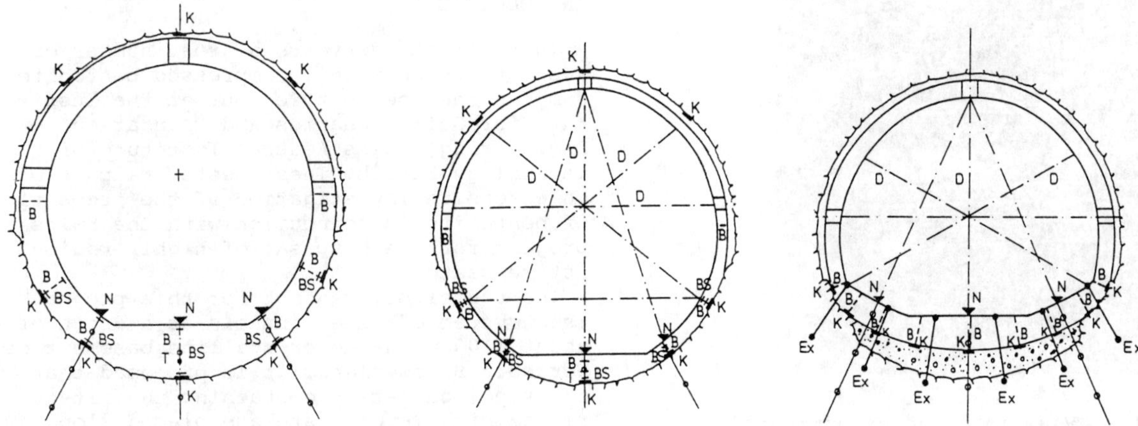

Figure 11. Different sections of the experimental cavern

Each of the sections is equipped with a big number of measuring devices (Fig. 11 and 12). Deformations of the ground below the tunnel are observed by means of sliding micrometers, which can localize even very small deformations in the beginning of swelling. From the side walls horizontal extensometers have been installed.

Deformations of the lining are measured by levelling and with distometers, curvometers and deformeters (Fig. 12).

Joint pressure between rock and lining respectively stress reduction zone and lining can be checked with Glötzl pressure cells. Such cells are also installed inside the lining to measure concrete stresses. Some reinforcement bars are equipped with strain transducers.

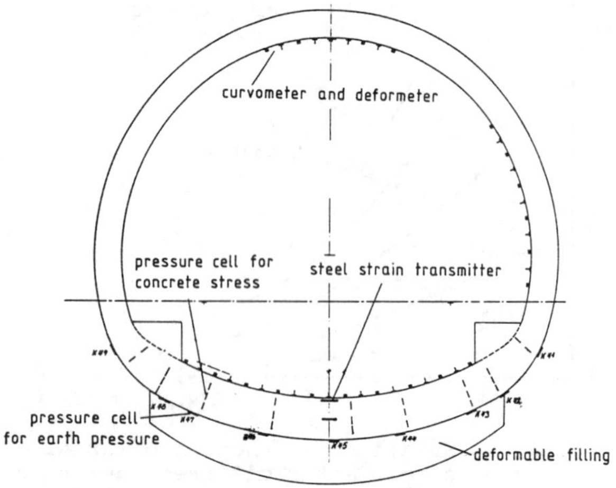

Figure 12. Layout of measuring devices in the tunnel lining

Figure 13. View into the experimental cavern with watering system on the invert

Most of the measuring devices can be read automatically, results are collected and evaluated in a central data station. Nevertheless, several technicians have to care for the tests continuously.

Swelling has been artificially initiated by means of an extensive watering system (Fig. 13) some months ago. Against to our expectations based on the laboratory tests, measuring systems indicated strong swelling reactions already some days after irrigation. Maximum joint pressure below the most rigid section has reached 1.6 MPa till now.

5 CONCLUSIONS

It seems to be clear, that swelling does actually highly threaten the stability of the Freudenstein-Tunnel. However, because of the fast start of the swelling process we have good hopes to find the optimum construction of a safe tunnel still in time.

INVESTIGATION OF SWELL PRESSURE OF HIGHLY COMPRESSED BENTONITE

F. Bucher (Switzerland)

ABSTRACT: The Swiss project for the ultimate storage of highly radioactive waste products involves the use of bentonite as a backfill and sealing material between the waste canisters and the host rock. For this purpose, the air-dry bentonite will be compressed by high isostatic pressures, cut into blocks and placed around the canisters.

One of the important properties of the bentonite, which had to be determined, was its swell pressure. As very high swell pressures were expected a new testing apparatus was designed and constructed which allows measuring swell forces up to 150 kN with vertical deformations of the sample of only a few tenths of a millimeter, varying the temperature of the sample in the range of 20 to 150°C and the pressure in the water supply system from 0 to 10 N mm^{-2}. In the tests sodium bentonite and calcium bentonite have been examined. The results show that the swell pressure is mainly determined by the final dry density of the bentonite. For values of e.g. 1.90 g/cm^3 swell pressures of 30 to 40 N mm^{-2} have been measured. Water pressures which are applied increase these values, but only by an amount of 60 to 70 % of the applied water pressure. The temperature of the sample is of minor influence on the swell pressure, as long as the water pressure is sufficiently high to avoid boiling of the water in the supply system.

INTRODUCTION

In the last few years a great number of swell tests on highly compressed bentonite samples has been carried out at the Institute for Soil Mechanics and Foundation Engineering, Swiss Federal Institute of Technology Zurich. These tests are part of an extensive investigation of the properties of bentonite in connection with the Swiss project for the disposal of highly radioactive waste [1].

The repository concept for this project is a system of mined tunnels at a depth of about 1200 m in the crystalline basement of Northern Switzerland. It is proposed that the steel canisters containing the vitrified high level waste are placed along the axis of the tunnels and are completely surrounded by blocks of bentonite. The bentonite forms a mechanical, hydraulic and chemical protection zone around the waste canisters.

From the many test results obtained in the laboratory investigations on bentonite only those from the swell tests are briefly presented in this paper [2, 3].

SWELL PRESSURE APPARATUS

As very high swell pressures had to be expected and in addition tests at elevated temperatures have been planned, a new test apparatus was constructed at the Institute. This apparatus had to fulfill the following requirements:

- measurement of swell forces up to 150 kN
- specimen temperatures up to 150°C
- water supply from top and/or bottom of specimen with pressures up to 10 N mm^{-2}
- only very small vertical deformation permitted (of the order of a few tenths of a millimeter)
- nearly complete prevention of radial strain by means of a robust cylinder

A section of the apparatus is given in fig. 1. At present more than 20 of these apparatuses are in use, not only for swell tests, but also for diffusion and corrosion studies.

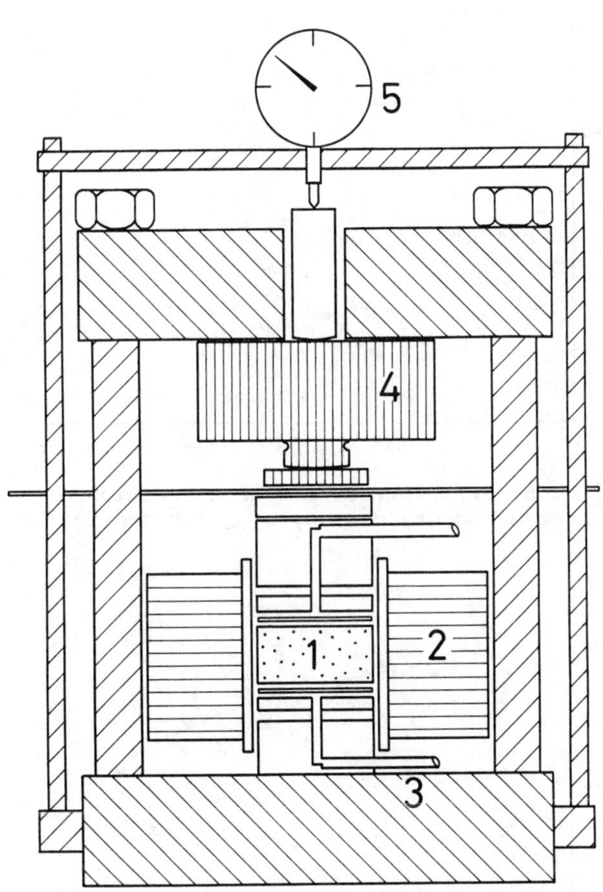

Fig. 1: Section of swell pressure apparatus 1. Specimen 2. heating element 3. water supply connection 4. load measuring cell 5. deformation dial gauge

EXAMPLES OF SWELL PRESSURE TESTS

A typical test result is shown in fig. 2. The test has been performed on MX-80, a sodium bentonite from Wyoming, USA, at a dry density of 1.92 Mg m^{-3}. During this test, only the temperature has been varied. The water pressure was kept constant. The vertical deformation of the sample was about 0.2 mm. The measured swell pressure was of the order of 35 N mm^{-2} and remained fairly constant over a period of 270 days with the exception of two rapid changes in temperature.

An example from a large series of tests, in which the influence of the initial dry density of the sample, the permitted volumetric strains, the temperature and the water pressure were investigated, is given in fig. 3. In this example a sample of Montigel, a calcium bentonite from Bavaria, was examined. The sample was allowed to expand in the early stages of the swell pressure; thus the initial dry density of 1.93 Mg m^{-3} was reduced to 1.70 Mg m^{-3}. The temperature was lowered from the initial value of 150°C to 90°C, and finally to 20°C, whereas the applied water pressure was either 0.6 N mm^{-2} or 10 N mm^{-2}.

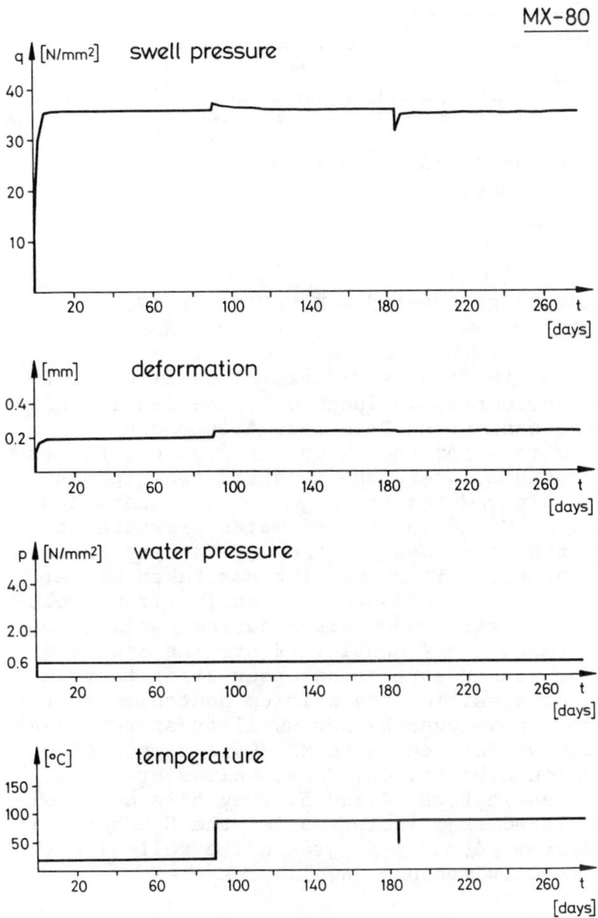

Fig. 2: Typical result of a swell pressure test (MX-80)

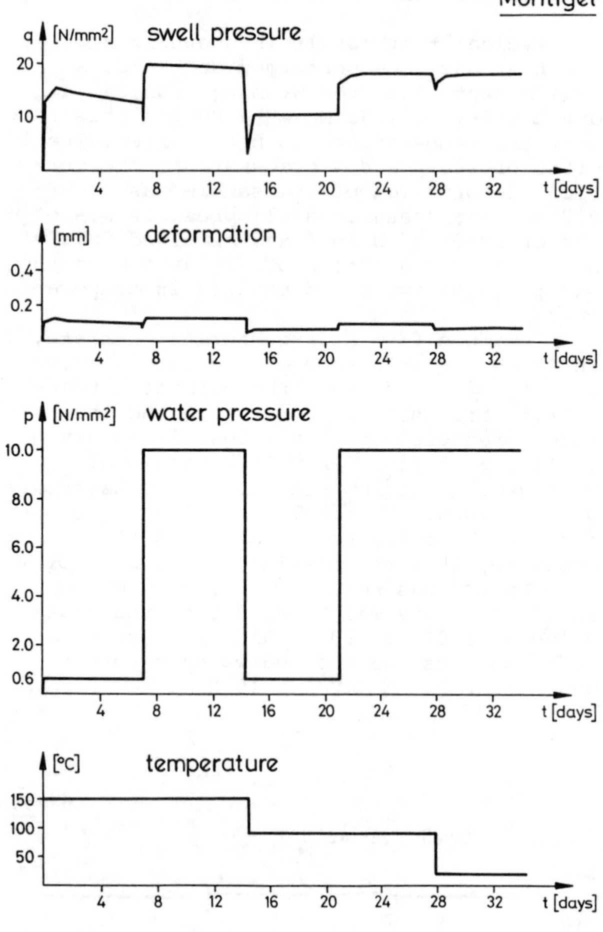

Fig. 3: Typical result of a swell pressure test (Montigel)

SWELL PRESSURE AS A FUNCTION OF THE DRY DENSITY

The tests showed that the swell pressure is primarily influenced by the density of the bentonite. This result is shown in figure 4 for the MX-80, in figure 5 for the Montigel, where the measured swell pressure is plotted against ρ_{red} for a temperature of 20°C and an applied water pressure of 0.6 N mm^{-2}. ρ_{red} is the dry density of the bentonite after swelling has taken place; i.e. the compressed dry density if no volumetric strains occurred during swelling and a reduced dry density if strains occurred. The rather surprising observation that at high densities the calcium bentonite Montigel shows even higher swell pressures than the sodium bentonite MX-80 is supported by calculated values. These values are also given in figs. 4 and 5. They have been calculated from isotherms for the H_2O/bentonite system [4, 5] and agree quite well with the directly measured swell pressures.

INFLUENCE OF THE APPLIED WATER PRESSURE ON THE SWELL PRESSURE

A water pressure which is applied on the sample increases the measured force in the swell apparatus. At high densities, however, there is no full response as is shown in fig. 6.

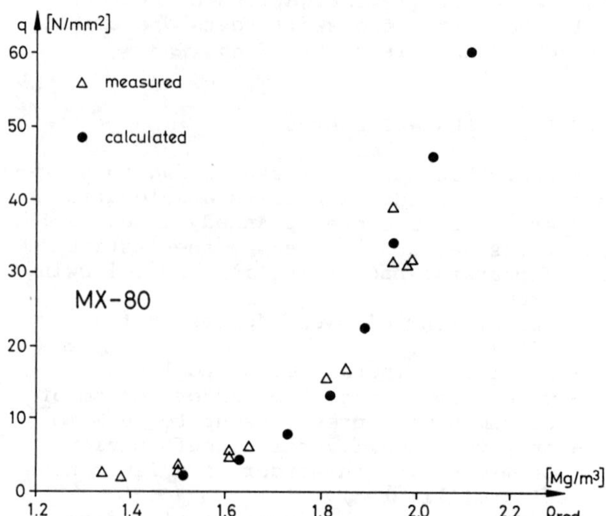

Fig. 4: Swell pressure q as a function of the reduced dry density ρ_{red} for MX-80

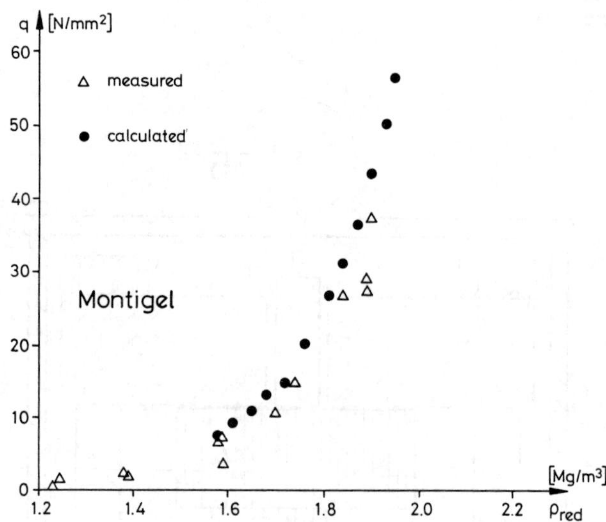

Fig. 5: Swell pressure q as a function of the reduced dry density ρ_{red} for Montigel

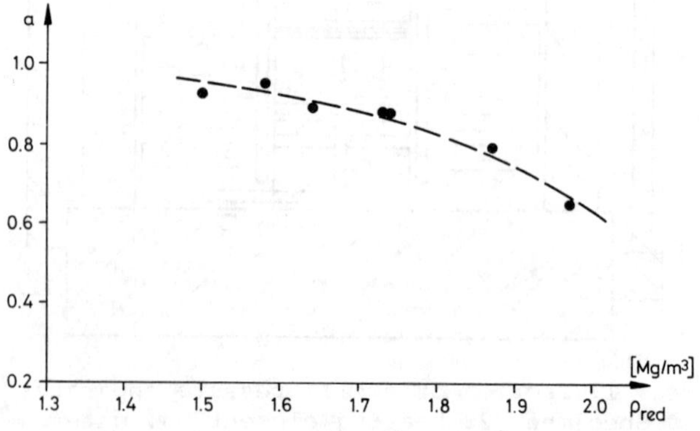

Fig. 6: Coefficient a as a function of the reduced dry density ρ_{red} of highly compressed bentonite

Using the formula

$$q = q_o + a \cdot p$$

where q : total swell pressure
q_o: swell pressure from a test without applied water pressure
p : applied water pressure

one obtains a coefficient a which is only 0.6 - 0.7 at densities of the order of 1.9 Mg m^{-3}. The reason for this reduced response is not yet fully known, but is certainly related to the extremely small porosity and the absence of free pore water in the highly compressed bentonite.

The influence of the temperature is not of great importance if a sufficiently high water pressure prevents the water from boiling.

An important aspect is the long-term stability of the swell pressure. To clarify this effect extensive investigations have been carried out which are still going on.

REFERENCES

[1] Nagra, 1985. Projekt "Gewähr 1985". Endlager für hochaktive Abfälle: das System der Sicherheitsbarrieren. NGB 85-04.

[2] BUCHER, F. and SPIEGEL, U., 1984. Quelldruck von hochverdichteten Bentoniten. Nagra Technischer Bericht 84-18.

[3] BUCHER, F. and MUELLER-VONMOOS, M., 1987. Bentonit als technische Barriere bei der Endlagerung hochradioaktiver Abfälle. Mitt. des Institutes für Grundbau und Bodenmechanik, ETH Zurich (in print).

[4] KAHR, G., KRAEHENBUEHL, F., MUELLER-VONMOOS, M. and STOECKLI, H.F., 1986. Wasseraufnahme und Wasserbewegung in hochverdichtetem Bentonit. Nagra Technischer Bericht 86-14.

[5] KRAEHENBUEHL, F., STOECKLI, H.F., BRUNNER, F., KAHR, G. and MUELLER-VONMOOS, M., 1987. Study of the water-bentonite system by vapour adsorption, immersion calorimetry and X-ray techniques: I. Micropore volumes and internal surface areas, following DUBININ's theory. Clay Minerals 22, 1-9.

WORKSHOP ON CONSTITUTIVE LAWS FOR SALT ROCK

Chairman: M.B. Dusseault (Canada)

The session goal was to present and discuss some current ideas on the constitutive behavior of salt rock, particularly salt. The interest in the behavior of salt rock comes from the nuclear waste isolation programs of several countries, specifically the United States and West Germany, as well as from the need for rational mine design in salt and potash. There is a deeper interest in the behavior of salt rock as well: understanding material behavior and appreciating the difficulties in making long-term predictions of material dominated by time-dependent behavior is of much broader interest as it verges on geotectonics, materials sciences, as well as other areas. Because the nuclear waste disposal agencies demand a considerable degree of "predictability", there is interest in reducing uncertainty, understanding the mechanisms of deformation, and appreciating the variability of natural materials.

There are other considerations beyond nuclear waste isolation. Increasingly, underground storage facilities in salt, usually created by solution mining, are being developed for storage of liquid products, for strategic stores of petroleum products, and for the disposal of toxic (but non-radioactive) liquid wastes. Particularly in the latter area, some knowledge of appropriate constitutive laws describing the rheological behavior of the salt would be of interest, because the organic liquid wastes have an indefinite life, and permitting them to enter into basin flow systems could have undesirable long-term effects.

Mining engineering traditionally has not been concerned with the physics of salt deformation. However, this is changing, particularly with the need to continue mining deep potash deposits and to develop salt mines at greater depths, or in more difficult conditions. For conventional shallow low-annual-volume salt or potash mines, a traditional approach may suffice, but for large scale exploitation at high extraction ratios, the long term behavior is important, and detailed numerical modelling of multiple entry systems consisting of barrier and yield pillars is important. For personnel safety as well as economics it no longer suffices to simply develop a mining approach purely by empirical means. New techniques which include backfilling of potash mine openings, either with straight waste halite or with some type of "enhanced" material, can not be evaluated without correct material constitutive laws, as there must be reasonable predictions as to when it is safe to return to previously mined area to effect second-pass extraction. This requires careful understanding of the behavior of viscoplastic backfill, as well as the behavior of the intact rocks.

The goal of research into constitutive laws for salt rock is to decrease the environmental risk of waste disposal, to increase the safety of miners in the underground environment, and to enhance the economics of mining in general. To these ends, a number of people participated in the the session on salt mechanics.

Dr. Darrol Munson of Sandia National Laboratories in Albuquerque, New Mexico in the USA presented a general statement of the physical and scientific requirements involved in repository work. The United States has two waste programs: the military wastes which are already being stored in salt repositories, and the civilian waste program, which generates must higher toxicity wastes with greater thermal output. The requirement for understanding the thermal properties of saltrock over a significant range of temperatures is a major difference between the repository work and the requirements of mining engineering, and adds a major dimension of complexity to the waste disposal problems. Also, the mining problem tends to be short-term, at least in comparison to the repository problem, which may last perhaps 50 years in an active storage state, then tens of thousands of years after backfilling and abandonment. The issue of fluid flow remains critical, and the mechanical behavior of the bedded salt repository location is intimately tied up to the flow question. The essence of the nuclear civilian program is that reasonable assurance must be given that safe disposal can be achieved in the target material (bedded salt in this case). Extensive performance criteria and guidelines, both scientific and legislative, exist to guide this process.

Dr. Wolfgang Wawersik, also of Sandia National Laboratories, presented information on the microstructure of salt deformed under various conditions. He emphasized that microscopy and the correct interpretation of the deformation mechanisms are closely tied together. Some of the mechanisms which could give rise to macroscopic deformation, but which must be examined on the microscopic scale, include various types of dislocations, vacancy migration, cross-slip, and intercrystalline processes such as Coble creep or superplasticity. Subgrain boundary development and alteration of fabric will be explored through a detailed study of samples taken from some creeping environments which have existed for many years, including domal mines and shallow salt mines in Poland which have been in production since the Middle Ages. Comparison of the microfabric of specimens from these environments to laboratory tested specimens should elucidate the mechanisms, and permit some confidence in extrapolations. It must be emphasized that extensive work in salt mechanics has gone on for a long period at Sandia, and these studies form a small part of a wide spectrum of activities designed to meet the goals of the civilian and military waste programs.

Dr. Chris Spiers from the Earth Sciences Institute at the University of Utrecht in The Netherlands presented information on the compaction behavior of granular salt gathered by him and his colleagues. The role of moisture seems very important to creep processes in salt. The postulated low-stress mechanism (corresponding to the "undefined mechanism" in the well-known Munson-Dawson deformation mechanisms map modelled after work by Ashby) is fluid-assisted diffusional creep. This is deformation behavior that arises due to mass transfer from more highly stressed regions of crystals to regions of lower stress. It can also be called "pressure-solution" or "solution-precipitation". Three possible rate-controlling factors were explored by Spiers and his colleagues: the rate of dissolution, the rate of precipitation, and the rate of diffusional mass transfer in the fluid phase. Their work shows that because salt is so soluble and also precipitates so readily from solution, the controlling factor on creep rate at low porosities is how fast the mass can transfer from one point to the other in the very thin films of moisture which are found between grains. This process is controlled by diffusion, and leads to an exponent of 1.0 on the plastic effective stress in a typical isothermal rate law of the form:

$$\dot{\varepsilon} = A\,(\sigma_1-\sigma_3)^n \qquad \text{(where } n \approx 1.0\text{)}$$

where A accounts for all factors other than stress, including the very important effects of moisture content and mean grain size. These two factors are critical, because the mass transfer rate is governed by them. However, the functional relationship to the moisture content is as yet unclear, and a macroscopic deformation law for intact salt has not arisen from this important work. It must be emphasized that the suggestion that the exponent is 1.0 applies only for the low stress regimes in intact salt: at higher shear stresses, other deformation mechanisms become dominant.

Dr. Michel Aubertin represented work that has been carried out in Montreal at Ecole Polytechnique with the guidance of Drs. Ladanyi and Gill. This work seeks to develop superior approaches to the description of the behavior of salt rock by using an internal state-variable approach based on the existence of an internal polycrystalline stress arising from non-homogeneous distributions of dislocations in the deformed crystals. By including this internal stress variable in the exponential power law of the form shown, a "natural" exponent of 3.0, postulated as being the correct exponent for dislocation, can work quite successfully. Matches to laboratory data for various natural salts from room temperature to 200°C have shown that the approach is very promising. Efforts are continuing on the development of evolutionary models (i.e. with some memory of stress history) to permit the modelling of strain-hardening and transient phase creep.

Dr. Paul Senseny of RE-SPEC (Rapid City, South Dakota), a major contractor to the US Department of Energy on salt testing, behavior, and modelling, outlined the major elements of the behavioral approach he and his numerous associates at RE-SPEC have taken recently. As with most other researchers in this area, they divide salt deformation into four categories: elastic deformation which occurs instantaneously on a change in stress and which is defined by essentially temperature independent elastic constants such as the Bulk Modulus; thermal deformation which is a volumetric change associated with a change in temperature and can be expressed by a coefficient "α", which is not necessarily linear; brittle deformation associated with explicit crystal debonding and dilational damage which occurs particularly at high shear stresses and low confining stresses; and finally, viscoplastic deformation which includes transient and steady-state components, the latter invariably governing the long term behavior of structures in salt rock. The viscoplastic component can be handled adequately by internal variable models such as the Munson-Dawson or Krieg model, which includes an explicit approach to the transient component of viscoplasticity. Appreciable success has recently been achieved in applying this approach in a continuum manner to trial excavations in salt. The major unknown in the whole process is the brittle field behavior and how brittle damage on a crystalline level affects the other processes. This area is worthy of careful investigation.

Dr. Manfred Wallner represented the German national effort on salt rock repository investigations, carried out largely by the German Federal Institute for Geosciences and Natural Resources under the direction of Dr. M. Langer. The major message of his talk is that our simplified models, incorporating the elements discussed above in Dr. Senseny's talk, are significant simplifications of a complex reality, and that it is not possible to incorporate all existing effects. In particular, the empirical formulations which are evolving tend to ignore second-order effects, and concentrate only on the major extrinsic factors such as normal stress, shear stress, temperature, and time. The second-order, more intrinsic factors such as moisture content and grain size, mineralogy and defect density, are handled empirically for each material by laboratory testing and back-fitting models to real room behavior in test mines. All these factors get lumped into a single parameter, such as the parameter A in the equation given. The major technical issue is the validation of these simplified empirical models for use as predictors of long-term behavior. Because they are empirical, validation must also be observational: a theoretical approach cannot validate an empirical equation. In particular, we must find out how conservative or non-conservative the simplified models are in predicting behavior over a period of decades: clearly a strong argument for continued instrumentation of test rooms in well-defined conditions.

Dr. Maurice Dusseault outlined the approach to isothermal creep laws which has evolved at the University of Waterloo (Ontario, Canada). This work has been oriented towards mining engineering rather than repositories, and has been carried out with Dennis Mraz, a consultant who applies it directly to field situations, and Leo Rothenburg, a physicist and modeller. They advocate a constitutive law which explicitly recognises multiple deformation mechanisms at different stresses, similar to the Munson-Dawson approach, but with some differences. Below a transition stress which is characteristic of each particular material (they call it the Prandtl Limit) the exponent on stress in the above equation is 3.0 to 1.0, depending on whether dislocation or diffusion is governing the creep mechanism. Above the Prandtl Limit, exponents higher than 4.0 are the rule and the mechanism is no longer simple dislocation or cross-slip, and a simple power-law rate equation on stress is inadequate. They showed data which suggests that the deformation mechanism at high shear stresses is associated with cracking and healing because the rate curves show a normal stress dependency. Therefore the exponent on the rate law is not a constant above the Limit, but is a function of normal confining stress, which suppresses the cracking process and provides energy to heal the cracks. The healing is also aided by moisture, likely by the fluid-assisted diffusional creep mechanism postulated by Spiers. Application of the results to isothermal field cases has given reasonable matches, even for five years of deformation in virgin ground openings.

The seminar went a long way to aid public exposition of the various approaches to salt rock deformation, and considerable debate took place after the session, and on into the evening over a delightful meal at an excellent French restaurant in Montreal. Clearly, the four main aspects of deformation are agreed on (brittle, elastic, thermal, viscous), and there is considerable agreement as to the general form of the rate equations for the four major processes. If there was an important new concensus that evolved, it was that it is important to know the explicit mechanisms involved in the creep process, and what are the more detailed effects of extrinsic and intrinsic variables such as stress, temperature, moisture content, grain size, and mineralogy. Everyone agrees that using the constitutive laws to apply to field data is critical, and will eventually prove the worth of the various approaches, which actually do not differ radically one from the other. It is clear that each of the approaches is part of a generally agreed-on view of deformation processes in salt, and that there are likely to be different mechanisms operating at different conditions. The state-of-the-art is still in somewhat a state of flux, but the achievements of the last decade have been impressive indeed: the rock mechanics of salt rock is on the way to a reasonable solution.

A major event has taken place in the United States since the 6th ISRM conference in Montreal: the United States Congress has cancelled all geological material repository studies other than those associated with the unsaturated zone tuff at the Nevada Test Site. This decision clearly solves a major political problem in that the Site is owned by the Department of Energy and is a low-population state, isolated from others. However, considering the major developments made in salt mechanics in the last few years, and considering the low hydraulic conductivity of salt rocks, the writer (probably accompanied by all the co-presenters in the seminar) wishes to emphasize that total abandonment of the salt repository concept is a premature decision, likely to be regretted in the future. The decision has come at a time when other countries (France, Great Britain) are beginning to look at salt more closely as a medium of superior qualities for waste disposal. The Germans already are well advanced in the salt repository area, and have a wealth of practical operational experience which is of considerable value.

At the very least, an intermediate level of activity should be maintained, and a smaller-scale demonstration version of a test facility is worthy of pursuit. Also, the use of low moisture content domal salt, abandoned in the early 1980's, should not be rejected purely in favor of bedded salt. The former has excellent low permeability characteristics, with low moisture contents, and without the problems associated with deep bedded strata which include non-salt stratification, higher moisture content, the presence of thick elastic strata in the roof, and so on. The identification of moisture-assisted diffusional creep as the likely dominant mechanism at low shear stresses suggests that dryer salt is better salt. It is to be hoped that the DOE does not totally abandon the valuable work which has gone on in the salt rock geomechanics area.

Submission received after the workshop

THE ACTIVE STRESS CONCEPT AND THE CREEP OF ROCK SALT

M. Aubertin (Canada)
D.E. Gill (Canada)
B. Ladanyi (Canada)

INTRODUCTION

The time-dependent plastic behavior of rocksalt as observed in laboratory testing has received much attention during the past two decades. As a result, the macroscopic and microscopic aspects of the latter seem to be now fairly well understood (see for instance reviews by: Handin et al., 1986; Wawersik and Zeuch, 1986; Aubertin et al., 1987a et 1987b).

The constitutive law used to define the behavior of rocksalt in relation with the analysis and the design of underground structures is often based on a power law equation of the following type (e.g., papers in these proceedings by: Preece, 1987; Côme, 1987; Borm, 1987):

$$\dot{\epsilon}_{ss} = A\,\sigma^n\,f(T) \tag{1}$$

where $\dot{\epsilon}_{ss}$ is the steady state deviatoric strain rate, σ is the deviatoric stress, A and n are constants and f(T) is a temperature function which is generally assumed to be the Arrhenius law. In this expression, the value of parameter n is usually close to 5.

In the following, equation (1) will be extended by introducing an internal state variable, $\bar{\sigma}_i$, defined as the mean long-range internal stress, so that the stress function becomes an active stress function, which is also called effective stress function by some other authors in the field of material sciences.

FUNDAMENTAL CONSIDERATIONS

From a micromechanistic point of view, the constitutive laws should account for a kinetic law and an evolution law. The former is often taken as the following function (Frost and Ashby, 1982; Poirier, 1985):

$$\dot{\epsilon} = g\,(\sigma,\ T,\ S_i\,,\ P_j) \tag{2}$$

where S_i is a variable associated with the microstructural internal state of the material (such as the dislocation density, the cell dimensions, the substructure evolution, etc.), and P_j is a nearly constant parameter related to the specific value of material properties (such as the Burgers vector, the stacking fault energy, the atomic volume, etc.); as before, σ denotes the stress state and T, the temperature.

The paremeter S_i, which is the key parameter in all internal state approaches, is affected by the plastic deformation, whether instanteneous or time-dependent. Its variation can be described by an evolution law defined by the following function:

$$\dot{S}_i = h\,(\sigma,\ T,\ S_i) \tag{3}$$

which assumes P_j to be a constant.

This evolution law is poorly understood, so that it is often postulated that S_i is a constant, or $\dot{S}_i = 0$; the kinetic law thus becomes:

$$\dot{\epsilon} = g'\,(\sigma,\ T) \tag{4}$$

This type of formulation, often used to describe the creep of metals, ceramics and other crystalline materials, implicitly admits that the mechanical behavior of the latter is not an explicit function of the microstructural state of the material (as it is the case with equation 1).

However, it has been demonstrated that there are many advantages in introducing an internal state variable, S_i, in the constitutive law of crystalline materials, especially when complex stress path and strain history are anticipated (McClintock and Argon, 1966; Hart et al., 1975). Viscoplastic laws making use of these advantages have recently been reviewed by Ohashi et al. (1982) and by James et al. (1987).

Many parameters can be used to characterize the microstructural state of the material. The present discussion paper suggests for this matter a particular formulation which makes use of the active stress concept presented by Solomon and Nix (1970) and Malinin and Khadjinsky (1972). This concept has been recently proposed for metals (Argon and Takeuchi, 1981; Oytana et al., 1982).

INTERNAL STRESS

Internal stresses are best explained by the composite structure model of Nix and Ilschner (1980) and Mughrabi (1983). The cellular structure observed in most plastically deformed crystals is formed by hard regions, where dislocation density is high, and soft regions, where dislocations density is low. The heterogeneous distribution of the dislocations gives rise to a long range internal stress, σ_i, as a consequence of the compatibility of the stress-strain behavior between hard and soft regions. Over the entire volume, we thus have an active stress σ_a which is responsible for the creep of the material. This active stress can be expressed as:

$$\sigma_a = \sigma - \overline{\sigma}_i \tag{5}$$

where $\overline{\sigma}_i$ is the average long-range internal stress over the entire volume. The increase of $\overline{\sigma}_i$ as a result of strain hardening mechanisms and its decrease as a result of recovery, explain transient phenomena such as isotropic and kinematic hardening and softening.

When σ_a is introduced into equation (1), one obtains:

$$\dot{\epsilon}_{ss} = A' \, (\sigma - \overline{\sigma}_i)^m \, f(T) \tag{6}$$

It has been observed in many cases that when such an equation is used to describe the creep of polycrystalline materials, the value of m is very close to 3, which is often considered to be the natural power law exponent of stress (Frost and Ashby, 1982; Langdon, 1985).

To apply the above formulation to the rocksalt data available in the literature, one may refer to the following relation, obtained by Pontikis (1977) for NaCl single crystals:

$$\frac{(\dot{\epsilon}_i)_2}{(\dot{\epsilon}_{ss})_1} = \left(\frac{\sigma_2}{\sigma_1}\right)^{7.3} \tag{7}$$

with $(\dot{\epsilon}_i)_2$ being the initial creep strain rate immediately following a stress increase from σ_1 (to which corresponds $(\dot{\epsilon}_{ss})_1$) to σ_2. Equation (7) is a constant-structure equation, independent of temperature and testing conditions.

Results of sample calculations are given in figure 1 in which the published data used are those for Avery Island Salt, taken from Hansen and Carter (1984) and Senseny (1984). The values of modulus of rigidity G used in these calculations have been corrected for temperature according to Eggeler and Blum (1981). The dotted line, corresponding to $\overline{\sigma}_i = \sigma$, represents the upper bound values of the internal stress. A linear regression analysis of these results indicates that

$$\overline{\sigma}_i = -0.05 + 0.72\sigma$$

with a coefficient of correlation of 0.97 (see the full line curve in figure 1). This implies that the average $\overline{\sigma}_i$ value is very close to 70% of the deviatoric stress in most cases. This percentage is about the same as that obtained by Hunsche (1984), who has performed experimental measurements of σ_i using dip test technique. A similar ratio of σ_i/σ has been frequently encountered, for other testing conditions, with different materials (Derby and Sherby, 1987).

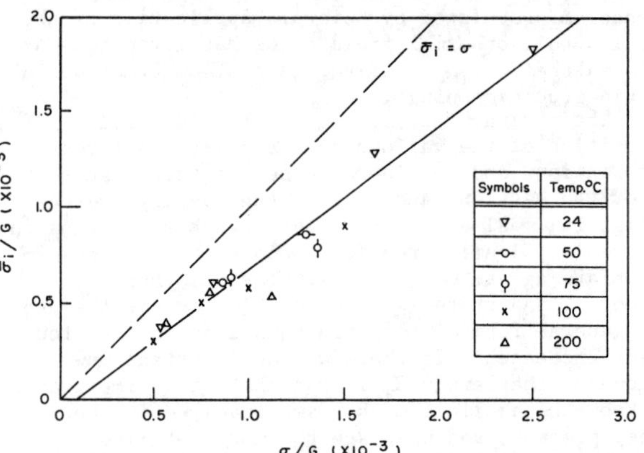

FIGURE 1: Results of sample calculations of the average long-range internal stress from data published by Hansen and Carter (1984) and by Senseny (1984); all are for Avery Island Salt.

REFERENCES

ARGON, A.S., TAKEUCHI, S. (1981): Internal stresses in power-law creep. Acta Metall., Vo. 29, pp. 1877-1884.

AUBERTIN, M., GILL, D.E., LADANYI, B. (1987a): Le comportement rhéologique du sel: Revue bibliographique - Tome I: Essais en laboratoire et modélisation empirique. Éditions de l'École Polytechnique de Montréal, Rapport EPM/RT-87/31, 204 pages.

AUBERTIN, M., GILL, D.E., LADANYI, B. (1987b): Le comportement rhéologique du sel: Revue bibliographique - Tome II: Mécanismes de déformation, modélisation physique, rupture, essais en place et comportement in situ. Éditions de l'École Polytechnique de Montréal, Rapport EPM/RT-87/32, 219 pages.

BORM, G. (1987): Bohrlochkonvergenz und spannungsrelaxation in steinsalzgebirge. Proc. 6th Cong. ISRM, Vol. 2, pp. 819-824.

CÔME, B. (1987): Benchmarking rock mechanics computer codes: the community project COSA. Proc. 6th Cong. ISRM, Vol. 1, pp. 55-60.

DERBY, B., ASHBY, M.F. (1987): A microstructural model for primary creep. Acta Metall., Vol. 35, pp. 1349-1353.

EGGELER, G., BLUM, W. (1981): Coarsening of the dislocation structure after stress reduction during creep of NaCl single crystals. Phil. Mag. A, Vol. 44, pp. 1065-1084.

FROST, H.J., ASHBY, M.F. (1982): Deformation-Mechanism Maps - The Plasticity and Creep of Metals and Ceramics. Pergamon Press.

HANDIN, J., RUSSELL, J.E., CARTER, N.L. (1986): Experimental deformation of rocksalt. Amer. Geophys. Union, Mono 36, pp. 117-160.

HANSEN, F.D., CARTER, N.L. (1984): Creep of Avery Island rocksalt. Proc. 1st Conf. Mechanical Behavior of Salt. Trans Tech. Pub., pp. 53-69.

HART, E.W., LI, C.Y., YAMADA, H., WIRE, G.L. (1975): Phenomenological theory: a guide to constitutive relations and fundamental deformation properties. Constitutive Equations in Plasticity, MIT Press, pp. 149-197.

HUNSCHE, V. (1984): Measurement of creep in rock salt at small strain rates. Preprint 2nd Conf. on

Mech. Behavior of Salt, Hannover, 10 pages.
JAMES, G.H., IMBRIE, P.K., HILL, P.S., ALLEN, D.H., HAISLER, W.E. (1987): An experimental comparison of several current viscoplastic constitutive models at elevated temperature. Journ. Engng. Mat. Techn., ASME, Vol. 109, (April), pp. 130-139.
LANGDON, T.G. (1985): Dislocations and creep. Dislocations and Properties of Real Materials. Institute of Metals, London, pp. 221-238.
McCLINTOCK, F.A., ARGON, A.S., (1966): Mechanical Behavior of Materials. Addison-Wesley.
MALININ, N.N., KHADJINSKY, G.M. (1972): Theory of creep with anisotropic hardening. Int. Journ. Mech. Sc., Vol. 14, pp. 235-246.
MUGHRABI, H. (1983): Dislocation wall and cell structures and long-range internal stresses in deformed metal crystal. Acta Metall., Vol. 31, pp. 1367-1379.
NIX, W.D., ILSCHNER, B. (1980): Mechanisms controlling creep of single phase metals and alloys. Strenght of Metals and Alloys, Vol. 3, Pergamon Press, pp. 1503-1530.
OHASHI, Y., OHNO, N., KAWAI, M. (1982): Evaluation of creep constitutive equations for type 304 stainless steel under repeated multiaxial loading. Journ. Engng. Mat. Techn., ASME, Vol. 104, (July), pp. 159-164.
OYTANA, C., DELOBELLE, P., MERMET, A. (1982): Constitutive equations study in biaxial stress experiments. Journ. Engng. Mat. Techn., ASME, Vol. 104, (January), pp. 1-11.
POIRIER, J.P. (1985): Creep of crystals - High temperature deformation processes in metals, ceramics and minerals. Cambridge University Press.
PONTIKIS, V. (1977): Phenomenological analysis of recovery - controlled transient creep supported by substructural observations in single crystals of sodium chloride and silver chloride. Acta Metall., Vol. 25, pp. 847-855.
PREECE, D.S. (1987): Borehole creep closure measurements and numerical calculations at the Big Hill, Texas SPR storage site. Proc. 6th Cong. ISRM, Vol. 1, pp. 219-224.
SENSENY, P.E. (1984): Creep properties of four rock salts. Preprint 2nd Conf. on the Mechanical Behavior of Salt, Hannover.
SOLOMON, A.A., NIX, W.D. (1970): Interpretation of high temperature plastic deformation in terms of measured effective stresses. Acta Metall., Vol. 18, pp. 863-876.
WAWERSIK, W.R., ZEUCH, D.H. (1986): Modeling and mechanistic interpretation of creep of rock salt below 200°C. Tectonophysics, Vol. 121, pp. 125-152.

WORKSHOP ON NUMERICAL METHODS AS A PRACTICAL TOOL

Chairman: B. Plischke (Germany)

The Workshop "Numerical Methods as a Practical Design Tool" was organized by the ISRM - Commission on Computer Programs. As there are various conferences dealing with the development and application of numerical methods in geotechnics, this workshop should not compete with these conferences. The limited time should rather be used to give some insight into the hardware and software developments, which enable to become numerical methods a valuable powerful and economic tool for design of structures in rock, not only to be handled by specialists.

As a starting-point, it was recalled that the design of structures in rock differs considerably from other engineering design tasks:

- The strength and deformability of the load bearing material cannot be chosen a priori by the engineer. Instead, these properties must be evaluated from a limited set of field and laboratory data.
- Also the loads acting on the stucture, as in situ stresses, are not known a priori.

Without doubt, numerical methods are principally suited best to tackle the problems:

- Practically any stress strain law can be incorporated.
- Complex geometries and threedimensional stress states can be handled.
- The influence of the uncertainty in the knowledge of parameters can be quantified by parametric studies.
- The evaluation of field measurements by means of numerical back analysis of the results yields realistic rock mass parameters.

Design concepts, as proposed for example, by WITTKE already in 1975, include numerical studies at various stages of design and construction. These concepts are nowadays more and more applied in practice.

Nevertheless, it is often stated that finite element studies are very expansive with respect to the improvement of the design gained by these calculations.This is true, as long:

- as the programs used are not userfriendly and require the employment of specialists,
- as a lot of time is spent with the preparation of the input, the interpretation of results and the translation of numerical results into those quantities which are of interest for the design, i.e. for example the amount of reinforcement steel in a tunnel lining,
- as programs do not exploit the high performance of nowadays computers and latest software developments,
- as computer costs are evaluated on the basis of costs of old mainframe computers and not on costs for workstations.

In this context it should be mentioned that most of the general purpose structural analysis FE-packages, which are commercialized worldwide and offer high performance, good support and continous improvement, are not suited sufficiently to meet the specific requirements of rock mechanics applications. Therefore, commonly rock engineers have developped a lot of specialized codes themselves.

The three presentations of this workshop attempted to demonstrate how the practice of application of numerical methods has changed during the last years. The first presentation was given by B. ALCOCK (IBM Canada Ltd., Toronto). He reported on "Increasing Performance and Decreasing Costs - Hardware Developments Enable Economic Use of Numerical Methods". He introduced the concept of a workstation on the engineer's desk and outlined its capability to solve even very complex problems. The engineer's schedule is not hampered by any limitations which are necessary for the operation of a large main frame computer which serves many users. Nevertheless, the workstation may be integrated into a network and thus connected to other computers and peripheral equipment. While the performance of new computers is growing from year to year, at the same time the prices decrease. As this trend will continue, the pure computation costs will no longer be a significant cost factor.

Subsequently, P. FRITZ (Swiss Federal Institut of Technology, Zurich) demonstrated that many standard finite element analyses of tunnels can today even be performed on a Personal Computer. The participants of the workshop watched on the screen projector how Dr. Fritz prepared the input for a 2D-analysis of a tunnel on a PC by means of an interactive mesh generator. The used program was a full implementation of the mainframe version of the RHEO STAUB code, and allows for up to 8000 triangular elements. The minimum requested hardware configuration is an IBM-PC XT, AT or compatible with 512 kB RAM and a hard disk of 10 MB. For small problems it takes only the duration of a coffee break to perform an elastic calculation. In his very enjoyable presentation, entitled: "Combination of Artificial Intelligence, Finite Elements and Graphics Editor on Personal Computers for the Design and Computation of Tunnels" Dr. Fritz also introduced a prototype of an "expert system" which may help the users to account for the various aspects of geological, topographical and rock mechanics conditions, which must be considered in a tunnel analysis. Unfortunately, time limitations did not allow for a discussion of the various aspects and implications of such a system.

The third presentation was given by B. BRADY (Itasca Consulting Group, Minneapolis). He introduced "Discontinuum Codes and Linked Boundary Element Continuum Codes in Rock Mechanics". Dr. Brady also took the opportunity to display some examples on the application of these codes, which run also on a personal computer. The discontinuum codes, mainly based on CUNDALL's developments, especially account for the dominant influence of joints and faults on the mechanical behaviour of rock mass. The easiness with which large displacements and rotations can be treated, for example in the analysis of blasting effects, will surely make these codes a valuable tool in rock mechanics, also in those countries, where discontinuum models are presently not yet well established.

Finally, it should be mentioned that the exhibition associated with the congress presented several other interesting applications of numerical methods in rock mechanics.

WORKSHOP ON ROCK FAILURE MECHANISMS AROUND UNDERGROUND OPENINGS

Chairman: V. Maury (France)

The schedule of the workshop was as follows:
- Introduction of the workshop by V. MAURY (15').
- Observations and classification of failures according to regional styles in the United States by E. CORDING (10').
- Observations and comments about failures in hard rocks under high stress in South Africa by D. STACEY and D. ORTLEPP (15').
- Failure in Finish mines by P. SARKKA (10').
- Laboratory test observations by N. BARTON and S. BANDIS (15').
- Experimental work on hollow cylinders by R. EWY (8').
- New Elastic stress distribution by F. SANTARELLI (6').
- Fundamentals of the theory of bifurcations to describe rock failure by I. VARDOULAKIS.
- Summary of the presentations and introduction to the discussion by V. MAURY (5').
- Discussion (35,).

Each of these items will be summarized in the text below.

A summary of the results of the national groups of the Commission on Rock Failure Mechanisms around Underground Excavation is given in Table 1. Note that the work of the Belgium, Dutch, Japanese and Australian groups which were not presented at the workshop were discussed during the pre-Congress meetings of the Commission on Rock Failure Mechanisms around Underground Openings and can be found in the minutes of these meetings. The first report of the Commission (1) is briefly presented and its main conclusions are summarized:
- In the case of deep boreholes (i.e. openings with an internal pressure), various modes of failures can occur depending on the stress distribution at the wall.
- The effects of thermal stresses can be important.
- A new approach in elasticity seems promising to explain various aspects of undergrounds excavation stability which were until now unexplained.

The scope of the Commission is reemphasized and consists in:
- A large collection of data.
- A definition of the various mechanisms of failure occuring in specific conditions (e.g. room and pillar mining, solution mined caverns, deep boreholes, long wall mining, etc...).

The presentations of the workshop will be made around these two axis - on the one hand field cases and on the other failure interpretation and prediction in the case of a single circular opening. The list of speakers is introduced and Pr. VARDOULAKIS is thanked for coming especially from Minneapolis to participate to and to address the workshop.

Cording presents a classification of failure cases observed around underground excavations in the USA. The first parameter of the classification is the part of the rock mass controlling the behaviour (intact rock, joint...) and the second parameter describes the environment (brittle failure, ductile failure, swelling rock, body force...). Cording points out the fact that many cases show combined environment with various part of the rock mass being affected by failure. This classification is illustrated by various case histories which are presented (e.g. a pressure water tunnel in Utah at 600 m depth).

Stacey and Ortlepp present observations made in deep mines of South Africa. They classify failures as corresponding to three different mechanisms:
- the most common is extension or indirect tensile fracture.
- direct tension also occurs and is often combined with gravity.
- Shear has not really been proved.

---

(1) (See paper, p. 1119 Proc. 6th ISRM Congr.)

NG - I.S.R.M. COMMISSION ROCK FAILURE MECHANISMS U.O. 30 AUGUST 87

COUNTRY	SPECIFIC TOPIC	STATUS, RESULTS, COMMENTS
AUSTRALIA B. BAMFORD J. ENEVER W. GALE J. ASKEW	Case histories record : - Coal mines - Civil Engineering - Hard metalliferous Mines	- Collection already done Civil Engineering (in progress). - Analytical data collected - In progress
BELGIUM B. BRYCH	- Single openings, rheological behaviour - Application to mining, Civ. Engng. - Application vert. horiz. drilling	- Meeting Louvain (CR), thesis (3). - Research program (University Industry) - Coordination EEC program
FRANCE V. MAURY	- Civ. Engng. Single openings.(shear, extention, failure modes). - Swelling rocks. - Failure in specific conditions	- 3d Edition of report - Claim of underground quarry, lab. test report issued - Report in preparation - Seminar lecture bifurcation
JAPAN P. SAKURAI	- Cases of failures in soft and very soft rock - Associated problems of construction	- Collect. case histories carried on - 1st Council Sydney 83 - New report in preparation
NORWAY N. BARTON (R. LIEN)	- Failure controled by in-filled weak material - Failure/single openings	- Rep. research undertaken : . Causes of caveins in water tunnels . Failures modes deep boreholes . Physical models of overstress rocks . Gouge material
S. AFRICA D. STACEY (D. ORTLEPP)	- Failure in hard rocks submitted to high stresses (single openings, mining).	- 4 cases histories analys. (Cambridge 84) - Reports on extension criteria - Discussion of 1st report of Commission
U.S.S.R. P. DMITRIEV	- General topic of failure	- Working group, probably active but great difficulty in relationship
USA E. CORDING	- Failure styles according to regional environment	- 1st classif. of failure according to main regional factors. - Interim report issued - Report in preparation
	OTHER CONTACTS	NO FORMAL GROUP
FINLAND P. SARKKA	- Some cases reported in hard rock, finish mines	- Draft report in preparation
NETHERLANDS P. ROEST P. GRAMBERG	- Theoretical and physical aspects of initiation and propagation of failure - Link with comment behaviour Griffith's theory	- Several reports issued. - Discussion, comments 1st report - Special report provided

They present many illustrations of remarkable quality. Ejection of fine material - at the grain scale - is compared to a micro rockburst. The periodic slabing leading to the formation of what is known as dog-earings is shown to take place at different scales. They indicate that the periodicity of these extension fractures is currently studied with great care. Another topic under investigation is the composition and structure of the gouge found in shear planes. This shear mechanism which affects the rock mass itself, shows mainly series of "en echelon" fractures. They insist on the fact that any useful case history must contain informations about the failure mechanism.

Sarkka presents briefly the rocks of the Finish bedrock before he describes failure around various mines. Buckling at all scales due to different elastic properties of hard and soft layers is observed in stratified media. Failure in room and pillar mines is presented alongside with F.E.M. computations. The lining method is also presented.

Bandis presents laboratory investigations carried out in Norway for oil companies on small scaled models of wells tested in a true triaxial cell. The artificial material used during his study is compared with the real material - rock -. The originality of the approach is a direct measurement of the stresses inside the rock itself. Stress distributions around the hole are therefore measured. Note that the rock is abnormally stable. This is interpreted by using a linear - strain hardening - strain softening - residual elasto-plastic model. The importance of the influence of the intermediate principal stress is emphasized.

Ewy presents experimental work performed at Berkeley with Pr. Cook on thick walled hollow cylinders of Berea Sandstone. The development of non axisymmetric failure zones - dog ears - from an axisymmetric specimen and loading is explained by the anisotropy of the tested material. The strength enhancement of the wall of the hollow cylinder is explained by the size effect and the influence of the intermediate principal stress. The importance of the stress path is emphasized.

Santarelli presents work performed at Imperial College with Pr. Brown. The influence of the variation of Young's modulus with confining pressure on the stress and strain distributions around the hole is studied. The possibility to explain at the same time, the "abnormal" curvature of the strain pressure curves of lab tests on small scaled models, the apparent strength enhancement at the wall of underground excavations, and the location of failure initiation inside the rock is emphasized. Photographs of tested hollow cylinders illustrate various aspects taken by the failed zone depending on the nature of the rock - a dolomite and a sandstone.

Vardoulakis presents the theory of bifurcations in a "down to earth" manner. A first important point in his talk is that the term failure must be defined more precisely. As far as rocks are concerned, two kinds of inhomogeneous deformations or past bifurcation paths can be defined. One is the axial splitting of the specimen or surface instability and the other is shear plane. This can in turn lead to two forms of instability around the hole. The behaviour law of the material must be able not only to predict when failure initiates but also how it forms (by surface instability or extension or by shear). The text of this communication is presented in Appendix.

V. MAURY summarizes the various points made during the presentations and opens the discussion by asking the following questions:

- What are the possible explanations of abnormal stability - in terms of conventional criteria?
- What are the true mechanisms of failures development - extension or true shear - ?
- Can the bifurcation theory explain the (pseudo) periodicity of the failure lines?
- How can we predict initiation and propagation of failure?
- What will the final shape of the excavation be?
- Can core discing give information about the periodicity of failure when it is experienced?

First of all, Heuze questions Ortlepp about the possibility of the determination of the velocity of fracture propagation from the observations of a purely tensile failure in the roof of a coal mine (see paper p. 1173 Proc. 6th ISRM Congr.). The similarity of this observation with old work of the 1900's is raised by Einstein. Finally, Gramberg proposes a mechanism to explain the purely tensile failure observed by Ortlepp.

Tanimoto raises the very important point of the boundary conditions and effects in the case of an already failed specimen tested by Bandis and Barton. Bandis replies that the hole is 3.5 cm diameter while the specimen is 30 cm wide. The ration R2/R1 = 8.6 should be enough to make the end effects negligeable. However, the size of the block will be increased up to 50 cm to improve this aspect of the tests.

Maury asks Vardoulakis which parameter can be hold responsible for the periodicity of fracture. Vardoulakis explains that this point is under intensive research at the moment.

Theories using flaw distributions and their critical wave length, theories introducing the size of the grains (sphere model by Mülhaus) are used. However, the most promissing approach consists in using a 2nd order Taylor development as a strain displacement relation instead of a 1st order development. The higher degree terms are then linked with the spacing of discontinuities or microcracks.

Pr. Habib uses a microscopic description to question the existence of extension fractures. Vardoulakis shows how the theory of damage can be used to explain why tensile stresses exist during uniaxial tests. Stacey reminds of triaxial tests showing such fractures.

Charlez questions the existence of true shear (mode II) in fracture mechanics i.e. the capacity for a crack to extend in a direction other than the maximum principal stress. Santiago Uriel goes in that sense by talking about the axial cleavage around mines - i.e. extension failure parallel to the maximum principal stress.

Plishke presents pseudo periodic failures in a soft layer between two harder observed around coal mines in Germany.

Finally, Maury concludes the meeting. Despite the ongoing discussion about failure mechanisms, it seems that indirect tensile failures exist around galery, were seen in the laboratory and correspond to the majority of the reported cases. However, no definitive conclusion can be made in this area yet. He emphasizes the challenge to the theory of bifurcation which consists in predicting properly the failure mode for a material whose volumetric behaviour is difficult to measure, and the difficulty to apply this theory to rock mechanics compared to soil mechanics where measurements of the geomechanical properties may be less difficult.

APPENDIX: Text of the Communication by I.G. VARDOULAKIS

UNDERGROUND INSTABILITIES AS BIFURCATION PHENOMENA

I.G. VARDOULAKIS, UNIVERSITY OF MINNESOTA
A. GUENOT, ELF AQUITAINE

1. Background

Failure in underground openings manifests itself in a great variety of mechanisms, like for example shear failure or slabbing of the wall. In order to predict these phenomena several continuum or fracture mechanics models have been suggested. These models use failure criteria that are usually determined experimentally; e.g. Mohr-Coulomb or Griffith criterion. However the aspect of instability is greatly overlooked. As it is well known from structural engineering, structures may fail even in the elastic regime of stresses through buckling instabilities (e.g. Euler buckling). One should thus distinguish between the strength of the material and the stability of the structural component or even the structure as a whole. This approach which is classified as bifurcation analysis, can be also used in the description of failure phenomena in underground openings (cf. Guenot, 1987).

2. Present State

Continuum bifurcation theory allows a unifying approach to the above stated problems. Bifurcation of a deformation process means that at some critical state the deformation process does not follow its "straight-ahead" continuation but it turns to an entirely different mode. Typical examples of such a spontaneous bifurcation are barrelling, buckling, necking, splitting and shear-band formation that are observed in laboratory tests. Bifurcation loads do not only depend on material properties but also on the geometry of the structure and the boundary conditions for loads and displacements. Typical example of this is the so-called shape effect in rock mechanics testing: i.e. the effect of the slenderness of the specimen on the measured failure load. A first application of these ideas in the context of the mechanics of rock bursting was done by Vardoulakis (1984) who found that shear-banding or surface instabilities may occur depending on the rock type. Surface instabilities are assumed to be the trigger mechanism for opening of latent preexisting, surface-parallel cracks, which may subsequently propagate. In this sense bifurcation theory is the link between continuum and fracture mechanics.

These ideas have been recently applied to study borehole stability (Vardoulakis and Papanastasiou, 1987; Vardoulakis et al., 1987). These studies demonstrated that the mode of failure depends strongly on the rock type. This means that predictions depend strongly on reliable experimental data on the pre-failure stress-strain behaviour of the material. Volume-change behaviour is found to be of critical importance. It was also found that for classical material description the stress gradient is not important.

3. Future Work

Bifurcation theory must be used in a thorough reexamination of procedures and interpretation of rock mechanics tests as it is already the case in the realm of soil mechanics. This eventually will result in the lay-out of new procedures and the design of new tests.

Theoretical work should focus on introducing appropriate constitutive laws for rock. In order to provide a better link between continuum and fracture mechanics, models with material damage should be developed and calibrated. Whereas in order to explain the various scale effects and/or deformation patterning observed in real structures, constitutive models with internal length (Mühlhaus and Vardoulakis, 1986) must be tested and calibrated. It is also of interest to study the coupling effect of temperature and pore-fluid pressure on the bifurcation load.

Both aspects, that is the experimental and the theoretical, must be carried out in conjunction with field observations and monitoring in order to achieve a proper and simple tool for future design of underground openings.

4. Literature

Guenot, A. (1987). Contraintes et ruptures autour des forages pétroliers. Proc. 5 th Congress ISRM, Montreal, 1987).

Vardoulakis, I.G., Sulem, J., Guenot, A. (1987). Borehole instabilities as bifurcation phenomena. Int. J. Rock Mech. & Min. Sc. & Geomech. Abstr., submitted.

Vardoulakis, I. (1984). Rock bursting as a surface instability phenomenon. Int. J. Rock Mech. & Min. Sc. & Geomech. Abstrat., Vol 21, 137-144.

Mühlaus, H.B. and Vardoulakis, I. (1986). Axially-symmetric buckling of the surface of a laminated half-space with bending stiffness. Mechanics of Materials. Vol. 5, 109-120.

Vardoulakis, I. and P. Papanastasiou (1987). Bifurcation analysis of deep boreholes. Int. Num. Anal. Meth. Geomechanics, accepted.

WORKSHOP ON ROCK TESTING AND TESTING STANDARDS

Chairman: J.A. Hudson (UK)
President, ISRM Commission on Testing Methods

EXECUTIVE SUMMARY

The Workshop on Rock Testing and Testing Standards was held on 2 September and attended by about 60 registrants. The meeting was chaired by Dr. Hudson of the UK and introduced by Dr. Franklin of Canada, the former President of the Commission on Testing Methods and new President of the ISRM. The main points made in discussion were recorded by Mr. Dyke of the Rock Mechanics Research Group at Imperial College. The objective of the Workshop was to provide a forum for discussion and for the contribution of ideas relating to rock testing and testing standards.

The Agenda items included the introduction to the Workshop together with an explanation of the activities of the ISRM Commission on Testing Methods, the rationale for choosing subjects for testing standards, the utility and style of testing standards documents, dissemination and translation of testing standards, and links with other relevant organisations (eg ASTM).

There was considerable discussion on the choice of subjects for Suggested Methods and the recommendation that the overall subject framework be reconsidered. It was also thought that some form of deadline system for the production of the Suggested Methods would be useful, although it was recognised that the system only succeeds through the professional goodwill of the rock mechanics experts involved in the Working Groups producing the documents.

The majority of those present thought that the Suggested Methods could be improved by the inclusion of a "commentary" section. This would give an overview of the purpose of the Suggested Methods and provide guidance on the meaning and interpretation of the testing procedures and values obtained. Other standards, especially those produced by the ASTM and in Australia, should be studied with a view to adding such a "commentary" in the future. (Since the Workshop, strong links have been established between the ISRM Commission on Testing Methods and ASTM Committee D18.12).

It was also generally agreed that a more "up-market" version of the Suggested Methods, as sold via the ISRM, would be in keeping with the enhanced international use of the documents. Associated with this could be a more commercial approach to dissemination.

The thorny question of translating the Suggested Methods was not resolved. It was agreed that maximum dissemination was the objective but that translation should be approved by the President of the Commission.

The group agreed that it would be helpful if the publication of the Suggested Methods had the status of a published paper for the members of the Working Group in each case.

Many other very helpful suggestions were made concerning the future production of the Suggested Methods and all these will be taken into account by the new Commission on Testing Methods in association with the ISRM President. Written suggestions for the Workshop were received from Dr. Zhou Simeng of China concerning the choice of subjects, matching projects and standards, and a variety of methods of upgrading the Suggested Methods. These comments were gratefully received and will also be used in forming the future Commission strategy.

The Workshop generated a great deal of interest and provided considerable support and enthusiasm for the concept of testing methods as developed by the ISRM. It is hoped that similar Workshops can be held at the main ISRM Conferences in future and already one is planned for the Madrid 1988 meeting.

WORKSHOP ON ROCK CUTTABILITY AND DRILLABILITY

Chairman: W.E. Bamford (Australia)

BACKGROUND :
During the 1979 Congress of the International Society for Rock Mechanics, in Montreux, Switzerland, a small group of researchers and consultants held informal discussions. It was agreed that it would be a good idea to co-operate, by exchanging details of testing procedures, comparing test results, and correlating measured field performances with the predictions available from different workers' tests. Engineers involved in constructing and using tunnel boring machines have also expressed interest in the rock mechanics fraternity developing a series of internationally-accepted test procedures, which they could apply to rock samples from any new project, whose results they could understand and correlate with their own machines' performance, and use as a basis for predicting future performance. It seemed to all present at these discussions that the rock mechanics profession has more to gain by friendly co-operation than by each worker attempting to "push" his own pet test procedures and to discredit those of his competitors.
It was agreed to form an International Working Group.

The I.S.R.M. Commission on Rock Boreability, Cuttability and Drillability was established in 1980, with the following Terms Of Reference :

The prediction of rock boreability (in full-face tunnelling or raise boring), cuttability (in part-face tunnelling or ripping) and drillability (in rotary or percussive drilling) is an important application of rock mechanics.
Ideally, we would wish to be able to predict the performance (advance rates, energy consumption, wear rates, etc.) of various rock excavation methods from a knowledge of rock properties and rock mass characteristics. In recent years, a considerable amount of effort has been expended in various parts of the world in attempts to define and measure the rock properties required for these predictions. As a result of this effort, a great number of rock testing and characterization procedures have been developed, but nonconsensus has been reached on the methods most appropriate for application to each general excavation method.

The aim of the Commission will be to develop a series of internationally accepted rock testing procedures which can be used to predict the performance of rock boring, cutting and drilling machines. The Commission will not carry out the basic research and development itself, but will seek to draw together the expertise and experience of its members. The rock testing and characterization methods used by various authorities will be set down and evaluated, and the results obtained using these methods compared. Particular attention will be paid to the correlation of predictions made using these procedures with observed field performance.

The ultimate aim of the Commission is that rock testing and characterization procedures recommended for the prediction of field performance in the various rock boring, cutting and drilling applications should be prepared for publication as "Suggested Methods of the Society".

The 1987 Workshop organised by the I.S.R.M. Commission on Rock Boreability, Cuttability and Drillability discussed the following material :

3 MAIN AREAS OF ACTION

A. Drilling
- Percussive Drilling
- Rotary Drilling

B. Surface Excavation
- Ripping
- Bucket Wheel Digging
- Dredging

C. Underground Excavation
- Roadheaders (using Picks)
- Tunnel Boring Machines (using Disc Cutters)

TASKS OF THE COMMISSION AND SUGGESTED CHAPTER ORGANISATION OF ITS REPORT

(1) Identification and description of the main drillability test procedures presently used, to encourage the accurate duplication of test apparatus.
Development of correlations and conversion factors between different test indices.

(2) Identification and description of the mechanisms of rock failure under the action of cutting and drilling tools, in the laboratory and in the field.

(3) Correlation of the performances of drills and excavation machines with rock physical properties and drillability test indices.
Assessment as to which of the properties and indices have the best correlations with performance, and consequently would appear to have the best predictive capability.

(4) Suggestion of standard test procedures for characterizing rock boreability cuttability and drillability.

The following draft outline of chapter 4 of the Commission's report was presented for comments :

SUGGESTED PROCEDURES FOR PREDICTING ROCK BOREABILITY, CUTTABILITY AND DRILLABILITY

Seven Parameters Are Relevant :

1. Hardness Of The Rock Substance;
2. Strength Of The Rock Substance;
3. Texture Of The Rock Substance;
4. Drillability Indices of Rock Substance;
5. Abrasiveness of the Rock Substance;
6. Geological Structure Of The Rock Mass;
7. Seismic Properties Of The Rock Substance And Rock Mass.

1. HARDNESS OF THE ROCK SUBSTANCE
Preferred Tests:
- Shore Scleroscope
- Roell & Korthaus Sklerograf

Acceptable Test:
Schmidt Rebound Hammer

2. STRENGTH OF THE ROCK SUBSTANCE
Preferred Tests:
"Brazilian" indirect tensile strength
Punch shear strength
Uniaxial compressive strength
Fracture toughness
Acceptable Test:
Point load strength index

3. TEXTURE OF THE ROCK SUBSTANCE
Preferred Test:
Quartz content
Acceptable Tests:
Texture Coefficient
Grain size and shapes

4. DRILLABILITY INDICES OF THE ROCK SUBSTANCE

Drillability tests "mimic" the 3 main modes of drilling and cutting :

(A) Translational (Rotary drilling or Linear cutting)
(B) Indentational or Penetrative
(C) Percussive

(A) **Translational**
Preferred Tests:
Goodrich Drillability
Sievers "J" Number
Acceptable Tests:
Core-Cutting Test Specific Energy (McFeat-Smith/Fowell/Roxborough)
VOEST-ALPINE Rock Cuttability Index
Taber Abradability

(B) **Penetrative**
Preferred Test:
NCB Cone Indenter Index
Acceptable Tests:
Morris Drillability
"Handewith" Test
O&K Wedge Test

(C) **Percussive**
Preferred Test:
Rock Impact Hardness Number
Acceptable Tests:
Coefficient Of Rock Strength
"Protodyakonov" Test

5. ABRASIVENESS TESTS OF THE ROCK SUBSTANCE
Preferred Tests:
Goodrich Wear Number
CERCHAR Abrasivity
Paddle Abrasiveness
F-coefficient
Acceptable Tests:
Core Cutting Test Abrasivity
L.C.P.C. Abrasimeter
Taber Abrasiveness

6. GEOLOGICAL STRUCTURE OF THE ROCK MASS
This can most conveniently be characterized by means of the most commonly used rock mass classification schemes.
Not every parameter in the classifications is necessarily relevant to rock boreability cuttability and drillability, so only the relevant parameters should be used for this purpose.

(A) The Rock Mass Rating System
(or Geomechanics Classification)

G_R = the sum of Parameters 3 and 4

where Parameter 3 describes the spacing of discontinuities
Parameter 4 describes the condition of discontinuities

(B) The Q System

$$G_Q = (RQD/J_N) * (J_R/J_A)$$

where RQD = Rock Quality Designation
J_N = joint set number
J_R = joint roughness number
J_A = joint alteration number

7. SEISMIC PROPERTIES OF THE ROCK SUBSTANCE AND ROCK MASS
Preferred Tests:
P-Wave and S-Wave velocities, measured in the rock substance in the laboratory
P-Wave velocity, measured in the rock mass
Acceptable Tests:
Modulus Ratio ($E_{dynamic}/E_{static}$)
Transmitted Amplitude Ratio

****

After the above material was presented and discussed by Commission members, a general discussion session ensued, during which the following facts and opinions which were stated :

Three distinctly different mechanisms of wear of drilling or cutting tools by rocks are observed : gouging, yielding, and erosion.

Acceptable tests for characterizing the cuttability and drillability of rock should be sensitive over the entire spectrum of rocks; they should be reliable; they should not be greatly affected by the cutting rate, or by the wear of the tool. They should be simple and quick and cheap to perform. The process of drillability testing should be physically relevant to the process of cutting in the field situation. Tests should be independent of machine geometry. They should have low sensitivity to the surface geometry of specimens.
Test apparatus should be easy to reproduce accurately. Tests should require only a small volume of rock.

In discussion of the CERCHAR Abrasivity test several users agreed that the test values tend to "flatten out" at a value of about 6. The test is insensitive at values of this order or higher, as the steel needle has such a large wear flat (by definition, greater than 0.6 mm diameter) ground upon its tip.

It was suggested that the scope of the Commission's activities could be widened to also include dredgeability. This topic was one of those of interest to the Permanent Association of International Navigation Conferences, and liaison with that organisation could prove useful.

The "Z-theory" of wear of ground - engaging tools was briefly discussed. The hardness ratio, or relative hardness of tool and rock is the most significant determinant of wear for one region, but the geometry of the tool is most significant in another region of the wear versus hardness curve.

The effect of water content on physical and mechanical properties was significant, so laboratory tests should be carried out on samples having undisturbed (or restored) field moisture content.

It is desirable and essential that suites of rocks from different tunnel projects be exchanged between the investigators of the different projects, so that "benchmark" test values can be obtained with different test apparatus.

The "Q-system", as modified by Kirsten, was recommended for rock mass classification, and as a moderator of laboratory determination of rock mass cuttability or drillability.

The "Total Hardness" technique, comprising 3 different tests, has been submitted to the A.S.T.M. for standardisation. The 3 component test methods are Shore Scleroscope, Schmidt (L-Type) Rebound Hardness, and the (modified) Taber Abrasion Hardness. It is crucial that the Schmidt Hammer made by the original manufacturer be used, and not an unauthorised copy made by another company. "Total Hardness" is being successfully used for the prediction of the performance of tunnel boring machines : Penetration rates; Utilization; Cutter wear and costs.

An experienced user of tunnel boring machines recommended that the geometry of cutters be varied, as a function of the type of rock being cut, and the position of a cutter on the head of the machine.

Among those present and contributing to the discussions were:

Bill Bamford	Australia
Don Banks	U.S.A.
Torgier Blindheim	Norway
Scott Burdick	U.S.A.
Dominique Fourmaintraux	France
Bob Fowell	England
Gerald Grainger	U.S.A.
Herbert Kutter	West Germany
Russ McFarlane	U.S.A.
Priscilla Nelson	U.S.A.
Peter Tarkoy	U.S.A.

Poster sessions

The Technical Program Committee invited contributors for the poster sessions which were to be held on each day of the Congress. These provided a venue for meeting with authors, as well as for the presentation of results that were not available when papers had to be submitted for publication in the Proceedings. This augmented the plenary sessions in a less formal manner

POSTER CATALOG

Fifty-eight delegates made use of the opportunity to display a poster and the following contribution were made.

Poster sessions were held in the Marquette room. Authors prepared their posters board Tuesday at 08:00 AM. Posters could remain in place for three days.

1. P.R. AGRAWAL (India) "Behaviour of Intra-Thrust Zone Along Water Conductor System of Yamuna Hydel Scheme, Stage II, Part II".
2. G. BALLIVY, B. BENMOKRANE, A. LAHOUD (Canada) "Integral Method for the Design of Grouted Rock Anchors".
3. K. BALTHASAR, M. HAUPT, CH. LEMPP, O. NATAU (Germany) "Stress Relaxation Behavior of Rock Salt: Comparison of In Situ Measurement and Laboratory Tests Results".
4. S. BANDIS, J. LINDMAN, N. BARTON (Norway) "Three-D Stress State Around Cavities in Overstressed Weak Rock".
5. B. BAMFORD (Australia) "Rock Borability, Cutability, Drillability".
6. W. BAWDEN, D. MILNE (Canada) "Geomechanical Mine Design Approach at Noranda Minerals Inc.".
7. B. BJARNASON (Sweden) "Non-Linear and Discontinuous Stress Variation with Depth in the Upper Crust of the Baltic Shield".
8. G. BORM (Germany) "Borehole convergence and Stress Relaxation in Rock Salt".
9. T.C. CHAN, N.W. SCHEIER (Canada) "Finite-Element Simulation of Groundwater Flow and Heat and Radionuclide Transport in a Plutonic Rock Mass".
10. F.H. CORNET, J. JOLIVET, J. MOSNIER (France) "Identification and Hydraulic Characterization of Fractures in Boreholes".
11. E. DETOURNAY, L. VANDAMME, A. H-D. CHENG (U.S.A.) "Propagation of a Vertical Hydraulic Fracture in a Poroelastic Formation".
12. C.H. DOWDING, K.M. O'CONNOR, M.B. SU (U.S.A.) "Quantification of Rock Mass Deformation Using Time Domain Reflectometry".
13. M.E. DUNCAN FAMA, L.J. WARDLE (Australia) "Numerical Analysis of Coal Mine Chain Pillar Stability".
14. DYWIDAG SYSTEMS INTERNATIONAL (Canada) "Dywidag Overload Indicator and Yield Anchor".
15. J. ENEVER (Australia) "Ten Years Experience with Hydraulic Fracture Stress Measurement in Australia".
16. G.S. ESTERHUIZEN (South Africa) "Evaluation of Stability of Mine Tunnels in Changing Field Stresses".
17. E. FJAER, R.K. BRATLI, J.T. MALMO, O.J. LOEKBERG, R.M. HOLT (Norway) "Optical Studies of Cavity Failure in Weak Sedimentary Rocks".
18. R.J. FOWELL, C.K. IP (England) "Mechanisms of Water Jet Assisted Drag Tool Cutting of Rock".
19. S. GENTIER (France) "Hydromechanical Behavior of a Single Natural Fracture Under Normal Stress".
20. D.E. GILL, B. LADANYI (Canada) "Time-Dependent Ground Response Curves for Tunnel Lining Design".
21. GAO HANG, JING ZIGANG, SHEN GUANGHAN (P.R. China) "Study on a Coal-Mining above Confined-Water in North China Coalfields".
22. T.F. HERBST (Germany) "Safety Increase by Complete Support Monitoring of Underground Excavations".
23. F. HEUZE, R.J. SCHAFFER, R.K. THORPE, A.R. INGRAFFEA, R.H. NILSON (U.S.A.) "Models of quasi-static and Dynamic Fluid-driven Fracturing in Jointed Rocks".
24. R.M. HOLT, J. BERGEN, T.H. HANSSEN (Norway) "Anisotropic Mechanical Properties of a Consolidated Sandstone".
25. D.R. HUGHSON, A.M. CRAWFORD (Canada) "Kaiser Effect Gauging: A Method for Determination of Rock Stress by Acoustic Emission".
26. O. HUNGR (Canada) "Examples of Wedge Stability Analysis Using the Method of Columns".
27. R.W. HUTSON, C.H. DOWDING (U.S.A.) "A New Method for Producing Identical Joints in Real Rock of all Hardnesses".
28. T. KYOYA, Y. ICHIKAWA, T. KAWAMOTO (Japan) "Deformation and Fracturing Process of Discontinuous Rock Mass and Damage Mechanics".
29. Y. KUDO (Japan) "Physical Properties and Microstructures in Granitic Rocks in Japan".
30. K. KIKUCHI (Japan) "Stochastic Estimation and Modelling Joint Distribution Based on Statistical Sampling".
31. V. LABUC, W. BAWDEN, F. KITZIGER (Canada)

"Seismic Monitoring System Using Fiber-Optic Signal Transmission".

32. LI XING CAI, SHI RU BIN, YU MU XING (China) "Accoustic Methods to Determine the Predominant Orientation of the Macrocracks in the Field".

33. LI XING CAI (China) "Zero or Negative Effective Pressure and its Effect on Physical Properties of Rock Mass in the Field".

34. J.C.S. LONG, D. BILLAUX, K. HESTIR, J.P. CHILES (U.S.A.) "Some Geostatistical Tools for Incorporating Spatial Structure in Fracture Network Modeling".

35. P.H. LU (U.S.A.) "Integrity Factor Approach for the Design of Longwall Interpanel Pillars".

36. J. MATHIS (Sweden) "Joint Statistics: Tunnel Mapping vs a Large Open Room - A Comparison".

37. KIKUO MATSUI, MASATOMO ICHINOSE, KENICHI UCHINO (Japan) "Deformational Behavior and its Prediction in Longwall Gate Roadway Under Pillar and Ribs".

38. P. MACDONALD (Canada) "Seismic and Numerical Modelling Research at Inco".

39. D.Z. MRAZ (Canada) "Proposed Design for High Pressure Isolation Structure in Saltrock".

40. H.-B. MUHLHAUS (Germany) "Stability of Deep Underground Excavations in Strongly Cohesive Rock".

41. T. MUTSCHLER, B. FROHLICH (Germany) "Analytical Calculation of the Strength Behavior of Interbedded Rockmass".

42. T. NISHIOKA, A. NARIHIRO, O. KYOYA (Japan) "A Numerical Analysis of Earth Anchor Retaining Wall Constructed on Soft Rock".

43. R.C.T. PALKANIS, H.D.S. MILLER, S. VONGPAISAL, T. MADIL (Canada) "An Empirical Approach to Open Stope Design".

44. D. PREECE (U.S.A.) "Borehole Creep Closure Measurements and Numerical Calculations of the Big Hill, Texaco SPR Storage Site".

45. J.P. JOISEN, J.P. PIQUET, R. REVALOR (France) "Contribution of Rock Mechanics to the understanding of Dynamic Phenomena in mines".

46. M. ROUSSIN, S. BURDICK (U.S.A.) "A New Tool for Rock Excavation: Impact Ripper Development".

47. S. SAKURAI, N. SHIMIZU (Japan) "Assessment of Rock Slope Stability by Fuzzy Set Theory".

48. J.C. SHARP, C.H. LEMAY, B. NEVILLE (Canada) "Observed Behavior of Major Open Pit Slopes in Weak Ultrabasic Rocks".

49. P.L. SWANSON (U.S.A.) "Resistance to Mode-I Fracture in Brittle Polycrystals: A Contribution from Crack-Interface Traction".

50. M. THIERCELIN, J.C. ROEGIERS, T.J. BOONE, A.N. INGRAFFEA (U.S.A.) "An Investigation of the Material Parameters that Govern the Behavior of Fractures Approaching Rock Interfaces".

51. SPE (France) "Rock at Great Depth: Rock Mechanics and Rock Physics at Great Depth".

52. KOT F. UNRUG (U.S.A.) "Blasting Effect on Roof of Underground Mine From Nearby Strip Operation Working Same Seam".

53. G. VENKATACHALAM, K. VINCENT PAUL (India) "Characterization of Rock Joints".

54. B.S. VERMA (Canada) "Suggested Techniques for Direct Determination of Tensile Strength of Competent Rocks".

55. B.S. VERMA (Canada) "Fracture Development Around Workings in Contiguous Mineral Formations".

56. T. VLADUT (Canada) "Retom LWD Logging System".

57. D. WULLSCHLAGER, O. NATAU (Germany) "The Bolted Rockmass as an Anisotropic Continuum Material Behavior and Design Suggestion for Rock Cavities".

QUANTIFICATION OF ROCK MASS DEFORMATION USING TIME DOMAIN REFLECTOMETRY

C.H. Dowding
Northwestern University, Evanston, Illinois, USA

K.M. O'Connor
New Mexico Institute of Mining and Technology, Socorro, New Mexico, USA

ABSTRACT: Time Domain Reflectometry (TDR) is an electrical pulse testing technique originally developed to locate breaks in power transmission cables. This technique has been adapted for use in monitoring the movement of rock during mining. Coaxial cables have been installed within the overburden over active and abandoned coal mines by drilling holes and grouting the cables into place. Rock mass movements cause damage to the cable and when electrical pulses are transmitted along the cable it is possible to locate rock mass movements by virtue of monitoring changes in pulse reflection signatures for the cable. Recent laboratory research has made it possible to not only characterize but also quantify rock mass movements that have caused changes in TDR signatures.

TDR Operating Principle and Installation

A TDR installation consists of a TDR cable tester and a cable grouted into the rock mass as shown in Figure 1. Ultra-fast rise time voltage pulses are sent down the cable from the tester. Cable defects such as crimps, short circuits or breaks cause reflected pulses which are detected by the tester. Pulse reflections from all changes in geometry along the cable are superimposed on the input pulse to form a reflected TDR signature. Consequently, the signature observed on a cable tester cathode ray tube (CRT) screen consists of many individual reflections associated with localized deformations along the cable. The characteristics of a TDR signature are determined not only by the magnitude of cable deformation but also by the type of cable defect (crimp, shear, break, abrasion, etc.).

Assuming a constant pulse propagation velocity, the distance to a cable defect is proportional to the elapsed time between initiation of the voltage step pulse and the arrival of the returned voltage. The cable tester measures this time in units of distance, so it is possible to determine the location of a cable defect and the type of cable defect by inspection of the recorded TDR signature.

Features of a TDR coaxial cable installation are shown schematically in Figure 1. Prior to installation, the cable is crimped at 6m(20 ft) intervals to optimize resolution and an appropriate connector (e.g., UHF) is attached to the top-of-hole end. The cable is connected to a TDR cable tester via a special 50-ohm connector cable and a reading is made to ascertain that all crimps will be recorded. The crimps must be deep enough to ensure a strong reflection but not so deep as to cause a short circuit in the cable. Signals attenuate over long cables especially when there is a large number of crimps. Panek and Tesch (1981) found it necessary to put as many as four closely-spaced crimps at the most distant locations on cables that were 100m (300 ft) long and this has also been found to be true in our field installations. After being adequately crimped, the cable is attached to an anchor, lowered down the hole, and bonded to the surrounding rock with an expansive cement grout that is tremied into the hole. Recommended installation procedures are presented by Dowding et al. (1986).

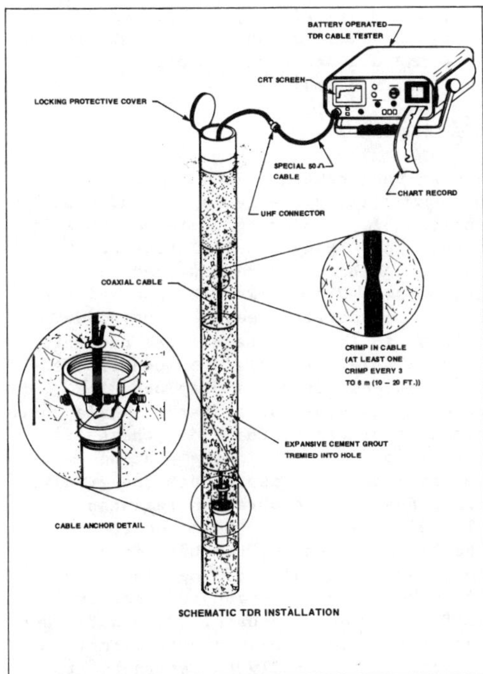

Figure 1. Schematic TDR Installation

When a cable is crimped at selected intervals, a set of reference reflections is recorded each time the cable is tested. The locations of cable defects which are created by rock mass movement can then be determined with respect to these references. If crimps are made in a cable every 6m (20 ft), the accuracy with which defects can be located is 12cm (5 in.).

Example Application

In one application, TDR monitoring cables were installed within the strata overlying two longwall coal mine panels at the Old Ben Mine No. 24 in Benton, Illinois (Wade and Conroy, 1980). The installations consisted of 22.2mm (0.875 in.) diameter coaxial cable (Cablewave Co., FXA 78-50) which was precrimped and grouted into vertical holes drilled from the surface to depths of up to 204m (670 ft) through the overburden and penetrating the coal seam being mined.

The TDR traces shown in Figure 2 were obtained from a monitoring cable installed approximately 244 m (800 ft) from the beginning of the second panel. The cable was subjected to rock strata response in front of the advancing longwall face as well as strata fracturing and caving after the cable location was undermined. The eight TDR traces describe cable deformation recorded over a period of approximately one year. The distance between cable and face is given for each trace with the convetion that distance is positive as the face approached the cable and negative as the face moved past the cable.

Su (1987) has developed a quantitative relationship between cable deformation and change in TDR reflection amplitude as well as distinguishing reflections associated with shear deformation from reflections associated with tensile deformation. Based on the results of this recent research, more information about the type and magnitude of rock movements has been extracted from existing TDR records.

Increasing local negative reflections, marked as (1) at a depth of 45 m (150 ft), indicate increasing shear deformation of the cable as the face approached. The negative reflection spike (2) on the plot at a depth of 105 m (350 ft) shows that shearing rather than extension severed the cable. On the other hand, tensile failure (3) at a depth of

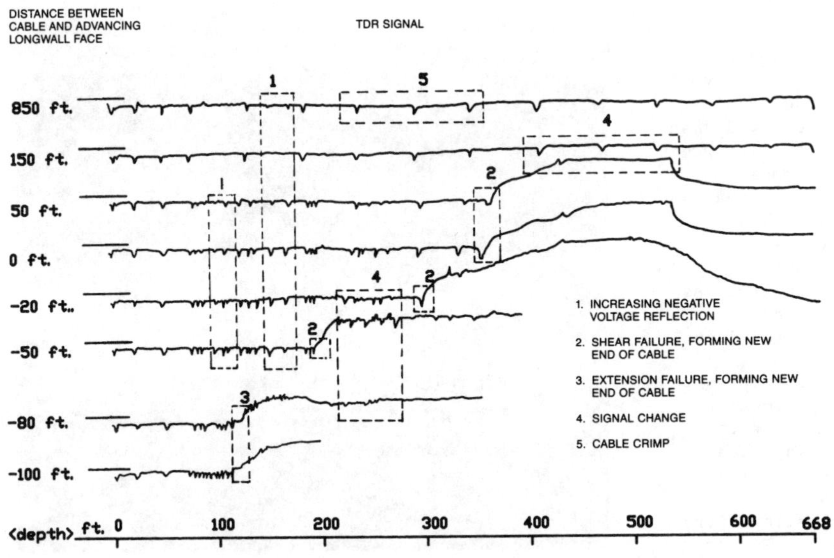

Figure 2. Interpretation of TDR Voltage Reflection Signatures Produced by Collapse of Strata Over A Longwall Panel

36 m (120 ft) shows only a slight negative reflection before the large positive open-circuit reflection. Signal changes (4) at a depth of 122 m (400 ft) show locations of secondary or even higher order reflections and do not present new data. Cable crimps (5) made at known distances during installation serve as references of distance along the cable.

REFERENCES

Cablewave Systems, 1985. Antenna and transmission line systems. Catalog 600, North Haven, Connecticut

Dowding, C.H., M.B. Su and K. O'Connor, 1986. Choosing coaxial cable for TDR monitoring. 2nd workshop on surface subsidence, Morgantown, West Virginia, June.

Panek, L.A. and W.J. Tesch, 1981. Monitoring ground movements near caving stopes-methods and measurement. RI8585, U.S. Bureau of Mines, Denver, Colorado.

Su, M.B., 1987. Quantification of cable deformation with time domain, reflectometry techniques. Ph.D. Dissertation, Northwestern University, June.

Wade, L.V. and P.J. Conroy, 1980. Rock mechanics study of a longwall panel. Mining Engineering, December.

A NEW METHOD FOR PRODUCING JOINTS IN REAL ROCK OF ALL HARDNESSES

R.W. Hutson and C.H. Dowding
Northwestern University, Evanston, IL., U.S.A.

ABSTRACT: This new computer numerical control method allows the experimenter to control joint geometry in real rock with sufficient precision to permit the production of multiple specimens with that produce repeatable results when tested in the same fashion. To extend this procedure to harder or more abrasion resistant materials, a custom diamond blade was designed and fabricated which has improved cutting stability, yielding joints with precision comparable to that previously obtained in softer Limestone.

1 INTRODUCTION

A basic goal of experimental work is to control all the variables while selectively changing only one at a time to observe its effect on the results. This "perfect" condition can rarely be achieved when measuring properties of natural materials, even when conducting relatively simple tests. For rock joints, the problem is further compounded in that no two natural joint specimens will ever be alike. Inherently, then most real rock joint testing programs involve at least one uncontrollable variable, joint geometry. This lack of control has led to stage testing, which is recognized as undesirable (Barton, 1982) yet is widely practiced out of necessity.

Historically, most investigators have attempted to circumvent the variability of natural specimens by casting identical joints from a model material. While widely practiced, this approach always leaves begging the question of how well the model material reflects the true behavior of real rock.

A new method has been developed which allows for the first time replication of joints in real rock with sufficient precision to permit multiple specimens that when tested will produce repeatable results (Hutson and Dowding, 1986 & 1987). A major difficulty was encountered when extending the procedure from limestone to harder or abrasive resistant rocks such as granite. The additional abrasiveness ultimately required the development of a new style of diamond blade.

2 COMPUTER CONTROL BAND SAW

The joints are produced with a specially constructed CNC (Computer Numerical Control) feed table mounted on a commercial size band saw. The only requirements are that the saw have drive wheels at least 24 in (61 cm) in diameter and that there be provisions for flood coolant. Potable water is preferred as a coolant to prevent contamination of the shear surfaces.

Control is provided by an inexpensive personal computer. This computer sends its signals to translators that interpret them and in turn energize the stepper motors to produce the desired motion. The complexity of the final shape is presently limited only by the minimum radius of curvature that a 0.25 in (6.4 mm) diamond blade is capable of cutting (0.375 in, 9.5 mm).

3 CUSTOM DIAMOND BLADE

When extending the technique from limestone to granite, the commercially available diamond blade became uncontrollable, wandering 10 mm and more from the desired cut line. Many possibilities, such as beam buckling of the blade, were examined to determine the cause. Ultimately, it appeared that some lateral cutting ability needed to be added.

The upper blade in figure 1 is the one currently commercially available. The blade is constructed by nickel plating diamonds from a slurry onto the leading edge of a stainless steel band. While there are a few diamonds plated to the side, these are really accidental or "errors" in the plating process. As this blade passes through the granite it encounters zones of more abrasion resistant material. These zones deflect the blade to one side. As the blade passes by, the hard zone continues to press on the side of the blade (the direction of weakest beam strength) deflecting it even further from the desired path. Once deflected in this manner the blade becomes uncontrollable because it cannot cut laterally to return to its true path.

The lower blade in figure 1 is the one specifically designed for this project. The diamond

Figure 1. Upper blade is commercially available. Lower custom blade has diamonds wrapped around side.

- nickel plating has been wrapped 1/16th of an inch (1.6 mm) around the side of the blade. These firmly attached side diamonds give the blade a lateral cutting action. This custome blade can also be deflected by an abrasive pocket; however, as the blade passes by these extra diamonds will continue to grind until the deflection pressure is relieved and the blade has once again returned to its desired track.

4 ANALYSIS OF RESULTING JOINTS

Surface finish, as a CLA (Center Line Average) in terms of microns, was measured using a Talysurf 4 both perpendicular and parallel to the direction in which a specimen would normally be sheared. There does not appear to be any significant directionality to the surface finish. Average CLAs were 2.4 microns perpendicular and 2.3 microns parallel for granite and 11.6 and 11.7 microns respectively for limestone. Surface finish appears to be more a function of the pores exposed to the surface by the passage of the saw rather than marks left by the saw in the solid material surrounding the pores.

Large scale geometry, parallel to the cut or in the direction in which a specimen would normally be sheared, was also measured to provide an indication of the degree to which the joints matched the theoretical cut line. These measurements were made with an LVDT (Linearly Variable Differential Transformer) profilometer. The surface coordinates of 3000 points were taken during an individual pass. The data were fitted to the theoretical cutting curve using a least squares technique. The mean deviation between the actual cut surface and the theoretical cut line was then calculated normal to the mean plane of the joint.

Table 1. Measured mean deviation of the cut surface from the theoretical cut line in the vertical direction based on a least squares fit for several profiles in different rock types.

Rock Type	Vertical Mean Deviation (mm)	(inches)
Indiana Limestone	0.033	0.0013
Charcoal Granite	0.056	0.0022
Coulton Sandstone	0.058	0.0023

Although the mean deviation is greater for the granite and sandstone than limestone, it is anticipated that actual test performance will be similar to that observed earlier for limestone (Hutson and Dowding, 1986).

5 CONCLUSIONS

A method employing a CNC feed table with a custom designed diamond band saw blade has been developed to cut replicate joints in real rock of all hardnesses. This technique allows for the first time the production of jointed specimens in real rock of all hardnesses with controllable joint geometry and sufficient precision as to be able to produce specimens which will test in a repeatable fashion. The custom blade developed in extending this procedure to harder or more abrasion resistant materials may also have commercial application for diamond sawing.

The future for this technique holds many intriguing possibilities. Potentially, average joint profiles could be measured or determined in the field. These shapes can then be digitized and reproduced in the same rock in the laboratory. The device can also be employed to cut individual pieces for a joint assembly. These pieces can then be assembled as a unit in the laboratory and tested to provide needed information on <u>rock mass</u> behavior.

BIBLIOGRAPHY

Barton, N. 1982. Modelling Rock Joint Behavior from an In Situ Block Test: Implications for Nuclear Waste Repository Design. Prepared for Battelle Memorial Institute, under Contract DE-AC06-76-RL01830-ONWI for the U.S. Department of Energy, pp. 96.

Hutson, R.W. and C.H. Dowding 1986. Identical Joints Artificially Produced in Real Rock. Proceeding of the 27th U.S. Symposium on Rock Mechanics, H. Hartman ed., pp. 213-218. American Institute of Mining Engineers: Denver Co.

Hutson, R.W. and C.H. Dowding 1987. A method for Producing Multiple Identical Joints in Real Rock for Laboratory Testing. Rock Mechanics and Rock Engineering, Vol. 20, pp. 39-56.

BLASTING EFFECT ON ROOF OF UNDERGROUND MINE FROM NEARBY STRIP OPERATION WORKING SAME SEAM

K.F. Unrug, University of Kentucky, Lexington, KY

ABSTRACT: Two companies were mining Indiana No. 6 seam coal on neighboring properties, one by underground operations and one by stripping. The property boundaries were such that the face of the surface operation was approaching the West Mains of the underground mine. These were of critical importance because the other three entries to the main reserves had been lost to roof falls. The management of the underground mine had observed that the roof falls occurred most frequently in the vicinity of the strip operations. The severity of blasting had been held to less than 1.2 cm/sec particle velocity, but the surface mine operators had asked for a variance of 5 cm/sec in view of reserves approaching exhaustion. The management of the underground mine believed that vibrations from the blasting were weakening the grip of the expansion heads of the roof bolts and opposed the variance request. A plan of field monitoring was devised to determine whether the blasting was indeed contributing to the roof falls.

The instrumentation included roof bolt dynamometers which were placed in critical areas and in a reference site considered to be outside the zone of blast influence. The dynamometers were of a proof-ring type which held two strain gauges in the region of maximum strain. The linearity range was 758 KN with accuracy +/- 2 KN. Several holes were bored in the roof for TV borescope observations.

Three types of roof falls had been observed in the mine:

1. Air slaking of weak shale strata causing small falls and progressive upward deteriorate by atmospheric weathering.
2. Falls associated with geological disturbances, which are types of defects in the roof, thus self-explanatory.
3. Massive falls unrelated to the above causes. Some were above anchorage level; most had flat tops with evidence of strata separation.

Bolts in the test area were installed and tightened similar to those in the reference area. Behavior was similar during the initial period of tension bleeding. Thereafter, the bolts in the test area continued to lose tension, whereas those in the reference area increased in tension. The detrimental effect of the nearby blasting was confirmed. Mechanisms of the vibration weakening effect are discussed.

FRACTURE DEVELOPMENT AROUND EXCAVATIONS IN CONTIGUOUS MINERAL FORMATIONS

B.S. Verma
Mining Consultant, Canada

ABSTRACT: Results of investigation on fracture development around multiple excavations, made in two contiguous mineral formations have been

discussed. First group of observations relate to drivages made in formations when either lying horizontally or vertically or when inclined gently or steeply. Next group of observations included relate to workings, placed one over the other and diagonally. The effect of relative advancement of workings has also been examined. In the first series only upper working advances on either side in stages; in the other upper and lower workings approach towards each other in stages. The experimental technique adopted, working geometries examined and the results obtained have been described and illustrated.

1. INTRODUCTION

Determination of stress concentration around mine excavations, essential for establishing the nature of strata movement and its ultimate failure, has drawn considerable attention but the studies on fracture development, though equally important, have not received the consideration they deserve.

Serious accidents mostly occur due to deep seated planes of fractures, leading to the dislodgement of strata from the roof, sides or in mass. Should it be possible to demonstrate experimentally, to the young mining engineers and more so to the miners and face workers, the origination of fracture planes, their propagation and as to how they would make the strata vulnerable for failure, it would go a long way in inculcating the ability to anticipate tendencies of failure of the mining ground and to plan timely the preventive measures. The experimental technique developed and the results obtained to fulfill these objectives have been presented in the Poster Sessions of the Congress.

2. EXPERIMENTAL TECHNIQUE

A suitable experimental technique, easy to repeat, economical for intensive experimental observations, dependable to give consistent and accurate results and capable of illustrating the phenomenon of fracture development and their extension was first developed. Plaster of Paris 6 inches (15.24 cm) cubes, simulating as physical models, nearly 300 ft. cubical block of mining ground were tried. They were casted, cured and tested for strength, uniformity of structure, isotropy and brittleness. They were found to show consistent behaviour and mostly failed along the anticipated planes of stress concentration. Material suitability thus having been established, trial models with circular hole drilled through, representing a round tunnel of a shaft, were tested under uniaxial loading. The tensile and shear fractures repeatedly originated and progressed along the well established positions and directions, confirming the suitability of the experimental technique.

The results of the first series of tests on single roadways of different shapes (square, rectangular, trapezoidal, cambered arch, arched, horseshoe and splay legged) representing the drivages in single mineral formations, mostly tested under uniaxial and some under bi-axial loading have already been discussed (Verma,1981).

3. WORKING GEOMETRIES EXAMINED

The working geometries as would be formed with the development of drivages and workings in two mineral formations, occurring contiguously and examined experimentally are described below and are shown in Figures 1,2.1,2.2,2.3,3 and 4.

3.1. Drivages

Two sets of roadways driven through each of the formation lying either

1. Horizontally or vertically - Figure 2.1,
2. Gentlly inclined - Figure 2.2,
3. Steeply inclined - Figure 2.3.

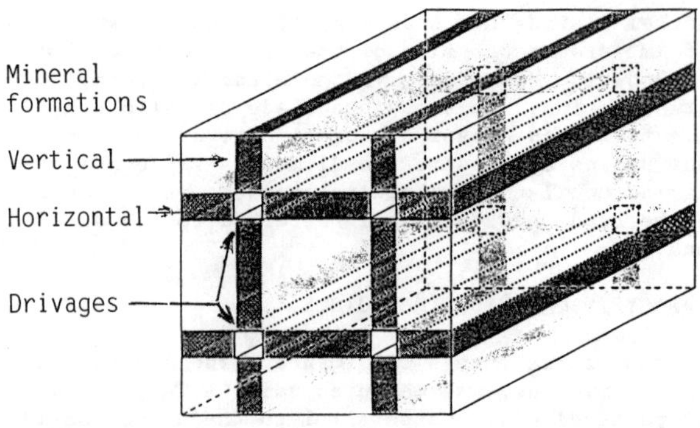

Figure 1. Experimental block with drivages in contiguous mineral formations, lying horizontally and vertically.

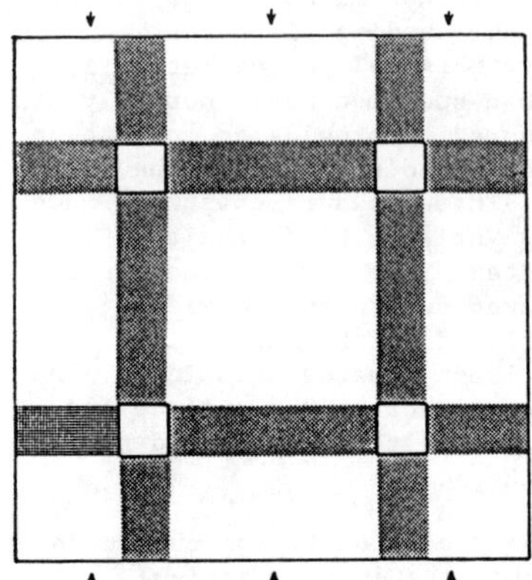

Figure 2.1. Set of two drivages in each of the two formations contiguous and lying horizontally and vertically.

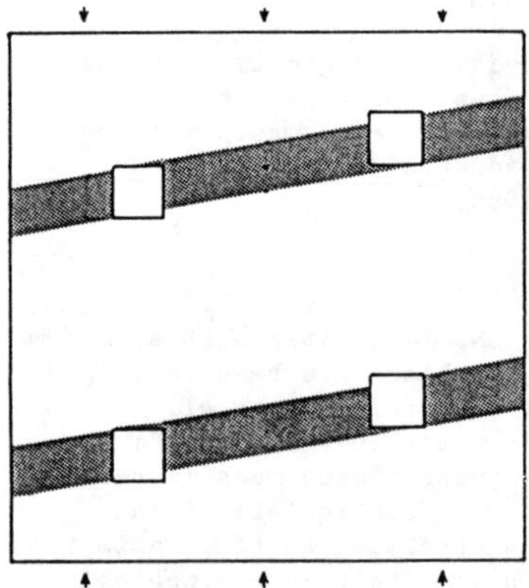

Figure 2.2. Set of two drivages in each of the two formations inclined gently.

3.2. Workings

Two series of geometries were examined representing the areas of ultimate extraction-rectangular and equal in size with width and height ratio 2:1, in the first series placed one over the other in each of the formation and in the second series the equal workings placed diagonally in each of the formation as shown in Figures 3 and 4 .

The effect of relative extension of workings in each of the series was also examined (extension of workings in different groups is shown by different shades in Figures 3 and 4.

The investigations described were restricted to workings being placed in two contiguous mineral formations lying only in horizontal plane.

3.2.1. Workings placed one over the other and
1. equal in size,
2. upper working extended on either side by ½ unit,
3. upper working extended on either side by 1 unit.

3.2.2. Workings placed diagonally and
1. equal in size,
2. both workings advanced towards each other by ½ unit,
3. both workings advanced towards each other by 1½ unit.

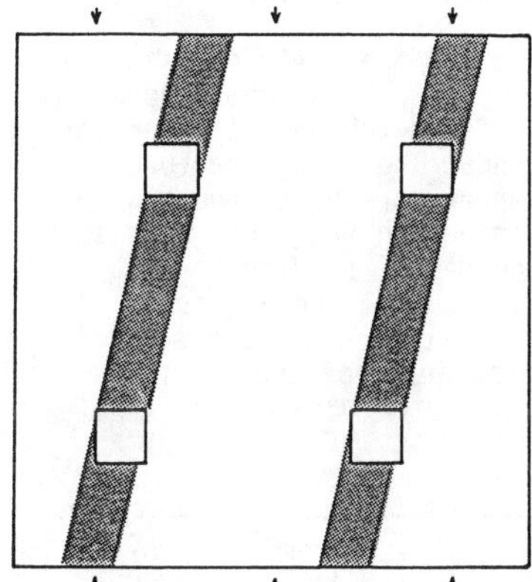

Figure 2.3. Set of two drivages in each of the two formations inclined steeply.

4. EXPERIMENTAL RESULTS

The fracture development around drivages and workings may be examined on a comparative basis taking each group at a time.

4.1. Drivages

Figures 5.a, 5.b and 5.c show fracture development around drivages made in mineral formations, their orientation changing from horizontal to vertical with observations on gently inclined and also in formations steeply inclined.

The tensile fractures in almost all cases originate first from the central parts of the top and bottom of the drivages. They mostly tend to extend vertically above and below the drivages in the body of the model. However, in horizontal and gently inclined formations the bottom fractures may not be so well pronounced as shown in Figures 2.a & 2.b. The fractures originating from the bottom edge of the upper formation drivages and the top of the drivages in the lower formation extend and get merged. But in the case of steeply inclined formations the tensile fractures tend to intersect the shear fractures that originate from the corners and extend sideways. The tensile fractures may occupy any or all the positions marked in the figures.

The shear fractures, in almost all cases mostly originate fromthe drivage corners and extend sideways and tend to join those originating from the other drivages. They tend to develop the zones of weakness about the tensile fractures. The shear fractures also tend to progress obliquely extending almost in all directions. With the change of inclination of the formations,the planes of shear fractures also change. The density of the shear fractures also increases as the inclination of the mineral formations changes and tends to be steeper.

The sides of the drivages experience crushing and wedge shaped parts are mostly dislodged, leaving V shaped cavities, their axis orienting at right angles to the direction of loading. This effect on the sides gradually diminishes as the inclination of the formations tends to be steeper.

4.2. Workings

The results of two series of working geometries developed in two contiguous formations lying horizontally and examined under uniaxial load are shown in Figures 6.a, 6.b, 7.a, 7.b & 7.c. These results are also examined and discussed on a comparative basis.

4.2.1. Workings placed one over the other

In this series three conditions were examined. Two equal and rectangular workings with width and height ratio of 2:1 and placed one over the other, the next with upper working extended on either side by ½ the unit of the width and the third with upper working extended by 1 unit of the width.

The tensile fractures dominate more prominently in all the three cases. They mostly originate from the middle part of the top and bottom edges of the workings and extend above and below. In the intervening

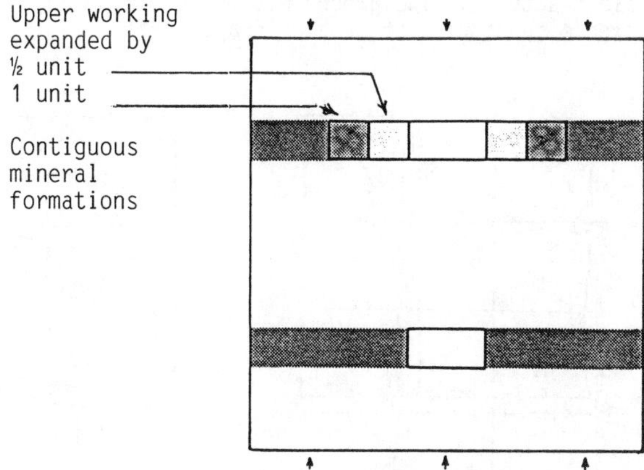

Figure 3. Workings formed in two contiguous horizontal formations, placed one over the other and upper working expanded relatively.

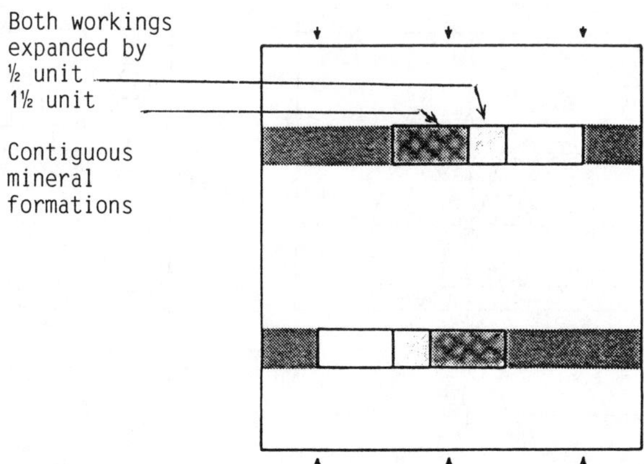

Figure 4. Workings formed in two contiguous horizontal formations, placed diagonally and both expanded towards each other.

ground between the formations, they merge to form continuous planes of tensile fractures. In the case of the workings extended by one unit, however, the tensile fractures do not occur in the lower part of the upper extended working and also do not appear in the upper and lower edges of the workings.

The shear fractures in all the three cases originate from the corners of the workings. They extend in both horizontal and angular directions. In the first case of equal workings, a set of shear fractures originate from near the corners of lower edges of the upper workings and extends downwards. A pair of shear fractures, identical to the first one, also originate from the upper edges of the lower working and extend upwards. These fracture planes merge and form a V shaped zone between the upper and lower formations as shown in Figures 6.a and 6.b. The tensile fractures also being present in this zone, the area would be having the tendencies of becoming weaker. The shear fractures are more prominent above the upper working and spread out in a greater area.

The crushing of sides is well defined and wedge shaped pieces fall out leaving V shaped cavities, their axis being at right angles to the direction of the load applied. With the extension of the upper working, this tendency of side crushing in the lower working is greatly reduced. However, the sides of the upper working show greater tendency of side crushing.

The pattern of fracture distribution within the adjoining ground changes considerably as the upper working extends. In the case of equal workings the shear fractures intersect almost halfway between the upper and lower workings. This zone is also traversed by tensile fractures. The ground within this zone may experience tendencies of greater fragmentation. This tendency within the intervening ground gradually reduces as the upper working extends. But the ground above the expanded working would be affected more adversely.

4.2.2. WORKINGS PLACED DIAGONALLY

Under this series also three groups were examind. In the first group rectangular workings of similar size were placed diagonally. In other groups both the workings were advanced appraching each other, in one by ½ unit of the width and in the other by 1½ unit. The expanded workings in the last group overlapped one over the other.

The tensile fractures in these cases also originate from the central portion of the upper and lower edges of the workings. They extend both above and below the workings. As the width of the workings increases, the dominance of the fracture planes becomes more pronounced. But in the case of overlapping workings, intrestingly, the tensile fractures merge with the shear fractures, originating from the lower working. Thus a zone is formed with fractures allaround. The block as a whole may tend to move if the circumstances favour.

The shear fractures in all the three cases originate from the corner of the workings, both upper and lower. They show well defined tendencies of orienting diagonally as they extend. The shear fractures from the inner side of the upper working extend towards the shear fractures originating from the inner side of the lower working and merge, tending to shear the space between the workings. With the extension of shear fractures from the outer corners of each of the working, tendencies of shearing of the entire body of the model are developed. In the case of overlapping

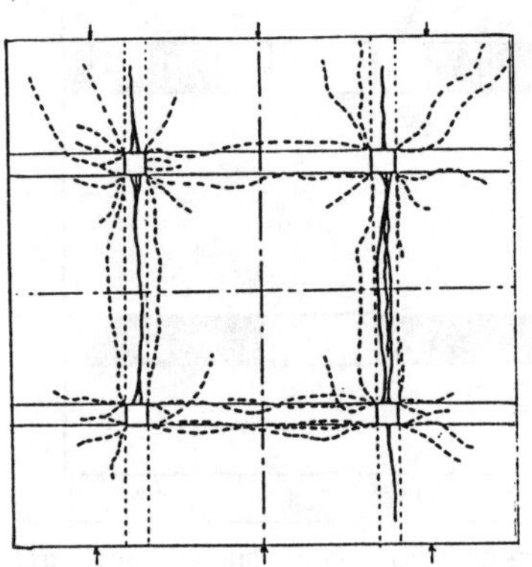

(a) Lying horizontally and vertically

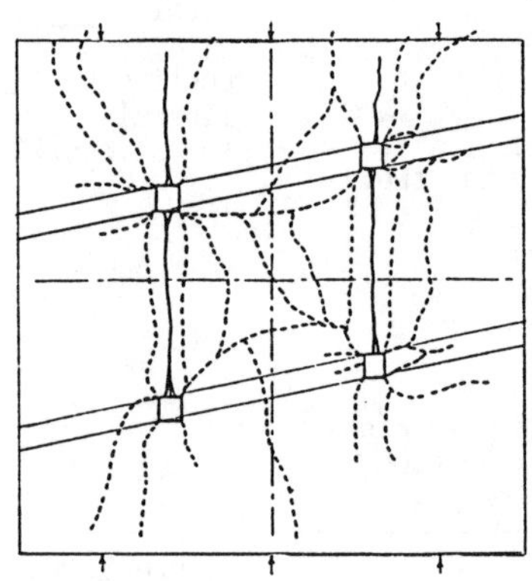

(b) Gently inclined

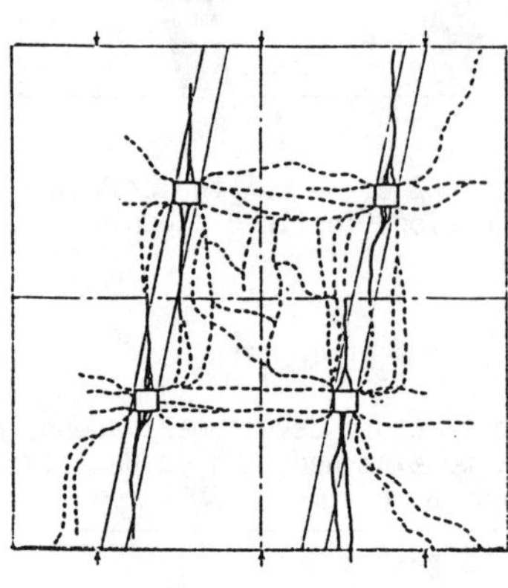

(c) Steeply inclined

Figure 5. Fracture development around drivages formed in two contiguous formations.

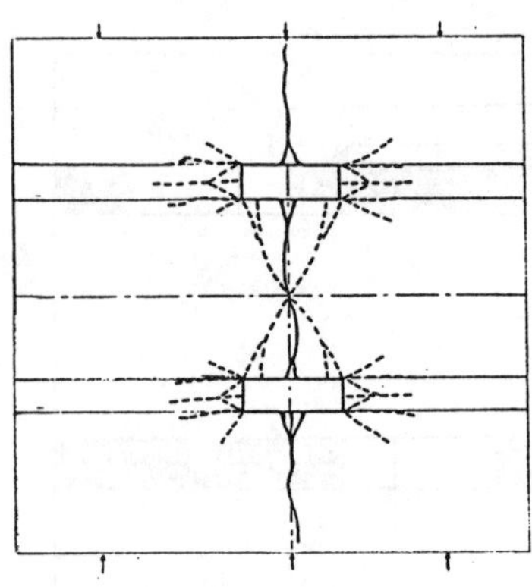

(a) Equal and placed one over the other

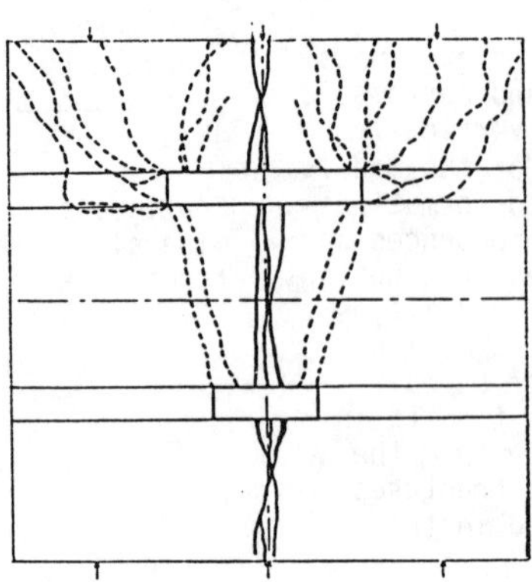

(b) Upper working extended by ½ unit

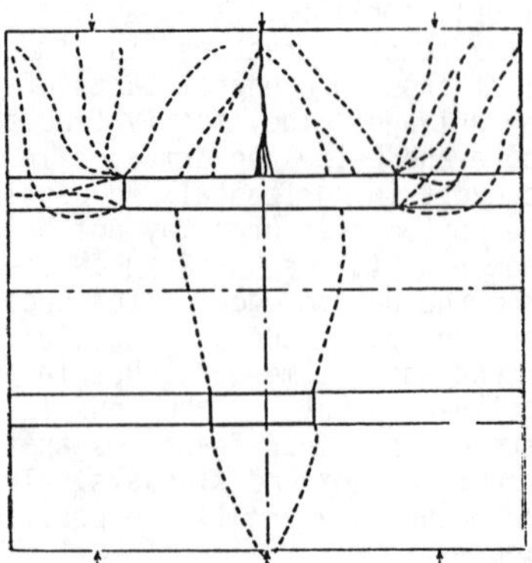

(c) Upper working extended 1 unit

Figure 6. Fracture development around workings formed in contiguous formations.

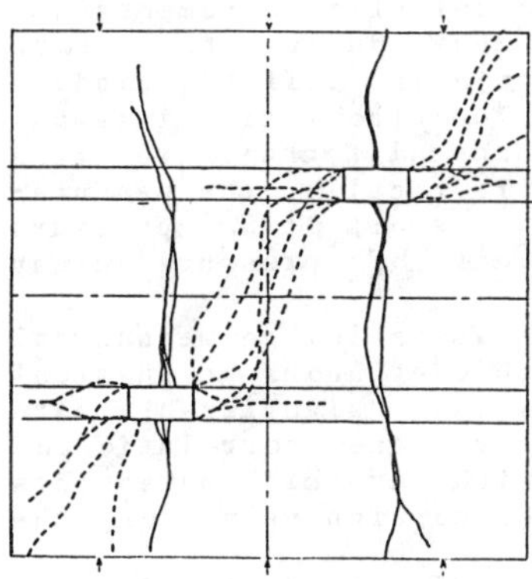

(a) Equal and placed diagonally

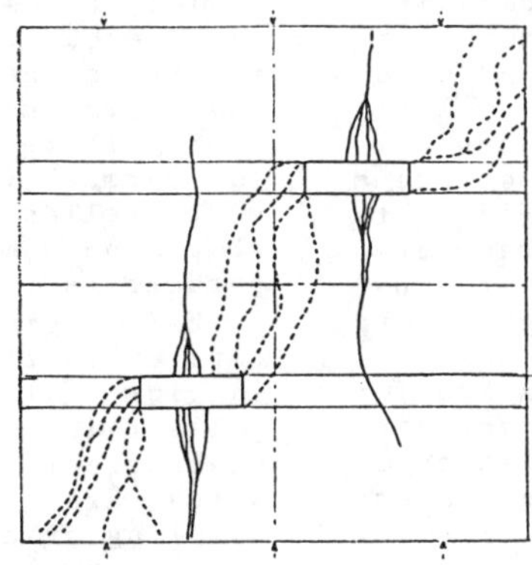

(b) Both advanced towards each other by ½ unit

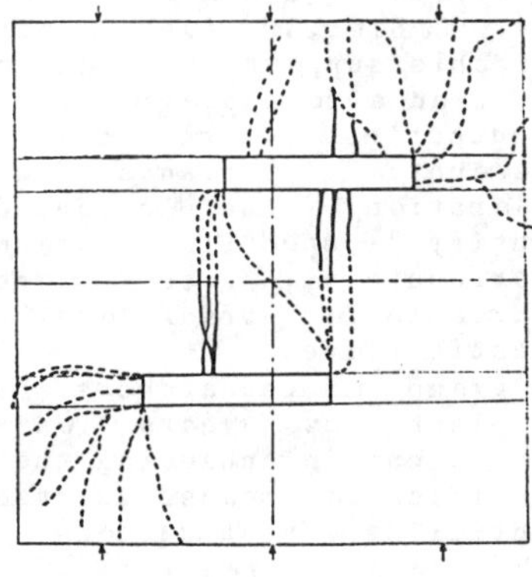

(c) Both advanced towards each other by 1½ unit

Figure 7. Fracture development around workings formed in contiguous formation.

workings merging of tensile fractures with the shear fractures would tend to weaken the intervening ground. The portions above the upper working appear to be affected more adversely.

The tendencies of crushing of the sides has been observed only in the first case. These tendencies have not been noticed in other geometries.

The intervening ground between the workings in the two formations seems to be most adversely affected and chances of failure of this ground under favourable circumstances are more than in other cases earlier discussed.

5. CONCLUSIONS

The experimental results seem to fulfill the objectives of yielding dependable, consistent and illustrative information about the origin of the fracture planes and their penetration within the adjoining space around the excavations as examined. They are interpretable in the terms of mining conditions that may be simulated.

The experimental technique is simple, easy to adopt in educational institutions and the training centres of the industry. The mechanics of fracture development and their extension within the mining ground adjoining the workings can be well demonstrated.

Underground it would be practically impossible to observe the penetration of fracture planes beyond the exposed surface of the workings. The technique and the results obtained thus have a very special significance to built the concepts of phenomenon of ground failure in mines. Even without the long exposure to the actual mining conditions and experience, these results can make the miners aware of the impending tendencies of origin of the failure planes, their direction of extension and merging. The knowledge thus gained and observations in the field may help to train the mine personnel to be able to anticipate the impending dangers from rock mass failure and adopt appropriate preventive measures well in time.

ACKNOWLEDGEMENTS

The support received from the Indian Council of Scientific and Industrial Research and the Department of Mining Engineering, Banaras Hindu University at investigating stages and the assistance received from the Technical University of Nova Scotia and its Department of Mining Engineering, particularly from Dr. Adorjan and Dr. Jones for analysis of results is duly acknowledged. Thanks are also due to the Organising Committee of the Congress and Dr. Herget, Congress Chairman for allowing the research results to be presented in the Congress Poster Sessions.

REFERENCES

1. Verma, B.S.1981.The Elements of Mechanics of Mining Ground, Vol. I, Varanasi, India Tuhin & Co.
2. Verma, B.S.1972.Charecteristics of ground damage due to workings in metalliferous mines, J. of Mines, Metals & Fuels-Proc. Underground Metal Mining Seminar.
3. Hoek, E.1964.Rock failure around mine excavations, Proc. Fourth Internl. Conf. on Strata Control and Rock Mechanics, University of Columbia.
4. Bernard,Y. & John, R.T.1953.An analysis of mine opening failures by means of models,American Institute of Mining Engineering, Vol.196.
5. Gale, W.J. & Blackwood, R.L.1986.Stress distribution and rock failure around coal mine roadways,Int. J. Rock Mech. & Min. Sci. Vol.24 No.3.
6. Aldorf, J. & Exner, K.1986 Mine Openings: Stability and Support,N.Y.Elsevier.

LOGGING WHILE DRILLING TECHNOLOGY FOR ROCK ENGINEERING

T. Vladut
Retom 1987 Geotechnology Ltd., Canada

Knowledge of ground environment is the fundamental of rock engineering. Several examples of failures are well known to be associated with luck of identification of soft thin beds, e.g., the Malpasse Dam failure. Similar problems are associated to big open pits in Alberta, where often weak thin beds in particular of bentonitic nature, are related to coal operations and oil sand surface mining. The oil sand operations concerns focuses to weak seams when affect the stability of benches on which big draglines operate.

Drilling procedures dominate the investigation means and provide data based on core analysis, testing and field tests are done sometimes after the completion of drilling programs. Even geophysics is most often desirable, cost and time (e.g., cost of rig immobilization) make utilization of such techniques an exception. Best example may be related to downhole geophysical logging which provide continuous logs but still the continuous usage is prohibitive because of cost relation.

Significant technical achievements are associated to property identification of very soft materials by continuous means of investigation mainly related to cone penetration (dynamic and static) technologies. Similar full continuous means are desirable support rock engineering procedures at an affordable cost.

Industries where drilling costs are paramount importance (e.g., petroleum exploration) use advanced logging to identify specific ground parameters. Unfortunately, such procedures are not affordable for traditional rock engineering investigations.

A group of researchers have embarked for the last two years on a concentrated development in enhancing the capabilities of identification means related to the process of drilling investigation.

The main direction of development is related to transfer of technology of logging technology from the oil industry to rock investigation. Such possibilities were identified associated to measurement while drilling technology (MWD). The MWD technology forefront in petroleum industry is associated mainly on the parameters of the drilling process, tool orientation and drilling parameters.

Retom´s approach is concentrated to continuous ground parameters identification along the drilling process by a specific Logging While Drilling technology.

The actual development produced a prototype which could be defined as a general carrier of interchangeable transducers. The LWD tool is an attachment on the drilling string, just behind the bit, in particular on the Kelly by a shell carrier host of specific transducers.

The shell carrier has four carrier slot units in which one is dedicated to a large downhole memory and one for the supply of power by lithium batteries. The other two units are dedicated to specific logging transducers, one for natural gamma and one for resistivity. Other shell carriers, four feet long, could be used to host other transducers as required.

The actual stage of development concentrated on the components associated to the electronics capability to survive the downhole conditions. The downhole conditions were evaluated by field measurements which indicated very high accelerations up to 150 times the gravity. The alternative for construction of electronic instruments rely only on military type of technology which are used by the space technology. The inital assembly, carrier, downhole memory (512k), natural gamma and resistivity transducers were tested in an API reference test hole in Fort Worth with good results.

Critical areas in which the future research will be concentrated on the necessity of development of a wireless communication between the downhole memory and the surface computer. Even with a large memory because of very large quantities of collected data a temporary solution for deeper holes is associated to intermediary wireline connection which may be used during the drilling breaks, e.g., during the addition of supplementary drilling connection pipes.

Further steps of development are concentrated on construction of downhole geomechanical testing unit in particular strength qualification by an indenter, hydrogeological parameters, particularly permeability and a sonic unit for qualification of the dynamic strength parameters, modulus through in hole velocity measurements.

Some of the geomechanical measurements are considered to be made during the short intervals when addition of drilling connection are made. The short time interval available requires unusual technologies like for permeability. Practically, the measurement relies on a short time pressure shot and measurement of pressure decay parameters.

The objective of parallel geomechanical data collection with continuous geophysical logging will provide significant data collection to improve the correlation of statistical data with dynamic parameters which are collected continuously along the drilling process.

The actual LWD tool (6 1/2" in diameter in size) represents the upper bound for geomechanical investigation drilling and the lower bound of size for petroleum exploration.

The future development concentrated on specific tools, slim hole type for mining and geomechanical exploration. Concentration on solving the downhole/surface communication will provide the opportunity to develop a logging instrument for oil exploration.

Several conceptual revisions may provide a simpler option for both domains of utilization. In a long term the Logging While Drilling technology is aiming to provide a continuous logging (geomechanical and geophysical) along the drilling process which will allow enhancement of the process of investigation in parallel with the drilling process. It must be recognized that the goals are significant and long development time will be associated up to the moment that an alternative like the cone investigation for soft grounds will be available for hard rock continuous monitoring.

Technical tours

Six tours were offered to provide a broad selection of engineering and mining projects located across Canada, from the high Arctic to Niagara Falls, and from the Rocky Mountains to the Atlantic. Locations are shown on the accompanying map.

Prices included road transportation, accommodation, technical visits and, at remote locations, meals and one-way air transportation to (pre-Congress tours) or from (post-Congress tours) Dorval airport, Montreal.

Pre-Congress Tours

Tour 1 - Deep Mines of Northern Ontario

August 25-28, 1987
Tour Leader: D.G.F. Hedley
Cost: $600.00 (meals not included)

Ontario's Sudbury Basin accounts for much of the total world supply of nickel, copper and platinum. Tours of Creighton and Strathcona nickel mines, 2000 m beneath the city of Sudbury, feature rockburst detection by a 64-channel microseismic network and seismic tomography, and also backfilling and cable bolting for ground control. Sudbury's "Science North" museum, a structure of architectural and rock mechanics interest, will be visited. The museum is built partly underground and the exposed rock displays geological features of the Precambrian shield.

At Elliot Lake, in the 1250 m deep Stanleigh Mine, the uranium ore body is mined using the room-and-pillar method with stabilizing barriers at 300 m centres. Horizontal stresses three times higher than vertical have been observed here. The shallower Denison Mine, with a similar layout, is using backfill to stabilize pillars. The effectiveness of this approach is being evaluated in a major instrumentation and testing program.

Denison Mine, Elliot Lake, Ontario (Tour 1)

Research projects on rockbursting, mine and regional stability are being carried out in the CANMET Mining Research Laboratory. The seismographic stations of the area are operated from here, and the data from the mines' seismic networks are analyzed by CANMET staff.

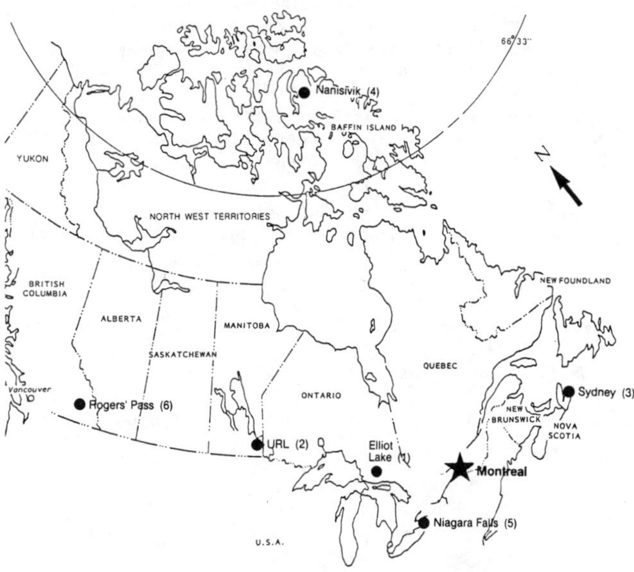

Tour destinations in Canada

Tuesday, August 25, 1987

The tour commenced in Sudbury.

07:00 Reception desk, Holiday Inn, Sudbury, Ontario.
a.m. Creighton Mine, Inco Ltd.
p.m. Science North Museum

Wednesday, August 26, 1987

a.m. Strathcona Mines, Falconbridge Ltd.
p.m. Bus to Elliot Lake.

Thursday, August 27, 1987

a.m. Stanleigh Mine, Rio Algom Ltd.
p.m. CANMET Mining Research Laboratory.

Friday, August 28, 1987

a.m. Denison Mine, Denison Mines Ltd.
p.m. Bus to Sudbury.
Travel to Montreal.
The tour terminates after arrival at Dorval Airport, Montreal.

Number of participants: 24

Tour 2 - Underground Research Laboratory (URL), Manitoba

August 26-28, 1987
Tour Leader: G.R. Simmons
Cost: $550.00 (some meals included)

The Underground Research Laboratory is a geotechnical research facility constructed by Atomic Energy of Canada (AECL) in the granitic Lac du Bonnet batholith. This unique facility is situated in a previously undisturbed pluton that was well characterized before construction. All potential testing areas are below the water table in saturated rock. The experimental activities focus on characterization of the environment surrounding the shafts, the 130 m and 240 m deep shaft stations, and the headings. The activities include stress determinations, geological and hydrogeological assessment, and rock core and fracture properties testing. In the next phases of development, the shaft will be extended and characterized to a depth of 465 m. The operating phase will involve complex multi-component experiments, which will help in the validation of mathematical models simulating repository conditions. The major components of this work are solute transport, thermal-mechanical-hydraulic tests, and engineering tests.

Participants will also tour AECL's Whiteshell Nuclear Research Establishment (WNRE) and Cold Spring Granite Company Limited's granite quarry. At WNRE participants will tour waste management facilities and Slowpoke Reactor System. At Cold Spring granite quarry a brief tour of the quarrying operations will be given by Cold Spring staff.

Wednesday, August 26, 1987

The tour commences in Winnipeg.
p.m. Arrive Winnipeg
p.m. Night at International Inn, Winnipeg.

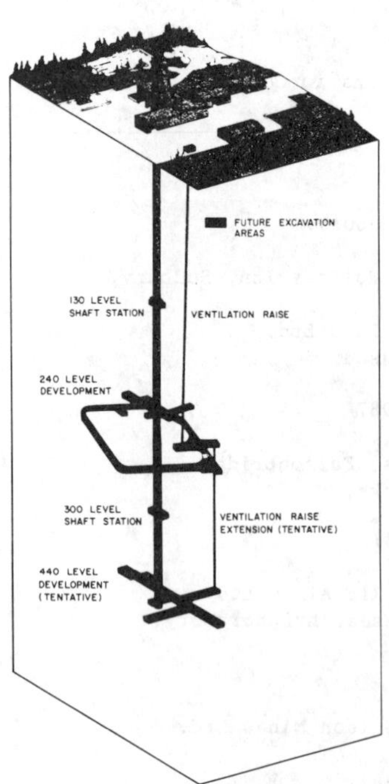

URL shaft extension and planned development (Tour 2).

Thursday, August 27, 1987

08:30 Meet at Reception desk, International Inn, Winnipeg.
a.m. Travel to URL. Presentation on the URL Technical Programs.
p.m. Tours and technical presentations in surface and underground facilities.
Dinner and accommodation in Pinawa.

Friday, August 28, 1987

a.m. Breakfast at Pinawa Motor Inn.
Travel to, and tours of Whiteshell Nuclear Research Establishement (Waste Management Facilities and Slowpoke Reactor System).
Lunch at WNRE (not included in tour).
p.m. Travel to, and tour of Granite Quarry owned and operated by Cold Spring Granite Company Limited.
Travel to Montreal.
The tour terminates after arrival at Dorval Airport, Montreal.

Number of participants: 28

Tour 3 - Coal Mining and tunnel Boring, Nova Scotia

August 27-29, 1987
Tour Leader: T.C.R. Aston
Cost: $600.00 (meals not included)

The Sydney Coalfield is the largest in eastern Canada, with an estimated 2.5 billion tons of coal reserves. The Cape Breton Development Corporation is operating two mines, Lingan and Prince, which produce nearly 2.5 million tons of saleable coal per annum. Two new mines, Phalen and Donkin-Morien, are under development. The tour will provide an opportunity to see and discuss undersea longwall mining operations in both the Lingan (advance longwall mining at a depth of 650 m) and Phalen (retreat longwall mining at a depth of 350 m) mines. At the Donkin-Morien site, a Lovat full-face tunnel boring machine has been used to drive two 7.6 m diameter tunnels 3.5 km long. Research programs in the CANMET laboratory include studies of strata mechanics, ventilation, mine dust and gases. Participants will inspect the rock mechanics preparation and testing facilities and hear presentations on current strata mechanics and environmental research programs.

A visit is also planned to the nearby Fortress of Louisbourg, which has been reconstructed together with part of the 18th-century town. Here the participants will have a chance to enjoy an authentic 18th-century lunch.

Thursday, August 27, 1987

The tour commences in Sydney, Nova Scotia
07:30 Meet at: Reception desk, Wandlyn Hotel, Sydney, Nova Scotia.
Lingan and Phalen Undersea Mines.

Friday, August 28, 1987

a.m. Donkin-Morien Mine.
p.m. CANMET Coal Research Laboratory.

Saturday, August 29, 1987

a.m. Louisbourg Fortress and Settlement.
p.m. Travel to Montreal.
The tour terminates after arrival at Dorval Airport, Montreal.

Number of participants: 5

Post-Congress Tours

Tour 4 - Mining in the High Artic, Baffin Island

September 4-7, 1987
Tour Leaders: J.E. Marshall, G. Clow, S.A. Luciani
Cost: $1,800.00 (meals and 3 nights accommodation included).

Visits to Polaris and Nanisivik Mines and the Eskimo settlement of Artic Bay.

Tour cancelled due to insufficient number of registrants. Unofficial tour conducted on individual basis; Number of participants: 3.

Tour 5 - Niagara Falls Power Plants and Rock Cliff Stabilization, Ontario

September 4-5, 1987
Tour Leader: E. McComb, D.A. McKay
Cost: $400.00 (some meals included)

The Niagara River is only 56 km long but has a powerful flow of 5700 cubic metres per second, half of which passes over Niagara Falls, and the remainder through three Ontario Hydro power plants and one hydropower plant in the State of New York. Downstream of the Falls a 100 m deep gorge has been cut in the limestones and shales. Ontario Power Generating Station, built in 1902, is at the base of the gorge, 250 m downstream from the Horseshoe Falls beneath an overhanging cliff. The overhang, stabilized by shotcrete, cable anchors and pressure relief holes, is being monitored by sliding micrometers and remotely-read computerized extensometers. Downstream, near the Sir Adam Beck No. 2 plant, differential weathering of the shales and limestones has resulted in some spalling and rock falls. The slope has been stabilized by rock bolts, dowels, pressure relief holes and cast concrete facing.

Before returning to Toronto, participants will visit the charming town of Niagara-on-the-Lake, the first capital of Upper Canada in the mid-18th century, which retains much of its original architecture and has excellent restaurants and gift stores.

Geotechnical investigations at Niagara Falls (Canada)

Friday, September 4, 1987

The tour commences at Dorval Airport, Montreal
09:00 Flight from Montreal to Toronto. Lunch at Niagara Parks Victoria Restaurant,
p.m. Ontario Power Generating Station: Geotechnical Situation, View of the Falls; Whirlpool Rapids; Sir Adam Beck Generating Stations, Presentation; Niagara-on-the-Lake, Prince of Wales Hotel; Toronto, L'Hôtel (Convention Centre)

Saturday, September 5, 1987

Meeting in the lobby of L'Hôtel
Information on Toronto
Free time - "Discover Toronto".

Number of participants: 43

Tour 6 - Railway Tunnels Through the Selkirk Mountains, British Columbia

September 4-6, 1987
Tour Leader: R.S. Tanaka
Cost: $1,500 (meals not included)

The tour was to visit Banff, Lake Louise, Rogers Pass, Spiral Railway tunnels, Kicking Horse Pass, Stoney Creek Viaduct, Rock cuts and retaining walls, Tour of Shaughnessy Tunnel, east portal, heading and shaft of Mount MacDonald Railway tunnel under construction.

Tour cancelled due to insufficient number of registrants.

Tour 7 - Jeffrey Pit of J.M. Asbestos, Quebec

September 4, 1987
Tour Leader: J.C. Sharp
Cost: $35.00

The Jeffrey Pit is an open pit asbestos mine located 250 km east of Montreal near Asbestos, Quebec and operated by J.M. Asbestos Inc. Visitors became acquainted with an extensive slope stability program which includes instrumentation as well as modelling. Bus left the Queen Elizabeth Hotel at 9:00 and returned at 16:30.

Number of participants: 35

Delegates at Jeffrey Pit

6th Congress Trade Exhibition

The trade exhibition at the 6th International Congress provided a rare opportunity for suppliers, distributors and service organizations to meet and discuss products and services with the key decision makers in rock mechanics, including civil and mining engineers, operating personnel, consulting engineers and executives. Hundreds of potential clients from North America and around the world were reached in person, at a fraction of the cost of most other sales alternatives.

Ribbon cutting ceremony at the trade exhibition. From left to right Dr. W. Bawden Trade Exhibition Chairman, The Honourable R. Savoie, Minister for Mines and Native Affairs - Province of Quebec, Mr. J. Gaydos, Trade Exhibition Manager - CIM.

The trade exhibition was held in the Foyer of the Conference floor for three days of the Congress. Booths measuring 2.4 m by 3.0 m (8 feet by 10 feet) were available at $900 Canadian and included a draped backwall, side railings and a booth sign measuring 18 cm by 112 cm (7 in. by 44 in.). Other display features were provided on request.

The Chairman of this Exhibition was Dr. W.F. Bawden and the exhibition manager was Mr. J. Gaydos, CIM Trade Exhibitions.

The exhibition was opened by the Honourable R. Savoie at 12:00, Monday, August 30, 1987 and remained open until Wednesday, September 1, 1987, 15:00. The exhibits were very well prepared and were visited with a lot of enthusiasm by many of the delegates. Fifty-five (55) booths were available in the foyer on the convention floor. A list of exhibitors is given below in alphabetical order:

BOOTH NO. 17
A.A. BALKEMA PUBLISHERS
P.O. Box 1675
NL 3000 BR Rotterdam, The Netherlands
Tel.: 01131104145822
Telex: 41605 TKOMNG
A.T. Balkema

Books and journals in the field of rock and soil mechanics, provision of services for publication.

BOOTH NO. 60-63
ATLAS COPCO
Mining & Construction Division
P.O. Box 745
Pointe Claire, Quebec, H9R 4S8
Tel.: (514) 631-5571
Charles E. Laws

Printed and video information on Atlas Copco's range of hard and soft rock drilling and haulage equipment.

BOOTH NO. 37
ATOMIC ENERGY OF CANADA LIMITED
Whiteshell Research
Pinawa, Manitoba, R0E 1L0
Tel.: (204) 345-8625
Telex: 07-57553
P.M. Thompson

AECL displayed the rock mechanics aspects of the geotechnical research at the Underground Research Laboratory (URL). Emphasis was on equipment, instruments, procedures, analysis and interpretation that have been used in the URL program. Information was also available on AECL's other geotechnical capabilities.

BOOTH NO. 18
C-I-L INC.
Explosives Division
P.O. Box 200, Station A
North York, Ontario, M2N 6H2
Tel.: (416) 229-8284
Telex: 06986505
C.L. Allen

The "SABREX" computer blasting model -- a blast optimization and analysis technique for both underground and surface mining operations. It provides blasting engineers with a valuable tool to improve the accuracy of their techniques and helps to control costs. It takes some of the guesswork out of blasting and so contributes to the profitable management of the mining operation.

BOOTH NO. 31
CME-BLASTING & MINING EQUIPMENT LTD.
36 Cameo Street
Oakville, Ontario, L6J 5Y1
Tel.: (416) 849-4530
Telex: 06982283 CME CAN OKVL
Robert Sjolander

CME program includes portable self-contained button bit grinding box -- compact and designed for threaded button bits and down-the-hole hammer bits. A super DIAROC grinds both carbide and steel in one step operation; DIAROC/BORROC has all sizes of grinding pins for grinding button bits; grinders (VS150 - 15 000 RPM, VS260 - 26 000 RPM, VS300 - 30 000 RPM) and grinding wheels for grindamatic B and abrasive products of all sizes and grades. It includes drill steel accessories such as bits, steel, couplings, DTH hammer bits; spare parts for all pneumatic hammers; tire treaded blast mats, 1 000 kg and more, drill hole plugs and connecting wire; and air hoses for low pressure, high pressure up to 600 psi. Safety equipment and accessories, such as head lights, torches, rainwear, etc. are included, as well as dust collectors and their accessories, electrical cable for pumps, and boom-mounted demolition tools.

BOOTH NO. 45
CROUCH RESEARCH, INC.
42 Island Road
St. Paul, Minnesota 55110, U.S.A.
Tel.: (612) 483-1829
Steven L. Crouch

Crouch Research, Inc. specializes in the development and application of personal computer software for stress analysis in mining and geotechnical engineering. A finite element package (FESOL) is available in addition to a variety of boundary element programs for both two- and three-dimensional modelling.

BOOTH NO. 50
DIASOL INC.
3120 rue Frechette
St-Jean Baptiste de Rouville, Québec,
J0L 2B0
Tel.: (514) 464-9152
Telex: 055 61006
Fax: (514) 464-9154
Carlo Brunner

Developed in Canada by ICOS, the slurry wall, DIASOL, uses jet grouting, a high-energy injection technology.

Literature that illustrates the evolution of this new technology was available at the booth or by mail request.

Jet grouting can be used for consolidation of soil columns, embankments, shaft or tunnel excavations, waterproofing cut-off for buildings, dams, etc.

Luncheon meeting during the trade exhibition (left to right): Mr. G. Reny, Assistant to the Honourable R. Savoie; Dr. W.F. Bawden, Trade Exhibition Chairman; Dr. E.T. Brown, President International Society for Rock Mechanics; the Honourable R. Savoie, Minister of Mines and Native Affairs - Province of Quebec; Dr. G. Herget, General Congress Chairman; Mr. J. Nantel, Session Secretary and Fundraising; Mr. J. Gaydos, Trade Exhibition Manager - CIM; Dr. B. Ladanyi, Honorary Congress Chairman; Dr. J.E. Udd, President, Canadian Rock Mechanics Association.

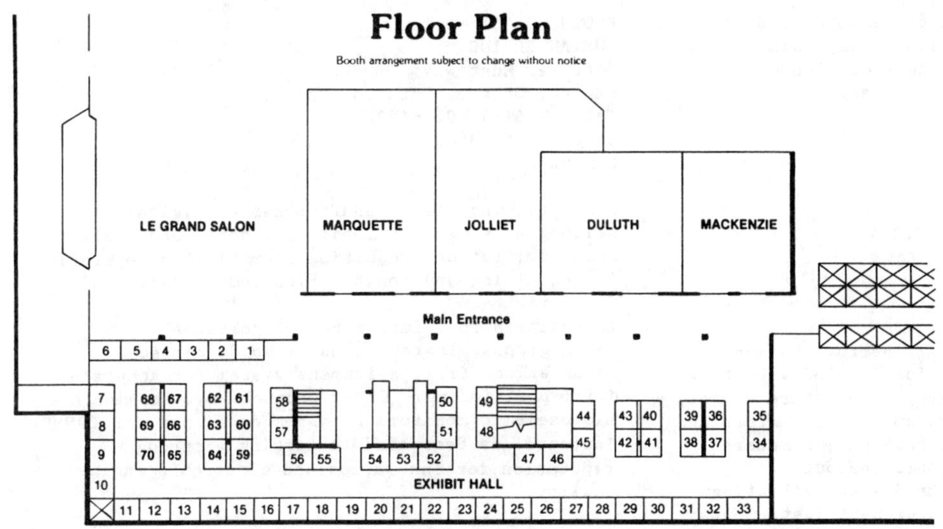

Layout of 6th ISRM Congress Trade Exhibition

BOOTH NO. 22
DUPONT CANADA INC.
Applied Technology and Blasting Services
Box 2200 - Streetsville
Mississauga, Ontario, L5M 2H3
Tel.: (416) 821-5157
Telex: 06-22304
Fax: (416) 821-5110
Brian Fawcett

Optimize productivity with advanced technology solutions from DuPont. Advanced technology is often required to solve problems and improve operations in the mining, construction and quarry industries. To help companies apply high technology for their special needs, DuPont Canada has formed a new business unit, the Applied Technology and Blasting Services Group.

BOOTH NO. 59
DYWIDAG SYSTEMS INTERNATIONAL CANADA LTD.
Eastern Canada Division
65 Bowes Road, Unit 8
Concord, Ontario, L4K 1H5
Tel.: (416) 669-4959
Telex: 06-964604
Adam Allan

Dywidag rock bolts are used as anchors for: (1) cavern construction, (2) tunnel and shaft construction, (3) securing embankment, (4) attachments, e.g. forms, and (5) mining. Dywidag rock and soil anchors are applied for: (1) anchorage external and uplift forces, (2) retaining walls, and (3) stabilization of slopes, rock walls and cuts. Anchor force can be transmitted by either cement grout or resin.

BOOTH NO. 42
EARTH TECHNOLOGY CORPORATION
Corporate Headquarters/Western Division
3777 Long Beach Blvd.
Long Beach, California, 90807, U.S.A.
Tel.: (213) 595-6611
Telex: 656338
Khosrow Bakhtar

The Earth Technology Corporation offers comprehensive rock mechanics services for both surface and underground projects. These services range from both in-house and on-site laboratory rock mass properties evaluation to field instrumentation before, during and after the project.

The rock mechanics staff has the geomechanical expertise necessary to perform sophisticated field and laboratory studies as well as numerical simulations, physical modelling, and engineering design. With comprehensive field instrumentation capabilities, accurate and immediate laboratory characterizations, and an experienced rock mechanics staff, The Earth Technology Corporation can design and implement rock mechanics testing programs to meet specific requirements.

BOOTH NO. 55
ÉCOLE POLYTECHNIQUE DE MONTRÉAL/UNIVERSITÉ DE SHERBROOKE
C.P. 6079, Succ. A
Montréal, Québec, H3C 3A7
Tel.: (514) 340-4220
Denis E. Gill

Short presentations of the various rock mechanics research projects that are carried out at both institutions were shown. These fall into the following fields: in situ rock testing, geostatistical rock mass characterization, stress measuring techniques, ground support and anchors, injections and rock bursts.

BOOTH NO. 32-33
ENERGY, MINES AND RESOURCES CANADA
Canada Centre for Mineral and Energy Technology (CANMET)
555 Booth Street
Ottawa, Ontario, K1A 0G1
Tel.: (613) 995-3065
Telex: 053-3395
Frank Jefferies

A display was shown of a variety of projects designed to illustrate CANMET's range of involvement in mineral and energy technology research and development.

BOOTH NO. 35
ENGINEERED INSTRUMENTS INC.
410 Jessop Avenue
Saskatoon, Saskatchewan, S7N 2S5
Tel.: (306) 955-0240
Telex: 074-2636
Fax: (306) 373-1979
H. Wayne Stoyko

Rock mechanics instrumentation - specializing in soft rock (uni-directional and radial pressure cells, extensometers, total closure meters, data acquisition systems). New products emphasized were "PAL" stressmeters and a rock bolt developed specifically for a soft rock application.

BOOTH NO. 43
GEOKON INC.
48 Spencer St.
Lebanon, New Hampshire 03766, U.S.A.
Tel.: (603) 448-1562
Telex: 4995473
Tony Simmonds

A complete line of geotechnical instrumentation including: the highly successful line of vibrating wire piezometers, strain gauges, temperature gauges, pressure cells, settlement systems, stressmeters and displacement transducers; borehole extensometers with mechanical and/or electronic readout capability; load cells; full bridge and vibrating wire types; in situ stress measurement instruments; USBM deformation gauge and CSIRO H.I. cell; dataloggers for remote unattended data acquisition and portable electronic readout instruments were presented.

BOOTH NO. 67
GEOPHYSICS G.P.R. INTERNATIONAL INC.
894 Front Street
Longueuil, Quebec, J4K 1Z7
Tel.: (514) 679-2400
France Goupil

G.P.R. provides all around the world within a short notice, the following services:
-- land and marine geophysical services and data interpretation;
-- pre-feasibility and feasibility studies;
-- research and development applied to specific problems;
-- technical assistance for geophysical specifications of tender and contract;
-- consultation and site investigation; and
-- training of personnel and transfer of technology.

BOOTH NO. 64
GROUND CONTROL (SUDBURY) LIMITED
1150 Kelly Lane Road
Sudbury, Ontario, P3E 5P4
Tel.: (705) 673-3020
Telex: 067-7536
Larry E. Alexander

Manufacturer and distributor of resin cartridges, rebar and rockbolts, cablebolts, mining hardware, Spedel-grouting equipment, Shotcrete equipment and rock mechanics instrumentation. Serving the mining and construction industry.

BOOTH NO. 2
HANDY GEOTECHNICAL INSTRUMENTS, INC.
P.O. Box 1200, Welch Avenue Station
Ames, Iowa 50010, U.S.A.
Tel.: (515) 292-6134
Telex: 283359 IASU UR
Richard L. Handy

No rock cores needed for the Rock Borehole Shear Tester, a hand-portable device for determining the cohesion, internal friction, and residual friction of soft to hard rocks tested in situ. Tests are performed inside 75 mm (3 in.) diameter boreholes, thereby avoiding data bias from core damage and poor core recovery. Developed at Iowa State University for U.S. Bureau of Mines.

BOOTH NO. 51
INSTANTEL INC.
8 Brewer Hunt Way
Kanata, Ottawa, Ontario, K2K 2B5
Tel.: (613) 592-4642
Telex: 053-3545
Ron Mask

Instantel Inc. is a manufacturer of digital seismographs used to monitor, record, and analyze ground vibrations resulting from blasting operations in the mining and construction industries.

In addition to a full range of portable seismographs, Instantel has also developed a computerized drill alignment system for accurate drill positioning. Other products/systems include a microseismic monitoring system developed for CANMET. Instantel is dedicated to forging a strong reputation for innovation and electronic engineering quality.

BOOTH NO. 27
INTERNATIONAL MINING SERVICES INC.
IMS (Electronics) Inc.
811-675 West Hastings Street
Vancouver, British Columbia, V6B 1N2
Tel.: (604) 689-1709
Telex: 04-54262 Aurorapac-VCR
Fax: (604) 689-1709
Mark Stoakes

IMS Inc. is a mining consulting company specializing in rock mechanics analysis and design and rock mechanics instrumentation. Products and services on display included stress measurement instrumentation, extensometers, rock mechanics monitors and remote monitoring utilizing underground radio communications equipment. Examples of analysis and design software, such as tunnel simulation, stress analysis modelling, etc. were also displayed.

BOOTH NO. 68
IRAD GAGE, A DIVISION OF KLEIN ASSOCIATES, INC.
Klein Drive
Salem, New Hampshire 03079, U.S.A.
Tel.: (603) 893-6131
Telex: 947439
William H. Key, Jr.

Manufacture and marketing of geotechnical instruments, including: the Smart Logger™ modular data acquisition system for vibrating wire, sonic probe, and thermistor instrumentation; piezometers, stress meters; strain gauges, extensometers, temperature measurement probes; total pressure cells; load cells; bentonite pellets; and other related instruments.

IRAD GAGE transducers are used to monitor stress, strain, load, pressure, deformation, and convergence in rock, soil, concrete and steel.

BOOTH NO. 15
IRECO INC.
Eleventh Floor, Crossroads Tower
Salt Lake City, Utah 84144, U.S.A.
Tel.: (514) 694-3040
Leo LeCouteur

Commercial explosives (bulk and packaged) and blasting accessories and up-to-date explosives technology were presented.

BOOTH NO. 44
ITASCA CONSULTING GROUP, INC.
P.O. Box 14806
Minneapolis, Minnesota 55414, U.S.A.
Tel.: (612) 623-9599
Telex: 201 682

Itasca Consulting Group is a team of geological, mining and civil engineers providing service in advanced geo-engineering practice. The exhibit featured samples of computational analyses for a variety of geo-engineering projects. PC software was demonstrated including FLAC (Fast Lagrangian Analysis of Continua), MUDEC (Micro Distinct Element Code) and 3DEC (3-D Distinct Element Code).

BOOTH NO. 40
LAROCQUE GEOTEK INC.
114 rue Point Langlois
Laval, Québec, H7L 3M5
Tel.: (514) 625-0717
Telex: 052-4807
Pierre Longval

The Larocque Group provides rock mechanic services and supplies to the mining industry. For all your drilling needs: acker drills, in-the-hole equipment, diamond bits, GS-550 biodegradable drilling fluids and BVS well screens.

BOOTH NO. 41
LYNX GEOSYSTEMS INC.
Geotechnical Software
800-1177 West Hastings Street
Vancouver, British Columbia, V6E 2K3
Tel.: (604) 669-5626
Telex: 04-55155
Don Plenderleith/Jay Wolfe

Lynx Geosystems displayed computer software for use by the tunnelling and geotechnical community. LYNX Geosystems offers the PC-based line of GEOLOG GEOTEC software for logging and displaying rock excavation parameters. This system also includes a stereonet package and rockmass Q and RMR computation. For engineering workstations, they offer finite element sectional and regional analysis programs for rock mechanics. On engineering workstations, they offer easy-to-use rock and soil slope analysis, wedge stability analysis, and a number of groundwater analysis programs. LYNX Geosystems addresses the software needs of the tunnelling, geotechnical, and mining enginner.

BOOTH NO. 56
McGILL UNIVERSITY/UNIVERSITÉ LAVAL
Dept. of Mining and Metallurgical Engineering
McGill University
3480 University Street
Montreal, Quebec, H3A 2A7
Tel.: (514) 398-4377
Telex: 05-268-510
Dr. F.P. Hassani
Université Laval
Sainte-Foy, Québec, G1K 7P4
Dr. Pierre Choquet

Recent and on-going research projects at McGill and Laval universities were displayed, including topics on rockbolting and backfilling research, novel instrumentation and monitoring application (cable-bolt support, backfill properties, drill and blast performances) and numerical modelling.

BOOTH NO. 39
MICROSEIS SYSTEMS INC.
105-902 Spadina Cres. East
Saskatoon, Saskatchewan, S7K 3H5
Tel.: (306) 244-3878
Neil Gendzwill

The Microseis System is a passive seismic monitoring system. It incorporates sophisticated event recognition, automatic first break picking and full wave recording.

Events are located using the Simplex iterative technique, with a tilted, layered velocity model. Geophones may be placed anywhere, any combination of velocities may be used and both critical and non-critical refractions are accounted for.

BOOTH NO. 16
MTS SYSTEMS CORPORATION
Box 24012
Minneapolis, Minnesota 55424, U.S.A.
Tel.: (612) 937-4000
Telex: 29-0521 MTS SYSTEMS ENPE
Fax: (612) 937-4515
Fred Begat

Calling attention at the 6th ISRM trade exhibition.

MTS Systems Corporation world-wide showed its rock mechanics testing system capabilities, through displays, posters and literature. These included test systems and accessories, including triaxial cells, extensometry, load cells and other test accessories unique to rock mechanics work. MTS manufacturers complete systems for rock mechanics testing.

BOOTH NO. 46
PITEAU ASSOCIATES ENGINEERING LTD.
Geotechnical and Hydrogeological Consultants
408, 100 South Park Royal
West Vancouver, British Columbia, V7T 1A2
Tel.: (604) 926-8551
Telex: 04 352896
Dennis Martin

Piteau Associates Engineering Ltd. offers a comprehensive range of geotechnical and hydrogeological services for numerous disciplines affecting the resource development, transportation, construction and industrial sectors. The staff of geological, mining, civil and groundwater engineers has developed specialist expertise in a wide range of disciplines including, engineering, geology, rock mechanics, hydrogeology, soil mechanics, environmental engineering, mining engineering, geophysics, remote sensing and northern engineering. Projects include open pit mines and quarries, underground mines, oil sands, coal mines, tunnels, highways, railways, pipelines, underground storage, dams, powerhouses, shafts, industrial plants, water supply, land fills, and waste disposal.

BOOTH NO. 54
QUEEN'S UNIVERSITY
Engineering Seismology Laboratory
Department of Geological Sciences
Queen's University
Kingston, Ontario, K7L 3N6
Tel.: (613) 545-6171
R.P. Young

Research work carried out by the rock physics and engineering seismology research group were displayed. The research involves the use of seismic tomographic imaging and microseismic techniques in the study of mining induced seismology.

BOOTH NO. 48-49
ROCTEST LTD.
665 Pine St.
St. Lambert, Quebec, J4P 2P4
Tel.: (514) 465-1113
Telex: 055-61134
Jacques Bourbonnais

Founded in 1967, Roctest Ltd. is a manufacturer of in situ testing and monitoring equipment used in soils, rocks and on structures. Roctest also provides complete services including instrument installation and maintenance as well as field personnel training. Its products have been sold worldwide in more than 60 countries, for civil and mining projects.

BOOTH NO. 20
ROCK ENGINEERING PTY LTD.

BOOTH NO. 21
ROMOR EQUIPMENT LTD.
1083 Britannia Rd. East
Mississauga, Ontario, L4W 3X1
Tel.: (416) 671-3209
Telex: 06-968828
Bob Morgan

OYO corporation products include MsSeis 1600 and McSeis 160 signal enhancement seismographs, the McOhm signal averaging resistivity meter and the Geologger 3030 borehole logging system. From DeRegt Special Cable B.V., the Seistec 100 Geophone Tester.

BOOTH NO. 38
R.S. TECHNICAL INSTRUMENTS LTD.
18-1780 McLean Avenue
Port Coquitlam, British Columbia, V3C 4K9
Tel.: (604) 941-4848
Telex: 04353577
Fax: (604) 941-4175
R.G. Straghan

Instrumentation for rock and soil mechanics including: CSIRO and ANZSI stress cells, extensometers, load cells, convergence monitors, data loggers, piezometers, etc.

BOOTH NO. 19
SHIMIZU CONSTRUCTION CO. LTD.
Technology Dept., Civil Engineering Division
Mita 43 Mori Bldg., No. 13-16, Mita 3-Chome,
Minato-Ku, Tokyo, 108, Japan
Tel.: Tokyo 03-451-6181, Ext. 432
Telex: 252-2373 SCC TKA-J
Fax: 03-798-2349
Takashi Oshii

The new TV System, i.e. 'Borehole Scanner', has been developed so that the surrounding (360°) inner wall of the borehole can be seen immediately as continuously and longitudinally scrolled images. With this system, investigation of the fracture system and the colour variation of deeper bedrocks is easy.

BOOTH NO. 36
SLOPE INDICATOR COMPANY
No. 240, 11300 River Road
Richmond, British Columbia, V6X 1Z5
Tel.: (800) 663-2374 or (604) 276-2545
Telex: 04-352848
Fax: (604) 276-0190
George A. Hancock

Manufacturer of instrumentation for mines, tunnels, and structures in rock, concrete and soil.

BOOTH NO. 23, 24 and 25
SOLEXPERTS LTD.
Ifangstr. 12, P.O. Box 230
CH-8630 Schwerzenbach, Switzerland
Tel.: (01) 825 29 29
Telex: 58273 sols ch
Fax: (01) 945 61 81 (att. Solexperts)
Arno Thut and Daniel Naterop

Solexperts manufactures and distributes worlwide the portable high-precision instruments developed at the Swiss Federal Institute of Technology, Zurich, Department of Rock Engineering and Tunnelling. Exhibited are: sliding micrometer - distribution of strain along borehole; TRIVEC - distribution of the three displacement vectors along vertical borehole; and PIEZODEX - piezometric head along borehole.

BOOTH NO. 6
SRP CONTROL SYSTEMS LTD./OPTIM ELECTRONIC CORP.
621 Guided Court
Rexdale, Ontario, M9V 4K6

The booth of Solexperts Ltd. won the first prize for best booth and product presentation.

tel.: (416) 746-0117
Telex: 065-27418
Joe Santo

High speed data acquisition systems, signal conditioning and recording systems; pressure transducers/transmitters, digital pressure indicators, calibrator/controllers; instrumentation amplifiers, transducer signal conditioning.

BOOTH NO. 26
TELEMAC
2, rue Auguste Thomas
92600 Asnières, France
Tel.: (1) 47 83 78 55
Telex: 610 448 F
J.L. Bordes

Telemac designs, builds and installs instrumentation for monitoring the performance and safety of dams, underground storage caverns for liquified gas and petroleum products, nuclear powerstation structures, road and railway tunnels and large excavation and cuts.

Fifty years ago, Telemac pioneered vibrating wire instruments and, more recently, linear differential induction sensor. Among some achievements, we may mention the vibrating wire pore pressure cell CL 1 and inductive long base borehole extensometer extensofor. Product line covers all instruments for soils and rock mechanics monitoring.

BOOTH NO. 29
TERRA TEK, INC.
Terra Tek Systems
360 Wakara Way
Salt Lake City, Utah 84108, U.S.A.
Tel.: (801) 584-2462
Telex: 910-9255284
Christopher F. Johnson

Terra Tek offers a complete line of rock mechanics and rock physics test equipment. Mechanical, thermal and physical properties can be measured under carefully controlled, simulated in situ conditions. Pressures to 60 000 psi and temperatures to 750°F can be simulated.

BOOTH NO. 57
UNIVERSITÉ DU QUÉBEC
UQAT, UQAM, UQAC
UQAT, C.P. 700, Rouyn, Québec, J9X 5E4;
UQAM, C.P. 8888, Montréal, Québec, H3C 3P8
UQAC, Chicoutimi, Québec, G7H 2B1
Tel.: (819) 762-0971
Jean-Yves Savard or Michel Aubertin

The three constituents of the Université du Québec presented some of their recent research work in the fields of geotechnique and engineering geology related to rock mechanics.

Université du Québec en Abitibi-Temiscamingue: study of the rockburst potential of mining exploitations in the Abitibi region -- a general approach.

Université du Québec à Montréal: urban geology for tunnelling and foundation engineering.

Université du Québec à Chicoutimi: geomechanical modelling and laboratory simulations of fractured rock mass behaviour.

BOOTH NO. 53
UNIVERSITY OF TORONTO
Rock Engineering
Department of Civil Engineering
Toronto, Ontario, M5S 1A4
Tel.: (416) 978-4611
Telex: 06-218915 UT ENG TOR
J. Curran

Posters and interactive computer graphics displays of recent research projects were presented. Details of the graduate program in rock engineering were also made available.

BOOTH NO. 47
UNIVERSITY OF WATERLOO
Earth Science Department
Waterloo, Ontario, N2L 3G1
Tel.: (519) 888-4590
Maurice B. Dusseault

Research and development facilities, experimental

laboratories, and computer software products and development were displayed.

An interactive geophysical log editing program on a PC and a rock fabric graphics analysis program were also demonstrated.

BOOTH NO. 28
U.S. BUREAU OF MINES
Branch of Technology Transfer
2401 E Street N.W., M.S. 1041
Washington, D.C. 20241, U.S.A.
Tel.: (202) 634-1225
Donald E. Ralston

The U.S. Bureau of Mines is a federal government agency which conducts research and gathers data related to minerals and mining. The Bureau's display at ISRM featured results of its rock mechanics related research efforts. Included were a newly developed borehole patented flatjack cell for measuring stress in coal pillars; a fundamental overview of techniques for use in predicting and controlling mountain bumps, cutter roofs and pillar failures; flow prediction and detection of leach fluid through fractured orebodies in preparation for in situ mining; and a computer model for use in making structural analyses and designing plans for longwall mines.

BOOTH NO. 1
WESTBAY INSTRUMENTS LTD.
507 E. Third St.
North Vancouver, British Columbia, V7L 1G4
Tel.: (604) 984-4215
Telex: 04-352606
Erik Rehtlane or Wendy Patton

Westbay Instruments Ltd. manufactures and installs the MP System, a modular multi-level groundwater monitoring system which provides high-quality, verifiable data from drillholes up to 4 000 ft in depth. The MP System is used for geotechnical investigations, groundwater pollution studies and nuclear waste disposal research.

BOOTH NO. 30
WILLIAMS FORM HARDWARE AND ROCK BOLT (CANADA) LTD.
P.O. Box 5
Ingersoll, Ontario, N5C 3K1
Tel.: (519) 485-2710
Telex: 06474151
Fax: (519) 485-4481
Chris Wilkinson

Rock reinforcement equipment was displayed including hollow core, groutable 'spin lock' rock bolts meeting ASTMA-615 requirements. This allows for prestressing by developing the full tensile capacity of ¾ in., 1 in, 1 1/8 in. and 2 in. hollow core rods by setting a mechanical cone shell assembly by torquing rod and full grouting by pumping through the core of the bolt. Solid bolts are available to working loads of 260 kips (1150 kN).

A congress provides many opportunities for the exchange of technical and cultural information. Here are from left to right:

Prof. W. Zhou, Tsinghua Univ., Beijing (China), Dr. A. Jakubick (Canada), Bus Driver (Canada), Ch. Laughton, CERN (Switzerland), Prof. Rongzun Huang, East China Petrol Institute, Beijing (China), Prof. Jingmin Zhu, Institute of Arch. & Engineering, Chongqing City (China).

2

La partie française

Cérémonie d'ouverture

Fig. 1.

Fig. 2.

Fig. 3.

Le lundi 31 août 1987, à 8 h 30, les congressistes se réunissaient dans le Grand Salon de l'Hôtel Reine-Élizabeth, Montréal. À 8 h 35, les invités de la table d'honneur pénétraient dans l'auditorium guidés par un groupe de soldats vêtus d'un uniforme français du 18e siècle. Ce groupe était composé d'un flûtiste, d'un tambour, de deux soldats armés d'un mousquet et d'un soldat de première classe (Figure 1).

Le président du Congrès, Dr. G. Herget, souhaite la bienvenue aux délégués dans les trois langues officielles et présente les invités de la table d'honneur qui sont de gauche à droite sur les figures 2 et 3:

N. Grossmann, Secrétaire général de la SIMR
R.P. Chapuis, Société canadienne de géotechnique
Pr D.E. Gill, Coprésident du 6e Congrès de la SIMR
Pr N.R. Morgensten, Programme technique
Pr B. Maldonado, Vice-président de la SIMR, Amérique du Nord
Pr B. Ladanyi, Président honoraire, 6e Congrès de la SIMR
L'Honorable R. Savoie, Ministre délégué aux Mines et aux Affaires indiennes, province de Québec
L'Honorable G. Merrithew, Ministre d'État aux Forêts et aux Mines, gouvernement du Canada
G. Herget, Président du 6e Congrès de la SIMR
Pr E.T. Brown, Président de la SIMR
Pr E. Hoek, Conférencier d'ouverture
J.E. Udd, Directeur des laboratoires de recherche minière, CANMET, ministère de l'Énergie, des Mines et des Ressources
L. Milne, Président de l'Institut canadien des mines et de la métallurgie
G. Miller, Président de l'Association minière du Canada

Le professeur Ladanyi souhaite la bienvenue aux congressistes et demande à E.T. Brown, président de la SIMR, d'adresser quelques mots aux congressistes au nom de la Société internationale de mécanique des roches.

Monsieur le Président, distingués invités, mesdames et messieurs

Au nom de la Société internationale de mécanique des roches et des nombreux congressistes en visite au Canada, j'aimerais remercier nos collègues canadiens pour la chaleureuse hospitalité qu'ils nous ont témoignée ces derniers jours, pour les mots de bienvenue qu'ils nous ont adressés ce matin et pour les dispositions qu'ils ont prises afin que ce congrès soit, nous en sommes sûrs, vraiment mémorable. Tous ceux d'entre nous qui viennent de l'extérieur du Canada ont également très hâte d'avoir la possibilité de visiter quelque peu cette fascinante ville de Montréal et d'apprécier les nombreuses et diverses attractions de ce grand et intéressant pays.

Pr E.T. Brown, Président de la SIMR

La Société internationale de mécanique des roches a été fondée à Salzbourg en Autriche en 1962 par un groupe d'enthousiastes prévoyants dirigé par Lépold Müller. Nous célébrons donc en 1987 le 25e anniversaire de la fondation de la société. Pendant ces 25 années, la mécanique des roches est devenue un sous-domaine bien établi et essentiel du génie civil, de l'exploitation minière et de la mise en valeur des ressources énergétiques, en plus de contribuer à la compréhension des processus géologiques. Il existe maintenant une imposante documentation dans ce domaine qui est devenue une spécialité dans les universités et dans certains organismes de consultation et de recherche. Elle est aussi suffisamment âgé pour que de nombreux chefs de file actuels en mécanique des roches y aient consacré toute leur carrière. C'est pour ces raisons, pour ne nommer que celles-là, que la mécanique des roches commence à présenter certaines caractéristiques qui appartiennent à une science mature.

Ce n'est donc pas sans raison que le 25e anniversaire de la société soit aussi l'année de la tenue du congrès qui a lieu tous les 4 ans et où il est de tradition de partager nos expériences et de passer en revue l'état de la science dans son ensemble. Il est également justifié, je crois, que ce congrès se tienne au Canada. Dans ce pays, de nombreux projets importants en exploitation minière et en mécanique des roches sont réalisés, sans compter que les Canadiens ont toujours été parmi ceux qui ont fait avancer la mécanique des roches, que ce soit sur le plan théorique ou pratique. Malgré cela, la société n'avait pas encore tenu de réunion internationale au Canada. Enfin, c'est en 1987 que nous corrigeons la situation.

Je suis heureux de mentionner que quelques-uns des membres fondateurs de la société et certains membres des anciens conseils ont pris des mesures spéciales pour participer à ce congrès anniversaire. Nous sommes donc très heureux de leur présence parmi nous et leur souhaitons de rapporter de bons souvenirs de ces quelques jours passés à Montréal.

J'aurais espéré que le père fondateur de la société, le professeur Leopold Müller, puisse se joindre à nous et nous adresser quelques mots ce matin. Malheureusement, des obligations de sa vie professionnelle encore active l'ont empêché de se libérer. Il nous a, cependant, envoyé un message que j'aimerais vous lire en son nom:

"Monsieur le Président, chers amis et collègues

J'avais sincèrement l'intention de participer à ce congrès anniversaire de notre société qui peut donner l'occasion de se remémorer ses années d'intenses activités, mais malheureusement je suis dans l'impossibilité d'y participer.

Vingt-cinq ans est en réalité une courte période de temps dans la vie culturelle du monde, mais à notre époque, tout se déroule beaucoup plus rapidement que dans le passé et il en est ainsi de la mécanique des roches. Au cours de ce quart de siècle, de nombreux concepts ont été maîtrisés et nos bibliothèques regorgent des résultats de recherche. Quiconque prenait la peine de déterminer quels termes et concepts actuellement utilisés ne l'étaient pas en 1962, en serait ébahi.

Il est regrettable de mentionner que ce sont surtout des catastrophes qui ont valu à cette jeune science de la mécanique des roches l'essor important qu'elle a connu particulièrement dans les années 60, alors qu'au cours des décennies précédentes, cette discipline ne suscitait qu'un faible intérêt.

Malheureusement, même si de nombreux travaux de recherche ont été réalisés au cours de ces 25 ans, un nombre incroyable d'entre eux sont tombés dans l'oubli. Je suppose qu'un petit pourcentage seulement de ce que nous, scientifiques, produisons sera utilisé en génie. Une petite quantité seulement des données de base est réellement utilisée dans les gros projets à travers le monde. D'un point de vue géomécanique, certains projets sont même plus mal conçus qu'il y a 20 ans.

Dans notre discipline, il est plus essentiel de comprendre les principes fondamentaux que les méthodes de calcul sophistiquées. Toutefois, les caractéristiques de la roche comme matériau, c'est-à-dire les particularités de cet élément qui réagit favorablement ou non favorablement à notre traitement, retient maintenant moins notre attention que les calculs élaborés.

Espérons qu'il ne sera pas nécessaire que de nouvelles catastrophes se produisent pour raviver cet intérêt.

Mon souhait le plus sincère est que ce congrès permette de réduire l'écart entre la théorie et la pratique en mécanique des roches.

Glück auf!

Leopold Müller"

Merci Monsieur le Président.

M. Herget remercie MM. Brown et Müller pour leur allocution et demande à C.G. Miller, président de l'Association minière du Canada, de prononcer quelques mots au nom de l'association.

"Monsieur le Président, mesdames et messieurs

Je suis heureux d'avoir la possibilité d'ajouter mes souhaits à ceux des autres conférenciers de ce matin. Il me fait grand plaisir, au nom de l'industrie minière du Canada, de souhaiter la bienvenue aux congressistes des nombreux pays.

Dr. G. Miller, Président de l'Association minière du Canada

Le choix du Canada pour ce 6e Congrès international de mécanique des roches est tout à fait approprié. Le Canada est l'un des principaux pays miniers dans le monde et notre industrie minière est redevable en grande partie à la science et à l'application de la mécanique des roches.

Le Canada extrait environ 60 minéraux de plus de 300 mines, le plaçant au premier rang des pays producteurs, compte tenu du nombre d'habitants. En 1986, notre pays occupait aussi la première place pour la production d'uranium et de zinc, la deuxième place pour la potasse, le nickel, le soufre élémentaire, l'amiante et le gypse, la troisième place pour les concentrés de titane, le cadmium, l'aluminium, l'or et les métaux du groupe du platine, la quatrième place pour le cuivre, le molybdène, le plomb et le cobalt, la cinquième place pour l'argent et la sixième pour le minerai de fer.

Le Canada est le plus important exportateur mondial de minéraux, plus de 80 % de sa production étant expédié vers les marchés étrangers à travers le monde. L'industrie minière donne plus de 100 000 emplois directs et les industries connexes, 300 000 emplois supplémentaires.

La science de la mécanique des roches contribue beaucoup et ne cesse de contribuer au bien-être de cette industrie importante et de ses travailleurs. L'extraction efficace et sécuritaire des gîtes minéraux dépend dans une large mesure d'une connaissance appropriée de la mécanique des roches.

Afin d'augmenter la productivité, de nouvelles méthodes d'extraction sont mises au point. Des modèles de mécanique des roches encore plus complexes doivent être élaborées pour appuyer la recherche de nouvelles méthodes qui permettront d'innover la conception de nouvelles mines. De plus, l'application de la mécanique des roches dans les mines existantes se traduit par des mesures plus élaborées et plus précises, combinées à des analyses de prévision, qui auront pour effet d'augmenter l'efficacité et la sécurité des opérations.

C'est pourquoi notre industrie est reconnaissante à l'Association canadienne de mécanique des roches et à l'Institut canadien des mines et de la métallurgie d'organiser cet important congrès. J'espère que vos délibérations feront rapidement progresser votre science et que les applications de votre art seront des plus efficaces!"

M. Herget remercie le conférencier pour ses mots de bienvenue et demande à l'Honorable R. Savoie, ministre délégué aux Mines et aux Affaires indiennes de la province de Québec de s'adresser aux délégués au nom de cette province.

Après l'allocution de l'honorable R. Savoie, l'honorable E. Merrithew, ministre d'État aux Mines, s'est adressé aux délégués au nom du gouvernement du Canada. Le texte de son allocution est reproduit ci-dessous :

Bonjour Mesdames et Messieurs

Au nom du gouvernement du Canada et à titre de ministre canadien responsable des mines, je vous souhaite tous la bienvenue et, en particulier, à ceux qui visitent notre pays pour la première fois. Je suis certain que vos séances plénières, vos ateliers de travail et autres réunions de discussion seront très productifs et que Montréal constituera un lieu incomparable pour mener vos activités sociales.

Nombre d'entre vous ont déjà vu certaines régions du centre et de l'est du Canada pour avoir participé à des tours guidés techniques qui ont eu lieu avant le congrès et d'autres iront bientôt contempler les chutes Niagara. Pour ceux qui peuvent saisir l'occasion, je les invite à passer du temps supplémentaire à explorer le Canada qui est un pays immense et spectaculaire et qui offre une grande variété de plaisirs scéniques et culturels. Et vous serez chaleureusement accueillis.

Comme toutes les disciplines qui traitent de la masse continentale et de son utilisation, la mécanique des roches est d'une importance vitale pour le Canada, pays axé sur les ressources. Tout progrès accompli en mécanique des roches a des répercussions importantes dans ce pays où nous nous efforçons d'utiliser nos ressources de façon aussi efficace et sûre que possible et de façon acceptable pour l'environnement.

Le Canada s'enorgueillit avec raison des travaux qu'il a accomplis en mécanique des roches. Il a toujours été un chef de file dans le secteur de l'exploitation minière en roche dure hautement productive et les récents perfectionnements apportés à l'exploitation par charges concentrées sont considérés comme une découverte importante qui combine productivité élevée et sécurité. La mécanique des roches joue un rôle central dans l'application de cette technique.

En outre, on utilise maintenant au Canada des mineurs à roche dure à des profondeurs dépassant souvent les 2000 mètres. Une exploitation aussi profonde ne pourrait pas être faite sans le concours de la mécanique des roches.

Je ne suis pas sans savoir que le sixième Congrès international de la mécanique des roches est un congrès spécial étant donné qu'il coïncide avec le 25e anniversaire de la Société internationale de mécanique des roches. Le Canada est par conséquent doublement honoré d'avoir été choisi comme pays hôte pour souligner cet événement important de l'histoire de votre organisme.

L'Honorable G. Merrithew, Ministre d'État aux Forêts et aux Mines, gouvernement du Canada

Je félicite la SIMR pour l'ensemble des projets remarquables qu'elle a accomplis depuis sa fondation. La mécanique des roches est une discipline relativement récente; les progrès accomplis dans ce domaine à travers le monde au cours des 25 dernières années ont pu l'être grâce en grande partie à la SIMR et à ses activités.

À titre d'exemple, j'aimerais citer la contribution de la SIMR à la normalisation des essais en laboratoire et sur le terrain, aux travaux de reconnaissance et aux méthodes d'analyse. On m'a indiqué qu'à ce jour 80 méthodes d'essai proposées ont été publiées, ce qui représente le fruit du travail d'environ 500 scientifiques.

Je suis en outre fier du fait que des spécialistes canadiens en mécanique des roches ont collaboré activement aux travaux de la SIMR depuis sa création. Nos spécialistes ont ainsi pu se tenir au courant des innovations qui se sont produites à travers le monde. Par la même occasion, ils ont été en mesure d'apporter leur propre contribution à la base de connaissances internationale de la mécanique des roches.

L'une des contributions canadiennes qui mérite d'être mentionnée est le Manuel sur les talus des mines à ciel ouvert préparé par l'un de nos pionniers remarquables en mécanique des roches, le regretté Donald Coates. Ce guide scientifique et technique exhaustif a eu des répercussions importantes sur la conception des mines à ciel ouvert dans tout le Canada et ailleurs.

Les quatre thèmes du congrès (écoulement de fluides et enfouissement de déchets dans les massifs rocheux; fondations et talus rocheux; sautage et excavation; et souterrains en massifs rocheux sous grandes contraintes) ont une grande portée sur les activités actuelles d'exploitation et d'ingénierie au Canada.

Au Canada, l'un des problèmes importants auxquels font face les sociétés d'exploitation minière est celui des coups de toit. De 1984 à 1985, pas moins de 217 coups de toit d'intensité variant de 1,5 à 4 à l'échelle Richter se sont produits dans ce pays. Le CANMET, organisme de recherche minière du gouvernement fédéral canadien, travaille en étroite collaboration avec l'industrie minière, des ministères provinciaux et des universités à mettre au point dans le cadre d'un programme de recherche quinquennal des technologies pour détecter les zones dangereuses dans les mines où des coups de toit sont les plus susceptibles de se produire.

Au cours du présent congrès, des exposés détaillés de ces travaux seront présentés, et ceux d'entre vous qui ont pu visiter les mines de Sudbury, mardi et mercredi dernier, auront eu un premier aperçu de la mise en oeuvre de ce programme de recherche.

En plus du rôle important qui leur est dévolu d'augmenter la sécurité et l'efficacité des opérations minières, les spécialistes canadiens de mécanique des roches participent à de nombreuses autres études et activités intéressantes, allant de l'élimination permanente des déchets nucléaires à la construction de fondations, de talus et de tunnels en roches. Quel que soit leur domaine d'intérêt particulier, je suis sûr que, comme vous tous, ils auront modifié à la fin de cette semaine certaines perceptions et qu'ils auront acquis de nouvelles connaissances.

Que ce soit des séances techniques aux tours guidés, des ateliers à l'exposition, la planification et la préparation de ce congrès ont été sans conteste réalisées de façon remarquable. Je félicite donc l'Institut canadien des mines et de la métallurgie, l'Association canadienne de géotechnique et l'Association canadienne de mécanique des roches pour les efforts qu'ils ont déployés.

Je déclare maintenant le sixième Congrès international de mécanique des roches officiellement ouvert.

M. Herget remercie le ministre d'avoir prononcé ces mots de bienvenue et d'avoir pris le temps de se libérer d'un horaire très chargé. Après l'ouverture du sixième Congrès international de mécanique des roches, le professeur N. Morgenstern, président du Comité du programme technique, présente le professeur E. Hoek qui doit faire un exposé sur l'ingénierie des roches au Canada. Le texte de l'exposé est reproduit ci-dessous:

INGÉNIERIE DES ROCHES AU CANADA

Evert Hoek

Département de génie civil, Université de Toronto

Toronto, Canada

INTRODUCTION

L'ingénierie des roches qui est l'application de la science de la mécanique des roches aux problèmes pratiques d'ingénierie ne date pas d'hier au

Canada. Dans la préparation du présent exposé, j'ai mis l'accent sur les cas pratiques et je tenterai d'illustrer la diversité des problèmes auxquels il a fallu et il faut encore faire face pour mettre en valeur cet immense pays.

Je ne prétends pas présenter un survol historique complet de l'évolution de l'ingénierie des roches au Canada; j'ai plutôt choisi de façon arbitraire des sujets que je considère importants et j'ai essayé de les traiter de façon suffisamment détaillée pour que les lecteurs intéressés puissent en tirer parti.

RÉSUMÉ HISTORIQUE

Le premier tunnel important qui a été construit au Canada a été un tunnel ferroviaire à Brockville (Ontario) dont la construction a commencé en 1854 pour se terminer en 1860. Selon Legget (1985), le traçage à grande échelle de tunnels au Canada aurait commencé en 1889 lorsqu'ont débuté les travaux de percement d'un tunnel sous la rivière Saint-Clair entre Sarnia (Ontario) et Port Huron (Michigan). Ce tunnel ferroviaire, tracé au bouclier, a été complété en 1890. Le traçage de tunnels en roche n'a pas commencé avant le début des années 1900, marquée par la fin de la construction des tunnels Spiral en 1909, et du tunnel Connaught du Canadien Pacifique, en 1916. Ce sont ces tunnels qui font traverser le chemin de fer à travers les Rocheuses canadiennes.

Le Répertoire des barrages du Canada, publié par le Comité national canadien de la Commission internationale des grands barrages (1984) énumère 613 barrages dont 43 dépassent 60 m de hauteur. Le premier barrage construit au Canada qui a figuré dans ce répertoire est le barrage-voûte en maçonnerie de Jones Falls mesurant 19 m de hauteur sur le canal Rideau et construit en 1832. Le barrage le plus haut au Canada est le barrage Mica de 243 m de hauteur en Colombie-Britannique dont la construction s'est terminée en 1972. Des 613 barrages terminés en 1983, 441 étaient associés à des projets hydro-électriques.

L'industrie minière a joué un rôle vital dans l'essor économique du pays et la technologie minière a évolué de façon à répondre aux besoins particuliers de cette industrie. Le premier programme de recherche en mécanique des roches appliqué aux mines a été entrepris à l'Université McGill au début des années 1950. Ce programme portait sur les ruptures provoquées par des contraintes et sur les coups de toit dans les mines souterraines. Le premier exposé sur les résultats de ces recherches a été présenté par Morrison et Coates à la réunion annuelle de l'Institut canadien des mines et de la métallurgie en 1955.

La recherche en mécanique des roches à la Direction des mines du ministère de l'Énergie, des Mines et des Ressources (renommée plus tard le Centre canadien de la technologie des minéraux et de l'énergie (CANMET)) a commencé en 1950. En 1964, un laboratoire de recherche en exploitation minière souterraine en roche dure s'ouvrait à Elliot Lake en Ontario et, en 1967, le Centre de recherche minière était mis sur pied à Ottawa sous la direction de D.F. Coates. Un relevé utile des travaux de recherche réalisés en mécanique des roches au Laboratoire de recherche minière de CANMET entre 1864 et 1984 a été compilé par Udd et Hedley (1985).

L'un des événements les plus importants marquant les débuts du développement de la mécanique des roches au Canada a été la publication par Coates en 1965 d'un manuel intitulé "Principes de mécanique des roches". Il s'agissait du premier manuel en langue anglaise sur le sujet et il est devenu un ouvrage de référence important pour les étudiants et les chercheurs au cours des années de défrichement du domaine.

Il est intéressant de noter que des travaux importants en mécanique des roches liés au génie civil et à la géologie ont été réalisés au début des années 1960. Ces travaux qui ont été publiés dans des revues scientifiques peu lues par les spécialistes en mécanique des roches sont moins connus que les travaux de recherche en exploitation minière mais leur apport à cette science n'en est pas moins important. Parmi ces travaux, mentionnons les essais au cisaillement de roches sédimentaires par Ripley et Lee (1961) et une étude très détaillée de diagrammes de texture tridimensionnels (projection sphérique) par Stauffer de l'Université de la Saskatchewan (1966).

Afin de procurer une tribune aux chercheurs en mécanique des roches pour qu'ils présentent leurs résultats de recherche et qu'ils en discutent, le premier Symposium canadien de mécanique des roches se tenait à Montréal en 1962. Il a été suivi par une série de symposia dont le prochain aura lieu à Toronto en septembre 1988. Depuis ses débuts, l'ingénierie des roches a acquis une certaine maturité qui en font maintenant une discipline de génie de plein droit. Des cours de premier et deuxième cycles sont offerts à un certain nombre d'universités canadiennes et des organismes de consultation, industriels et gouvernementaux à travers le pays recourent à des techniques de mécanique des roches pour résoudre des problèmes pratiques de génie.

LABORATOIRE DE RECHERCHE SOUTERRAIN DE L'ÉNERGIE ATOMIQUE DU CANADA LTÉE

L'Énergie atomique du Canada Limitée (EACL) est chargée des travaux de recherche et développement de technologies qui permettront d'éliminer de façon sûre et permanente les déchets de combustion nucléaire au Canada.

Dans le cadre de ce vaste programme, l'EACL construit actuellement un laboratoire de recherche souterrain (LRS) près de Lac du Bonnet (Manitoba) pour évaluer certains aspects de l'enfouissement des déchets dans des formations géologiques stables. Aucun déchet nucléaire ne sera utilisé dans le programme du LRS.

Le LRS est excavé dans le batholite granitique du Lac du Bonnet, massif rocheux d'excellente qualité, que l'on considère représentative de nombreuses roches intrusives granitiques du bouclier précambrien au Canada. Avant que ne débute l'excavation, la massif rocheux a d'abord été entièrement caractérisé par une cartographie détaillée de la géologie de surface, une diagraphie de trous de sondage forés au diamant et par percussion, des levés géophysiques en surface et en sondages ainsi que des mesures de contraintes par fracturation hydraulique (Lang, 1985). Un programme hydrogéologique global a aussi été réalisé afin de caractériser le site du LRS avant le début de l'excavation (Davison, 1984).

En 1984 et 1985, un puits de section rectangulaire 2,8 m sur 4,9 m a été foncé par forage et sautage jusqu'à une profondeur de 255 m. Des recettes de puits ont été excavées aux profondeurs de 130 m et 240 m et les laboratoires ont été excavés au même niveau que les recettes.

Au cours du fonçage du puits, un certain nombre de programmes expérimentaux ont été réalisés. Ils consistaient, entre autres, à mesurer la convergence en puits, à déterminer les contraintes in situ par surcarottage et fracturation hydraulique et à analyser les comportements géomécanique et hydrogéologique combinés de la roche entourant le puits (Lang et Thompson, 1985), (Kuzyk, Lang et Peters, 1986).

Pr E. Hoek, Conférencier d'ouverture

Au cours de l'abattage de certaines chambres dans la recette de fond, on a eu recours à des techniques d'excavation par forage et sautage minutieusement contrôlées pour déterminer l'endommagement induit dans ce type de roche par le sautage de haute qualité utilisé. Ce type de dommage peut avoir des effets importants sur l'élimination des déchets nucléaires étant donné qu'il peut créer un trajectoire de fuite parallèle à la limite d'excavation. Kuzyk, Lang et Le Bel (1986) et Lang, Humphreys et Ayotte (1986) ont décrit le type de sautage utilisé et les excellents résultats obtenus dans le cadre de ce programme d'excavation.

Divers programmes expérimentaux se poursuivent au LRS et quelques résultats importants ont déjà été obtenus. Davison (1986) a comparé les données observées et prévues en se fondant sur un modèle hydrogéologique mis au point par Guvanasen (1984). Le volume d'infiltration observé dans le puits est inférieur au volume prévu par le modèle; cependant, les estimations par le modèle du rabattement piézométrique en fonction de données spatio-temporelles, dans 171 piézomètres entourant le puits, se sont avérées relativement exactes. En général, on a constaté que la précision des prévisions du modèle était moins élevé près du puits mais qu'elle augmentait à mesure qu'on s'en éloignait. D'autres travaux sont réalisés afin de réduire les différences entre les observations et les prévisions et afin de mettre à l'essai d'autres modèles hydrogéologiques.

L'un des aspects intéressants des travaux réalisés en hydrogéologie au LRS est la mise au point, en collaboration avec la Westbay Instruments Ltd. de Vancouver, d'instruments spécialisés pour effectuer des mesures piézométriques et prélever des échantillons d'eau (Rehtlane et Patton, 1982). Pour mettre au point cette instrumentation, on s'est basée sur une série très réussie d'instruments semblables conçus par la Westbay au cours de la dernière décennie. Grâce à ces instruments, on peut mesurer les pressions piézométriques et prélever des échantillons d'eau à intervalles rapprochés et à des profondeurs précises et, si le volume d'échantillonnage est maintenu très petit, minimiser la perturbation du régime des eaux souterraines.

Lang et al. (1986) ont présenté en détail les résultats qu'ils ont obtenus dans le cadre d'un programme initial de mesure des contraintes in situ dans le cadre duquel 42 essais ont été effectués selon la technique élaborée par le Bureau of Mines des États-Unis. Dans des essais ultérieurs, on a eu recours aux cellules de mesure des contraintes triaxiales de la CSIRO d'Australie et aux cellules modifiées de la CSIR d'Afrique du Sud. Les résultats préliminaires indiquent que les contraintes horizontales maximales et minimales subissent un changement marqué à une profondeur de 185 m. Cette anomalie est associée à une zone de fracture située à cette profondeur et l'on espère que d'autres mesures des contraintes triaxiales permettront de mieux comprendre la corrélation qui existe entre les contraintes mesurées et les conditions géologiques observées à différentes profondeurs. Le rapport de la contrainte horizontale moyenne à la contrainte verticale correspond bien à une relation établie par Herget (1980) pour des mesures de contraintes réalisées dans le Bouclier canadien.

En plus de revêtir une certaine importance pour le programme d'élimination des déchets nucléaires au Canada, les travaux réalisés dans le Laboratoire de recherche souterrain de l'EACL intéressent beaucoup les spécialistes canadiens de la mécanique des roches en général. Les informations qui auront été recueillies dans les expériences très particulières réalisées dans le LRS contribueront à augmenter considérablement notre compréhension du comportement des massifs rocheux.

EXPLOITATION DES SABLES PÉTROLIFÈRES PAR LA SYNCRUDE DANS LE NORD DE L'ALBERTA

La mine et l'usine de la Syncrude Canada Ltd. sont situées à environ 40 km au nord de Fort McMurray dans le nord de l'Alberta. L'exploitation des sables pétrolifères, qui nécessite quatre draglines de 65 mètres cubes accompagnées de quatre excavateurs de reprise et de convoyeurs, a commencé en juin 1977. Avec une capacité de production de 129 000 barils de pétrole synthétique par jour (le niveau du permis actuel), la production quotidienne de la mine, pendant les 25 années de vie prévues, s'élèvera à environ 250 000 tonnes par jour (Fair et Lord, 1984).

Les sables pétrolifères de l'Athabasca dans le nord de l'Alberta contiennent l'une des plus grandes accumulations de bitume visqueux dans le monde. La formation de McMurray qui contient le gisement de sable pétrolifère résulte du remplissage sédimentaire d'un bassin formé sur des carbonates dévoniens au cours de la transgression qui a eu lieu avant le Crétacé. Les différents faciès sédimentaires témoignent de dépôts successifs de sédiments fluviaux, estuariens et marins qui ont produit les plaines de marée faiblement inclinées du niveau des hautes eaux au niveau des basses eaux tels qu'ils existaient à l'époque de la sédimentation (List, Lord et Fair, 1985).

Le drainage des plaines intertidales s'est accompli suivant un réseau de chenaux de marée à méandres interconnectés qui ont sans cesse remanié les plaines de marée et qui ont déposé des couches inclinées de sable, de limon et d'argile. Les couches d'argile dont l'inclinaison peut dépasser 20 degrés, forment des surfaces de rupture qui sont à l'origine des éboulements qui se sont produits dans les talus de 60 mètres de hauteur créés par les draglines (Brooker et Khan, 1980).

Si un éboulement était survenu du côté du talus où se trouvait une dragline, l'exploitation en aurait été gravement perturbée. En général, ces éboulements prennent tout au plus une minute ou deux à se développer et, étant donné qu'il faut environ 15 minutes pour qu'une dragline puisse s'éloigner de la crête de la muraille, il est essentiel d'identifier les éboulements possibles et de prendre les mesures qui s'imposent. Fair et Lord (1984) ont décrit les techniques utilisées pour identifier les zones d'éboulement possibles et pour les surveiller durant l'exploitation.

L'une des techniques innovatrices qui a été utilisée pour stabiliser les talus est le sautage dont on se sert pour briser la continuité des couches d'argile inclinées le long desquelles se forment les surfaces de rupture. Ce sautage est réalisé avant que la dragline ne soit utilisée dans les zones où des éboulements sont susceptibles de se produire. En plus d'accroître l'angle de frottement des matériaux le long des surfaces de rupture possibles, le sautage augmente la perméabilité de ces matériaux et contribue à dissiper les pressions interstitielles présentes dans les sables pétrolifères. Ces modifications améliorent la stabilité des zones d'éboulement possibles.

Pour obtenir les résultats souhaités, on déplace les sables pétrolifères verticalement en utilisant des charges sphériques ou cabrées dans le sautage par charges concentrées. La conception des trous de mine se fonde sur les travaux réalisés en collaboration avec le professeur Alan Bauer de l'Université Queen's et ils sont décrits en détail par List, Lord et Fair (1985). Il s'avère que la méthode la plus économique de sautage comme mesure de stabilisation consiste à utiliser du nitrate d'ammonium et du fuel (ANFO) dans les trous de mine de grand diamètre.

Dans le sautage courant, on utilise des trous de mine mesurant 0,76 m de diamètre et environ 15 m de profondeur. Le poids de l'ANFO par trou de mine est d'environ 1180 kg et la vitesse de pointe des particules créées par les tirs est limitée en contrôlant la charge de chaque détonateur à retard et le retard entre les trous de mine successifs.

En plus des éboulements susmentionnés, plusieurs autres types de rupture de pente ont été identifiés (Brooker et Khan, 1980). Seules les zones d'éboulement, cependant, peuvent être stabilisées par la technique de sautage préalable décrite. Comme toutes les ruptures de muraille peuvent avoir des répercussions graves sur l'exploitation minière, des mesures ont été prises pour identifier et traiter tous les types de ruptures possibles. Fair et Lord (1984) et Fair et Isaac (1985) ont décrit les programmes de surveillance de grande envergure qui sont mis en oeuvre pour mesurer les pressions piézométriques et les déplacements de talus pendant l'exploitation.

L'interprétation des mesures recueillies dans ce programme de surveillance a facilité la prise de décisions sur la stratégie d'exploitation à court terme et a permis de comprendre les mécanismes de rupture de pente possibles dans la zone exploitée par la Syncrude.

EXPLOITATION MINIÈRE SOUTERRAINE PROFONDE EN ROCHE DURE

Le Canada est riche en ressources minérales qui sont extraites d'un grand nombre de mines souterraines et à ciel ouvert. Dans le bouclier précambrien, les mines souterraines d'or et de métaux communs sont situées à des profondeurs atteignant 2300 m et, dans nombre de ces mines, les coups de toit peuvent constituer un danger pour le personnel et nuire à la production.

Le premier coup de toit enregistré a eu lieu dans la région de Sudbury en 1929. Des coups de toit sont survenus régulièrement dans les mines de Kirkland Lake durant les années 1930 et, en 1939, l'Association minière de l'Ontario retenait les services du professeur Morrison pour qu'il la conseille sur les mesures à prendre pour réduire les risques de coups de toit (Morrison, 1942). Les coups de toit n'ont pas cessé de constituer un défi pour l'industrie minière et pour les groupes de recherche en mécanique des roches qui tentent d'approfondir ce phénomène.

Les règlements de l'Ontario sur les mines et les usines d'exploitation (Regulations for Mines and Mining Plants) définissent un coup de toit comme une rupture instantanée d'un massif rocheux causant une explosion de matériaux à la surface d'une ouverture ou un ébranlement sismique dans une mine à ciel ouvert ou souterraine. En 1984, on a enregistré, dans les mines de l'Ontario, un total de 105 coups de toit dont l'intensité était supérieure à 1,5 à l'échelle Richter (Scoble, 1986). Ce chiffre peut être considéré comme représentatif des coups de toit qui se sont produits ces dernières années. Bien que de nombreux coups de toit soient trop faibles pour avoir des conséquences importantes, plusieurs ont causé des dommages notables (gouvernement de l'Ontario, 1986).

En 1985, CANMET a entrepris un important projet de recherche quinquennal sur les coups de toit (CANMET, 1986). Ce projet comporte une surveillance microsismique de base dans plusieurs petites mines et des études détaillées, par l'utilisation de systèmes de surveillance microsismique de pointe, dans deux mines de Sudbury. De plus, des études seront réalisées dans une grande mine de l'Ontario où l'on recourra à un système microsismique prototype perfectionné permettant des analyses détaillées des événements sismiques. En association avec ce projet, un certain nombre d'activités parallèles, telles que la mise au point d'un modèle numérique tridimensionnel sophistiqué (McDonald, Wiles et Villeneuve, 1988), sont actuellement entreprises par des groupes de recherche financés en partie par CANMET.

Deux commissions d'enquête ont préparé des rapports, l'un publié en 1981 et l'autre en 1986, sur la sécurité dans les mines souterraines du nord de l'Ontario. Comme l'ingénierie des roches, c'est-à-dire les questions relatives aux coups de toit, à l'élimination des contraintes, au remblayage, à la récupération des piliers, au minage et au soutènement, a reçu une attention spéciale des deux commissions, leurs conclusions et recommandations devraient être lues par quiconque s'intéresse à l'exploitation souterraine. Comme conséquence directe du rapport de 1986 préparé pour le gouvernement de l'Ontario par un comité présidé par M. Trevor Stevenson, une direction de recherche minière a été mise sur pied en Ontario pour coordonner les travaux de recherche portant sur l'exploitation minière profonde en roche dure. Ces travaux porteront d'abord sur la mécanique des roches appliquée mais il est prévu que les activités de cet organisme s'étendront graduellement à d'autres domaines de recherche liés à l'exploitation minière. La direction de recherche fonctionnera parallèlement aux organismes de recherche existants, soit les universités, CANMET et les groupes de recherche qui travaillent dans les sections de services techniques des grandes sociétés minières.

PERCEMENT DE TUNNELS DANS DES TERRAINS INSTABLES ET CONVERGENTS

Pendant que les coups de toit constituent un problème dans l'exploitation minière des roches dures du bouclier précambrien, la fermeture graduelle ou la convergence est l'un des problèmes auxquels doivent faire face les tunnelliers dans d'autres régions du pays. Dans le sud de l'Ontario, où des contraintes horizontales élevées ont été observées (Lo, 1978), les schistes argileux relativement instables que l'on trouve couramment dans cette région sont particulièrement sujets à la convergence (Lo, Cooke et Dunbar, 1986), (Lo, 1986).

Le percement de tunnels en roche tendre crée aussi des problèmes dans d'autres régions du pays. Certains de ces problèmes ont d'ailleurs été décrits par Kaiser et McKay (1983), Kaiser, Guenot et Morgenstern (1985) et Boyd, Yuen et Marsh (1986). Ce dernier document est particulièrement intéressant du fait qu'il y est question d'un tunnel sous-marin de 4 km de long foré mécaniquement dans des roches sédimentaires relativement instables de la zone houillère située au large des côtes de la Nouvelle-Écosse.

Des études théoriques sur le comportement des tunnels dans des roches convergentes ont été entreprises par Ladanyi (1974, 1976), Ladanyi et Gill (1984), Lo et Yuen (1981), Emery, Hanafy et Franklin (1979), et Brown et al. (1983). Dans des travaux de recherche réalisés sous la direction du professeur K.Y. Lo de l'Université Western en Ontario, Ogawa (1986) a mis au point une solution tridimensionnelle des contraintes et des déplacements dans la roche autour d'un tunnel et l'application de cette solution à l'interprétation du comportement d'un tunnel a été décrite par Lo, Lukajic et Ogawa (1984) et par Lo (1986). Une autre analyse tridimensionnelle des contraintes et des déplacements qui s'exercent dans la roche entourant le front de taille avançant d'un tunnel a été traitée par Eisenstein, Heinz et Negro (1984).

PROJETS HYDRO-ÉLECTRIQUES AU CANADA

En raison de ses nombreuses régions montagneuses et de ses abondantes ressources hydrologiques, le Canada ne manque pas de possibilités pour la construction de centrales hydro-électriques. Un certain nombre de projets hydro-électriques ont été réalisés mais relativement peu d'entre eux sont connus à l'extérieur du pays étant donné que les données relatives à ces projets ont été, soit non publiés, soit publiés dans des revues scientifiques qui ne sont pas habituellement lues par les spécialistes de la mécanique des roches.

Parmi les projets les moins connus mentionnons celui de Spray en Alberta (Eckenfelder, 1952). L'élément de ce projet qui retient particulièrement l'attention est la conception de la galerie en pression qui comporte une section inclinée de 45° et qui traverse des massifs de calcaire, decalcaire dolomitique, d'argilite massive et quelques rares bandes de schiste argileux. Selon les normes actuelles, le rapport de la couverture rocheuse à la charge statique qui sert à déterminer la longueur du soutènement métallique de la centrale de Spray ne serait pas suffisant à moins que des essais d'acceptation de pression soient réalisés dans le cadre du processus de conception (Humphreys et Hoek, 1987). Le fait que cette galerie ait fonctionné avec succès pendant de nombreuses années laisse supposer que, dans certaines circonstances, les critères actuels de profondeur de la couverture peuvent être jugés prudents et que d'autres recherches devraient être effectuées dans le domaine de la conception des galeries en pression.

Huber (1952, 1953) et Wise (1952) ont publié des données détaillées sur la construction de la centrale de Kitimat-Kemano par l'Aluminium du Canada Ltée dans le nord de la Colombie-Britannique. Lors de sa construction, cette centrale souterraine était considérée comme la plus grande du monde.

Les aspects de la géologie de l'ingénieur et de l'ingénierie des roches qui ont été traités lors de la conception des centrales hydro-électriques de W.A.C. Bennett (mont Portage) et de Mica en Colombie-Britannique ont été présentés par Imrie et Jory (1969) et Imrie et Campbell (1976). Les problèmes d'instabilité souterraine associés aux cisaillements de foliation observés dans les centrales hydro-électriques de Bersimis et de Chute-des-Passes au Québec sont décrits dans un document publié par Benson et al. (1974). Humphreys et Lang (1986) et Humphreys et Hoek (1987) ont présenté des données détaillées de certains aspects de l'ingénierie des roches appliquée à la centrale hydro-électrique de Cat Arm à Terre-Neuve. Ce projet est inhabituel du fait qu'il comprend une galerie en pression élevée non soutenue et minée avec soin dans un granite de très bonne qualité.

L'ingénierie des roches a joué un rôle important dans l'élaboration du projet hydro-électrique des chutes Churchill au Labrador qui était considéré à l'époque comme un projet souterrain de grande envergure et très ambitieux. Les analyses géotechniques qui ont été effectuées et la conception de l'usine souterraine ont été décrites dans une série de documents publiés par Benson et al. (1970a, 1970b, 1971, 1974) et certains détails de construction ont été présentés par Mamen (1969) et Gagné (1972).

En raison de son éloignement et de ses grandes dimensions, le projet de la Grande dans la région de la baie James (nord du Québec) a suscité beaucoup d'intérêt. Ce projet vise à mettre en valeur le potentiel hydro-électrique de la rivière La Grande et de certaines parties des bassins hydrographiques adjacents de Caniapiscau et d'Eastmain-Opinaca. Le projet global comprend neuf centrales d'une puissance prévue de 13 753 MW et d'une capacité de production d'électricité de 81,6 terawatt-heures. Tout le complexe se situe dans le Bouclier canadien et la plupart des emplacements du projet reposent sur des roches granitiques de texture allant de massive à gneissique. En se fondant sur l'expérience acquise dans de nombreux projets d'exploitation minière et de construction qui ont été réalisés dans d'autres zones du bouclier canadien, on peut résumer la conception et la philosophie de la construction adoptée pour l'aménagement de La Grande de la façon suivante (Murphy et Levay, 1982):

1. À moins d'indication contraire, la qualité de la roche est supposée d'être de bonne à excellente.

2. Les techniques de forage et de sautage utilisées sont en général plus importantes que les conditions géologiques du terrain pour minimiser la surfragmentation et pour s'assurer la stabilité des profils d'excavation.

3. Pour évaluer les besoins de traitement de la roche (renforcement et protection en surface), il est habituellement plus valable de faire preuve de jugement en se fondant sur les observations effectuées sur le terrain que d'adopter une méthode basée uniquement sur des données théoriques.

Même s'ils ont laissé entendre que la mécanique des roches a peu servi dans la conception du projet hydro-électrique de La Grande, Murphy et Levay poursuivent en décrivant l'utilisation qui est faite des photographies aériennes, des levés structuraux et des sondages au diamant combinés à l'utilisation d'appareils de prise de vue dans les sondages et à des essais de mise en pression. Dans le cas de la centrale souterraine de LG-2, on a creusé un puits d'essai et des galeries le long de la voûte de la centrale afin de mesurer les modules et les contraintes in situ; les résultats de ces analyses ont été utilisés dans des études par la méthode des éléments finis sur les dimensions des piliers et la géométrie des cavernes. Je trouve fascinant que toutes ces analyses soient qualifiées "de jugement fondé sur l'expérience et d'observations sur le terrain" par opposition à "une méthode basée uniquement sur des données théoriques". Cela laisse entendre que dans l'esprit des gestionnaires du projet de La Grande, l'ingénierie des roches s'est acquis le statut d'un instrument de conception pratique, ce que je considère comme un compliment pour les nombreux chercheurs en mécanique des roches qui se sont acharnés pendant de nombreuses années sur leur "méthode basée uniquement sur des données théoriques" pour atteindre ce but.

La centrale souterraine de LG-2 est, prétend-on, la plus grande du monde. Creusée dans du granite massif à une profondeur de 160 m, la salle des machines a une portée de 28,4 m tandis que celle de la cheminée d'équilibre parallèle est de 23,6 m. Ces deux excavations mesurent environ 47 m de hauteur et la salle des machines contient 16 turbogénérateurs de 333 MW. Dans la centrale, deux grues de 405 tonnes reposent sur des corniches taillées dans une roche de qualité élevée en appliquant des techniques d'excavation minutieusement contrôlées.

L'excellente qualité de la roche a également permis d'utiliser des galeries de dérivation et de fuite non revêtues permettant à l'eau de couler à une vitesse pouvant atteindre 14 m/s. Plus de 8000 m de galerie atteignant jusqu'à 14 x 20 m de dimensions ont été excavés mais on n'a eu recours au revêtement en béton que dans les portails et pour les vannes ou les batardeaux.

Le déversoir de LG-2 est une autre pièce d'ingénierie intéressante. Il s'agit d'un déversoir en cascade conçu pour faire face à un débit de pointe de 16 280 mc/s. La longueur totale du chenal du déversoir est de 2200 m et il comprend 10 niveaux ou gradins variant de 9,1 à 12,2 m de hauteur et de 127 à 200 m de longueur. La largeur nominale du chenal est de 122 m mais comme tous les déblais d'excavation ont été utilisés pour remblayer le barrage principal, sa largeur a été augmentée à 183 m pour les derniers 1100 m afin d'accroître la quantité de remblais (Levay et Aziz, 1978). Les gradins du chenal en cascade ont été excavés en rocher et quelques bordures seulement où les fissures étaient peu espacées ont été renforcées avec des boulons d'ancrage et du béton projeté. Le déversoir fonctionne depuis 1979 et n'a montré aucun signe de détérioration important malgré le fait que les vitesses de l'eau aient atteint la limite nominale. La vitesse limite est de 11 m/s à la fin du premier gradin et de 26 m/s au dernier gradin.

CONSTRUCTION DE TUNNELS DANS LES MONTAGNES DE L'OUEST CANADIEN

Le col Rogers est situé dans une belle région sauvage des monts Selkirk en Colombie-Britannique. Il atteint une altitude maximale de 1200 m au-dessus du niveau de la mer, au milieu de cimes dépassant les 3300 m. Au cours d'un hiver moyen, il y tombe plus de 9 m de neige.

Le premier chemin de fer à traverser le col Rogers a été construit par le Canadien Pacifique en 1885. Malgré de graves problèmes, notamment un glissement où 62 travailleurs ont trouvé la mort en 1910, cette ligne de chemin de fer a continué d'être exploitée pendant de nombreuses années. Un programme visant à augmenter la capacité de la ligne du CP à travers le pays s'est traduit par le percement d'un tunnel dans la base du mont Macdonald de la chaîne Selkirk. La construction de ce tunnel, appelé au début tunnel Selkirk et, plus tard, tunnel Connaught, a commencé en 1913 et s'est terminée en 1916. Il mesure environ 8 km de long et a été excavé à partir d'un petit tunnel-pilote parallèle au tunnel principal, ce qui permettait d'avoir accès à un certain nombre de fronts de taille à partir de petits tunnels de communication qui étaient percés à intervalle périodique le long du tunnel. Ce tunnel, qui est encore utilisé aujourd'hui, a permis non seulement de raccourcir la ligne et de réduire la déclivité mais également d'éviter certaines des zones d'avalanches les plus dangereuses.

Pour éliminer les embouteillages dans la ligne de chemin de fer principale reliant Calgary et Vancouver, CP Rail travaille à un nouveau projet d'amélioration. L'un des principaux éléments de ce projet est un nouveau tunnel, appelé tunnel du mont Macdonald, qui a été creusé à travers le mont Macdonald et le mont Cheops. Ce tunnel de 14,7 km de longueur est le plus long tunnel ferroviaire en Amérique du Nord. La construction a commencé en 1984 et les deux fronts de taille avançant dans des directions opposées se sont rejoints au début de l'année. Ce tunnel en forme de fer à cheval revêtu de béton aura une portée de 5,18 m et une hauteur, du sommet de la voie à la couronne, de 7,87 m.

L'excavation du tunnel à partir du portail de l'ouest a été effectuée par méthode de forage et sautage à pleine section, avant d'être revêtu de béton coulé sur place (Stewart, 1987). Un tunnelier a été utilisé pour creuser la galerie d'avancement à partir du portail de l'est avant que le reste du tunnel soit abattu à l'explosif par gradins jusqu'à sa pleine hauteur (Knight, 1987). Le soutènement utilisé durant l'excavation du tunnel était composé de tirants ainsi que de treillis et de béton projeté. Il est intéressant de noter que, selon les études de classification des massifs rocheux, le tunnel aurait dû être soutenu sur des longueurs importantes par des boulons d'ancrage, du béton projeté et des treillis.

Basées sur l'expérience acquise depuis la construction du tunnel parallèle Connaught, qui est demeuré sans soutènement pendant 70 ans, les exigences en soutènement ont diminué pour atteindre le niveau minimal actuel. Un revêtement complet en béton est actuellement placé dans le tunnel et sa capacité portante est suffisante pour assurer une stabilité à long terme du massif rocheux entourant le tunnel.

Une brève description du tunnel creusé dans le col Rogers a été présentée par Gadsby (1985) et un survol des éléments techniques généraux de ce projet a été publié par Tanaka (1987). L'établissement de cartes géologiques et le traitement assisté par ordinateur des données réalisés dans le cadre de ce projet sont traités en détail dans un document de Klassen et al. (1987).

Dans une autre région de la Colombie-Britannique, on évalue actuellement les données relatives à un certain nombre de tunnels qui seraient creusés pour améliorer la voie ferroviaire du Canadien National à travers les montagnes. Benson, Hostland et Charlwood (1985) ont énuméré les exigences en matière de géotechnique et de ventilation qu'il faudra satisfaire dans ces tunnels.

Kaiser et Gale (1985) ont publié les résultats d'une étude très complète portant sur l'utilisation des systèmes de classification des massifs rocheux pour déterminer les besoins en soutènement des tunnels Tumbler Ridge de la BC Rail en Colombie-Britannique. Un certain nombre de classifications populaires sont comparées et les avantages et inconvénients de chacune sont présentés. Selon moi, ce document constitue un apport important à la documentation de plus en plus nombreuse sur l'utilisation des classifications des massifs rocheux pour évaluer les besoins en soutènement des excavations souterraines.

L'EXPLOITATION MINIÈRE DANS LE GRAND NORD CANADIEN

Située sur la petite île Cornwallis, à 77 degrés au nord de l'Équateur, la mine Polaris de la Cominco se trouve être la mine la plus septentrionale du monde. Le gîte de plomb-zinc est contenu dans une zone pergélisolée qui, à cette latitude, s'étend jusqu'à une profondeur d'environ 450 m. Les températures de surface varient de -55 °C, en hiver, à 15 °C, en été. Même pendant les étés les plus chauds, quelques centimètres seulement de la surface du sol dégèlent (Leggatt, 1982).

Étant donné la teneur élevée en glace du minerai, il est important de garder la masse rocheuse gelée afin d'assurer la stabilité des excavations souterraines. Pour ce faire, la mine est ventilée avec de l'air froid qui permet de garder la température constante à environ -7 °C. Un mélange de neige et de gravier est déversé dans les cheminées pour former dans les chantiers un remblai que le gel rend très dur et efficace.

Depuis le début de l'exploitation de la mine en 1982, il est à remarquer que très peu de problèmes géotechniques se sont posés. Malgré le fait que le personnel doit s'habiller un peu plus chaudement que dans une mine ordinaire, l'exploitation de la mine, la conception des chantiers et des piliers ainsi que le choix des systèmes de soutènement sont remarquablement semblables à ceux que l'on trouverait dans une mine située dans un climat plus tempéré.

INGÉNIERIE DES TALUS ROCHEUX

Les voies de transport qui traversent les montagnes et les grandes mines à ciel ouvert ont en commun d'être caractéristisées par des talus rocheux abrupts mais stables. Compte tenu de la nature montagneuse de certaines régions du Canada et de l'importance de son industrie minière, il n'est pas surprenant que l'ingénierie des talus rocheux joue un rôle important dans la mise en valeur du pays. Comme la documentation canadienne sur le sujet est trop étendue pour être passée en revue, je me limiterai à ne mentionner que quelques faits saillants.

L'un des premiers praticiens de l'ingénierie des talus rocheux a été C.O. Brawner, actuellement professeur de géomécanique à l'Université de la Colombie-Britannique. Plus praticien que théoricien, Brawner a consacré ses travaux aux problèmes des talus rocheux associés aux grandes routes, aux chemins de fer et aux mines à ciel ouvert du monde entier et ses contributions à ce domaine ont été très importantes (Brawner, 1986).

Coates, dont les travaux en exploitation minière souterraine ont déjà été mentionnés, s'est également vivement intéressé à la stabilité des talus rocheux. Le point culminant de son intérêt a été la mise sur pied en 1972 d'un projet du CANMET qui s'est traduit par la publication d'un ouvrage en plusieurs volumes intitulé "Manuel des pentes des mines à ciel ouvert" (CANMET, 1977). Cet imposant ouvrage de 27 volumes couvre tous les aspects de la conception des talus rocheux, des dépôts de stériles et des digues de retention de rejets associés aux mines à ciel ouvert en plus de constituer un ouvrage de référence important dans le domaine de l'ingénierie des talus rocheux.

Le professeur Peter Calder et ses collègues à l'Université Queen's à Kingston ont contribué à l'analyse de la stabilité des talus, à la conception du minage dans les mines à ciel ouvert et à la surveillance des talus (Calder, 1969). Douglas Piteau et ses collègues de Vancouver ont utilisé des techniques probabilistiques pour optimiser la conception des talus rocheux (Piteau et al., 1985).

L'analyse, la surveillance et la stabilisation des ruptures par basculement des talus ont été traitées par un certain nombre de chercheurs, notamment par Choquet et Tanon (1985), Wyllie (1980) et Brawner, Stacey et Stark (1975). Le recours à un renforcement à l'aide de câbles sous tension et de tirants injectés a été décrit par Martin (1987); ces techniques devraient jouer un rôle grandissant dans l'élaboration des talus rocheux à mesure que seront mises au point des techniques de renforcement plus efficaces et plus économiques.

AMÉLIORATIONS DU BÉTON PROJETÉ

Pour répondre à la nécessité de protéger les talus contre l'érosion dans les milieux où les chutes de pluie sont élevées et pour effectuer des réparations rapides dans les tunnels ferroviaires achalandés, des travaux importants ont été consacrés à l'amélioration de la résistance et de l'aptitude au façonnage du béton projeté.

Les améliorations apportées comportent l'utilisation de gros agrégats (MacRae, 1973), l'ajout de fibres d'acier (Little, 1985 et Morgan, 1985) et l'ajout de microsilice (Anon, 1987), (Morgan et al., 1987). Aujourd'hui, dans l'Ouest canadien et de plus en plus dans le reste du pays, on utilise couramment du béton projeté à microsilice armé aux fibres d'acier. L'ajout de microsilice améliore de façon importante l'aptitude au façonnage et la résistance du béton projeté et, de ce fait, le rend utilisable dans des situations qui, il y a quelques années, nécessitaient le recours à des cadres en acier ou des revêtements en béton.

CONCLUSION

J'ai tenté d'illustrer par quelques exemples pratiques la situation actuelle de l'ingénierie des roches au Canada. L'évolution de cette discipline dont la base théorique relève de la mécanique des roches a été lente et parfois frustrante. Même si une ou deux personnes seulement ont joué un rôle important dans l'impulsion de départ donnée à l'ingénierie des roches, l'évolution générale de cette science au Canada est redevable au travail à long terme d'un grand nombre de chercheurs travaillant dans de nombreux groupes de recherche publics et privés, universités, et organismes de consultation de toutes les régions du pays. Il n'existe pas de centre reconnu dans le domaine de l'ingénierie des roches au Canada mais, comme j'ai tenté de le montrer dans cet exposé, un certain nombre de groupes d'ingénieurs et de géologues s'intéressent aux applications de ce domaine pour résoudre des problèmes pratiques. Selon moi, cette évolution parallèle est saine du fait qu'elle produit une diversité de méthodes et d'opinions et que les solutions adoptées se fondent plutôt sur leur valeur pratique que sur la réputation des personnes ou des organismes qui les ont proposées.

REMERCIEMENTS

Au cours de la préparation de cet exposé, j'ai recherché et obtenu l'aide de nombreuses personnes de toutes les régions du Canada. Il me serait impossible de les nommer toutes mais j'aimerais leur exprimer ma sincère gratitude pour les notes, les documents et les photographies (que je n'ai pas tous utilisés) qu'ils m'ont envoyés.

BIBLIOGRAPHIE (voir le texte anglais)

M. Udd remercie le professeur E. Hoek pour son exposé à la fois bien documenté et bien présenté et en guise de remerciement, lui offre une sculpture esquimaude de la part du comité organisateur.
M. Herget termine ensuite la cérémonie d'ouverture et demande au président de la première séance, B. Bjurstrom (Suède) de convoquer les délégués pour les séances techniques après la pause café.

Thème 1
Ecoulement de fluides et enfouissement de déchets dans les massifs rocheux

Reconnaissance des sites et caractérisation des massifs rocheux; écoulement des eaux souterraines; transport des contaminants; isolation des déchets; contrôle des écoulements; contraintes thermiques et effets dommageables sur les excavations; surveillance et rétro-analyse

Président: S.A.G. Bjurström (Suède)
Secrétaire: R. Chapuis (Canada)
Conférencier thématique: T.W. Doe (États-Unis)

CONCEPTION DES PROGRAMMES D'ESSAIS DE PUITS POUR L'ENFOUISSEMENT DES DÉCHETS DANS LES ROCHES CRISTALLINES, avec John Osnes, Michael Kenrick, Joel Geier, Scott Warner (États-Unis)

RÉSUMÉ

L'étude de programmes d'essais de puits pour une roche fissurée à faible perméabilité comprend deux aspects: le choix des méthodes d'essais de puits et l'étude du programme d'échantillonnage.

Les méthodes d'essais de puits ont été développées indépendamment en quelque sorte à partir de deux sources, l'une étant le génie civil et l'ingénierie minière, l'autre étant l'ingénierie des pétroles et l'hydrogéologie. Les essais de puits par le génie civil ont été concentrés sur l'essai d'injection à pression constante ou essai Lugeon qui présuppose un régime permanent. Les essais pétroliers et de nappes d'eau souterraines ont principalement utilisé des méthodes de pompage à débit constant pour les roches à perméabilité élevée et des essais par impulsions pour une perméabilité faible. Ces essais utilisent des méthodes de régime transitoire qui incluent les conditions d'emmagasinement, ce dernier reflétant la compressibilité et la déformabilité de l'eau et de la roche.

L'emploi de méthodes de régime permanent plutôt que transitoire, n'aboutit pas à des erreurs capitales concernant la détermination de la conductivité hydraulique; cependant, les méthodes d'écoulement transitoire fournissent en pratique plus d'informations, y compris les effets de compressibilité dans le puits (accumulation dans le forage du puits), les effets d'écoulement non linéaire, les effets de déformation des fractures, l'influence des frontières (telles que les intersections de fractures s'éloignant du trou ou les terminaisons des acculs) et l'influence du sondage sur la perméabilité à proximité du trou de sonde. Des solutions tenant compte de ces facteurs ont été amplement développées pour les essais en régime permanent et peuvent être adaptées aux essais à pression constante, là où l'écoulement transitoire est analysé. En ce qui concerne les roches à faible perméabilité, l'essai à pression constante présente des avantages précis, y compris l'applicabilité à une grande variété de types de perméabilité, l'élimination de l'accumulation dans le forage du puits et une vitesse relative surtout si des appareils de mesure du débit à très faible gamme peuvent être employés.

Il y a deux façons distinctes d'aborder l'étude des programmes d'échantillonnage. La première, l'essai de zone discrète, sélectionne des intervalles spécifiques d'importance. Cette méthode est employée lorsqu'une zone spécifique telle qu'une zone de fracture ou couche aquifère constitue le point essentiel de l'essai. La second, l'essai à longueur d'intervalle fixe (FIL), prélève des échantillons du trou dans les zones adjacentes ayant une longueur fixe. La méthode à longueur d'intervalle fixe est employée quand l'essai doit fournir un échantillonnage statistique des propriétés hydrologiques de la roche. La répartition des valeurs de perméabilité résultant d'un programme d'essai à longueur d'intervalle fixe reflète à la fois la fréquence des fractures et les répartitions de perméabilité. Les données résultant d'un essai FIL peuvent être analysées pour fournir des informations sur ces deux points.

1 INTRODUCTION

Le présent article décrit une approche pour les essais hydrologiques dans la roche cristalline. L'expression roche cristalline comprend les roches tant plutoniques que fortement métamorphisées; cependant, les méthodes examinées dans le présent article sont généralement applicables à toute roche dont la matrice a une perméabilité très faible et où l'écoulement s'effectue principalement dans un réseau de fissures.

Dans la conception d'un programme d'essais de puits dans la roche cristalline, il faut tenir compte du choix de la méthode d'essai et de la démarche d'échantillonnage. Les roches où l'écoulement s'effectue principalement dans des fissures peuvent avoir des propriétés hydrologiques extrêmement hétérogènes. Par conséquent, les méthodes d'essai choisies doivent être capables de fournir des données pour une vaste gamme de conductivités hydrauliques. L'hétérogénéité signifie également que le programme d'échantillonnage doit être conçu pour vérifier directement ou déduire les propriétés des voies d'écoulement dans la masse rocheuse, qu'elles soient ou non recoupées par les trous de sonde.

Le présent article est divisé en quatre sections. La première traite d'expressions et de principes de base. Les principaux facteurs qui influent sur les essais de puits sont définis en termes d'hydrologie et de mécanique. Cette section aborde également les distinctions qui existent entre roche poreuse et fissures en recourant à des définitions fondamentales des propriétés hydrologiques.

La deuxième section décrit les quatre principales méthodes d'essai dans la roche cristalline (à débit constant, à pression constante, par remontée de la pression et par impulsions/bouchons) et les compare en termes de leur fiabilité pour une gamme de valeurs de perméabilité, de leur rapidité et de l'information qu'on peut tirer de chacune d'elle.

La troisième section traite en détail de l'essai d'injection à pression constante et des différents facteurs qui influent sur cet essai : effets des frontières, déformation des fissures, écoulement non linéaire et perturbations près de la paroi du puits.

La dernière section de cet article porte sur quelques stratégies d'échantillonnage dans la roche cristalline, notamment sur la déduction de l'espacement des fissures et la distribution des transmissivités à partir d'une série d'essais de puits. Le problème de l'écoulement canalisé est aussi abordé.

L'emphase est mise sur les essais en puits unique parce qu'il est probable que la reconnaissance des masses rocheuses se fera à l'aide d'un nombre limité de sondages. Pour mieux utiliser les ressources de forage, il faudra vraisemblablement répartir les sondages sur de grandes surfaces plutôt que de les concentrer dans des zones limitées qui conviennent à des essais dans plusieurs trous de sonde.

2 QUESTIONS FONDAMENTALES TOUCHANT LES ESSAIS DE PUITS EN MILIEU PEU PERMÉABLE

Cette section traite de la terminologie des principes de base et définit les principaux facteurs qui influent sur les essais de puits, tant en termes d'hydrologie qu'en termes de mécanique. Lorsqu'on définit des propriétés hydrologiques, il est important d'insister sur la distinction entre roche poreuse et fractures.

2.1 L'équation de diffusion appliquée aux essais de puits

L'équation de diffusion est le fondement de toutes les méthodes d'essai de puits portant sur des écoulements laminaires saturés. En termes simples, la plupart des méthodes d'essai de puits, y compris celles décrites dans le présent article, sont basées sur des solutions de l'équation :

$$\nabla^2 h = \frac{S}{T}\frac{\partial h}{\partial t} \tag{1}$$

où : h = charge hydraulique [L]
t = temps [T]
T = transmissivité [L^2/T]
S = coefficient d'emmagasinement [-]
∇ = opérateur de divergence

Cette équation est exprimée sous la forme linéaire; elle peut être transformée directement sous une forme radiale ou sphérique pour d'autres géométries d'écoulement. Les deux paramètres fondamentaux du milieu, que les essais de puits permettent d'obtenir, sont la transmissivité T qui décrit la capacité d'écoulement de l'aquifère ou des voies d'écoulement recoupées par la zone d'essai, et le coefficient d'emmagasinement S qui décrit la capacité d'emmagasinement des voies d'écoulement.

2.2 Transmissivité, conductivité hydraulique et perméabilité

La théorie des écoulements laminaires saturés est fondée sur la loi de Darcy selon laquelle la vitesse d'écoulement et le flux sont des fonctions linéaires du gradient hydraulique :

$$v = K\,\partial h/\partial l \tag{2}$$

où : v = vitesse [L/T]
K = conductivité hydraulique [L/T]
l = longueur [L]

La conductivité hydraulique est une propriété de la roche poreuse et du fluide. K dépend des propriétés de la roche et du fluide suivant l'équation :

$$K = \frac{k\rho g}{\mu} \tag{3}$$

où : k = perméabilité [L^2]
μ = viscosité dynamique [M/LT]
g = accélération de la pesanteur [L/T^2]
ρ = masse volumique du fluide [M/L^3]

Les paramètres du fluide, sa masse volumique et sa viscosité, sont fonction de la température. Dans les essais de puits, il faut connaître particulièrement bien les conditions de température et les maintenir constantes pour éviter toute estimation erronée des propriétés de la roche.

Dans l'étude de la capacité d'écoulement d'un milieu aquifère poreux, la transmissivité T est en général définie par :

$$T = Kb \tag{4}$$

où : b = épaisseur de l'aquifère [L]

La transmissivité mesure la capacité d'une voie d'écoulement ou d'un milieu aquifère à transporter l'eau, tandis que la conductivité hydraulique se rapporte aux propriétés de conduction de la matière contenue dans la voie d'écoulement ou dans le milieu aquifère.

Les essais de puits visent à mesurer la transmissivité. La conductivité hydraulique est calculée à partir de la transmissivité en posant b comme étant l'épaisseur de la nappe aquifère ou des nappes aquifères recoupées par le puits. Dans les milieux aquifères stratifiés, la valeur de b est simplement l'épaisseur du milieu stratifié qui contient de l'eau. Pour les essais de puits dans la roche cristalline massive, b n'est pas aussi bien défini. En général, la valeur utilisée comme épaisseur pour le calcul de la conductivité hydraulique est simplement la longueur de la zone d'essai, malgré le fait que seule une petite partie de la zone d'essai permet le mouvement du fluide. À moins que la longueur de la zone d'essai ne soit très grande par rapport à l'espacement des fissures, de telles conductivités ne sont probablement pas représentatives de l'ensemble de la masse rocheuse. Aucune méthode de calcul de la conductivité des masses rocheuses n'est reconnue. Carlsson et al. (1983) ont proposé d'utiliser la moyenne géométrique ou harmonique des mesures de conductivité dans un puits comme mesure de la conductivité de la masse rocheuse. Des estimations plus complexes peuvent être établies à partir de modèles de réseau de fissures (par exemple Long et al., 1982; Dershowitz, 1985).

La distinction entre transmissivité et conductivité hydraulique est importante dans l'étude des écoulements dans des fissures. Une fissure simple peut être considérée comme une nappe aquifère ou une voie d'écoulement simple. La transmissivité de la fissure simple peut être exprimée de la même façon que celle d'une nappe aquifère poreuse :

$$T = K_f e \tag{5}$$

où : K_f = conductivité de la fissure [L/T]
e = ouverture de la fissure [L]

Si la fissure est remplie, K_f est la conductivité hydraulique du matériau de remplissage. Si la fracture est ouverte, la conductivité est une mesure de la résistance de l'écoulement le long des parois de la fracture et de la tortuosité des voies d'écoulement à l'intérieur de la fissure. La composante frottement de la conductivité de la fissure peut être exprimée par la loi dite cubique qui décrit un écoulement entre des plaques parallèles :

$$Q \propto T = \frac{\rho g e^3}{12\mu} \tag{6}$$

où : Q = débit [L^3/T]

Cette équation cubique détermine la valeur maximale de la transmissivité d'une fissure d'ouverture donnée. Les fissures simples ouvertes ne constituent en général pas des plaques parallèles idéales, et les transmissivités mesurées donnent habituellement des ouvertures apparentes qui sont beaucoup plus petites que l'écartement réel des parois des fissures. Les distinctions entre les définitions de K, S et T pour des milieux fissurés et poreux sont représentées schématiquement dans la figure 1.

2.3 Coefficient d'emmagasinement

Dans une nappe aquifère poreuse, le coefficient d'emmagasinement se rapporte au volume de fluide par unité de surface de la nappe aquifère qui est ajouté dans le milieu d'emmagasinement ou qui en est retiré, quand la charge hydraulique varie d'une unité. Le coefficient d'emmagasinement est un paramètre global sans dimension qui découle de l'analogie entre un écoulement souterrain et la diffusion thermique (Theis, 1935). Dans un écoulement confiné saturé, le coefficient d'emmagasinement dépend de la compressibilité de la roche et du fluide. Dans les milieux poreux, il est lié aux compressibilités de la roche et de l'eau par l'équation suivante :

$$S = \rho g b(C_r + nC_f) \qquad (7)$$

où : C_r = compressibilité de la roche $[LT^2/M]$
C_f = compressibilité du fluide $[LT^2/M]$
n = porosité [-]

Pour une fracture simple, cette équation devient (Doe et Osnes, 1985) :

$$S = \rho g(\frac{1}{k_n} + eC_f) \qquad (8)$$

où : k_n = rigidité normale de la fissure $[M/L^2T^2]$

Comme dans un milieu poreux, le coefficient d'emmagasinement a deux composantes, une due à la compressibilité du fluide et l'autre due aux variations de l'ouverture. Dans les milieux poreux, la composante de compressibilité de la roche est habituellement beaucoup plus petite que la composante de compressibilité du fluide et est en général négligeable. Par contre, dans les fractures simples, les volumes de fluide sont souvent très petits et la composante de rigidité peut alors être dominante. La déformation pose un problème dans ces définitions. En particulier, dans le calcul de l'équation de diffusion, les termes de perméabilité qui dépendent de la pression sont exclus, et la perméabilité est supposée constante (Snow, 1968). Par conséquent, il y a incohérence inhérente dans la définition du coefficient d'emmagasinement des fissures : l'emmagasinement doit être dû principalement à des variations d'ouverture, mais la perméabilité varie en fonction du carré de l'ouverture.

Le coefficient d'emmagasinement est une propriété d'une nappe aquifère ou d'une voie d'écoulement. L'emmagasinement spécifique, obtenu en divisant le coefficient d'emmagasinement par l'épaisseur de l'aquifère, est une propriété de la matière contenue dans la voie d'écoulement. Pour les fractures simples ouvertes, l'emmagasinement spécifique, qui est une propriété de la matière, ne correspond à aucune réalité physique.

2.4 Emmagasinement dans le trou de sonde

Le but premier des essais de puits est de déterminer les propriétés du réservoir de fluide dans la roche autour du trou de sonde. La réponse transitoire dans les essais de puits tient au fait qu'on retire ou qu'on ajoute du fluide dans ce réservoir. Cependant, le trou de sonde et le matériel d'essai agissent aussi comme réservoirs de fluide. Ces réservoirs additionnels ont des propriétés d'emmagasinement qui, dans certaines circonstances, peuvent dominer le comportement transitoire de l'essai. Comme l'emmagasinement dans le trou de sonde n'a rien à voir avec les propriétés de la roche réservoir, il est important de reconnaître son effet pour que le comportement du trou de sonde ne soit pas interprété comme un comportement de roche-réservoir. Il est aussi important de concevoir les essais de façon à limiter ou éliminer l'effet d'emmagasinement dans le trou de sonde.

L'emmagasinement dans les trous de sonde a été étudié en hydrogéologie par Papadopulos et Cooper (1967) et pourrait être défini par le volume de fluide ajouté dans le trou de sonde (ou retiré du trou) quand la pression varie d'une unité (Earlougher, 1977) :

$$C = \frac{\Delta V}{\Delta P} \qquad (9)$$

Les termes utilisés pour décrire mathématiquement l'emmagasinement dans un trou de sonde sont analogues aux termes qui définissent le coefficient d'emmagasinement d'une nappe aquifère poreuse. Ces termes comprennent :

- C_1, la variation d'emmagasinement due à un changement du niveau d'eau
- C_2, la variation d'emmagasinement due à la compressibilité du fluide, et
- C_3, la variation d'emmagasinement due à la déformation du trou de sonde, des tubes et à d'autre matériel d'essai de puits.

L'emmagasinement dans un trou de sonde est la somme des composantes, ou :

$$C = C_1 + C_2 + C_3 \qquad (10)$$

Le terme du niveau d'eau, C_1, n'apparaît que dans les essais en trou ouvert, tel que les essais de soutirage à débit constant (ou essais de pompage). Le cas du trou ouvert est analogue à celui des nappes aquifères confinées où l'emmagasinement est principalement le résultat des fluctuations du niveau d'eau. Dans un essai en trou ouvert dans une nappe aquifère non confinée, C_1 sera beaucoup plus grand que les autres termes, de sorte que C_2 et C_3 deviennent en général négligeables. Le terme d'emmagasinement dans le trou de sonde, C_1, est simplement le volume d'eau emmagasiné par unité de longueur du trou, divisé par la pression par unité de hauteur de charge hydraulique, ou :

$$C_1 = A/\rho g \qquad (11)$$

où : A = section transversale à travers laquelle le niveau du fluide fluctue $[L^2]$

Pour les essais où il n'y a aucune fluctuation du niveau d'eau, l'emmagasinement dans le trou de sonde ne tient qu'aux termes de compressibilité, C_2 et C_3. La plupart des essais avec packers, notamment les essais d'injection sous pression dans des sections isolées par des packers, entrent dans cette catégorie. Dans ces essais, l'emmagasinement dans le trou de sonde est analogue au coefficient d'emmagasinement d'une nappe aquifère confinée. Le terme de compressibilité du fluide, C_2, peut être défini ainsi :

$$C_2 = V_f C_f \qquad (12)$$

où : V_f = volume d'eau sous pression (ou sous dépression) dans le trou de sonde $[L^3]$

Le dernier terme d'emmagasinement dans le trou de forage, C_3, se rapporte à la déformabilité de la paroi du trou et du matériel d'essai. Pour les besoins du présent article, nous utiliserons les termes compressibilité du matériel et compressibilité du trou de sonde, la compressibilité étant simplement une variation de volume pour une variation donnée de la pression. La compressibilité du matériel est due à la déformation des packers, des tubes conduisant le fluide dans le trou et de toute autre composante du système d'essai. La compressibilité du trou de sonde est due à la déformation du trou ouvert dans la zone d'essai. Le terme C_3 est la somme des compressibilités du trou de sonde et du matériel. La compressibilité des tubes d'essai et du trou de sonde peut être calculée si les propriétés élastiques des matériaux sont bien connues. La compressibilité des tubes non rigides et des packers devrait être mesurée directement dans un essai de laboratoire en conditions contrôlées.

2.5 Effets pariétaux

La couche pariétale d'un puits est une zone mince près de la paroi du puits où la perméabilité de la roche a été modifiée par les travaux de forage ou par des traitements visant à augmenter la productivité comme l'indique la figure 2 (Earlougher, 1977; Agarwal et al., 1970, Almen et al., 1986). L'infiltration de boue et de débris de forage dans la formation produit un effet pariétal positif. Une augmentation de la perméabilité près de la paroi du puits due à des dommages causés par le forage ou au lavage des matériaux de remplissage des pores semble avoir un effet pariétal négatif. Des effets naturels associés à l'hétérogénéité de l'aquifère ou à des fissures peuvent aussi être considérés comme des effets pariétaux. Dans une fissure où l'effet de canal est important, le passage du trou de sonde dans un îlot de faible perméabilité peut sembler produire un effet pariétal positif dans l'essai du puits, tandis que le recoupement d'une petite fissure ou d'un canal à l'intérieur d'une fissure peut sembler produire un effet pariétal négatif.

La couche pariétale peut être considérée comme une zone de conductivité modifiée, K_s, dans un rayon r_s autour du trou, et dont le facteur pariétal ξ est défini comme suit (Earlougher, 1977) :

$$\xi = (K/K_s - 1) \ln (r_s/r_w) \qquad (13)$$

où : K = conductivité hydraulique de la formation
K_s = conductivité hydraulique de la couche pariétale
r_w = rayon du puits

La section 3 traite de l'estimation des facteurs pariétaux dans les essais d'injection et de pompage.

2.6 Effets thermiques dans les essais de puits

Les variations de température peuvent avoir un effet important sur les résultats des essais, notamment dans les roches peu perméables. Les effets thermiques tiennent au fait que la viscosité et la masse volumique du fluide dépendent de la température ainsi qu'aux effets d'expansion thermique.

Les variations de température modifient la viscosité et la masse volumique du fluide, ce qui entraîne un changement de la conductivité hydraulique. Des variations de température dans la roche autour du puits peuvent entraîner des variations de conductivité hydraulique, même lorsque la roche est entièrement homogène. À moins qu'elles ne soient décelées, les variations de température peuvent être prises pour des variations de perméabilité à une certaine distance du puits.

L'expansion thermique peut être un facteur particulièrement important dans les essais de puits en milieu peu perméable. Dans un essai d'injection ou un essai par impulsions de pression, où le système est rendu étanche, toute variation de température peut entraîner une augmentation de la pression du fluide. La variation de pression dans un système fermé est donnée par :

$$P = \frac{\Delta\theta\alpha}{C_f} \qquad (14)$$

où : $\Delta\theta$ = variation de température
α = coefficient d'expansion thermique du fluide

De faibles variations de température peuvent entraîner de fortes variations de pression à cause de la compressibilité relativement faible et du coefficient d'expansion thermique élevé de l'eau, comme le souligne Grisak et al. (1985). Dans un essai d'injection à faible débit et à pression constante, l'augmentation de volume par expansion thermique peut entraîner une sous-estimation du débit ou encore un jaillissement au lieu d'une injection de fluide. La régularisation de la température est sans aucun doute un facteur important dans la conception des essais de puits en milieu peu perméable.

2.7 Géométrie d'écoulement

Les méthodes classiques d'interprétation des essais de puits portent en général sur des aquifères (ou des fissures) homogènes et sur des géométries très simples. Les trois principales géométries d'écoulement sont l'écoulement radial, l'écoulement linéaire et l'écoulement sphérique (figure 3). Dans le présent article, nous traiterons principalement de l'écoulement radial, car c'est le genre d'écoulement qui se produit dans les fissures ou les nappes aquifères qui gisent perpendiculairement au trou de sonde. Dans un essai donné, la réponse hydraulique peut prendre une ou toutes les formes suivantes. Pendant un essai, la progression se fait habituellement de la dimension la plus petite (linéaire) à la dimension la plus élevée (sphérique).

L'écoulement linéaire présente un certain intérêt car cette géométrie peut être observée lorsque les fissures sont parallèles au trou de sonde (c'est-à-dire des fissures verticales dans des trous verticaux) ou lorsque l'écoulement s'effectue dans un nombre restreint de canaux. Karasaki (1986) a mis au point un modèle dans lequel l'écoulement est linéaire près du puits et radial dans un réseau équivalent à un milieu poreux loin du puits. Ce modèle pourrait être une bonne analogie pour les écoulements canalisés. Almen et al. (1986) rapportent que la plupart des essais dans le cadre d'un programme de recherche sur les méthodes hydrologiques ont montré que l'écoulement était radial ou linéaire et non sphérique. Ershagi et Woodbury (1985) ont réalisé une analyse intéressante des effets de la géométrie des écoulements sur les essais à débit constant.

2.8 Effets des limites extérieures

En plus des influences des limites intérieures (tels que le facteur pariétal et l'emmagasinement dans le trou de sonde), un essai de puits peut être influencé par des limites extérieures. Il existe deux types fondamentaux de limites extérieures : des limites fermées ou sans écoulement et des limites ouvertes ou à pression constante (figure 4). Une limite fermée s'entend d'une limite sur laquelle se termine une fissure dans un matériau imperméable. Une limite ouverte ou à pression constante se dit d'une limite assurant avec un réservoir une liaison qui est suffisamment développée et conductrice pour que l'essai n'influe aucunement sur sa pression. Une limite à pression constante dans la roche cristalline comporterait des fissures suffisamment plus conductrices et plus développées que les fissures recoupées par le puits, y compris d'importantes zones de fissuration. Des ouvrages artificiels peuvent établir des limites à pression constante. Tout écoulement vers une ouverture souterraine ou tout écoulement retournant dans le trou de sonde lui-même peut constituer une limite à pression constante.

Les fuites dans la roche encaissante peuvent être considérées comme ayant un effet aux limites extérieures si la perméabilité de la matrice rocheuse est suffisamment élevée. Hantush (1956, 1959) a mis au point des solutions pour les fuites dans les essais à pression constante et les essais à débit constant. Les solutions de Hantush ne tiennent pas compte du coefficient d'emmagasinement de la roche encaissante; ce facteur a été analysé par Neuman et Witherspoon (1969). La roche cristalline intacte, malgré sa faible conductivité, peut être le siège de fuites si la surface des fissures est assez grande.

3 MÉTHODES D'ESSAI PAR PUITS UNIQUE

Dans cette section, on décrit les principaux types d'essai par puits unique qui sont habituellement utilisés dans la roche cristalline et on analyse leur utilité relative. Les principales caractéristiques qui sont recherchées dans une méthode d'essai de puits dans la roche cristalline sont (1) la capacité à contrôler une fourchette étendue de conductivités et (2) la rapidité relative de la méthode.

La conductivité de la roche cristalline peut varier en général d'environ 10^{-5} m/s à près de 10^{-12} m/s (Brace, 1980). Une vérification de toute la fourchette des conductivités n'est pas toujours nécessaire : les zones de conductivité élevée sont nettement plus importantes que les zones de faible conductivité dans les problèmes de transport. Cependant, si l'objectif est de déterminer formellement la distribution des conductivités à des fins d'analyse statistique, la méthode devrait permettre de mesurer les conductivités inférieures à la médiane et, de préférence, les conductivités jusqu'au 25ème percentile.

Quant à la rapidité de la méthode, il est nettement préférable que les essais soient les plus courts possibles. Si l'objectif est d'évaluer la variabilité de la conductivité d'une masse rocheuse par un échantillonnage d'un grand nombre de zones, l'essai choisi devrait permettre de vérifier les faibles conductivités en moins d'un jour.

3.1 Définitions des types d'essai

L'essai par puits unique mesure les propriétés hydrauliques de la roche par observation de la réponse du puits à une perturbation de la pression. Dans le présent article, nous traiterons de quatre méthodes de collecte de données hydrauliques :

- essais à débit constant
- essais à pression constante
- essais par impulsions
- analyses de remontée

La figure 5 illustre des variations typiques de la pression et du débit en fonction du temps pour chacun de ces essais.

Dans les essais à débit constant, on perturbe le puits en injectant ou en prélevant du fluide pendant un certain temps à débit constant. Les caractéristiques hydrauliques de la roche sont évaluées à partir de variations de la pression dans le puits. Dans les essais à pression constante, on maintient la pression dans le puits constante et on détermine les caractéristiques hydrauliques à partir du débit qui entre dans le puits ou qui en sort.

Dans les méthodes par remontée de la pression, on observe comment, après une perturbation à pression ou débit constant, la pression dans le puits revient à la valeur hydrostatique d'équilibre de la formation. Les méthodes par remontée de la pression reposent essentiellement sur la même théorie que les essais à débit constant.

Dans les essais par impulsions, on provoque une variation instantanée de la pression dans le puits et on détermine les caractéristiques hydrauliques de la roche à partir de la courbe de rétablissement de la pression.

Le système à l'essai est qualifié de fermé si la zone d'essai est isolée des surfaces de fluide libres dans le puits. Dans tous les essais d'injection, à pression ou débit constant, les sytèmes sont fermés dans la mesure où la pompe d'injection est branchée directement sur la zone d'essai. Les essais de prélèvement s'effectuent rarement dans des systèmes fermés à moins qu'il ne s'agisse d'un puits qui jaillit naturellement, c'est-à-dire que la charge totale dans la nappe aquifère est supérieure à la charge au point de prélèvement du fluide.

Les essais par impulsions peuvent s'effectuer dans un système tant ouvert que fermé. Dans le premier cas, l'essai est appelé essai par "bouchons"; dans le deuxième cas, il s'agit d'un essai par "impulsions de pression" ou simplement d'un essai par "impulsions".

La principale différence entre les essais en système ouvert et les essais en système fermé tient à la nature de l'emmagasinement dans le trou de sonde. Dans les essais en système ouvert, l'emmagasinement dans le trou de sonde tient essentiellement à la variation du niveau d'eau dans le trou ouvert, et les effets de la compressibilité de l'eau sont secondaires. Dans les essais en système fermé, l'emmagasinement dans le trou de sonde n'est attribuable qu'à la compressibilité de l'eau et du matériel, composantes qui sont négligeables par rapport à la composante liée au niveau d'eau dans un trou ouvert.

Des essais à débit constant peuvent s'effectuer en circuit tant fermé qu'ouvert. Les essais traditionnels de pompage (essais de prélèvement à débit constant) s'effectuent presque toujours dans un système ouvert. Les essais à pression constante s'effectuent habituellement en système fermé comme les essais d'injection ou comme les essais de prélèvement dans des puits jaillissants.

3.2 Essais d'injection à pression constante

L'essai d'injection à pression constante, aussi appelé l'essai d'injection, l'essai aux packers et l'essai Lugeon, est très répandu dans les applications du génie pour évaluer la transmissivité d'une roche cristalline peu conductrice. Les données d'essai peuvent être analysées par diverses méthodes analytiques, y compris la méthode classique du débit en régime permanent, la méthode de la courbe type de pression en fonction du débit en régime transitoire (Jacob et Lohman, 1952; Hantush, 1959), ou la méthode par remontée de la pression en fonction du débit en régime transitoire (Uraiet et Raghavan, 1980; Ehlig-Economides et Ramey, 1981).

Dans les essais à pression constante, les ingénieurs utilisent le débit permanent observé pour obtenir la transmissivité :

$$T = C_s Q/\Delta h \tag{15}$$

où : C_s = facteur de forme qui varie selon la géométrie présumée de l'écoulement

Ziegler (1976) et Hoek et Bray (1974) ont établi des listes de facteurs de forme courants. Un des facteurs de forme les plus répandus est basé sur l'équation de l'écoulement permanent appliquée à des écoulements laminaires radiaux à une distance limite R du trou où la pression est constante, ou :

$$C_s = \frac{1}{2\pi} \ln(R/r_w) \tag{16}$$

Le rayon limite (ou rayon d'influence) n'est jamais connu avec une grande précision. Cependant, comme l'analyse porte sur le logarithme de ce paramètre, de fortes variations de R ne produisent que de faibles variations de T. Par exemple, si le rayon d'un trou de forage r_w est de 38 mm, une variation de R de 20 à 500 m entraîne une variation de la transmissivité calculée d'à peine 1/3. Cette erreur est petite comparativement au grand intervalle de variation des mesures de transmissivité dans la roche fissurée.

Dans les essais à pression constante, une expression du débit transitoire d'un écoulement a été établie par Jacob et Lohman (1952) pour des nappes aquifères infinies et, plus tard, développée par Hantush (1959) :

$$Q = 2\pi \Delta h T Q_D(t_D) \qquad (17)$$

et

$$T_D = \frac{Tt}{S r_w^2} \qquad (18)$$

où : $Q_D(t_D)$ = débit sans dimension
t_D = temps sans dimension

La courbe de $Q_D(t_D)$, en coordonnées logarithmiques, est appelée une courbe type.

Les données transitoires peuvent être analysées au moyen de courbes types ou d'une droite semi-log d'approximation **(figure 6)**. L'application de la méthode semi-logarithmique aux essais à débit constant est abordée à la section 3.3; la méthode peut aussi être appliquée aux essais à pression constante si on remplace Δh par 1/Q (Ehlig-Economides, 1979; Uraiet, 1979).

L'équation de la droite semi-logarithmique est donnée dans la figure 7; la transmissivité est donnée par la pente de cette droite. L'équation tient compte des effets pariétaux et de l'emmagasinement. Dans la pratique, il n'est possible de déduire qu'un seul de ces termes des données, et il faut attribuer à l'autre terme une valeur arbitraire. En géohydrologie traditionnelle, la pratique veut qu'on attribue au facteur pariétal une valeur nulle ($\xi = 0$) pour ainsi obtenir une valeur d'emmagasinement. Par ailleurs, les ingénieurs de réservoirs pétroliers calculent le facteur pariétal en attribuant une valeur arbitraire d'emmagasinement basée sur la porosité et la compressibilité du fluide.

En comparant des calculs d'écoulement transitoire et d'écoulement permanent, Doe et Remer (1981) ont conclu que les erreurs associées à l'hypothèse d'un écoulement permanent pour calculer T sont relativement petites.

La limite inférieure de transmissivité qui peut être évaluée au moyen de l'essai d'injection à pression constante dépend du débit le plus faible qui peut être mesuré (voir Section 5.3). Les débitmètres les plus sensibles ont une résolution d'environ 0,1 mL/min (Almen et al., 1986). Pour une charge de 100 m et un rayon d'influence arbitraire de 30 m, cela signifie que la limite inférieure de mesure de transmissivité est d'environ 10^{-10} m^2/s. La limite supérieure est habituellement déterminée par les pertes de frottement dues à la capacité limitée du matériel; cependant, on peut habituellement réduire ces pertes si on utilise une différence de charge plus petite.

3.3 Essais à débit constant (y compris les essais par remontée)

Les essais à débit constant comprennent les essais classiques de pompage utilisés en hydrogéologie et l'essai de base utilisé en génie des réservoirs de pétrole. L'essai à débit constant consiste à injecter dans un puits ou à en extraire un débit constant de fluide. Les caractéristiques hydrauliques de la formation sont déterminées à partir de la réponse transitoire de la pression. Les méthodes standard d'analyse des essais à débit constant et par remontée de la pression après de tels essais sont bien documentées dans Earlougher (1977) et dans Kruseman et De Ridder (1979).

La solution de base au problème de l'écoulement radial dû à un prélèvement de fluide à pression constante est l'intégrale exponentielle (Ei). Si nous définissons le temps sans dimension comme dans l'équation 18, la pression sans dimension dans le puits est donnée par :

$$P_D = -Ei(-1/4t_D)/2 \qquad (19)$$

et

$$P_D = 2\pi T \Delta h/Q \qquad (20)$$

En appliquant l'approximation logarithmique à l'intégrale exponentielle pour de grandes valeurs de t_D, on obtient :

$$P_D = (0{,}809 + \ln t_D)/2 \qquad (21)$$

L'équation 19 définit la courbe bien connue de Theis à laquelle des courbes log-log de la variation de pression en fonction du temps peuvent être ajustées (Theis, 1935). La forme exponentielle produit une droite semi-log bien connue sur un diagramme de la pression en fonction du logarithme du temps (figure 7). Idéalement, la courbe de Theis est celle d'un puits de diamètre infinitésimal; des corrections pour des puits de diamètre fini sont données dans Mueller et Witherspoon (1965).

La remontée de la pression après pompage est modélisée en superposant une source égale de signe opposé (Earlougher, 1977) et en introduisant dans l'approximation logarithmique (l'équation 21) le temps de "Horner", t_H :

$$t_H = \frac{t_p + \Delta t}{\Delta t} \qquad (22)$$

où : t_p = temps de pompage
Δt = temps de remontée

La principale contrainte dans les essais à débit constant et les essais de remontée de la pression dans la roche peu perméable est l'effet d'emmagasinement dans le puits, qui masque la réponse de la masse rocheuse en début d'essai. Les effets d'emmagasinement dans le puits sont progressivement plus marqués dans la roche de faible conductivité.

Les effets d'emmagasinement dans le puits peuvent être décelés dans les données de début d'essai à partir d'une droite log-log, de pente unitaire, de la pression en fonction du temps. Le temps nécessaire pour compenser les effets d'emmagasinement dans le puits est le principal facteur qui détermine la limite inférieure de résolution en transmissivité dans les essais à débit constant. La durée d'emmagasinement dans le puits varie considérablement selon que l'essai est effectué dans un système ouvert ou dans un système fermé. La durée approximative d'emmagasinement dans le puits, en temps sans dimension, est donnée par Earlougher (1977) :

$$T_{wD} = (60 + 3{,}5\xi)C_D) \qquad (23)$$

où la capacité d'emmagasinement sans dimension dans le puits, C_D, est définie par :

$$C_D = \frac{C \pi g}{2\pi S r_w^2} \qquad (24)$$

Si on néglige les effets pariétaux, la durée d'emmagasinement dans le puits peut être exprimée en termes du coefficient d'emmagasinement décrit à la section 2.4 :

$$t_w = 30C\rho g/\pi T \qquad (25)$$

Comme indiqué à la section 2.4, l'emmagasinement dans le puits, C, pour un essai en système ouvert, est donné par le facteur C_1, qui, lorsqu'introduit dans l'équation 23, donne :

$$t_w = 30A/T \qquad (26)$$

où : A = section transversale du puits à travers laquelle le niveau du fluide fluctue

La figure 8 donne la durée d'emmagasinement dans le puits en fonction de T pour un puits de section transversale efficace de 45 cm^2. Il faudrait compter près de 5 jours dans une zone d'essai où $T = 10^{-7}$ m^2/s pour que l'effet d'emmagasinement dans le puits soit compensé. Il est clair que les essais de pompage en trou ouvert ne sont pas efficaces dans la plupart des essais de puits en milieu peu perméable.

La durée d'emmagasinement dans le puits pour un essai en système fermé (tel qu'un essai d'injection à débit constant et sous pression) peut être calculée à partir de l'équation 23 et des coefficients C_2 et C_3. Si les effets de compressibilité du matériel sont négligeables, la durée d'emmagasinement est alors :

$$t = \frac{30V_wC_f\rho g}{\pi T} \qquad (27)$$

Le terme d'emmagasinement dans le puits dans le cas d'un système fermé est beaucoup plus faible que celui d'un essai en système ouvert, ce qui se traduit par une durée d'emmagasinement plus courte dans le cas d'un essai en système fermé. L'introduction d'un terme de compressibilité du matériel aurait pour effet d'augmenter la durée d'emmagasinement. Par exemple, la durée d'emmagasinement dans le puits pour un essai où le volume de fluide est de 0,3 m^3, (ce qui correspond à une série de tubes de 20 mm de diamètre descendus dans un trou de 1000 m) est représentée à la figure 9. Les essais en système fermé permettent de mesurer approximativement des transmissivités de deux ordres de grandeur inférieures pour des durées comparables d'emmagasinement. La limite inférieure de résolution de la transmissivité, eu égard aux effets d'emmagasinement dans un puits, peut être abaissée davantage en réduisant le volume de fluide dans le système d'essai.

3.4 Essais par impulsions et par bouchons

L'essai par bouchons a été introduite pour évaluer la transmissivité d'une nappe aquifère en observant la réponse du niveau d'eau à une variation instantanée de la charge dans un puits ouvert. Des modifications pour tenir compte du diamètre fini du puits ainsi que des courbes types pour évaluer la transmissivité à partir de données charge-réponse ont été proposées par la suite (Cooper et al., 1967). Des modifications plus récentes ont été apportées à l'analyse des données des essais par bouchons pour tenir compte de l'effet possible des couches pariétales d'épaisseur finie sur la réponse du puits (Faust et Mercer, 1984) et pour adapter l'analyse des essais par bouchons à des nappes aquifères fissurées (Black, 1985; Barker et Black, 1983.

Karasaki (1986) a mis au point différentes méthodes par bouchons pour des systèmes qui sont dominés par des fissures près des parois du puits, mais qui se comportent comme des milieux poreux à une certaine distance. Moench et Hsieh (1985) ont étudié l'effet pariétal sur les essais par bouchons. Les essais par impulsions et par bouchons se sont imposés partout au cours des dernières années et ont été utilisés de façon répandue dans les programmes de gestion des déchets radioactifs tant canadiens (Davison et Simmons, 1983) que suisses (Grisak et al., 1985).

L'estimation du coefficient d'emmagasinement n'est souvent pas réaliste parce que le volume perturbé de la nappe aquifère peut être relativement petit. Il faut aussi être prudent dans l'application de l'analyse standard des écoulements radiaux homogènes à un milieu fissuré ou à une nappe aquifère lorsque le rapport de la perméabilité horizontale à la perméabilité verticale approche l'unité (Papadopulos et al., 1973). Dans le cas d'une nappe aquifère fracturée ou fissurée, l'expression quantitative de la réponse en charge doit tenir compte des variations de charge non seulement dans le puits, mais aussi dans la matrice rocheuse et dans les fissures (Black, 1985; Barker et Black, 1983).

L'essai par bouchons dure habituellement peu de temps (de quelques minutes à quelques jours) et est plus souvent appliqué à des formations dont la transmissivité varie entre 10^{-2} et 10^{-9} m^2/s. Dans les nappes aquifères caractérisées par une transmissivité beaucoup plus élevée, le bouchon a tendance à s'amortir si rapidement qu'il est impossible de recueillir des données significatives. Si la transmissivité est trop faible (par exemple dans des formations de faible perméabilité comme le granite), la variation instantanée de charge peut durer beaucoup trop longtemps.

Le pourcentage d'amortissement du bouchon jugé nécessaire pour obtenir une bonne correspondance entre les données et les courbes types standard de Cooper et al. (1967) (voir figure 10) est de 70 % (Wilson et al., 1979). Aucune indication n'est donnée quant au pourcentage d'amortissement nécessaire pour obtenir une bonne correspondance entre les données et les courbes types de "fissures" modifiées de Black (1984), même si un pourcentage de 70 % peut être considéré comme suffisant. La figure 10 indique le temps nécessaire pour qu'un bouchon soit amorti de 70 % dans une nappe aquifère homogène en fonction de la transmissivité de la nappe aquifère. Pour fin de comparaison, la courbe de 95 % d'amortissement est aussi indiquée dans la figure. Les courbes se rapportent à un puits de 76 mm de diamètre dans lequel descendent des tubes de 25 mm de diamètre. Une plage raisonnable de temps écoulé de 1 minute à 1 jour est aussi représentée. Dans ces conditions, une zone aquifère ayant une transmissivité de 10^{-8} m^2/s et un coefficient d'emmagasinement 10^{-5} pourrait être vérifiée en un peu moins de 24 heures si on utilisait la courbe d'amortissement de 70 %. Pour une zone d'essai de 10 m, la conductivité hydraulique résultante serait de 10^{-9} m/s.

Les résultats ci-dessus se rapportent à une situation idéale dans une nappe aquifère homogène. Plusieurs facteurs, dont le manque d'homogénéité du milieu, l'état du puits et les effets pariétaux, les fluctuations de température et la compressibilité du matériel peuvent accroître la marge d'incertitude dans l'analyse des essais par bouchons. Une analyse de nappe aquifère fissurée, basée sur les courbes types standard pour un milieu homogène, peut aussi comporter des erreurs. Barker et Black (1983) ont supposé qu'il était possible, lorsqu'on applique les courbes types de Cooper et al. (1967) aux données d'une nappe aquifère fissurée, de surestimer la transmissivité par un facteur de 2 à 3 et de surestimer ou de sous-estimer le coefficient d'emmagasinement par un facteur pouvant atteindre 10^5.

La méthode d'essai par impulsions est essentiellement une version modifiée de l'essai classique par bouchons qui permet de mesurer rapidement la transmissivité dans des fissures simples de faible ouverture (Wang et al., 1978). Au lieu de noter la chute lente de charge dans une colonne ouverte à la suite d'une variation instantanée du niveau d'eau, on emprisonne un petit volume d'eau sous pression dans une section isolée et on observe la chute de pression (ou la remontée de la pression si la section isolée est mise en dépression) en fonction du temps (Bredehoeft et Papadopulos, 1980). La formulation mathématique de l'analyse de l'essai par impulsions dans un milieu présumé poreux est semblable à celle de l'essai par bouchons et n'en diffère que par le fait que la vitesse à laquelle l'eau fuit hors du puits est équivalente à la vitesse à laquelle le volume d'eau, initialement sous pression, augmente à mesure que la pression diminue (Grisak et al., 1985). Les mêmes courbes types utilisées pour estimer la transmissivité et le coefficient d'emmagasinement dans l'essai par bouchons, sont utilisées dans l'analyse des données de l'essai par impulsions sauf que, dans ces paramètres hydrologiques, il faut tenir compte de la compressibilité et de la masse volumique de l'eau à la température et à la pression de la formation auxquelles l'essai est mené. Si on suppose que l'écoulement est transitoire dans un milieu fissuré, la continuité des vitesses d'écoulement dans la fissure et à l'interface fissure/paroi du puits devient une condition aux limites, c'est-à-dire à l'extrémité du rayon du puits (Wang et al., 1978). Si on suppose que la roche et le matériel d'essai sont parfaitement rigides, la chute de pression dans l'essai par impulsions est une fonction des propriétés hydrauliques de la roche et de la compressibilité de l'eau sous pression qui est emmagasinée dans le puits (Wilson et al., 1979). Dans des conditions idéales, la transmissivité peut être déterminée après une chute de 10 % de la pression de l'impulsion par rapport à sa valeur initiale. La figure 11 montre les fourchettes de transmissivité qui peuvent être déterminées pendant une période d'essai pouvant aller jusqu'à un jour, après des chutes de pression de 10 % et de 90 % à une température et à une pression de la formation de 40 °C et de 100 kPa respectivement. Les courbes indiquées se rapportent à un puits ayant un diamètre de 76 mm sur une profondeur sous pression de 100 m (l'intervalle d'essai n'est pas pris en compte dans le calcul de la transmissivité). Dans le cas d'une chute de pression de 10 %, une transmissivité de moins de 10^{-12} m^2/s peut être décelée (conductivité hydraulique de 10^{-13} m/s pour une intervalle d'essai de 10 m en milieu poreux), valeur qui est de beaucoup inférieure à celle qui peut être décelée dans un essai classique par bouchons dans une période comparable. Selon l'hypothèse du milieu fissuré de Wang et al. (1978), une fissure simple de 1 µm d'ouverture peut, en théorie, être vérifiée en moins de 2 heures.

Un autre avantage de l'essai par impulsions est que les temps de réponse plus longs peuvent être utilisés pour évaluer la géométrie de la fissure au-delà de la paroi du puits. Des courbes d'amortissement théoriques, établies par Wang et al. (1978), indiquent que, pour des limites tant imperméables que parfaitement conductrices, la chute de pression est une fonction de l'ouverture de la fissure, de la distance entre packers, du volume de la fissure et du puits de forage.

Plusieurs difficultés rencontrées dans l'analyse de l'essai par impulsions doivent être examinées. Une d'entre elles porte sur l'hypothèse selon laquelle la dissipation d'une impulsion de pression appliquée instantanément n'est possible que si du fluide s'écoule de la section d'essai isolée (Wilson et al., 1979). Comme l'impulsion est en réalité appliquée dans un intervalle de temps bref, mais fini, une partie de l'énergie transmise pendant l'application d'une impulsion est dépensée pour comprimer la roche et le matériel descendu dans le trou. Neuzil (1982) a étudié le problème de l'évaluation de la compressibilité du matériel dans un puits fermé et a conclu que le coefficient de compressibilité du matériel d'essai peut en fait être considérablement plus élevé (d'un facteur de 6) que celui de l'eau. Le remplacement du coefficient de compressibilité de l'eau par celui du matériel dans le calcul de la transmissivité mènerait à des valeurs proportionnellement plus élevées de la transmissivité et déplacerait la courbe d'amortissement de la figure 12 vers la droite.

Un examen plus poussé des effets de la compressibilité du matériel a révélé qu'une forte augmentation de la pression dans un trou de sonde pendant un temps relativement court pourrait avoir un effet important sur la compressibilité des packers, ce qui donnerait lieu à un effet marqué sur la réponse en pression (Thackston et al., 1984) et, finalement, sur le calcul de la transmissivité et du coefficient d'emmagasinement. Toutefois, l'énergie dépensée pour comprimer le matériel dans le trou (et la matrice rocheuse) est récupérable dans une certaine mesure. Cela donne lieu à un retour d'énergie dans le fluide, ce qui se traduit par le maintien d'une pression élevée pendant un temps plus long (la courbe d'amortissement est déplacée vers la gauche). S'il n'est pas corrigé, un tel effet peut introduire des erreurs dans le calcul de l'ouverture des fractures. Autant Wilson et al. (1979) que Neuzil (1982) ont proposé d'analyser, avant les essais, les effets possibles de compressibilité du matériel pour établir des courbes types propres au matériel.

3.5 Comparaison des méthodes d'essai

La fourchette générale des transmissivités qui peuvent être calculées par les différentes méthodes d'essai décrites ci-dessus est illustrée à la figure 12. Les fourchettes de transmissivités mesurables indiquées dans la figure se rapportent à un milieu poreux équivalent ayant un coefficient d'emmagasinement faible (10^{-5}). Les effets d'emmagasinement dans le trou de sonde ont été pris en compte dans les essais à débit constant pour représenter plus fidèlement les transmissivités qui peuvent être mesurées dans des intervalles de temps donnés.

Les résultats indiqués concordent avec les fourchettes de conductivité hydraulique jugées mesurables par les différentes méthodes d'essai, décrites par Wilson et al. (1979) et calculées dans l'étude de caractérisation de sites suédois par Almen et al. (1986). L'essai d'injection à pression constante et l'essai par impulsions semblent applicables à l'intérieur d'une vaste gamme de transmissivités, y compris des valeurs inférieures à 10^{-8} m^2/s (valeur typique de la roche cristalline), tandis que l'essai à débit constant et l'essai par bouchons ne devraient pas être utilisés pour étudier des formations dont les transmissivités sont inférieures à 10^{-5} et 10^{-8} m^2/s respectivement. Les segments pointillés de la figure 12 représentent les fourchettes de transmissivité à l'intérieur desquelles un coefficient d'emmagasinement plus élevé justifierait l'application de l'essai particulier dans des fourchettes de transmissivité plus larges.

À la lumière de l'analyse qui précède et compte tenu des données hydrauliques qui peuvent être recueillies par chaque méthode d'essai, l'injection à pression constante est une méthode appropriée d'essai hydrologique en puits unique dans la roche cristalline peu perméable. Les avantages de cette méthode sont la grandeur de la plage de transmissivités qui peuvent être mesurées, la possibilité d'établir la géométrie de l'écoulement et les conditions aux limites, les effets négligeables d'emmagasinement dans le trou de sonde et le volume relativement élevé de roche qui peut être vérifié en peu de temps. Le programme suédois (Almen et al., 1986) renferme aussi des recommandations quant à l'utilisation de l'essai par remontée de la pression après la phase d'injection pour confirmer les données obtenues pendant l'essai d'injection, même si l'essai par remontée de la pression a un champ d'application limité à cause de l'effet d'emmagasinement dans le trou de sonde.

L'essai de prélèvement à débit constant ne convient pas à une évaluation rapide de la roche de faible perméabilité à cause de la persistance de l'emmagasinement dans le trou de sonde. L'essai d'injection à débit constant, quoique fondamentalement semblable à l'essai d'injection à pression constante, est aussi très sensible à l'emmagasinement dans le trou de sonde.

En outre, à moins que les transmissivités aient été estimées avec précision avant l'essai dans la formation à l'étude, l'utilisation d'un débit d'injection autre que prévu pendant une période prolongée peut entraîner une déformation des fissures ou même la fissuration hydraulique de la roche.

Les essais par impulsions conviennent également à l'étude de la roche cristalline, particulièrement lorsqu'ils sont complétés par des essais par bouchons qui permettent de mesurer les transmissivités plus élevées dans la formation à l'étude. Les inconvénients de l'essai par impulsions par rapport à l'essai à pression constante sont qu'il porte en général sur un volume de roche plus petit et qu'il dure parfois plus longtemps dans les zones de faible conductivité.

Un des avantages des essais par impulsions est la facilité relative de l'interprétation lorsque le gradient hydraulique de fond varie pendant un essai. L'hypothèse d'une pression statique initiale n'est parfois pas respectée dans les essais de puits en milieu peu perméable. Le problème est particulièrement apparent dans les essais en trous de sonde effectués à partir d'installations d'essai souterraines qui subissent de fortes variations de pression lorsque le trou est ouvert et que l'écoulement est libre. Un autre avantage des essais par impulsions est le petit volume d'eau utilisé, ce qui minimise les effets sur les autres travaux d'essai comme l'échantillonnage géochimique.

Quand on compare des essais, il faut noter que toutes les méthodes décrites ici sont basées sur des solutions de l'équation de diffusion, et que les différences résident simplement dans les conditions présumées aux limites et dans le type de perturbation. Il n'est pas étonnant que les comparaisons révèlent en général que les diverses méthodes produisent des valeurs de transmissivité comparables pourvu que le matériel utilisé et les durées des essais conviennent à la gamme des transmissivités de la zone à l'étude (Almen et al., 1986). Les facteurs principaux de différenciation des méthodes sont avant tout pratiques et tiennent notamment à la simplicité du matériel, à la facilité des contrôles et au temps requis pour effectuer les mesures.

4. ESSAIS D'INJECTION À PRESSION CONSTANTE

Cette section traite de l'application de l'essai d'injection à pression constante dans la roche cristalline. L'accent est mis sur l'interprétation des résultats d'essais en régime transitoire et permanent.

4.1 Bref historique

La théorie de l'essai d'injection à pression constante a été mise au point de façon quelque peu indépendante dans les domaines du génie civil, de l'hydrogéologie et du pétrole. L'approche du génie civil remonterait à Lugeon (Serafim, 1968) et est basée sur les essais d'injection.

En hydrologie, les essais à pression constante ont porté sur des puits dans des formations où le niveau de la nappe aquifère se situe au-dessus de la tête du puits, c'est-à-dire dans des puits jaillissants (Jacob et Lohman, 1951; Hantush, 1959). Dans ce cas, la pression constante est la pression atmosphérique.

Les méthodes à pression constante ont été étudiées en génie pétrolier en partie pour les cas où le niveau piézométrique a été abaissé au niveau de la pompe, - de sorte que le débit dans le puits est maintenu constant (Ehlig-Economides, 1979; Uraiet, 1979) -, et en partie parce que les ingénieurs de réservoirs n'avaient pas d'autres problèmes à résoudre.

En génie civil, l'utilisation de plus en plus répandue des essais à pression constante est révélatrice du matériel couramment utilisé et des types de roche rencontrés. Les essais traditionnels à débit constant ont été mis au point principalement parce que des producteurs travaillant dans de la roche perméable exploitaient des fluides souterrains au moyen de pompes à débit plus ou moins constant. Par contre, les études de sites pour des ouvrages de génie, situés notamment dans de la roche dure, étaient généralement basées sur des essais d'injection au moyen de pompes de forage fonctionnant à pression plus ou moins constante. En outre, les trous de sonde standard pour les études de terrain sont trop petits pour recevoir des pompes immergées, et la capacité des pompes était trop élevée pour étudier la roche peu perméable.

Les analyses des essais à pression constante se distinguent surtout du fait qu'elles utilisent des méthodes d'écoulement transitoire plutôt que permanent. Les calculs de génie civil sont essentiellement basés sur des analyses qui supposent que l'écoulement est permanent, tandis que dans d'autres disciplines, les méthodes sont basées sur un débit transitoire ou sur la remontée de la pression. Les méthodes basées sur un régime permanent supposent l'existence d'une frontière à pression constante qui peut exister ou pas. Même si cette hypothèse est inexacte, les calculs de transmissivité et de conductivité hydraulique ne sont pas perturbés outre mesure (Doe et Remer, 1980). Dans les méthodes basées sur un régime permanent, on perd de l'information sur les propriétés d'emmagasinement (et, par conséquent, sur la déformabilité), sur les effets aux limites et sur les effets pariétaux.

4.2 Méthodes basées sur un écoulement permanent

Les essais d'injection constituent la méthode la plus répandue pour évaluer la roche cristalline à des fins de génie. L'exécution des essais basés sur un écoulement permanent est un sujet qui a été approfondi par Zeigler (1976), Snow (1966, 1968), Louis et Maini (1970) et Banks (1972), entre autres. La méthode standard d'essai aux packers consiste à isoler une section du trou de forage au moyen de packers et à y injecter de l'eau sous diverses pressions d'injection. On maintient chaque niveau de pression pendant plusieurs minutes jusqu'à obtenir un écoulement qui semble permanent. Dans le cas d'un écoulement laminaire, les différents débits permanents devraient être linéairement proportionnels à la pression d'injection. Les écarts de linéarité peuvent être interprétés soit comme des écoulements turbulents, soit comme des déformations de fissures, selon la nature de l'écart, comme l'indique la figure 13.

Dans l'analyse des écoulements permanents, la zone d'essai est définie comme étant une nappe aquifère confinée qui est entièrement traversée par le puits (une fissure simple dans de la roche imperméable est une approximation raisonnable de cette condition). On suppose qu'il existe une limite ouverte ou à pression constante à un rayon R du trou. L'écoulement permanent est une conséquence de cette limite. La transmissivité pour une telle géométrie est définie simplement par :

$$T = \frac{Q \ln(R/r_w)}{2\pi\Delta h} \qquad (28)$$

La position exacte de la limite à pression constante (ou rayon d'influence) n'est jamais connue, mais comme R figure dans le terme logarithmique, tout écart par rapport à sa valeur exacte influe peu sur la valeur de la transmissivité.

Des facteurs de correction ont déjà été établis pour d'autres géométries d'écoulement, par exemple des écoulements sphériques ou elliptiques plutôt que radiaux. Ils sont résumés dans Zeigler (1976) et dans Hoek et Bray (1981). L'application de facteurs de correction pour d'autres géométries d'écoulement ne modifie généralement pas la valeur calculée de la transmissivité de plus de 30 % environ (Zeigler, 1976), écart qui est faible car la transmissivité peut varier de plusieurs ordres de grandeur d'une fissure à l'autre dans une masse rocheuse.

4.3 Analyse des essais transitoires

L'analyse en régime transitoire des données d'essai à pression constante est semblable à l'analyse en pression transitoire utilisée dans les essais classiques de pompage, la seule différence étant que c'est la pression dans le trou de sonde plutôt que le débit qui est maintenu constant.

4.3.1 Historique

Des solutions aux écoulements transitoires ont été élaborées de façon indépendante dans les domaines de l'hydrologie et du génie des réservoirs.

En hydrologie, Jacob et Lohman (1952) ont introduit la courbe de débit transitoire pour des nappes aquifères infinies comme méthode d'analyse des données sur les puits jaillissants. Ils ont aussi noté que (1) la courbe de débit transitoire correspond approximativement à l'inverse de la courbe de pression transitoire (Theis) lorsque la pression est remplacée par le débit et que (2) la remontée de la pression après production à pression constante pouvait être analysée de la même façon que dans le cas d'une production à débit constant. Hantush (1959) a établi les équations pour les essais à pression constante, avec frontières ouvertes (à pression constante) ou fermées (sans écoulement), ainsi que pour les fuites dans un aquitard où il n'y a pas d'emmagasinement. Plus récemment, Jaiswal et Chuhan (1978) ont établi des courbes pour un écoulement sphérique autour d'un puits qui pénètre partiellement une nappe aquifère.

Les analyses d'écoulement transitoire peuvent couvrir une vaste gamme de phénomènes : déformations de fissures (Noorishad et Doe, 1982), écoulements non linéaires (Elsworth et Doe, 1986), effets pariétaux (Uraiet, 1979) et influences des frontières.

Un effet très peu étudié dans la documentation est celui de l'emmagasinement dans le trou de sonde. Dans les essais à pression constante, il n'y a effectivement aucun emmagasinage dans le trou de sonde parce qu'une fois l'essai commencé, il ne se produit aucune variation de pression dans le trou ou dans le matériel, de sorte que les seuls effets d'emmagasinement se produisent dans la nappe aquifère ou dans les fissures. La masse d'eau dans le trou et dans le matériel est constante, empêchant ainsi toute variation de température.

L'absence d'emmagasinement dans le trou de sonde est le principal avantage de l'essai à pression constante dans la roche peu perméable. Contrairement aux essais à débit constant, les données sont utiles dès le début de l'essai.

Dans le domaine du génie des réservoirs, le même problème des écoulements transitoires a été étudié par Moore et al. (1933) et par Hurst (1934). Fetkovich (1980) a utilisé un système à limite fermée pour décrire comment les réservoirs s'épuisent dans les dernières étapes de la production. Ehlig-Economides (1979) et Uraiet (1979) ont formulé la théorie de l'analyse des débits transitoires ainsi que celle de l'analyse de la remontée de la pression après une production à pression constante. Le premier auteur a élaboré des solutions analytiques, le deuxième, des solutions numériques. Les deux auteurs ont tenu compte des effets aux frontières et ont conclu que l'analyse de la remontée de la pression était identique pour les essais à pression constante et pour les essais à débit constant.

4.3.2 Effets aux frontières

Les effets aux frontières peuvent jouer un rôle important dans l'établissement des courbes de débit transitoire. Si on assimile les fissures simples à des aquifères confinés, l'extrémité des fissures devrait constituer des limites hydrauliques. L'extrémité d'une fissure dans la roche solide constitue effectivement une limite sans écoulement. Les frontières imperméables ou fermées donnent des formes distinctives aux graphiques des essais à pression constante. Comme l'indique la figure 14, la réponse du débit transitoire suit la courbe d'une nappe aquifère infinie jusqu'à ce que le front de pression d'injection atteigne le bord de la fissure. À ce point, le débit devrait chuter et atteindre tôt our tard un débit relativement faible qui dénote la présence de fuites dans la roche encaissante peu perméable. Des courbes types pour des conditions aux limites fermées ont été établies au moyen d'une inversion numérique des solutions en espace laplacien. Les données sont présentées à l'appendice 1.

Les frontières sont ouvertes ou à pression constante lorsque la fissure dans laquelle est injecté du fluide à partir du puits recoupe une autre fissure plus perméable. Comme l'indique la figure 15, cette situation donne ultimement lieu à un écoulement permanent. Comme dans le cas des limites fermées, le débit transitoire devrait diminuer le long de la courbe d'un aquifère infini jusqu'à ce que le front de pression depuis le point d'injection atteigne le point de recoupement. Cette condition définit l'écoulement permanent qui est souvent observé dans les essais à pression constante.

La distance entre la frontière et le puits est le rayon d'influence mal défini utilisé dans les calculs d'écoulements permanents. Cette grandeur peut être estimée en faisant correspondre les observations de débit transitoire à la courbe type. Le temps requis pour observer les effets aux limites peut être estimé à partir de l'expression suivante (Uraiet, 1979) :

$$t_D = 0{,}809\ R_D^{\,1.99} \qquad (29)$$

où : $t_D = Tt/Sr_w^2$
$R_D = R/r_w$

Dans la pratique, il est peu probable que la frontière à pression constante pour une fissure simple soit uniforme dans un rayon donné. Strictement, il existe plutôt une frontière ouverte seulement à l'intersection et non sur tout le reste de la périphérie de la fissure où il n'y a pas d'écoulement. Cette condition mixte aux limites n'a pas été étudiée analytiquement, mais peut être traitée numériquement par modélisation de l'écoulement dans un aquifère circulaire autour d'un puits où seuls quelques noeuds de la périphérie sont maintenus à pression constante. La figure 16 montre les résultats d'un modèle aux éléments finis où les conditions à la limite extérieure sont telles. Les résultats pour un système infini à frontière fermée sont indiqués à titre de vérification générale. Comme la courbe de débit pour une frontière partiellement ouverte l'indique, il y a un début de chute rapide du débit associé à une frontière extérieure fermée, suivi de l'établissement d'un débit permanent dû à la frontière à pression constante.

Dans ces exemples, on a fait varier sur la frontière le nombre de noeuds à pression constante entre 1 et 4 (maximum 15 % de la périphérie). Dans chaque cas, le débit devient permanent à des niveaux progressivement plus élevés, et l'effet de la limite fermée est de plus en plus difficile à discerner. Cela montre que, si une partie importante des limites d'une fissure recoupe des fissures plus conductrices, le comportement transitoire de l'écoulement approchera celui d'un écoulement pour une frontière circulaire à pression constante; c'est-à-dire que la chute transitoire de débit devrait évoluer de la courbe d'un aquifère infini vers celle d'un écoulement permanent.

4.3.3 Écoulement non linéaire - turbulence et déformation des fissures

Les effets d'écoulement non linéaire dus à des turbulences sont traités dans Elsworth et Doe (1986) et ne sont pas détaillés ici. Un écoulement turbulent se reconnaît soit au tracé de la pression en fonction du débit permanent (figure 13), soit à l'horizontalité de la partie initiale de la courbe de débit (indiquée schématiquement dans la figure 17). Un écoulement sera vraisemblablement turbulent au début d'un essai lorsque les gradients de pression près du trou de sonde sont les plus élevés; l'écoulement peut devenir laminaire plus tard au cours de l'essai lorsque les gradients diminuent. Elsworth et Doe (1986) ont aussi noté que la distance entre un puits et une frontière à pression constante peut être calculée à partir des résultats d'un essai à pression en paliers dans lequel on peut observer des écoulements laminaires et turbulents.
On peut aussi reconnaître des déformations de fissures dans une courbe pression-débit par l'examen d'une série de paliers de débit permanent (figures 13 et 18). Par ailleurs, la déformation d'une fissure peut avoir des effets discernables sur le débit transitoire lui-même. Noorishad et Doe (1982) ont étudié ces derniers effets en utilisant un modèle combiné contrainte-débit pour montrer que la déformation des fissures entraîne une augmentation de débit en fonction du temps plutôt qu'une diminution. Cet effet est représenté schématiquement à la figure 17. Noorishad et Doe ont remarqué que les effets de la déformation d'une fissure ne se manifestent qu'après que le front de pression s'est propagé sur une certaine distance dans la fissure, constatant ainsi que la fissure se déforme non pas sous l'effet de la pression dans le trou de sonde, mais sous l'effet de la pression exercée sur toute la surface de la fissure, c'est-à-dire sous l'effet de la charge totale. Par ailleurs, Sabarly (1968) a indiqué que la courbe pression-débit pourrait comporter une composante de puissance quatre si la déformation de la fissure est linéaire dans la fourchette d'injection.

4.3.4 Effets pariétaux

Les effets pariétaux (changements dans les propriétés de la roche ou des fissures sur le bord ou à proximité du trou de sonde) peuvent être pris en compte dans les courbes types à pression constante. Les effets pariétaux peuvent être interprétés soit à partir de la courbe type logarithmique, soit à partir de courbes semi-logarithmiques. Des courbes types logarithmiques illustrant des effets pariétaux sont représentées dans les figures 19 à 21 pour une nappe aquifère infinie et deux conditions aux limites. Malheureusement, les courbes types avec effets pariétaux ne diffèrent pas considérablement des courbes sans effets pariétaux. Les données des essais comportant des effets pariétaux peuvent facilement s'aligner sur des courbes sans effets pariétaux.

L'autre méthode d'interprétation des effets pariétaux est fondée sur l'analyse semi-logarithmique comme cela se fait couramment dans le domaine du génie pétrolier. Le programme suédois de gestion des déchets radioactifs (Almen et al., 1987) s'appuie sur cette méthode pour l'interprétation des effets pariétaux dans les essais effectués à pression constante dans la roche cristalline. Malheureusement, comme il en est question à la section 3.1, cette méthode ne permet pas d'isoler le coefficient d'emmagasinement des effets pariétaux, de sorte qu'on ne peut interpréter un effet pariétal que si on attribue une valeur arbitraire au coefficient d'emmagasinement.

4.4 Résumé des essais à pression constante

La figure 16 résume les différents effets qui peuvent être observés dans la courbe de débit transitoire d'un essai à pression constante.
Au début, il peut y avoir augmentation ou diminution du débit à cause des effets pariétaux ou d'un écoulement turbulent. Par la suite, le débit peut être sujet à des effets aux limites ou à une déformation des fissures. Une fissure qui n'en recoupe pas d'autres peut donner lieu à une brusque diminution du débit à mesure qu'on progresse vers la limite fermée. Si la fissure recoupe une fissure plus perméable, l'écoulement devrait devenir permanent. La déformation des fissures devrait produire une augmentation caractéristique du débit en fonction du temps.

5 MATÉRIEL D'ESSAI DES TROUS DE SONDE

Il est essentiel que le matériel soit bien conçu et bien choisi pour s'assurer que les essais à pression constante produisent des données utiles. Dans la section qui suit, on passe en revue le matériel utilisé pour les essais effectués dans les trous de sonde.

5.1 Isolation de la zone d'essai - packers

La plupart des essais menés dans la roche peu perméable sont effectués dans des trous ouverts au moyen de packers en caoutchouc gonflables qui permettent d'isoler temporairement la zone d'essai. Les packers permettent en général d'isoler des sections de 1 à 2 m de long.

Les packers peuvent être gonflés soit d'air, soit d'eau. Le gonflage à l'air est une méthode très pratique, surtout lorsqu'on dispose de bouteilles d'air comprimé. Cependant, les packers gonflés d'air sont plus compressibles que les packers gonflés d'eau, et peuvent contribuer à l'emmagasinement dans le trou de sonde. Le gonflage à l'air ne peut être utilisé qu'à des profondeurs où la pression de l'air comprimé peut compenser la pression de la colonne de fluide dans le trou de sonde. Dans la pratique, le gonflage à l'air est impossible à des profondeurs de plus de 500 m environ.

Le gonflage à l'eau rend les packers beaucoup plus rigides. Cette méthode n'est véritablement désavantageuse que dans des puits où la nappe phréatique se trouve à plus de 20 à 30 m sous la surface. Dans de tels cas, la charge statique dans la canalisation de gonflage des packers peut être suffisante pour maintenir les packers en place après que la pression de gonflage des packers a été coupée en surface. Ce problème peut être résolu en installant un détendeur de pression dans le trou ou en remplissant ce dernier d'eau pendant que les packers sont dégonflés et déplacés dans le trou.

Dans les formations denses, les packers, en se gonflant, peuvent comprimer l'eau dans la zone d'essai, exerçant sur cette dernière une pression qui peut approcher la pression de gonflage des packers. Ce problème peut être résolu soit en mettant la conduite d'injection à l'atmosphère pendant le gonflage des packers, soit en installant une soupape dans le trou pour empêcher que la zone d'essai ne soit comprimée à cause du gonflage.
Il existe plusieurs possibilités pour acheminer la pression de gonflage vers les packers. La solution la plus répandue et la plus simple consiste à installer une canalisation de gonflage séparée de la canalisation d'injection. Pour les systèmes d'essai

à tuyauterie rigide, il existe dans le commerce des outils d'essai aux tiges qui comportent une soupape mécanique pour acheminer le fluide soit dans les packers, soit dans la zone d'essai.

La plupart des systèmes à packers ne comportent que deux packers; d'autres packers peuvent être ajoutés pour isoler des zones de garde. Ces dernières servent parfois à isoler des zones d'injection distinctes au-dessus et au-dessous de la zone principale d'injection de façon à réaliser un écoulement radial (Maini et al., 1971). Les zones de garde peuvent aussi servir à contrôler les fuites dans les packers. Des variations de pression dans les zones de garde peuvent indiquer des phénomènes autres que des fuites dans les packers, notamment un retour de fluide dans le puits ou des variations de pression dûes à l'isolation de zones qui vraisemblablement communiquaient entre elles par le trou ouvert.

L'aménagement de systèmes à packers multiples est extrêmement complexe. Il faut contrôler la pression non seulement dans chaque zone isolée, mais aussi dans les sections au-dessus et au-dessous du jeu de packers. En outre, il faut parfois installer des détendeurs de pression dans le trou pour éviter que la pression n'augmente dans la zone d'essai lors du gonflage des packers.

5.2 Canalisation d'injection du fluide

Traditionnellement, les essais de puits sont effectués au moyen de tuyaux d'injection rigides en acier ou de tiges de forage creuses montées par sections. À moins qu'il ne s'agisse de tiges de très grande qualité, les raccords dans un tel système peuvent fuire, et ces fuites sont presque impossibles à localiser.

Au cours des dernières années, l'usage de systèmes à câbles métalliques tels que le système d'essai ombilical suédois s'est répandu (Almen et al., 1986). Les systèmes à câbles métalliques sont constitués de boyaux hydrauliques ou de tubes en acier souples, combinés en faiseaux à des câbles électriques et à un élément de tension pour supporter le poids du système. Malgré les premières appréhensions quant aux possibilités de perte ou de coincement d'outils dans le trou de sonde, l'application de ces systèmes a donné d'assez bons résultats. Un des avantages du système à câbles métalliques est la rapidité d'exécution des essais. En outre, leur coût est moins élevé car elles réduisent les temps de forage et de main d'oeuvre.

5.3 Mesure et régularisation du débit

Il existe deux types fondamentaux de débitmètres qui peuvent être qualifiés d'hydrodynamiques et de volumétriques (figure 22). Les débitmètres hydrodynamiques évaluent le débit directement en mesurant un paramètre de l'énergie du fluide en mouvement. Les débitmètres volumétriques enregistrent simplement le volume ou la masse de fluide injectée ou prélevée; le débit est alors obtenu en divisant le volume débité par le temps écoulé pendant la mesure du volume.

Parmi les compteurs hydrodynamiques capables de mesurer de faibles débits, mentionnons les rotamètres, les débitmètres à turbine et les compteurs à effet de Coriolis. Les rotamètres sont des tubes transparents contenant une bille. Le rotamètre est orienté verticalement de façon que l'écoulement puisse soulever la bille sur une hauteur proportionnelle au débit. Ces appareils sont en général capables de mesurer des débits à l'intérieur d'une gamme d'environ un ordre de grandeur. Les rotamètres sont peu coûteux et se vendent avec diverses échelles allant des centièmes de millilitres par minute à plusieurs litres par seconde. Les rotamètres nécessitent en général un relevé manuel des données et ne peuvent donc être utilisés dans des systèmes automatisés de saisie des données.

Les débitmètres à turbine assurent l'écoulement à partir d'une turbine montée dans la canalisation. Un capteur magnétique ou électrique détecte le passage des pales de la turbine et produit un signal dont la fréquence est proportionnelle au débit. Le signal en fréquence peut être enregistré directement ou peut être converti en un signal analogique de courant ou de tension. Tout comme le rotamètre, le débitmètre à turbine est efficace pour des fourchettes de débit d'un à deux ordres de grandeur seulement. En outre, aux débits élevés, la perte de charge due à la présence du débitmètre peut être importante. Les pertes de charge élevées nécessitent l'installation de capteurs ou de régulateurs de pression en aval du débitmètre. Le seuil de détection des débitmètres à turbine est d'environ 4 mL/min, valeur qui est encore trop élevée pour une grande proportion des essais de puits effectués dans la roche peu perméable.

Dans les débitmètres à effet de Coriolis, un tube en U est branché sur le tuyau d'écoulement pour mesurer le débit massique. Le tube est mis en vibration à une fréquence établie. L'accélération du fluide dans la dérivation du tube se traduit par une déformation du tube vibrant. Cette déformation est proportionnelle à la masse de fluide se déplaçant dans le tube. Comme d'autres compteurs hydrodynamiques, les compteurs à effet de Coriolis sont précis dans une fourchette de débits d'environ un ordre de grandeur. Si l'erreur acceptable est d'environ 5 %, ces appareils peuvent être utilisés pour mesurer des débits aussi faibles que 0,1 mL/min dans des fourchettes de plus de quatre ordres de grandeur (Almen et al., 1986).

Dans les débitmètres volumétriques, on détermine le débit en mesurant le volume de fluide injecté pendant un temps donné. Le débit est donné par le rapport du volume sur la durée de mesure. Le débitmètre volumétrique est en général intégré au système de pompage. Dans les systèmes volumétriques simples, le déplacement d'un piston unique peut être contrôlé au moyen d'un transducteur à déplacement variable linéaire (TDVL). Si le piston est actionné par un moteur électrique, la vitesse de rotation des engrenages ou du moteur lui-même peut alors être contrôlée.

Un modèle courant de débitmètre volumétrique est le débitmètre à déversoir. Il s'agit d'un réservoir d'eau sous pression, de diamètre intérieur constant. La partie supérieure du réservoir communique avec une source d'air comprimé. Pour les essais à pression constante, la pression de l'air est régularisée pour obtenir la charge d'injection désirée. Le volume de fluide débité est donné par la baisse du niveau d'eau dans le réservoir. La mesure peut être effectuée manuellement par observation visuelle ou automatiquement au moyen d'un transducteur de pression différentielle qui détecte la différence de pression entre le dessus et le fond du réservoir. Un transducteur de pression différentielle peut détecter des dénivellations d'eau de moins de 1 mm. Avec des systèmes constitués d'une batterie de réservoirs de différents diamètres, on peut mesurer une vaste gamme de débits. Des réservoirs ont été fabriqués à partir de tubes en acier inoxydable de petit diamètre (Haimson et Doe, 1983).

Les débitmètres volumétriques peuvent avoir une résolution presque infinitésimale pourvu que le débit puisse être mesuré sur une longue période. La précision de la méthode diminue avec la fréquence des mesures, de sorte que les débitmètres volumétriques conviennent le mieux aux mesures de faible débit permanent et sont moins efficaces pour mesurer les débits qui peuvent varier rapidement. Ils ne sont en général pas indiqués pour mesurer des débits en début d'essai d'injection transitoire.

5.4 Mesure et régularisation de la pression

Il est essentiel de mesurer et de régulariser la pression dans tous les essais de puits. Comme les méthodes d'analyse font intervenir la pression dans la zone d'essai, il n'est pas indiqué de mesurer la pression à l'entrée du puits : il faut prendre soin d'étalonner le système d'essai pour tenir compte des pertes de charge entre le point de mesure et la zone d'essai. Il est donc devenu souhaitable de mesurer la pression dans le trou pendant les essais en milieu peu perméable.

De nombreux systèmes à transducteur de pression installés dans le trou ont été construits au cours des dernières années; tous nécessitent un câble électrique indépendant. Lorsque plusieurs instruments sont descendus dans le trou, le câble électrique peut devenir encombrant à moins de recourir au multiplexage. Pour protéger les transducteurs de pression installés dans un trou, il convient de les loger dans un récipient étanche à l'eau. Par ailleurs, il existe des transducteurs de pression étanches qui peuvent supporter de fortes pressions d'eau. Pour optimiser les mesures de pression, tous les transducteurs devraient avoir une compensation de température et être couplés avec des transducteurs de température tels que des thermistors ou des dispositifs thermométriques à résistance.

5.5 Installation de soupapes dans les trous de sonde - Méthodes de production d'impulsions

Dans les essais de puits de faible perméabilité, il est essentiel de descendre une soupape d'arrêt d'injection au fond du trou. Cette soupape est utilisée pour isoler la zone d'essai du tube d'injection avant l'essai. En fermant la soupape et en mettant le tube d'injection sous pression avant l'essai, on peut éliminer la plupart des effets de compressibilité du matériel. La soupape facilite aussi la mise en route rapide de chaque essai et le rétablissement des vitesses en agissant directement dans la zone d'essai.

Une telle soupape est aussi essentielle dans les essais par impulsions. Le moyen le plus simple d'effectuer un tel essai consiste à mettre le tube d'injection sous pression en maintenant la zone d'essai isolée, puis à établir instantanément une impulsion de pression dans la zone d'essai en ouvrant et en fermant rapidement la soupape.

6 STRATÉGIES D'ÉCHANTILLONNAGE

Le succès d'un programme d'essais de puits dépend et de la précision de chaque essai, et de l'efficacité du programme d'échantillonnage. La présente section décrit deux grandes méthodes d'échantillonnage dans des trous de sonde : l'échantillonnage par intervalles de longueur fixe (ILF) et l'échantillonnage par zones discrètes (ZD). La section se termine par une analyse de l'effet de canal sur l'exactitude des conclusions tirées d'essais de puits.

6.1 Essais par ILF et par ZD

L'échantillonnage par ILF consiste à étudier un trou de sonde par intervalles successifs séparés par des packers (obturateurs) régulièrement espacés. La longueur des intervalles choisis doit être assez petite pour fournir un minimum de détails sur la variabilité de la roche, mais assez grande pour que le trou puisse être étudié en un temps raisonnable. Si les obturateurs sont trop rapprochés, il peut arriver qu'un grand nombre d'intervalles ne contiennent aucune voie d'écoulement, et que l'essai soit vain. Néanmoins, comme l'a montré Snow (1970), le nombre d'essais sans écoulement est un paramètre utile pour évaluer la fréquence des voies d'écoulement. Idéalement, un programme d'essais devrait comporter un petit nombre de zones sans écoulement, peut-être 20 % environ, pour être efficace. D'autres méthodes d'analyse des données d'essais par ILF sont traitées dans la section qui suit.

Le but principal des essais par ILF est de fournir des données destinées au traitement statistique des propriétés hydrauliques de la roche. Chaque trou de sonde devient essentiellement une ligne d'échantillonnage pour fins d'essais hydrauliques. Le programme d'essais par ILF produit une distribution de valeurs de conductivité qui est habituellement log-normale. Un exemple d'une distribution log-normale est donné à la figure 23. Cette distribution renseigne sur la variabilité de la conductivité et sur la géométrie des fissures conductrices. La méthode par ILF est décrite dans la figure 24.

L'autre grande méthode d'essai est celle de l'échantillonnage par zones discrètes (ZD). La méthode par ZD est utilisée lorsqu'il s'agit d'étudier des caractéristiques géologiques particulières. Par exemple, on pourrait effectuer un essai par ZD pour sonder les propriétés hydrauliques d'une zone de cisaillement particulière le long d'un alignement du tunnel. Dans ce cas, seules des zones particulières sont étudiées; les propriétés générales de la masse rocheuse ne sont pas importantes. La méthode par zones discrètes peut être utilisée statistiquement si des données sont recueillies dans plusieurs trous de sonde forés dans la même formation géologique, ce qui permet d'obtenir une distribution de valeurs qui représentent bien la variabilité de la formation. La méthode par ZD est illustrée à la figure 25.

6.2 Analyse statistique des données d'essais par ILF

La distribution des résultats d'un essai par ILF représente fidèlement la géométrie du réseau des fissures d'écoulement. On peut donc en tirer de l'information sur les distributions de fréquence et de transmissivité des fissures ainsi que sur l'effet de canal. La présente section traite de l'aspect théorique de ces méthodes.

L'utilisation des données recueillies par ILF pour en déduire la géométrie des fissures a été proposée par Snow (1970).

La base de la méthode de Snow est que le débit dans un essai de puits est égal à la somme des débits dans chacune des fissures conductrices recoupées par l'intervalle d'essai.

Par conséquent, selon le même raisonnement, on peut établir des estimateurs de la moyenne et de la variance des transmissivités d'un ensemble de fissures conductrices recoupées lors d'une série d'essais de puits par intervalles de longueur fixe :

$$T = \sum_{i=1}^{N_f} T_i \qquad (30)$$

où : T = transmissivité mesurée
T_i = transmissivité de la $i^{\text{ème}}$ fissure conductrice recoupée
N_f = nombre de fissures conductrices recoupées

À noter que le nombre de fissures conductrices recoupées, ainsi que leur transmissivité et la mesure résultante de transmissivité sont des variables intrinsèquement aléatoires.

Snow (1970) a fait les hypothèses suivantes :

- La masse rocheuse caractérisée par une série d'essais de puits est statistiquement homogène;
- Les transmissivités et les espacements des fissures conductrices sont statistiquement et mutuellement indépendants;
- L'espacement des fissures conductrices a une distribution exponentielle, ce qui implique que le nombre de fissures conductrices recoupées a une distribution de Poisson.

Selon l'équation 30 et les hypothèses d'homogénéité et d'indépendance statistique, les moments (la moyenne et la variance) des mesures de transmissivité peuvent être exprimés en termes des moments du nombre de fissures coupées et de leur transmissivité. Selon l'hypothèse de la distribution de Poisson, la moyenne et la variance du nombre de fissures conductrices recoupées par essai et la probabilité d'un essai "sans écoulement" dans lequel aucune fissure conductrice n'a été recoupée peuvent être exprimées explicitement en termes de la fréquence moyenne des fissures conductrices et de l'espacement des packers.

Par intuition, la fréquence relative des essais "sans écoulement" dans une série d'essais de puits devrait être une estimation de la probabilité des essais "sans écoulement". En outre, la moyenne arithmétique et la variance des transmissivités mesurées dans une série d'essais de puits peuvent être utilisées pour estimer la moyenne et la variance des mesures de transmissivité, respectivement. En introduisant ces estimateurs dans les expressions des moments des mesures de transmissivité et du nombre de fissures conductrices recoupées par essai, on peut définir des estimateurs de la fréquence moyenne des fissures conductrices et de la moyenne et de la variance des transmissivités des fissures conductrices. Ces estimateurs sont :

$$\hat{\lambda}_c = \frac{\ln(N/n_o)}{b} \qquad (31)$$

$$\hat{\mu}_T = \frac{\hat{\mu}_T}{\ln(N/n_o)} \qquad (32)$$

$$\hat{\sigma}_T^2 = \frac{1}{\ln(N/n_o)}\left[\hat{\sigma}_T^2 - \frac{\hat{\mu}_T}{\ln(N/n_o)}\right] \qquad (33)$$

où : $\hat{\lambda}_c$ = estimateur de la fréquence moyenne des fissures conductrices
N = nombre d'essais de puits dans la série
n_o = nombre d'essais "sans écoulement" dans la série
b = espacement des packers
$\hat{\mu}_T$ = estimateur de la transmissivité moyenne des fissures conductrices
$\hat{\mu}_T$ = moyenne arithmétique des transmissivités mesurées dans une série d'essais de puits

$\hat{\sigma}_T^2$ = estimateur de la variance des transmissivités des fractures conductrices
$\hat{\sigma}_T^2$ = variance des transmissivités mesurées dans une série d'essais de puits.

Les estimateurs définis par les équations 31, 32 et 33 sont essentiellement équivalents aux estimateurs obtenus par Snow (1970), quoique Snow ait travaillé en termes de débit plutôt qu'en termes de transmissivité.

Carlsson et al. (1984) ont proposé une autre méthode qui peut être utilisée pour estimer la fréquence moyenne des fissures conductrices. Comme Snow (1970), ces chercheurs ont fait la double hypothèse de l'homogénéité et de l'indépendance statistique et ont utilisé la fréquence des essais "sans écoulement" dans leur technique d'estimation. Toutefois, ils n'ont fait aucune hypothèse quant à la distribution du nombre de fissures conductrices recoupées. Les résultats des essais de puits sont plutôt complétés par des données sur le nombre réel de fissures recoupées par chaque essai d'après l'étude des carottes prélevées dans les zones d'essai.

Les données d'étude de carottes servent à répartir les essais de puits en m groupes, chaque essai dans le i[ème] groupe recoupant i fissures. Pour chaque groupe d'essais, il est possible de définir, selon le principe du maximum de vraisemblance, un estimateur de la probabilité qu'une fissure ne soit pas conductrice :

$$\hat{p}_i = \left[\frac{n_{oi}}{N_i}\right]^{1/i}, \quad i = 1, \ldots, m \qquad (34)$$

où : $\hat{p}_i$ = estimateur de la probabilité qu'une fissure ne soit pas conductrice d'après les essais qui ont recoupé i fissures
n_{oi} = nombre d'essais "sans écoulement" qui ont recoupé i fissures
N_i = nombre total d'essais qui ont recoupé i fissures

L'équation 34 définit m estimateurs de la probabilité qu'une fissure ne soit pas conductrice. Le nombre d'estimateurs réellement évalués est choisi en fonction de la fréquence totale de fissures dans la région examinée. Un estimateur global de la probabilité qu'une fissure ne soit pas conductrice est défini par la somme pondérée des m estimateurs, les facteurs de pondération étant choisis de façon à minimiser la variance de l'estimateur global. Cette somme et les coefficients de pondération sont définis par les équations 35 et 36 respectivement.

$$\hat{p} = \sum_{i=1}^{m} a_i \hat{p}_i \qquad (35)$$

$$a_i = \left[\mathrm{Var}(\hat{p}_i)\sum_{j=1}^{m}\frac{1}{\mathrm{Var}(\hat{p}_j)}\right]^{-1}, \quad i = 1, \ldots, m \qquad (36)$$

où : $\hat{p}$ = estimateur de la probabilité qu'une fissure ne soit pas conductrice

$\mathrm{Var}(\hat{p}_i)$ = variance de l'estimateur $\hat{p}_i$

La variance de l'estimateur $\hat{p}_i$ peut être représentée par l'expression approchée suivante :

$$\mathrm{Var}(\hat{p}_i) \approx \frac{p^{2-i}(1-p^i)}{n_i i^2}, \quad i = 1, \ldots, m \qquad (37)$$

où : p = probabilité qu'une fissure ne soit pas conductrice

L'estimation de la probabilité qu'une fissure ne soit pas conductrice d'après les équations 35, 36 et 37 est un processus itératif puisque la probabilité qu'on cherche à évaluer dans l'équation 35 est la même probabilité nécessaire au calcul des variances dans l'équation 37.

La fréquence moyenne des fissures conductrices est le produit de la fréquence moyenne des fissures et de la probabilité qu'une fissure soit conductrice. Par conséquent, la fréquence moyenne des fissures conductrices peut être estimée à partir d'une estimation de la fréquence moyenne des fissures provenant des données d'étude des carottes et d'une estimation de la probabilité qu'une fissure soit non conductrice :

$$\hat{\lambda}_c = (1 - \hat{p})\left[\sum_{i=1}^{m} n_{oi} i\right] \Big/ \left[\sum_{i=1}^{m} n_{oi} b\right] \qquad (38)$$

Dans la pratique, la difficulté d'obtenir des estimateurs qui sont basés sur le nombre d'essais "sans écoulement" tient au fait que le seuil de mesure dans les essais de puits, quoique très bas en termes de valeurs mesurables de la transmissivité (par exemple 10^{-10} m^2/s pour des sections d'essai de 25 m) n'est néanmoins pas nulle. Les essais effectués dans des sections où la transmissivité est inférieure au seuil de mesure doivent être traités comme des essais "sans écoulement" même si les sections peuvent contenir quelques fissures conductrices. Cette ambiguïté est particulièrement critique dans l'évaluation de l'estimateur de Snow de la fréquence moyenne des fissures conductrices à cause de la relation logarithmique qui existe entre la fréquence relative des essais "sans écoulement" et l'estimateur.

6.3 Modèle probabiliste ILF

Les valeurs moyennes et les variances dont les estimateurs ont été traités dans la section précédente caractérisent les tendances centrales et les dispersions des propriétés hydrauliques des fissures recoupées dans une série d'essais de puits par intervalles fixes. Un modèle probabiliste fournit des renseignements complets en matière de distribution, c'est-à-dire qu'il permet d'obtenir non seulement la valeur des variables telles que la moyenne et la variance, mais aussi la probabilité que la valeur d'un propriété dépasse une valeur donnée. En outre, les caractéristiques statistiques telles que l'erreur systématique et l'erreur quadratique moyenne des estimateurs peuvent être calculées à partir d'un modèle probabiliste qui décrit la distribution implicite des estimateurs. Ces caractéristiques peuvent être utiles dans la conception de programmes optimaux d'essais de puits qui permettent de réduire ou de minimiser les erreurs quadratiques moyennes des estimations, à partir des essais effectués dans le cadre des programmes.

En plus des hypothèses faites par Snow (1970) dans son calcul des estimateurs des propriétés hydrauliques des fissures recoupées par des essais de puits, il faut faire une hypothèse quant à la distribution des transmissivités des fractures conductrices pour élaborer un modèle probabiliste qui décrit la distribution des transmissivités mesurées dans les essais de puits. Plusieurs chercheurs ont observé que la distribution normale représente bien les données de transmissivité des essais de puits (Snow, 1970). Cependant, il n'existe pas de fonction analytique pour représenter la somme des variables ayant une distribution log-normale, fonction qui est nécessaire pour représenter la somme des transmissivités des fissures conductrices de l'équation 30.

La distribution gamma a une forme semblable à celle de la distribution log-normale. Cette ressemblance a été relevée dans les données sur la longueur des fissures, même si la distribution log-normale convient mieux à de telles données (Baecher et al., 1977). Contrairement à la distribution log-normale, la distribution gamma est régénérative sous l'addition. Cette propriété additive est précisément ce qu'il faut pour établir la distribution de la somme des transmissivités des fissures conductrices. Par conséquent, on suppose que les transmissivités de fissures conductrices peuvent être représentées fidèlement par une distribution gamma parce que la forme de la distribution gamma ressemble à la forme de la distribution log-normale, mais elle est plus souple sur le plan mathématique.

Selon les hypothèses énoncées ci-dessus, le modèle probabiliste suivant peut être élaboré en appliquant le théorème de la probabilité totale à l'équation 30 :

$$Pr[T \leq \tau] = \exp(-\lambda b) \left[1 + \sum_{n=1}^{\infty} \frac{(\lambda_c b)^n \gamma(n\nu_T, \lambda_T t)}{n!\Gamma(n\nu_T)} \right] \quad (39)$$

où :

ν_T = paramètre de forme de la distribution gamma représentant les transmissivités de fissures conductrices

$Pr[T \leq \tau]$ = probabilité que la transmissivité mesurée est inférieure ou égale à τ

$\Gamma(n\nu_T)$ = fonction gamma

λ_T = paramètre d'échelle pour la distribution gamma représentant les transmissivités de fissures conductrices

$\gamma(n\nu_T, \lambda_T t)$ = fonction gamma incomplète

6.4 Effets de canal

L'effet de canal est un facteur géologique qui pourrait avoir un effet important sur les mesures de transmissivité effectuées dans les essais d'injection dans un puits unique (Tsang et Tsang, 1987). On a conclu en particulier que, si des sections ouvertes dans les fractures peuvent être représentées par des canaux étroits très espacés, les essais d'injection dans un trou unique constituerait un moyen très inefficace d'échantilloner les propriétés hydrauliques du réseau de fissures. Dans la présente section, un modèle conceptuel simple pour les réseaux de canaux est défini. Ce modèle permet de quantifier les effets de canal sur les essais d'injection dans un trou unique.

Par définition, un modèle est une représentation abstraite d'une entité ou d'un phénomène naturel. Par conséquent, un modèle d'un réseau de canaux devrait représenter la distribution des surfaces ouvertes et des surfaces de contact à l'intérieur de fissures. Cependant, la géométrie des canaux dans les fissures n'a jamais été observée sur le terrain parce que, pour ce faire, il faudrait pouvoir mesurer la distribution de la surface de l'ouverture d'une fissure sans perturber ni séparer au préalable la fissure, tâche qui est presque impossible à réaliser dans la pratique. Par conséquent, tout modèle d'un réseau de canaux est tout à fait spéculatif parce qu'il doit être basé sur des observations indirectes de la géométrie des canaux.

Rasmusson et Neretnieks (1986) ont proposé le modèle illustré à la figure 25. Comme il est indiqué dans la référence, leur "idée est la suivante : l'eau se déplace dans des canaux qui sont assez espacés et qui peuvent, sur des distances assez considérables, ne pas recouper d'autres canaux". Ce concept est corroboré par des observations et des mesures minutieuses de la distribution de l'écoulement naturel dans une excavation (Neretnieks, 1985), effectuées à partir de la distribution irrégulière de traceurs adsorbés sur les surfaces des fissures lors d'une expérience sur le terrain (Abelin et al., 1985) et à partir d'analyses d'expérience effectuées au moyen de traceurs sur des échantillons de fissures naturelles (Moreno et al., 1985). Dans ce concept, les canaux sont plus ou moins parallèles et sont recoupés par endroits par des canaux transversaux.

Tel qu'illustré à la figure 25, les zones délimitées par les canaux sont appelées "îles". Une île est simplement une zone dans laquelle le débit est négligeable comparativement au débit dans les canaux. Par conséquent, les îles comprennent non seulement les zones de contact, mais aussi les zones dans lesquelles les ouvertures sont beaucoup plus petites que celles des canaux.

L'effet des canaux sur les mesures de transmissivité effectuées au moyen d'essais d'injection dans un trou unique est une question d'échelle : quelle est la largeur des canaux et leur séparation par rapport au diamètre du trou de sonde? Par exemple, la figure 26 pourrait représenter une photomicrographie des voies d'écoulement tortueuses entre les aspérités sur les surfaces d'une fissure. Si cela était le cas, un trou de sonde recouperait en général un grand nombre de canaux, et un essai d'injection révélerait avec exactitude les caractéristiques de transmission du fluide de la fissure même si les canaux sont extrêmement étroits. Cet exemple montre que l'effet de canal est tolérable lorsque la probabilité de recoupement d'un canal est relativement élevée. Si cette probabilité peut être estimée, l'effet de canal peut être évalué.

Représentons des canaux par un réseau régulier de voies d'écoulement orthogonales de largeur constante égale à $\underline{d}$. Les îles formées par les canaux sont rectangulaires en plan et mesurent $\underline{w}_o$ sur $\underline{w}_x$, où $\underline{w}_x$ (la distance entre les canaux transversaux) est supérieure ou égale à $\underline{w}_o$. La fraction de la surface de la fissure qui est composée d'îles et qui est essentiellement exempte de tout écoulement est :

$$\theta = \frac{w_o w_x}{(w_o+d)\ (w_x+d)} \tag{40}$$

où : θ = fraction de la surface de la fissure composée d'îles

Le calcul de la probabilité qu'un canal soit recoupé par un trou de sonde de rayon r_w est direct. Supposons que tout point de la surface de la fissure peut être pris comme le centre du trou de sonde; alors la probabilité Pr[I] qu'une partie du trou recoupera un canal est :

$$Pr[I] = \begin{cases} 1 - \theta \left[1 - \dfrac{2r_w}{w_o}\right]\left[1 - \dfrac{2r_w}{w_x}\right], & \dfrac{2r_w}{w_o} < 1 \\[2ex] 1, & \dfrac{2r_w}{w_o} \geq 1 \end{cases} \tag{41}$$

L'équation 41 représente quantitativement les énoncés qualitatifs précités : la probabilité qu'un trou de sonde recoupe au moins un canal est fonction de la largeur des canaux et de leur espacement par rapport au diamètre du trou de sonde. L'équation 41 indique qu'un trou de sonde recoupera certainement un canal (Pr[I]=1) si le diamètre du trou de sonde est supérieur à l'espacement des canaux ($2r_w/\underline{w}_o>1$) peu importe la largeur des canaux.

D'après l'équation 41, la probabilité qu'un trou de sonde recoupe au moins un canal est représentée à la figure 27 en fonction du diamètre du trou de sonde par rapport à l'espacement des canaux. Comme l'indique cette figure, une enveloppe renfermant toutes les géométries possibles d'îles rectangulaires peut être définie pour une valeur donnée de θ par la courbe inférieure $\underline{w}_o = \infty$ (aucun canal transversal) et par la courbe supérieure $\underline{w}_x = \underline{w}_o$ (îles carrées). L'écart maximal de probabilité de recoupement entre ces deux extrêmes est θ/4 pour $2r_w/\underline{w}_o = 0,5$. Par conséquent, la géométrie des îles a un effet secondaire sur la probabilité qu'un puits recoupe au moins un canal.

Deux valeurs de θ sont prises en compte dans la figure 26. Iwai (1976) a réalisé des expériences de laboratoire sur l'écoulement dans des fissures et observé que de 10 à 20 % de la surface d'une fissure était en contact avec son vis-à-vis.
En outre, la valeur supérieure n'a été approchée que lorsque des contraintes moyennement élevées (20 MPa) ont été exercées en travers de la fissure. Par conséquent, θ = 0,15 pourrait être une valeur représentative de la surface des îles que comportent de nouvelles fissures planes dont les lèvres se marient bien telles que les fissures de traction étudiées par Iwai. Des valeurs beaucoup plus élevées de θ ont été proposées par Rasmusson et Neretnieks (1986). Leurs observations de l'écoulement dans des excavations dans le socle suédois indiquent que ce n'est que de 5 à 20 % de la surface d'une fissure qui transporte la plus grande quantité de l'eau. Dans un modèle de transport à canaux parallèles, Rasmusson et Neretnieks ont supposé que chaque canal avait des zones perméables de 0,2 m de large séparées par des zones imperméables de 1 m de large, ce qui est équivalent à θ = 0,83. Par conséquent, θ = 0,85 serait une valeur représentative d'anciennes fissures qui ont subi une importante érosion mécanique et géochimique, comme c'est le cas des fissures du socle précambrien de la Suède.

L'équation 41 indique que la probabilité minimale de recoupement d'un canal est 1 - θ, une fonction de la surface des îles seulement. Par conséquent, si la surface des îles représente une petite fraction de la surface de la fissure, la probabilité de recoupement d'au moins un canal par un trou de sonde est élevée, peu importe combien petit est le diamètre du trou par rapport à l'espacement des canaux. Le cas de θ = 0,15 dans la figure 27 illustre ce point. Cependant, si la surface des îles représente une fraction importante de la surface de la fissure, comme dans le cas de θ = 0,85 dans la figure 27, la probabilité de recoupement d'un canal est étroitement liée au diamètre du trou par rapport à l'espacement des canaux.

En résumé, le modèle des canaux illustré à la figure 26 indique que :

- la forme des îles entre les canaux a un effet secondaire sur la probabilité de recoupement d'un canal;
- si les îles constituent une petite fraction de la surface de la fissure, la probabilité de recoupement d'un canal est élevée, peu importe l'espacement des canaux;
- si les îles constituent une fraction importante de la surface de la fissure, la probabilité de recoupement d'un canal est étroitement liée au diamètre du trou de sonde par rapport à l'espacement des canaux.

À noter que le modèle duquel les conclusions qui précèdent sont tirées est une représentation abstraite simplifiée, basée sur l'observation indirecte de la géométrie des canaux. Même si la méthode de quantification de l'effet de canal qui s'exerce sur les mesures de transmissivité en termes de la probabilité de recoupement peut être utile en général, le modèle particulier présenté ici et les conclusions qui en découlent doivent être considérés comme spéculatifs.

La présente section n'a porté que sur un des effets de canal qui peuvent affecter les mesures de transmissivité. L'analyse particulière n'a porté que sur l'efficacité de l'échantillonnage des propriétés hydrauliques des fissures avec effet de canal dans le cadre d'essai d'injection dans un puits unique. Même si la probabilité de recoupement de canaux est relativement élevée de sorte qu'un essai d'injection indiquera en général si une fissure conductrice est effectivement conductrice (échantillonnage efficace), il reste à interpréter les conditions de l'essai suivant un modèle théorique quelconque du régime d'écoulement pour déterminer la transmissivité.

REMERCIEMENTS

Le présent article est la somme de plusieurs années d'efforts axés sur la recherche de la vérité en matière d'essais au moyen de packers. Un grand nombre de personnes ont contribué à cet article lors de discussions, y compris d'anciens collègues et des collègues actuels, notamment Derek Elsworth, Jahan Noorishad, Gen-Hua Shi, Janet Remer, Charles Wilson et William Dershowitz. Nous remercions tout particulièrement nos collègues suédois Leif Carlsson, Karl-Erik Almen, Jan-Erik Andersson et Bjorn Rossander avec qui nous partageons un certain nombre de préjugés.

Liste de figures

Pour les figures, voir le texte anglais. Les légendes apparaissent en français à la fin.

Figure 1. Illustration de la transmissivité, de la perméabilité (conductivité hydraulique) et du coefficient d'emmagasinement dans une roche poreuse et une roche fissurée.

Figure 2. Illustration des effets pariétaux dans une roche poreuse et des fissures.

Figure 3. Géométries d'écoulement dans les essais de puits.

Figure 4. Illustration des deux types de limites extérieures.

Figure 5. Comportement type de la pression et du débit dans les essais à débit constant, à pression constante et par impulsions/bouchons.

Figure 6. Illustration d'une superposition entre courbes types logarithmiques et d'une analyse semi-logarithmique.

Figure 7. La méthode d'analyse semi-logarithmique.

Figure 8. Durée des effets d'emmagasinement dans le puits pendant un essai à débit constant en système ouvert.

Figure 9. Durée des effets d'emmagasinement dans le puits pendant un essai à débit constant en système fermé.

Figure 10. Temps nécessaire pour qu'un bouchon soit amorti de 90 % et de 70 % dans les conditions indiquées en médaillon.

Figure 11. Temps nécessaire pour qu'une impulsion soit amortie de 90 % et de 10 % dans les conditions indiquées en médaillon.

Figure 12. Fourchettes pratiques des transmissivités qui peuvent être mesurées au moyen des méthodes d'essai de puits dans la roche cristalline.

Figure 13. Courbe pression-débit pour une série théorique d'injections à débit constant (d'après Louis et Maini, 1970).

Figure 14. Courbes types de débit transitoire pour des frontières imperméables.

Figure 15. Courbes types de débit transitoire pour des frontières à pression constante (ouvertes).

Figure 16. Courbe de débit transitoire pour une fissure simple avec une frontière extérieure partiellement ouverte.

Figure 17. Courbes théoriques de débit transitoire pour les essais à pression constante.

Figure 18. Pression et débit en fonction du temps pour un essai d'injection par paliers. À noter l'augmentation transitoire de débit dans le palier de pression le plus élevé.

Figure 19. Courbe type de débit transitoire pour un aquifère infini avec effets pariétaux.

Figure 20. Courbe type de débit transitoire pour un aquifère avec frontière extérieure à pression constante et effets pariétaux.

Figure 21. Courbe type de débit transitoire pour un aquifère avec frontière extérieure imperméable et effets pariétaux.

Figure 22. Illustration de divers types de débitmètres.

Figure 23. Courbe logarithmique de probabilité des données d'essais par intervalles de 25 m, provenant d'un puits foré dans la roche cristalline en Suède (d'après Carlsson et Winberg, 1983).

Figure 24. Illustration de la méthode d'essai par intervalles de longueur fixe (ILF).

Figure 25. Illustration de la méthode d'essai par zones discrètes (ZD).

Figure 26. Modèle conceptuel d'un réseau de fissures représenté par des canaux.

Figure 27. Probabilité qu'un trou de sonde recoupe des canaux dans une fissure en fonction de la surface de contact relative, θ.

RÉFÉRENCES (voir section anglais)

APPENDICE 1

Les données pour les courbes types adimensionnelles d'essais à pression constante sont présentées ci-dessous pour trois conditions aux limites (nappe aquifère infinie, milieu sans écoulement et milieu à pression constante) et pour de légers effets pariétaux tant positifs que négatifs. Les données ont été produites par un programme de Golder Associates pour un PC, CHQP, qui effectue une inversion numérique des solutions spatiales par transformation de Laplace des équations différentielles à pression constante. Les solutions spatiales par transformation de Laplace sont données dans Ehlig-Economides (1979).

Pour les tableaux voir texte anglais.

Tableau 1. Effets de limites ouvertes et fermées. Pour des limites situées aux rayons indiqués, des données sont produites à partir du moment où les courbes types commencent à s'écarter du cas de la nappe infinie, jusqu'à ce que les débits sans dimension pour le cas d'une limite fermée atteignent 10^{-3}.

Tableau 2. Données de courbes types indiquant les effets pariétaux pour le cas d'une limite à l'infini.

Tableau 3. Données de courbes types indiquant des effets pariétaux pour le cas d'une limite ouverte à un rayon sans dimension de 1000.

Tableau 4. Données de courbes types indiquant des effets pariétaux pour le cas d'une limite fermée à un rayon sans dimension de 1000.

11:15 Conférences bref

Elsworth, D., Piggott, A.R., États-Unis
Analogies physiques et numériques pour l'écoulement en milieu fissuré

Azevedo, A.A., Corrêa Fo, D., Quadros, E.F., da Cruz, P.T., Brésil
Conductivité hydraulique dans les discontinuités des roches de fondation du barrage de Taquaruçu-Brésil

By, T.L., Norvège
Utilisation de la géotomographie pour la caractérisation du rocher et la prédiction de problèmes d'écoulement pour deux tunnels routiers majeurs sous la ville d'Oslo

Charo, L., Zélikson, A., Bérest, P., France

Risques de formation de dômes après stockage de déchets nucléaires dans une couche de sel

12:00 - 13:30 Pause déjeuner et sessions d'affichage

13:30 Duluth-Mackenzie

Chen Gang, Liu Tian-Quan, Chine
Analyse détaillée des effets et des mécanismes des venues d'eau soudaines du plancher d'exploitations houillères

Crotogino, F., Lux, K.-H., Rokahr, R., Schneider, H.-J., Allemagne-Ouest
Aspects géotechniques du stockage des déchets dangereux dans les cavités de sel

Helm, D.C., Australie
Prédiction d'affaissements causés par le retrait des eaux souterraines dans la vallée de Latrobe en Australie

Johansson, S., Finlande
Excavation de grandes cavernes souterraines pour le stockage de pétrole à la raffinerie de Porvoo de la compagnie finlandaise Neste Oy

Kojima, K., Ohno, H., Japon
Relation entre les étendues proches et éloignés dans le contexte des fractures et/ou des chemins d'écoulement pour l'isolation géologique des déchets radioactifs

Maury, V., Sauzay, J.-M., Fourmaintraux, D., France
Approche géomécanique de la production pétrolière. Problèmes essentiels, premiers résultats

Nakagawa, K., Komada, H., Miyashita, K., Murata, M., Japon
Étude de conservation de l'air comprimé dans des poches rocheuses non revetues

Pahl, A., Liedtke, L., Kilger, B., Heusermann, S., Bräuer, V., Allemagne-Ouest
Recherche sur la permeabilité et sur l'état des contraintes dans la roche crystalline pour l'estimation de son rôle de barrière

Pyrak-Nolte, L.J., Myer, L.R., Cook, N.G.W., Witherspoon, P.A., États-Unis
Propriétés hydrauliques et mécaniques de fractures naturelles dans une roche peu perméable

Watanabe, K., Ishiyama, K., Asaeda, T., Japon
Instabilité de l'interface entre gaz et liquide dans le modèle d'une fracture ouverte

15:00 - 15:30 Pause-café (Foyer)

15:30 - 17:00 Discussion (Duluth/Mackenzie)

Modérateur: N. Barton (Norvège)

Panélistes: M. Langer (Allemagne de l'ouest)
J.A. Hudson (Angleterre)
P. Rissler (Allemagne de l'ouest)

M. Langer (Allemagne de l'ouest):

CALCUL DE L'EFFICACITÉ D'UNE BARRIÈRE ROCHEUSE POUR L'ISOLATION DES DÉCHETS

RÉSUMÉ: Selon le principe de plusieurs barrières l'environ géologique d'un stockage des déchets est un des plus importants facteurs pour l'isolation des déchets contre la biosphère. L'évaluation de la sûreté de la barrière géologique est démontrer seulement par des calculs avec des modèles géomécaniques et hydrogéologiques valides. La bonne idéalisation de la roche dans le modèle de calcul est la base pour des calculs réalistes des contraintes thermiques et des fissurations. L'analyse d'écoulement le long des fractures est nécessaire pour évaluer l'efficacité d'une roche cristalline.

Discussion pour le Thème I: de gauche à droite: P. Rissler, J.A. Hudson, N. Barton, M. Langer, G. Herget, W. Bawden.

P. Rissler (Allemagne de l'ouest):

RÉALITÉ ET MODÈLES HYDROGÉOLOGIQUES

RÉSUMÉ: Au cours des vingt dernières années, des modèles conceptuels de systèmes d'abattage et de forage hydraulique ont été développés. Ils reflètent l'approche de l'ingénieur qui doit toujours simplifier les procédés ou conditions complexes afin d'être en mesure de trouver des solutions dans des délais acceptables. Toute simplification se fait au dépens de l'attention au détail. À l'occasion, cela peut donc donner lieu à des discussions entre ceux qui croient que la simplification occasionne une perte des éléments essentiels et que la nature dans toute sa complexité ne peut être réduite à l'échelle d'un modèle.

Résumé du modérateur: N. Barton (Norvège)

Les articles soumis au Thème I sont regroupés en six sous-thèmes: modèles de réseaux (problèmes des tenseurs de perméabilité), modèles discrets de joints (comportement couplé), imperméabilisation des massifs rocheux (problèmes d'apport d'eau), problèmes de fluage et techniques de mesure (hydrauliques et géophysiques). Les sujets d'importance majeure soulevés dans chaque sous-thème sont énumérés. Une brève intervention de la part du modérateur est faite sur les thèmes de modèles discrets de joints, comportement couplé hydro thermo mécanique, et problèmes de validation des modèles.

1 INTRODUCTION

Des cinquante-deux articles soumis sous le thème I, seuls quatre ont fait l'objet de présentations orales. Inévitablement, le choix de ces articles intéressants s'est fait au détriment de plusieurs autres thèmes. La lacune la plus remarquée a été l'exclusion d'articles sur les modèles numériques couplés, domaine de grand intérêt lorsqu'il s'agit de résoudre des problèmes d'isolement des déchets nucléaires. Pour établir quelque peu la situation, le modérateur fait les commentaires suivants qui portent particulièrement sur certains des problèmes identifiés dans ce domaine.

2 LISTE DE SOUS-THÈMES ET D'AUTEURS

La liste d'auteurs et de sous-thèmes suivante éclairera les lecteurs des comptes-rendus (volumes 1 et 2). Quoique incomplète, elle indiquera au lecteur en quête d'articles sur des sujets particuliers à orienter ses recherches.

2.1 Modélisation de réseaux

* Elsworth et Piggot
* Dershowitz et Einstein
* Long, Billaux, Hestir, Chiles

Tenseurs de perméabilité

* Brown et Boodt
* Barbraux, Cacas, Durand, Fenga
* Oda, Hatsuyama, Kamemura

2.2 Modélisation/essais de joints discrets

* Pyrak-Nolte, Myer, Cook, Witherspoon
* Gentier
* Nakagawa Komada, Miyashita, Murata
* Watanabe, Ishiyama, Asaeda

2.3 Imperméabilisation des masses rocheuses (problèmes d'apport d'eau)

A. Roche fissurée

* Azevedo, Correa, Quadros, Cruz
* Gang, Tian, Quan
* Matsumoto, Yamaguchi
* Mineo, Hori, Ebara
* Semprich, Speidel, Schneider
* Johansson

B. Sel/potasse

* Charo, Zelikson, Berest
* Langer
* Crotogino, Lux, Rokahar, Schneider
* Schwab, Fischle
* Mraz

2.4 Modèles numériques couplés

* Chan, Scheir
* Curran, Carvalho
* Izquierdo, Romana
* Ohnishi, Kobayashi, Nishigaki
* Sato, Ito, Shimiza
* Shaffer, Heuze, Thorpe, Ingraffea, Nilson
* Thiercelin, Roegiers, Boone, Ingraffea
* Kuriyagawa, Matsunaga, Yameguchi, Zyvdloshi, Kelkar
* Charo, Zelikson, Berest
* Côme
* Crotogino, Lux, Rokahr, Schneider
* Langer

2.5 Problèmes de fluage

(i) Fluage du sel

* Balthasar, Haupt, Lempp, Natau
* Loanger
* Preece
* Charo et coll.
* Crotogino et coll.
* Mraz

(ii) Problèmes de stabilité des puits

* Maury, Sauzay, Fourmaintraux
* Rongzun, Zuhuri, Jingen
* Geunot

2.6 Méthodes de recherche (hydrauliques, géophysiques, etc.)

* By
* Pahl, Liedtke, Kilger, Heusermann, Bräur
* Barbreau, Cacas, Durand, Fenga
* Brown, Boodt
* Cornet, Jolivet, Mosnier
* Andrade
* Haimson
* Heystee, Raven, Belanger, Semec
* Azevedo, Correa, Quadros, Cruz
* Matsumoto, Yamaguchi
* Mineo, Hori, Ebara
* Agrawal, Jain

3 CONTRIBUTIONS DES PARTICIPANTS

1. Peter Rissler (Association des réservoirs de la Ruhr)
"Felshydraulische Modelle und Wirklichkeit, Ansätze für die weitere Forschung" (Les modèles hydrauliques de la roche et la réalité, base des recherches futures)

2. Prof. Michael Langer (BGR, Hanovre)
"Berechung der Barrierenwirksamkeit von Fels zur Abfallisolierung"
(Évaluation de l'efficacité des masses rocheuses comme barrières pour l'isolement des déchets)

3. John Hudson (Imperial College, Londres)
"Coupled mechanisms - the matrix approach to the overall engineering objective"
(Mécanismes couplés - l'approche matricielle dans l'atteinte de l'objectif global du génie)

4 PROBLÈMES IMPORTANTS ET QUESTIONS-CLÉS

Une courte liste de questions-clés est fournie ci-dessous pour attirer l'attention sur certains grands problèmes. Cette liste est inévitablement incomplète et subjective, mais aborde certains des problèmes soulevés lors de la lecture des comptes-rendus.

4.1 Modélisation de réseaux de joints (bidimensionnels et tridimensionnels)

a) Quelle est l'importance de l'erreur que nous introduisons lorsque nous ne modélisons pas des joints déformables?

b) Pouvons-nous modéliser les chenaux sur les arêtes de blocs de roche (intersections de joints)?

c) Est-il possible de détecter et de modéliser des super-conducteurs?

d) Quelle est l'utilité des tenseurs de perméabilité pour des milieux équivalents?

e) Qu'entendons-nous par volume d'essai représentatif?

4.2 Modélisation de joints discrets

a) Les modèles approchés de rigidité des joints linéaires sont-ils acceptables?

b) Quelle est l'importance de modéliser le cisaillement?

c) Combien de définitions d'ouverture de joints sont-elles nécessaires?

d) Surface de contact → effets de canal → effets d'échelle → loi cubique?

4.3 <u>Imperméabilisation des masses rocheuses</u>

a) À quel point les mesures de perméabilité indiquent-t-elles si une fondation fissurée peut être colmatée avec du coulis de ciment?

b) À quel point est-il possible d'imperméabiliser les cloisons dans le sel gemme?

c) À quel point est-il possible d'imperméabiliser les tunnels dans la roche fissurée?

d) Problèmes de migration de la saumure?

4.4 <u>Modèles numériques couplés</u>

a) Des études hydrothermomécaniques sont justifiées si les simplifications apportées à un processus introduisent des erreurs plus grandes que l'effet total du fait de les ignorer?

b) Quels sont les paramètres d'entrée clés et comment les évaluer?

4.5 <u>Problèmes de fluage</u>

a) Obturation des trous de sonde et effets de redistribution des contraintes?

b) Qu'est-ce en fait qu'une zone "plastique"?

c) Prévisions du fluage du sel - pourquoi ne sont-elles pas plus exactes?

4.6 <u>Techniques de mesure</u>

a) Essais hydrauliques - à quel point le résultat est-il représentatif de l'ensemble de la masse rocheuse?

b) Essais géophysiques - quelle est la résolution des détails géologiques obtenus?

5 DIFFICULTÉS DE MODÉLISATION

La modélisation de l'écoulement de fluides dans des masses rocheuses au moyen de réseaux rigides de plaques parallèles également espacées qui se recoupent représente malheureusement mal la réalité. L'ouverture, l'orientation et la continuité variables des joints, l'écoulement par chenaux, l'écoulement canalisé (le long d'intersections, en direction normale du plan bidimensionnel) et le couplage efficace entre les contraintes et les ouvertures sont parmi les facteurs qui rendent la simulation difficile.

Pour des raisons évidentes, il semble qu'il y ait divergence des tendances en matière de modélisation des joints dans la roche. Les hydrologues et les spécialistes en migration des radionucléides d'une part ont mis au point des modèles fascinants dont certains sont maintenant capables de simuler plusieurs des caractéristiques réelles qui sont jugées importantes, par exemple l'écoulement par chenaux et les super-conducteurs pour n'en nommer que deux. Les spécialistes en mécanique des roches d'autre part ont mis au point des modèles discontinus ou à éléments discrets qui tiennent compte de l'effet des variations des contraintes sur l'ouverture des joints dans la roche, qu'il s'agisse de contraintes de compression, de traction ou de cisaillement.

Comme l'ouverture (e) de plaques parallèles équivalentes à parois lisses est directement liée à la conductivité du joint (k) par la relation bien connue

$$k = e^2/12,$$

il semblerait en principe possible d'établir un lien direct entre k et la fermeture, l'ouverture et le cisaillement d'un joint en fonction des contraintes de compression, de traction et de cisaillement.

Il ne faut pas beaucoup d'imagination pour comprendre les limites actuelles de l'informatique lorsqu'on tente de coupler les meilleurs modèles hydrologiques de réseaux rigides tridimensionnels aux meilleurs modèles déformables de la mécanique des roches. Heureusement, il convient parfois de ne pas tenir compte d'un des processus dans un système couplé lorsqu'il peut être démontré que les variations (par exemple les variations thermiques) peuvent être faibles dans l'intervalle de temps considéré.

6 MODÉLISATION DE JOINTS DÉFORMABLES

Un exemple d'un modèle perfectionné de processus unique, actuellement exploité à l'ING, est donné à la figure 1. La simulation est réalisée au moyen du programme à éléments discrets UDEC (Cundall, 1980) dans sa version pour micro-ordinateur (μDEC). Ce programme comporte un nouveau sous-programme BB qui définit les caractéristiques non linéaires du comportement des joints à partir du couplage de la résistance au cisaillement, de la déformation et de la conductivité des joints (Barton, Bandis et Bakhtar, 1985). Le programme peut être utilisé pour des écoulements dans des joints couplés à condition de l'exécuter sur un gros ordinateur.

L'ouverture conductrice des joints représentés pour le tunnel de droite dans la figure ci-dessous est basée sur la surveillance continue des ouvertures mécaniques (E). Ces ouvertures de joints varient en fonction de la fermeture, de l'ouverture, du cisaillement et de la dilation causés par les travaux d'excavation. Les ouvertures mécaniques (E) sont converties en ouvertures conductrices (e) à partir des ensembles de données produits dans la figure 2. Ces inégalités entre (e) et (E) sont causées par la tortuosité des voies d'écoulement et par l'écoulement par chenaux autour des surfaces de contact dans le plan des joints; facteurs qui accentuent l'inégalité aux contraintes élevées.

Le programme comprend un sous-programme de calcul de l'écoulement d'un fluide dans les joints. Ce sous-programme n'a pas été utilisé dans la présente étude sur l'effet tunnel à cause du prégunitage et de la projection de béton renforcé de fibres; les premières mesures visaient à limiter les infiltrations d'eau (et les variations ultérieures de la pression du fluide) à de très petites valeurs.

Le but ultime sera d'intégrer un programme réaliste de calcul des joints non linéaires dans un programme à éléments discrets tridimensionnel qui modélise l'écoulement de l'eau et les gradients thermiques dans les joints. Ce but ne sera atteint que dans quelques années, principalement à cause du manque de données d'entrée générales (ou particulières).

7 PROBLÈMES DE COUPLAGE TOTAL

Le comportement des joints dans des conditions de couplage hydro thermo mécanique total est peu connu actuellement à cause du nombre limité d'études effectuées sur le sujet. L'essai réalisé sur un bloc chauffé à la mine expérimentale de la Colorado School of Mines par Terra Tek (Hardin et coll., 1982) a révélé que les lèvres des joints peuvent s'ajuster plus étroitement les unes aux autres lorsqu'on élève la température.

On ne sait si cet effet augmente aussi la résistance au cisaillement, mais cela semble probable. Le comportement pourrait être tel que la résistance au cisaillement augmente avec la température sous une charge purement normale, en conditions de conductivité réduite. Si les différences de contraintes dues au chauffage sont importantes au point de causer un cisaillement, la résistance au cisaillement, après un maximum, pourrait retourner à des valeurs ambiantes, ou à des valeurs inférieures (à cause de la perte de verrouillage et d'une diminution de la résistance des parois des joints à haute température). Les programmes d'essais in situ et les études de modélisation constitutives sont loin de produire les résultats escomptés, mais des étapes importantes ont été franchies dans la réalisation de tels essais de blocs.

8 PROBLÈMES DE VALIDATION DES PROGRAMMES

De récentes expériences de validation des codes effectuées à l'ING ont fait ressortir un dilemme. Lorsqu'un objet d'essai tel qu'un essai in situ doit être modélisé au moyen d'un nouveau code, et que l'objectif est de valider le code par comparaison à un ensemble de mesures, la question est de savoir jusqu'à quel point il est possible de simplifier l'objet d'essai et de le connaître au préalable?

La figure 4 illustre un modèle µDEC-BB simplifié de l'essai de bloc de la CSM dans lequel on utilise la pression du fluide pour solliciter le bloc fissuré et simuler ainsi la charge réelle exercée par un vérin plat. La simplification adoptée consistait à modéliser quatre blocs comportant seulement trois joints chacun. Il ne s'agit là que d'une fraction du nombre total de joints observés dans la surface des blocs. Cependant, les quatre blocs ont été identifiés par Richardson (1986) comme étant ceux qui comportaient des déplacements différentiels importants.

Dans les prévisions qui suivent, il est impossible de savoir quels blocs sont "actifs" dans une masse rocheuse donnée. Le problème exige donc que l'on trouve des méthodes améliorées pour distinguer à distance les joints continus des joints discontinus, ainsi que les joints lisses des joints rugueux. La combinaison des caractéristiques "continues" et "lisses" permet sans doute de mieux définir les joints actifs (susceptibles d'être cisaillés) que la combinaison "discontinue" et "rugueux".

Le besoin pressant de méthodes de recherches in situ qui permettront de distinguer les joints importants des joints non importants doit être manifeste, et y répondre constitue un véritable défi autant pour les géophysiciens que pour les hydrologues.

Liste de figures

Pour les figures, voir le texte anglais. Les légendes apparaissent en français à la fin.

Figure 1. Modèle à éléments discrets µDEC-BB des effets tunnels sur les déformations de blocs et les ouvertures conductrices de joints. (Makurat et coll., 1987).

Figure 2. Écart entre de l'ouverture réelle d'un joint (E) et l'ouverture conductrice théorique (e), dû à la tortuosité de la voie d'écoulement, aux surfaces de contact et à l'écoulement par chenaux. (Barton et coll., 1985).

Figure 3. Couplage hydro thermo mécanique dans un essai sur un bloc chauffé. Le comportement de l'ouverture conductrice d'un joint (e) montre que les parois du joint se marient mieux à 75 °C sous une charge normale pure. (Hardin et coll., 1982).

Figure 4. Contraintes principales qui s'exercent sous une charge biaxiale et sous une charge uniaxiale sur un bloc de roche fissurée. Barton et coll., 1987.

RÉFÉRENCES (voir section anglais)

Thème 2
Fondations et talus rocheux

Reconnaissance des sites et caractérisation des massifs rocheux; stabilité de talus et exploitations minières en fosse; fondations de ponts, d'édifices et de barrages; approche probabiliste à la conception; prévision, surveillance et rétro-analyse

Président: F.H. Tinoco (Vénézuela)
Secrétaire: P. Rosenberg (Canada)
Conférencier thématique: M. Panet (France)

Renforcement des fondations et des talus à l'aide d'ancrages actifs et passifs

RÉSUMÉ

Le renforcement des massifs rocheux par tirants précontraints est une technique bien connue et éprouvée. Les essais préliminaires et le contrôle des mises en tension provisoire et définitive permettent de s'assurer du bon fonctionnement de chaque tirant. Concurremment, le renforcement par barre passives est une technique qui s'est largement répandue grâce à un coût plus faible et qui a prouvé son efficacité sur de nombreux chantiers. Toutefois les conditions de l'interaction entre l'armature et le massif rocheux sont plus difficiles à analyser, et les errements actuels en matière de calcul ne sont pas entièrement satisfaisants. Par ailleurs, il n'existe pas de méthode de contrôle qui donne toute garantie sur la bonne exécution, le bon fonctionnement et la pérennité du renforcement. Il conviendrait de poursuivre les efforts de recherche dont l'objectif serait de définir les caractéristiques d'un massif de roche armée.

La stabilité des massifs rocheux ne peut être assurée dans bien des cas avec un coefficient de sécurité convenable que grâce à des mesures correctives telles que : modifications de la géométrie des déblais, drainage du massif, réalisation d'ouvrages de soutènement, ou renforcement de la masse rocheuse.

Le choix de la meilleure solution résulte d'une analyse de stabilité dans laquelle les paramètres de comportement des surfaces de discontinuité ont un poids essentiel. Dans les massifs rocheux, la rupture s'amorce par des décollements et des glissements au niveau des surfaces de discontinuité. Aussi toute méthode de renforcement qui tend à rétablir la continuité d'une masse rocheuse permet d'améliorer de manière très significative sa stabilité. Cet objectif peut être atteint par la mise en oeuvre de barres ou de câbles d'acier, et ce type de renforcement tend à se généraliser. Deux techniques sont en concurrence :

- le renforcement par aciers actifs ou précontraints
- le renforcement par aciers passifs.

L'examen des pratiques dans divers pays montre que les conceptions sont diverses et favorisent l'une ou l'autre de ces deux techniques. Elles présentent évidemment chacune des avantages et des inconvénients que nous nous proposons d'examiner à la lumière d'une analyse détaillée de leur fonctionnement dans un massif rocheux discontinu.

1 RENFORCEMENT PAR ARMATURES ACTIVES

Le renforcement des masses rocheuses par des tirants ou des barres précontraintes est une technique déjà ancienne. Des tirants de forte capacité (jusqu'à 10.000 kN) ont été mis en oeuvre par André Coyne et Eugène Freyssinet au cours des années trente.

Un tirant d'ancrage précontraint comprend (fig. 1) :

- la zone d'ancrage où le tirant est scellé au terrain
- un bouchon permettant d'isoler la zone de scellement de la partie libre en évitant les remontées de coulis le long du forage au moment de l'injection du scellement
- la longueur libre du tirant
- la tête d'ancrage et le massif d'appui avec le dispositif de blocage.

Les recommandations françaises distinguent :

- la traction limite T_1 qui correspond soit à la rupture de l'armature du tirant, soit à la rupture du scellement
- la traction T_G correspondant à la limite élastique de l'armature du tirant
- la traction admissible en service T_a

On doit observer :

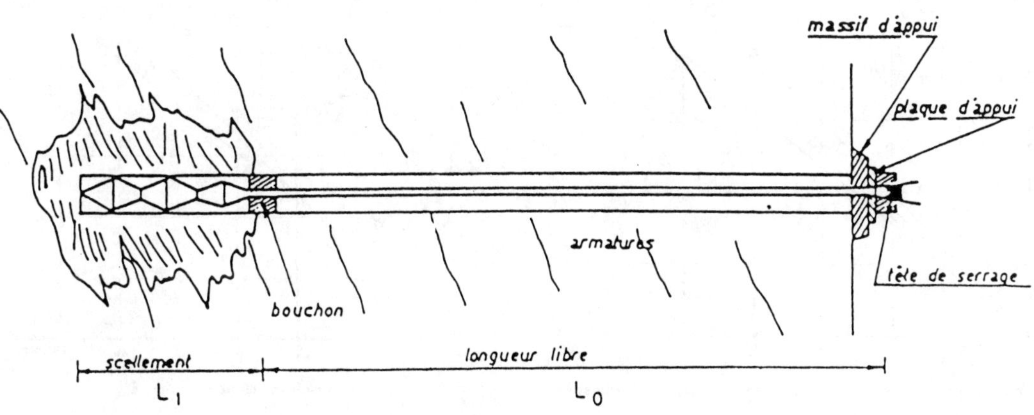

Fig. 1 - TIRANT D'ANCRAGE PRECONTRAINT - SCHEMA

$T_a \leqslant$ 0,6 T_G pour les tirants permanents

$T_a \leqslant$ 0,75 T_G pour les tirants provisoires d'une durée inférieure à 18 mois.

Il n'existe toujours pas de théorie convaincante donnant la traction qui conduit à la rupture d'un scellement. Mais les essais de conformité et les procédures de mise en tension permettent de vérifier le bon comportement des tirants.

Selon les recommandations françaises, la mise en tension s'effectue par paliers successifs jusqu'à la traction d'épreuve T_e telle que

T_e = 1,2 T_a pour les tirants provisoires

T_e = 1,3 T_a pour les tirants permanents.

Par ces essais on peut également contrôler la traction critique de fluage mais également la longueur libre équivalente.

L'effort apporté par la précontrainte s'oppose à l'ouverture des discontinuités ou augmente leur résistance au cisaillement ; dans le premier cas, il agit par sa composante normale et doit donc être orienté perpendiculairement aux discontinuités correspondantes ; et dans le second cas, la composante normale et la composante tangentielle interviennent ; il convient alors de déterminer son orientation optimale en tenant compte évidemment des possibilités réelles de mise en oeuvre (W. Wittke).

Les calculs de stabilité des massifs rocheux renforcés par des tirants précontraints sont simples dans leur principe puisqu'ils consistent en une analyse à l'équilibre limite dans laquelle on introduit les forces de précontrainte comme des forces extérieures de direction et d'intensité connues et constantes.

Mais il faut vérifier que ces hypothèses sont effectivement satisfaites dans la réalisation. Dire que les forces sont extérieures au volume à stabiliser signifie que les zones de scellement sont au-delà des surfaces de rupture potentielles, et en particulier que les longueurs libres des tirants ont été respectées. Des remises en tension de tirants montrent trop souvent que les longueurs libres sont beaucoup plus courtes qu'on ne le pensait du fait de scellements locaux intempestifs dûs à une maîtrise insuffisante des circulations de coulis.

De même les forces de précontraintes peuvent le plus souvent être considérées comme constantes puisque les déformations du massif postérieures à la mise en tension n'induisent qu'un déplacement relatif entre la zone d'ancrage et la tête d'appui très faible vis-à-vis de la longueur libre du tirant. Mais ce n'est pas toujours le cas, un déplacement de 1 cm pour une longueur libre de 20 m induit une variation de la contrainte de 100 MPa dans le tirant. Aussi lorsque des tirants sont mis en place avant que les travaux de terrassement ne soient totalement exécutés, il faut évaluer les déformations qui peuvent se produire afin de décider si les tirants doivent être mis provisoirement à une tension inférieure à la tension de service définitive (fig. 2). Une auscultation par extensomètres en forage est alors très utile. La remise en tension permet de vérifier une nouvelle fois la longueur libre du tirant, et d'évaluer l'évolution de la traction dans le tirant entre les deux phases de mise en tension. Vis-à-vis du comportement dans le temps des tirants précontraints, les ingénieurs ont deux préoccupations majeures : les pertes ou augmentation de tension par fluage relaxation, et la corrosion. Nous disposons désormais de suffisamment de recul pour affirmer que des tirants précontraints réalisés avec soin constituent des ouvrages sûrs et durables. A cet égard, il paraît intéressant de citer les constatations faites à Hong Kong et rapportées par Brien-Boys et Howells. Elles ont porté sur 214 ancrages correspondant à 17 projets et ont donné les principaux résultats suivants :

- la perte de tension moyenne sur 4 ans est de 9 %
- 16 % des ancrages ont une tension égale ou supérieure à 1,2 fois la tension de service.

L'observation des normes et recommandations existant dans plusieurs pays permet de réaliser une protection efficace contre la corrosion.

Il nous apparaît que la technique de renforcement par tirants précontraints peut être complètement maîtrisée par les ingénieurs compétents tant du point de vue de la conception que de la réalisation.

Cependant lorsque les masses rocheuses à stabiliser deviennent importantes, cette technique a un coût élevé. En effet pour améliorer de manière significative le coefficient de sécurité conventionnel issu des analyses de stabilité classiques, par exemple pour le faire passer d'une valeur de 1,1 à une valeur de 1,5, l'effort de précontrainte est souvent de l'ordre de 10 % à 25 % du poids de la masse rocheuse à stabiliser, ce qui est considérable.

2 RENFORCEMENTS PAR ARMATURES PASSIVES

Par renforcement passif, il faut entendre une armature qui est scellée au massif sur toute sa longueur et qui n'est pas mise en tension lors de sa mise en place. S'il ne se produit pas de déformation dans le massif renforcé, les aciers passifs ne sont pas sollicités.

A notre connaissance, un des premiers exemples d'utilisation de cette technique de renforcement de massif rocheux, est celui des appuis du barrage de Chaudanne en France.

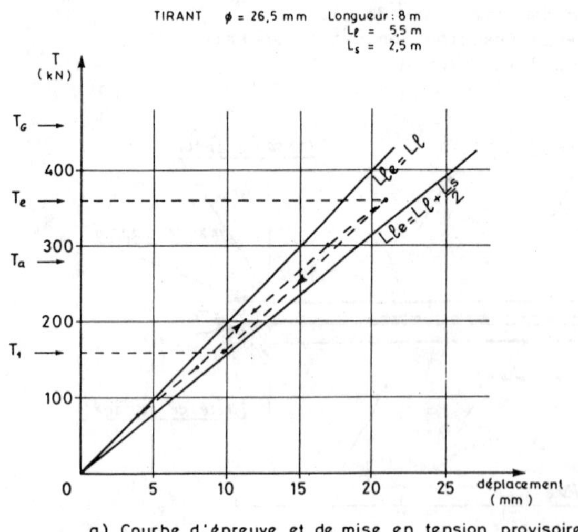

a) Courbe d'épreuve et de mise en tension provisoire

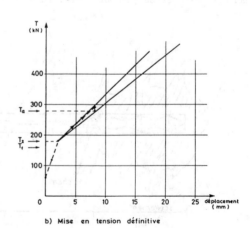

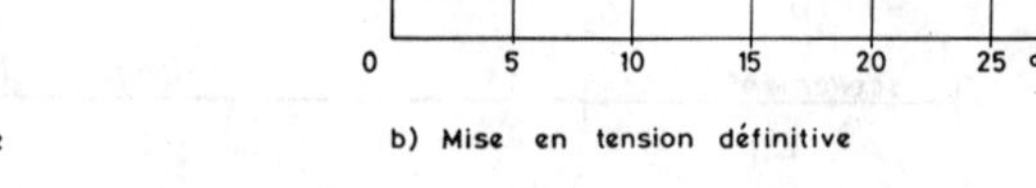
b) Mise en tension définitive

Fig. 2 - COURBES DE MISE EN TENSION D'UN TIRANT PRECONTRAINT EN DEUX PHASES

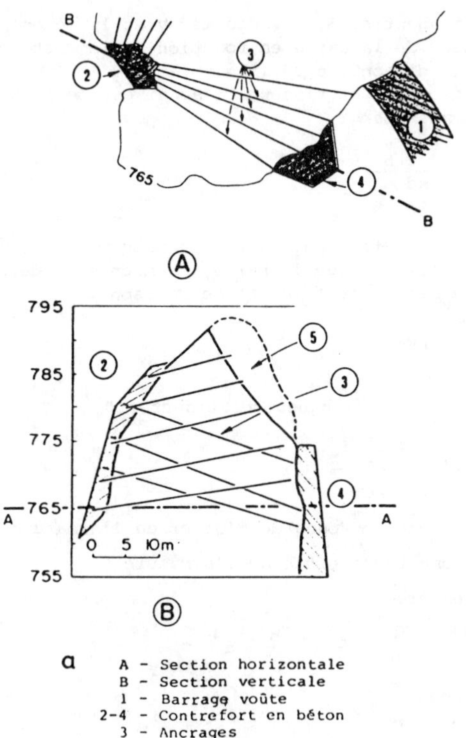

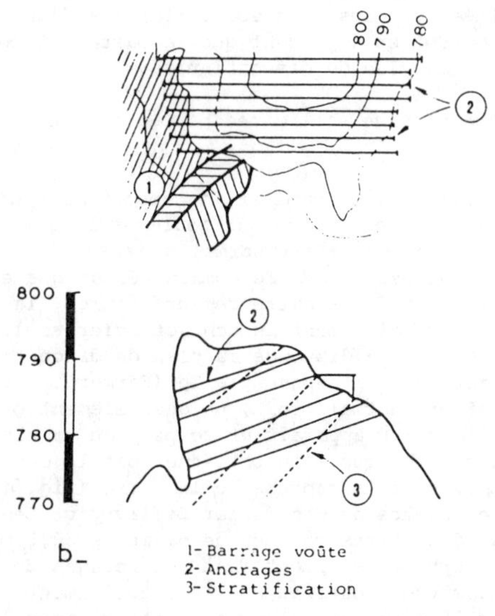

Fig. 3 - CONFORTEMENT DES APPUIS DU BARRAGE DE LA CHAUDANNE PAR ARMATURES PASSIVES

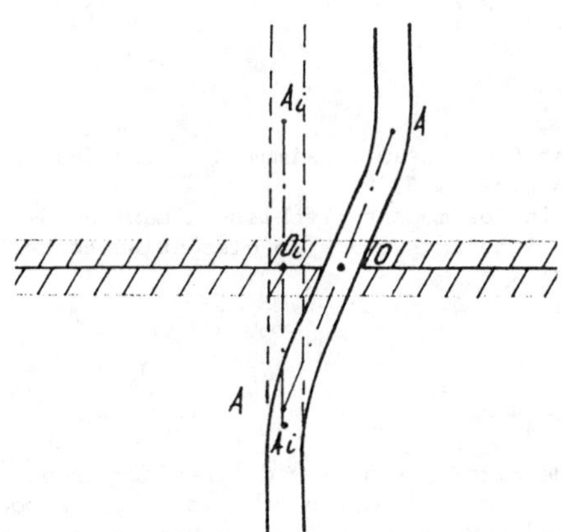

FIG. 4 - LOCALISATION DU POINT D'INFLEXION À 0 AVEC DÉPLACEMENT TANGENTIEL

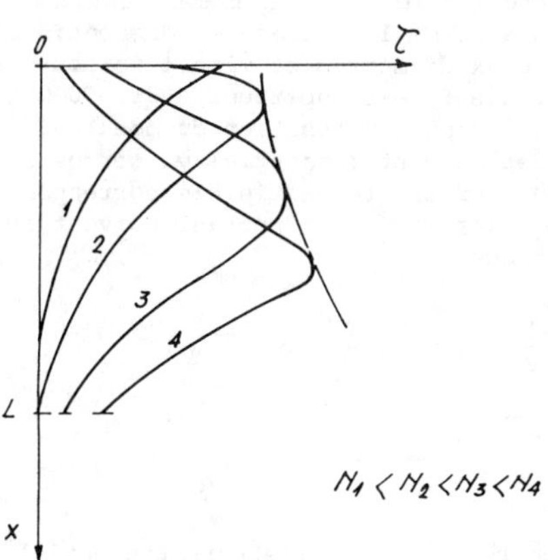

Fig. 5 - DESCELLEMENT PROGRESSIF D'UNE BARRE SCELLEE SOUMISE A UN EFFORT NORMAL DE TRACTION

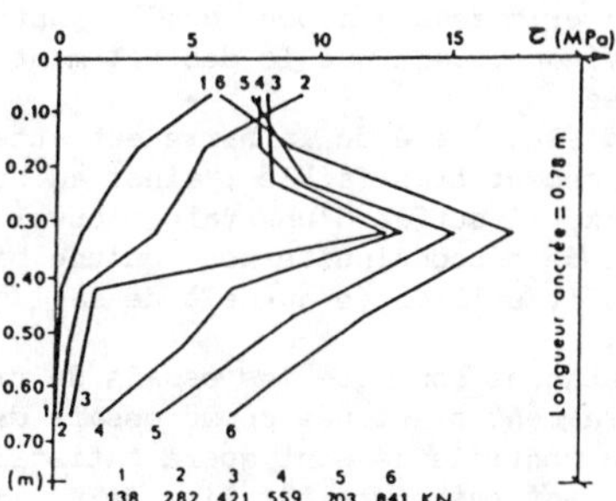

Fig. 6 - ESSAI DE TRACTION D'UNE BARRE SCELLEE DE 0,78 m DE LONGUEUR (CHARGE DE RUPTURE 950 kN) (d'après Ballivy et Benmokrane)

Le barrage de Chaudanne est une voûte mince de 70 m de hauteur construite entre 1949 et 1952 sur le Verdon dans le Sud de la France. La voûte s'appuie sur un massif calcaire affecté par des plans de discontinuité ayant un pendage de 45° dirigé vers l'amont et vers la rivière. En rive droite le confortement à l'aval de l'appui a été réalisé par des barres horizontales Ø 50 mm de 25 à 30 m de longueur et des contreforts en béton. En rive gauche, le massif a été consolidé par des câbles composés de 115 à 185 fils de Ø 5 mm inclinés à 15° sur l'horizontale. Ces cables n'ont pas été tendus mais injectés avec un coulis de ciment sur toute leur longueur. (fig. 3)

Le fonctionnement des armatures passives est beaucoup plus difficile à analyser que celui des armatures précontraintes, car ce type de renforcement crée un milieu composite avec des interactions locales entre la roche et les armatures, notamment au niveau des intersections avec les surfaces de discontinuité. Considérons le torseur des sollicitations d'une barre passive au point d'intersection 0 avec un plan de discontinuité soit :

No, la composante normale de la force au plan de discontinuité

So, la composante tangentielle de la force au plan de discontinuité

Mo, le moment fléchissant.

On peut faire l'hypothèse que Mo est nul car s'il se produit un déplacement tangentiel, le point d'inflexion de la barre est en O (fig.4). Aussi nous allons examiner successivement le comportement d'une barre scellée sous les deux sollicitations élémentaires No et So, en supposant que la barre est perpendiculaire au plan de discontinuité.

2.1 Armature passive sollicitee par un effort normal

Nous nous proposons d'examiner le comportement d'une barre de diamètre d scellée sur toute sa longueur et sollicitée par un effort normal N.

Tant que l'on reste dans le domaine élastique et que le scellement de la barre demeure intact, la contrainte de cisaillement au contact acier-scellement décroît avec la distance au plan de discontinuité suivant une loi exponentielle (Farmer). Mais lorsque l'effort normal croît, un descellement progressif de la barre apparaît et se propage en profondeur jusqu'à ce que l'on obtienne soit le descellement total, soit la rupture de la barre (fig.5)

Les études faites au Canada par Ballivry et Ben Mokrane sur des barres courtes de diamètre 36,8 mm scellés au mortier de ciment dans des forages de 76 mm de diamètre, montrent que le descellement local apparaît pour une valeur de l'effort normal qui n'est que de 25 % à 30 % de l'effort maximum (fig.6).

Pour des barres de diamètre compris entre 20 mm et 40 mm et un scellement correct de ciment ou à la résine, il suffit généralement d'une longueur de scellement inférieure à 2 m pour que la rupture de la barre survienne avant que le descellement total ne se produise.

La limite d'élasticité de la barre est atteinte pour un déplacement très faible ; ainsi au sein d'un massif rocheux, il suffit d'une faible ouverture d'une surface de discontinuité pour qu'une barre qui la traverse soit sollicitée au-delà de la limite d'élasticité.

Ces circonstances font que les essais de traction qui sont couramment pratiqués comme essais de conformité ou de contrôle ne sont guère satisfaisants. Ils ne permettent pas en particulier de vérifier si le scellement de la barre est effectif sur toute sa longueur. Aussi les essais ne donnent pas de garantie sur le bon fonctionnement de la barre au sein du massif, ni sur sa perennité dans le temps vis-à-vis de la corrosion pour les barres simplement scellées. Il y a là un inconvénient majeur en particulier dans les massifs à fracturation ouverte où les pertes de produit de scellement sont difficilement contrôlables. Il est alors souvent nécessaire de procéder à une injection préalable permettant d'obturer les discontinuités ouvertes.

2.2 Armature passive sollicitée par un effort tangentiel

Le fonctionnement d'une armature passive sollicitée par un effort tangentiel au plan de discontinuité est beaucoup plus difficile à analyser. Il a été l'objet de plusieurs études tant théoriques qu'expérimentales (Haas - Bjurstrom - Azuar - Dight- Egger)

Le mécanisme de l'interaction fait intervenir le comportement en flexion de l'armature et la résistance que la roche et le produit de scellement oppose au déplacement de celle-ci. Dans cet exposé, nous supposerons pour simplifier que la roche et le produit de scellement ont les mêmes caractéristiques mécaniques, et que la discontinuité est fermée sans produit de remplissage.

Si l'effort tangentiel est suffisamment faible pour que l'on reste dans le domaine élastique du milieu dans lequel la barre est scellée, une analyse classique permet de déterminer en fonction de l'effort tangentiel S, la déformée y et le moment fléchissant de la barre en fonction de la distance au plan de discontinuité (fig.7).

Le déplacement tangentiel au niveau de la surface est donnée par :

$$yo = \frac{2\ S}{kd\ \lambda}$$

où

k est le coefficient de réaction du milieu dans lequel est scellée la barre, de module de déformation E_m et de coefficient de Poisson ν_m

On peut poser
$kd = mE_m$

λ est appelé "longueur de transfert"

$$\lambda^4 = \frac{4\ E_b\ I_b}{kd}$$

E_b module de Young de l'acier de l'armature

I_b moment d'inertie de l'armature

soit encore

$$\lambda^4 = \frac{\pi}{16}\ \frac{1}{m}\ \frac{E_b}{E_m}\ d^4$$

En posant

$$\alpha^4 = \frac{\pi}{16}\ \frac{1}{m}\ \frac{E_b}{E_m}$$

on obtient

$$\frac{yo}{d} = \frac{2\ S}{mE_m \alpha\ d^2}$$

et

$$\lambda = \alpha\ d$$

Le moment fléchissant maximum M_{max} est égal à
$M_{max} = 0,322 \alpha\ dS$
et le point de moment fléchissant maximum est à une distance l de la surface de discontinuité telle que :

$$l = \frac{\pi}{4} \lambda$$

soit

$$\frac{l}{d} = \frac{\pi}{4} \alpha$$

Afin de fixer les ordres de grandeur, nous allons examiner le cas de barres en acier type précontrainte de module de Young $E_b = 2 \times 10^5$ MPa et dont la limite d'élasticité $\sigma_e = 840$ MPa. Les barres ont des diamètres compris entre 20 mm et 50 mm.

La figure 8 donne la variation du rapport $\frac{l}{d}$ en fonction de E_m en coordonnées semi-logarithmiques. Elle montre que le point de moment maximum se trouve à une distance de la surface de discontinuité inférieure à deux diamètres et même inférieure à un diamètre lorsque E_m est supérieur à 38.000 MPa.

La fig. 9 donne en fonction du module E_m la variation du déplacement tangentiel yo correspondant à une valeur de l'effort tangentiel Se correspondant à la limite d'élasticité en supposant suivant le critère de Tresca que :

$$Se = \frac{1}{2}\ Ne = \frac{\pi}{8}\ d^2\ \sigma_e$$

d'où

$$\frac{yo}{d} = \frac{\pi}{4}\ \frac{\sigma_e}{\alpha\ mE_m}$$

Pour cette valeur de l'effort tangentiel, le déplacement yo reste inférieur à $\frac{d}{4}$ et même inférieur à $\frac{d}{10}$ lorsque E_m devient supérieur à 10.000 MPa.

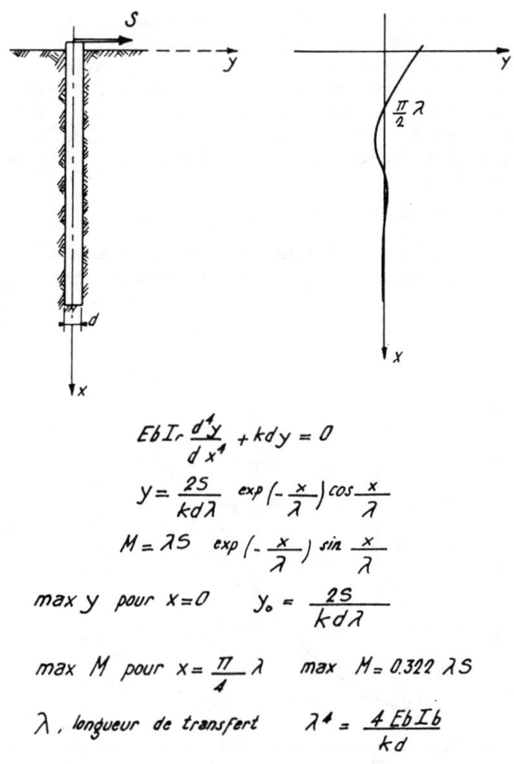

S
y
x
d
π/2 λ
$EbI_r \frac{d^4y}{dx^4} + kdy = 0$
$y = \frac{2S}{kd\lambda} \exp(-\frac{x}{\lambda}) \cos\frac{x}{\lambda}$
$M = \lambda S \exp(-\frac{x}{\lambda}) \sin\frac{x}{\lambda}$
max y pour x=0 $y_o = \frac{2S}{kd\lambda}$
max M pour $x = \frac{\pi}{4}\lambda$ max $M = 0.322\,\lambda S$
λ, longueur de transfert $\lambda^4 = \frac{4EbIb}{kd}$

Fig. 7 - COMPORTEMENT D'UNE BARRE PASSIVE DANS UN MILIEU ELASTIQUE SOUMISE A UN EFFORT TANGENTIEL

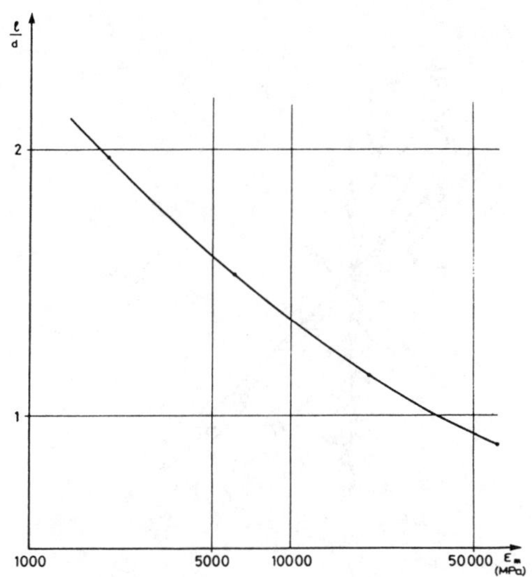

FIG. 8 - VARIATION DU RAPPORT 1/d EN FONCTION DU MODULE DE DÉFORMATION DE LA BARRE, Em

La fig.10 représente la variation du moment fléchissant maximum correspondant à la valeur de l'effort tranchant limite Se.

Le moment fléchissant maximum M_{max} est rapporté au moment plastique Mp de la barre d'acier.

Pour le moment plastique Mp, nous avons adopté l'expression proposée par Dight

$$Mp = 1,7 \frac{d^3}{32} \sigma_e$$

d'où

$$\frac{Mmax}{Mp} = 0,76\,\alpha \quad \text{pour } S = Se$$

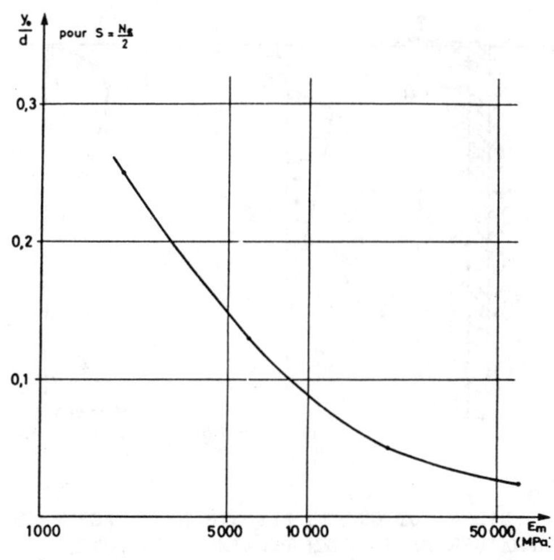

FIG. 9 - VARIATION DU DÉPLACEMENT TANGENTIEL Yo EN FONCTION DU MODULE Em CORRESPONDANT À UNE VALEUR DE L'EFFORT TANGENTIEL Se À LA LIMITE D'ÉLASTICITÉ

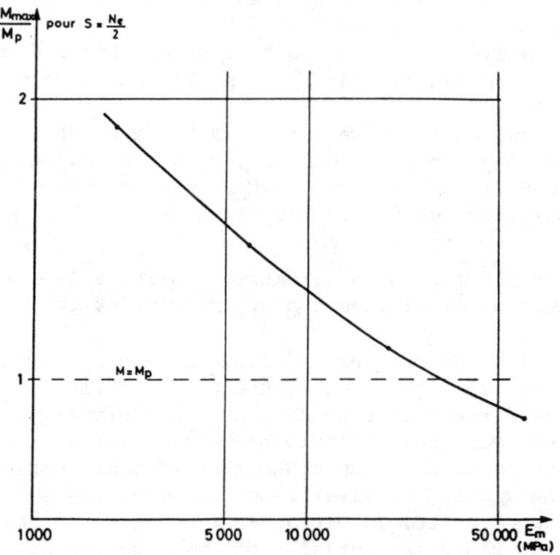

FIG. 10 - VARIATION DU MOMENT FLÉCHISSANT MAXIMUM CORRESPONDANT À LA VALEUR DE L'EFFORT TRANCHANT LIMITE Se

lorsque E_m > 32.000 MPa, M_{max} < Mp ; cela signifie que la valeur de l'effort tranchant est atteinte avant que n'apparaisse une rotule plastique.

Dight a étudié le comportement d'une armature passive dans le domaine plastique, et il suppose une distribution des efforts ainsi qu'il est indiqué sur la fig. 11

Le moment plastique Mp est atteint pour un effort tangentiel à la surface de discontinuité S_1 tel que

$$S_1 = \frac{d^2}{4} \sqrt{1,7\,\pi\,\sigma_e\,p_u}$$

la distance 1 du point du moment fléchissant maximum à la surface de discontinuité est donnée par :

$$\frac{1}{d} = \frac{1}{4} \sqrt{\frac{1,7\,\pi\,\sigma_e}{p_u}}$$

La détermination de la pression p_u entre la barre et le milieu dans lequel est scellée a été discutée par Dight. On peut admettre en première approximation la relation suivante entre p_u et la résistance en compression simple

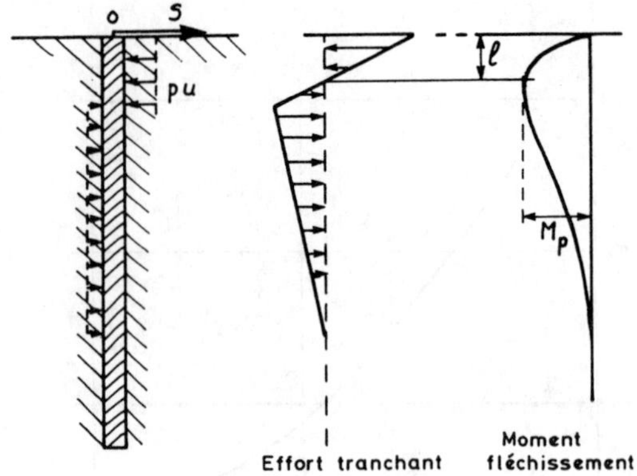

FIG.11 - DISTRIBUTION DES EFFORTS ET COMPORTEMENT D'UNE ARMATURE PASSIVE EN DOMAINE PLASTIQUE

$p_u = 6,5\ \sigma_c$

Dans ces conditions, la figure 12 donne la variation de $\frac{l}{d}$ en fonction d'une variation de la résistance en compression simple entre 6 et 200 MPa. Ce rapport varie entre 3 et 0,5.

Sur la figure 13 on a porté les variations du rapport $\frac{S_l}{Sa}$ en fonction de σ_c, dans le même domaine de variation. On remarque qu'au-dessus de 60 MPa, ce rapport devient supérieur à 1, ce qui signifie que l'effort tranchant limite est atteint avant que n'apparaisse une rotule plastique.

2.3 Contribution d'une armature passive à la résistance au cisaillement d'une discontinuité

Dans les paragraphes précédents, nous avons pu analyser le comportement d'une barre scellée dans une masse rocheuse soumise à des sollicitations simples. Mais les sollicitations d'une armature passive au droit d'une discontinuité sont complexes. Au cours d'un cisaillement suivant une surface de discontinuité, le déplacement n'a pas seulement une composante tangentielle Ut, mais également une composante normale Un à la discontinuité du fait des caractéristiques de dilatance de cette dernière.

Considérons une armature passive intersectant une discontinuité au sein d'un massif rocheux et faisant un angle θ avec la normale à la surface de discontinuité ; la résultante $\vec{R}$ des actions s'exerçant sur la section de la barre au droit de la discontinuité fait un angle β avec la direction de la barre. On peut décomposer $\vec{R}$ en un effort normal N parallèle à la barre et un effort tranchant S perpendiculaire à la direction de la barre (fig. 14a).

On admet généralement que la contribution de la barre à la résistance au cisaillement est donnée par l'expression :

$$C_b = R\cos(\theta+\beta)\ \mathrm{tg}\varphi + R\sin(\theta+\beta)$$

En pratique, les méthodes les plus couramment utilisées consistent à faire des hypothèses très simples.

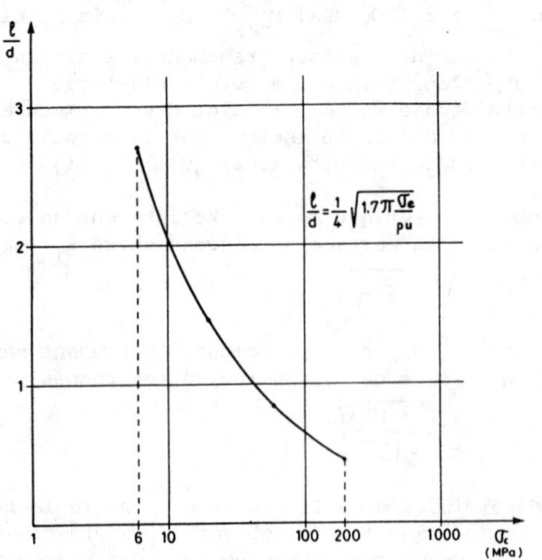

FIG.12 - VARIATION DU RAPPORT l/d EN FONCTION DE LA VARIATION σ_c ENTRE 6 ET 200 MPa

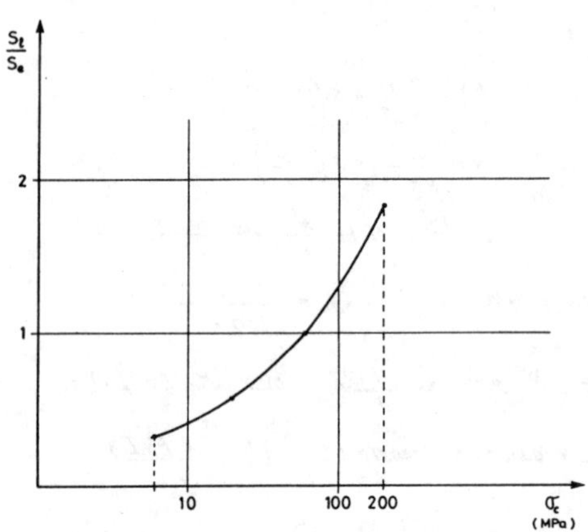

FIG. 13 - VARIATION DU RAPPORT Sl/Sa EN FONCTION DE σ_c

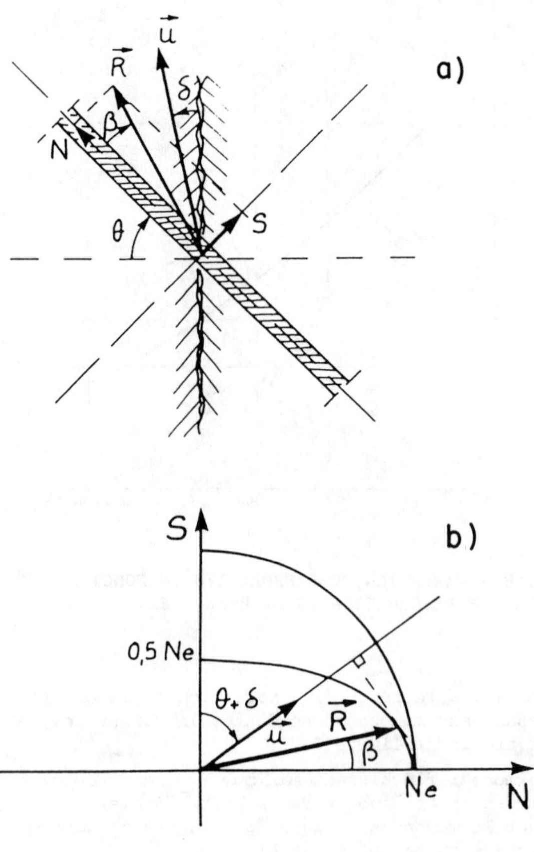

FIG. 14 - DÉCOMPOSITION DE LA RÉSULTANTE DES ACTIONS R EN EFFORT PARALLÈLE ET UN EFFORT PERPENDICULAIRE À LA DIRECTION DE LA BARRE

Dans une première méthode on suppose :

$\beta = 0$

$R = Ne$

La barre est sollicitée à sa limite élastique en traction parallèlement à sa direction (Gaziev). On prend souvent un coefficient de sécurité sur l'acier en prenant comme contrainte admissible, les 2/3 de la limite élastique (Colombet - Bonazzi).

La valeur de C_b est maximale si

$$\theta = \frac{\pi}{2} - \varphi$$

Dans une deuxième méthode, les hypothèses sont les suivantes :

$$\beta = \frac{\pi}{2} - \theta$$

$$R = \frac{Ne}{2}$$

Cette méthode est surtout utilisée lorsque θ est très faible ; la barre n'intervient que par sa résistance à l'effort tranchant et l'effet de dilatance n'est pas pris en compte. Cette méthode est a priori pessimiste, cependant on obtient des résultats voisins ou très peu supérieurs lorsqu'on fait des essais de cisaillement directs sur des discontinuités peu dilatantes avec des armatures ayant une direction normale au plan de cisaillement (Azuar - Groupe Français). La barre est effectivement cisaillée (fig. 15). Cela suppose également que les épontes rocheuses soient en contact sans interposition d'un matériau de remplissage. A cet égard, Haas a suggéré l'image d'un ciseau qui a trop de jeu et qui coupe mal.

Pour un déplacement de direction $\vec{u}$ faisant un angle δ par rapport à la surface de discontinuité, le principe du travail maximum nous dit que pour toute valeur licite $\vec{R}^*$ l'effort limite R doit vérifier la relation

$$(\vec{R} - \vec{R}^*).\vec{u} > 0$$

Par conséquent l'extrémité du vecteur R est le point de l'ellipse dont la tangente est perpendiculaire à $\vec{u}$

On en déduit :

$$\frac{R}{Ne} = \left[\frac{1 + \frac{m^2}{16}}{1 + \frac{m^2}{4}}\right]^{\frac{1}{2}}$$

$$\mathrm{tg}\,\beta = \frac{m}{4}$$

$$m = \mathrm{cotg}\,(\theta + \delta)$$

Nous avons tracé sur la figure 16 des abaques donnant $\frac{R}{Ne}$ et β pour des valeurs de l'angle δ : 0°, 5°, 10°, 20° et θ variant entre 0° et 60°. Elles montrent clairement que si θ est très faible, selon la valeur de δ, donc de la dilatance, $\frac{R}{Ne}$ est compris entre 0,5 et 0,8, ce qui est conforme aux résultats expérimentaux.

Par contre, si $\theta \geqslant 45°$, l'angle β est inférieur à 15°, et $\frac{R}{Ne}$ est supérieur à 0,9.

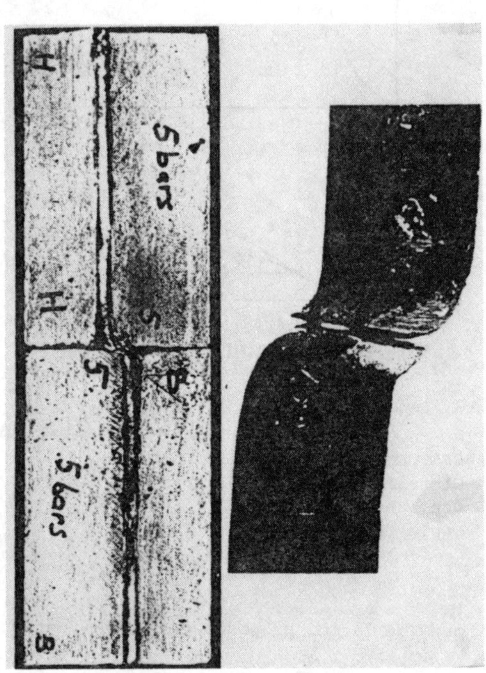

Fig. 15 - RUPTURE D'UNE BARRE PASSIVE PAR CISAILLEMENT AU COURS D'UN ESSAI DE CISAILLEMENT DIRECT

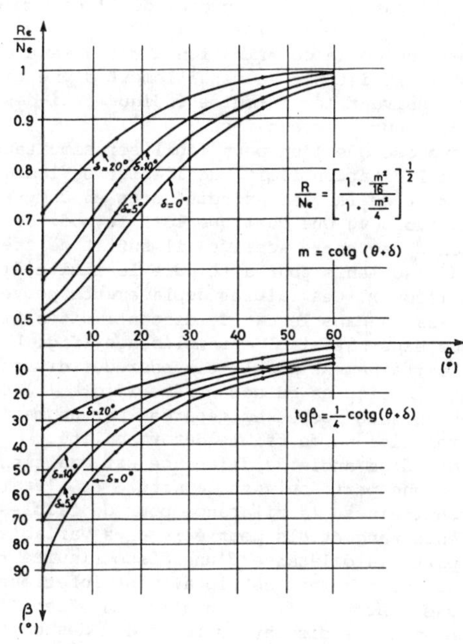

FIG. 16 - VALEUR DES RAPPORTS Re/Ne ET β AVEC δ = 0°, 5°, 10°, 20° ET θ = 0° À 60°

Une troisième méthode nous est fournie par l'application simple du principe du travail maximum (fig. 14b). Elle a été introduite dans certains programmes de clouage en mécanique des sols (Schlosser).

La sollicitation limite de l'armature R a son extrémité située sur l'ellipse

$$\left(\frac{N}{Ne}\right)^2 + \left(\frac{S}{0,5\ Ne}\right)^2 = 1 \quad \text{(critère de Tresca)}$$

Il est intéressant de déterminer l'orientation des barres θ qui conduit à la valeur maximale de C_b. Sur la figure 17 sont portées les variations de C_b en fonction de θ dans trois cas.

Cas A $\varphi = 30°$ $\delta = 0$

Cas B $\varphi = 40°$ $\delta = 5°$

Cas C $\varphi = 50°$ $\delta = 10°$

Dans ces trois cas, le maximum de C_b se situe

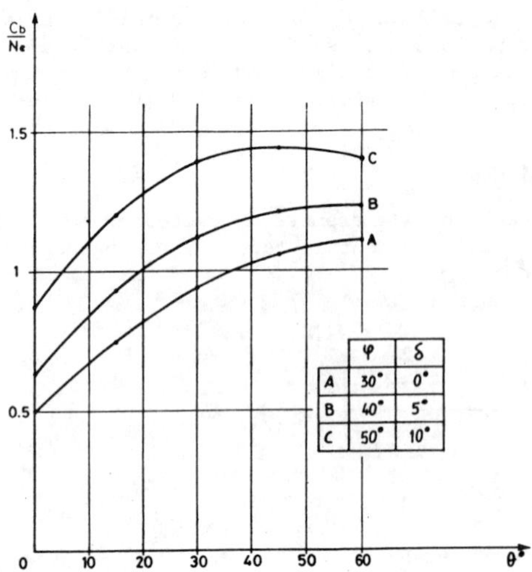

FIG. 17 - VALEUR DU RAPPORT Cb/Ne EN FONCTION DE θ POUR TROIS CAS DE γ - δ

respectivement pour $\theta > 60°$, $\theta = 60°$, $\theta = 45°$

Plus δ et φ sont faibles, plus l'angle θ correspondant au maximum de C_b est élevé.

Mais il convient de remarquer que cette optimisation ne correspond pas nécessairement à une optimisation économique dans laquelle intervient également la longueur des barres et qui doit tenir compte des possibilités pratiques de mise en oeuvre. Il s'agit de cas d'espèces. Mais dans de nombreux cas, un angle $\theta = 45°$ est une valeur proche de l'optimum économique.

L'évaluation de la contribution d'une armature passive à la résistance au cisaillement d'une discontinuité suivant les formules indiquées ci-dessus appelle plusieurs commentaires :

a) elle suppose que l'on peut mobiliser simultanément la sollicitation limite de l'acier et l'angle de frottement de la discontinuité pris en compte.

Or nous avons vu que quel que soit le mode de sollicitation d'une barre scellée, il suffit de très faibles déplacements pour atteindre la limite élastique de l'acier. Ces faibles déplacements peuvent être suffisants dans le cas de discontinuités très imbriquées sans matériau de remplissage (fig. 18a), ce n'est certainement pas le cas pour des discontinuités mal imbriquées ou desserrées (fig.18b) ou ayant les épontes rocheuses séparées par un matériau de remplissage de faible déformabilité.

b) le choix d'un angle de dilatance est également délicat. D'une part, suivant la nature et l'état des discontinuités, la dilatance pour de faibles déplacements tangentiels peut être très variable. D'autre part, la dilatance d'une discontinuité renforcée par une armature est localement inférieure à celle d'une discontinuité naturelle. En effet, la dilatance dn d'une discontinuité dépend des contraintes normales qui s'exercent sur la discontinuité.

Barton admet la loi

$$d_n = 0{,}5\ K \log \frac{\sigma c}{\sigma n}$$

$$\text{si} \quad 1 < \frac{\sigma c}{\sigma n} < 100$$

K - coefficient lié à la nature de la discontinuité et à la géométrie des épontes (K varie entre 0 et 20)

σc - résistance en compression simple de la roche constituant les épontes

σn - contrainte normale

Pour des valeurs faibles de σ_n, $\frac{\sigma c}{\sigma n} > 100$

$dn = dn° = K$

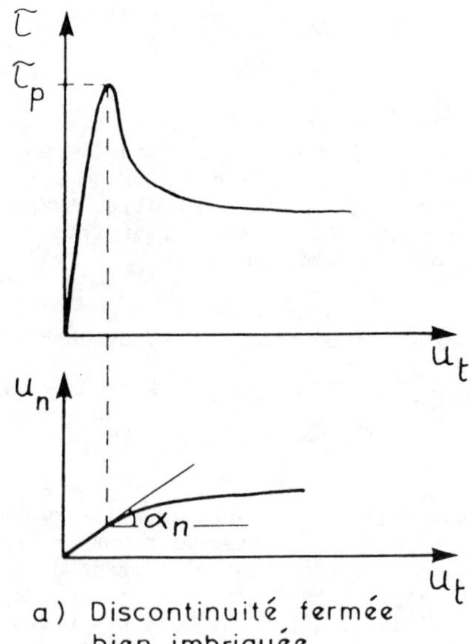

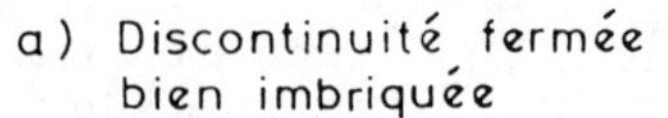

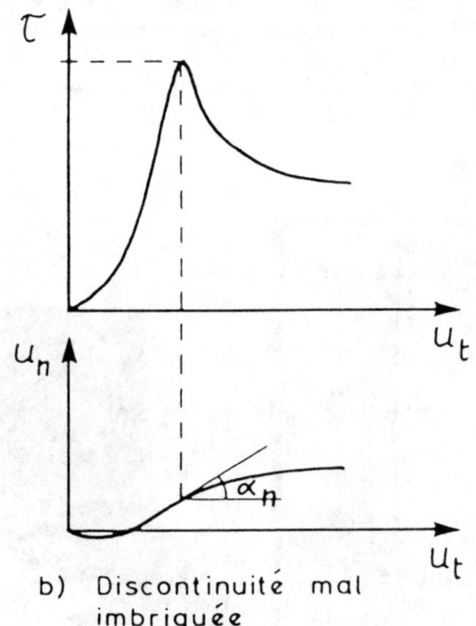

Fig. 18 - COURBES DE CISAILLEMENT DE DISCONTINUITES ROCHEUSES CONTRAINTE NORMALE CONSTANTE

Contrainte de cisaillement

Déplacement tangentiel

Déplacement normal

La forme de la loi proposée par Gaziev est différente

$$d_n = d°_n \left[1 - \frac{\sigma_n}{\sigma_c}\right]^{K}$$

Etant donné l'influence du paramètre δ dans les analyses faites ci-dessus, il convient d'être très circonspect et il est recommandé de choisir une valeur de δ inférieure à l'évaluation qui peut être faite de la valeur d_n avec les méthodes usuelles.

c) enfin les analyses ci-dessus supposent que le scellement de la barre est efficace. Si le scellement est insuffisant, il se produit un descellement progressif de la barre qui se déforme peu à peu avec deux rotules plastiques s'éloignant l'une de l'autre (fig. 19). Dans ce cas, la contribution de la barre à la résistance au cisaillement devient beaucoup plus faible (Azuar & al).

Fig. 19 - DESCELLEMENT PROGRESSIF D'UNE BARRE PASSIVE AU COURS D'UN ESSAI DE CISAILLEMENT DIRECT

2.4 Conception générale du renforcement d'un massif rocheux par armatures passives

A la lumière des analyses présentées rapidement ci-dessus, les principes de renforcement d'un massif rocheux apparaissent clairement.

a) Il suffit de faibles déplacements au droit des discontinuités pour que des tirants passifs soient localement sollicités à leur limite élastique. Ces déplacements sont d'autant plus faibles que le massif est plus rigide, les discontinuités plus serrées et par conséquent plus dilatantes. Il est donc possible en armant un massif rocheux au fur et à mesure d'une excavation de bloquer pratiquement les déformations au niveau des discontinuités.

b) Une analyse de la cinématique des déformations des modes de rupture potentiels et une appréciation des caractéristiques des discontinuités permettent d'orienter correctement les armatures et d'optimiser un projet de renforcement.

c) Il convient de veiller tout particulièrement au bon scellement des barres de part et d'autre des discontinuités ; cela est d'autant plus difficile que les discontinuités sont ouvertes et le massif décomprimé. Si les essais de traction permettent de juger de la qualité du coulis et du mode de scellement, ils ne donnent pas de garantie sur la totalité du scellement.

d) Dans de grandes excavations où les déformations ne peuvent pas être négligées, les sollicitations des barres passives peuvent dépasser la limite élastique et entraîner des descellements locaux. Si ce mode de fonctionnement est tout à fait acceptable pour des ouvrages temporaires, il convient d'être beaucoup plus prudent pour les ouvrages permanents du fait des risques de corrosion. Une auscultation par extensomètres en forage du massif renforcé constitue la meilleure manière d'analyser son comportement pendant les travaux d'excavation, la mise en service et la vie de l'ouvrage. En fin de travaux, il peut s'avérer nécessaire de compléter le renforcement de manière à assurer la sécurité définitive de l'ouvrage. Cependant, dans de grandes excavations qui atteignent parfois quelque cent mètres de hauteur, les parties les plus hautes sont quasiment inaccessibles à un coût raisonnable, si les accès n'ont pas été maintenus au fur et à mesure de l'excavation.

e) Pour des raisons qui tiennent au comportement de l'ouvrage : meilleure homogénéïté de la répartition des contraintes et à la sécurité de l'ouvrage : plus faible probabilité de ruine, il convient de préférer, à poids égal d'acier, un nombre plus élevé de barres, à des barres de section plus forte, même si cette disposition conduit à un prix plus élevé du fait de l'accroissement du coût de foration.

Le succès de l'emploi des boulons à ancrage répartis dans le domaine des ouvrages souterrains a gagné celui des ouvrages superficiels où l'emploi de renforcement par barres passives est de plus en plus fréquent. En 1974, au congrès de Denver, P. Londe et D. Bonazzi introduisent le concept de "roche armée" Depuis cette date des progrès notables ont été réalisés dans l'étude du fonctionnement d'une barre passive au sein du massif rocheux. Toutefois le concept n'est pas complètement éclairé. Le praticien souhaiterait pouvoir effectivement disposer d'une méthode pour déterminer les caractéristiques mécaniques globales : déformabilité, conditions de rupture, d'un massif renforcé qu'il pourrait par exemple introduire dans un modèle numérique.

Les recherches sur les matériaux composites sont très actives dans de nombreux domaines. A ma connaissance, les recherches faites sur les massifs renforcés ne sont pas suffisantes, et pourtant les quelques tentatives qui ont été faites à partir de modèles de blocs sont très séduisantes.

3 CONCLUSIONS

Dans leur conception, leur mode de fonctionnement, dans la manière d'aborder les problèmes de sécurité et de pérennité, les renforcements des massifs rocheux par tirants précontraints et barres passives sont très différentes.

Les opinions exprimées dans cette conférence sont évidemment celles de son auteur, mais il tient à exprimer ses remerciements à tous ses collègues des centres de recherche et des bureaux d'études français avec lesquels il a travaillé depuis de nombreuses années sur ce thème et qui lui ont apporté de nombreuses réflexions tellement pertinentes qu'il les a fait siennes.

BIBLIOGRAPHIE

Azuar, J.J. 1977 Stabilisation des massifs rocheux fissurés par barres d'acier scellées - Rapport de Recherche n° 73 - Laboratoire Central des Ponts et Chaussées Paris

Azuar, J.J., Panet, M. 1978 Le comportement au cisaillement des aciers passifs dans les massifs rocheux. Industrie Minérale. Document SIM B4 pp 93-98

Azuar & al 1979 Le renforcement des massifs rocheux par armatures passives. Proc. 4th ISRM Cong. Montreux. Vol. 1 pp. 23-29

Ballivy, G., Benmokrane, B. 1984 Les ancrages injectés et la consolidation des parois rocheuses Int. Conf. Renforcement en place des sols et des roches. ENPC. Louisiana State University Paris pp. 213-218

Bjurstrom, S. 1974 Shear strength of hard rock joints reinforced by grouted untensioned bolts Proc. 3rd Cong. ISRM. Denver, Vol. 2B, pp 1194-1199

Bonazzi, D., Colombet, G. 1984 Réajustement et entretien des ancrages de talus. Int. Conf. Renforcement en place des sols et des roches ENPC. Louisiana State University. Paris pp. 225-230

Descoeudres, F., Gencer, M. 1979 Rupture progressive en versant rocheux stratifié et fissuré. Proc. 4th ISRM Cong. Montreux, Vol. 1, pp 613-619

Dight, P.M. 1983 Improvements to the stability of rock slopes in open pit mines. Ph.D Thesis. Monash University Australia, 348 p.
Dight, P.M. 1985 The theoretical behaviour of full contact bolts subject to shear and tension. Int. Symp. on the Role of Rocks Mechanics. Zacatecas Mexico, pp 290-297
Egger, P., Fernandes, H. 1983 Nouvelle presse triaxiale. Etude de modèles discontinus boulonnés. Proc. 5th ISRM Cong. Melbourne, pp A171-A175
Gaziev, E.G., Lapin L.V. 1983 Passive anchor to shearing stress in a rock joint. Proc. Int. Symp. on Rock. Abisko, pp 101-108
Haas, C.J., & al 1978 An investigation of the interaction of rock and types of rock bolts for selected loading conditions. University of Missouri Rolla. US Bureau of Mines PB 293988
Heuzé, F.E. 1981 Analysis of bolt reinforcement in rock slopes. 3rd Int. Conf. on Stability in Surface Mining. Vancouver
Hoek, E., Londe, P. 1974 The design of rock slopes and foundations. General Report. Proc. 3rd ISRM Cong. Denver
Londe, P., Bonazzi, D. 1974 La roche armée. Proc. 3rd ISRM Cong. Denver
Panet, M., 1978 Stabilisation et renforcement des massifs rocheux par aciers précontraints et aciers passifs. Atti del Seminario sur consolidamento di terreni e roce in posto nell' ingegneria civile. Stresa pp 33-62
Schlosser, F., Jacobsen, H.M., Juran, I. 1983 Soil reinforcement. General Report 8th European Conf. on Soil Mechanics and Foundation Engineering. Helsinki, pp 83-104
Wittke, W. 1965 Verfahren zur Berechnung der Standsicherheit belasteter und unbelasteter Felsböshungen. Rock Mechanics and Engineering Geology. Vol. 30, Suppl. II, pp 52-79

9:20 Conférences brefs (Le Grand Salon)

Carrere, A., Nury, C., Pouyet, P., France
Un modèle tridimensionnel non linéaire aux éléments finis contribue à évaluer la convenance du site géologiquement difficile de Longtan en Chine au support d'une voûte de 220 mètres de hauteur.

Egger, P., Spang, K., Suisse
Études de stabilité d'une fondation de barrage améliorée par ancrages passifs

Irfan, T.Y. Koirala, N., and Tang, K.Y., Hong Kong
Défaillance complexe d'une pente dans un massif rocheux très altéré

Marchuk, A.N., Khrapkov, A.A., Zukerman, Y.N., Marchuk, M.A., Russie
Effets de contact à l'interface fondation-barrage en béton avec une centrale située au pied du barrage

10:00. FOYER, Pause-Café

10:30.

Martin, D.C., Canada
Utilisation du renforcement de roches et du soutènement artificiel dans les mines à ciel ouvert

Nilsen, B., Norvège
Flambement flexural de roche dure - un genre de rupture en pentes abruptes?

Rosengren, K.J., Friday, R.G., Parker, R.J., Australie
Le pré-placement de câble d'ancrage pour le renfort des talus dans les mines à ciel ouvert

Rowe, R.K., Armitage, H.H., Canada
Méthode de conception pour l'emplacement de piliers forés dans une masse rocheuse molle évaluée pour trois cas documentés

Savely, J.P., États-Unis
Analyse de probabilité de masses rocheuses soumises à de fortes fractures

Serra de Renobales, T., Espagne
Flambage des strates dans l'éponte inférieure d'une mine de charbon

Sharp, J.R. Bergeron, M., Ethier, R., U.K., Canada
Excavation, renforcement et surveillance d'un mur d'ardoise d'une hauteur de 300m

Sueoka, T., Muramatsu, M., Torii, Y., Shinoda, M., Tanida, M., Narita, S., Japon
Étude à coût réduit et construction d'un grand talus excavé dans la pélite

Widmann, R., Promper, R., Autriche
Les effets de la roche de fondation sur les contraintes agissant près de la base d'un barrage en voûte

12:00 - 13:30 Pause déjeuner

13:30 - 15:00 Discussion (Le Grand Salon)

Modérateur: K. Kovari (Suisse)

Panélistes: B.K. MacMahon (Australie)
R. Ribacchi (Italie)
R. Yoshinaka (Japon)

Résumé du modérateur: K. Kovari (Suisse)

Prise de décisions en génie des talus et des fondations

Les talus et les fondations aménagées dans des massifs rocheux sont très variables quant à leur taille et aux risques qu'ils présentent. Les tranchées creusées pour la construction des routes et des voies ferrées ne dépassent en général pas quelques dizaines de mètres, sinon il serait plus rentable de creuser un tunnel. Les grands barrages peuvent atteindre 300 m de hauteur, mais le risque que comporte un tel choix dépend de la capacité des réservoirs. En outre, l'exploitation minière à ciel ouvert se fait à des échelles de plus en plus grandes, ce qui se traduit par des talus de plusieurs centaines de mètres de hauteur. L'enjeu dans ce dernier cas est toutefois complètement différent de celui dont il est question dans l'aménagement d'un barrage. Enfin, les glissements de terrain, comme celui de Vajont, peuvent aussi poser un problème important, particulièrement s'ils se produisent en bordure du réservoir d'un barrage. Les dimensions et les risques en jeu déterminent en grande partie la démarche de l'ingénieur ainsi que les conditions d'acceptation de tels ouvrages de génie. Il est évident que plus les masses rocheuses sont grandes, plus il est difficile d'en modifier le comportement et, dans certains cas, la seule solution consiste à abandonner un emplacement ou à éviter de modifier les caractéristiques de la masse rocheuse, par exemple le niveau d'un réservoir. L'efficacité d'un ouvrage de surface peut être exprimée quantitativement au moyen de paramètres tels que : facteur de sécurité, probabilité de rupture, contraintes exercées dans le béton et fuites d'eau. Les fuites d'eau peuvent aussi influer sur la stabilité du barrage.

Mesures pour l'aménagement de talus rocheux

Il faut d'abord tenir compte de l'angle de la pente qui peut avoir d'importantes conséquences en matière d'économie et de sécurité. Pendant la construction, les travaux d'excavation doivent progresser sans retard indû et sans effondrement; dans le cas de gros talus, la seule mesure corrective qu'on puisse prendre consiste à contrôler l'eau en surface et dans la masse rocheuse. Il est possible d'abaisser le niveau de la nappe phréatique ou de diminuer la pression de l'eau dans les zones potentielles de glissement de terrain en forant des trous et en aménageant des galeries de drainage. Le contrôle de la hauteur piézométrique indiquera si ces mesures sont efficaces. En matière de renforcement, la tendance est aux boulons d'ancrage entièrement cimentés qui sont relativement peu coûteux et qui résisteraient mieux à la corrosion que les systèmes actifs. L'effet statique des boulons d'ancrage actifs est simple à comprendre; chaque boulon exerce une paire de forces dont la grandeur peut être réglée en tout temps. Les boulons d'ancrage cimentés ne sont efficaces qu'une fois la masse rocheuse déformée. Si les déformations ne sont concentrées que dans quelques joints, les contraintes se concentreront en ces points dans des boulons d'ancrage cimentés, tandis que les contraintes dans le boulon d'ancrage actif n'auront pas augmenté sensiblement. Il est définitivement nécessaire de pousser la recherche sur le comportement des boulons d'ancrage passifs. Il arrive souvent qu'on doive écarter l'aménagement d'un talus au profit du creusement d'un tunnel.

Mesures pour la construction de fondations de barrage

Ces mesures consistent à éliminer la roche tendre, ce qui entraîne souvent des problèmes de stabilité des talus dans les vallées à versants abrupts. La cimentation de consolidation vise à augmenter la rigidité et la résistance de la masse rocheuse, tandis que l'application de couches de ciment vise à diminuer l'infiltration dans la roche. Une des plus grandes opérations de cimentation que l'histoire ait connues a été réalisée récemment au barrage d'El Cajon, au Honduras, qui a comporté l'aménagement de 11 km de galeries de cimentation et le forage de trous d'une longueur totale d'environ 500 km. Il est extrêmement difficile de prévoir les besoins de cimentation dans les zones karstiques parce que, dans certains cas, il faut effectuer 60 % plus de travaux que prévu. Aux problèmes de stabilité des talus s'ajoute souvent celui du glissement de la masse de béton elle-même qu'il faut à ceux sûr prévenir. Dans certains cas, la pose de boulons d'ancrage entièrement cimentés peut être une solution. Dans d'autres cas, un système de galeries est aménagé et rempli de béton de façon à établir des clés de cisaillement. Ces dernières visent parfois à augmenter la rigidité de la masse rocheuse.

Prise de décisions

Dans les ouvrages en surface aménagés dans la roche, il y a tant de décisions de grande portée à prendre qu'il vaut la peine d'examiner de plus près le processus de prise de décisions. Il arrive souvent que la situation soit considérée si "évidente" que les décisions sont prises simplement en fonction de l'expérience sans que le raisonnement n'intervienne. Tout à l'opposé, certaines situations sont si "complexes" que l'expérience seule ne semble pas suffire et qu'il faut recourir à ce qu'on appelle la "boîte à outils" de la mécanique des roches. Avant d'examiner de plus près le contenu de cette boîte, il faut souligner qu'une situation est souvent considérée "évidente" seulement parce qu'on ne possède pas les connaissances nécessaires en mécanique des roches. De toute façon, les notions de situations "évidentes" et de situations "complexes" sont plutôt subjectives et les avantages que présentent cette "boîte à outils" sont aussi déterminés en grande partie par l'expérience des ingénieurs responsables des travaux. Les "outils" peuvent aider à évaluer

- les propriétés pertinentes de la masse rocheuse
- les conditions des eaux souterraines
- le facteur de sécurité
- la déformation des ouvrages.

Il a déjà été mentionné que, malgré la disponibilité d'outils puissants, il faut faire preuve de bon jugement à tous les niveaux de la démarche ainsi que dans l'évaluation des résultats. Idéalement, toutes les décisions devraient être prises avant la construction pour que les documents contractuels soient préparés en bonne et due forme, pour que le choix se porte sur du matériel approprié et pour que les calendriers soient au point. Cette démarche ne peut souvent pas être suivie de sorte que certains prônent la philosophie de la "planification au fur et à mesure" qui est certes applicable à un grand nombre d'exploitations à ciel ouvert, notamment lorsque le propriétaire et l'entrepreneur font partie de la même organisation. Pour en revenir à la "boîte à outils" de la mécanique des roches, ne considérons qu'un seul aspect du comportement des matériaux. Pour les problèmes de stabilité des talus ou des barrages contrôlés par les joints, le potentiel de déformation des discontinuités peut être un paramètre crucial. Lorsque ce potentiel est faible, la rupture se fera en mode fragile et des déplacements peu importants au départ pourront entraîner une perte importante de résistance et de sécurité. Dans ce cas, les boulons d'ancrage avec précontrainte sont vraiment plus efficaces que les boulons passifs, et il est à conseiller d'introduire dans les calculs des valeurs de résistance résiduelle plutôt que des maximums. Par ailleurs, un grand potentiel de déplacement propice à des déformations élastiques, donc à des ruptures progressives, pourrait nécessiter la pose de boulons d'ancrage entièrement cimentés pour permettre l'étude des paramètres de résistance maximale. Un aspect peu connu de cette question porte évidemment sur l'évaluation des propriétés des matériaux à l'échelle des ouvrages à l'étude! Un autre problème persistant en mécanique des roches se rapporte à l'effet des discontinuités sur la déformabilité d'une masse rocheuse. Une façon d'aborder ce problème expérimentalement consiste à effectuer des essais de charge à grande échelle (par exemple, un essai d'arrachement de câble) et à observer le profil de déformation le long d'un trou de sonde dans la direction de la charge appliquée. Les pics distinctifs dans la courbe de distribution de déformation indiquent clairement la présence de joints actifs et de couches peu résistantes qui recoupent le trou de sonde. Considérons les méthodes de calcul actuellement utilisées dans l'analyse des talus et des fondations de barrage. Tout examen statique repose sur des modèles, notamment des modèles d'analyse du comportement des matériaux dans un système statique et des modèles de calcul des charges. La modélisation consiste à ramener un problème réel à sa plus simple expression. Le principe des corps rigides est le plus utilisé dans l'étude de comportement et de sécurité finale. Dans l'analyse des contraintes et des déformations, on suppose que la roche a un comportement élastique linéaire. Il arrive qu'on suppose que la masse rocheuse a un comportement non linéaire dans certains calculs, mais de tels calculs n'ont pas pour but d'évaluer les conditions finales (effondrement).

En réalité, les ouvrages sont rarement bidimensionnels, mais il arrive parfois que des modèles bidimensionnels soient tout à fait justifiés comme dans le cas de la rupture d'un talus ou de l'analyse des contraintes d'un barrage-gravité. Les charges sont statiques ou dynamiques, et les spécialistes ne s'entendent toujours pas sur les questions suivantes : quand faut-il effectuer une analyse de la réponse dynamique et quand suffit-il de ne considérer que la charge statique pour simuler un séisme.

Les analyses techniques reposent dans une grande mesure sur une compréhension poussée des mécanismes qui régissent le comportement structural des talus et des fondations. Il faut observer certaines règles pratiques avant de commencer les calculs et d'accepter un ensemble de résultats : il faut souvent étudier un grand nombre de comportements différents pour déterminer le pire cas possible.

Le prochain point de discussion est le suivi, qui vise à

- vérifier un état d'équilibre
- déceler les comportements inattendus pendant la phase initiale
- expliquer des comportements inattendus de façon à prendre les mesures correctives appropriées
- aider à évaluer la sécurité des ouvrages existants et
- prévoir les ruptures.

Il arrive souvent qu'on considère ces buts de façon combinée. Les paramètres physiques les plus importants sont les déformations (telles que les déplacements absolus ou relatifs, les déformations de volume et les changements d'inclinaison) ainsi que la hauteur piézométrique et les forces d'ancrage.

Dans la plupart des cas, les déformations sont mesurées de façon ponctuelle. Le mouvement de certains points sur une surface inclinée peut indiquer si la structure est stable ou instable. Pour être beaucoup mieux informé, on procède à des "observations le long de lignes", c'est-à-dire qu'on observe comment les mesures d'un paramètre sont distribuées le long d'une ligne droite ou d'une ligne courbée. Par exemple, à partir de la distribution des déplacements horizontaux dans un trou de sonde sur une pente instable, on peut évaluer la surface de glissement qui est la cause du comportement indésirable. De même, à partir de la distribution des mesures des déformations de volume dans un trou de sonde dans le roc sur lequel repose un barrage, on peut déterminer où des joints existants deviennent actifs, c'est-à-dire que les parois s'ouvrent et se ferment lorsque le niveau d'eau du réservoir fluctue. Les pics très nets dans le profil des déformations révèlent que le trou de sonde recoupe de tels joints actifs. Il est important de savoir s'il existe des joints actifs pour fins d'évaluation en matière d'infiltration et de sécurité.

Un autre aspect de la surveillance des talus se rapporte à la prévision des ruptures. Dans les talus dont le potentiel de déformation et de fluage est élevé, les ruptures seront surtout progressives, ce qui permettra, dans de nombreux cas, de prévoir les ruptures de talus, mais aussi l'efficacité des mesures correctives. Par ailleurs, si la cause d'une rupture est une forte pluie ou un séisme intense, la surveillance des déformations sera peu utile. Dans l'analyse, les hypothèses quant à la nature des discontinuités, aux propriétés de la masse rocheuse et aux conditions des eaux souterraines comportent des incertitudes inhérentes. En outre, des données ne sont pas disponibles. Les résultats des calculs sont donc, en principe, hypothétiques; cependant, il sont explicatifs puisqu'ils indiquent les causes et les effets. Par ailleurs, le suivi comporte des lacunes: nombre limité d'observations, périodes sans aucune observation et aussi des techniques de mesure déficientes. Les résultats des observations décrivent la réalité, mais ne permettent pas nécessairement de la comprendre. En l'absence d'efforts particuliers, la surveillance pourrait continuer d'informer sans expliquer. Cependant, s'il est possible de combiner analyse et surveillance, on finira par décrire, et comprendre, la réalité.

Pour tirer le plus grand profit possible de l'analyse, il faudrait adapter les modèles de calcul au problème au moyen d'un nombre minimal de paramètres qui devraient tous être mesurables, du moins en principe. Un modèle très perfectionné mène à des interprétations complexes qui ne facilitent pas nécessairement la prise de décisions pour l'ingénieur. À noter que la surveillance n'est utile que si elle est basée sur des principes bien définis. En d'autres termes, "aucune observation sans théorie".

C'était l'intention du modérateur de terminer ce bref examen en soulignant certaines lacunes existantes dans notre approche aux problèmes. Nos ingénieurs en mécanique des roches continueront de réaliser des progrès sur les plans tant théorique que pratique. Cependant, n'est-il pas vrai qu'en mécanique des roches l'ingénieur doit prendre des décisions et assumer de lourdes responsabilités même en l'absence de faits et de connaissances suffisantes? Le présent congrès a définitivement contribué à notre effort en vue de réduire les incertitudes et de mettre à jour plus de faits lorsque des décisions doivent être prises.

Thème 3
Sautage et excavation

Vibrations de sautage: analyse, effets et contrôle; technologie et contrôle du forage; fragmentation; performance des tunnelliers et des traceuses; forage de puits à grand diamètre

Président: Tan Tjong Kie (Chine)
Secrétaire: W. Comeau (Canada)
Conférencier thématique: C.K. McKenzie (Australie)

SAUTAGE EN ROCHE DURE: TECHNIQUES DE DIAGNOSTIC ET MODÉLISATION DE L'ENDOMMAGEMENT ET DE LA FRAGMENTATION

RÉSUMÉ : Des modèles visant à prévoir la réaction des roches au sautage ont été mis au point afin d'évaluer de nouvelles configurations de trous de mine qui seraient utilisées dans des applications et des types de roche spécifiques. Pour optimiser la configuration, il faudrait cependant effectuer d'autres travaux de surveillance et de mesure afin de régler la séquence des détonations et identifier les facteurs les plus importants influant sur la réaction des roches.

1. INTRODUCTION

Un modèle de la fragmentation des roches par le sautage a été mis au point par le Centre de recherche minérale Julius Kruttschnitt d'Australie. Il se fonde sur les résultats de calcul et de surveillance recueillis dans plus de 25 mines et correspondent à plus de 200 semaines de travail sur place pendant les trois dernières années. Ce modèle a permis de définir les dommages causés à la roche par le sautage et on l'utilise actuellement pour optimiser la configuration des trous de mine afin d'atteindre les objectifs fixés.

Ce modèle a comme principale caractéristique d'utiliser directement les données sur les fractures tirées de cartes et de stéréominutes figurant les fractures. De plus, la fracturation naturelle de la roche y est considérée comme le principal facteur de fragmentation et d'endommagement. Probablement tout aussi importante est l'élaboration des techniques de surveillance et de mesure utilisées pour recueillir les données nécessaires au modèle. Le recours à ces techniques de surveillance a mis en lumière certaines lacunes de base du sautage, notamment le manque de maîtrise de l'amorçage, la performance variable des explosifs et des initiateurs, le manque de méthodes d'évaluation détaillée des configurations et une connaissance insuffisante des mécanismes et des facteurs d'endommagement. Dans la plupart des applications courantes, la modélisation est considérée comme étant moins importante pour le réglage de base que les méthodes de chargement, la configuration des charges et les caractéristiques des produits.

Les résultats de certains travaux de surveillance de base du sautage sont présentés pour souligner dans quelle mesure la surveillance permet d'optimiser le rendement du sautage. L'augmentation de la productivité résultant de la surveillance a été importante et dépasse la portée actuelle de la modélisation. Les sociétés minières ont enregistré des économies dépassant 2 millions de dollars par année en explosifs et en travaux de foration seulement ainsi qu'une réduction de plus de 50 % de la densité de chargement en plus d'améliorer la fragmentation, de limiter les dommages et de contrôler les ébranlements du sol.

2. MODÉLISATION DE LA FRAGMENTATION

Le modèle de fragmentation et d'endommagement du Centre de recherche minérale Julius Kruttschnitt (CRMJK) se fonde sur les observations suivantes :

1. la fissuration est le facteur le plus important de la fragmentation et de l'endommagement;

2. le degré de fragmentation d'une roche dépend de la dimension des particules et de l'énergie transmise;

3. au cours du sautage, la fragmentation est principalement causée par l'onde de choc ou la brisance et les roches sont déplacées par les pressions dues au dégagement gazeux.

L'hypothèse de base de ce modèle mécanisme de fragmentation peut, par conséquent, s'énoncer de la façon suivante : si les dimensions de la particule et l'énergie libérée sont connues, il est possible de prévoir l'importance du fractionnement de la particule. Cet énoncé découle d'observations portant sur le fractionnement de milliers de particules rocheuses soumises à des essais de fractionnement en laboratoire. La principale caractéristique du modèle est peut-être le fait qu'il se fonde sur des mesures plutôt que sur des propriétés théoriques de la roche ou des fractures et que l'exactitude de toutes les prévisions du modèle peuvent être vérifiée par des mesures.

2.1 Mesure des fractures

La densité des fractures est établie par balayage linéaire ce qui permet de localiser les points d'intersection de toutes les fractures naturelles et de mesurer leur pendage et leur direction. Ces fractures sont ensuite représentées sur un stéréogramme à projection équivalente et une simulation Monte Carlo est utilisée pour calculer la distribution granulométrique des blocs de trois configurations d'espacement de joints différentes. La roche est par conséquent décrite comme un système de forme irrégulière dont la répartition dimensionnelle des blocs est définie par la macrofracturation naturelle.

La mesure des fractures renseigne sur la dimension maximale des particules qui pourrait être observée dans les déblais et le pourcentage des particules dépassant une certaine dimension. Si l'on peut définir la distribution granulométrique des fragments de déblais souhaitée, on peut comparer les deux distributions de la façon indiquée à la figure 1. Cette comparaison permet d'obtenir une indication immédiate de l'importance de la fragmentation produit par explosif et permet de définir un "indice de fragmentation".

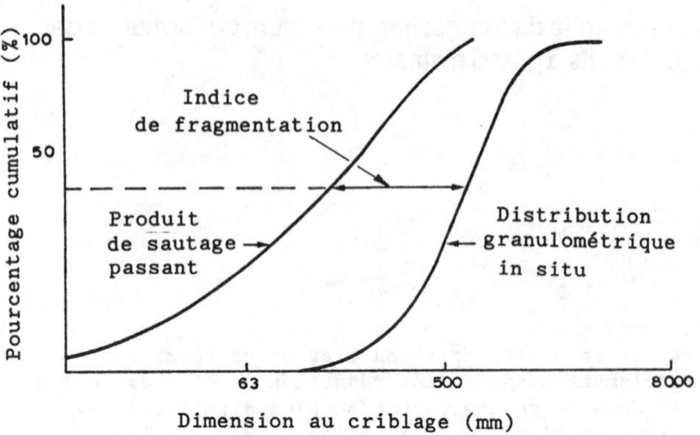

Figure 1. Distribution granulométrique avant et après l'abattage à l'explosif, montrant l'étendue de la fragmentation.

2.2 Déplacement de l'onde de choc

L'onde de choc correspond à l'ébranlement du terrain mesuré au moyen de vélocimètres (géophones). Des vélocimètres sont disposés en réseau autour du trou de mine et les informations sont enregistrées sous forme numérique ou analogique. On utilise des formes d'onde de vibration complète pour calculer l'énergie dégagée à un point particulier en se fondant sur la densité de la roche, la vitesse de l'onde p dans la roche et la variation temporelle de la vitesse des particules au point en question. Les mesures de vitesse des particules sont effectuées près du trou de mine (de 5 m à 25 m) afin d'éviter que de trop grandes erreurs soient faites dans le calcul par extrapolation de l'énergie de vibration.

À partir de plusieurs enregistrements de vitesse des particules, on peut définir les équations de propagation des vibrations en fonction de la combinaison particulière roche-explosif du trou de mine. Les équations utilisée au CRMJK tiennent compte des mécanismes d'atténuation géométrique et de frottement des ondes sphériques et permettent de faire des extrapolations détaillées sur la forme et l'amplitude des ébranlements du sol à tout endroit. Au moyen de mesures et de calculs, il est par conséquent possible de déterminer l'onde de choc ou l'énergie de vibration à tout endroit dans un trou de mine ou autour de celui-ci et d'établir les contours de l'ébranlement du sol autour d'une configuration de trous de mine. C'est cette énergie de vibration que l'on croit être à l'origine de la formation de nouvelles fractures ou de fractionnement d'un massif rocheux.

2.3 Mesure de la fragmentation

Tout programme de prévision de la fragmentation à l'explosif doit être vérifié en mesurant la fragmentation résultante. L'une des techniques utilisées au CRMJK est un procédé qui consiste à prendre de nombreuses photographies d'un tas de déblais à divers stades d'excavation; toutes les fragments dicernables sur la photographie sont ensuite numérisées manuellement afin d'obtenir les dimensions de base telles que la surface, le périmètre, les longueurs de corde maximales et minimales prévues. Par des étalonnages poussés, on établit un lien entre les mesures numérisées et la dimension des particules, ce qui permet de calculer la distribution granulométrique. Cette technique décrit avec exactitude les changements observés dans la distribution granulométrique mais elle exige beaucoup de travail : il faut plus d'une heure pour numériser plusieurs centaines de particules d'une seule photographie.

De nos jours, la mesure de la fragmentation est de plus en plus nécessaire dans le sautage étant donné que des objectifs précis de fragmentation sont visés pour permettre l'application de techniques d'exploitation plus rentables, telles que le nettoyage continu, l'utilisation de concasseurs mobiles et le transport par convoyeur des matériaux abattus à l'explosif. Même s'il est possible de définir la fragmentation nécessaire à ces applications, il ne semble pas pratiquement faisable de mesurer la fragmentation pour que les ingénieurs miniers puissent s'en servir pour optimiser davantage les procédés de minage et de concassage primaire. Il s'agit sans aucun doute d'un secteur où l'analyse assistée par ordinateur pourrait être utile mais à la condition que les problèmes de discrimination des limites soient résolus.

2.4 Prévisions au moyen d'un modèle

En utilisant toutes les techniques susmentionnées, il serait possible de modéliser le processus de fragmentation à l'explosif en intégrant les données d'observation et de mesure recueillies sur le terrain pour déterminer les paramètres du modèle. Plusieurs sautages ordinaires ont été surveillés pour ce qui est de la réaction de la roche aux vibrations, la fragmentation des déblais et la répartition des fractures avant l'explosion. Les résultats obtenus ont été utilisés pour déterminer la valeur des paramètres critiques. La conception du schéma de tir a alors été modifiée et le modèle a été utilisé pour prévoir les changements de fragmentation des déblais selon deux configurations différentes. Dans la première configuration, le diamètre des trous de mine a été réduit de 165 mm à 114 mm et la densité de chargement a été maintenu constante en réduisant la charge et l'espacement de façon proportionnelle. Dans la seconde configuration, la densité de chargement a été réduite de 20 % en augmentant la charge et l'espacement et en utilisant des trous de mine courant de 165 mm de diamètre. La figure 2 présente la concordance entre la fragmentation observée et prévue en utilisant les paramètres déterminés à partir de mesures de sautages antérieurs.

2.5 Détermination et définition de l'endommagement

La méthode de modélisation mécanisé peut sans conteste permettre de prévoir avec exactitude la

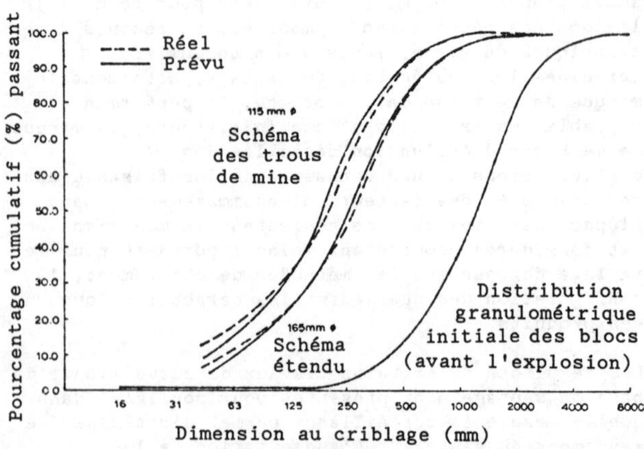

Figure 2. Distributions granulométriques prévue et observée à partir du modèle de fragmentation appliquée à des conceptions de sautage réels

fragmentation et peut facilement être étendue à la prévision de l'endommagement que l'on peut maintenant définir comme la modification de la distribution granulométrique des blocs en place par suite d'un sautage voisin et que l'on peut lier à l'"indice de fragmentation" déjà présenté, c'est-à-dire le déplacement de la courbe granulométrique des blocs après le coup de mine. Selon l'application qui est faite des mesures de l'endommagement, cette définition peut ne pas être appropriée, soulignant le besoin d'une définition plus explicite de l'endommagement. La définition utilisée ci-dessus convient si l'on compare les effets des différentes configurations de trous de mine sur les hors profils mais donne peu d'indications sur la perte de stabilité résultante qui peut se traduire par une rupture, une dilution du minerai et une perte de récupération.

3. LES LIMITES DE LA MODÉLISATION

Dans la modélisation de prévision, contrairement à la rétro-analyse, il faut énoncer des hypothèses sur le comportement des explosifs et sur la pertinence des caractéristiques des trous de mine, tels que leur diamètre, leur emplacement, les dimensions de leur configuration, etc. Sans une quelconque surveillance, le rendement réel d'un trou de mine ne peut pas être déterminé, ni les causes réelles du rendement observé.

Dans les modèles actuels de fragmentation des roches à l'explosif, on suppose une correspondance entre le rendement réel et prévu de la charge. Cette hypothèse, comme on le constatera, constitue une limite importante à la modélisation et un obstacle notable au programme d'optimisation des conceptions de sautage. Les progrès accomplis en modélisation et en calcul ne comptent pas si l'amorçage ne peut pas être totalement maîtrisé. La simple surveillance du rendement des charges et la capacité de détecter les défauts d'amorçage peuvent permettre d'augmenter davantage la rentabilité du sautage que la modélisation de prévision. Bien qu'il soit mathématiquement possible de varier le diamètre des trous de mine, les charges, les espacements, la force des explosifs et les longueurs de bourrage de façon indépendante sur de grandes étendues, en pratique, cette façon de procéder peut facilement provoquer l'amorçage de la charge par influence, la désensibilisation de la charge ou la dislocation de la charge, situations qui peuvent ne pas être détectées sans une surveillance minutieuse. Les intervalles de retard peuvent aussi être théoriquement modifiés sur une grande étendue mais la modélisation ne tient pas compte des effets d'un intervalle inapproprié sur l'inversion de la séquence, l'accroissement des vibrations ou la dislocation des charges adjacentes.

4. TECHNIQUES DE SURVEILLANCE DES TROUS DE MINE

S'ils sont situés près d'un trou de mine, les instruments peuvent enregistrer avec netteté la détonation des charges explosives. La figure 3 montre la réponse d'une charge unique, enregistrée à partir d'un géophone situé à environ 20 mètres de la charge et cimenté en permanence à la roche dans un trou de sonde. À partir de cet enregistrement, on peut calculer avec précision le temps précis de l'amorçage, l'erreur d'amorçage et l'amplitude de la vibration. Le chronométrage de l'amorçage fournit des informations sur la mise en séquence, les ratés d'allumage, le décalage des retards, tandis que l'amplitude de la vibration fournit des informations sur le rendement de l'explosif et l'énergie dégagée pour la fragmentation.

La Figure 4 présente la réponse d'un sautage à plusieurs retards ainsi que la conception du sautage. Dans cet exemple, les charges sont superposées, c'est-à-dire que plusieurs charges sont contenues dans un seul trou de mine et les temps d'amorçage nominaux et réels de chaque charge sont indiqués. Le décalage d'amorçage des deux charges de 300 ms est clairement indiqué. Malgré les erreurs de retard, ces charges réagissent tel que prévu; l'étalement des retards observés ne dépassent pas les limites prévues et toutes les charges ont produit à l'allumage une forte vibration.

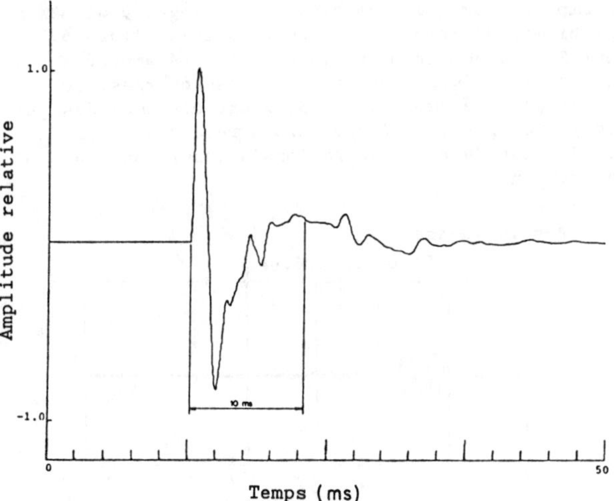

Figure 3. Vibration causée par une seule charge explosive et enregistrée à moins de 20 m d'un trou de mine.

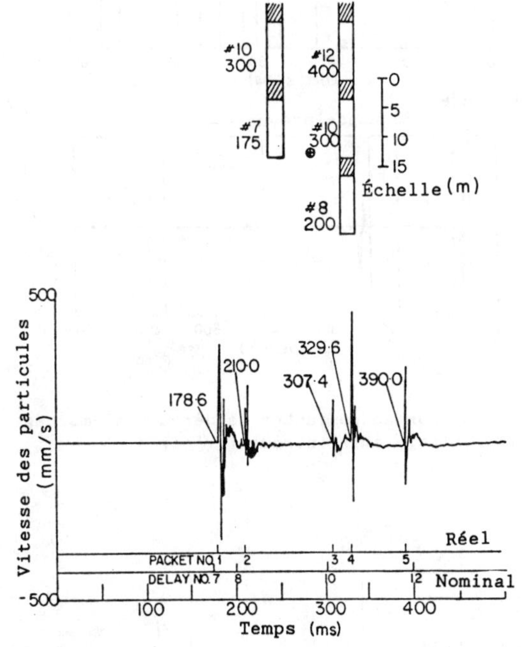

Figure 4. Vibration causée par un sautage à plusieurs retards.

4.1 Amorçages défectueux

La figure 5 présente trois des formes les plus courantes d'amorçages défectueux qui ont été observées en Australie, au Canada, aux États-Unis et au Chili. L'importance des défauts d'allumage dépend fortement de la conception des trous de mine, du nombre de charges différées, de la durée du tir et de la géologie du terrain. La surcharge est la cause la plus courante des problèmes d'amorçage, d'une fragmentation insuffisante et des hors profils excessifs. Lorsque la fragmentation est insuffisante, on augmente, en général, la densité de chargement, ce qui peut aggraver la situation en favorisant les ratés d'allumage et en amplifiant les hors profils qui ont tendance à suivre les fractures naturelles, donc à être moins bien définis.

L'ampleur possible des ratés d'allumage et d'autres problèmes d'amorçage est indiquée à la figure 6. Sur 20 charges, neuf seulement ont été amorcées tel que prévu. Sur le nombre total des charges, 20 % n'ont pas été amorcées et 35 % ont été amorcées par influence par les détonateurs à retard 5 et 8, produisant deux concentrations d'impulsions de vibration.

Par influence

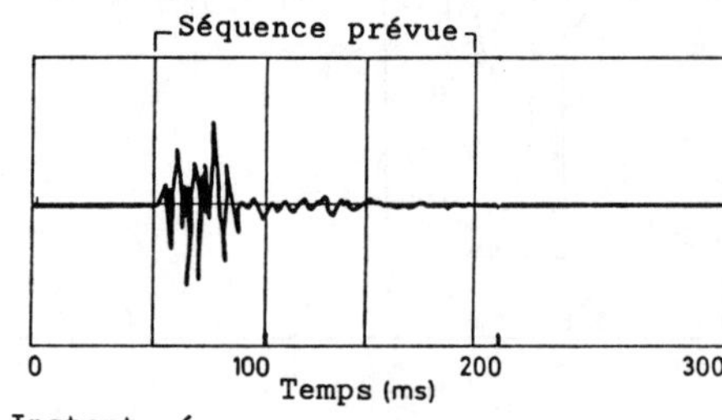

Instantanée

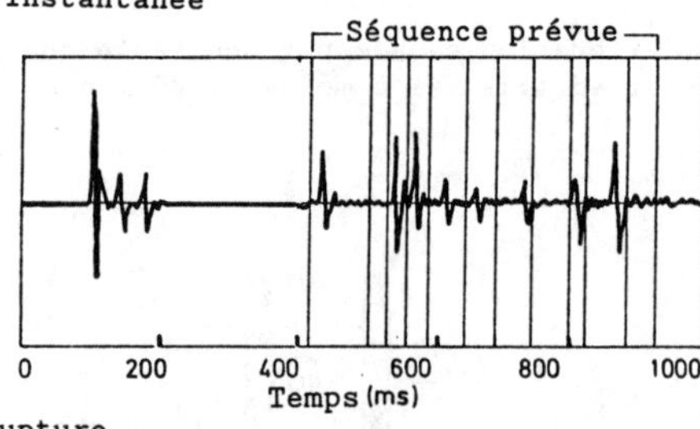

Rupture

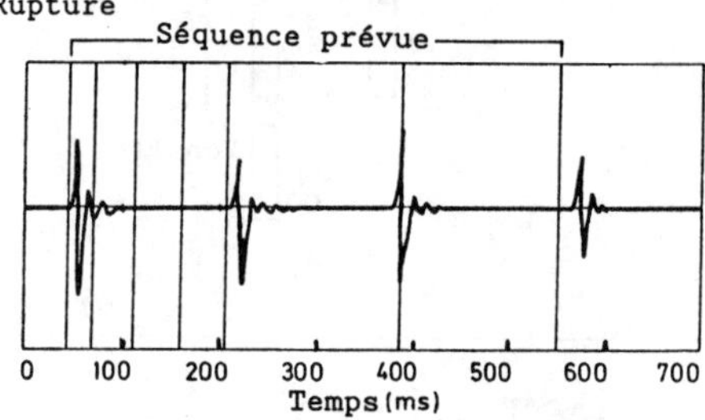

Figure 5. Formes courantes de défauts d'amorçage des explosifs: enregistrement réel.

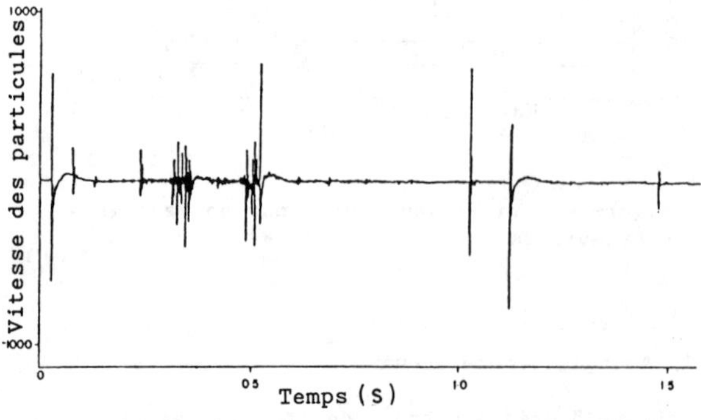

Figure 6. Enregistrement montrant dans quelle mesure les ratés d'allumage peuvent se produire - enregistrement réel de tir souterrain.

Après avoir été identifiés, les amorçages défectueux peuvent être éliminés en rétablissant le contrôle de l'amorçage et en diminuant invariablement la quantité d'explosifs. La figure 7 présente un allumage bien contrôlé obtenu après avoir modifié la conception du sautage utilisé dans le diagramme de la figure 6. Les ratés d'allumage ont été éliminés, le nombre de petites charges superposées a été réduit, la densité de chargement a été diminuée d'environ 50 % et les niveaux de vibration, l'endommagement et les hors profils ont été réduits.

Un quatrième défaut courant d'amorçage est le rendement réduit des charges qui est surtout associé aux explosifs à émulsion. Ces explosifs produisent fréquemment des niveaux de vibration variables, une fragmentation variable et un déplacement variable des déblais en plus de faire couramment l'objet de ratés d'allumage. Ces caractéristiques laissent supposer que ces explosifs ne détonent pas toujours au taux maximal ou prévu et que l'énergie libérée pour le fractionnement est, par conséquent, beaucoup moins élevée. Pour être rentable, les explosifs à émulsion doivent toujours atteindre des niveaux de rendement nominaux.

À la lumière des données concernant la portée des problèmes d'amorçage dans les travaux de production et de préparation ou le minage des galeries, il y aurait plus d'avantages pratiques à optimiser la conception du sautage en recourant à des instruments qui permettraient d'identifier rapidement les problèmes qu'en élaborant des modèles. La surveillance du rendement des tirs s'est fréquemment traduite par une réduction d'environ 50 % de la quantité d'explosifs et une maîtrise améliorée de 25 % en moyenne de la fragmentation, de la vibration et de l'endommagement.

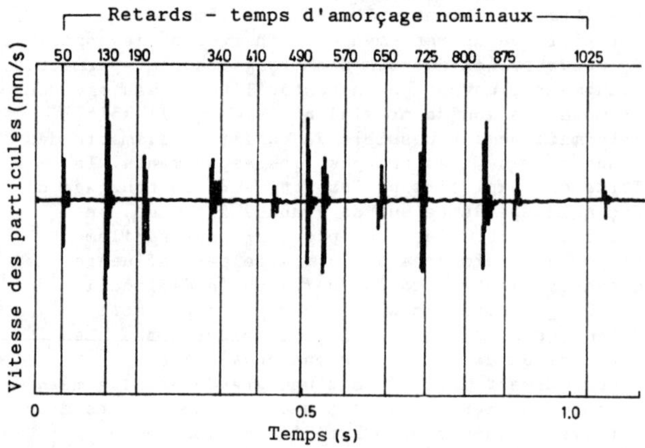

Figure 7. Amorçage contrôlé des charges, sans ratés d'allumage - enregistrement réel de tir souterrain.

5. ÉVALUATION DE LA CONCEPTION DES TROUS DE MINE

La surveillance du rendement des trous de mine a amplement démontré l'importance d'exercer un contrôle serré de la répartition et de la concentration des charges explosives. Cependant, même une densité de chargement moyennement faible peut causer des défauts d'amorçage localisés qui peuvent indiquer un surabattage. De simples algorithmes informatiques peuvent être mis au point pour permettre aux concepteurs des trous de mine d'effectuer une meilleure évaluation d'un trou de mine et de son mode de chargement.

La figure 8(a), par exemple, montre comment la concentration d'explosifs dans une partie de trou de mine a été accrue de façon à doubler la concentration moyenne. Dans ce cas, on a augmenté la concentration par suite d'un élargissement du diamètre du trou de mine de 70 mm à 115 mm afin d'allonger le trou de mine sans perte de précision en matière de foration. Dans cette partie du trou de mine, il devrait se produire une détonnation par influence des charges causant un hors profil et un endommagement accru. La figure 8(b) montre comment cette concentration des charges peut être réduite en utilisant un explosif de force inférieure dans des trous de plus grand diamètre.

En appliquant des méthodes semblables à des configurations en éventail de trous de mine souterrains, on a réduit de plus de 10 % la densité de chargement en réglant la longueur des colonnes d'explosifs dans les trous de mine.

Des techniques d'évaluation des conceptions aussi simples que celles-là offrent en outre des possibilités d'optimisation en matière de conception des trous de mine supérieures aux modèles actuels. Les modèles devront dans l'avenir permettre de déterminer le point auquel la détonation par influence des charges et l'amorçage des charges devraient se produire.

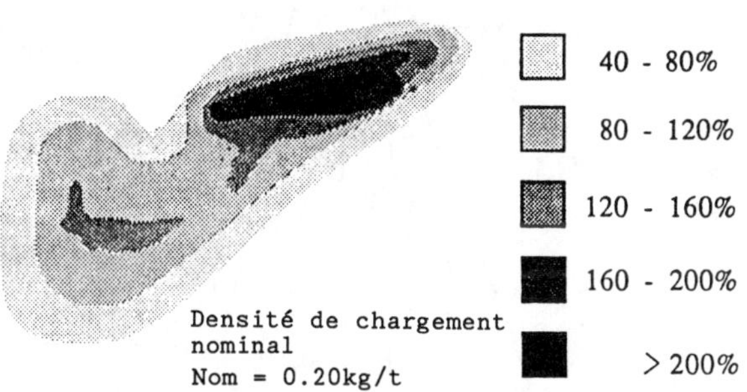

Figure 8(a). Concentration de l'explosif au cours du chargement par suite d'une augmentation du diamètre du trou de mine.

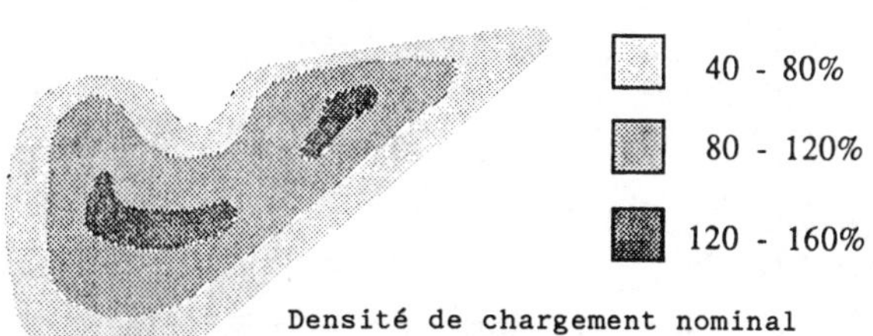

Figure 8(b). Distribution égale de l'explosif obtenue en ajustant la force de l'explosif dans des trous de mine à grand diamètre.

6. DÉCALAGE DES RETARDS

Le décalage des retards est une caractéristique bien acceptée des détonateurs pyrotechniques mais dans les programmes et les modèles de conception des trous de mine, on tient rarement compte, semble-t-il, de ce phénomène. Plusieurs milliers de détonateurs à retard non électriques ont été analysés pour déterminer avec précision le décalage de leur temps de retard en les chronométrant tous électroniquement et en prenant des photographies ultra rapides d'un grand nombre d'entre eux. Les détonateurs analysés provenaient tous de l'Australie et leur retard se situait pour la plupart entre 25 ms et environs 10 s.

Les résultats indiquent que, bien que le décalage à l'intérieur d'un même lot soit très faible (de 3 à 4 % environ), le décalage entre des lots est beaucoup plus élevé. Il faut en conclure que lorsqu'on utilise un seul lot de détonateurs de même retard, le décalage de retard peut être considéré comme minime soit de 3 ou 4 % environ tandis que si on utilise des détonateurs de retards ou de lots différents, le décalage des retards sera beaucoup plus élevé, pouvant atteindre 13 % selon le détonateur. La figure 9 montre le décalage des retards de deux lots de détonateurs à retard de 500 ms. Le décalage total lorsque ces deux lots sont mélangés est d'au moins deux fois supérieur au décalage de chaque lot individuel. Les résultats d'essai de plus de 40 lots différents indiquent que si on mélange les lots, le décalage peut s'accroître d'un facteur de plus de 3, bien que cela dépende du type de détonateur utilisé. Pour les fins de cet exposé, un décalage de retard de ±13 % sera utilisé pour décrire la distribution de l'erreur de retard, c'est-à-dire que 95 % des détonateurs à retard détoneront avant ±13 % du retard nominal ou prévu. On a démontré expérimentalement que cette définition du décalage des retards était réaliste.

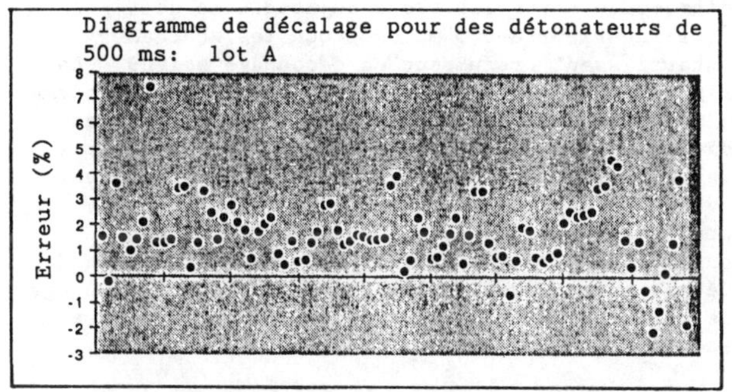

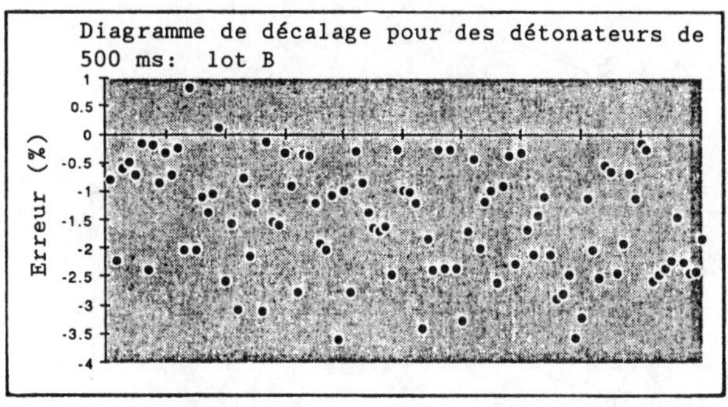

Figure 9. Décalage des retards pour différents lots de détonateurs.

6.1 Influence du décalage des retards sur la conception des trous de mine

Le décalage des retards étant connu, on peut évaluer les probabilités de désordre dans la séquence des retards. Une évaluation supplémentaire du contrôle de l'amorçage d'une configuration de trous de mine est ainsi réalisée.

Les détonateurs à retard sont utilisés dans le minage pour que l'un des deux objectifs suivants puisse être atteint:

1. s'assurer que les charges ne sont pas amorcées en même temps;
2. promouvoir l'amorçage instantanné des charges.

En utilisant les données décrivant le décalage des retards, on peut évaluer les possibilités d'atteindre ces objectifs dans toute situation. Prenons l'exemple d'une volée dans un tunnel ou dans des travaux préparatoires, telle que montrée à la figure 10 qui présente la configuration des trous de mine et la séquence des retards utilisés dans une volée souterraine typique. Dans l'analyse, on tiendra compte, en particulier, des probabilités suivantes:

1. les trous en périphérie se partageant un détonateur à retard vont interagir pour favoriser un découpage soigné et des murs de galerie bien découpés;
2. les trous en périphérie seront mis à feu après que les trous intérieurs auront été allumés et dégagés, minimisant ainsi les dommages causés aux murs et à l'arrière-taille.

Pour que des trous de mine adjacents interagissent pour faciliter un découpage soigné, ils doivent être amorcés dans un intervalle critique déterminé par la vitesse de propagation des fissures dans la roche. Si l'on suppose une vitesse de l'onde p d'environ 6000 m/s, une vitesse de propagation des fissures égale au tiers de la vitesse de l'onde p et une distance entre les trous de mine ne dépassant pas un mètre, l'intervalle d'amorçage entre les trous adjacents ne doit pas être supérieur à 0,5 ms pour que les charges interagissent produisent un découpage soigné. La figure 11 montre la probabilité d'interaction comme une fonction du retard d'environ 13 %. Pour les détonateurs puissants situés autour de la galerie, la probabilité que des trous adjacents détoneront avant 0,5 ms est effectivement nulle et la probabilité qu'ils détoneront avant 5 ms est inférieure à 2 %. C'est pourquoi l'utilisation de détonateurs puissants à macroretard dans le sautage de galeries ne permet pas d'obtenir un découpage soigné.

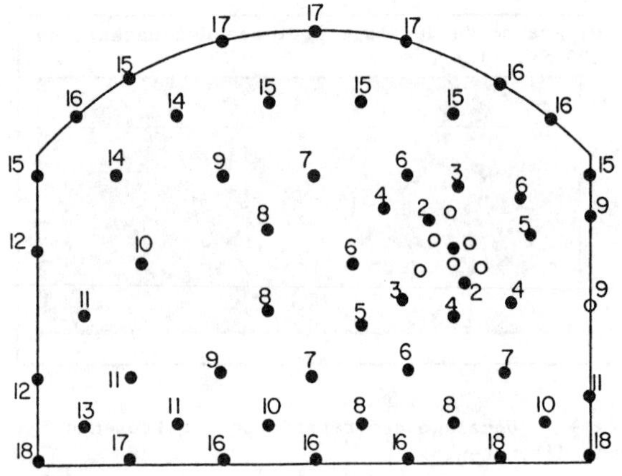

Figure 10. Configuration d'un sautage préparatoire et mise en séquence de trous de mine de 52 mm de diamètre.

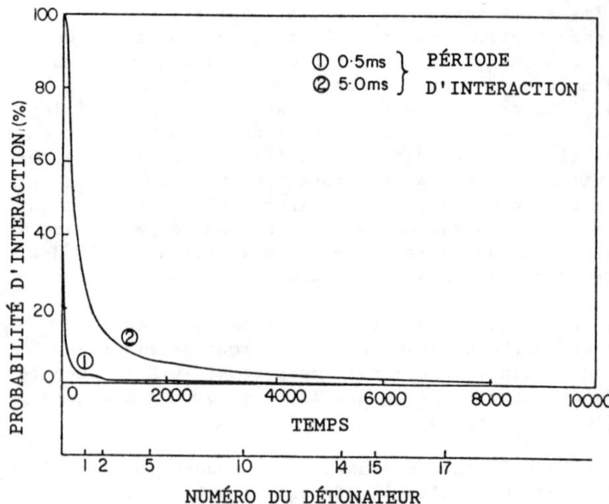

Figure 11. Fonction de probabilité décrivant le potentiel d'interaction constructive entre des trous périphériques adjacents dans un minage préparatoire.

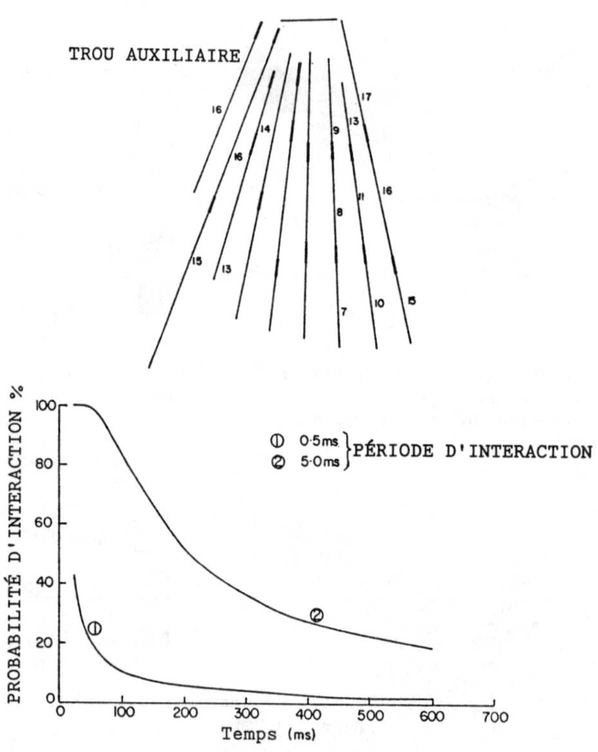

Figure 12. Fonction de probabilité décrivant le potentiel d'interaction constructive entre des trous périphériques adjacents dans un minage par chambres vides.

La figure 12 montre une analyse semblable de découpage soigné réalisé le long des murs d'un chantier en utilisant des détonateurs à retard de l'ordre du millième de seconde. L'avant-trou dans ce cas représente un trou intermédiaire entre les rangées, en périphérie du chantier. Idéalement, cet avant-trou devrait détoner en même temps que le trou du mur dans la rangée qui se trouve derrière pour que le mur soit découpé avec soin. La probabilité d'obtenir cet effet est d'environ 3 % pour des détonateurs à retard de 600 ms et ne dépasse pas 10 % pour des détonateurs à retard de 100 ms.

La probabilité qu'il se produise un désordre de la séquence et que des trous en périphérie soient amorcés avant les trous adjacents intérieurs a aussi été analysée. La figure 13 montre la fonction de probabilité de recouvrement des détonateurs à retard utilisés dans le sautage si le décalage des retards est d'environ 13 %. Par probabilité, on entend les possibilités de recouvrement entre des détonateurs à retard successifs. Pour des retards d'environ 6 secondes, on y indique une probabilité de recouvrement maximal d'environ 14 %. Il est par conséquent fort probable qu'un trou en périphérie sera amorcé avant un trou intérieur, causant des dommages aux murs, produisant des excavations de forme irrégulière et augmentant les coûts de soutènement. En s'assurant qu'au moins deux intervalles séparent la détonation des trous en périphérie des trous intérieurs, la probabilité de recouvrement diminuerait de façon à devenir effectivement nulle.

Le fait de relier les charges à des détonateurs distincts n'indique pas nécessairement que l'amorçage sera réalisé séparément ou selon la séquence souhaitée, en particulier si les intervalles courants ont été réduits en combinant des détonateurs à retard. Les modèles de sautage doivent d'abord tenir compte de l'influence des retards sur la fragmentation et ensuite de l'influence des amorçages désordonnés sur la fragmentation et l'endommagement.

6.2 Effet du décalage des retards sur le contrôle des vibrations

L'une des principales raisons pour lesquelles on utilise de nos jours des détonateurs à retard dans le minage est que l'on vise à minimiser les vibrations et les dommages qu'elles causent. C'est pourquoi dans de nombreux dépilages par longs trous, on utilise plusieurs charges dans un seul trou de mine, chaque charge étant séparée par un bourrage inerte et amorcée séparément. Il faut, par conséquent, un nombre plus grand de détonateurs pour fragmenter par sautage les quantités souhaitées (tonnes) de roches. Étant donné que le nombre de détonateurs à retard est limité à environ 40, il est courant d'augmenter ce nombre en combinant des détonateurs, c'est-à-dire en utilisant un détonateur au sommet du trou de mine pour amorcer la ligne descendante vers un second détonateur dans le fond du trou. Cette façon de procéder nécessite un plus grand nombre de détonateurs mais comme l'intervalle entre les détonateurs diminue, la probabilité de recouvrement des retards successifs risque d'augmenter encore plus.

Les équations de prévision des vibrations couramment utilisées portent sur le poids maximal de la charge par détonateur à retard et mettent peu l'accent sur la détermination du retard minimal. En pratique, ce dernier dépend entièrement du type de roche, étant relativement court pour les roches élastiques à module élevé et beaucoup plus longs pour les roches plastiques à module faible. On peut établir un lien entre le retard et la réponse de la roche aux vibrations induites. L'intervalle minimal des vibrations est le temps qui s'écoule après la détonation mais avant que la roche retourne à un état de vibration presque nul, tel que montré à la figure 3. Les vibrations produites par des détonations de charges individuelles ne subissent pas les effets des autres charges et l'intensité des vibrations est bien maîtrisée.

Pour prévoir les niveaux de vibration en intégrant les concepts susmentionnés, on peut considérer la probabilité d'amélioration des vibrations provoquées par différentes charges en déterminant la probabilité de mise à feu des charges dans un intervalle particulier causée par le décalage des retards, à partir de données de surveillance des vibrations enregistrées par des instruments capables d'enregistrer toutes les vibrations dans le sol quel que soit l'endroit. Cette méthode a été, entre autres, utilisée pour simplifier la conception d'un sautage où un grand nombre de charges superposées sont utilisées. C'est par une analyse probabiliste qu'on a comparé des configurations qui comportaient un grand nombre de charges superposées séparées par des microretards à d'autres caractérisées par un nombre moins élevé de charges superposées (charges de poids supérieur) à macroretard. Dans bien des cas, lorsque les charges sont de poids supérieur et les retards plus longs, on a constaté que les vibrations étaient plus faibles, qu'elles étaient moins variables (amélioration du contrôle des vibrations) et que le chargement était plus facile et plus rapide.

7. DOMMAGES CAUSÉS PAR LA PÉNÉTRATION DES GAZ

Les données recueillies sur le sautage indiquent que les gaz explosifs sous pression élevée pénètrent dans les fractures qui se dirigent dans toutes les directions à partir d'un trou de mine. La pénétration des gaz vers l'avant favorise le déplacement et le gonflement des déblais tandis que la pénétration des gaz vers l'arrière cause le gonflement des roches derrière le trou de mine avant de se traduire par une perte de frottement le long des surfaces de fracture, donc d'augmenter l'instabilité. Les effets causés par les gaz dépendent fortement de la nature des fractures présentes dans le massif rocheux.

Des expériences simples réalisées pour montrer l'écoulement des gaz derrière les trous de mine ont permis de receuillir des informations sur leur vitesse de propagation, de chronométrer le mouvement de la roche et de déterminer l'importance des variables de conception du sautage. Dans ces expériences, des trous de sondage sont forés derrière les trous de mine que l'on scelle avec un bouchon de béton après y avoir inséré des indicateurs de pression. Près du trou de sondage, ou même à l'intérieur du trou, on utilise un vibromètre pour déterminer le temps exact de l'amorçage de la charge explosive située près de l'indicateur de pression.

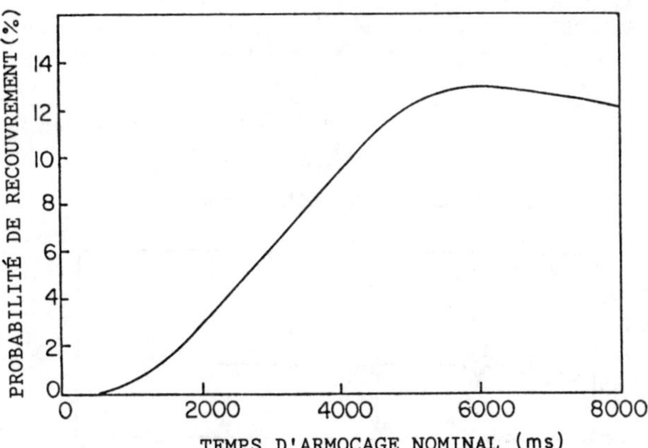

Figure 13. Fonction de probabilité décrivant le potentiel de recouvrement entre des détonateurs successifs, susceptibles de produire un amorçage non séquentiel de charges en mur dans un minage de galerie.

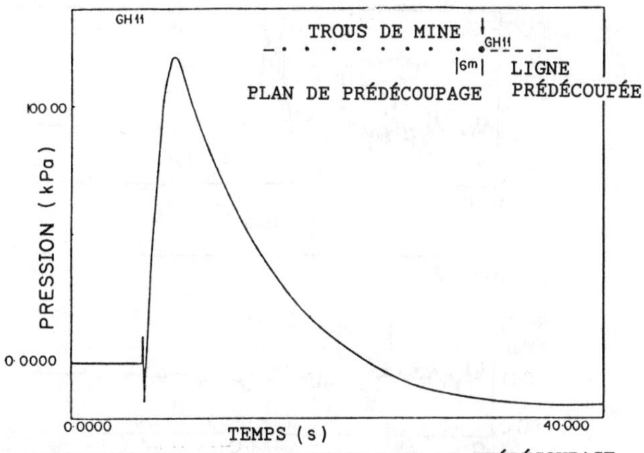

Figure 14. Réponse à la pression des gaz d'un trou de sondage non chargé situé à 6 m de l'extrémité d'une ligne prédécoupée de trous de mine.

La figure 14 montre la réponse à la pression d'un trou de sondage situé à 6 mètres de l'extrémité d'une ligne de prédécoupage par trous de mine. L'indicateur de pression, qui a été compensé pour tenir compte des vibrations, réagit encore aux détonations des trous de mine les plus proches et affiche une augmentation soudaine de la pression suivie d'une diminution graduelle. Dans cette expérience, la trajectoire des gaz sous haute pression a clairement suivi la ligne de prédécoupage qui s'étendait au-delà du dernier trou de mine plutôt que les fractures naturelles.

Dans une seconde expérience, la surveillance de la pression des gaz et des vibrations a été réalisée derrière un tour de mine régulier foré dans le terrain de couverture d'une mine de charbon à ciel ouvert. Les résultats sont présentés à la figure 15. Les charges ont été amorçées suivant cinq rangées parallèles au front libre et la réponse aux vibrations de chacune des rangées est clairement indiquée sur l'enregistrement des vibrations par l'indicateur de pression radiale situé dans le trou BM 11. Ces pointes de vibration peuvent aussi être repérées sur les enregistrements de pression des gaz des deux autres signaux, indiquant ainsi que les piézographes n'ont pas été complètement compensés pour tenir compte des vibrations. La réponse à la pression des gaz du trou GM 7 est également facilement discernable: il y a d'abord augmentation puis diminution de la pression, ce qui a pour effet de créer un vide dans le trou de sondage. Enfin, le retour à la pression ambiante a lieu une seconde environe après le début de l'écoulement des gaz dans le trou. La vitesse de propagation des gaz calculée dans de nombreuses expériences varie de 8 à 12 mètres/seconde.

La création d'un vide indique que le volume du trou de mine augmente, à cause probablement d'une relaxation de décharge à mesure que la roche qui se trouve en face du dernier trou de mine s'éloigne du front du gradin. Les temps de déplacement de la charge mesurés en fonction de cette hypothèse indiquent qu'il faut environ 90 ms après la détonation des charges explosives pour que la roche soit projetée.

La pénétration des gaz qui a été mesurée à des distances de 8 mètres derrière un trou de sondage constituerait une cause principale d'endommagement et de perte de stabilité. On peut minimiser la pénétration des gaz par les moyens suivants qui consistent à :

1. réduire le temps d'action de la pression des gaz (maximiser la vitesse de déplacement de la roche);

2. minimiser la pression de pointe du trou de mine;

3. prévoir un évent des gaz tel qu'une ligne prédécoupée.

Parmi ces techniques, la réduction du temps pendant lequel la pression des gaz exerce ses effets devrait être le moyen le plus efficace de limiter leur pénétration. La vitesse de déplacement de la roche peut être maximisée sans augmenter la pression de pointe dans le trou de sondage, en minimisant les dimensions de la roche à abattre. Les conceptions de sautage permettant de minimiser les dommages devraient par conséquent posséder les caractéristiques suivantes:

1. ne comporter que quelques rangées de trous de mine, afin de ne pas limiter le déplacement de la roche;

2. réduire les dimensions de la roche à abattre;

3. prédécouper le long des limites critiques.

Les modèles de minage devront éventuellement être capables de prévoir l'endommagement et la fragmentation. Ces deux facteurs dépendent fortement de l'état des fractures présentes dans la roche. La pression des gaz joue un rôle crucial dans la détermination de la forme finale des déblais et sur l'étendue des dommages dus aux gaz. Les modèles doivent donc produire une prévision de l'écoulement des gaz ainsi que du volume et de la pression des gaz produits par le sautage.

8. CONCLUSION

L'utilisation de modèles mathématiques a permis de mieux comprendre certains mécanismes mis en jeu durant le sautage. Cependant, contrairement à la plupart des autres modèles mathématiques, les modèles de sautage ont été élaborés sans que les effets ne soient surveillés et mesurés, ce qui

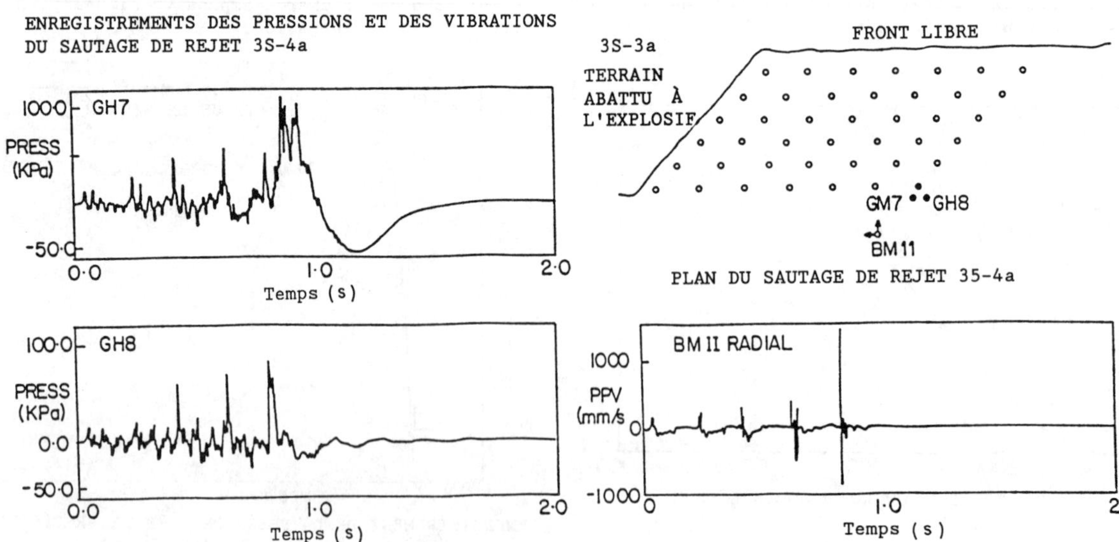

Figure 15. Réponse à la pression des gaz dans un trou de sondage non chargé situé derrière un trou de mine de production dans une roche très fracturée.

aurait permis de valider les prévisions ou même de confirmer les principaux mécanismes. On suppose que les explosifs et les détonateurs fonctionnement exactement comme prévu de sorte que l'on accorde peu d'importance à l'évaluation des facteurs qui peuvent les empêcher de fonctionner de façon optimale. La fragmentation et l'endommagement, principaux facteurs dont il faut tenir compte dans les exercices de modélisation, ne peuvent pas être mesurés avec précision à grande échelle. L'absence de techniques et d'instruments de mesure de base qui pourraient être utilisés dans l'industrie minière est un obstacle important à l'application et à l'élaboration d'autres modèles de sautage. Il existe actuellement des instruments pour surveiller le rendement des charges explosives et il est essentiel de les utiliser si l'on veut s'assurer que le minage conçu produise les résultats souhaités ou prévus.

9:20 Conférences brefs (Le Grand Salon)

Bonapace, B., Autriche
Effet et limites de l'avancement par tunnelier en suite de l'excavation d'une galerie en charge de 22 km

Brych, J., Ngoi Nsenga, Ziao Shan, Belgique, Zaire, Chine
Contribution à l'étude de destructibilité des roches en forage rotatif

Farmer, I.W., Garrity, P., Angleterre
Prédiction de la performance d'une machine à attaque ponctuelle en fonction de la résistance à l'extension d'une fracture

Ginn Huh, Kyung Won Lee, Han Uk Lim, Corée
Détermination empirique de l'équation de la vitesse de particules des vibrations des tirs à l'explosif

10:00. FOYER, Pause-Café

10:30.

Goto, Y., Kikuchi, A., Nishioka, T., Japon
Comportement dynamique du revêtement d'un tunnel lors d'un sautage effectué à proximité

Grant, J.R., Spathis, A.T., Blair, D., Australie
Investigation de l'influence de la longueur d'une colonne d'explosifs sur les vibrations de tirs

Heraud, H., Rebeyrotte, A., France
Essais de prédécoupage et mesures des arrière-bris dans les roches granitiques du massif central Français

Inazaki, T., Takahashi, Y., Japon
Évaluation de la qualité de la masse rocheuse au moyen de la tomographie séismique

Maerz, N.H., Franklin, J.A., Rothenburg, L., Coursen, D.L., Canada, États-Unis
Analyse photographique par méthode digitale de la fragmentation induite par explosifs

Miranda, A.M., Mendes, F.M., Portugal
L'altération des roches comme paramètre de forabilité

Mitani, S., Iwai, T., Isahai, H., Japon
Relations entre les conditions du massif rocheux et la faisabilité de TBM

Singh, R.D., Virendra Singh, Khare, B.P., Inde
Étude pour l'optimisation de la fragmentation par sautage dans une couche de houille épaisse jointée

Wilson, W.H., Holloway, D.C., États-Unis
Études de fragmentation sur des modèles en béton instrumentés

12:00 - 13:30 Pause déjeuner

13:30 - 15:00 Discussion (Le Grand Salon)

Modérateur: P.A. Lindquist (Suède)

Panélistes: R.F. Favreau (Canada)
O.T. Blindheim (Norvège)
C.D. de Gama (Brésil)

Voir la version anglaise pour discussions dans la salle et contribution de la table ronde

Résumé du modérateur: P.A. Lindquist (Suède)

1. INTRODUCTION

Le troisème thème du sixième Congrès international de mécanique des roches s'intitule Sautage et excavation. Trente exposés sur le forage, l'excavation mécanique des roches et le sautage ont été présentés. Le tableau 1 est un résumé des domaines techniques traités dans les exposés.

Tableau 1 - Domaines techniques traités dans le cadre du thème III

Forage	Nbre d'exposés
Forage - généralités	3
Forage percutant	2
Forage rotatif	1
Foration de puits de pétrole	2
Excavation mécanique	
Tunnelier	3
Machine de traçage	4
Foreuse de monterie	1
Démolisseuse statique	1
Sautage	
Caractérisation de la masse rocheuse	1
Sautage - généralités	4
Sautage - mine à ciel ouvert	3
Prédécoupage	1
Percement de tunnels - aspects divers	4

Comme on peut le constater, la répartition des exposés entre les domaines du forage, de l'excavation mécanique et du sautage est bien équilibrée.

Il est peut-être plus intéressant de se pencher sur les problèmes traités par les différents auteurs. Dans le tableau 2, les exposés sur le forage et l'excavation mécanique ont été regroupés sous quatre en-têtes principales.

Tableau 2 - Problèmes choisis - Forage et excavation mécanique

Prévision de la vitesse d'avancement/rendement/faisabilité	Nbre d'exposés
Bernaola & Oyanguren	
Howarth & Rowlands	
Miranda & Mendes	
Farmer & Garrity	
Bonapace	
Mitani, Twai & Isahai	6

Conception des taillants/fleurets

Brysch, Nsenga & Shan
Brighenti & Mesini
Nishimatsu, Okubo & Jinno
Xu, Tang & Zou 4

Stabilité des forages

Kaiser & Maloney 1

Mécanismes de taillage/jets d'eau

Hood, Geier & Xu
Iihoshi, Nakao, Torii & Ishii
Ip & Fowell
Kutter & Jütte
Sekula, Krupa, Koci, Krepelka & Olos
Fukuda, Kumasaka, Ohara & Ishijima 6

Enfin, dans le tableau 3, les problèmes de sautage traités par les auteurs ont été répartis en six catégories. À remarquer qu'aucun exposé ne porte sur la modélisation informatique du sautage ni sur la modélisation empirique de la fragmentation.

Tableau 3. Problèmes choisis - Sautage

Théorie du sautage	Nbre d'exposés
Singh & Sastry	
Wilson & Holloway	2
Conception du sautage	
R D Singh, V Singh & Khare	1
Critères de vibration	
Fadeev, Glosman, Kartuzov & Safonov	
Huh, Lee & Lim	
Goto, Kikuchi & Nishioka	
Grant, Spathis & Blair	
Siskind	5
Prédécoupage	
Heraud & Rebeyrotte	1
Mesure de la fragmentation	
Maerz, Franklin, Rothenburg & Coursen	1
Caractérisation de la masse rocheuse/modèle de coûts	
Inazaki & Takahashi	
Pöttler & John	
Mikura	3

En résumé, les exposés du troisième thème indiquent que des travaux de recherche et de développement fructueux sont actuellement réalisés à travers le monde dans ce domaine. Depuis le dernier Congrès de mécanique des roches, quelques réalisations importantes doivent être mentionnées, notamment l'utilisation accrue de jets d'eau à moyenne haute pression pour seconder les machines de traçage et l'établissement généralisé de critères sur la vibration du sautage fondés sur des données locales mais tenant souvent compte de l'expérience acquise à l'échelle internationale.

Les exposés du troisième thème comprennent également quelques études de cas importantes qui contribueront au développement à long terme de la technologie du sautage et de l'excavation.

2. FAITS SAILLANTS

Compte tenu de la grande qualité des exposés, il est difficile d'en sélectionner quelques-uns pour les commenter. Certains, toutefois, méritent d'être soulignés.

L'exposé "Études de fragmentation sur des modèles en béton instrumentés" de W.H. Wilson et D.C. Holloway porte sur l'utilisation d'une combinaison de techniques d'observation différentes qui permettent de tirer des conclusions nouvelles et solides. Il s'agit d'un exposé clé qui, à mon avis, contribue largement à la compréhension de la mécanique des roches préalable à la fragmentation.

Dans leur exposé, intitulé "Investigation de l'influence de la longueur d'une colonne d'explosifs sur les vibrations de tirs" J.R. Grant et ses co-chercheurs, indiquent qu'il est possible de détecter les interruptions de détonation qui constituent actuellement un grave problème dans l'exploitation souterraine.

L'exposé "Analyse photographique par méthode digitale de la fragmentation induite par explosifs" de N.H. Maerz et de ses co-chercheurs illustre de façon instructive comment on utilise les ordinateurs pour trouver de nouvelles façons de résoudre des problèmes techniques difficiles.
J. Brych et ses co-chercheurs présentent les résultats de recherche impressionnants obtenus dans le cadre de plusieurs thèses de maîtrise et de doctorat dans un exposé intitulé "Contribution à l'étude de destructibilité des roches en forage rotatif".

3. RÉALISATIONS FUTURES

Lorsqu'on envisage l'avenir, il y a plusieurs aspects à considérer. L'un de ces aspects est la période de temps à prévoir. Au cours de la discussion portant sur les besoins de recherche futurs, j'ai plutôt adopté une perspective à court terme. Que les scientifiques doivent considérer le proche avenir est discutable. Selon moi, cependant, il se produit actuellement ou il se produira bientôt des changements qui auront des répercussions importantes sur le choix des domaines de recherche.

Avant d'entreprendre une recherche, il faut d'abord choisir le domaine technique dans lequel travailler et choisir ensuite le problème scientifique à résoudre et, enfin, déterminer le type de méthodes à utiliser (essais en laboratoire, essais sur le terrain ou modalisation informatique). En situation idéale, il faudrait recourir à toutes ces techniques pour solutionner un problème mais ce n'est certainement pas toujours le cas.

Dans la section suivante, quelques réalisations techniques importantes sont présentées avec les conséquences prévues pour chacune. Les besoins de recherche futurs seront ensuite énumérés.

Réalisations futures	Conséquences
Nouveaux matériaux	L'équipement de forage à grande vitesse et à grande puissance réduira le coût de percement des trous.
Jets d'eau à moyenne haute pression	Des tunneliers et des machines de traçage seront utilisés pour les travaux préparatoires et l'extraction dans les mines en roche dure.
Équipement de grande capacité	La grande disponibilité de l'équipement sera un critère très important.
Exploitation et excavation souterraine à plus	Des trous plus grands (de 50 à 100 mm de diamètre, de 10 à 50 m de longueur) seront même

grande échelle	utilisés dans le cadre d'exploitations à petite échelle.
Augmentation de l'automatisation et de la robotisation	Précision accrue des opérations de forage et de sautage.
Surveillance et diagraphie accrue des forages	Augmentation spectaculaire de l'acquisition, de l'analyse, du stockage et de la présentation de données géotechniques.
Manipulation des données	
Amorces de précision et explosifs de force variable et injectables par pompage.	Augmentation spectaculaire des possibilités de conception du sautage.

4. BESOINS EN MATIÈRE DE RECHERCHE

Compte tenu des réalisations mentionnées et de leurs conséquences, les projets de recherche suivants sont proposés.

4.1 Forage et excavation mécanique

Études de cas mettant l'accent sur de nouvelles expériences.

Élaboration de modèles d'élimination des débris de forage.

Élaboration de modèles de prévision appliqués aux machines de traçage en roche dure.

Mise au point de méthodes rapides pour mesurer la longueur des forages et leur déviation, de modèles et de techniques pour réduire la déviation des forages.

Mise au point de nouvelles méthodes de cueillette de données géotechniques:

Mesure en cours de forage avec les marteaux perforateurs, y compris l'élaboration de modèles.

Échantillonnage de débris de forage et analyses chimiques et physiques, y compris l'élaboration de modèles.

Foration au marteau suivie de diagraphie géophysique du forage, y compris l'élaboration de modèles.

Élaboration de modèles de forage et de fragmentation mécanique.

Mise au point de systèmes intégrés d'acquisition, d'analyse, de stockage et de présentation efficace de données géotechniques.

4.2 Sautage

Cueillette de données et élaboration d'autres modèles empiriques et conceptions de sautage en mettant l'accent sur les problèmes réels et les nouveaux explosifs.

Mise au point de méthodes rapides de surveillance du rendement du sautage relativement à la durée du retard, à la détonation de la charge totale, aux vibrations et à la fragmentation.

Mise au point de nouvelles méthodes de prédécoupage et de tir périphérique avec de nouveaux explosifs.

Perfectionnement de modèles de sautage théorique (principe), tenant compte de l'influence notamment du type de roche, du comportement du massif rocheux et des explosifs.

Thème 4
Souterrains en massifs rocheux sous grandes contraintes

Reconnaissances des sites et caractérisation des massifs rocheux; méthodes de conception; soutènement et renforcement; coups de charge et séismicité; mécanismes de rupture des excavations souterraines; venues de gaz et autres catastrophes éventuelles; prévision; surveillance et rétroanalyse

Président: W. Bamford (Australie)
Secrétaire: J. Nantel (Canada)
Conférencier thématique: H. Wagner (Afrique du Sud)

CALCUL ET SOUTÈNEMENT DES EXCAVATIONS SOUTERRAINES DANS LA ROCHE SOUMISE À DE FORTES CONTRAINTES

RÉSUMÉ

Dans l'exploitation minière et, dans une moindre mesure, dans les travaux de génie civil, il arrive souvent que la roche se fracture autour des excavations souterraines sans qu'on n'y puisse rien. Dans ces circonstances, l'accent en matière de calcul des excavations doit être mis sur l'étude de la zone de fracturation et sur le soutènement de la roche fracturée. Le manque général de données factuelles sur la résistance et, en particulier, sur les propriétés après rupture des masses rocheuses dénote que des méthodes de calcul semi-empiriques s'imposent. Dans la roche dure et cassante, une analyse d'élasticité, combinée à une connaissance des phénomènes fondamentaux de fracturation de la roche,constitue une base utile pour les calculs d'excavation. Le contrôle de la déformation après rupture de la roche fracturée est le principe des modèles de soutènement des masses rocheuses soumises à de fortes contraintes car il permet non seulement à la roche de conserver sa résistance intrinsèque, mais limite aussi l'étendue de la zone de fracturation. Le moyen le plus direct pour respecter ces conditions est de recourir à des systèmes de soutènement intégré dans lesquels on utilise des armatures de précontrainte, des grillages métalliques et du béton projeté. Les forces sismiques jouent un rôle négligeable dans le calcul et le soutènement des excavations soumises à de fortes contraintes, sauf lorsque les excavations sont situées très près des épicentres.

INTRODUCTION

L'homme, dans son effort pour extraire des minéraux, mettre sur pied des infrastructures de transport et répondre aux besoins énergétiques de la société moderne, s'enfonce de plus en plus profondément dans la croûte terrestre. Il a réussi à forer des trous de sonde jusqu'à des profondeurs de plus de 12 000 m, à percer des montagnes de tunnels de plus de 3000 m de longueur et à exploiter des "reefs" aurifères jusqu'à des profondeurs de plus de 3800 m. L'excavation de la roche à ces profondeurs pose des problèmes technologiques uniques et présente des dangers uniques.

C'est probablement en Afrique du Sud qu'on peut se procurer le plus d'information sur l'excavation de la roche en profondeur, car on y abat chaque année plus de 100 millions de tonnes de minerai aurifère dans des reefs à une profondeur moyenne de 1600 m. Les niveaux de travail les plus profonds sont situés à 3800 m sous la surface, et on a entrepris, pour extraire la roche de reef, la réalisation de réseaux de puits qui s'enfoncent à des profondeurs dépassant de beaucoup les 4000 m. Chaque année, plus de 800 km de puits, de cheminées à minerai, de tunnels et de galeries de roulage sont aménagés dans les mines d'or d'Afrique du Sud pour faciliter l'extraction de minerai de reefs étroits à ces profondeurs. Même si les problèmes de mécanique des roches liés aux reefs tabulaires étroits sont uniques, beaucoup d'autres problèmes de pression exercée par les roches se posent dans ces mines profondes qui sont d'un intérêt et d'une importance plus grands pour les ingénieurs en mécanique des roches.

Le but de cet article général est de faire état de certains des résultats les plus importants concernant le comportant des excavations dans des terrains soumis à de fortes contraintes et de dégager certains critères généraux de calcul et de soutènement de ces excavations. Une grande partie de l'information sur le comportement des excavations soumises à de fortes contraintes a trait à des excavations situées dans de la roche quartzitique relativement résistante. Cependant, de l'information sur le comportement des excavations dans d'autres formations géologiques soumises à de fortes contraintes a été incluse à des fins de comparaison et pour permettre de tirer des conclusions plus générales.

Comme l'expérience sur le comportement des excavations dans la roche soumise à de fortes contraintes a été principalement acquise dans des mines profondes, l'accent sera mis sur cet aspect de l'ingénierie des roches. En exploitation minière, de nombreux aspects du comportement d'excavations soumises à de fortes contraintes peuvent être étudiés, lesquels, pour diverses raisons, ne pourraient être observés dans des travaux souterrains de génie civil.

ASPECTS TECHNIQUES DES EXCAVATIONS SOUMISES À DE FORTES CONTRAINTES

Dans le contexte du présent article, l'ingénierie des masses rocheuses soumises à de fortes contraintes se rapporte à des situations où les contraintes qui résultent de la construction d'un souterrain sont supérieures à la résistance de la masse rocheuse. Par conséquent, il y a aura fracturation de la masse rocheuse autour de l'excavation. Dans une telle situation, l'objectif de l'ingénierie des roches comporte deux volets : calculer la forme et la taille de l'excavation de façon à minimiser l'étendue de la zone de fracturation et choisir un système de soutènement fonctionnel et économique pour contrôler la zone de fracturation autour des excavations.

Il s'ensuit que la fracturation de la roche est une conséquence inévitable des travaux d'excavation souterraine dans une masse rocheuse soumise à de fortes contraintes. Par conséquent, une connaissance détaillée des propriétés de résistance de la masse rocheuse et des processus de fracturation de la roche sont des aspects essentiels du calcul de l'excavation. Il est aussi important de connaître l'ensemble des contraintes auxquelles sont soumises la masse rocheuse dans laquelle l'excavation sera réalisée. En exploitation minière, un autre aspect mérite qu'on s'y attarde, notamment les variations des contraintes qui ont lieu après l'excavation à cause d'autres travaux miniers.

Un aspect important du calcul et du soutènement des excavations soumises à de fortes contraintes est la formation de zones de fracturation distinctes autour du souterrain. L'objectif du processus de l'ingénierie des roches est d'établir la séquence des travaux d'excavation et la géométrie du souterrain de façon que la zone de fracturation soit la plus favorable possible sur le plan du soutènement. Par contre, le calcul des excavations dans une roche soumise à des contraintes faibles à moyennes est fortement lié aux discontinuités géologiques existantes le long desquelles il peut se produire des mouvements dus à la gravité.

Toute analyse du comportement des excavations soumises à de fortes contraintes doit tenir compte de la formation de nouvelles fractures ainsi que des facteurs qui contrôlent ces fractures, et ne peut se limiter à la seule analyse des discontinuités géologiques existantes. Même si ces dernières sont importantes dans le calcul des excavations souterraines, elles ont tendance à être moins déterminantes dans une roche soumise à de fortes contraintes. Néanmoins, ces discontinuités et, notamment, les différences marquées entre les propriétés mécaniques de différentes couches rocheuses peuvent avoir un effet majeur sur le mécanisme de formation des fractures dans une roche soumise à de fortes contraintes. Une telle situation peut se présenter dans une mine de charbon profonde où il existe des différences considérables entre les propriétés mécaniques de la couche de charbon et celles des couches environnantes. Cependant, même des différences relativement faibles peuvent avoir un effet marqué sur la formation de fractures autour des excavations dans les mines de roche dure profondes.

RÉSISTANCE ET FRACTURATION DE LA ROCHE

Propriétés de résistance

Il est juste de dire que de plus grands efforts de recherche ont été consacrés à la clarification des propriétés de résistance qu'à tout autre aspect du comportement de la roche. Il n'en reste pas moins difficile d'établir des prévisions fiables sur l'effondrement de la roche autour des excavations souterraines.

Cette difficulté tient à plusieurs raisons. Premièrement, les propriétés de résistance de la roche sont habituellement déterminées à partir de petits échantillons de fragments intacts de roche qui sont soumis à des essais dans des conditions de sollicitation artificielles. Aux fins des calculs, il est difficile, voire impossible, d'extrapoler les résultats obtenus de cette façon à cause de la présence de diaclases, de fissures et d'autres plans de faiblesse qui peuvent avoir un effet important sur la résistance de la masse rocheuse. Des efforts remarquables ont été déployés au cours des dernières années pour expliquer ces effets. Une des thèses les plus prometteuses est celle du critère de rupture empirique défini par Hoek et Brown (1982, page 137).

$$\sigma_1 \leq \sigma_3 + \sqrt{(m\sigma_c\ \sigma_3 + s\sigma_c^2} \qquad (1)$$

où σ_1 et σ_3 sont les contraintes principales, majeure et mineure, au point de rupture, σ_c est la résistance à la compression uniaxiale de la masse rocheuse intacte, et "m" et "s" sont des constantes qui sont fonction des propriétés de la roche et de l'étendue de la zone de rupture avant l'apparition des contraintes σ_1 et σ_3. Pour une masse rocheuse intacte, s = 1 et m varie entre 7 et 25 selon le type de roche.

La deuxième source de difficulté est la définition de la résistance elle-même, comme l'a souligné Salamon en 1974. La définition la plus répandue s'appuie sur l'inégalité suivante :

$$\sigma_1 \leq a\sigma_3{}^b + \sigma_c \qquad (2)$$

où le paramètre "a" représente l'augmentation de résistance de la roche en fonction de la contrainte σ_3; l'exposant est une mesure de la linéarité de la relation et est en général voisin de l'unité. Dans de nombreux cas, il convient donc de simplifier l'équation (2) ainsi :

$$\sigma_1 \leq a\sigma_3 + \sigma_c \qquad (2a)$$

Ce qui est important de retenir de cette inégalité, et incidemment aussi du critère de rupture de Hoek et Brown, l'équation (1), c'est qu'ils définissent la limite supérieure de de résistance de la roche pour une contrainte donnée, mais ne permettent de tirer aucune conclusion en ce qui a trait à la résistance ou à l'état de la roche lorsque cette inégalité est satisfaite.

Une troisième source de difficulté est la modification importante des propriétés de la roche lorsque ces inégalités sont satisfaites. Pour tenir compte de ce changement de propriété et, notamment, du niveau réduit de confinement qu'assure la masse rocheuse fracturée autour d'une excavation soumise à de fortes contraintes, Hoek et Brown (1982) ont modifié leur critère initial, l'équation (1), en introduisant une cote pour la masse rocheuse effondrée ou fracturée. Conformément à ce principe, les valeurs de m et de s de la masse rocheuse fracturée sont réduites selon le degré de fracturation et de diaclasage.

D'après Hoek et Brown, les valeurs de m_r et de s_r pour une masse rocheuse fracturée et diaclasée peuvent être estimées à partir des systèmes de classification des masses rocheuses NGI ou CSIR. La figure 1 met en relation la cote CSIR et le facteur de réduction des valeurs de m et de s qu'il faut appliquer pour tenir compte du degré de diaclasage et d'altération de la masse rocheuse. Le critère de Hoek-Brown pour les masses rocheuses fracturées et diaclasées a été appliqué avec beaucoup de succès à l'analyse des excavations dans les mines profondes. Toutefois, il faut faire bien attention lorsqu'on introduit de telles corrections car elles sont étroitement liées aux géométries de confinement et aux effets de soutènement. Par exemple, des études de Wagner et de Schümann (1971) ont montré que l'étendue réelle de la zone de fracturation au-dessus et au-dessous de piliers soumis à de fortes contraintes avait tendance à être inférieure à celle calculée au moyen des équations 1 ou 2. La raison en est que ces équations ne tiennent pas compte des contraintes locales de confinement qui résultent de la dilatation de la roche fracturée.

Quatrièmement, l'espacement des diaclases et les autres faiblesses structurales tels que les plans de stratification et de séparation, par rapport à la dimension linéaire critique de l'excavation, sont des facteurs importants qui déterminent l'étendue de la zone de fracturation. Cet aspect est d'une importance particulière dans le calcul des excavations dans des formations rocheuses sédimentaires, par exemple des galeries dans des houillères profondes ou de grandes ouvertures dans des quartzites ou des formations de schiste argileux et de grès feuilleté (Jacobi, 1976; Piper, 1984).

Une des grandes lacunes de la plupart des systèmes de classification des masses rocheuses en usage est qu'ils ne tiennent pas compte de l'effet des dimensions de l'excavation (Piper, 1985).

Dans le roc massif, une relation utile est celle établie par Hoek et Brown (1982) entre la résistance à la compression uniaxiale et le diamètre de l'échantillon.

$$\sigma_c = \sigma_{c50}(^{50}/d)^{0,18} \qquad (3)$$

où σ_{c50} est la résistance à la compression uniaxiale d'un échantillon de roche de 50 mm de diamètre et "d" est le diamètre de l'échantillon.

L'équation 3 a été utilisée avec succès pour estimer la fracturation dans des tunnels de 2 à 3 m de diamètre creusés dans ces quartzites massifs (figure 2).

Cinquièmement, la présence de couches de roche tendre dans les parois des excavations soumises à de fortes contraintes peut avoir un impact majeur sur le comportement de l'excavation et sur le mode de fracturation (Jacobi, 1976; Kersten et coll., 1983; Reuther, 1987). La présence de couches de roche tendre influe de plusieurs façons sur le comportement des excavations. Premièrement, si la couche de roche tendre est relativement épaisse, la roche se déformera surtout dans cette couche. Deuxièmement, comme la roche tendre a un faible module de déformation et un coefficient de Poisson généralement élevé, il s'établit de fortes contraintes de cisaillement à l'interface des couches de roche dure et de roche tendre lorsque les conditions de charge varient comme c'est le cas pendant une excavation (Brummer, 1984), (figure 3).

Comme la roche est beaucoup moins résistante en traction qu'en compression, il se développent des fractures de traction dans les couches de roche dure qui sont prises en sandwich entre des couches de roche tendre, et les parois rocheuses s'effondrent sous des contraintes beaucoup plus faibles que prévu par les essais de compression effectués sur des échantillons de la couche de roche dure. Laubscher (1984) a mis au point un nomogramme pour estimer la résistance des parois rocheuses constituées de couches de roche dure et de roche tendre (figure 4).

Fracturation de la roche

La figure 5 montre les différents modes de fracturation de la roche qui sont observés dans les excavations soumises à de fortes contraintes verticales. Dans la roche massive et cassante de la figure 5a), des fractures d'extension se forment dans les murs de l'excavation. L'étendue et l'espacement de ces fractures dépendent de la grandeur des contraintes et de la fragilité de la roche. La déformation de la roche autour des excavations situées dans des masses rocheuses diaclasées est contrôlée par les contraintes de cisaillement et les forces de gravité, comme l'indique la figure 5b). La présence de couches de roche tendre dans les murs des excavations favorise la formation de fractures d'extension dans les couches de roche cassante qui sont prises en sandwich entre les couches de roche tendre. Ces fractures d'extension ont tendance à se produire sous des contraintes beaucoup plus faibles que celles observées en l'absence des couches de roche tendre. En outre, ces fractures se propagent dans les murs beaucoup plus loin que cela n'est le cas dans la roche massive cassante, comme l'indique la figure 5c). Dans la figure 5d), des plans de stratification caractérisés par un faible coefficient de frottement favorisent la formation de fractures en coins qui peuvent s'étendre loin dans les murs de l'excavation. De plus, de fortes contraintes latérales sont induites dans les couches environnantes. Selon la nature du milieu environnant, des fractures d'extension ou des phénomènes de flambage et de plissement de couches feuilletées sont observés, d'après Jacobi (1976) et Reuther (1987).

En conclusion, il est raisonnable d'affirmer que, malgré les nombreux travaux qui ont été effectués sur les propriétés de résistance de la roche et des masses rocheuses et sur la fracturation de la roche, il y a beaucoup d'autres avenues à explorer. Du point de vue de l'ingénieur d'études, la situation actuelle est loin d'être satisfaisante, et il faudra s'engager dans de nouvelles avenues pour réaliser vraiment des progrès. La mise au point de modèles numériques perfectionnés et l'accès à des ordinateurs puissants indiquent que l'analyse rétrospective de cas réels est la route à suivre. La puissance inhérente de cette méthode a été démontrée par Salamon (1967) lorsqu'il a mis au point une méthode de calcul des aménagements par chambres et piliers dans des mines de charbon d'Afrique du Sud, méthode basée sur l'analyse de 125 aménagements par chambres et piliers, certains étant restés intacts, d'autres ayant cédé aux contraintes. A noter à cet égard les efforts déployés par des techniciens en mécanique des roches dans des mines d'or d'Afrique du Sud pour relier le comportement des murs de tunnel à la résistance à la compression uniaxiale de la formation rocheuse et aux contraintes qui agissent sur ces tunnels (Cook, 1972; Ortlepp et coll., 1972; Wiseman, 1979; Hoek et Brown, 1982; Piper, 1984).

CALCUL DES EXCAVATIONS SOUTERRAINES

Généralités

La démarche traditionnelle de l'ingénieur mécanicien et de l'ingénieur civil est de s'assurer que la structure soit stable en la calculant de façon que les contraintes dans chaque élément de la structure soient toujours inférieures à la résistance de cet élément. Selon l'importance de la structure et les conséquences d'une rupture structurale, on définit un facteur de sécurité approprié. Les objectifs du calcul consistent alors à choisir des matériaux appropriés pour la structure, en prévoyant des sections suffisantes pour les membres de la structure soumise à des contraintes et en évitant que ne s'exercent des conditions de charge non favorables. Pour l'ingénieur en mécanique des roches, cette liberté de choix est toutefois fortement limitée par le but particulier de l'excavation souterraine, but qui détermine souvent le lieu d'excavation et, partant, la formation géologique dans laquelle elle sera située. En outre, le champ de contraintes dans lequel l'ouverture devra être creusée est souvent prédéterminé par les contraintes locales, et la seule liberté dont jouit l'ingénieur est celle de choisir la forme de l'excavation, le système de soutènement et la séquence des travaux d'excavation.

Stabilité des excavations et fracturation de la roche

Dans le calcul et le soutènement des excavations soumises à de fortes contraintes, il appert immédiatement que l'ingénieur ne peut adopter une démarche traditionnelle car, en termes de la définition susmentionnée, les contraintes qui agissent sur la roche autour de l'excavation sont supérieures à la résistance du matériau rocheux, et la fracturation est inévitable. Cependant, cela ne signifie pas nécessairement que la stabilité de l'excavation est menacée. En effet, il arrive souvent que la roche se fracture par endroits autour des excavations souterraines et que ces excavations demeurent stables même sans être soutenues. Le critère voulant que les contraintes doivent toujours être inférieures à la résistance de la roche peut donc être une condition suffisante pour assurer que la structure soit stable, même si cette condition n'est pas nécessaire. Cela s'explique du fait que la plupart des roches après rupture sont encore capables de supporter certaines contraintes. D'après Jaeger et Cook (1979, p. 80), un matériau est dit ductile tant qu'il peut supporter une déformation permanente sans perdre sa capacité de résister à une charge. Un matériau est dit fragile ou cassant lorsque sa capacité de résister à une charge décroît lorsque la déformation augmente.

Dans les excavations souterraines, et notamment

dans les masses rocheuses soumises à de fortes contraintes, il arrive souvent qu'une partie de la roche encaissante devienne cassante. Par conséquent, il faut tenir compte du comportement après rupture de la roche, tel que défini par la partie décroissante de la courbe contrainte-déformation, si l'on veut bien comprendre le mécanisme des excavations souterraines.

La figure 6a) montre un diagramme idéal de compression en fonction de la charge pour une roche cassante. Ce diagramme indique aussi la relation force-déplacement du système de charge tel que défini par la pente, k, de la courbe de charge.

La figure 6b) montre la variation de la pente, λ, de la courbe force-déplacement de la roche et la valeur de la pente, k, du système de charge. D'après Salamon (1974), le système est en équilibre stable si et seulement si

$$k + \lambda > 0 \qquad (4)$$

Comme, par définition, k est positif, l'inégalité (4) indique que l'équilibre sera stable, peu importe la rigidité du système de charge, tant que la résistance de la roche à la déformation augmente en fonction de la déformation (région OB). Dans la région après rupture, BC, l'équilibre sera stable jusqu'au point auquel la courbe de charge du système devient tangentielle à la courbe de force-déplacement, $P = P_S$. A ce point, il y aura rupture violente de la roche, mais cette rupture violente ne se produira pas tant que la pente minimale de la courbe force-déplacement de la roche, λ_{min}, satisfait l'inégalité $k + \lambda_{min} > 0$.

Il découle de cette analyse que le calcul des excavations dans la roche soumise à de fortes contraintes ne doit pas être confiné à l'analyse des zones de fracture autour des excavations. Il faut aussi tenir compte de la rigidité de la masse rocheuse encaissante qui, ajoutée au comportement après rupture de la roche, détermine s'il y aura vraisemblablement rupture violente de la roche.

En regard de ce dernier facteur, Jaeger et Cook (1979, p. 472-474) ont examiné le cas idéal d'une ouverture circulaire, sujette à une pression hydrostatique, p, telle que les contraintes exercées dans un anneau mince de roche d'épaisseur, t, autour de l'ouverture de rayon R, dépassent à peine la résistance de la roche. Si, pour r > R, la roche autour de l'ouverture se comporte linéairement et élastiquement, la rigidité de la contrainte radiale exercée sur l'anneau par la roche encaissante est donnée par

$$k_R = \sigma_R/\varepsilon_R = 2G \qquad (5)$$

où G est le module de rigidité de la roche. Il s'ensuit l'inégalité suivante

$$\left|\frac{t}{R}\frac{d\sigma}{d\varepsilon}\right| < |\,2G\,| \qquad (6)$$

qui devient le critère de stabilité de l'anneau de roche après rupture en fonction des contraintes exercées sur lui par la roche encaissante. Comme dans la plupart des cas t/R 1, il s'ensuit qu'un anneau mince de roche fracturée est considéré comme "mou" par rapport à la rigidité de la roche encaissante, et donc comme stable.

Quoique très simplifié, cet exemple illustre néanmoins que la fracturation de la roche autour de tunnels creusés dans une masse rocheuse non perturbée sur le plan géologique sera vraisemblablement stable. Ce résultat est corroboré par un grand nombre d'observations souterraines qui montrent que la fracturation autour des tunnels et des cavernes serait un processus stable.

Nous attirons votre attention sur le travail de Deist (1965) et d'Ulgudur (1973) qui ont étudié dans les fins détails le problème de la formation de fractures stables ou instables autour de tunnels soumis à de fortes contraintes.

Il est important de noter que ces observations ne s'appliquent pas aux excavations souterraines dont la géométrie diffère beaucoup de celle des tunnels. En particulier, la rupture de piliers de soutien dans des exploitations de gisements tabulaires peut entraîner des effondrements régionaux catastrophiques de roche instable qui libèrent de grandes quantités d'énergie sismique. De même, la fracturation de la roche autour d'excavations dans des formations tabulaires soumises à de fortes contraintes peut être instable, particulièrement s'il y a des perturbations géologiques (Heunis, 1980).

Facteurs énergétiques

Au début des années 1960, Cook (1963) a attiré l'attention sur les fortes variations énergétiques qui ont cours pendant l'exploitation minière. Il affirme notamment que "un excès d'énergie potentielle entraîne des dommages appelés coups de toit". Depuis ce temps, diverses grandeurs énergétiques, et en particulier le taux de libération d'énergie, sont devenues des paramètres de calcul très répandus dans l'exploitation des gisements profonds.

En 1974 et 1984, Salamon a examiné certains des aspects fondamentaux des variations énergétiques produites lors de l'élargissement d'une excavation. Salamon souligne qu'à cause de la taille ou du nombre accru des excavations minières, il se produit des déplacements dans la roche encaissante. Par ces déplacements, les forces extérieures et les forces exercées dans la masse effectuent un certain travail, W. Ce travail est généralement qualifié de variation d'énergie "gravitationnelle" ou "potentielle".En outre, une certaine quantité d'énergie de déformation est stockée dans la roche, énergie qui est libérée pendant l'exploitation minière, U_m. La somme $(W + U_m)$ est l'apport d'énergie dont il faut tenir compte et qui doit être dépensé d'une manière ou d'une autre.

Cette énergie est partiellement prise en compte par une augmentation du contenu en énergie de déformation de la masse rocheuse environnante, U_c. Lorsque les excavations sont remblayées ou supportées, une certaine quantité de travail est effectuée pour déformer les supports, W_s. Si la masse rocheuse est un milieu élastique, aucune énergie n'est dissipée dans la fracturation ou dans une déformation non élastique. Par conséquent, l'énergie dépensée pendant une opération minière est $(U_c + W_s)$.

Salamon conclut que ces processus simples n'expliquent pas la totalité des apports en énergie et qu'une certaine quantité d'énergie

$$W_r = (W + U_m) - (U_c + W_s) > 0 \qquad (7)$$

doit être libérée ou dissipée d'autre façon. Le seuil d'énergie libérée est donc donné par

$$W_r \geq U_m > 0 \qquad (8)$$

Dans le cas d'excavations dans des formations tabulaires, l'énergie U_m et, par conséquent, la limite inférieure de W_r peuvent extimées à partir de la moyenne des composantes de contrainte, $\sigma_{k3}^{(p)}$, et de la convergence, $S_k^{(i)}$, avant et après l'enlèvement du minéral sur la petite surface ΔA où $\sigma_{k3}^{(p)}$ (k = 1, 2, 3,) sont les composantes du tenseur de contrainte agissant sur la surface A avant l'exploitation de cette surface et $S_k^{-(i)}$ sont les composantes du vecteur de déplacement relatif après l'exploitation de la même région.

$$\Delta U_m = \int_{\Delta A} \tfrac{1}{2} S_k^{(i)} \sigma_{k3}^{(p)} \, \Delta A \qquad (9)$$

Ces composantes peuvent être calculées directement au moyen d'un réseau analogue de résistances électriques ou d'un des nombreux simulateurs numériques d'exploitation de formations tabulaires qui ont été mis au point au cours des dernières années (Ryder, 1986).

L'expérience acquise dans les mines d'or profondes montre que le taux $\Delta U_m/\Delta A$ est un paramètre important pour la prévision des problèmes de maintien d'un contrôle satisfaisant des couches dans les chantiers d'exploitation profonds, d'après Hodgson et Joughin (1966), Heunis (1980), Salamon et Wagner (1979). Un exemple de la relation qui existe entre $\Delta U_m/\Delta A$ et le nombre de coups de toit par 1000 m^2 de surface exploitée est donné à la figure 7.

Cependant, peu de travaux ont été faits concernant les variations d'énergie dans des excavations souterraines non tabulaires. Salamon (1984) a étudié les variations énergétiques résultant d'une augmentation soudaine du rayon d'un tunnel circulaire d'une valeur a à une valeur c. L'expression suivante du taux de libération d'énergie, W_r, a été obtenue pour un tunnel sans soutènement, soumis à une pression hydrostatique de grandeur p

$$W_r = \frac{2(1 - \nu^2)p^2}{E} V_m \qquad (10)$$

où V_m est le volume de roche à abattre par unité de longueur de tunnel

$$V_m = \pi(c^2 - a^2) \qquad (11)$$

L'énergie cinétique potentielle qui est libérée lors d'une augmentation soudaine du rayon du tunnel est donnée par

$$W_k = \frac{(1 + \nu)p^2}{E} (1 - \frac{a^2}{c^2}) V_m \qquad (12)$$

Ce résultat est d'une importance considérable car il indique qu'une quantité importante d'énergie cinétique peut être libérée lorsque le diamètre d'un tunnel est augmenté soudainement (figure 8).

Finalement, on compare ci-dessous l'énergie libérée pendant l'extraction d'un gisement tabulaire situé à une profondeur de 3000 m avec l'énergie libérée lors du creusement d'un tunnel à la même profondeur.

La limite supérieure du taux de libération d'énergie la plus probable lors d'un abattage extensif est donnée par

$$\Delta U_m/\Delta A = S_m \sigma_v \qquad (13)$$

où S_m est la hauteur d'abattage et σ_v est la composante verticale du tenseur de contrainte vierge. Si la hauteur d'abattage est S_m = 1 m et la contrainte verticale est σ_v = 80 MPa, la valeur de $\Delta U_m/\Delta A$ est de 80 MJ/m^2.

Par contre, le taux de libération d'énergie résultant de l'excavation d'un tunnel de 2 m de rayon dans un champ de contrainte de 80 MPa est de l'ordre de 0,2 MJ/m^3, soit plus de deux ordres de grandeur inférieur à celui observé dans un chantier d'abattage.

Cet exemple montre que les facteurs énergétiques sont probablement moins importants dans le calcul des tunnels profonds et des cavernes aménagées dans la roche qu'ils ne le sont pour le calcul des excavations pour fins d'exploitation minière. Cependant, à noter que le fait que les taux de libération d'énergie résultant de l'aménagement de tunnels profonds sont du même ordre de grandeur que ceux résultant de la rupture d'un mètre cube de roche dure sous l'effet de la compression.

Forme de l'excavation et contraintes

Le choix de la forme de l'excavation est l'une des décisions les plus importantes que doit prendre l'ingénieur en mécanique des roches, comme il a été mentionné plus tôt. Cela est particulièrement le cas lorsque les excavations sont soumises à de fortes contraintes.La forme de l'excavation et son orientation par rapport aux contraintes locales non seulement détermine les contraintes aux limites, mais aussi régit la formation de la zone de fractures et influe sur les décisions en matière de soutènement.

Dans un milieu où les contraintes sont faibles, ce sont normalement les discontinuités géologiques et structurales qui contrôlent le comportement de l'excavation et qui déterminent les exigences en matière de soutènement. Dans ces circonstances, la forme de l'excavation est habituellement choisie de façon à offrir la géométrie la plus favorable du point de vue du contrôle des mouvements le long des diaclases et de la préservation de la résistance naturelle des formations rocheuses.Par exemple, dans la roche stratifiée, il peut être très profitable d'utiliser des plans de stratification ou de séparation bien définis comme toit de l'excavation ou de choisir une des couches les plus fortes et les plus compétentes comme toit naturel même si elle n'offre pas la géométrie la plus favorable du point de vue des contraintes.

À mesure que les contraintes augmentent par rapport à la résistance de la masse rocheuse dans laquelle l'excavation est réalisée, il faut accorder plus d'attention au choix de la géométrie la plus favorable du point de vue des contraintes. L'expérience acquise dans l'aménagement de tunnels dans des mines profondes de l'Afrique du Sud montre que dans le cas de tunnels à section carrée de 3 ou 4 m de largeur, les murs s'effritent légèrement lorsque la composante verticale, σ_v, du tenseur de contraintes dépasse une valeur d'environ 0,2 fois la résistance à la compression uniaxiale, σ_c, de la roche, telle que déterminée en laboratoire. On observe un effritement marqué des murs lorsque $\sigma_v/\sigma_c > 0{,}3$, et il faut prévoir d'importants ouvrages de soutènement lorsque ce rapport dépasse une valeur d'environ 0,4 (Ortlepp et coll., 1972).

Quoique très utile dans les mines d'or, le simple critère des contraintes exercées dans le terrain, qui est basé sur la composante verticale du tenseur de contraintes ne tient pas compte de l'effet des autres composantes de contrainte sur le comportement des excavations souterraines. Wiseman (1979) en a tenu compte en définissant un facteur de concentration des contraintes exercées sur les murs, SF,

$$SF = \frac{3\sigma_1 - \sigma_3}{\sigma_c} \qquad (14)$$

où σ_1 et σ_3 sont les contraintes principales, maximale et minimale, agissant sur l'excavation, et σ_c est la résistance à la compression uniaxiale de la roche telle que déterminée en laboratoire. Dans une étude détaillée sur plus de 20 km de tunnels typiques aménagés dans des mines d'or, Wiseman a observé que l'état des tunnels non soutenus se détériorait considérablement lorsque le facteur de concentration des contraintes dans les murs atteignait une valeur d'environ 0,8 (figure 9).

Ces exemples montrent que dans une formation de roche massive, les contraintes aux limites, lorsque normalisées par rapport à la résistance uniaxiale de la formation rocheuse, peuvent être de bons indicateurs de problèmes d'excavation éventuels. Il faut attirer l'attention sur les difficultés que posent une estimation raisonnable de la résistance à la compression uniaxiale de la masse rocheuse. Ces difficultés ont été abordées précédemment, mais doivent être soulignées de nouveau.

Une première étape dans le calcul consiste donc à

déterminer les contraintes qui agissent à la limite de l'excavation projetée. Une condition essentielle est une connaissance des contraintes locales exercées à l'emplacement prévu de l'excavation. Dans une exploitation minière, il convient de modéliser les effets généralement très étendus des excavations, sur les contraintes qui s'exercent à l'emplacement prévu de l'ouverture. Un grand nombre de programmes d'analyse par éléments aux limites et par éléments finis ont été mis au point à cette fin. Dans le cas d'excavations de génie civil, il peut s'avérer nécessaire de mesurer les contraintes sur place. Là où ces mesures sont impossibles à effectuer, il faudrait relever les tendances régionales dans le champ des contraintes lorsqu'on veut évaluer l'état des contraintes au lieu d'excavation prévu. En règle générale, la composante verticale du tenseur de contraintes est voisin de la contrainte exercée par les morts-terrains, telle que calculée à partir de l'épaisseur de la couche de morts-terrains et de la densité des formations rocheuses. Toutefois, il arrive souvent que même la direction et la grandeur des contraintes horizontales sont incertaines, selon Hoek et Brown (1982).

Une première approximation, souvent suffisante, des contraintes aux limites autour d'une ouverture souterraine peut être rétablie à partir des relations suivantes de Brady et Brown (1985, p. 195).

$$\sigma_A = p(1 - K + \sqrt{\frac{2W}{\rho A}}) \qquad (15)$$

$$\sigma_B = p(K - 1 + K\sqrt{\frac{2H}{\rho B}}) \qquad (16)$$

où σ_A et σ_B sont les contraintes circonférentielles aux limites dans le mur (A) et dans le toit (B) de l'excavation, ρ_A et ρ_B sont les rayons de courbure aux points A et B et K est le rapport des contraintes horizontales aux contraintes verticales (figure 10).

Formations de roche massive

Dans les formations de roche massive, les problèmes d'excavation liés aux contraintes sont peu probables lorsque les valeurs des contraintes aux limites, qui ont été normalisées par rapport à la résistance à la compression de la masse rocheuse, sont bien inférieures à l'unité. Dans les cas où les contraintes aux limites normalisées approchent l'unité, il faudrait envisager des changements dans la forme de l'excavation de façon à réduire les contraintes aux limites. Cependant, il arrive souvent que les contraintes sur le terrain sont si élevées que même la géométrie la plus favorable ne suffit pas à empêcher la rupture de la roche à la limite de l'excavation. Cela survient lorsque la valeur maximale de la contrainte principale subsidiaire approche une valeur d'environ la moitié de la résistance à la compression uniaxiale de la roche, $\sigma_1 \approx 0,5\ \sigma_c$. La forme de l'excavation doit alors être calculée de façon à minimiser l'étendue de la zone de fracture et à faciliter le soutènement de l'excavation.

L'expérience acquise dans les tunnels soumis à de fortes contraintes et situés dans des quartzites massifs montre que dans ces cas, la géométrie d'excavation la plus favorable s'écarte considérablement de celle qui découle d'une analyse des contraintes élastiques. En général, dans les tunnels de section équidimensionnelle, la fracturation de la roche commence avec la formation de fractures d'extension dans les murs du tunnel, lesquels sont sujets aux contraintes de compression les plus élevées. À mesure que des pans de roc se forment de plus en plus à l'intérieur des murs du tunnel, leur longueur a tendance à diminuer et, après un certain temps, il s'établit un équilibre où il n'y a plus de fracturation. Ce processus a été décrit en détail par Fairhurst et Cook (1966). La géométrie résultante du tunnel est une ellipse allongée dont le grand axe est orienté perpendiculairement à la direction de la contrainte de compression maximale, c'est-à-dire dans le sens opposé à la forme élastique la plus favorable.

L'analyse la plus détaillée d'un tunnel situé dans un champ de contraintes très élevées est probablement celle d'un tunnel expérimental qui a été aménagé dans le roc solide au-delà d'un front de chantier dans une mine d'or profonde; cette étude de sismicité dans les mines profondes a été effectuée par Ortlepp et Gay (1984). Ce tunnel a été aménagé dans un champ de contraintes qui variait d'environ 50 MPa dans la zone d'abattage à 140 MPa dans le roc solide au-delà de la zone d'abattage. Après une intensification de l'abattage, les contraintes agissant sur le tunnel ont augmenté jusqu'à une valeur d'environ 230 MPa. Ces contraintes correspondent à une profondeur équivalente de plus 8000 m. Le tunnel lui-même était situé dans des quartzites massifs dont la résistance à la compression uniaxiale était d'environ 350 MPa. Cette valeur, une fois corrigée quant à la taille de l'excavation au moyen de l'équation (3), a été réduite à environ 200 MPa. La figure 11 montre la variation des contraintes suivant l'axe du tunnel au moment de l'excavation, ainsi que la section transversale prévue dans le plan de tir et la section résultante, en plusieurs points le long du tunnel.

Dans les zones soumises à de fortes contraintes, une bande bien définie de fracturation intense a été observée à mi-hauteur environ du front du tunnel. Cette zone d'effritement s'est transformée en une "entaille" donnant lieu à des contraintes qui s'est rapidement enfoncée plus en profondeur dans le mur à mesure que le front progressait, suivant un processus de flambage et d'éjection de "micro-pans" à la pointe de l'entaille. Ce processus essentiellement non violent de fracturation s'est poursuivi jusqu'à ce que, après un certain temps, un profil elliptique caractéristique, aigu aux extrémités, se soit formé avec une largeur se stabilisant à quelque 3 à 4,5 m. À noter que, ultimement, le tunnel a atteint une géométrie finale stable avec un minimum de soutènement. Dans une partie du tunnel, des travaux de soutènement ont été effectués très tôt dans le but premier de contenir la roche fracturée dans les murs du tunnel. Il est aussi important de noter l'absence d'une fracturation extensive dans le toit et la sole du tunnel et l'état remarquablement stable du toit.

Ces observations sont caractéristiques de tunnels creusés dans des roches cassantes massives qui sont soumises à des contraintes verticales élevées. À noter qu'un mode semblable de fracturation de la roche caractérise aussi les tunnels qui sont situés dans des zones de fortes contraintes horizontales. Dans ce dernier cas, la fracturation se produit dans le toit et la sole.

En général, on peut donc affirmer que la fracturation de la roche dans les tunnels situés dans de la roche cassante massive où existent de fortes contraintes se limite principalement aux parois du tunnel qui sont parallèles à la direction de la contrainte de compression subsidiaire maximale. Après fracturation de la roche de ces parois, la section du tunnel a tendance à prendre la forme d'une ellipse dont le grand axe est orienté perpendiculairement à la direction de la contrainte de compression maximale. Le rapport de forme de l'ellipse a tendance à augmenter lorsque la valeur de la contrainte de compression maximale augmente par rapport à la résistance de la roche. Du point de vue du contrôle de la fracturation de la roche, la forme la plus favorable des excavations situées dans des roches cassantes soumises à de fortes contraintes est donc l'opposé de celles basées sur l'analyse des contraintes s'exerçant à la limite d'une ouverture dans un milieu élastique.

On peut obtenir une première estimation de l'étendue de la zone de fracture dans les murs des tunnels soumis à de fortes contraintes en effectuant une analyse des contraintes élastiques et en établissant des courbes de calcul où les contraintes autour de l'excavation satisfont un des critères de rupture de la roche bien connus, l'équation 1 ou l'équation 2. Un des désavantages de cette méthode est qu'elle ne tient pas compte des variations des propriétés de la roche dans les zones de "rupture". Néanmoins, l'expérience a montré que cette méthode idéale permet de comparer les mérites relatifs de différentes formes d'excavation dans un environnement rocheux soumis à de fortes contraintes. La figure 12 montre l'étendue de la zone de fracture, ΔW_r, dans les murs de quatre tunnels de formes différentes en fonction du rapport de la composante verticale du tenseur de contraintes, σ_v, à la résistance à la compression uniaxiale de la roche. Les résultats indiqués correspondent à un rapport des contraintes horizontales aux contraintes verticales de $\sigma_h/\sigma_v = 0,5$. On est à même de constater que, dans la gamme des faibles contraintes, $\sigma_v/\sigma_c < 0,5$, les formes de tunnel les plus favorables sont l'ellipse allongée à la verticale et le carré. Cependant, après fracturation des parois du tunnel, $\Delta W_R/w > 0$ l'ellipse allongée à l'horizontale et la section circulaire sont les formes les plus favorables, notamment si $\sigma_v/\sigma_c > 1$.

La figure 13 montre que les formes de tunnel les mieux adaptées à l'excavation sont celles où les murs sont très courbes; on y voit la distribution des contraintes principales σ_1 et σ_3 suivant l'axe horizontal de la section du tunnel. Il ressort de ces distributions que le gradient $\sigma_1/\Delta x$ est presque nul pour les tunnels dont les murs ont un rayon de courbure très grand, notamment les tunnels à section elliptique debout et à section carrée. La valeur de la contrainte principale minimale, σ_3, près des murs des quatre tunnels, est très variable. Plus le rayon de courbure est petit, plus la contrainte principale minimale est élevée près des murs du tunnel. Comme la résistance de la roche augmente rapidement en fonction de la contrainte d'étreinte, il s'ensuit que la fracturation des murs autour des excavations présentant de faibles rayons de courbure sera confinée dans une zone très étroite.

Un autre facteur qui joue en faveur des tunnels dont les murs ont un faible rayon de courbure est l'effet des gradients de contrainte sur la résistance de la roche. Cet effet a été bien étudié par Cook et Jaeger (1979, p. 197-198) dans des essais visant à mesurer indirectement la résistance à la traction de la roche; ils ont découvert que la résistance à la traction de la roche augmente en fonction du gradient de contrainte. Un effet semblable a été observé par Wagner et Schümann (1979) qui ont étudié la résistance de la roche sous des contraintes de contact élevées. Ces auteurs ont observé un effet marqué du gradient de contrainte sur la résistance de la roche cassante. Ces observations et celles de Johnson (1985) indiquent qu'il faut tenir compte des effets des gradients de contrainte dans le calcul des excavations en profondeur.

Couches laminées

L'alternance de couches de roche dure et de roche tendre et l'existence de plans de faiblesse bien définis sont des caractéristiques souvent observées dans les formations géologiques sédimentaires. Comme ces caractéristiques peuvent avoir un effet marqué sur le comportement des excavations, il faut en tenir compte dans le calcul et le soutènement des excavations. Les connaissances les plus poussées sur le comportement des excavations dans les couches laminées sont de loin celles acquises dans les mines de charbon, et il serait des plus profitables d'étudier les travaux effectués par Jacobi (1976) et Reuther (1987) sur les houillères profondes d'Europe.

Toute excavation a pour effet de modifier la distribution des contraintes dans la masse rocheuse et de déformer la roche autour de l'ouverture. Dans le cas d'excavations dans des formations de roche cassante massive, l'accent a été mis sur la formation de zones de fracture sous l'effet de contraintes excessives exercées par la roche. Par ailleurs, dans les couches laminées, on observe que la présence de plans de stratification bien définis le long desquels la roche peut se déplacer a un effet marqué sur le comportement des excavations. Ce problème a été étudié par un grand nombre d'auteurs à l'aide de modèles de laboratoire constitués de matériaux équivalents, notamment par Everling (1962) et Buschmann (1964). Plus récemment, Purrer (1984) a mis au point des modèles numériques pour analyser les effets de couches de charbon tendre et de plans de faible résistance au frottement sur le comportement des excavations.

À partir d'une simple analyse des contraintes, Reuther (1987) a identifié des zones autour d'une galerie de mine de charbon typique où il peut y avoir glissement avec frottement le long de plans de stratification horizontaux (fig. 14). D'après cet exemple, qui suppose l'existence d'un champ de contraintes hydrostatiques, le glissement le long des plans de stratification peut se produire dans la partie supérieure de la galerie et peut s'étendre loin dans le toit ou dans les reins de voûte de l'excavation si le coefficient de frottement, μ, est faible. Dans le cas de couches épaisses, tout glissement et toute fissuration au-dessus des reins de voûte peut entraîner une rupture par cisaillement de la couche. Dans les couches finement laminées, le problème intrinsèque tient à la stabilité des couches du toit sous l'effet de la force de la pesanteur et surtout de la poussée latérale exercée sur les couches, laquelle peut entraîner un important flambage ou plissement des couches, phénomène qui est souvent observé dans les houillères profondes.

La contrainte axiale σ_a sous laquelle un pan de roc d'épaisseur t et de longueur ℓ flambera est donnée par

$$\sigma_a = \frac{\pi^2 E}{12\zeta^2 (\ell/t)^2} \qquad (17)$$

où E est le module d'élasticité de la roche, ℓ/t est le rapport de forme du pan de roc et ζ est une constante qui dépend de l'état des extrémités du pan et qui varie entre 0,5 pour un pan encastré aux deux extrémités et 1 pour un pan dont les deux extrémités peuvent être considérées comme reposant sur des rotules.

L'équation 17 montre que la contrainte axiale que peut supporter un pan de roc avant de flamber est inversement proportionnelle au carré du rapport de forme. Par conséquent, des pans minces flamberont plus facilement que des pans épais.

Du point de vue des calculs, il est souhaitable que les couches du toit de l'excavation soient épaisses pour éviter autant que possible leur flambage. Lorsque cela n'est pas le cas, il faudrait rendre le toit artificiellement compétent en boulonnant le toit et en renforcant la roche le plus possible.

Une méthode possible pour réduire la possibilité de flambage et de plissement des couches finement laminées consiste à forer des trous de relaxation des contraintes dans les reins de voûte du tunnel de façon à permettre un certain glissement des plans de stratification sans exercer de poussée sur la couronne du toit. Les couches partiellement relaxées sont ensuite boulonnées aux couches supérieures et forment un toit compétent. Les avantages possibles de cette méthode ont été démontrés dans des essais sur modèles et méritent une attention plus poussée. La fig. 15 montre un exemple de trous de relaxation des contraintes dans une galerie de mine de charbon, tiré de Roest et

Gramberg (1981).

Même si la plupart des dommages observés dans les excavations situées dans des couches laminées horizontalement se produisent dans le toit et la sole de l'excavation, la hauteur des murs de l'excavation est un paramètre de calcul critique car elle contrôle dans une large mesure la profondeur de fracturation des murs. Cela influe ensuite sur les exigences en matière de portée efficace et de soutènement du toit. De ce point de vue, la hauteur des excavations dans des couches laminées horizontalement devrait être aussi petite que possible. Cette exigence est plus importante lorsqu'un glissement peut se produire le long de plans de stratification où le frottement est faible. Lorsqu'il n'y a pas de plan de faible frottement, la hauteur des murs est moins critique.

Jusqu'ici, l'analyse a porté sur le calcul des excavations dans des couches laminées horizontalement, sujettes à de fortes contraintes verticales. Dans le cas de fortes contraintes qui s'exercent horizontalement, l'épaisseur des premières couches du toit et de la sole de l'excavation est un facteur critique car elle détermine la portée à laquelle il y a flambage. Dans les schistes charbonneux en couches minces hoirizontales, soumis à des contraintes horizontales relativement élevées, on a souvent vu le toit de galeries de six mètres de large s'effondrer sur des longueurs de dix mètres ou plus dans les houillères d'Afrique du Sud. Ces effondrements s'arrêtent en général à la première couche de grès compétente.

SOUTÈNEMENT DES EXCAVATIONS DANS LA ROCHE SOUMISE À DE FORTES CONTRAINTES

Dans l'analyse des facteurs généraux de calcul des excavations dans des masses rocheuses soumises à de fortes contraintes, on a souligné l'importance d'un choix judicieux de la géométrie de l'ouverture souterraine. Dans la figure 12, on voit que le choix de la forme de la section d'un tunnel est particulièrement critique dans tous les cas où la valeur de la contrainte de compression maximale est égale ou supérieure à la résistance à la compression de la masse rocheuse. L'ampleur du problème de soutènement dans les masses rocheuses soumises à de fortes contraintes est donc étroitement liée aux calculs de la géométrie de l'excavation, laquelle détermine non seulement la forme mais aussi la profondeur de la zone de fracture autour de l'ouverture.

La fonction première du soutènement dans des masses rocheuses soumises à de fortes contraintes est de mobiliser et de préserver autant que possible la résistance de la masse rocheuse pour que cette dernière devienne autoportante. On y arrive en limitant la déformation après rupture de la roche fracturée autour de l'excavation. Cela non seulement préserve la résistance inhérente de la roche fracturée, mais aussi limite la croissance non contrôlée de la zone de fracture en confinant la roche à l'extérieur de cette zone. Pour y arriver de façon efficiente, il faut que le système de soutènement devienne efficace le plus tôt possible après les travaux d'excavation.

Idéalement, le système de soutènement devrait être actif, c'est-à-dire qu'il devrait pouvoir confiner la roche dans la zone de fracture sans qu'il faille la dégager. En outre, le système de soutènement devrait pouvoir résister à la déformation de la roche autour de l'excavation en développant rapidement une résistance de soutènement. Enfin, le système de soutènement devrait comporter un point de rupture bien défini de façon à éviter qu'il ne soit surchargé par suite soit d'une expansion de la zone de fracture, soit d'une charge dynamique qui s'établit parfois dans les mines souterraines profondes.

Comme la zone de fracture autour des ouvertures souterraines soumises à de fortes contraintes ne se forme pas instantanément, mais se développe en un certain temps, l'ouvrage de soutènement initial ou primaire doit être capable de supporter un certain niveau de déformation de la roche sans perte de capacité de soutènement. Cet aspect revêt une importance particulière dans le cas du soutènement des excavations minières car le nombre, la taille et la position relatives des excavations dans une mine changent continuellement. Par conséquent, les contraintes qui s'exercent sur une excavation minière évolueront pendant toute la durée utile de l'excavation.

Dans les ouvrages souterrains de génie civil, où il est peu probable que surviennent des changements dans les contraintes une fois l'excavation terminée, il est devenu pratique courante de retarder l'installation de la structure ou du revêtement final de soutènement jusqu'à ce que les parois de l'excavation aient cessé de se déplacer, d'après Rabcewicz et coll. (1972).

Interaction roche-soutènement

Le calcul du soutènement des excavations est difficile à établir parce que la structure, composée du soutènement et de la roche encaissante, est en général statistiquement indéterminée, c'est-à-dire que la charge exercée sur le soutènement est déterminée par la déformation des deux composantes de la structure. Cependant, cette interaction complexe n'a été résolue numériquement que pour quelques situations idéales par Salamon (1974); et par Brady et Brown (1985).

Facteurs pratiques

Dans la pratique, un grand nombre de décisions relatives au soutènement des excavations sont basées sur l'expérience et des mesures de déformation. On peut toutefois éviter de graves erreurs d'évaluation des besoins de soutènement dans le cas d'excavations souterraines soumises à de fortes contraintes en suivant quelques règles simples.

Premièrement, il est recommandé de commencer l'évaluation par une analyse de la distribution des contraintes autour de l'excavation. Dans de nombreux cas, et particulièrement dans le cas d'excavations dans de la roche cassante massive, cette analyse peut être basée sur la théorie de l'élasticité. En comparant les contraintes résultantes et les propriétés de résistance de la roche, on est en mesure de prévoir la possibilité d'un effondrement sous contrainte autour de l'excavation et l'étendue probable de cet effondrement. Dans le cadre de cette analyse, il faudrait examiner la sensibilité de l'étendue de la zone de fracture à des incertitudes quant aux propriétés de résistance de la masse rocheuse (figure 16).

Deuxièmement, il faudrait comparer avec d'autres excavations dont le comportement dans des masses rocheuses semblables est déjà connu. Pour être significative, cette comparaison devrait être fondée sur une analyse des contraintes et de la zone de rupture. Lorsqu'on compare les exigences de soutènement d'excavations qui sont situées dans la même formation rocheuse mais qui ont des dimensions différentes, il faut tenir compte du fait que la résistance de la masse rocheuse diminue avec la taille de l'excavation d'après Piper (1984), par exemple en ajustant l'espacement des diaclases dans le système de classification géomécanique CSIR en fonction de la hauteur de l'excavation (figure 17).

Troisièmement, il faudrait évaluer l'effet des discontinuités géologiques connues sur le comportement des parois d'excavation au moyen de techniques d'analyse des contraintes ou de modèles à base de blocs ou de coins. Il faudrait enfin élaborer des modèles conceptuels de rupture possible des parois d'une excavation et évaluer le plan de soutènement en fonction de ces modèles.

Méthodes de soutènement éprouvées

L'expérience acquise dans les mines d'or du Witwatersrand et dans les tunnels profonds a montré qu'un système de soutènement intégré où l'on utilise des boulons d'ancrage, du grillage métallique et des treillis de câble satisfait à la plupart des exigences de soutènement pour les excavations soumises à de fortes contraintes, qui sont situées dans de la roche cassante massive. Ce système de soutènement est parfois complété par la projection d'une fine couche de béton.

Dans le calcul des systèmes de soutènement destinés à un tel usage, les grandes lignes directrices suivantes se sont avérées utiles :

(i) La longueur des boulons d'ancrage ou des ancrages de treillis de câbles ne devrait pas être inférieure à la moitié de la largeur de l'excavation dans le cas des boulons de toit ou à environ la moitié de sa hauteur dans le cas des boulons de mur. Cette règle générale peut être modifiée à la lumière des résultats de l'analyse des zones de fracture autour de l'excavation qui pourraient indiquer qu'une longueur de boulon égale à la moitié de la hauteur de l'excavation est insuffisante dans le cas de tunnels soumis à de très fortes contraintes.

(ii) L'espacement maximal des boulons devrait, de préférence, ne pas dépasser la moitié de leur longueur. Dans le cas de boulons ou d'ancrages de treillis de câbles très longs, il est recommandé d'ajouter des boulons plus courts dans le système de soutènement pour réduire l'espacement des boulons à la surface des parois de l'excavation.

(iii) Dans le soutènement du toit, la résistance et l'espacement des boulons devraient assurer un soutènement pour une charge totale d'environ le double du poids mort des couches traversées par les boulons.

(iv) Dans une masse rocheuse laminée ou stratifiée, les boulons devraient, de préférence, couper les couches suivant un angle d'au moins 45 degrés.

(v) Les éléments de précontrainte employés dans le soutènement des excavations soumises à de fortes contraintes devraient être complétés par des grillages de fils entrelacés pour empêcher des pans de roc de se détacher entre les éléments de précontrainte. Dans le cas d'un soutènement plus permanent, une fine couche de béton devrait être projetée pour limiter l'altération et agir comme clé de voûte.

Grâce au système de soutènement décrit ci-dessus, des tunnels mesurant 4 m sur 4 m ont résisté à des champs de contraintes dépassant 120 MPa. Des études effectuées par Hepworth et Gay (1986) ont montré que la convergence dans les tunnels soumis à de fortes contraintes peut être réduite de beaucoup en augmentant la résistance de soutènement dans le tunnel. Ce dernier paramètre est exprimé en kN/m^2 de paroi de tunnel et est déterminé par la limite d'élasticité des éléments de précontrainte et par la densité du soutènement. Pour tenir compte des variations des contraintes qui se sont produites après l'installation du soutènement, ces auteurs ont normalisé la résistance de soutènement par rapport aux variations des contraintes. La figure 18 montre une relation évidente entre les valeurs normalisées de la résistance de soutènement et la convergence des parois du tunnel. Ces résultats indiquent qu'en deçà d'une résistance de soutènement critique, la convergence dans les tunnels de mine augmente rapidement, tandis qu'au-delà de la valeur critique, les améliorations ne sont que marginales. Des résultats semblables ont été obtenus par Brady et Brown (1985) qui ont analysé l'effet de la pression de soutènement sur la déformation des parois d'un tunnel circulaire soumis à un champ de contraintes hydrostatiques.

Wiseman (1979) a évalué l'efficacité de différents systèmes de soutènement des tunnels utilisés dans les mines profondes de l'Afrique du Sud. Les résultats de son étude sont donnés à la figure 19. À noter l'efficacité plutôt médiocre des cadres et des cintres coulissants en acier par rapport à celle des différents types de soutènement à base d'éléments de précontrainte. À la lumière de ces résultats, il n'est pas étonnant de contaster qu'on n'utilise plus les cadres et les cintres coulissants en acier dans les mines profondes de roche dure.

La projection de béton combinée à la pose de boulons d'ancrage scellés au coulis sur toute leur longueur et de grillages métalliques s'est avérée la méthode de soutènement la plus efficace lorsque les contraintes sont élevées, d'après Hepworth et Gay (1986).

Dans la roche tendre stratifiée, notamment dans la plupart des houillères européennes, l'usage des cintres coulissants est répandu selon Jacobi (1976). Ce système de soutènement s'est avéré efficace lorsque des galeries d'exploitation par longues tailles sont soumises à des contraintes élevées après l'extraction des couches de charbon. Dans de telles galeries, on observe souvent des convergences de plus de 50 pour cent. Cependant, même lorsque les galeries se trouvent à l'extérieur de la zone d'influence des contraintes exercées dans les murs du soutènement des fronts de longues tailles, la convergence dépasse souvent 30 pour cent. Cela peut être attribué à la difficulté de réaliser un bon contact entre les cintres et la masse rocheuse encaissante et à la faible rigidité radiale des cintres. Ces facteurs favorisent un développement relativement illimité de la zone de fracture autour des galeries (figure 20).

Les premiers essais effectués sur les boulons d'ancrage scellés à la résine dans des houillères profondes d'Allemagne ont été réussis sur les plans technique, sécurité et économique, selon Reuther (1987). Ce type de soutènement connaîtra donc un champ d'applications plus étendu dans les houillères et remplacera massivement les cintres en acier.

COMPORTEMENT DES EXCAVATIONS SOUMISES À DES CHARGES SISMIQUES

Les secousses sismiques qui s'exercent sur les excavations souterraines soumises à de fortes contraintes constituent un phénomène souvent observé dans les mines profondes, mais sont aussi importantes dans beaucoup d'autres situations. La sismicité peut influer sur les excavations souterraines de plusieurs façons. Premièrement, lorsque des ondes sismiques se propagent dans la roche, elles interagissent avec l'excavation et peuvent donner lieu à des augmentations dynamiques des contraintes. Deuxièmement, les mouvements du terrain peuvent déformer l'excavation et exercer des contraintes sur l'enveloppe intérieure de l'excavation. Troisièmement, les mouvements du terrain peuvent entraîner l'accélération de pans de roches lâches et exercer des forces additionnelles sur les ouvrages de soutènement.

Variations dynamiques des contraintes

Mow et Pao (1971) ont étudié le comportement d'ondes P en régime permanent dans des cavités cylindriques lorsque le front d'onde progresse suivant l'axe du tunnel. Ces auteurs ont conclu que la concentration maximale des contraintes due à des charges harmoniques peut dépasser les valeurs statiques d'un pourcentage atteignant 10 à 15 pour cent pour les ondes P et 5 pour cent pour les ondes S. Ces maximums se produisent à des longueurs d'onde relativement grandes, de l'ordre de 10 à 20 diamètres de tunnel. Les valeurs maximales des contraintes normales, σ_p, et de cisaillement, τ_{xy}, dues à une onde P sont données par

$$\sigma_p = \frac{(1 - \nu) E}{(1 + \nu)(1 - 2\nu)} \cdot \frac{V_p}{c_p} \qquad (18)$$

et

$$\tau_{xy} = \frac{G V_p}{2 c_p}$$

où E et G sont les modules d'élasticité et de cisaillement du matériau, ν est le coefficient de Poisson, V_p est la vitesse maximale des particules et c_p est la vitesse de l'onde P.

Les contraintes σ_p et τ_{xy} donnent une indication de la variation des contraintes due au passage de l'onde P. Si $E = 50$ GPa, $\nu = 0{,}2$, $V_p = 1\ ms^{-1}$ et $c_p = 5.10^3\ ms^{-1}$, la variation de la contrainte de compression est d'environ 5 MPa, valeur qui est faible par rapport aux contraintes qui s'exercent en profondeur. A noter que la variation dynamique de la contrainte est proportionnelle à la vitesse maximale des particules, V_p. D'après McGarr et coll. (1981), la vitesse maximale des particules V_p des secousses sismiques est donnée par

$$\log R\, V_p = 3{,}95 + 0{,}57\, M_L \qquad (19)$$

où R est la distance par rapport au foyer de la secousse et M_L est l'intensité de la secousse.

D'après l'équation 19, la vitesse maximale des particules est inversement proportionnelle à la distance entre l'excavation et la source de la secousse et, dans la plupart des situations réelles, est inférieure à 1 m/s. Il s'ensuit que les variations dynamiques des contraintes dues à une secousse sismique lointaine sont négligeables à toutes fins pratiques. Cependant, les variations dynamiques des contraintes peuvent être importantes si l'excavation est située près de la source d'une secousse intense. La déformation maximale à la compression ε_p, due à une onde P approchant une excavation, est donnée par

$$\varepsilon_p = \frac{V_p}{c_p} \qquad (20)$$

Comme V_p est de l'ordre des ms^{-1} et que c_p est de l'ordre des km^{-1}, il s'ensuit que les déformations à la compression dues à des charges sismiques ε_p, sont inférieures à 10^{-3} sauf pour les excavations situées à proximité du foyer de secousses intenses.

Exigences de soutènement

Wagner (1984) a examiné les exigences de soutènement dans des tunnels soumis à des charges sismiques. Il a basé son modèle de soutènement sur l'hypothèse selon laquelle tout pan de roche encaissante autour des tunnels, qui a une masse $\overline{m}$, est accéléré jusqu'à la vitesse maximale des particules, V_p, et que le soutènement peut être capable d'absorber l'énergie cinétique, W_k, de ce pan de roche sans risque de rupture.

$$W_k = \tfrac{1}{2}\, \overline{m}\, V_p^2$$

Si la limite d'élasticité des éléments de précontraite, F_y, est de 100 kN et que l'allongement, $\Delta\ell$, est de 20 mm (valeurs normales pour des éléments de précontraite utilisés dans des tunnels de mine d'or), l'épaisseur des pans de roche qui peuvent être supportés dans des conditions de charge dynamique peut être estimée à partir de la figure 21. Ce diagramme montre que les boulons d'ancrage classiques sont un moyen de soutènement plus efficace pour supporter des pans de roche épais. Wagner (1984) a conclu que le seul moyen pratique d'optimiser lessystèmes de soutènement en condition dynamique consiste à augmenter leur élasticité, soit en utilisant des éléments de précontrainte souples enrobés de coulis, soit en intégrant des éléments élastiques dans le système de précontrainte.

CONCLUSIONS

L'expérience acquise dans les mines profondes a montré qu'il est possible de calculer les excavations et leur structure de soutènement pour des niveaux de contrainte qui approchent la résistance à la compression de la masse rocheuse. Pour l'ingénieur concepteur, la plus grande inconnue reste l'évaluation de la résistance de la masse rocheuse et de ses variations en fonction de la taille de l'excavation. Le seul moyen pratique de réaliser des progrès réels à cet égard est de s'engager dans un programme international d'analyse rétrospective d'excavations soumises à de fortes contraintes. Beaucoup de travaux ont été effectués dans des mines profondes de roche dure, mais ces travaux devront être étendus à d'autres types de roche.

Dans le calcul des excavations réalisées dans des masses rocheuses soumises à de fortes contraintes, l'accent doit être mis sur le calcul de la zone de fracture autour de ces excavations de façon à faciliter le soutènement de cette zone de fracture.

Les systèmes de soutènement, actifs dès la pose et comportant des éléments relativement rigides, sont les plus prometteurs car ils conservent à la roche fracturée ses propriétés de résistance inhérentes en minimisant la déformation après rupture et permettent en outre l'établissement d'un état triaxial de contrainte à proximité des parois de l'excavation.

Les charges sismiques exercées sur les excavations souterraines ne constituent pas en général un problème grave. Cependant, dans les exploitations minières où des secousses peuvent se produire très près d'excavations, de fortes contraintes dynamiques peuvent s'exercer et le système de soutènement peut être surchargé. Dans ce cas, la vitesse maximale des particules, V_p, est le paramètre le plus critique car il détermine autant les variations dynamiques des contraintes que l'efficacité du soutènement.

Liste de figures

Pour les figures, voir le texte anglais. Les légendes apparaissent en français à la fin.

Figure 1. m et s en fonction de la cote de la masse rocheuse

Figure 2. Effet du diamètre de l'excavation sur la résistance de la roche (d'après Johnson, 1986)

Figure 3. Effet d'une couche de roche tendre sur les contraintes horizontales exercées au centre d'un échantillon de roche cylindrique.

Figure 4. Diminution de la résistance de la roche par bandes de roche tendre (d'après Laubscher, 1984)

Figure 5. Mode type de rupture de la roche dans les tunnels soumis à de fortes contraintes

Figure 6. Courbe idéale de compression en fonction de la charge pour une roche cassante (d'après Salamon, 1974)

Figure 7. Relation entre l'énergie libérée ERR et le nombre de coups de toit par 1000 m^2 exploités

Figure 8. Répartition de l'énergie totale libérée, W_r en énergie cinétique (potentielle) totale, W_k, et en énergie totale de déformation (libérée au moment de l'excavation de la roche) (d'après Salamon, 1984)

Figure 9. État des tunnels sans soutènement dans des mines d'or en fonction du facteur de concentration des contraintes dans les murs (D'après Wiseman, 1979)

Figure 10. Définition de la géométrie et des contraintes autour d'une ouverture elliptique.

Figure 11. Distribution des contraintes principales subsidiaires suivant l'axe de tunnels expérimentaux, et sections transversales types

Figure 12. Largeur de la zone de fracture dans les murs de tunnels de différentes formes en fonction de σ_v/σ_c

Figure 13. Variation des contraintes principales σ_1 et σ_3 en fonction de la distance dans le mur de l'excavation

Figure 14. Zones de glissement potentiel autour d'un tunnel dans des couches laminées horizontales

Figure 15. Trous de relaxation des contraintes dans une galerie de mine de charbon

Figure 16. Effet de la résistance de la roche sur la zone de fracture autour d'un tunnel carré (d'après Piper, 1984)

Figure 17. Effet des dimensions de l'excavation et des contraintes sur le terrain sur la profondeur de la zone de fracturation autour de l'excavation

Figure 18. Effet de la résistance de soutènement sur la convergence des tunnels (d'après Hepworth et Gay, 1986)

Figure 19. Efficacité de divers systèmes de soutènement dans des conditions de fortes contraintes (d'après Wiseman, 1979)

Figure 20. Effet de contraintes dans la roche et de la résistance de soutènement sur le rayon de la zone de faible cohérence autour d'un tunnel dans une mine de charbon (d'après Jacobi, 1976)

Figure 21. Efficacité des systèmes de soutènement des tunnels dans les mines d'or sous des charges sismiques

BIBLIOGRAPHIE (voir section anglaise)

09h20 : Conférences brèves (Le Grand Salon)

Barla, G., Scavia, C. Antonellis, M., et Guarascio, M., Italie
Caractérisation du massif rocheux par analyse géostatistique à la mine Masua

Borg, T., Holmstedt, A., Suède
Investigation de mécanique des roches d'une ouverture longitudinale à la mine Kiirunavaara

Budavari, S., Croeser, R.W., Afrique du Sud
Effets de l'extraction des dépôts tabulaires autour des puits verticaux dans les mines à grande profondeur

Ewy, R.T., Kemeny, J.M., Zheng Ziqiong, Cook, N.G.W., États-Unis
Génération et analyse de formes d'excavations stables sous fortes contraintes

10h00 : Foyer, Pause-Café
10h30 : Le Grand Salon

Gale, W.J., Nemcik, J.A., Upfold, R.W., Australie
Application de méthodes de contrôle des contraintes à la conception de mines de charbon souterraines où existent de fortes contraintes latérales

Guenot, A., France
Contraintes et ruptures autour des forages pétroliers

Ishijima, Y., Fujii, Y., Sato, K., Japon
Microséismicité causée par une activité d'extraction de charbon en profondeur

Jager, A.J., Piper, P.S., Gay, N.C., Afrique du Sud
Aspect de mécanique des roches de remblayage dans les mines profondes sud-africaines

Kersten, R.W.O., Greeff, H.J., Afrique du Sud
L'influence de la géologie et fatigue sur l'excavation d'un puits : une étude de cas

Kusabuka, M., Kawamoto, T., Ohashi, T., Yoichi, H., Japon
Évaluation intégrée de la stabilité d'une caverne souterraine

Kyung Won Lee, Ho Yeong Kim, Hi Keun Lee, Corée
Interprétation des mesures de déformations pour la détermination de la zone de relâchement en périphérie d'un tunnel dans une mine de charbon

Matsui, K., Ichinose, M., Uchino, K., Japon
L'influence de piliers rémanants sur la déformation des galeries de panels de longue taille sousjacents

Petukhov, I.M., Russie
Prédiction des coups de toit et leur prévention : développements récents

12h00 : Marquette
13h30 : Le Grand Salon

Tanimoto, C., Hata, S., Fujiwara, T., Yoshioka, H., Michihiro, K., Japon
Relation entre la déformation et la pression sur le soutènement en creusant un tunnel au travers des roches trop contraintes

Udd, J.E., Hedley, D.G.F., Canada
Recherche sur les coups de toit au Canada - 1987

Van Sint Jan, M.L., Valenzuela, L., Morales, R., Chili
Revêtement souple pour des chambres souterraines dans une mine par foudroyage

Vervoort, A., Thimus, J.-F., Brych, J., Crombrugghe, O. de, Lousberg, E., Belgique
Vérification par la méthode des éléments finis de l'influence du soutènement marchant sur les conditions de toit dans les exploitations "longwall"

Yuen, C.M.K., Boyd, J.M. Aston, T.R.C., Canada

Étude de l'interaction massif-supports d'un tunnel foncé au moyen d'un tunnelier à la mine Donkin, Nouvelle-Écosse

Zhu Weishen, Lin Shisheng, Zhu Jiaqiao, Dai Guanyi, Zhan Caizhao, Chine
Cas pratiques de rétro-analyse de la déformabilité de cavités souterraines en fonction du temps et du milieu

14:30: Foyer, Pause-café

15:00 - 16:30: Le Grand Salon, Discussion

Modérateur: B.N. Whittaker (Angleterre)

Panélistes: V.M. Sharma (Inde)
P.K. Kaiser (Canada)
V. Maury (France)

Texte des panélistes, voir section anglaise.

Résumé du modérateur: B.N. Whittaker (Angleterre)

RÉSUMÉ: Les principaux problèmes et sujets couramment traités dans le domaine des souterrains en massifs rocheux sous contraintes excessives sont passés en revue. Certains exposés du congrès sont cités dans le résumé où l'accent est mis sur les secteurs vers lesquels il est recommandé d'orienter les travaux de développement futurs.

INTRODUCTION

Le terme "contrainte excessive" correspond à la moyenne des contraintes qui sont appliquées à un massif rocheux dans des conditions normales et qui, étant supérieures à sa résistance naturelle, se traduisent par une certaine forme de rupture du massif. Des exemples particuliers sont présentés pour mettre en relief les conséquences des contraintes excessives: d'abord, les voies dans les houillères profondes peuvent traverser un massif rocheux soumis à des contraintes excessives à cause de la faiblesse des couches environnantes, ce qui provoque l'éboulement des roches autour de la voie et, habituellement, la réduction graduelle de l'excavation. Au contraire, un pilier de charbon sous contrainte excessive peut soudainement se rompre si son coefficient d'élancement favorise une telle instabilité bien que la plupart des piliers de charbon soient conçus de façon que s'ils subissent des contraintes excessives, la déformation résultante soit limitée. Les excavations en roche dure réalisées dans des mines profondes subissent les effets de surcharge d'amplitude variable, allant de dommages localisés du toit ou des parements à une fermeture soudaine qui peut être causée par des coups de toit. La nature des propriétés de la roche, en particulier sa capacité de travail de déformation, la géologie du terrain et le champ de contraintes, est un facteur important.

Lorsqu'on creuse dans un massif rocheux surchargé, il est courant de vouloir prévoir la fermeture résultante et le degré d'instabilité et d'être obligé de prendre des mesures spéciales pour régler les problèmes associés aux conditions de contraintes excessives. Un éventail d'instruments de prévision peuvent être utilisés pour analyser le champ de contraintes prévu autour des excavations et l'usage de ces instruments a été étendu à l'étude de l'influence des différentes formes de soutènement visant à promouvoir la stabilité. Les excavations effectuées dans des roches profondes sous contraintes excessives laissent encore planer une grande incertitude face à la nature de certains problèmes, notamment la rupture soudaine des massifs ou même la production de coups de toit. Des mesures spéciales ont été prises pour perfectionner l'utilisation de modèles dans les travaux de conception. La modélisation des fractures joue un rôle important et son utilisation sera sans doute plus grande dans l'avenir du fait que les concepteurs devront mieux connaître la zone possible de fissuration. La modélisation des fractures permet de mieux prévoir les conséquences de la surcharge sur les excavations et leur soutènement.

PROBLÈMES DE BASE ET SOLUTIONS RECHERCHÉES

Le tableau 1 présente un résumé des principaux problèmes de base et des solutions recherchées. Pour trouver des solutions satisfaisantes à ces problèmes, il faut d'abord, en ingénierie des roches, connaître les contraintes qui se créent au sein du massif rocheux. Il est également important de connaître les effets de telles contraintes sur la réaction des roches liée à la présence de l'excavation.

Le tableau 1 présente cinq domaines dans lesquels il est nécessaire d'effectuer des recherches. Les remarques ci-dessous concernent chacun de ces domaines.

1. Il est nécessaire d'identifier tous les facteurs géologiques, tels que les principaux défauts et discontinuités de la roche, sinon les évaluations du champ de contraintes peuvent s'avérer trompeuses. Il est urgent d'évaluer de façon appropriée l'importance des diverses formes de défauts et de discontinuités géologiques sur le champ de contraintes général.

2. La stratification joue un rôle important sur la configuration de la fissuration et sur la déformation associée aux excavations souterraines. Il faudrait examiner la façon dont la stratification peut arrêter la propagation des fractures et provoquer des déformations asymétriques.

3. La rupture de la roche autour des souterrains est un phénomène en général graduel et d'autres recherches devront être réalisées pour déterminer son importance.

4. Comme un excès de contraintes se traduit par une rupture, le comportement du matériau à divers stades de post-rupture soulève souvent des problèmes.

5. Les effets de l'altération et du vieillissement de la relation contrainte-déformation des roches sont limités dans le temps. La question, à savoir comment les propriétés des roches évoluent avec le temps, se pose donc? Dans la conception des excavations en massif rocheux, il ne faut pas dépasser les limites acceptables qui sont fonction de certains facteurs que sont la forme, les dimensions et le soutènement. Il est donc important de déterminer l'effet de ces facteurs dans les conditions de massifs sous contraintes excessives.

Tableau 1. Importants problèmes de base non résolus et solutions recherchées relativement aux souterrains en massif rocheux sous contraintes excessives

---

Certains problèmes de base:

a) la prévision des contraintes créées dans le massif rocheux par la présence de l'excavation;

b) la prévision exacte, à partir de l'analyse des contraintes, des déplacements produits autour de l'excavation.

Réponses requises aux questions suivantes:

1. Quels sont les effets du champ de contraintes initial sur le comportement de l'excavation?

2. De quelle façon la stratification influe-t-elle sur la redistribution des contraintes et le comportement de l'excavation?

3. Quelle est l'importance de la rupture de la masse rocheuse autour de l'excavation dans certains champs de contraintes donnés?

4. Quelles sont les propriétés de plasticité-déformation de la roche fragmentée et quels effets ont-ils sur la conception et le rendement de l'excavation?

5. Quels sont les effets du temps sur le massif rocheux sous contraintes, notamment sur sa déformation, sa rupture et sa vitesse de déplacement?

RÉSUMÉ DU MODÉRATEUR DES PRINCIPAUX DOMAINES TRAITÉS PAR DIFFÉRENTS PARTICIPANTS

Un large éventail de problèmes, associés à la conception et à la stabilité des excavations souterraines, ont été présentés par divers congressistes. Le résumé suivant met l'accent sur des points considérés comme particulièrement importants pour les principales questions du quatrième thème. Tous les exposés des auteurs cités sont contenus dans le deuxième volume des Comptes rendus du sixième Congrès international de mécanique des roches de la SIMR. Les lecteurs qui veulent prendre plus ample connaissance des exposés sont donc priés de se reporter au deuxième volume des Comptes rendus. Les principaux domaines traités par les différents participants du quatrième thème peuvent se subdiviser en quatre parties qui sont chacune commentées ci-dessous. Il y a eu de nombreux exposés remarquables au cours du congrès, mais quelques-uns seulement ont été sélectionnés pour identifier les principaux secteurs problèmes liés à l'application de la mécanique des roches aux excavations réalisées dans des massifs rocheux sous contraintes excessives.

1. CONTRÔLE DES TERRAINS ET MÉTHODES DE CONCEPTION POUR APPROFONDIR LA STABILITÉ DES EXCAVATIONS

Les chercheurs canadiens Bawden et Milne démontrent de façon efficace comment la modélisation (numérique et empirique) peut être appliquée à la conception des mines et en améliorer la stabilité. Il est également important de signaler la contribution de Duncan Fama et Wardle d'Australie en ce qui concerne l'utilisation de la modélisation numérique pour mieux comprendre le comportement des piliers de charbon.

Del Greco et ses collègues d'Italie ont montré comment il est possible d'évaluer la stabilité du toit d'une excavation en fonction de la voûte de pression.

Vervoort et ses collègues de Belgique ont démontré avec succès le rôle utile de la modélisation par éléments finis pour prévoir les conditions du toit dans les excavations minières. Cunha du Portugal a aussi montré de façon efficace comment les modèles peuvent constituer des instruments de conception importants pour évaluer la stabilité des tunnels.

Ewy et ses collègues des États-Unis ont fait un exposé digne de mention sur leurs travaux de simulation réussie de la création de formes stables pendant une rupture progressive.

Un exposé important a été donné par Herget et Mackintosh du Canada dans lequel ils montrent comment l'instrumentation peut servir à évaluer la convergence et les contraintes, donc à améliorer la conception des excavations.

À ce chapitre, il faut mentionner l'exposé de Maury sur la stabilité des galeries uniques, en particulier sur le choix de la théorie appropriée pour expliquer la stabilité.

2. ASPECTS ASSOCIÉS AU SOUTÈNEMENT DES EXCAVATIONS DANS DES CONDITIONS DE CONTRAINTES EXCESSIVES

Depuis de nombreuses années, la conception des stots de protection de puits et leur abattage subséquent préoccupent des ingénieurs du monde entier. L'exposé de Budavari et Croeser d'Afrique du Sud sur l'abattage des stots de protection de puits tombe à point et leurs travaux démontrent la faisabilité d'augmenter l'efficacité de l'abattage en appliquant les principes de la mécanique des roches. L'exposé de McKinnon, également d'Afrique du Sud, traite d'autres aspects des stots de protection de puits.

Le soutènement continue d'intéresser énormément les ingénieurs du monde entier. Tan de Chine a fait un exposé intéressant sur l'application de la NATM dans des conditions difficiles. Le renforcement des massifs rocheux continue d'être une fonction de soutènement importante et des exposés dignes de mention ont été donnés par Thompson d'Australie, par Aydan et ses collègues du Japon et par Ballivy du Canada.

Selon Jager et ses collègues d'Afrique du Sud, le remblayage est un moyen très efficace de réduire les contraintes s'exerçant sur les piliers, donc d'augmenter la sécurité.

Des exposés intéressants sur le soutènement des terrains ont aussi été donnés par Van Sint Jan du Chili, Barroso et Lamas du Portugal, Fine de France et Korini d'Albanie ainsi que par Yen et ses collègues du Canada qui ont traité de l'interaction roche-soutènement.

3. COMPORTEMENT DE LA ROCHE À L'ABATTAGE, ÉVALUATION ET CONTRÔLE DES CONDITIONS SOUS CONTRAINTES EXCESSIVES

Les travaux de Barlow et Kaiser du Canada ont clairement montré l'importance de l'interprétation des mesures de convergence pour améliorer l'évaluation de l'excavation et du soutènement des tunnels. Les conditions changeantes du champ de contraintes en fonction de leurs effets sur le comportement des tunnels peuvent être évaluées, tel que montré par Esterhuizen d'Afrique du Sud.

Le comportement d'un tunnel est caractérisé par des courbes de réponse en fonction du temps et Gill et Ladanyi du Canada ont montré comment on peut tenir compte du temps pour prévoir le comportement d'un tunnel. Le facteur temps sur le comportement d'un tunnel est également traité de façon efficace par Minh et Rousset de France. La stabilité des tunnels continue de retenir une attention spéciale comme en fait foi l'exposé de Sharma et de ses collègues de l'Inde.

La formation de fractures au cours de l'abattage a été analysée de façon exhaustive par Ortlepp et Moore (Afrique du Sud et G.-B.) dont les conclusions sur la vitesse, la nature et les caractéristiques générales des déplacements associés constituent une conclusion importante. L'interprétation de la zone de déplacements plastiques est traitée par Lee et Kim de Corée qui font de judicieux commentaires sur les configurations de zones de détente autour des tunnels. Ishijima et ses collègues du Japon attirent l'attention sur de nouveaux procédés microsismiques pour faciliter l'interprétation des phénomènes de fractures autour des excavations.

4. CARACTÉRISATION, COMPORTEMENT ET CONTRÔLE DES CONTRAINTES DES ROCHES DANS LES EXCAVATIONS SOUTERRAINES

La caractérisation des massifs rocheux par Barla et Scavia d'Italie attire l'attention sur le rôle important que peut jouer l'analyse géostatistique dans l'interprétation des discontinuités rocheuses et leurs effets sur le comportement des massifs rocheux dans les excavations minières.

L'un des éléments importants des travaux de Bandis et de ses collègues de Norvège est leur explication du comportement aux contraintes et aux fractures des cavités situées dans un terrain de mauvaise qualité. Les résultats valables qu'ils ont obtenus justifient sans aucun doute l'examen approfondi d'autres conditions rocheuses.

Gale et ses collègues de l'Australie ont démontré la valeur du contrôle des contraintes pour améliorer la conception des mines, en particulier pour optimiser la conception de l'exploitation et l'orientation de l'abattage afin de réduire les effets des contraintes.

Les coups de toit continuent d'intéresser fortement les ingénieurs et des progrès importants ont été accomplis par Udd et Hedley du Canada sur la compréhension du phénomène et son élimination. Tout aussi important ont été les travaux de Sugawara et de ses collègues du Japon, pour expliquer les mécanismes et les aspects d'élimination des projections de charbon.

L'interaction de la géologie et du minage est soulignée par Kersten et Greef d'Afrique du Sud dont les travaux montrent clairement comment une meilleure compréhension devient possible en regard de la stabilité lors des travaux de fonçage d'un puits dans des conditions de minage en grande profondeur. Leur interprétation et les principales observations qu'ils ont faites sur l'interaction de la géologie et de l'exploitation minière méritent une attention spéciale relativement à leur application aux problèmes connexes de stabilité dans d'autres situations d'excavation souterraine sous contraintes excessives. Notre conférencier principal, Horst Wagner d'Afrique du Sud souligne également l'importance de la forme de l'excavation sur l'amélioration de la stabilité dans des conditions de contraites excessives dans la roche.

En conclusion, je reviens à la liste des réponses nécessaires indiquées au tableau 1 du présent rapport. Ces cinq points nécessiteront de notre part une attention spéciale dans l'avenir.

ANNEXE

ANALYSE DE LA STABILITÉ DANS LES TUNNELS PERCÉS EN ROCHE TENDRE

La présente annexe fait partie du rapport du modérateur. Elle vise à montrer comment certains problèmes mentionnées font l'objet de recherche dans un centre de recherche en mécanique des roches au Royaume-Uni.

Définition du problème

La mécanique des roches, comme tout autre branche de la mécanique, traite du mouvement. Elle peut, par conséquent, être subdivisée en deux principes fondamentaux:

i) la différence de propriété d'un milieu entre deux points, cause une inégalité dans ce milieu, appelée "potentiel moteur", et

ii) le déplacement du milieu, du potentiel supérieur vers le potentiel inférieur, a lieu jusqu'à ce qu'un état d'équilibre soit atteint.

En mécanique des roches, le potentiel moteur est l'état de contrainte dans le massif qui est produit par une excavation, l'équilibre des contraintes étant rétablie après déplacement des roches vers cette excavation qu'elle soit soutenue ou non extérieurement.

La prévision du comportement des tunnels peut comporter deux aspects distincts:

a) Il faut prévoir les contraintes dans la masse rocheuse produite par la présence de l'excavation et déterminer s'il y aura rupture;

b) Il faut prévoir avec exactitude les déplacements autour de l'excavation après analyse des contraintes.

Plusieurs facteurs influent sur la stabilité des tunnels percés dans des roches tendres de terrains houillers. Ce sont:

i) l'état des contraintes dans le terrain avant l'excavation,

ii) la résistance du massif rocheux au stade pré-rupture qui contrôle la formation d'une zone de roches fragmentées (zone de déplacements plastiques) et la résistance du massif rocheux dans la zone de déplacements plastiques, laquelle correspond au comportement post-rupture du massif rocheux;

iii) la séquence de strates dans laquelle le tunnel est percé et la résistance relative des types de roches qui s'y trouvent;

iv) la forme et les dimensions du tunnel ainsi que le type et les caractéristiques de résistance du système de soutènement qui est utilisé dans le tunnel;

v) les déplacements élastiques (pré-rupture) et plastiques (post-rupture) causés par le changement de contraintes dans la masse rocheuse. Le déplacement élastique est le résultat de l'effet de Poisson et son importance est relativement moins grande que le déplacement plastique de la roche fragmentée qui résulte de l'augmentation de son volume causée par le changement d'état (Whittaker et Frith, 1987).

Pour être réaliste, il est souhaitable de tenir compte de tous ces facteurs lorsqu'on applique une méthode de conception. Mais pour ce faire, il faut effectuer de nombreux travaux sur le terrain pour exprimer les lois de comportement de chacun de ces facteurs. Très souvent, les lois théoriques se sont avérées inappropriées pour décrire un tel comportement et les lois empiriques fondées sur des mesures restent à établir. C'est pourquoi le modèle de prévision mis au point par l'Université de Nottingham est fortement biaisé.

Un autre aspect de la mise au point de modèles de conception est l'utilisation de données réelles sur les dimensions de la zone de convergence et de la zone de déplacements plastiques pour évaluer leur précision. Pour recueillir ces données, il faut consacrer de nombreuses heures de surveillance souterraine. C'est à cette fin que l'Université de Nottingham a participé activement à des projets réalisés dans plusieurs houillères du Royaume-Uni

(Whittaker et al., 1981).

État actuel du domaine à l'Université de Nottingham

Des travaux antérieurs ont été réalisés pour élaborer des méthodes et des normes d'essai post-rupture sur des roches de couches de terrains houillers qui serviront à recueillir des données utilisables dans des travaux de prévision. Une grande variété de types de roches ont été soumis à des essais selon ces normes et l'accent a été mis sur les résistances pré-rupture et post-rupture et sur leurs caractéristiques de déformation volumétrique. À partir des résultats de ces essais, on a établi des lois de comportement pour la résistance des roches et pour les déplacements plastiques associées au massif fragmenté (Whittaker et al., 1985). Ces deux lois ont été instrumentées dans le modèle de conception qui en découle, et dont il sera maintenant question.

Le modèle de conception, mis au point par l'Université de Nottingham, se fonde actuellement sur les hypothèses suivantes :

i) Il s'applique à un tunnel circulaire de toutes dimensions;

ii) Il tient compte des conditions de contraintes hydrostatiques initiales;

iii) Le soutènement installé dans le tunnel exerce une pression constante sur la masse rocheuse, quelle que soit sa déformation;

iv) Il tient compte de la stratification en stockant informatiquement ces informations, en divisant la massifrocheux entourant le tunnel en de petits éléments indépendants et en leur attribuant des propriétés qui sont fonction du type de roche dans laquelle ils se trouvent;

v) La résistance de la roche est caractérisée par l'équation $\sigma_1 = A\sigma_3{}^B + C$, où A, B et C sont uniques à un type de roche et à l'état de rupture; cette équation se fonde sur les résultats du programme d'essai sur les roches, mentionné précédemment;

vi) la déformation volumétrique de la roche fragmentée est donnée par l'équation: $V_e = ae^{-\sigma_3/b} + c$ où a, b et c sont uniques à un type de roche. Cette équation correspond aux déformations volumétriques maximales qui devraient se produire et représente par conséquent la pire situation possible. Il est souhaitable d'en tenir compte dans la conception initiale (Whittaker et al., 1984).

L'analyse des contraintes est réalisée en utilisant l'équation différentielle pour un cylindre symétrique dans des conditions d'équilibre. Cette équation est une équation de contrainte simplifiée qui ne s'applique qu'à un tunnel circulaire dont les conditions de contraintes initiales sont hydrostatiques. Il est donc nécessaire d'énoncer les hypothèses (i) et (ii). L'hypothèse (iii) résulte de la nécessité de supposer une condition limite pour obtenir une solution unique à l'équation des contraintes résultantes. Les hypothèses (iv) à (vi) nécessitent l'utilisation d'un micro-ordinateur étant donné qu'il n'est pas possible de trouver une solution analytique en appliquant ces équations. Cette technique est par conséquent numérique plutôt qu'analytique et on l'a dénommée la méthode des éléments indépendants.

L'un des problèmes courants soulevés par ces méthodes de prévision réside dans la classification des massifs rocheux. Il est généralement admis que la résistance des roches intactes mesurée en laboratoire est plus élevée que celle des roches in situ, étant donné que l'échantillon choisi provient d'une massif rocheux fracturé. Diverses méthodes ont été proposées pour modifier les valeurs de résistance pour qu'elles se rapprochent de celles du massif rocheux in situ. Dans la plupart des cas, cependant, il faut obtenir des informations géologiques détaillées qui ne sont pas toujours disponibles. Par conséquent, il faut procéder autrement. C'est ainsi que l'on a trouvé une solution possible en supposant que le massif rocheux intact est aussi caractérisé par la résistance résiduelle ou post-rupture de la roche. Cette valeur représente une fois de plus la pire situation possible; on a cependant constaté, en évaluant les convergences mesurées et prévues dans le tunnel, que les corrélations établies étaient très satisfaisantes (fig. 1) (Kapusniak et al., 1984). L'hypothèse est ainsi indirectement justifiée et c'est la raison pour laquelle cette méthode est actuellement utilisée à l'Université de Nottingham dans les travaux de prévision du comportement des roches dans les tunnels.

On a donc élaboré un modèle de conception relativement simple fondé sur des lois de comportement des roches stables établies à partir de mesures d'essais qui tiennent compte de la stratification du massif rocheux. La validité de cette méthode a donc été confirmée en comparant les convergences prévues et mesurées d'un tunnel circulaire. Cependant, ce modèle ne s'applique qu'aux voies circulaires dans des conditions de contraintes hydrostatiques et les effets du soutènement y sont représentés sous forme d'une pression de soutènement unique. C'est dans ces secteurs que des travaux de développement devront être réalisés dans l'avenir étant donné que les fondements sont maintenant posés.

1) Charbon
2) Profondeur (m)
3) Remblai
4) Remarque: Déplacements radiaux exprimés en cm

Figure 1. Calcul de la déformation radiale et de la zone de déplacements plastiques d'un tunnel circulaire dans des roches stratifiées.

Réalisations futures

Le soutènement d'un tunnel relève de la mécanique des roches et nécessite une analyse approfondie. En résumé, tout type de soutènement comporte ses propres caractéristiques de contrainte-déformation qui devraient être incluses dans le modèle de conception étant donné que le soutènement ne peut pas dépasser en toute sécurité certaines limites finies de déformation. Toute déformation au-delà de cette limite est considérée nuisible à la stabilité du tunnel et il peut être nécessaire dans de telles situations de recourir à un type de soutènement différent. À l'heure actuelle, des recherches visent à obtenir des données exactes sur ces caractéristiques de soutènement par des essais sur modèle et à les incorporer dans le modèle de conception en les stockant dans la mémoire de l'ordinateur et en les rappelant au besoin. Il sera ainsi possible de prévoir avec exactitude la déformation d'un tunnel lorsque le soutènement subira une perte partielle ou totale de sa résistance après rupture, situation relativement courante dans les couches de charbon profondes.

Le progrès le plus important à accomplir est d'être en mesure de considérer différentes formes de tunnels dans diverses conditions de contraintes initiales. Il faut pour ce faire délaisser les équations de contrainte actuelles et utiliser une autre technique. Le concept actuellement formulé est d'utiliser la méthode des éléments finis et de

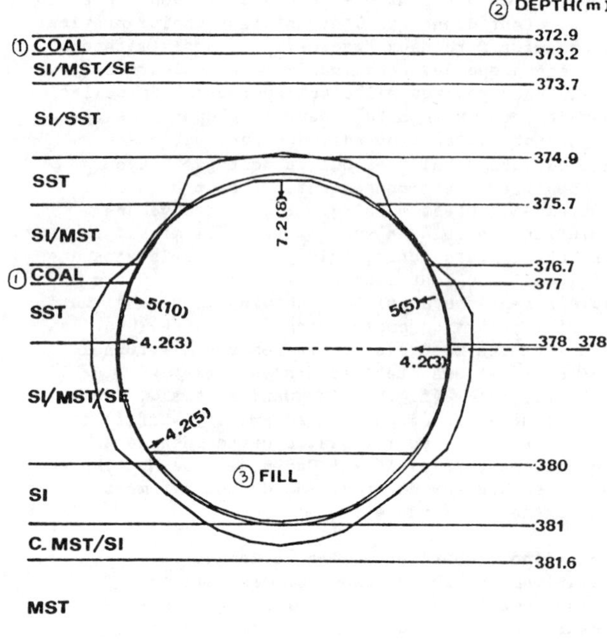

Remarque : Déplacements radiaux exprimés en cm

1) Charbon
2) Profondeur (m)
3) Remblai

Figure 1. Calcul de la déformation radiale et de la zone de coulissement d'un tunnel circulaire dans des roches stratifiées.

l'adapter à l'équation de la résistance des roches déjà établie. Cette façon de procéder permettra d'effectuer une analyse acceptable des contraintes dans des tunnels de différentes formes soumis à diverses conditions de contraintes initiales. Cependant, la méthode des éléments finis comporte des limites lorsqu'on ne tient compte que de la déformation élastique qui est négligeable lorsqu'une zone de déplacements plastiques importante se forme. C'est pourquoi il faudrait quand même recourir à l'analyse des déplacements en appliquant la loi de déformation volumétrique dont la validité a été confirmée mais qui est basée sur la méthode améliorée de l'analyse des contraintes. Cette méthode "hybride" constituera certainement un instrument de conception très précis.

Il est particulièrement important de prévoir diverses formes de tunnels et de conditions de contraintes. L'utilisation de tunnels circulaires est encore relativement rare dans les mines de charbon de la Grande-Bretagne et les conditions de contraintes hydrostatiques ne sont considérées que comme une règle générale. Récemment, on a attribué à des conditions de contraintes anisotropiques la rupture d'un tunnel, conditions qui résultent de la présence de failles et de discontinuités géologiques. Par conséquent, il est essentiel que ces travaux portent fruit dès que possible pour que d'autres soutènements puissent être conçus avec assurance à partir des résultats.

Conclusion

Les travaux de recherche réalisés à l'Université de Nottingham ont permis de mettre sur pied une base concrète de solutions aux problèmes de mécanique des roches en ce qui concerne les lois de comportement des massifs rocheux et les techniques de calcul. Le modèle de conception actuel comporte des limites à cause des équations de contraintes qui y sont utilisées; il a tout de même permis d'établir une excellente corrélation entre les données de convergence prévues et mesurées. Nombre de perfectionnements futurs restent à réaliser et, s'ils le sont, la technique de conception des tunnels qui en résultera sera des plus appropriées.

RÉFÉRENCES (voir section anglais)

16h30 - 17h00

CÉRÉMONIE DE CLÔTURE - 3 SEPTEMBRE 1987

Lors de la clôture des séances techniques, le jeudi 3 septembre 1987, le Président du congrès, G. Herget, demande au Président honoraire, B. Ladanyi, de remercier au nom du comité organisateur canadien, les présidents, les conférenciers, les modérateurs et les participants aux tables rondes pour leur excellente participation aux séances techniques. Le professeur Ladanyi remercie tous les conférenciers et offre à tous les représentants officiels supérieurs une sculpture esquimaude.

Après ces quelques mots de remerciement prononcés par le professeur Ladanyi, le professeur Wittke de l'Allemagne de l'Ouest est prié de faire un compte rendu personnel des apports dont a bénéficié la mécanique des roches depuis la fondation de la Société internationale de mécanique des roches, il y a 25 ans, ainsi que des lacunes qui restent à combler.

Le professeur E.T. Brown, président de la Société internationale de mécanique des roches prononce le mot de la fin en ces termes :

Monsieur le Président, Mesdames et Messieurs

Lors de la cérémonie d'ouverture, j'avais remercié au nom de la Société internationale et des délégués, nos collègues canadiens pour leur chaleureuse hospitalité et pour leurtravail de préparation du congrès. Maintenant que les séances techniques sont terminées, j'ai une fois de plus le plaisir de les remercier en votre nom pour le travail qu'ils ont investi dans l'organisation de cette excellent congrès.

Comme nous l'a fait remarqué notre distingué ancien président, le professeur Wittke, notre domaine a réalisé des progrès considérables au cours des 25 années qui se sont écoulées depuis la création de la Société internationale même s'il reste beaucoup encore à accomplir. Au cours de la conférence, j'ai parlé à un certain nombre de vieux routiers, c'est-à-dire ceux qui sont de ma génération ou ceux qui sont juste un petit peu plus âgés. Nous sommes émerveillés devant les capacités et les réalisations des nombreux jeunes qui nous ont suivi dans ce domaine, notamment certains candidats de haut calibre à la médaille Manuel Rocha de la société. Comme la mécanique des roches est beaucoup plus un domaine de haute technologie que dans le passé et qu'elle compte dans ses rangs de jeunes gens talentueux, particulièrement en Amérique du Nord, je suis confiant que d'autres progrès intéressants auront été accomplis lorsque nous nous réunirons de nouveau à Aix-la-Chapelle en 1991. Je vous convie donc tous à faire un effort spécial pour participer à notre prochain congrès.

Vous vous rendez tous compte, j'en suis sûr, qu'un congrès comme celui-ci doit être planifié et organisé de façon très minutieuse par un groupe de personnes enthousiastes et dévouées comme le sont les membres du comité organisateur et leurs nombreux collaborateurs. Nous sommes particulièrement reconnaissants à leur parrain, le professeur Branko Ladanyi, qui a entrepris l'ensemble du projet et qui a surveillé les activités d'une façon qui pourrait paraître superficielle, mais je n'en suis pas complètement sûr. Plusieurs membres du comité organisateur ont contribué de façon spéciale à l'organisation des aspects techniques du congrès. Le professeur Norbert Morgenstern a été le cerveau derrière le Programme technique qui a atteint un très haut niveau de qualité. Tous ceux avec lesquels je me suis entretenu sur le sujet étaient d'accord pour dire que les ateliers spécialisés ont remporté un franc succès.

L'excellente exposition commerciale qui a été l'une des mieux réussies qu'il m'a été donné de voir à une réunion sur la mécanique des roches est le fruit du travail de Bill Bawden. Soit dit en passant, Bill est celui à qui nous devons la présence du joueur de cornemuse qui a accompagné le ministre lors de l'ouverture de l'exposition lundi.

Bill s'est aussi vu confié l'organisation détaillée des séances techniques qui a consisté à donner des instructions au président des séances et des conférenciers et à s'assurer qu'elles soient bien exécutées. Le choix de Bill pour remplir cette tâche a été des plus judicieux: il était la personne toute indiquée pour ce travail.

Le professeur Denis Gill a occupé le poste de coprésident du congrès spécialement chargé des nombreux détails de ce qui était considéré comme des fonctions locales mais qui, selon mes observations, touchait à peu près à n'importe quoi. Denis s'est montré un soutien précieux et nous lui en sommes tous des plus reconnaissants.
C'est au président du congrès, Gerhard Herget, l'homme qui n'a de gauche que la main gauche,que nous devons la cohésion de l'ensemble. Le travail de Gerhard et des membres de son comité a commencé il y a quelques années et il est loin d'être terminé. Ceux qui le connaissent, comme le secrétaire général et moi-même, et qui ont été en étroite relation avec lui pendant une période de temps prolongée savent combien il a investi de temps et d'énergie pour que ce congrès soit réussi.

Branko, Gerhard et les autres membres du comité organisateur, nos salutations et tous nos remerciements!

Notre société est une société vraiment internationale. Trois langues officielles et plusieurs non officielles y sont en usage, y compris mon propre anglais australien. C'est pourquoi il nous faut compter beaucoup sur la qualité de la traduction simultanée pour que la communication soit efficace au cours des séances techniques. La contribution des traducteurs qui accomplissent leurs fonctions exigeantes avec talent passe souvent inaperçue. Mais pas cette fois-ci Mesdames et Messieurs les traducteurs, nous vous remercions donc chaleureusement.

Ceci met fin à cette séance et aux comptes rendus techniques du congrès. Je vous donne rendez-vous au prochain colloque parrainé par la SIMR à Minneapolis (É.-U.), en juin 1988, et à Madrid (Espagne), en septembre 1988, et il va s'en dire à notre prochain congrès à Aix-la-Chapelle en Allemagne de l'Ouest dans quatre ans.

Mesdames et Messieurs, cette séance est terminée. Merci.

Dîner soulignant le 25e anniversaire

Ont participé au dîner organisé pour commémorer le 25e anniversaire de fondation de la Société internationale de mécanique des roches 575 personnes pour lesquelles il a fallu réservé le Grand Salon, les salles Marquette et Jolliette de l'hôtel Reine Élizabeth. La soirée a débuté par la dégustation de cocktails servis dans une salle attenante. Les congressistes se sont ensuite dirigés vers les tables et lorsque les invités de la table d'honneur eurent été assis, le président du congrès, G. Herget, a souhaité la bienvenue aux délégués dans les trois langues officielles et leur a souhaité un agréable repas. Le menu est reproduit ci-dessous et le service à table par l'hôtel Reine Elizabeth s'est effectué rondement sous la direction compétente de J. Druda, directeur des banquets de l'hôtel.

Pendant qu'étaient servis le café et les digestifs, le Président, G. Herget, a présenté les invités de la table d'honneur qui étaient de droite à gauche:

P. Michaud, Montréal, Canada, Directeur exécutif de l'Institut canadien des mines et de la métallurgie
N. Grossmann, Lisbonne, Portugal, Secrétaire général de la SIMR
Pr Langer, Hanovre, Allemagne de l'Ouest, ancien président de la Société internationale de géologie de l'ingénieur
Pr Bello M., Mexico, Mexique, Vice-président de la SIMR pour l'Amérique du Nord
M.D. Everell, Ottawa, Canada, Sous-ministre adjoint, Ministère de l'Énergie, Mines et des Ressources
Mme J. Udd, Ottawa, Canada
Pr Tan T.-K., Beijing, Chine, Vice-président de la SIMR pour l'Asie
Mme L. Gill, Montréal, Canada
Pr B. Ladanyi, Montréal, Canada, Président honoraire du congrès, ancien président de l'Association canadienne de mécanique des roches
Mme U. Herget, Ottawa, Canada
Pr E.T. Brown, Londres, Royaume-Uni, Président de la Société internationale de mécanique des roches
G. Herget, Ottawa, Canada, Président général du congrès
Mme N. Ladanyi, Montréal, Canada
Pr D.E. Gill, Montréal, Canada, Coprésident du congrès
Mme Tan T.-K., Beijing, Chine
J.E. Udd, Président de l'Association canadienne de mécanique des roches
Mme A. Reid, Londres
Pr B. Bamford, Melbourne, Australie, Vice-président de la SIMR pour l'Océanie et l'Australie
Pr Tinoco, Caracas, Vénézuela, Vice-président de la SIMR pour l'Amérique du Sud
Pr Wagner, Johannesburg, Afrique du Sud, Vice-président de la SIMR pour l'Afrique du Sud
M. J.A. Franklin, Orangeville, Canada, Président élu de la SIMR
M. Bozozuk, Ottawa, Canada, Président de la Société canadienne de géotechnique

G. Herget remercie ensuite tous les membres du comité organisateur qui n'ont pas ménagé leur temps pendant trois ans pour préparer ce congrès. Au cours de ces années, ils ont tous consacré du temps supplémentaire en plus d'accomplir leurs tâches habituelles. Les membres du comité organisateur sont chaudement applaudis lorsqu'ils se rendent à tour de rôle à la table d'honneur recevoir leur prix:

B. Ladanyi, Président honoraire et chargé de la levée de fonds (sculpture esquimaude);
D.E. Gill, Coprésident du congrès et chargé des fonctions locales (plaque);
N.R. Morgenstern, Programme technique (plaque);
J.A. Franklin, publicité (plaque);
T. Carmichael, finances (plaque);
W.F. Bawden, expositions (plaque);
J. Gaydos, ICM, exposition commerciale (sculpture esquimaude);
A.T. Jakubick, visites techniques (plaque);
L. Geller, traduction (plaque);
J. Robertson, programme pour les conjoints (plaque);
D. Grégoire, inscription, ICM (sculpture esquimaude);
J. Bourbonnais, activités sociales (plaque);
J. Nantel, levée de fonds (plaque);
S. Vongpaisal, éditeur adjoint (plaque).

Le Président souligne que ces personnes sont les chefs des groupes de personnes qui ont travaillé très fort, en particulier au cours des derniers mois, pour préparer le congrès.

Après avoir présenté et remercié les membres du comité organisateur, le Président donne la parole au président sortant de la Société internationale de mécanique des roches, le Pr E.T. Brown:

Merci Gerhard, bonsoir Mesdames et Messieurs

Et bien, ce congrès tire à sa fin et le moment de nous quitter approche. C'est la dernière fois que je vous adresse la parole à titre de président de la Société internationale. C'est sans doute un moment de grand soulagement pour nous tous, pour nos collègues canadiens qui ont organisé le congrès et qui ne sont pas mécontents que tout soit terminé, pour moi qui peux maintenant confier mes responsabilités à quelqu'un d'autre et pour vous tous qui serez délivrés dans quelques minutes de ne plus avoir à écouter tous ces discours, en l'occurence le mien.

Avant de me taire et de me rasseoir, j'aimerais, au nom des visiteurs de l'étranger, remercier à l'instar de Gerhard, les membres du comité organisateur. J'aimerais remercier, en particulier ce soir, les personnes qui ont organisé le Programme des activités sociales et le Programme pour les personnes qui ont accompagné les congressistes, notamment Mme Joan Robertson et

Jacques Bourbonnais. Nous avons apprécié votre hospitalité, nous avons pris du plaisir à participer aux activités spéciales que vous avez organisées pour nous : les excursions, les réceptions et le spectacle offert par Les Ballets Jazz de Montréal lundi soir, en particulier le pas de deux exécuté par Gerhard Herget et Jacques Bourbonnais, et nous nous amusons encore certainement ce soir. Cette rencontre à Montréal a été particulièrement appréciée. C'est une ville des plus attrayantes, donc un excellent choix pour la tenue d'un congrès.

À la cérémonie de clôture, cet après-midi, j'ai remercié le président honoraire, le professeur Branko Ladanyi et le président, Gerhard Herget, pour le rôle essentiel qu'ils ont joué dans l'organisation de ce congrès. Maintenant que les circonstances sont moins officielles, j'aimerais les remercier une fois de plus au nom de vous tous. Lorsque je suis entré dans la salle ce soir, Gerhard était déjà là avec son épouse à s'affairer de sa manière méthodique et discrète habituelle pour que tout soit prêt pour le banquet. Je suis heureux de noter qu'il s'est accordé un peu de répit.

Comme vous le savez, en 1987, nous célébrons le 25e anniversaire de la société et le dîner de ce soir est le banquet d'anniversaire. Pour souligner cet événement important, nous avons demandé à tous les membres des anciens conseils de la société qui ont pu participer au congrès de s'assoir ensemble ce soir avec leur épouse à une table spéciale. Bien qu'ils constituent un corps distingué d'hommes qui ont apporté des contributions très importantes à l'existence et au bien-être de notre société, je dois admettre qu'ils ne constituent pas un spectacle particulièrement impressionnant. Néanmoins, je vais leur demander de se lever pendant quelques secondes pour que nous puissions leur souhaiter la bienvenue et reconnaître leurs contributions spéciales à notre domaine et à notre société. Messieurs, s'il-vous-plaît.

Au cours des 20 dernières années, la société a été administrée par un secrétariat situé à la LNEC à Lisbonne au Portugal. La société a été particulièrement chanceuse de compter dans ses rangs

Dr. J. Franklin, nouveau président-élu de la Société Internationale pour les Mécaniques des Roches, remercie l'ancien vice-président de l'Amérique du Nord, Prof. Bello-Maldorado, pour ces services distingués rendus à la société. À l'avant Mme. U. Herget et Prof. Dr. E.T. Brown.

des secrétaires généraux remarquables, dont deux d'entre eux, Fernando Mello Mendes et Ricardo Oliviera, font partie du groupe que vous venez d'applaudir. Ces quatre dernières années, le Secrétaire général a été Nuno Grossmann. Tous ceux qui ont participé aux réunions pendant cette période auront appris à connaître ce personnage trapu et affairé, qui organise différentes choses et s'assure que nous tous, moi en particulier, agissons correctement et dans les meilleurs intérêts de la société. À titre de président, je crois être le mieux placé parmi nous pour dire avec quel dévouement, vaillance, loyauté il a travaillé pour la société. Nuno, je vous remercie personnellement et au nom de la société.

Dr. J. Franklin remercie l'ancien vice-président de l'Australasie, Prof. Dr. W.M. Bamford, pour ces services distingués rendus à la société. À l'avant Dr. G. Herget.

Au cours de ces vingt années, l'autre membre important du secrétariat a été la Secrétaire, Mme Maria de Lurdes Eusébio. Si les secrétaires généraux ont maintenu les présidents et les conseils au pas, c'est Maria de Lurdes qui, de façon douce et charmante, a maintenu les secrétaires généraux au pas. L'une des plus grandes satisfactions à occuper un poste de direction dans cette société est d'avoir la possibilité de travailler avec Maria de Lurdes. Elle ne manque jamais de tact en plus d'être charmante, élégante et des plus efficaces. Nous lui sommes tous très attachés et nous ne pourrions tout simplement pas nous passer d'elle. Pour souligner ses 20 ans de services exceptionnels dans la société et lui signifier quelque peu notre appréciation sans mesure, j'aimerais lui remettre un présent au nom de la société.

Les membres du conseil qui est composé des vice-présidents de chacune des six régions ont également un rôle important à jouer dans la société. Je suis personnellement très reconnaissant aux membres du conseil sortant pour l'appui et les conseils sans réserve qu'ils m'ont accordés au cours des quatre dernières années. Je suis particulièrement reconnaissant à notre premier vice-président, Sten Bjurström de la Suède. Sten a dû repartir pour Stockholm pour négocier des questions importantes; il n'a donc pas pu, malheureusement, nous honorer de sa présence ce soir.

Certains d'entre vous ont remarqué qu'au cours de cette semaine passée à Montréal, il y a eu, pourrait-on dire des discussions sur l'élection du nouveau conseil qui a eu lieu à la réunion qui s'est tenue plus tôt ce soir. Pour clore mon mandat de président, il me fait grand plaisir de vous présenter les membres du nouveau conseil qui dirigeront les affaires de la société à partir de maintenant jusqu'à la fin du 7e congrès à Aix-la-Chapelle, en septembre 1991. Ce sont:

Vice-président pour l'Amérique du Sud - C. Dinis da Gamma (Brésil)
Vice-président pour l'Amérique du Nord - Jim Coulson (États-Unis)
Vice-président pour l'Europe - Marc Panet (France)

Dr. J. Franklin remercie l'ancien vice-président de l'Afrique, Dr. H. Wagner, pour ces services distingués rendus à la société. À l'avant de gauche à droite, Dr. G. Herget et Mme. N. Ladanyi.

Vice-président pour l'Australasie - Ian Johnston (Australie)
Vice-président pour l'Asie - S.L. Mokhashi (Inde) (absent)
Vice-président pour l'Afrique - Oscar Steffen (Afrique du Sud)

Enfin, et fait très important, j'aimerais vous présenter mon successeur comme président de la société que je félicite et à qui je cède les rênes de la société: John Franklin du Canada.

Le nouveau président de la Société internationale de mécanique des roches, J.A. Franklin, est ensuite prié de s'adresser aux congressistes.

Après l'allocution de J.A. Franklin, le Président, G. Herget, prononce les mots de la fin du 6e congrès de la SIMR. Il remercie les congressistes de leur venue à Montréal et les invite à regarder le spectacle donné par Les Sortilèges, groupe canadien de danse folklorique, dont le programme est reproduit ci-dessous:

L'ensemble national de folklore canadien, Les Sortilèges.

Après ce spectacle, très apprécié par tous les membres, les danseurs du groupe demandent aux congressistes de les rejoindre sur la piste de danse qui est rapidement envahie par une foule joyeuse. Le 6e Congrès international de mécanique des roches se termine ainsi et la plupart des congressistes repartent vers 23 h.

Séances d'affichage

La Comité technique a invité les congressistes à participer aux séances d'affichage qui eurent lieu quotidiennement durant le congrès. Ces séances ont constitué un lieu de rencontre avec les auteurs ainsi qu'un lieu de présentation des résultats qui n'étaient pas disponibles lorsque les documents ont été présentés pour publication dans les comptes rendus. Les séances plénières furent ainsi prolongées d'une façon moins structurée

CATALOGUE DES AFFICHES

1. P.R. ARGAWAL (Inde) "Comportement de l'intrazone de chevauchement le long du tunnel d'amenée d'eau de l'aménagement Yamuna Hydel, stade II, partie II".

2. G. BALLIVY, B. BENMOKRANE, A. LAHOUD (Canada) "Méthode intégrale de dimensionnement d'ancrage injectés en rocher".

3. K. BALTHASAR, M. HAUPT, CH. LEMPP, O. NATAU (Allemagne) "Relaxation des contraintes dans le sel gemme : Comparaison des résultats obtenus par mesure in situ et en laboratoire".

4. S. BANDIS, J. LINDMAN, N. BARTON (Norvège) "État des contraintes tridimensionnelles autour de cavités en massif de roche faible sous grandes contraintes".

5. B. BAMFORD (Australie) "Capacité au sondage, au havage, au forage de la roche".

6. W. BAWDEN, D. MILNE (Canada) "Approche géomécanique pour la conception des mines à Noranda Minerals Inc.".

7. B. BJARNASON (Suède) "Variation non linéaire et discontinue des contraintes en fonction de la profondeur dans la partie supérieure du bouclier baltique".

8. G. BORM (Allemagne) "Fluage et relaxation autour d'un trou de sonde perçé dans un massif de sel gemme".

9. T.C. CHAN, N.W. SCHEIER (Canada) "Simulation par la méthode des éléments finis de l'écoulement de l'eau souterraine et du transfert de la chaleur et des radionucléides dans un massif rocheux plutonique".

10. F.H. CORNET, J. JOLIVET, J. MOSNIER (France) "Identification et caractérisation hydraulique des fractures recoupées par un forage".

11. EMMANEL DETOURNAY, L. VANDAMME, A. H-D CHENG (É.-U.) "Propagation d'une fracture hydraulique verticale dans une formation poro-élastique".

12. C.H. DOWDING, K.M. O'CONNOR, M.B. SU (É.-U.) "Quantification de la déformation d'une masse rocheuse par échométrie".

13. M.E. DUNCAN FAMA, L.J. WARDLE (Australie) "Analyse numérique de la stabilité des piliers interchambres des mines de charbon".

14. DYWIDAG SYSTEMS INTERNATIONAL (Canada) "Ancrage déformable et indicateurs de surcharge Dywidag".

15. J. ENEVER (Australie) "Dix ans d'expérience dans la mesure des contraintes par fracturation hydraulique en Australie".

16. G.S. ESTERHUIZEN (Afrique du Sud) "Évaluation de la stabilité des tunnels de mine dans des champs de contraintes changeants".

17. E. FJAER, R.K. BRATLI, J.T. MALMO, O.J. LOEKBERK, R.M. HOLT (Norvège) "Études optiques de la rupture d'une cavité dans les roches sédimentaires de faible résistance".

18. R.J. FOWELL, C.K. IP (Angleterre) "Mécanisme de taillabilité des roches par pics et jets d'eau".

19. S. GENTIER (France) "Comportement hydromécanique d'une fracture naturelle sous contrainte normale".

20. D.E. GILL, B. LADANYI (Canada) "Courbes caractéristiques différées pour le calcul du revêtement des souterrains".

21. GAO HANG, JING ZIGANG, SHEN GUANGHAN (R.P. de Chine) "Étude d'une mine de charbon au-dessus d'une nappe d'eau captive dans les bassins houillers du nord de la Chine".

22. T.F. HERBST (Allemagne) "Accroissement de la sécurité par la surveillance complète du soutènement".

23. F. HEUZE, R.J. SCHAFFER, R.K. THORPE, A.R. INGRAFFEA, R.H. NILSON (É.-U.) "Modèles quasi statiques et dynamiques de fracturation produite par des fluides dans un milieu rocheux fracturé".

24. R.M. HOLT, J. BERGEN, T.H. HANSSEN (Norvège) "Anisotropie des propriétés mécaniques d'un grès peu consolidé".

25. D.R. HUGHSON, A.M. CRAWFORD (Canada) "Mesure de l'effet Kaiser: Une méthode pour déterminer les contraintes dans les roches par émission acoustique".

26. OLDRICH HUNGR (Canada) "Exemples d'analyse de stabilité d'un coin par la méthode des colonnes".

27. R.W. HUTSON, C.H. DOWDING (É.-U.) "Nouvelle méthode de production de fractures identiques dans de la roche réelle de n'importe quelle dureté".

28. T. KYOYA, Y. ICHIKAWA, T. KAWAMOTO (Japon) "Procédé de déformation et de fracture d'une masse rocheuse discontinue et mécanique d'endommagement".

29. YOZO KUDO (Japon) "Propriétés physiques et microstructures de roches granitiques au Japon".

30. KOKICHI KIKUCHI (Japon) "Calcul probabiliste et modélisation de la répartition des diaclases selon l'échantillonnage statistique".

31. V. LABUC, W. BAWDEN, F. KITZIGER (Canada) "Réseau d'écoute sismique utilisant la fibre optique pour la transmission des signaux".

32. LI XING CAI, SHI RU BIN, YU MU XING (Chine) "Méthodes acoustiques pour déterminer sur le terrain l'orientation prédominante des macrofissures".

33. LI XING CAI (Chine) "Pression effective nulle ou négative et ses effets sur les propriétés physiques d'une masse rocheuse in situ".

34. J.C.S. LONG, D. BILLAUX, K. HESTIR, J.P. CHILES (É.-U.) "Techniques géostatistiques permettant d'utiliser la structure spatiale pour modéliser un réseau de fractures".

35. PAUL H. LU (É.-U.) "Facteur d'intégrité appliqué à la conception des piliers interpanneaux des longues tailles".

36. J. MATHIS (Suède) "Statistique des fissures : Comparaison entre la cartographie d'un tunnel et d'une grande chambre vide".

37. KIKUO MATSUI, MASATOMO ICHINOSE, KENICHI UCHINO (Japon) "Prévision de la déformation des voies par de longues tailles sous piliers et parements".

38. P. MACDONALD (Canada) "Recherche en modélisation sismique et numérique à l'INCO".

39. D. MILNE (Canada) "Approche géomécanique pour la conception des mines à Noranda Minerals Inc.".

40. D.Z. MRAZ (Canada) "Conception proposée pour des structures isolantes à haute pression dans le sel".

41. H.-B. MUHLHAUS (Allemagne) "Stabilité des excavations souterraines profondes dans des roches à forte cohésion".

42. T. NISHIOKA, A. NARIHIRO, O. KYOYA (Japon) "Analyse numérique d'un mur de retenue ancré construit dans une roche tendre".

43. R.C.T. PALKANIS, H.D.S. MILLER, S. VONGPAISAL, T. MADIL (Canada) "Approche empirique à la conception des chantiers ouverts".

44. D. PREECE (É.-U.) "Mesures de convergence par fluage de forage et calculs numériques au site de stockage SPR de Big Hill, Texas".

45. J.P. JOSIEN, J.P. PIQUET, R. REVALOR (France) "Apports de la mécanique des roches à la maîtrise des phénomènes dynamiques dans les mines".

46. M. ROUSSIN, S. BURDICK (É.-U.) "Nouvel appareil d'excavation rocheuse : Mise au point d'un scarificateur à impact".

47. S. SAKURAI, N. SHIMIZU (Japon) "Détermination de la stabilité des talus rocheux par la théorie des ensembles flous".

48. J.C. SHARP, C.H. LEMAY, B. NEVILLE (Canada) "Observation du comportement de grands talus de fosse dans les roches tendre ultrabasiques".

49. P.L. SWANSON (É.-U.) "Résistance à la fracturation en mode-I dans des polycristaux fragiles: Apport de la traction à l'interface".

50. M THIERCELIN, J.C. ROEGIERS, T.J. BOONE, A.N. INGRAFFEA (É.-U.) "Investigation des paramètres d'un matériau affectant le comportement de fractures à l'approche d'interfaces rocheuses".

51. SPE (France) "Roche à grande profondeur : Mécanique des roches et physique des roches à grande profondeur".

52. K.F. UNRUG (É.-U.) "Effet sur le toit d'une mine souterraine du minage réalisé dans une mine à ciel ouvert voisine exploitant la même couche".

53. G. VENKATACHALAM, K. VINCENT PAUL (Inde) "Caractérisation des fractures rocheuses".

54. B.S. VERMA (Canada) "Techniques proposées de détermination directe de résistance à la traction de roches compétentes".

55. B.S. VERMA (Canada) "Formation de fissures autour des chantiers dans des formations minérales contiguës".

56. T. VLADUT (Canada) "Système de diagraphie LWD".

57. D. WULLSCHLAGER, O. NATAU (Allemagne) "Le massif rocheux boulonné en tant que continuum anisotrope - Comportement et proposition de dimensionnement pour les excavations".

58. T.H. MUTSCHLER, B. FROHLICH (Allemagne) "Calcul analytique du comportement de résistance de massifs rocheux à couches alternées".

3

Deutscher Teil

Eröffnungsveranstaltung

Abb. 1

Abb. 2

Abb. 3

Am Montag, dem 31. August 1987, um 8.30 Uhr versammelten sich die Kongreßteilnehmer im Grand Salon des Hotels Queen Ezliabeth, im Montreal. Um 8.35 Uhr wurden die Ehrengäste von Soldaten in französischen Uniformen des 18. Jh. - Flötenspieler, Trommler, zwei Soldaten mit Musketen, Gefreiter - zu ihrem Tisch geleitet.

Der Kongreßvorsitzende, Dr. G. Herget, begrüßte die Teilnehmer in den drei Kongreßsprachen und stellte die Ehrengäste vor (Abb. 2 und 3, von rechts nach links):

N. Grossmann, Generalsekretär der IGFM
Dr. Chapuis, Kanadische Gesellschaft für Geotechnik
Prof. Dr. Gill, 2. Kongreßvorsitzender
Prof. Dr. N.R. Morgenstern, Wissenschaftliches Programm
Prof. B. Maldonado, Vizepräsident der IGFM (Nordamerika)
Prof. Dr. B. Ladanyi, Ehrenvorsitzender des 6. Internationalen Kongresses der IGFM
The Hon. R. Savoie, Minister der Provinz Quebec für Bergbau und Angelegenheiten der Ureinwohner
The Hon. G. Merrithew, kanadischer Staatsminister für Forstwirtschaft und Bergbau
Dr. G. Herget, Vorsitzender des 6. Internationalen Kongresses der IGFM
Prof. Dr. E.T. Brown, Präsident der IGFM
Prof. Dr. E. Hoek, Eröffnungsreferat
Dr. J.E. Udd, Leiter des Bergbau-Forschungsinstituts
L. Milne, Präsident des Kanadischen Instituts für Bergbau und Metallurgie
Dr. G. Miller, Präsident der Kanadischen Bergbauvereinigung

Prof. Ladanyi hieß die Kongreßteilnehmer willkommen und bat Prof. Dr. E.T. Brown, Präsident der IGFM, sie im Namen der Internationalen Gesellschaft für Felsmechanik zu begrüßen.

"Herr Vorsitzender, verehrte Gäste, meine Damen und Herren!

Im Namen der Internationalen Gesellschaft für Felsmechanik und der vielen Delegierten, die zu diesem Kongreß nach Kanada gekommen sind, möchte ich unseren kanadischen Kollegen danken für die Gastfreundschaft, die uns in den letzten Tagen zuteil geworden ist, für die liebenswürdige Begrüßung heute morgen und für ihre Vorbereitungen zu dieser Tagung, die uns sicher unvergeßlich bleiben wird. Diejenigen unter uns, die nicht in Kanada leben, freuen sich auch über die Gelegenheit, Montreal, diese faszinierende Stadt, und Kanada überhaupt mit seiner herrlichen Landschaft und seinen vielen Sehenswürdigkeiten näher kennenzulernen.

Professor Dr. E.T. Brown, England

Die Internationale Gesellschaft für Felsmechanik wurde im Jahre 1962 in Salzburg von einer Gruppe weitblickender, von ihrer Wissenschaft erfüllter Felsmechaniker um Prof. Leopold Müller gegründet. Wir feiern daher in diesem Jahr, 1987, das 25. Gründungsjubiläum der Gesellschaft. Im Laufe dieser 25 Jahre ist die Felsmechanik zu einer anerkannten Disziplin geworden, die aus der Praxis des Ingenieurbaus, des Bergbaus und der Rohstoffgewinnung nicht mehr wegzudenken ist und auch zum größeren Verständnis geologischer Vorgänge beiträgt. Die Felsmechanik besitzt heute eine umfangreiche Literatur, wird an den Universitäten und Hochschulen sowie bei Beratenden Ingenieurbüros und Forschungsorganisationen als eigenständiges Wissensgebiet betrachtet und besteht als selbständige Disziplin schon so lange, daß viele der führenden Felsmechaniker ihr ganzes Berufsleben diesem Fachgebiet gewidmet haben. In dieser und mancher anderer Hinsicht wird die Felsmechanik langsam aber sicher zu dem, was oft als ausgereifte Wissenschaft bezeichnet wird.

Es ziemt sich daher, daß im Jahr des silbernen Gründungsjubiläums unserer Gesellschaft auch einer unserer vierjährlichen Kongresse stattfindet, die dem Erfahrungsaustausch und einer wissenschaftlichen Bestandsaufnahme gewidmet sind. Ebenso, meine ich, ziemt es sich, daß der Tagungsort dieses Jubiläumskongresses Kanada ist, denn hier laufen zur Zeit einige bedeutende berg- und felsbauliche Projekte, und kanadische Wissenschaftler haben schon immer zum Fortschritt in Theorie und Praxis der Felsmechanik beigetragen. Trotzdem hat bisher noch kein Felsmechanikkongreß in Kanada stattgefunden, und es ist wohl höchste Zeit, daß wir dieses Versäumnis 1987 nachholen.

Ich kann Ihnen auch mitteilen, daß einige Gründungsmitglieder und frühere Beiratsmitglieder der IGFM heute anwesend sind. Wir freuen uns, daß sie sich dieser Mühe unterzogen haben, begrüßen sie herzlich in unserer Mitte und hoffen, daß sie schöne Erinnerungen an diese Tagung und an den Aufenthalt in Montreal mit nach Hause nehmen werden.

Ich hatte gehofft, daß auch der eigentliche Gründer der Gesellschaft, Herr Prof. Leopold Müller, in der Lage sein würde, an dem Kongreß teilzunehmen, um bei dieser Gelegenheit zu uns zu sprechen. Leider ist Prof. Müller durch die Pflichten eines noch äußerst regen Berufslebens daran verhindert; er hat mir jedoch Grußworte an Sie übermittelt, die ich Ihnen jetzt vorlesen möchte:

"Sehr geehrter Herr Vorsitzender,
liebe Freunde und Kollegen!

Ich hatte die ernstliche Absicht, mich zur Feier des silbernen Jubiläums unserer Gesellschaft einzufinden und mit Ihnen einen gemeinsamen Rückblick auf 25 Jahre intensiver wissenschaftlicher Tätigkeit zu tun. Ich bin nun leider doch verhindert und kann also nicht bei Ihnen sein.

25 Jahre sind im kulturellen Leben der Welt nur eine kurze Zeitspanne. In unserer Zeit entwickelt sich jedoch alles viel schneller als je zuvor, auch die Felsmechanik. In diesem Vierteljahrhundert haben wir viel erreicht, und unsere gefüllten Bibliotheken sind der beste Beweis dafür. Wenn man bedenkt, wieviele Begriffe und Ausdrücke, die uns heute vollkommen geläufig sind, im Jahre 1962 noch unbekannt waren, so kann man nur staunen.

Bedauerlicherweise hat die junge Wissenschaft der Felsmechanik hauptsächlich durch Katastrophen und Unglücksfälle den stärksten Auftrieb erhalten, besonders in den sechziger Jahren, während in den Jahrzehnten davor wenig Interesse für dieses Gebiet bestand.

Leider muß ich feststellen, daß in diesen 25 Jahren die felsmechanische Forschung einerseits unser Wissen sehr bereichert hat, daß aber andererseits in derselben Zeit erstaunlich viel auch in Vergessenheit geraten ist. Wahrscheinlich findet nur ein Bruchteil der neuen felsmechanischen Erkenntnisse Eingang in die Praxis des Felsbaus. Nur ein kleiner Teil dieser grundlegenden theoretischen Arbeit kommt bei praktischen Großprojekten zur Anwendung, und wenn wir ehrlich sein wollen, so müssen wir zugeben, daß es um die geomechanische Praxis heute manchmal schlechter bestellt ist als vor 20 Jahren.

In der Felsmechanik spielen nicht so sehr ausgeklügelte Computerverfahren, sondern fundierte theoretische Kenntnisse eine entscheidende Rolle, vor allem aber das Interesse am Fels, am Gebirge selbst, an den Eigenschaften dieses Partners in unserer Arbeit, der je nachdem, wie wir ihn behandeln, Gelingen oder Mißerfolg in unseren Unternehmungen bewirken kann. Es ist jedoch gerade dieses Interesse, das nun vielfach hinter den eindrucksvolleren mathematischen Verfahren zurückstehen muß.

Wir wollen hoffen, daß es nicht neuer Katastrophen bedarf, um dieses Interesse wieder zum Erwachen zu bringen.

Es ist mein aufrichtiger Wunsch, daß dieser Kongreß dazu beitragen möge, die Kluft zwischen Theorie und Praxis in der Felsmechanik wieder zu schließen.

Glückauf!

Leopold Müller"

Ich danke Ihnen, Herr Vorsitzender.

Dr. Herget dankte Dr. Brown und Dr. Müller und bat hierauf Dr. C.G. Miller, den Präsidenten der Kanadischen Vereinigung für Bergbau, im Namen seiner Vereinigung das Wort an die Versammelten zu richten.

"Herr Vorsitzender, meine Damen und Herren!

Ich freue mich, daß auch ich Gelegenheit habe, Sie nach den übrigen Sprechern dieses Vormittags begrüßen zu dürfen. Es bereitet mir großes Vergnügen, die Delegierten zahlreicher Länder im Namen der kanadischen Bergbauindustrie herzlich willkommen zu heißen.

Dr. C.G. Miller, Kanada

Die Wahl Kanadas als Tagungsort für den Sechsten Internationalen Kongreß der Felsmechanik ist durchaus angemessen. Kanada ist eines der wichtigsten Bergbauländer der Welt, und unsere Bergbauindustrie hat der Theorie und Praxis der Felsmechanik sehr viel zu verdanken.

Kanada produziert etwa 60 mineralische Rohstoffe in mehr als 300 Abbaubetrieben und steht damit pro Kopf der Bevölkerung an erster Stelle unter den Förderländern. 1986 war es außerdem Spitzenproduzent von Uran und Zink, stand an zweiter Stelle in der Produktion von Kali, Nickel, gediegenem Schwefel, Asbest und Gips, an dritter Stelle mit Titankonzentraten, Kadmium, Aluminium, Gold und Platinmetallen, an vierter Stelle mit Kupfer, Molybdän, Blei und Kobalt, an fünfter Stelle mit Silber und an sechster Stelle mit Eisenerz.

Kanada ist mit Exporten bis zu 80 % der Gesamtproduktion das größte Ausfuhrland von mineralischen Rohstoffen. Die Bergbauindustrie beschäftigt in Kanada mehr als 100 000 Arbeitskräfte; weitere 300 000 sind in angeschlossenen Industriezweigen tätig.

Die Felsmechanik leistet einen großen, ständig wachsenden Beitrag zur Förderung dieses wichtigen Industriezweiges und zur Sicherheit der darin Beschäftigten. Der wirtschaftliche, gefahrlose Abbau der mineralischen Lagerstätten hängt in hohem Maße von einer ausreichenden Kenntnis der Felsmechanik ab. Im Zuge der Bestrebungen nach gesteigerter Produktivität werden laufend neue Abbauverfahren entwickelt. Die bergbauliche Forschung erfordert immer kompliziertere gebirgsmechanische Modelle zur Entwicklung neuer Verfahren für die Konzipierung moderner Abbaubetriebe. Darüber hinaus stützt man sich bei der neuer Verfahren für den Entwurf moderner Abbaubetriebe. Darüber hinaus stützt man sich bei der praktischen Anwendung der Felsmechanik in bereits bestehenden Bergwerken auf umfassendere und genauere Messungen sowie auf Voraussagen, um den Betrieb rationeller und sicherer gestalten zu können.

Unsere Industrie betrachtet es daher als besonders anerkennenswert, daß die Kanadische Vereinigung für Felsmechanik und das Kanadische Institut für Bergbau und Metallurgie die Organisation dieses wichtigen Kongresses übernommen haben. Mögen Ihre Diskussionen schnellere Fortschritte in Ihrer Wissenschaft und eine wirksamere Anwendung Ihres Könnens in der Praxis zur Folge haben!"

Dr. Herget dankte Dr. Miller für seine freundlichen Worte und bat dann Herrn R. Savoie, Minister der Provinz Quebec für Bergbau und Angelegenheiten der Ureinwohner, im Namen seiner Provinz das Wort an die Teilnehmer zu richten. (Offizieller Wortlaut nicht vorhanden). Im Anschluß daran sprach der kanadische Staatsminister für Bergbau, The Hon. G. Merrithew, im Namen der Bundesregierung (Wortlaut der Rede wie folgt):

"Guten Morgen, meine Damen und Herren!

Im Namen der kanadischen Bundesregierung und als Staatsminister für das Bergbauressort heiße ich Sie alle herzlich willkommen, besonders diejenigen, die zum ersten Mal in Kanada weilen. Ich bin überzeugt, daß sich die Plenarsitzungen, Workshops und anderen Diskussionen sehr fruchtbar gestalten werden, und ich weiß, Sie werden mir beistimmen, wenn ich sage, daß Montreal die ideale Stadt für den gesellschaftlichen Teil der Tagung ist.

Viele von Ihnen haben bei den Fachexkursionen vor dem Kongreß schon einen Teil von Zentral- und Ostkanada gesehen, und manche haben sich für eine Fahrt zu den Niagarafällen angemeldet. Wer jedoch Zeit hat, sollte sich diese Gelegenheit nicht entgehen lassen, auch das übrige Kanada kennenzulernen - ein riesiges Land voll landschaftlicher Schönheit, das Ihnen auch kulturell vieles zu bieten hat. Man wird Sie überall herzlich willkommen heißen.

Wie alle Disziplinen, die sich mit dem Land und seiner Nutzung befassen, spielt auch die Felsmechanik in Kanada eine entscheidende Rolle. Kanada lebt zu einem großen Teil von seinen Bodenschätzen; so hat jeder Fortschritt in der Gebirgsmechanik hier nachhaltige Auswirkungen, da unsere Bodenschätze so effizient, gefahrlos und umweltfreundlich wie möglich genutzt werden sollen.

Kanada ist stolz auf seine felsmechanischen Leistungen, und mit Recht. Seit jeher steht es im förderungsintensiven Hartgesteinabbau an führender Stelle, und die Entwicklungen der letzten Zeit im Bereich des Senkrechttrichter-Rückwärtsbaus stellen eine bedeutende Pionierleistung dar, da sie hohe Förderleistungen mit gefahrloser Arbeit verbinden. Die Felsmechanik spielt bei diesem Abbauverfahren eine zentrale Rolle.

Auch dringt der Bergbau in Kanada in immer größere Teufen vor; Abbautiefen bis zu 2000 Meter und mehr sind nichts Ungewöhnliches. Ohne den fortdauernden Beitrag der Felsmechanik wäre eine derartige Entwicklung nicht möglich.

Auf diesem Sechsten Internationalen Kongreß der Felsmechanik gilt es aber auch, etwas Besonderes zu feiern - das silberne Gründungsjubiläum der Internationalen Gesellschaft für Felsmechanik. Kanada fühlt sich daher doppelt geehrt, an diesem wichtigen Meilenstein in der Geschichte der IGFM das Gastland für Ihren Kongreß sein zu dürfen.

The Hon. G. Merrithew, kanadischer Staatsminister für Forstwirtschaft und Bergbau

Ich beglückwünsche die IGFM zu ihren hervorragenden Leistungen in den letzten 25 Jahren. Die Felsmechanik ist eine verhältnismäßig junge Wissenschaft; die im letzten Vierteljahrhundert international in der Felsmechanik erzielten Fortschritte sind in beträchtlichem Maße der IGFM und ihrer Tätigkeit zu verdanken.

Gestatten Sie mir, dafür ein Beispiel zu geben, nämlich den Beitrag der IGFM zur Vereinheitlichung der Methoden für Labor- und Feldversuche sowie in Baugrunderkundung und Analytik. Im Rahmen dieser Bemühungen sind 80 Prüfverfahren veröffentlicht worden, die sich auf die Arbeit von ca. 500 Wissenschaftlern stützen.

Ich bin auch stolz darauf, daß kanadische Felsmechaniker seit der Gründung der IGFM aktive Mitglieder der Gesellschaft sind. Durch diese intensive Mitarbeit halten sich unsere Wissenschaftler über die neuesten Entwicklungen in aller Welt auf dem laufenden und tragen zur Erweiterung und Vertiefung des internationalen felsmechanischen Wissens bei.

Eine kanadische Leistung, der eine kurze Erwähnung gebührt, ist das Handbuch der Grubenböschungen von Donald Coates, dem leider inzwischen verstorbenen Pionier auf dem Gebiet der Felsmechanik. Dieser umfassende Leitfaden durch die Theorie und Praxis dieses Themas hat den Tagebau in Kanada und anderswo entscheidend beeinflußt.

Alle vier Themenkreise des Kongresses - Flüssigkeitsbewegung und Abfallisolierung im Fels, Felsgründungen und Böschungen, Sprengen und Ausbruch, Untertägige Hohlräume im überbeanspruchten Gebirge - sind von ausschlaggebender Bedeutung für den Berg- und Felsbau in Kanada.

So sind z.B. Gebirgsschläge ein schwieriges Problem im kanadischen Bergbau. In den Jahren 1984-1985 ereigneten sich nicht weniger als 217 Felsschläge in Kanada, deren Stärke 1,5 bis 4 auf der Richter-Skala betrug. CANMET, die auf Bundesebene für Bergbauforschung zuständige Behörde, hat in enger Zusammenarbeit mit der Bergbauindustrie, mit den Bergbauressorts der Provinzregierungen und den Hochschulen ein fünfjähriges Forschungsprogramm erstellt, welches Technologien zur Feststellung von Gebirgsschlag-Gefahrenzonen entwickeln soll.

Sie werden im Laufe des Kongresses Näheres über die vorläufigen Ergebnisse dieses Programms erfahren, und wer von Ihnen am vergangenen Dienstag und Mittwoch an der Exkursion zu den Gruben im Sudbury-Becken teilgenommen hat, kann sich bereits aus eigener Erfahrung ein Bild machen.

Die kanadischen Felsmechaniker tragen nicht nur entscheidend zu größerer Sicherheit und höheren Förderleistungen im Bergbau bei, sondern sind auch in zahlreichen anderen Bereichen tätig, von der Endlagerung radioaktiver Abfälle bis zu Felsgründungen, Böschungs- und Tunnelbau. Ich bin sicher, daß sie wie auch alle anderen Teilnehmer nach Beendigung dieser Tagung mit vielen neuen Erkenntnissen nach Hause zurückkehren werden.

Zum Schluß möchte ich dem Kanadischen Institut für Bergbau und Metallurgie, der Kanadischen Gesellschaft für Geotechnik und der Kanadischen Vereinigung für Felsmechanik meine Anerkennung für die offenbar hervorragende Planung und Vorbereitung der Diskussionen, Fachexkursionen, Workshops und Ausstellungen dieses Kongresses aussprechen.
Hiermit erkläre ich nun den Sechsten Internationalen Kongreß der Felsmechanik für eröffnet!"

Dr. Herget dankte dem Sprecher für seine Worte und dafür, daß er trotz vielfacher anderer Verpflichtungen gekommen war. Da nun der 6. Internationale Kongreß der Felsmechanik eröffnet worden war, erteilte Prof. Dr. N.R. Morgenstern, Vorsitzender des Wissenschaftlichen Programmausschusses, Prof. Dr. E. Hoek das Wort für sein Eröffnungsreferat über den Felsbau in Kanada.

DER FELSBAU IN KANADA

Evert Hoek

Department of Civil Engineering, University of Toronto, Toronto, Canada

EINLEITUNG

Der Felsbau, also die technisch angewandte Felsmechanik, kann in Kanada auf eine lange Geschichte zurückblicken. Ich habe mich in diesem Übersichtsreferat auf die felsbauliche Praxis konzentriert und möchte so einen Einblick in die Vielzahl der verschiedenartigen Probleme bei der Erschließung unseres riesigen Landes geben.

Ich kann für meinen Rückblick über den Werdegang des Felsbaus in Kanada nicht den Anspruch der Vollständigkeit erheben, sondern muß mich auf eine willkürliche Auswahl von Themen beschränken, die meiner Ansicht nach wichtig sind; ich hoffe, daß meine

Schilderung genügend ausführlich ist, um dem interessierten Leser eine weitere Beschäftigung mit dem jeweiligen Thema zu ermöglichen.

KURZER GESCHICHTLICHER RÜCKBLICK

Der erste größere kanadische Tunnel war ein Eisenbahntunnel in Brockville/Ontario, der 1854 begonnen und 1860 fertiggestellt wurde. Nach Legget (1985) beginnt in Kanada der Tunnelbau in großem Stil im Jahr 1889 mit der Untertunnelung des St. Clair River zwischen Sarnia/Ontario und Port Huron/Michigan (USA). Dieser im Schildvortrieb aufgefahrene Eisenbahntunnel wurde 1890 fertiggestellt. Der Felstunnelbau begann erst Anfang des 20. Jh. mit der Fertigstellung der von der Eisenbahngesellschaft Canadian Pacific Railway gebauten Spiral-Tunnels 1909 und des Connaught-Tunnels 1916, der durch die kanadischen Rockies führenden Verbindung zwischen den Eisenbahnstrecken zu beiden Seiten des Gebirgszugs.

Das vom Kanadischen Nationalausschuß der Internationalen Kommission für Großdämme 1984 herausgegebene Kanadische Damm-Register nennt 613 Dämme, von denen 43 höher als 60 m sind. Der erste kanadische Damm, der in dieses Register aufgenommen wurde, war der 1832 gebaute, 19 m hohe Mauergewölbe-Damm der Jones-Wasserfälle am Rideaukanal in der Provinz Ontario. Der höchste Damm in Kanada ist der Mica-Damm (243 m) in der Provinz British Columbia, der 1972 fertiggestellt wurde. Von den 613 im Jahre 1983 bestehenden Dämmen dienten 441 der Stromerzeugung aus Wasserkraft.

Die Bergbauindustrie spielt seit jeher eine maßgebende Rolle in der wirtschaftlichen Entwicklung Kanadas, und die Bergbautechnik ist der aus den Bedürfnissen dieser Industrie entstandene Zweig der Technik. Mit dem ersten felsmechanischen Forschungsprogramm für bergmännische Zwecke wurde Anfang der fünfziger Jahre an der McGill University in Montreal begonnen. Es befaßte sich mit spannungsbedingten Brüchen und Gebirgsschlägen im Untertagebau, und die ersten Ergebnisse der Untersuchungen wurden von Morrison und Coates auf der Jahresversammlung 1955 des Kanadischen Instituts für Bergbau und Metallurgie mitgeteilt.

Die felsmechanische Forschung auf öffentlicher Ebene nahm 1950 bei der Abteilung Bergbau des kanadischen Ministeriums für Energiewirtschaft, Bergbau und Bodenschätze (dem heutigen CANMET - Kanadisches Institut für Rohstoff- und Energietechnik) ihren Anfang. 1964 wurde in Elliot Lake/Ontario ein Laboratorium für Untertagebau im Hartgestein geschaffen, 1967 das Bergbauforschungsinstitut in Ottawa unter Dr. D.F. Coates gegründet. Udd und Hedley (1985) geben einen informativen Überblick über die in den Bergbau-Forschungslaboratorien von CANMET in den Jahren 1964-1984 durchgeführten felsmechanischen Forschungsarbeiten.

Zu den Meilensteinen der frühen Entwicklung der Felsmechanik in Kanada zählt das von Coates 1965 veröffentlichte Lehrbuch mit dem Titel "Rock Mechanics Principles" (Grundlagen der Felsmechanik). Es war das erste seiner Art in englischer Sprache und wurde in den entscheidenden Entwicklungsjahren der Felsmechanik zu einem unentbehrlichen Nachschlagewerk für Studenten und Wissenschaftler.

Es ist bemerkenswert, daß schon zu Beginn der fünfziger Jahre bedeutende felsmechanische Arbeit im Zusammenhang mit dem Ingenieurbau und der Geologie geleistet wurde. Die Arbeitsergebnisse wurden in selten von Felsmechanikern gelesenen Zeitschriften veröffentlicht und sind somit weniger bekannt als die bergmännisch ausgerichtete felsmechanische Forschung, was jedoch den Wert dieser Beiträge nicht mindert. Typisch für diese Arbeit sind die von Ripley und Lee durchgeführten Scherversuche an Sedimentgesteinen (1961) sowie eine sehr ausführliche Untersuchung dreidimensionaler Gefügediagramme (sphärische Projektionen) von Stauffer an der Universität von Saskatchewan (1966).

Um der Felsmechanik-Forschung ein Forum für die Mitteilung und die Diskussion von Forschungsergebnissen zu schaffen, wurde 1962 in Toronto das erste Kanadische Symposion der Felsmechanik abgehalten. Weitere Symposien folgten, und das nächste soll im September 1988 in Toronto stattfinden. Aus diesen bescheidenen Anfängen entwickelte sich der Felsbau zu einer ausgereiften Wissenschaft, die als selbständige Ingenieurdisziplin anerkannt wird. Eine Reihe kanadischer Universitäten bieten Ausbildungsgänge für Studenten und Graduierte, und felsmechanische Verfahren werden von Beratenden Ingenieurbüros, Industriefirmen und Behörden in ganz Kanada zur Lösung praktischer Bauprobleme eingesetzt.

UNTERIRDISCHES FORSCHUNGSLABORATORIUM DER ATOMIC ENERGY OF CANADA LTD.

Atomic Energy of Canada Limited (AECL) ist für die Erforschung und Entwicklung geeigneter Technologien zur sicheren Endlagerung radioaktiver Abfälle in Kanada zuständig. Im Rahmen dieser Aufgabe wird zur Zeit in der Nähe von Lac du Bonnet/Manitoba ein Untertage-Forschungslaboratorium (URL) gebaut, das dem Studium verschiedener Aspekte der Endlagerung in tiefen, stabilen geologischen Formationen gewidmet ist. Radioaktive Abfälle selbst werden dabei nicht verwendet.

Das Laboratorium wird im Granitbatholit von Lac du Bonnet angelegt, einem äußerst guten Gestein, das als repräsentativ für viele Intrusivgranite des Präkambrischen Schilds gilt. Dem Ausbruch ging eine ausführliche Baugrunderkundung mit obertägiger geologischer Kartierung, Schlag- und Drehbohruntersuchungen, geophysikalischen Oberflächen- und Bohrlochmessungen sowie hydraulischen Bruchspannungsmessungen voraus (Lang, 1985). Ebenso fand im Rahmen eines umfassenden hydrogeologischen Programms eine eingehende Standortcharakterisierung statt (Davison, 1984).

Prof.Dr. E. Hoek, Kanada

Nach der Abteufung eines rechteckigen Schachts (2,8 x 4,9 m) mit Sprengmethoden bis zu einer Tiefe von 255 m (1984 und 1985) erfolgten die Ausbrüche für das Laboratorium von Schachtansatzpunkten in 130 m und 240 m Tiefe.

Bei der Schachtabteufung wurden eine Reihe von Versuchen durchgeführt, darunter Schachtkonvergenzmessungen, In-Situ-Spannungsmessungen durch Überbohren und hydraulische Aufweitung sowie Untersuchungen des geomechanischen und hydrogeologischen Verhaltens des Nebengesteins (Lang und Thompson, 1985; Kuzyk, Lang und Peters, 1986).

Für den Ausbruch einiger Kammern am unteren Schachtansatzpunkt wurden teilweise besonders präzise Bohr- und Sprengverfahren angewandt, um den beim Einsatz solcher Verfahren hervorgerufenen Schadensumfang zu bestimmen. Letzterer kann bei der Endlagerung radioaktiver Abfälle von entscheidender Bedeutung sein, da sich dadurch ein parallel zum Ausbruchsrand verlaufender Sickerweg bilden kann. Kuzyk, Lang und Le Bel (1986) sowie Lang, Humphreys und Ayotte (1986) beschreiben den dabei verwendeten Sprengplan und die damit erzielten ausgezeichneten Ergebnisse.

Im URL laufen z.Z. verschiedene Versuchsprogramme, von denen einige schon wichtige Resultate geliefert haben. Davison (1986) hat die auf einem von Guvanasen (1984) entwickelten hydrogeologischen Modell beruhenden Vorhersagen mit Beobachtungen in der Praxis verglichen. Dabei hat sich herausgestellt, daß das Modell zu hohe Werte für die Sickerung im Schacht, aber für den räumlichen und zeitlichen Verlauf der piezometrischen Senkung des Wasserspiegels wesentlich genauere Werte lieferte. Allgemein kann man sagen, daß die Modellgenauigkeit in unmittelbarer Nähe des Schachtes am niedrigsten war und mit wachsender Entfernung vom Schacht zunahm. Weitere Arbeiten sind im Gange mit dem Ziel, die Diskrepanz zwischen empirischen Beobachtungen und Prognosen zu beseitigen und andere hydrogeologische Modelle zu testen.

Ein weiterer interessanter Aspekt der im URL geleisteten hydrogeologischen Arbeit ist die Entwicklung von Spezialinstrumenten für piezometrische Messungen und Wasserproben-Entnahme in Zusammenarbeit mit Westbay Instruments Ltd. aus Vancouver (Rehtlane und Patton, 1982). Es handelt sich dabei um die erfolgreiche Weiterentwicklung einer Reihe ähnlicher, von Westbay im Laufe der letzten 10 Jahre entwickelten Instrumente. Piezometrische Druckmessungen und die Entnahme von Wasserproben können in kurzen Intervallen in genau festgesetzten Tiefen erfolgen, und durch das geringe Probenvolumen wird die Störung des Grundwasserhaushalts so klein wie möglich gehalten.

Lang et al. (1986) haben die Ergebnisse einer ersten Reihe von In-Situ-Spannungsmessungen veröffentlicht, darunter 42 Versuche mit den Verfahren des US Bureau of Mines. Spätere Versuchsreihen bedienten sich des australischen CSIRO-Verfahrens und eines modifizierten südafrikanischen CSIR-Dreiaxialgeräts. Die vorläufigen Resultate deuten darauf hin, daß sich in einer Tiefe von 185 m sowohl die maximale wie auch die minimale Horizontalspannung sprunghaft ändern. Diese Anomalie wird auf eine Bruchzone in dieser Tiefenlage zurückgeführt, und man hofft, daß eine genauere Analyse der im Dreiaxialversuch erhaltenen Meßwerte eine Deutung der Korrelation zwischen den gemessenen Spannungen und der geologischen Gebirgsbeschaffenheit in verschiedenen Tiefen ermöglichen wird. Es zeigte sich eine gute Übereinstimmung des Verhältnisses von horizontaler zu vertikaler Spannung mit einer von Herget (1980) aufgestellten Beziehung für Spannungswerte im Kanadischen Schild.

Die im unterirdischen AECL-Laboratorium geleistete Arbeit ist nicht nur für die nukleare Entsorgung in Kanada von Bedeutung, sondern sie ist auch für die kanadische Felsmechanik im allgemeinen von großem Interesse. Die Ergebnisse der eingehenden URL-Versuche werden unser Verständnis des In-Situ-Gebirgsverhaltens um etliches bereichern.

AUSBEUTUNG DER ÖLSANDVORKOMMEN IN NORD-ALBERTA DURCH DIE FIRMA SYNCRUDE

Die Abbau- und Aufbereitungsanlagen der Firma Syncrude Canada Ltd. liegen rd. 40 km nördlich von Fort McMurray im Norden der Provinz Alberta. Die Ausbeutung der Ölsande mit vier riesigen Schürfkübelbaggern (65 Kubikmeter) sowie vier Schaufelradbaggern und Transportbandsystemen begann im Juni 1977. Bei einer Förderleistung von 129 000 Barrels synthetischen Rohöls pro Tag (die derzeit genehmigte Grenze) beträgt die Tagesleistung des Bergwerks über die geplante Lebensdauer von 25 Jahren etwa 250 000 Tonnen (Fair und Lord, 1984).

Die Athabasca-Ölsande im nördlichen Alberta bilden eines der reichsten Vorkommen schwerflüssigen Bitumens in der Welt. Die McMurray-Formation, in der die Lagerstätte liegt, besteht aus den sedimentären Ablagerungen eines Beckens, das im Laufe vorkretazischer Transgressionen auf abgetragenen devonischen Karbonatgesteinen entstanden ist. Die einzelnen sedimentären Fazies spiegeln die fluvialen, ästuarinen und marinen Ablagerungsfolgen wider, aus denen Watte entstanden, die von den damals herrschenden Hochwasserständen zu den Niedrigwasserständen hin sacht abfallen (List, Lord und Fair, 1985).

Die Entwässerung der Watte erfolgte über ein Netz miteinander verbundener, sich schlängelnder Wattrinnen, wobei die Wattengebiete einer ständigen Wandlung unterworfen waren und geneigte Sand-, Schlamm- und Tonschichten abgelagert wurden. Die geneigten Tonschichten mit einem Einfallwinkel von 20 Grad oder mehr bilden Bruchflächen für Blockrutschungen, die sich in den 60 m hohen, von den Schürfkübelbaggern ausgehobenen Böschungen ereignet haben (Brooker und Khan, 1980).

Blockrutschungen auf Böschungen, auf denen die Schürfkübelbagger arbeiten, können schwerwiegende Folgen für den Abbaubetrieb haben. Da eine solche Rutschung innerhalb von 1-2 Minuten abläuft und das Rückziehen eines solchen Baggers von der Böschungskante mindestens 15 Minuten erfordert, ist es unerläßlich, mögliche Rutschzonen zu identifizieren und zu stabilisieren. Fair und Lord (1984) schildern die Methoden, die zur Identifizierung solcher Gefahrenzonen und zu ihrer Beobachtung während des Abbaus verwendet werden.

Eine neuartige Methode der Böschungsstabilisierung benutzt Sprengungen zur Destabilisierung der geneigten Tonschichten, in denen die Bruchflächen entstehen. Die Sprengungen werden vor dem Beginn des Abbaubetriebs dort vorgenommen, wo die Möglichkeit von Blockrutschungen festgestellt worden ist. Die Sprengungen vergrößern nicht nur den Reibungswinkel des Gesteins an den potentiellen Bruchflächen, sondern erhöhen auch die Gesteinsdurchlässigkeit und tragen zur Auflösung des Porenwasserdrucks in den Ölsanden bei. Beide Folgeerscheinungen fördern die Standfestigkeit in den Gefahrenzonen.

Die erwünschte Wirkung der Sprengmaßnahmen beruht auf der vertikalen Verschiebung der Ölsande durch Kratersprengungen mit kugeligen oder kompakten Ladungen. Der Sprengplan stützt sich auf die Ergebnisse von Untersuchungen, die in Zusammenarbeit mit Prof. Alan Bauer von der Queen's Universität durchgeführt wurden; sie werden von List, Lord und Fair (1985) ausführlich beschrieben. Die Verwendung von Ammoniumsalpetersprengstoffen mit Sprengöl (ANFO) in großkalibrigen Sprenglöchern hat sich als die kostengünstigste Methode der Stabilisierungssprengung erwiesen. Bei den derzeit gebräuchlichen Sprengplänen haben die Sprenglöcher einen Durchmesser von 0,76 m und sind rd. 15 m tief. Das Gewicht der ANFO-Ladungen pro Sprengloch beträgt etwa 1180 kg, die von den Sprengungen verursachte Teilchenhöchstgeschwindigkeit wird durch die Zahl von Ladungen pro Zündintervall und die Dauer des Zündintervalls zwischen den aufeinanderfolgenden Sprenglöchern in angemessenen Grenzen gehalten.

Neben den oben erörterten Blockrutschungen kennt man noch eine Reihe anderer Mechanismen des Böschungsbruchs (Brooker und Khan, 1980). Als einzige lassen sich Blockrutschungen durch die eben beschriebene Sprengmethode stabilisieren. Da alle Böschungsbrüche nachteilige Folgen für den Abbaubetrieb haben können, geht das Bestreben dahin, alle potentiellen Arten des Bruchs zu identifizieren und wirksam auszuschalten. Fair und Lord (1984) sowie Fair und Isaac (1985) erörtern umfassende Meßprogrammme zur Beobachtung des piezometrischen Drucks und der Böschungsverschiebung während der Abbautätigkeit.

Die Auswertung der dabei erhaltenen Daten hilft bei kurzfristigen bergmännischen Entscheidungen und trägt zum besseren Verständnis potentieller Mechanismen des Böschungsbruchs im Syncrude-Abbaubetrieb bei.

DER HARTGESTEIN-TIEFBAU IN GROSSEN TEUFEN

Kanada ist reich an mineralischen Rohstoffen, die in zahlreichen Tage- und Untertagebaubetrieben gewonnen werden. Untertägige Gold- und Nichtedelmetallbergwerke im präkambrischen Schild liegen in Tiefen bis zu 2300 m, und in vielen dieser Betriebe können Felsschläge die Arbeiter gefährden und die Förderleistungen beeinträchtigen.

Von einem Gebirgsschlag im Gebiet von Sudbury wird das erste Mal 1929 berichtet. In den dreißiger Jahren kam es in den Abbaubetrieben Kirkland Lake regelmäßig zu Felsschlägen, und 1939 wurde Prof. Morrison von der Bergbauvereinigung der Provinz Ontario beauftragt, über mögliche Vorbeugemaßnahmen zu berichten (Morrison, 1942). Auch heute noch stellen Gebirgsschläge eine Herausforderung für die Bergbauindustrie und für die felsmechanische Forschung dar, die sich um ein besseres Verständnis dieser Erscheinung bemüht.

Die Verordnung der Provinz Ontario über Abbau- und Bergbaubetriebe definiert den Gebirgsschlag als "einen plötzlichen Felsbruch, der zu einer Gesteinsexplosion an der Oberfläche eines Hohlraums oder zu einer seismischen Störung in einem Tage- oder Untertagebau führt". 1984 wurden in Bergwerken der Provinz Ontario insgesamt 105 Gebirgsschläge mit einer Magnitude von mehr als 1,5 auf der Richter-Skala registriert (Scoble, 1986). Diese Zahl kann als typisch für die Gebirgsschlagtätigkeit in den letzten Jahren gelten. Während die meisten Felsschläge so klein sind, daß sie ohne nennenswerten Folgen bleiben, haben einige doch beträchtlichen Schaden angerichtet (Ontario Government, 1986).

1985 leitete CANMET ein großangelegtes Fünfjahresprogramm in die Wege. Im Rahmen dieses Projekts werden in verschiedenen kleinen Bergwerken einfachere mikroseismische Untersuchungen und in zwei Bergwerken in Sudbury mit Hilfe modernster Meßinstrumente detaillierte Untersuchungen durchgeführt. Ferner werden in einem großen Bergwerk in Ontario Untersuchungen mit dem Prototyp eines weiterentwickelten mikroseismischen Systems unternommen, mit dessen Hilfe mikroseismische Vorgänge mit großer Genauigkeit aufgezeichnet werden können. Parallel dazu läuft eine Reihe ähnlicher, teilweise von CANMET mitfinanzierter Projekte - z.B. die Entwicklung eines ausgefeilten dreidimensionalen numerischen Modells (McDonald, Wiles & Vielleneuve, 1988) - im Gange.

1981 und 1986 legten zwei Untersuchungskommissionen Berichte über die Sicherheit im Untertagebau im nördlichen Ontario vor. Diese Berichte schenkten dem Felsbau und seiner Bedeutung für Gebirgsschläge, Entspannungs-, Ausbau-, Versatz- und Sprengmaßnahmen sowie für die Pfeilergewinnung besondere Aufmerksamkeit; die Schlußfolgerungen und Empfehlungen der Kommissionen sind für alle, die mit dem Untertagebau befaßt sind, besonders lesenswert. Als unmittelbare Folge des von einem Ausschuß unter der Leitung von Trevor Stevenson für die Provinzregierung Ontario ausgearbeiteten Berichts 1986 wurde in Ontario ein Bergbauforschungsinstitut zur Koordinierung der Forschungstätigkeit auf dem Gebiet des Untertagebaus in großen Teufen gegründet. Anfänglich wird sich die Arbeit an diesem Institut hauptsächlich auf die angewandte Felsmechanik konzentrieren; eine Erweiterung des Tätigkeitsbereichs auf andere bergmännische Probleme ist jedoch geplant. Das Institut wird mit bereits bestehenden Forschungseinrichtungen wie den Hochschulen, CANMET und den Forschungsabteilungen führender Bergbauunternehmen zusammenarbeiten.

DER STOLLENBAU IM SCHWACHEN UND DRUCKHAFTEN GEBIRGE

Im Gegensatz zur Abbautätigkeit im Hartgestein des präkambrischen Schilds, wo Gebirgsschläge ein ernstes Problem darstellen, muß sich der Stollenbau in anderen Regionen Kanadas mit der Erscheinung der Druckhaftigkeit des Gebirges auseinandersetzen. In Süd- Ontario, einem Gebiet mit starken Horizontalspannungen im Gestein (Lo, 1978), zeigen besonders die verhältnismäßg schwachen, dort sehr häufig auftretenden Schiefer eine starke Neigung zur Druckhaftigkeit (Lo, Cooke & Dunbar, 1986; Lo, 1986).

Der Stollenbau in weichem Gestein ist auch anderswo in Kanada mit Schwierigkeiten verbunden, die zum Teil von Kaiser und McKay (1983), Kaiser, Guenot und Morgenstern (1985) sowie von Boyd, Yuen und Marsh (1986) geschildert worden sind. Das letztgenannte Referat ist von besonderem Interesse, da es sich mit einem 4 km langen, in Maschinenbohrweise aufgefahrenen Unterwasserstollen in relativ schwachem Sedimentgestein im Kohlenrevier vor der Küste von Nova Scotia befaßt.

Theoretische Untersuchungen zum Stollenverhalten in druckhaftem Gebirge sind von Ladanyi (1974, 1976), Ladanyi und Gill (1984), Lo und Yuen (1981), Emery, Hanafy und Franklin (1979) und Brown et al. (1983) angestellt worden. An der University of Western Ontario hat Ogawa (1986) unter der Leitung von Prof. K.Y. Lo ein dreidimensionales Modell der Gebirgsspannungen und -verlagerungen um ein Stollenmundloch entwickelt. Die Anwendung dieser Modelle auf die Deutung des Stollenverhaltens ist von Lo, Lukajic und Ogawa (1984) und von Lo (1986) beschrieben worden. Eisenstein, Heinz und Negro (1984) schildern ein ähnliches Modell.

KANADISCHE WASSERKRAFTWERKE

Die Gebirge und die reichlichen Wasservorräte in vielen Teilen Kanadas können zur Stromerzeugung eingesetzt werden. Obwohl in Kanada bereits eine ganze Reihe von Wasserkraftwerken in Betrieb ist, weiß man im Ausland davon nur verhältnismäßig wenig, da entweder nichts über sie veröffentlicht worden ist oder eventuelle Berichte in Fachzeitschriften erschienen sind, die normalerweise nicht von Felsmechanikern gelesen werden.

Zu diesen weniger bekannten Projekten zählt das Spray-Wasserkraftwerk in der Provinz Alberta (Eckenfelder, 1952). Von besonderem Interesse bei diesem Kraftwerk ist die Konstruktion des Druckstollens, der einen um 45 Grad geneigten Abschnitt umfaßt, der durch massiven Kalkstein, Dolomitkalkstein, Tonschiefer und einige Schiefermittel führt. Nach heutigen Maßstäben würde das zur Festlegung der Länge der Stahlauskleidung benutzte Verhältnis von Felsüberdeckung zur Druckhöhe als nicht ausreichend bezeichnet werden, es sei denn, daß in der Entwurfsphase Druckfestigkeitsversuche durchgeführt worden sind (Humphreys und Hoek, 1987). Die Tatsache, daß dieser Druckstollen schon seit vielen Jahren ohne Probleme in Betrieb steht, scheint darauf hinzudeuten, daß unter gewissen Umständen die heutigen Überlagerungskriterien eher vorsichtig formuliert sind und daß der Druckstollenbau noch nicht voll erforscht ist.

Huber (1952, 1953) und Wise (1962) berichteten über das Kitimat-Kemano-Projekt der Aluminium Company of Canada im nördlichen British Columbia. Als diese Anlage gebaut wurde, galt sie als das größte unterirdische Kraftwerk in der Welt.

Die ingenieurgeologischen und felsbaulichen Aspekte des Entwurfs der Wasserkraftwerke W.A.C. Bennett (Portage Mountain) und Mica in British Columbia sind von Imrie und Jory (1969) und Imrie und Campbell (1976) erörtert worden. Ein Referat von Benson et al. (1974) befaßt sich mit untertägigen Festigkeitsproblemen, die bei den Kraftwerken Bersimis und

Chute-des-Passes in Quebec im Zusammenhang mit schieferungsparallelen Störungen aufgetreten sind. Humphreys und Lang (1986) und Humphreys und Hoek (1987) berichten ausführlich über die felsbaulichen Aspekte des Cat-Arm-Wasserkraftwerks in Newfoundland, das insofern ungewöhnlich ist, als der durch schonendes Sprengen in sehr gutem Granitgestein aufgefahrene Hochdruck-Stollen keine Auskleidung hat.

Eine wichtige Rolle spielte der Felsbau bei dem Churchill-Falls-Wasserkraftwerk in Labrador, das zu seiner Zeit ein kühnes unterirdisches Großprojekt darstellte. Die geotechnischen Untersuchungen und der Entwurf des unterirdischen Krafthauses schildert eine Artikelreihe von Benson et al. (1970a, 1970b, 1971, 1974); Einzelheiten über den Bau sind zum Teil von Mamen (1969) und Gagne (1972) veröffentlicht worden.

Das La Grande-Projekt im Gebiet der James Bay hat wegen seiner Abgelegenheit und seiner Größe viel Aufmerksamkeit auf sich gelenkt. Es soll die Wasserkraft des La Grande River und teilweise der angrenzenden Einzugsgebiete der Flüsse Caniapiscau und Eastmain-Opinaca nutzen. Insgesamt besteht die Anlage aus neun Krafthäusern mit einer geplanten Leistung von 13 753 MW und einer Jahresleistung von 81,6 TWh. Der gesamte Komplex liegt im Bereich des kanadischen Schilds, und die meisten Teilanlagen sind auf Granitgestein mit massiver bis gneissartiger Struktur gebaut. Unter Berücksichtigung der in anderen Gebieten des kanadischen Schilds gewonnenen umfassenden berg- und ingenieurbaulichen Erfahrung wurden für das La Grande-Projekt Konstruktionsprinzipien angewandt, die sich wie folgt zusammenfassen lassen (Murphy und Levay, 1982):

1. Die Gebirgsqualität gilt als gut bis ausgezeichnet, wenn nichts Gegenteiliges bewiesen werden kann.
2. Bohr- und Sprengmaßnahmen sind im allgemeinen im Hinblick auf die Reduzierung von Mehrausbrüchen und die Standfestigkeit der Ausbruchprofile wichtiger als die geologischen Verhältnisse.
3. Bei der Feststellung der erforderlichen felsbaulichen Maßnahmen (Stützung und Oberflächenschutz) sind auf Beobachtungen im Gelände abgestützte Erfahrungswerte gewöhnlich wertvoller als rein theoretische Ansätze.

Obwohl Murphy and Levay in ihrem Artikel die Vermutung äußern, daß die Felsmechanik bei der Planung des La Grande-Projekts nur eine untergeordnete Rolle spielte, befassen sie sich doch eingehend mit den beim Bau dieses Kraftwerks eingesetzten Methoden der Luftbildauswertung, Klüftungsvermessungen, Diamantbohrungen mit Bohrlochkameraeinsatz und der Wasserdruckmessung. Für das unterirdische Kraftwerk LG-2 wurden zu In-Situ-Messungen des Deformationsmoduls und der Spannungen entlang der Wölbung des Krafthauses ein Versuchsschacht abgeteuft und Stollen aufgefahren, und die Meßergebnisse wurden mit der Methode der finiten Elemente für Pfeilerdimensionierung und Kavernengeometrie ausgewertet. Ich finde es, gelinde gesagt, erstaunlich, daß bei diesen Untersuchungen von "Urteilsbildung auf Erfahrungsgrundlage und Beobachtungen im Gelände" im Vergleich zu "rein theoretischen Methoden" gesprochen wird. Man kann daraus schließen, daß die Erbauer des La Grande-Projekts die Felsmechanik als praktisches Entwurfswerkzeug betrachteten, ein Umstand, den ich als ein Kompliment für die vielen Felsmechaniker auffasse, die sich jahrelang mit ihren "rein theoretischen Methoden" abgemüht haben, um dieses Ziel zu erreichen.

Das unterirdische Krafthaus LG-2 gilt als das größte der Welt. Die in massivem Granit in einer Tiefe von 160 m liegende Maschinenhalle hat eine Spannweite von 28,4 m; die Spannweite des parallel dazu gelegenen Ausgleichsbeckens beträgt 23,6 m. Beide Ausbrüche sind rd. 47 m hoch, und in der Maschinenhalle sind 16 333-MW-Turbogeneratorsätze untergebracht. Im Krafthaus ruhen zwei 405-t-Kräne auf Felsvorsprüngen, die das Ergebnis sorgfältiger Ausbrucharbeiten in gutem Gestein sind.

Aufgrund der ausgezeichneten Gesteinsqualität konnte auch in den für Strömungsgeschwindigkeiten bis zu 14 m/s konzipierten Umleitungsstollen und Abflußkanälen auf eine Auskleidung verzichtet werden. Die Gesamtlänge der bis zu 14 m x 20 m großen Stollen beträgt über 8000 m; Betonauskleidungen wurden nur für Stollenmundlöcher sowie für Schleusen- und Verschlußeinrichtungen verwendet.

Von technischem Interesse ist weiterhin der Kaskadenüberlauf des Krafthauses LG-2, der für einen maximalen Durchsatz von 16 280 m^3/s ausgelegt ist. Der 2200 m lange Überlaufkanal hat zehn 9,1 m - 12,2 m hohe und 127 m - 200 m hohe Stufen. Die Nennbreite des Kanals beträgt 122 m; auf den letzten 1100 m wurde er jedoch auf 183 m verbreitert, um zusätzlich zu dem beim Ausbruch anfallenden Material auch hier noch Füllmaterial für den Hauptstaudamm zu gewinnen (Levay and Aziz, 1978). Die Felsstufen des Kaskadenüberlaufkanals sind nur an Stellen mit enger Felsverklüftung durch Anker und Spritzbeton abgestützt. Der Überlauf steht seit 1979 in Betrieb und zeigt trotz fallweisen Strömungsgeschwindigkeiten bis an die zulässige Grenze - 11 m/s am Ende der ersten Stufe, 26 m/s bei der letzten Stufe - keine nennenswerten Abnutzungserscheinungen.

DER TUNNELBAU IM GEBIRGE WESTKANADAS

Der Rogers Pass befindet sich in dem landschaftlich reizvollen Wildnisgebiet der Selkirk-Berge (Provinz British Columbia). Sein Scheitel liegt 1200 m über dem Meeresspiegel, umgeben von mehr als 3300 m hohen Gipfeln. Hier fallen in einem normalen Winter über 9 m Schnee.

Die erste Eisenbahnstrecke über den Rogers Pass wurde 1885 von der Eisenbahngesellschaft Canadian Pacific gelegt. Trotz erheblicher Schwierigkeiten - z.B. kamen 1910 bei einem Lawinensturz 62 Streckenarbeiter ums Leben - wurde diese Strecke jahrzehntelang benutzt. Im Zuge einer geplanten Intensivierung des transkanadischen Bahnverkehrs der Canadian Pacific wurde in den Jahren 1913-1916 durch den Fuß des Mount Macdonald in den Selkirk-Bergen ein Tunnel gebaut, der ursprünglich Selkirk-Tunnel genannt, später jedoch auf Connaught-Tunnel umgetauft wurde. Zum Auffahren dieses rd. 8 km langen Tunnels wurde zuerst ein kleiner Richtstollen parallel zur Hauptbohrrichtung angelegt, von dem der Tunnelvortrieb dann durch mehrere kurze, in regelmäßigen Abständen gebaute Verbindungstunnel erfolgte. Dieser auch heute noch befahrene Tunnel verkürzte nicht nur die Fahrstrecke, sondern verringerte auch die Steigungen und ermöglichte die Umfahrung der gefährlichsten Lawinengebiete.

Zur Beseitigung von Engpässen auf der Hauptstrecke zwischen Calgary und Vancouver plant Canadian Pacific nun weitere Ausbaumaßnahmen. Dazu gehört ein neuer Tunnel, der Mount Macdonald-Tunnel, der durch den Mount Macdonald und Cheops Mountain führt. Dieser 14,7 km lange Tunnel wird der längste Eisenbahntunnel in Nordamerika sein. Mit dem Bau wurde 1984 begonnen, und die von den beiden entgegengesetzten Enden des Tunnels aufeinander zuschreitenden Vortriebsfronten trafen vor kurzer Zeit zusammen. Der hufeisenförmige, mit Beton ausgekleidete Tunnel hat eine Spannweite von 5,18 m und eine Höhe (vom Schienenstrang bis zum Tunnelscheitel) von 7,87 m.

Das Auffahren des Tunnels vom westlichen Mundloch erfolgte mit einer Vollschnitt-Tunnelbohrmaschine und Sprengungen, wobei die Ortsbetonauskleidung der voranschreitenden Front unmittelbar folgte (Steward, 1987). Die Kopfstrecke vom östlichen Mundloch wurde mit einer Tunnelbohrmaschine vorgetrieben, und die volle Höhe des Tunnels wurde durch Strossensprengung erzielt (Knight, 1987). Während des Auffahrens erfolgte der Tunnelausbau mittels Nagelungen, Drahtmatten und Spritzbeton. Hier ist bemerkenswert, daß nach den Ergebnissen von Gebirgsklassifikationsstudien für den Ausbau langer Strecken des Tunnels Anker, Spritzbeton und Matten erforderlich gewesen wären. Da jedoch der parallel verlaufende Connaught-Tunnel seine Standfestigkeit 70 Jahre lang ohne Stützbauten bewahrt hatte, entschied man sich für weniger umfassende Ausbaumaßnahmen. Die geplante volle Betonauskleidung des Tunnels soll dem umgebenden Gestein die erforderliche Langzeit-Festigkeit verleihen.

Gadsby (1985) hat eine kurze Beschreibung des Rogers Pass-Tunnels, Tanaka (1987) einen guten Überblick über die allgemeinen bautechnischen Eigenschaften dieses Projekts veröffentlicht. Eine ausführliche Erörterung der geologischen Kartierung und der computergestützten Datenverarbeitung für das Projekt stammt von Klassen et al. (1987).

Anderswo in British Columbia werden zur Zeit im Rahmen von Maßnahmen zur Streckenmodernisierung der Canadian National Railways eine Reihe von Tunneln geprüft. Benson, Hostland und Charlwood (1985) haben die geotechnischen Bedingungen und Belüftungserfordernisse dieser Tunnels besprochen.

In einem Bericht von Kaiser und Gale (1985) über die Ergebnisse eingehender Untersuchungen zur Verwendung von Gesteinsklassifikationssystemen bei der Ausbauplanung für die Tumbler Ridge-Tunnels (British Columbia) der Eisenbahngesellschaft BC Rail vergleichen die Autoren eine Reihe gebräuchlicher Klassifikationssysteme und erörtern die jeweiligen Vor- und Nachteile. Ich betrachte diesen Bericht als einen wesentlichen Beitrag zur ständig wachsenden Literatur über die Bedeutung der Gesteinsklassifikation bei der Planung von Ausbaumaßnahmen in untertägigen Hohlräumen.

BERGBAU IM HOHEN NORDEN KANADAS

Das nördlichste Bergwerk der Welt - die Polaris-Grube der Firma Cominco - liegt auf Little Cornwallis Island, 77 Breitegrade nördlich vom Äquator. Das Blei- und Zinkvorkommen befindet sich im Permafrost, der sich in dieser geographischen Breite bis zu einer Tiefe von rd. 450 m erstreckt. Die Oberflächentemperaturen schwanken zwischen -55 Grad Celsius im Winter und 15 Grad Celsius im Sommer. Selbst im wärmsten Sommer tauen jedoch nur wenige Zentimeter des Bodens auf (Leggatt, 1982).

Wegen des hohen Eisgehalts im Erz muß die Gesteinsmasse gefroren bleiben, um ein Einstürzen der untertägigen Hohlräume zu verhindern. Das erreicht man durch Kaltluftbewetterung, bei der die Bergwerkstemperatur konstant um -7 Grad Celsius liegt. Als Füllmaterial für die Abbaufronten wird ein über schwebende Strecken hinunterbefördertes Gemisch aus Schnee und Kies verwendet, welches gefriert und so zur gewünschten Festigkeit führt.

Seit der Inbetriebnahme dieses Bergwerks im Jahre 1982 haben sich nur sehr wenige geotechnische Probleme ergeben. Abgesehen davon, daß die Bergleute etwas wärmere Kleidung als in herkömmlichen Bergwerken benötigen, unterscheiden sich Abbaubetrieb, Entwurf der Abbaufronten und Pfeiler sowie der Ausbau wenig von südlicheren Bergwerken.

FELSBÖSCHUNGSBAU

Sowohl Gebirgstunnel wie auch große Tagebaubetriebe erfordern steile, dabei aber standfeste Felsböschungen. Angesichts des teilweise gebirgigen Charakters der kanadischen Landschaft und der Größe der kanadischen Bergbauindustrie ist es nicht verwunderlich, daß der Felsböschungsbau in der Entwicklung des Landes seit jeher eine wichtige Rolle spielt. Da die einschlägige kanadische Literatur für eine Besprechung in diesem Rahmen zu umfangreich ist, möchte ich mich auf einige wichtige Punkte beschränken.

C.O. Brawner, derzeit Professor für Geomechanik an der University of British Columbia, ist einer der Pioniere des Felsböschungsbaus. In seiner mehr praktisch als theoretisch ausgerichteten Arbeit hat sich dieser Wissenschaftler weltweit mit Problemen des Felsböschungsbau im Hinblick auf Straßen-, Eisenbahn- und bergmännischen Tagebau beschäftigt und hat so einen wesentlichen Beitrag zum Felsbau geleistet (Brawner, 1986).

Coates, dessen Arbeiten über den Untertagebau bereits erwähnt worden sind, befaßte sich eingehend mit dem Problem der Felsböschungsfestigkeit. Sein diesbezügliches Interesse führte zu einem von ihm 1972 in die Wege geleiteten CANMET-Projekt, dessen Ergebnisse als "Pit Slope Manual" veröffentlicht worden sind (CANMET, 1977). Dieses umfangreiche, 27 Bände umfassende Werk behandelt ausführlich alle Aspekte des Baus von Felsböschungen, Halden und Berge-Böschungen für den bergmännischen Tagebau und stellt somit ein grundlegendes Nachschlagewerk für den Felsböschungsbau dar.

Prof. Peter Calder und seine Mitarbeiter an der Queen s Universität in Kingston/Ontario haben Beiträge zur Analyse der Böschungsfestigkeit, zum Entwurf tagebaulicher Sprengungen und zur Böschungsüberwachung veröffentlicht (Calder, 1969). Dr. Douglas Piteau und seine Mitarbeiter in Vancouver haben über ihre Erfahrungen mit Wahrscheinlichkeitsmethoden bei der Optimierung von Felsböschungen berichtet (Piteau et al., 1985).

Die Analyse, Beobachtung und Stabilisierung von Böschungsüberkippungen sind von einer Reihe von Autoren behandelt worden, darunter Choquet und Tanon (1985), Wyllie (1980) und Brawner, Stacey und Stark (1975). Martin (1987) hat sich mit der Verwendung von Spannkabeln und einbetonierten Dübeln bei der Felsverbesserung beschäftigt, und es ist zu erwarten, daß diese Mittel bei der Entwicklung von wirksameren und wirtschaftlicheren Festigungsmethoden eine immer größere Rolle spielen werden.

SPRITZBETON ALS AUSBAUMATERIAL

Die Verwendung von Spritzbeton kann in regenreichen Gebieten zum Schutz von Böschungen gegen Erosion und zur schnellen Vornahme von Reparaturen in vielbefahrenen Eisenbahntunnel dienen. Man hat sich daher darauf konzentriert, die Festigkeit und Bearbeitbarkeit von Spritzbeton laufend zu verbessern.

Derartige Verbesserungen sind unter anderem die Verwendung grober Zuschläge (MacRae, 1973) sowie der Zusatz von Stahlfasern (Little, 1985; Morgan, 1985) und "microsilica" (Anon, 1987). Im westlichen Kanada ist stahlfaserverstärkter "microsilica"-Spritzbeton bereits gebräuchlich und ist dabei, auch im übrigen Kanada Fuß zu fassen. Insbesondere der Zusatz von "microsilica" macht Spritzbeton viel leichter zu bearbeiten und wesentlich fester, so daß er heute auch dort eingesetzt werden kann, wo noch vor wenigen Jahren ein starrer Stahlausbau oder Betonauskleidungen erforderlich gewesen wären.

SCHLUSSBEMERKUNG

An Hand einiger praktischer Beispiele habe ich versucht, den gegenwärtigen Stand des Felsbaus in Kanada darzustellen. Seine Entwicklung auf der theoretischen Grundlage der Felsmechanik ist nur langsam vorangegangen und war vielfach von Enttäuschungen begleitet. Während die Gründung des Felsbaus als wissenschaftliche Disziplin im wesentlichen ein oder zwei Einzelpersonen zu verdanken ist, hat sich der Felsbau in Kanada im allgemeinen dank des unermüdlichen Einsatzes verschiedener Universitäten, Behörden und privater Forschungsteams und Beratender Ingenieurbüros in ganz Kanada entwickelt. In Kanada gibt es kein anerkanntes Zentrum des Felsbaus, sondern es arbeiten, wie ich versucht habe zu zeigen, Ingenieure und Geologen in einer Reihe von Arbeitsgruppen an der praktischen Anwendung der felsbaulichen Theorie. Meiner Meinung nach ist diese Parallelentwicklung von Vorteil, da sie die verschiedensten Betrachtungsweisen und Ansichten hervorbringt und so zu Lösungen führt, die nicht nach dem Ruf dessen, der sie vorgeschlagen hat - ob Einzelperson oder Organisation -, sondern nach ihrem praktischen Wert beurteilt werden.

LITERATURVERZEICHNIS (siehe englischer Teil)

DANK

Bei der Zusammenstellung dieser Übersicht sind mir viele Kollegen in ganz Kanada mit Rat und Tat zur Seite gestanden. Ich kann sie nicht alle mit Namen nennen, möchte aber allen herzlich für die mir übersandten Notizen, Referate und Fotografien (die ich bedauerlicherweise nicht alle verwenden konnte) danken.

Dr. Udd dankte Prof. E. Hoek für sein ausgezeichnetes, sachlich fundiertes Referat und überreichte ihm im Namen des Organisationsausschusses als kleine Anerkennung eine Inuit-Kleinplastik. Dr. Herget erklärte hierauf die Eröffnungsveranstaltung als beendet und bat den Vorsitzenden der I. Plenarsitzung, Dr. B. Bjurstrom (Schweden), die Teilnehmer nach der Pause zur ersten Fachsitzung einzuberufen.

Thema 1
Flüssigkeitsbewegung und Abfallisolierung im Fels

Standortuntersuchungen und Charakterisierung des Gebirges; unterirdische Wasserbewegung; Wanderung von Giftstoffen; Abfallisolierung; Wasserzuflußkontrolle; Rißbildung durch Thermalbelastung und Druckspannung; In situ Messungen und Rückrechnungen

Vorsitzender: S.A.G. Bjurström (Schweden)

Schriftführer: R. Chapuis (Kanada)

Sprecher: T.W. Doe (USA)

Bohrlochuntersuchungen für Abfallagerung im kristallinen Gestein (mit Beiträgen von John Osnes, Michael Kenrick, Joel Geier, Scott Warner (USA))

ZUSAMMENFASSUNG:

Bohrungen werden auf niedrige Wasserdurchlässigkeit und auf das Vorhandensein von rissigem Fels überprüft. Bohrungsprüfmethoden und die Art der Probenahme werden beschrieben.

Bohrungsprüfmethoden stammen aus zwei unabhängigen Quellen, die eine vom Bergwerks- und Bauingenieurwesen, die andere von der Erdölforschung und Wasserkunde. Die Prüfmethode beim Bauingenieurwesen beschränkt sich hauptsächlich auf den sogenannten Lugeon Test, bei dem Wasser mit konstantem Druck eingepumpt wird, und wo mit stetigem Fliessen gerechnet wird. In der Erdölforschung werden meistens bei den hochdurchlässigen Gesteinen konstante Pumpmethoden eingesetzt. Bei geringen Durchlässigkeiten werden "puls tests" eingesetzt. Diese Prüfungen benutzen "transient flow" mit einer Kapazitätsangabe. Die Kapazität spiegelt die Zusammendrückbarkeit und die Verformbarkeit von Gestein und Wasser wider.

Die Annahme von stetigem gegenüber unterbrochenem Fliessen ergibt keine grösseren Rechnungsfehler in bezug auf die Wasserdurchlässigkeit. Unterbrochenes Fliessen kann aber mehr Aussagen liefern, zum Beispiel, Zusammendrückbarkeit (Bohrschacht-kapazität); nicht lineare Fliesseffekte; Rissverformungseffekte, Einfluss durch Rissgrenzen (vom Bohrschacht entfernte, sich kreuzende Risse, oder aufhörende Risse). Auch kann die Auswirkung vom Bohren auf Wasserdurchdringlichkeit in der Umgebung eines Schachtes festgestellt werden.

Für solche Fälle sind Lösungen für die Prüfungen mit konstantem Fliessen weitgehend entwickelt worden, und diese lassen sich auch bei den Prüfungen anwenden, wo unterbrochenes Fliessen analysiert wird. Beim Gestein von niedriger Durchdringlichkeit hat die Gleichdruck-Methode bestimmte Vorteile, zum Beispiel Anwendbarkert für einen weiten Messbereich, Kompensation des Bohrlochinhaltes; und relative Geschwindigkeit, besonders wenn Niedrigbereich-Fliessmeter gebraucht werden.

Für die Probenahme gibt es zwei Ansätze. Beim ersten Verfahren, dem diskreten Zonenprüfen, werden bestimmte Abschnitte gewählt. Dieses Verfahren wird gewählt, wenn eine bestimmte Zone, zum Beispiel eine Bruchzone oder eine Wasserader, von Interesse ist. Das zweite Verfahren mit der festgelegten-Abstandslänge (FIL), wird gewählt, wenn ein representativer Wert der Wasserdurchdringlichkeit des Gesteins geliefert weren soll. Die Verteilung der verschiedenen Durchdringlichkeitswerte bei einem solchen Verfahren spiegeln sowohl die Anhäufung der Gesteinsrisse, als auch die Verteilung der Durchdringlichkeitseigenschaften wider. FIL- Daten können so analysiert werden, dass Aussagen über Bruchflächenhäufigkeit und Fliessgeschwindigkeiten gemacht werden können.

11:15 Kurzvorträge
Elsworth, D., Piggott, A.R., U.S.A.
Physikalische und numerische Analogien der Grundwasserbewegung im geklüfteten Medium

Azevedo, A.A., Corea Fo, D., Quadros, E.F., da Cruz, P.T., Brasilien
Hydraulische Konduktivität in Klüften des basaltischen Untergrundes des Taquaruçu Dammes

By, T.L., Norwegen
Seismische Tomographie zur Felscharakterisierung und Vorhersage von Sickerproblemen für zwei Autobahntunnel unter der Stadt von Oslo

Charo, L., Zélikson, A., Bérest, P., Frankreich
Salzdombildungsrisiko in Folge von HLW Einlagerung in einem Salzlager

12:00 - 13:30 Mittagspause

12:00 Eröffnung der Fachausstellung

13:30 Kurzvorträge
Chen Gang, Liu Tian-Quan, China
Eine umfassende Analyse der Einflüsse und Mechanismen der Wasserzuführung aus der Sohle von Kohleflözen

Crotogino, F., Lux, K.H., Rokahr, R., Schneider, H.J.,West-Deutschland
Geotechnische Aspekte der Endlagerung von Sonderabfällen in Salzkavernen

Helm, D.C., Australien
Vorhersage einer durch Grundwassersenkung verursachten Senkung im Latrobetal, Australien

Johansson, S., Finland
Auffahrung von Grosskavernen im Fels der Neste Oy Porvoo Werke, Finland

Kojima, K., Ohno, H., Japan
Zusammenhang zwischen Fern- und Nahfeld in Bezug auf Brüche und/oder Fliessbahnen für die geologische Isolierung radioaktiver Abfälle

Maury, V., Sauzay, J.-M., Fourmaintraux, D., Frankreich
Felsmechanische Annäherung der Erdoel- und Gasgewinnung; erste Ergebnisse

Nakagawa, K., Komada, H., Miyashita, K., Murata, M., Japan
Untersuchung über Druckluftlagerung in Felskavernen ohne Auskleidung

Pahl, A., Liedtke, L., Kilger, B., Heusermann, S., Bräuer, V., West-Deutschland
Untersuchungen von Durchlässigkeit und Spannungszustand im kristallinen Gebirge zur Beurteilung der Barrierewirkung

Pyrak-Nolte, L.J., Myer, L.R., Cook, N.G.W., Witherspoon, P.A., U.S.A.
Hydraulische und mechanische Eigenschaften der natürlichen Bruchflächen in einem wenig durchlässigen Fels

Watanabe, K., Ishiyama, K., Asaeda, T., Japan
Die Instabilität der Zwischenphase zwischen Gas und Flüssigkeit an Hand eines Beweismodells

15:00 - 15:30 Pause (Foyer)

15:30 - 17:00 Diskussion (Duluth/MacKenzie)

Diskussionsleiter: N. Barton (Norwegen)

Podiumsdiskussion: M. Langer (West-Deutschland)
J.A. Hudson (England)
P. Rissler (West-Deutschland)

M. Langer (West-Deutschland)

Berechnung der Barrierenwirksamkeit zur Abfallisolierung

ZUSAMMENFASSUNG: Entsprechend dem Mehrfach-Barrierenprinzip muß die geologische Umgebung eines Abfallendlagers wesentlich zur Isolierung des Abfalls von der Biosphäre beitragen. Die Bewertung der Integrität der geologischen Barriere kann nur durch Berechnungen mit validierten geomechanischen und hydrogeologischen Modellen erfolgen. Die richtige Idealisierung des Wirtsgesteins im Rechenmodell ist die Grundlage für realistische Berechnungen von Thermospannungen und Rißbildungen. Die Bestimmung der Wasserdurchlässigkeit entlang Trennflächen ist notwendig, um die Barrieren-wirksamkeit von Kristallingebirge zu berechnen.

P. Rißler (West-Deutschland)

Felshydraulische Modelle und Wirklichkeit

Die Felshydraulik entwickelt seit zwei Jahrzehnten Modellvorstellungen. Sie sind Ausdruck der Arbeitsweise des Ingenieurs, der immer gezwungen ist, komplexe Vorgänge oder Sachverhalte zu vereinfachen, um überhaupt in annehmbarer Zeit zu Lösungen zu kommen. Vereinfachung geht immer auf Kosten der Betrachtungsschärfe. Deshalb besteht zuweilen ein Disput mit denjenigen, welche meinen, durch die Vereinfachung ginge Entscheidendes verloren - und - man könne die komplexe Natur nicht in Modelle pressen. Lassen Sie uns diese Kontroverse aufgreifen, weil sie zur Standortbestimmung geeignet ist.

Modelle sind nicht Selbstzweck sondern Hilfsmittel für die Lösung von Aufgaben. Der Wert eines Modells hängt generell, damit auch in der Felshydraulik davon ab

- ob trotz Vereinfachung die wesentlichen Charakteristiken der Natur erfaßt werden,

- von den Anforderungen der Aufgabenstellung.

Diskussion zum Thema I: v.l.n.r.: P. Rissler, J.A. Hudson, N. Barton, M. Langer, G. Herget, W. Bawden

Wirklichkeitsnähe:

Die Frage nach der Wirklichkeitsnähe kann nur im Einzelfall und für die jeweilige Fragestellung beantwortet werden. Verschiedene Merkmale sind Maßgebend:

- Wir unterscheiden das kontinuierliche und das diskontinuierliche Modell. Bei ersterem ist unterstellt, daß die gegenseitigen Abstände d der Fließwege im Vergleich zu den Abmessungen a des Bauwerks sehr klein sind, so daß von homogener Durchströmung ausgegangen werden kann. Als Grenze zum diskontinuierlichen Modell wird $d/a = 1/10$ bis $1/20$ angenommen.

- Zu unterscheiden ist weiterhin, ob die Sickerströmung in offenen Poren des Gebirges stattfindet, im Trennflächengefüge oder in beiden. Dann ist die Durchlässigkeit nämlich entweder bereichsweise von vornherein isotrop, die Strömung laminarer. Bei Beteiligung des Trennflächengefüges kann sie u.U. anisotrop und/oder in Teilbereichen auch turbulent sein.

- Bilden hauptsächlich die Trennflächen die Strömungswege, so ist im Einzelfall zu fragen, ob diese mehr flächig oder mehr linienförmig ausgebildet sind.

- Vielfach wird - der einfacheren Behandlung wegen und weil entsprechende Daten fehlen - davon ausgegangen, daß die Raumstellungen der Trennflächen streng nach Scharen geregelt, die Öffnungsweiten konstant, die Wandrauhigkeiten unveränderlich und die Fließwege großräumig gleichförmig sind. Dies ist offenbar der Haupteinwand der Kritiker. Sie verkennen dabei, daß es sich um eine vereinfachende Modellierung handelt, die im ersten Schritt erforderlich war, um die Aufgabe einer Berechnung überhaupt zugänglich zu machen. Ich bin glücklich, daß wir inzwischen einen Schritt weiter sind. Neuerdings liegt mit der Arbeit von Koppelberg (RWTH Aachen) nämlich eine sehr detaillierte Untersuchung vor, welche gestattet, den Fehler dieser Vereinfachung abzuschätzen. Ich halte es für außerordentlich bemerkenswert, daß diese Arbeit von einem Geologen stammt. Für mich ist es der Idealzustand, daß ein zunächst notwendigerweise sehr vereinfachtes Modell im nächsten Schritt an die Wirklichkeit angepaßt wird, bzw. daß im nächstein Schritt darüber Rechenschaft abgelegt wird, wie groß der Fehler des Modells äußerstenfalls sein kann.

Aufgabenstellung:

Der Felshydraulik stellen sich die Aufgaben:

1. Bestimmung der Durchlässigkeit
2. Prognose der Potential-, Druck- und Gradientenverteilung
3. Prognose der Kraftwirkungen aus dem Sickerwasser
4. Prognose von Sickerwassermengen
5. Prognose der Verpreßbarkeit

Die Aufgaben 2 bis 4 bauen auf der Durchlässigkeit des Gebirges auf. Die Hilfsgrößen Spaltweite und Wandrauhigkeit sind zur Bestimmung der Durchlässigkeit nicht erforderlich. Mithin sind Unzulänglichkeiten bei der Beschreibung bzw. Abschätzung von Spaltweite und Wandrauhigkeit bei der Bestimmung der Wasserdurchlässigkeit unerheblich und es entfällt auch ein Teil der Einwände der Kritiker.

Zur Prognose der Verpreßbarkeit wären allerdings bessere Informationen über Spaltweiten, Wandrauhigkeiten und die Durchtrennung wünschenswert.

Insgesamt wird hier die These vertreten, daß die Felshydraulik inzwischen hinreichend wirklichkeitsnahe Modelle für die meisten Erscheinungsformen der Wasserbewegung im Gebirge bereit stellt. Für die üblichen Aufgabenstellungen des Talsperrenbaus, für welche ich hier in erster Linie spreche, dürfte dies zunächst keine weitere Forschung vordringlich machen. Zur Prognose der Verpreßbarkeit wären allerdings gesicherte Informationen über die Ausbildung der Wasserleiter wünschenwert. Auch ist das alte Problem der Beziehung zwischen Wasseraufnahme und Injektionsgutaufnahme noch nicht gelöst.

Das wesentliche Problem bei der Beurteilung der Durchlässigkeit liegt gewöhnlich weniger in der mangelnden Vollkommenheit der felshydraulischen Modellvorstellungen, als vielmehr in unserem mangelnden Wissen über die Gebirgseigenschaften in den großen Bereichen, welche nicht untersucht worden sind. Dies verdeutlicht folgendes Beispiel: In der Herdmauerebene eines 470 m langen Dammes wurden 16 bis zu 50 m lange Untersuchungsbohrungen abgeteuft. Die durchgehend ausgeführten WD-Versuche erfaßten das Durchlässigkeitsverhalten lediglich in 1000 - 2000 m^2 der gesamten Dichtungsfläche von 30000 m^2. Das sind 3 bis 6% der Gesamtfläche. Über 90% des Gebirges sind damit nicht erkundet, sondern nur im Analogieschluß eingeschätzt. Hier vorhandene Inhomogenitäten stören die Schlußfolgerungen wesentlich mehr als Unvollkommenheiten bei den Modellen. Meines Erachtens muß hier die Forschung verstärkt ansetzen.

Ein weiteres Problemfeld, das m.E. noch weitgehend ungelöst ist, stellt das langzeitliche Durchlässigkeitsverhalten dar. Es ist wichtig für die Talsperrensicherheit und für die Langzeitlagerung von Abfällen. Kalke im Untergrund lösen sich durch weiches Talsperrenwasser. Kluftfüllungen werden bei hohen Gradienten u.U. in Jahrzehnten ausgewaschen. Chemische Substanzen lösen Dichtungsmaterialien. Über die Rolle des Wassers im Zusammenhang mit Salzstöcken wird sicher Herr Professor Langer referieren. Auch die Auswirkungen thermischer Belastungen auf das Durchlässigkeitsverhalten ist nicht klar. Mir scheint, daß hier ein außerordentlicher Bedarf für Forschung besteht.

Schrifftum:

Koppelberg, W.: Numerische und statistische Untersuchungen zur Durchlässigkeit geklüfteter geologischer Körper und ihrer Bestimmung durch Wasserdruckversuche. Mitt. zur Ingenieurgeologie und Hydrogeologie, RWTH Aachen, Heft 23, Februar 1986

Diskussionsbeiträge der Sitzungsteilnehmer sind nur im englischen Teil aufgeführt.

Flüssigkeitsbewegung und Abfallisolierung - Thema I

Zusammenfassung des Diskussionsleiters

N. Barton (Norwegen)

Die zu Thema 1 eingereichten Referate wurden in sechs Gruppen unterteilt: Modelle für Kluftsysteme (Problematik der Durchlässigkeitstensoren); Modelle für Einzelklüfte; Gesteinsabdichtung (Wassereinbruch); Kopplung numerischer Modelle; Problematik des Kriechens; und hydraulische und geophysikalische Meßverfahren. Die wichtigsten in jeder Gruppe besprochenen Probleme werden angeführt. Der Diskussionsleiter nimmt kurz zu folgenden Themen Stellung: Modelle für Einzelklüfte; Verhalten unter Berücksichtigung aller hydro-thermomechanischen Vorgänge; Nachweis der Modellgültigkeit.

1. EINLEITUNG

Von den 52 zu Thema I eingereichten Referaten kamen lediglich 14 zum Vortrag. Bei der Auswahl der hochinteressanten Beiträge konnten leider nicht alle Themen berücksichtigt werden. Als besonders gravierend wurde die Tatsache empfunden, daß sich keines der vorgetragenen Referate mit der Frage der Kopplung numerischer Modelle beschäftigte, denn diese Problematik ist für die Endlagerung radioaktiver Abfälle von großem Interesse. Um diese Lücke etwas zu schließen, werden in der Zusammenfassung des Diskussionsleiters gezielt einige der auf diesem Gebiet aufgeworfenen Probleme angesprochen.

2. THEMENGRUPPEN UND REFERENTEN

Die folgende Gliederung der Referenten nach Themengruppen soll die Benutzung des Berichtwerks (Band 1 und 2) vereinfachen. Obwohl nicht alle Referenten angeführt sind, erleichtert diese Liste doch das Auffinden bestimmter Beiträge.

2.1 Modelle von Kluftsystemen

- Elsworth und Piggot
- Dershowitz und Einstein
- Long, Billaux, Hestir, Chiles

Durchlässigkeitstensoren

- Brown und Boodt
- Barbraux, Cacas, Durand, Fenga
- Oda, Hatsuyama, Kamemura

2.2 Modelle für Einzelklüfte/Versuche

- Pyrak-Nolte, Myer, Cook, Witherspoon

- Gentier
- Nakagawa, Komada, Miyashita, Murata
- Watanabe, Ishiyama, Asaeda

2.3 Gesteinsabdichtung (Wassereinbruch)

A. Geklüfteter Fels

- Azevedo, Correa, Quadros, Cruz
- Gang, Tian, Quan
- Johansson
- Matsumoto, Yamaguchi
- Mineo, Hori, Ebara
- Semprich, Speidel, Schneider

B. Steinsalz/Kali

- Charo, Zelikson, Berest
- Crotogino, Lux, Rokahar, Schneider
- Langer
- Mraz
- Schwab, Fischle

2.4 Kopplung numerischer Modelle

- Chan, Scheir
- Curran, Carvalho
- Izquierdo, Romana
- Ohnishi, Kobayashi, Nishigaki
- Sato, Ito, Shimiza
- Shaffer, Heuze, Thorpe, Ingraffea, Nilson
- Thiercelin, Roegiers, Boone, Ingraffea
- Kuriyagawa, Matsunaga, Yameguchi, Zyvdloshi, Kelkar
- Charo, Zelikson, Berest
- Côme
- Crotogino, Lux, Rokahr, Schneider
- Langer

2.5 Problematik des Kriechens

(i) Kriechen des Steinsalzes

- Balthasar, Haupt, Lempp, Natau
- Langer
- Preece
- Charo et al.
- Crotogino et al.
- Mraz

(ii) Bohrlochstandfestigkeit

- Maury, Sauzay, Fourmaintraux
- Rongzun, Zuhui, Jingen
- Preece
- Geunot

2.6 Untersuchungsverfahren (hydraulische, geophysikalische u.a.)

- By
- Pahl, Liedtke, Kilger, Heusermann, Bräur
- Barbreau, Cacas, Durand, Fenga
- Brown, Boodt
- Cornet, Jolivet, Mosnier
- Andrade
- Haimson
- Heystee, Raven, Belanger, Semec
- Azevedo, Corrêa, Quadros, Cruz
- Matsumoto, Yamaguchi
- Mineo, Hori, Ebara
- Agrawal, Jain

3. BEITRÄGE DER DISKUSSIONSTEILNEHMER

1. Dr. Peter Rissler (Ruhrtalsperrenverein): "Felshydraulische Modelle und Realität im Vergleich: Ansätze zur weiteren Forschung"

2. Prof. Michael Langer (BGR, Hannover): "Berechnung der Barrierenwirksamkeit von Fels zur Abfallisolierung"

3. Dr. John Hudson (Imperial College, London): "Gekoppelte Mechanismen - Beitrag der Matrixmethode zur Lösung ingenieurtechnischer Probleme"

4. WICHTIGE THEMEN UND SCHLÜSSELFRAGEN

Die folgende Aufführung einiger Schlüsselfragen soll die Aufmerksamkeit auf einige der wichtigsten Themen lenken. Sie erhebt keinen Anspruch auf Vollständigkeit und völlige Objektivität, sondern greift lediglich einige der beim Studium des Berichtwerks aufgeworfenen Fragen auf.

4.1 Zwei- und dreidimensionale Modelle für Kluftsysteme

a) Wie groß sind die durch die Nichtberücksichtigung verformbarer Klüfte in den Modellen entstehenden Fehler?
b) Können die Strömungsrinnen an den Kluftkörperrändern (Kluftschnittpunkten) modellmäßig behandelt werden?
c) Können Gesteine mit hohen Flüssigkeitsgeschwindigkeiten erkannt und erfaßt werden?

d) Wie relevant sind die Durchlässigkeitstensoren gleichwertiger Medien?

e) Wie groß ist ein repräsentatives Prüfvolumen?

4.2 Modelle für Einzelklüfte

a) Sind lineare Näherungslösungen bei der Bestimmung der Steifheit von Klüftungen zulässig?
b) Wie wichtig ist die Modellbildung von Schervorgängen?
c) Wie viele Definitionen der Kluftöffnung sind erforderlich?
d) Kontaktzone → Strömungsrinnen → Maßstabseffekte → kubisches Gesetz?

4.3 Gesteinsabdichtung

a) Wie genau läßt sich aus Durchlässigkeitsmessungen die Verpreßbarkeit geklüfteter Gründungen feststellen?
b) Wie wirksam lassen sich Abschlußbauwerke im Salzgestein abdichten?
c) Wie wirksam lassen sich Stollen im geklüfteten Gestein abdichten?
d) Welche Probleme verursacht die Salzlaugenwanderung?

4.4 Kopplung numerischer Modelle

a) Wie sinnvoll sind hydro-thermomechanische Untersuchungen, wenn im Zuge der Vereinfachungen eines Vorgangs Fehler entstehen, die größer sind als der durch die Vernachlässigung eines Vorgangs bewirkte Gesamteffekt?
b) Welches sind die wichtigsten Eingabeparameter und wie kann man sie ermitteln?

4.5 Problematik des Kriechens

a) Bohrlochkonvergenz und Spannungsumlagerungseffekte?
b) Was versteht man eigentliche unter einer "plastischen" Zone?
c) Warum läßt sich das Kriechverhalten im Salzgestein nicht genauer vorhersagen?

4.6 Meßverfahren

a) Wie repräsentativ für die gesamte Gesteinsmasse sind die Ergebnisse hydraulischer Versuche?

b) Wie viele geologische Einzelheiten lassen sich durch geophysikalische Versuche bestimmen?

5. SCHWIERIGKEITEN BEI DER MODELLBILDUNG

Modelle, die die Flüssigkeitsbewegungen im Gestein durch ein Netzwerk starrer, in regelmäßigen Abständen verlaufender und sich überschneidender, paralleler Gesteinslagen darstellen, sind leider sehr realitätsfern. Veränderungen in Öffnungsweite, Ausrichtung und Kontinuität der Klüfte, Strömungsrinnen, Röhrenströmungen (entlang der Schnittpunkte der zweidimensionalen Ebene) und die Wechselwirkung zwischen effektivem Druck und Kluftweite zählen zu den Faktoren, die eine wirksame Simulierung erschweren.

Aus leicht ersichtlichen Gründen scheint es bei der Modellbildung von Felsklüftungen zu einer Arbeitsteilung gekommen zu sein: Die Hydrologen und die Spezialisten auf dem Gebiet der Radionuklidwanderung haben ihrerseits interessante Modelle entworfen, mit deren Hilfe sich schon mehrere als wichtig geltende Faktoren - z.B. Strömungsrinnen und Gesteine mit hoher Flüssigkeitsgeschwindigkeit, um nur zwei davon zu nennen - wirklichkeitsnah simulieren lassen. Die Felsmechaniker wiederum haben Diskontinuum-Modelle entwickelt, in denen die Auswirkungen von druck-, zug- oder scherungsbedingten Spannungsänderungen auf die Felskluftweite berücksichtigt werden können.

Da die äquivalente glattwandige Kluftweite paralleler Schichten (e) durch die bekannte Gleichung

$$k = e^2/12$$

unmittelbar mit der Flüssigkeitsgeschwindigkeit (k) in den Klüften verknüpft ist, scheint es im Prinzip gerechtfertigt zu sein, (k) zu der durch Änderungen der Druck-, Zug- oder Scherspannung verursachten Konvergenz, Öffnung und Scherung der Klüfte in Beziehung zu setzen.

Auch ohne große Phantasie läßt sich unschwer vorstellen, wie schnell die Leistungsgrenze eines Computers erreicht ist, wenn die derzeit besten dreidimensionalen Modelle von Kluftsystemen in Gestalt starrer Netzwerke mit den derzeit besten felsmechanischen Deformationsmodellen kombiniert werden. Glücklicherweise kann unter gewissen Umständen bei einer solchen Kombinierung einer der Vorgänge vernachlässigt werden, wenn man beweisen kann, daß in dem betreffenden Zeitraum wahrscheinlich nur geringfügige Änderungen (z.B. Erhitzen) auftreten werden.

6. MODELLE VERFORMBARER KLÜFTE

Abbildung 1 veranschaulicht ein ausgefeiltes Einprozeß-Modell, das zur Zeit am NGI verwendet wird. Die Simulierung erfolgt unter Benutzung des Diskrete-Elemente-Codes UDEC (Cundall, 1980) in der Mikromaschinen-Fassung (uDEC). Dieser Code kann mit einer neuen Subroutine BB die nichtlinearen Einzelheiten des Kluftverhaltens durch die Koppelung von Scherfestigkeit, Deformation und Flüssigkeitsbewegung der Klüfte darstellen (Barton, Bandis & Bakhtar, 1985). Bei Einsatz eines Großrechners läßt sich der Code mit kombinierter Flüssigkeitsbewegung in den Klüften verwenden.

Die durchströmbare Kluftöffnung des im unteren Teil von Abbildung 1 dargestellten rechten Tunnels stützt sich auf kontinuierliche Messungen der mechanischen Öffnungsweiten (E), die ihrerseits von der durch die Ausbrucharbeiten verursachten Schließung, Öffnung, Scherung und Dehnung abhängen. Die mechanischen Öffnungsweiten (E) werden mittels der in Abbildung 2 dargestellten Datensätze in durchströmbare Kluftöffnungen (e) umgewandelt. Die Diskrepanz zwischen (e) und (E) ist auf die Tortuosität der Flüssigkeitsbewegung und auf Strömungsrinnen um die Kontaktzonen in der Kluftebene zurückzuführen - Faktoren, die bei höheren Spannungen die Diskrepanz vergrößern.

Der Code enthält, wie bereits erwähnt, eine Subroutine für die Darstellung der Flüssigkeitsbewegung an den Klüften. Bei der hier besprochenen Tunneluntersuchung wurde diese Subroutine jedoch nicht verwendet, da als erste Maßnahmen zur Minimierung des Wassereinbruchs (und der dadurch verursachten Änderungen des Flüssigkeitsdrucks) eine Vorbehandlung des Gesteins erfolgte und stahlfaserverstärkter Spritzbeton eingesetzt wurde.

Das endgültige Ziel dieser Bemühungen ist es, einen realistischen nichtlinearen Code in einen dreidimensionalen Diskrete-Elemente-Code einzubauen, mit dem sich die Flüssigkeitsbewegung in Klüften und geothermischen Gradienten simulieren läßt. Hauptsächlich wegen der sehr geringen Menge an allgemeinen und spezifischen Eingabedaten wird es bis zur Erreichung dieses Endziels jedoch noch einige Jahre dauern.

7. BERÜCKSICHTIGUNG MEHRERER VERÄNDERUNGEN

Das Kluftverhalten bei der Berücksichtigung aller hydro-thermomechanischen Veränderungen ist zur Zeit noch weitgehend unerforscht, da bisher nur sehr wenige einschlägige Untersuchungen vorliegen. Der in der Versuchsgrube der Colorado School of Mines von Terra Tek durchgeführte Versuch mit geheizten Kluftkörpern hat gezeigt, daß Klüfte bei höheren Temperaturen enger zusammenrücken können (Hardin et al., 1982).

Ungewiß, wenn auch wahrscheinlich, ist es noch, ob dieser Effekt auch die Scherfestigkeit erhöht. Es ist durchaus denkbar, daß bei reiner Normalbelastung die Scherfestigkeit bei steigenden Temperaturen unter gleichzeitigem Absinken der Durchströmbarkeit ansteigt. Sind die durch den Temperaturanstieg verursachten Änderungen des Wirkdrucks so groß, daß sie zu Abscherungen führen, dann kann die Scherfestigkeit nach ihrem Maximum als Folge einer Entzahnung der Schichten voneinander und des durch den Temperaturanstieg verursachten Festigkeitsverlustes der Kluftwände wieder auf die früheren Werte oder noch tiefer absinken. In-Situ-Messungen und grundlegende Untersuchungen über die Modellierung der Festkörpergesetze im Gestein sind heute noch weit von ihrem endgültigen Ziel entfernt, doch sind mit Kluftkörperversuchen bereits wesentliche Fortschritte erzielt worden.

8. PROBLEME BEIM CODE-GÜLTIGKEITSNACHWEIS

Kürzlich am NGI beim Code-Gültigkeitsnachweis gemachte Erfahrungen haben folgende Fragen aufgeworfen: Wie weit darf die Vereinfachung gehen und wie viel darf von dem Versuchsobjekt bekannt sein, wenn z.B. für einen In-Situ-Versuch mittels eines neuen Code eine Modellbildung zum Nachweis der Übereinstimmung des Code mit tatsächlichen Meßergebnissen erfolgen soll?

Abbildung 4 zeigt das vereinfachte μDEC-BB-Modell eines CSM-Versuchs, bei dem der geklüftete Block durch Flüssigkeitsdruck ähnlich wie in einem hydraulischen Druckkissenversuch beaufschlagt wird. Die Vereinfachung bestand darin, daß der Block lediglich drei Klüfte (und somit vier Teilblöcke) enthielt - einen Bruchteil der tatsächlich an der Blockoberfläche beobachteten Kluftzahl. Richardson (1986) konnte jedoch nachweisen, daß die größten relativen Verschiebungen in diesen vier Blöcken zu beobachten waren, weshalb sie als aktive Blöcke betrachtet werden.

Bei Prognosen weiß man leider im vorhinein nicht, welche Blöcke in einer bestimmten Felsmasse "aktiv" sein werden. Es bedarf daher besserer Verfahren zur Fernunterscheidung zwischen kontinuierlichen und nicht kontinuierlichen Klüften sowie zwischen glatten und rauhen Kluftflächen. Aktive (scherfähige) kontinuierlich" und "rauh" definiert.

Wir ersehen hieraus, wie dringend die Entwicklung von In-Situ-Verfahren zur Unterscheidung zwischen wichtigen und unwichtigen Klüftungen ist und welche Herausforderungen sie an Geophysiker und Hydrologen stellt.

LITERATURVERZEICHNIS(Siehe englischer Teil)

Abbildung 1. µDBE-BB-Diskrete-Elemente-Modell der Auswirkungen des Tunnelbaus auf Kluftkörperverformung und durchströmbare Kluftweiten (Makurat et al., 1987).

Abbildung 2. Diskrepanz zwischen tatsächlicher (E) und theoretisch durchströmbarer (e) Kluftöffnungsweite - eine Folge der Tortuosität der Flüssigkeitsbewegung, der Kontaktzonen und der Strömungsrinnen (Barton et al., 1985).

Abbildung 3. Berücksichtigung aller hydro-thermomechanischen Veränderungen in einem Erwärmungsversuch mit Kluftkörpern. Die durchströmbare Kluftöffnungsweite (3) zeigt eine bessere Ausrichtung der Kluftwände bei 75 °C und rein lotrechter Beaufschlagung (Hardin et al., 1982).

Abbildung 4. Wichtigste Spannungen bei zweiachsiger und einachsiger Kraftbeaufschlagung eines In-Situ-Kluftkörpers (Barton et al., 1987)

Abbildungen sind im englischen Text wiedergegeben.

Thema 2
Felsgründungen und Böschungen

Standortuntersuchungen und Charakterisierung im Fels; Standsicherheit von Böschungen im Tagebau; Brücken-, Bau- und Dammgründungen; Wahrscheinlichkeitsberechnung in der Planung und bei Rückrechnungen

Vorsitzender: F.H. Tinoco (Venezuela)

Schriftführer: P. Rosenberg (Kanada)

Sprecher: M. Panet (Frankreich)

Felsverbesserungen für Gründungen und Böschunen mit vorgespannten und nicht vorgespannten Ankern

KURZFASSUNG: Die Bewehrung von Felsgründungen mit Hilfe vorgespannter Anker ist ein weitverbreitetes und bewährtes Verfahren. Durch Vorversuche und durch Überwachung der vorläufigen und endgültigen Spannung wird der feste Sitz jedes Ankers sichergestellt. Auch die aufgrund ihrer Kostengünstigkeit weitverbreitete Bewehrung mittels ungespannter Stangen hat ihre Wirksamkeit auf vielen Baustellen bewiesen. Die Wechselwirkungsbedingungen zwischen Bewehrung und Gestein sind jedoch schwieriger zu untersuchen, und die derzeit angewandten Rechenverfahren sind nicht voll zufriedenstellend. Im übrigen gibt es kein Verfahren, um das sachgemäße Setzen, das einwandfreie Funktionieren und die Langlebigkeit des Gesteinsankers sicherzustellen. Künftige Forschungsarbeiten sollten sich mit der Definition von Kenndaten für bewehrtes Gestein befassen.

In vielen Fällen kann die Festigkeit einer Gesteinsmasse mit der erforderlichen Sicherheitsspanne nur mit Hilfe korrektiver Maßnahmen gewährleistet werden, z.B. durch Veränderung der Hohlraumgeometrie, Entwässerung des Gesteins, Bau von Stützmauern oder durch Bewehrung des Gesteins.

Für das geeignetste Verfahren entscheidet man sich nach Durchführung einer Standfestigkeitsuntersuchung, bei der die Verhaltensparameter der Diskontinuitätsflächen eine entscheidende Rolle spielen. Brüche beginnen im Gestein mit Ablösungen und dem Abgleiten entlang der Diskontinuitätsflächen. Jedes Bewehrungsverfahren, das auf eine Wiederherstellung des Kontinuums abzielt, bewirkt deshalb eine erhebliche Verbesserung der Standfestigkeit des Gesteins. Zu diesem Zweck können Stahlstangen oder Stahlkabel verwendet werden; beide Verfahren setzen sich immer mehr durch. Es gibt zwei konkurrierende Bewehrungsverfahren:

Einsatz aktiver (vorgespannter) Stahlstangen;
Einsatz passiver (ungespannter) Stahlstangen.

Eine Untersuchung zeigt, daß die Auffassungen von Land zu Land verschieden sind und daß jeweils einem der beiden Verfahren der Vorzug gegeben wird. Offensichtlich hat jedes Verfahren seine Vor- und Nachteile, die im folgenden im Rahmen einer eingehenden Untersuchung der Funktionsweise jedes Verfahrens in einem Diskontinuum ermittelt werden sollen.

(Figure 1)

Abb. 1 - SCHEMATISCHE DARSTELLUNG EINES VORGESPANNTEN ANKERS

1. AKTIVE VERSTÄRKUNG

Die Bewehrung von Gesteinsmassen mit vorgespannten Ankern oder Stangen ist bereits ein relativ altes Verfahren. Hochleistungsanker (bis zu 10 000 kN) wurden schon in den dreißiger Jahren von André Coyne und Eugène Freyssinet eingesetzt.

Vorgespannte Anker (Abb. 1 (siehe englischer Teil)) bestehen aus folgenden Funktionselementen:

- der Verankerungsstrecke, d.h. dem Ankerabschnitt, der eingebettet wird;
- einer Dichtungsmanschette, die die Verankerungsstrecke vom freien Stangenabschnitt trennt und verhindert, daß bei der Einzementierung des Ankers das Verfüllmaterial in das Bohrloch überläuft;
- der freien Länge der Ankerstange;
- dem Ankerkopf mit Auflagermasse und Arretierungsvorrichtung.

In Frankreich werden folgende Unterscheidungen empfohlen:

- maximale Zugbelastung T_l , die der Dehnungsgrenze der Ankerstange bzw. der Verankerungsstrecke entspricht;
- Zugspannung T_G, die der Elastizitätsgrenze der Ankerstange entspricht;
- T_a, die höchstzulässige Betriebsspannung.

Zu beachten ist:

$T_a \leqq 0{,}6\ T_G$ bei Daueranker

$T_a \leqq 0{,}75\ T_G$ bei Ankern für vorübergehende Zwecke mit einer Nutzungsdauer von weniger als 18 Monaten.

Es gibt noch immer keine befriedigende theoretische Erklärung dafür, welche Spannung ein Ausreißen der Verankerung verursacht. Mit Hilfe von Abnahmeprüfungen und den Spannverfahren kann jedoch die einwandfreie Ankerwirkung geprüft werden.

In Frankreich wird empfohlen, die Vorspannung eines Ankers stufenweise bis zur Prüfspannung T_e zu erhöhen, und zwar dergestalt, daß:

$T_e = 1{,}2\ T_a$ für Anker für vorübergehende Zwecke

$T_e = 1{,}3\ T_a$ für Daueranker.

Im Rahmen dieser Tests können auch die kritische Kriechspannung sowie die entsprechende freie Ankerstangenlänge geprüft werden.

Die durch die Vorspannung eingeleitete Kraft wirkt einem Öffnen der Unstetigkeitsflächen entgegen oder erhöht ihre Scherfestigkeit; im ersten Falle wirkt die Kraft durch ihre Normalkomponente und muß deshalb rechtwinklig zu den entsprechenden Unstetigkeitsflächen aufgegeben werden; im zweiten Fall kommt die senkrechte und auch die Tangentialkomponente zur Geltung, und folglich muß die optimale Ausrichtung unter Berücksichtigung der in der Praxis möglichen Gegebenheiten ermittelt werden (W. Wittke).

Stabilitätsberechnungen für Gestein, das mit vorgespannten Ankern bewehrt wurde, sind im Prinzip verhältnismäßig einfach, denn sie bestehen aus einer Analyse im Stabilitätsbereich, in die die Vorspannungskräfte, deren Richtung und Stärke bekannt und konstant sind, als von außen wirkende Kräfte bekannter und konstanter Richtung und Stärke eingebracht werden.

Diese Hypothesen müssen jedoch noch auf ihre Übereinstimmung mit der Praxis hin überprüft werden. Die Behauptung, daß es sich um Kräfte handele, die außerhalb der zu stabilisierenden Gesteinsmasse wirksam sind, bedeutet, daß die Verankerungsstrecken außerhalb der möglichen Bruchflächen liegen, und insbesonders, daß die freie Länge der Ankerstangen auch wirklich den angegebenen Werten entspricht. Das Nachspannen von Ankern hat nur allzu oft gezeigt, daß die freien Ankerlängen wesentlich kürzer sind als angenommen, weil es durch unkontrollierte Verfüllmaterial-Bewegungen zu unerwünschten Verankerungseffekten gekommen ist.

Vorspannungskräfte können in den meisten Fällen ebenfalls als konstant betrachtet werden, weil die nach der Verankerung stattfindenden Gesteinsverformungen auf die freie Stangenlänge bezogen nur noch äußerst geringfügige Verschiebungen zwischen der Verankerungsstrecke und dem Ankerkopf verursachen. Dies ist jedoch nicht immer der Fall; eine Verschiebung um 1 cm bewirkt beispielsweise bei einem 20 m langen freien Stangenabschnitt eine Änderung der Ankerspannung um 100 MPa. Deshalb muß in Fällen, in denen Anker eingebracht werden, bevor der Ausbau abgeschlossen ist, die Verformung ermittelt werden, um entscheiden zu können, ob die Anker zunächst mit einer vorläufigen Spannung beaufschlagt werden sollen, die geringer als die endgültige Betriebsspannung ist (Abb. 2). In diesem Fall kann eine Untersuchung des Bohrlochs mit Hilfe eines Dehnungsmessers sehr nützlich sein. Das Nachspannen bietet die Gelegenheit, den freien Stangenabschnitt erneut zu prüfen und den Verlauf der Zugspannung innerhalb der beiden Spannvorgänge festzustellen. Ingenieure haben hinsichtlich des zeitabhängigen Verhaltens vorgespannter Anker im wesentlichen zwei Bedenken: Ab- bzw. Zunahme der Spannung durch Kriechrelaxation und Korrosion. Aufgrund der Erfahrungen, die wir inzwischen gewonnen haben, kann festgestellt werden, daß sorgfältig eingebrachte vorgespannte Anker sicher und zuverlässig sind. Von Interesse in diesem Zusammenhang sind die in Hongkong gemachten Beobachtungen, von denen Brien-Boys und Howells berichten. Die Beobachtungen beziehen sich auf 214 über 17 verschiedene Projekte verteilte Anker; die wichtigsten Ergebnisse lassen sich wie folgt zusammenfassen:

- der durchschnittliche Zugspannungsverlust betrug in einer Zeitspanne von 4 Jahren 9 %;
- bei 16 % der Anker war die Spannung mindestens 1,2mal so groß wie die Betriebsspannung.

Die Einhaltung der Normen und Empfehlungen ist der beste Korrosionsschutz.

Wir sind der Meinung, daß die Technik der Planung und Durchführung des Ankerausbaus mit vorgespannten Ankern von qualifizierten Ingenieuren vollkommen beherrscht werden kann.

Wenn das zu stabilisierende Gesteinspaket jedoch sehr mächtig ist, wird dieses Verfahren recht kostspielig. Um nämlich den mittels herkömmlicher Stabilitätsanalyse ermittelten konventionellen Sicherheitsfaktor spürbar - z.B. von 1,1 auf 1,5 - zu erhöhen, muß die Vorspannkraft häufig 10 % bis 25 % des Gewichts der zu stabilisierenden Gesteinsmasse betragen, was recht erheblich ist.

(Figure 2)

a) Kennlinien für die Prüfspannung und die vorläufige Ankerspannung

b) Endgültige Ankerspannung

Abb. 2 - KENNLINIEN FÜR DIE ZWEISTUFIGE SPANNUNGSBELASTUNG EINES ANKERS

2. PASSIVE BEWEHRUNG

Unter passiver Bewehrung versteht man den Einsatz einer Bewehrung, die auf ihrer gesamten Länge mit dem Fels verbunden ist und beim Einbringen nicht verspannt wird. Wenn im bewehrten Gestein keine Verformungen auftreten, werden die passiven Ankerstangen auch keinerlei Spannungen ausgesetzt.

Eines der frühesten Beispiele für den Einsatz dieser Ankertechnik ist unseres Wissens das Auflager der Chaudanne-Staumauer in Frankreich.

Die Chaudanne-Talsperre ist eine geringmächtige, siebzig Meter hohe Bogenstaumauer, die 1949 bis 1952 im Verdon in Südfrankreich errichtet wurde. Die Bogenmauer steht auf Kalkstein, der Diskontinuitätsflächen aufweist, die in einem Winkel von 45° stromaufwärts und zur Wasserseite hin einfallen. Am rechten Ufer wurde die stromabwärtige Seite mit horizontal eingebrachten, 20 bis 30 m langen Ankerstangen mit einem Durchmesser von 50 mm und mit Betonauflagern verstärkt. Am linken Ufer wurde das Gestein mittels Seilen konsolidiert, die aus 115 bis 185 Drähten von 5 mm Stärke bestehen und in einem Winkel von 15° gegen die Horizontale eingebracht wurden. Diese Seile wurden nicht verspannt, sondern auf ihrer gesamten Länge mit Zement vergossen (Abb. 3).

(Figure 3)

A - Horizontaler Querschnitt
B - Vertikaler Querschnitt
1 - Bogenstaumauer
2 - 4 Betonpfeiler
3 - Anker
4 - Rutschung

1 - Bogenstaumauer
2 - Anker
3 - Schichtung

Abb. 3 - PASSIVE BEWEHRUNG DES AUFLAGERS DER CHAUDANNE-STAUMAUER

Die Wirkungsweise passiver Anker ist schwieriger zu beschreiben als die vorgespannter Anker, weil passive Anker komplexe Umgebungsbedingungen und örtliche Wechselbeziehungen zwischen Anker und Gestein schaffen, insbesondere an den Schnittpunkten mit Diskontinuitätsflächen. Betrachten wir nun das Drehmoment der Spannungen einer passiven Ankerstange im Schnittpunkt 0 einer Diskontinuitätsfläche; es sei:

No die Normalkomponente der Kraft auf der Diskontinuitätsfläche;

So die tangential zur Diskontinuitätsfläche verlaufende Komponente der Kraft;

Mo der Biegemoment.

Es kann die Annahme gemacht werden, daß Mo gleich Null ist, denn im Falle einer Tangentialverschiebung befindet sich die Biegestelle im Punkt 0 (Abb. 4). Daher werden wir das Verhalten einer fest eingebetteten Ankerstange untersuchen, die durch die beiden Elementarspannungen No und So belastet wird, wobei von der Annahme ausgegangen wird, daß die Stange senkrecht auf der Diskontinuitätsfläche steht.

2.1 Belastung passiver Anker mit einer Normalkraft

Untersucht werden soll im folgenden das Verhalten einer Stange mit Durchmesser d, die auf ihrer gesamten Länge eingebettet ist und mit einer Normalkraft N beansprucht wird.

Solange wir im elastischen Bereich bleiben und die Zementierung der Stange unversehrt ist, nimmt die Scherspannung am Berührungspunkt zwischen Stahl und Zementierung mit zunehmendem Abstand von der Diskontinuitätsfläche exponentiell ab (Farmer). Wenn jedoch die Normalkraft zunimmt, kommt es zu einer progressiven Lockerung der Stange, die sich so lange in die Tiefe fortsetzt, bis sich die Stange auf ihrer gesamten Länge abgelöst hat oder bis sie bricht (Abb. 5).

In Kanada haben Ballivy und Benmokrane Untersuchungen mit kurzen Ankerstangen mit 36,8 mm Durchmesser durchgeführt, die mit Mörtelzement in Bohrlöchern mit 76 mm Durchmesser einzementiert wurden; dabei hat sich gezeigt, daß es zu einer örtlichen Ablösung kommt, wenn die Normalkraft lediglich 25 % bis 30 % der Maximalkraft beträgt (Abb. 6).

Bei Stangen mit einem Durchmesser zwischen 20 und 40 mm, die ordnungsgemäß in Zement oder Harz eingebettet sind, genügt im allgemeinen schon eine Verankerungsstrecke von weniger als 2 m, damit die Stange bricht, bevor es zur vollständigen Ablösung kommt.

Die Elastizitätsgrenze der Stange ist schon bei sehr geringen Bewegungen erreicht; so genügt beispielsweise schon die geringfügige Öffnung einer Diskontinuitätsfläche in der Gesteinsmasse, um die Stange, die diese Diskontinuitätsfläche durchquert, über ihre Elastizitätsgrenze hinaus zu beanspruchen.

Angesichts dieser Tatsachen sind die Ziehversuche, die in der Praxis als Abnahme- und Prüftests durchgeführt werden, kaum zufriedenstellend. Insbesondere kann anhand der Ziehversuche nicht geprüft werden, ob die

(Figure 4)

Abb. 4 - BIEGESTELLE EINES PASSIVEN ANKERS IM FALL EINER TANGENTIALVERSCHIEBUNG

(Figure 5)

Abb. 5 - PROGRESSIVE ABLÖSUNG EINER EINGEBETTETEN STANGE BEI NORMALKRAFT-BEAUFSCHLAGUNG

(Figure 6)

Abb. 6 - ZIEHVERSUCH MIT EINER 0,78 M LANGEN STANGE (BRUCHSPANNUNG 950 KN) (nach Ballivy und Benmokrane)

Stange auf ihrer gesamten Länge eingebettet ist. Daher sind diese Versuche keine Garantie für das einwandfreie Funktionieren der Stange im Gestein und auch nicht für die langfristige Korrosionsfestigkeit von Stangen, die lediglich eingebettet wurden. Dies ist ein erheblicher Nachteil, besonders in Gestein mit geöffneter Klüftung, in dem der Verlust von Verfüllmaterial schwer festzustellen ist. In diesem Fall ist es oft erforderlich, zunächst Verfüllmaterial in die offenen Klüfte zu injizieren.

2.2. Tangentialbelastung passiver Anker

Die Wirksamkeit passiver Anker, die mit einer tangential zur Diskontinuitätsfläche wirkenden Kraft belastet werden, läßt sich wesentlich schwieriger analysieren. Dieses Thema war Gegenstand mehrerer theoretischer und praktischer Untersuchungen (Haas, Bjurstrom, Azuar und Dight-Egger).

In den Wechselwirkungsmechanismus geht auch das Biegeverhalten der Ankerstange ein sowie der Widerstand, den Gestein und Verfüllmaterial der Bewegung der Ankerstange entgegensetzen. In der vorliegenden Untersuchung wird der Einfachheit halber davon ausgegangen, daß Gestein und Verfüllmaterial dieselben mechanischen Eigenschaften haben, und daß die Diskontinuität ohne Zuhilfenahme von Verfüllmaterial geschlossen werden kann.

Wenn die Tangentialkraft so gering ist, daß sie innerhalb des Elastizitätsbereichs des Mediums bleibt, in das der Stab eingebettet ist, kann mit Hilfe einer Standardanalyse die Verformung y als Funktion der Tangentialkraft S sowie das Biegemoment der Stange als Funktion des Abstandes von der Diskontinuitätsfläche ausgedrückt werden (Abb. 7).

Die Tangentialverschiebung an der Fläche wird durch folgende Gleichung gegeben:

$$yo = \frac{2\ S}{kd\lambda}$$

in der

k der Reaktionskoeffizient des Mediums ist, in dem die Stange mit einem Deformationsmodul E_m und der Poissonzahl ν_m eingebettet ist.

Es kann folgendes gesetzt werden:

$$kd = mE_m$$

λ wird als "Längenübertragung" bezeichnet

$$\lambda^4 = \frac{4\ E_b\ I_b}{kd}$$

E_b ist der Elastizitätsmodul des Ankerstahls

I_b ist das Trägheitsmoment der Ankerstange

und somit

$$\lambda^4 = \frac{\pi}{16}\ \frac{1}{m}\ \frac{E_b}{E_m}\ d^4$$

Setzt man

$$\alpha^4 = \frac{\pi}{16}\ \frac{1}{m}\ \frac{E_b}{E_m}$$

so erhält man

$$\frac{yo}{d} = \frac{2\ S}{mE_m \alpha\ d^2}$$

und

$$\lambda = \alpha_d$$

Das maximale Biegemoment M_{max} ist gleich:

$M_{max} = 0{,}322\,\alpha\,dS$,

und der maximale Biegemomentpunkt befindet sich in einem Abstand l von der Diskontinuitätsfläche, so daß

$$l = \frac{\pi}{4}\lambda$$

und somit

$$\frac{l}{d} = \frac{\pi}{4}\alpha$$

Um die Größenordnungen festzulegen, wird der Fall vorgespannter Stahlstangen mit einem Elastizitätsmodul $E_b = 2 \times 10^5$ MPa und mit einer Elastizitätsgrenze $\sigma_e = 840$ MPa untersucht. Der Durchmesser der Stangen liegt zwischen 20 mm und 50 mm.

Abbildung 8 zeigt die Änderung des Verhältnisses in halblogarithmischen Koordinaten als Funktion von E_m. Sie zeigt, daß sich der maximale Biegemomentpunkt in

(Figure 7)

$$E_bI_r\frac{d^4y}{dx^4} + kdy = 0$$

$$y = \frac{2S}{kd\lambda}\exp\left(-\frac{x}{\lambda}\right)\cos\frac{x}{\lambda}$$

$$M = \lambda S\ \exp\left(-\frac{x}{\lambda}\right)\sin\frac{x}{\lambda}$$

max y bei $x=0$ $\quad y_o = \frac{2S}{kd\lambda}$

max M bei $x = \frac{\pi}{4}\lambda$ $\quad$ max $M = 0.322\ \lambda S$

λ, Längenübertragung $\quad \lambda^4 = \frac{4 E_b I_b}{kd}$

Abb. 7 - VERHALTEN EINER PASSIVEN STANGE IN EINEM ELASTISCHEN MEDIUM BEI TANGENTIALER KRAFTBEAUFSCHLAGUNG

(Figure 8)

Abb. 8 - ÄNDERUNG DER 1/d BEZIEHUNG IN ABHÄNGIGKEIT VOM DEFORMATIONSMODUL, Em

einem Abstand von der Diskontinuitätsfläche befindet, der geringer als zwei Durchmesser ist, und der sogar weniger als ein Durchmesser beträgt, wenn E_m größer als 38 000 MPa ist.

Abbildung 9 zeigt die Tangentialverschiebung yo bei einer Tangentialkraft Se, die der Elastizitätsgrenze entspricht, als Funktion des Moduls E_m, wobei nach der Trescaschen Fließbedingung davon ausgegangen wird, daß:

$$Se = \frac{1}{2} \quad Ne = \frac{\pi}{8} d^2 \tau_e$$

und damit:

$$\frac{yo}{d} = \frac{\pi}{4} \frac{\tau_e}{\alpha mE_m}$$

(Figure 9)

Abb. 9 - TANGENTIALVERSCHIEBUNG Yo MIT DER TANGENTIALSPANNUNG Se IN ABHÄNGIGKEIT VON Em

Bei diesem Tangentialkraftwert bleibt die Verschiebung yo geringer als d/4 und sogar geringer als d/10, wenn E_m größer als 10 000 MPa ist.

Abbildung 10 zeigt die Änderung des maximalen Biegemomentes, die dem Wert der Grenzscherkraft Se entspricht.

Das maximale Biegemoment Mmax wird dann auf das Plastizitätsmoment Mp der Stahlstange bezogen.

Für das Plastizitätsmoment Mp wurde der von Dight vorgeschlagene Ausdruck benutzt:

$$Mp = 1,7 \frac{d^3}{32} \tau_e$$

folglich

$$\frac{Mmax}{M_p} = 0,76 \alpha \quad \text{bei } S = Se$$

wenn Em > 32 000 Mmax < Mp; das bedeutet, daß der Scherkraftwert vor dem Auftreten eines plastischen Knickpunktes erreicht wird.

Dight hat das Verhalten einer passiven Bewehrung im plastischen Bereich untersucht, und er geht davon aus, daß die Spannungsverteilung Abb. 11 entspricht.

Das Plastizitätsmoment Mp ist bei einer tangential zur Diskontinuitätsfläche aufgegebenen Kraft S_1 dergestalt erreicht, daß

$$S_1 = \frac{d^2}{4} \sqrt{1,7 \pi \sigma_e p_u}$$

Der Abstand l vom Punkt des maximalen Biegemomentes auf der Diskontinuitätsfläche ist gegeben durch:

$$\frac{l}{d} = \frac{1}{4} \sqrt{\frac{1,7 \pi \sigma_e}{p_u}}$$

Abb. 10 - ÄNDERUNG DES MAXIMALEN BIEGEMOMENTES MIT DEM HÖCHSTWERT DER TANGENTIALSPANNUNG Se

(Figure 11)

Abb. 11 - SPANNUNG UND VERHALTEN EINER PASSIVEN BEWEHRUNG IM PLASTISCHEN BEREICH

Die Bestimmung des Drucks p_u zwischen der Stange und dem Einbettungsmilieu der Stange wurde von Dight untersucht. Es kann in erster Annäherung davon ausgegangen werden, daß zwischen p_u und der einfachen Druckfestigkeit folgende Beziehung besteht:

$$p_u = 6,5 \tau_c$$

Für diesen Fall zeigt Abb. 12 den Verlauf von l/d für einfache Druckfestigkeits-Werte von 6 bis 200 MPa. Dieses Verhältnis liegt zwischen 3 und 0,5.

In Abb. 13 wurde der Verlauf des Quotienten Sl/Sa als Funktion von τ_c für denselben Wertebereich eingetragen. Man stellt fest, daß der Quotient ab 60 MPa größer als 1 wird, was bedeutet, daß der Grenzscherkraftwert erreicht ist, bevor ein plastischer Knickpunkt auftritt.

2.3 Verbesserung der Scherfestigkeit einer Diskontinuität durch passive Anker

Im vorhergehenden Abschnitt wurde das Verhalten einer in das Gestein eingebetteten Stange untersucht, wenn diese mit einfachen Kräften beansprucht wird. Die Kräfte, denen ein passiver Anker an einer Diskontinuität ausgesetzt ist, sind jedoch komplex. Bei einer Scherung entlang einer Diskontinuitätsfläche hat die Bewegung aufgrund der Expansionscharakteristiken der Diskontinuität außer der Tangentialkomponente Ut auch eine senkrechte Komponente Un.

Betrachten wir den Fall einer passiven Ankerstange, die eine Diskontinuität in einer Gesteinsmasse schneidet und zwar in einem Winkel θ zur Senkrechten auf der Diskontinuitätsfläche; die Resultierende $\vec{R}$ der an der Diskontinuität auf den Stabquerschnitt wirkenden Kräfte bildet auf die Stabrichtung bezogen einen Winkel β. $\vec{R}$ kann in eine parallel zur Stange verlaufende Normalkraft

(Figure 12)

Abb. 12 - VERÄNDERUNG DER 1/d BEZIEHUNG MIT σ_c IM BEREICH VON 6 BIS 200 MPa

(Figure 13)

Abb. 13 - VERÄNDERUNG DER S1/Sa BEZIEHUNG MIT σ_c

N und in eine senkrecht auf der Stabrichtung stehende Schwerkraft S aufgelöst werden (Abb. 14).

Im allgemeinen wird davon ausgegangen, daß die von der Stange bewirkte Bewehrung der Scherfestigkeit durch folgende Gleichung ausgedrückt wird:

$$C_b = R \cos(\theta + \beta) \, tg \varphi + R \sin(\theta + \beta)$$

Bei den in der Praxis am häufigsten verwendeten Methoden werden sehr einfache Annahmen gemacht.

(Fig. 14 a and 14 b)

Abb. 14 - AUFTEILUNG DER RESULTIERENDEN R IN EINE KOMPONENTE PARALLEL UND IM RECHTEN WINKEL ZUR SPANNUNGSRICHTUNG

Bei der ersten Methode werden folgende Annahmen gemacht:

$\beta = 0$
$R = Ne$

Die Stange wird parallel zu ihrer Richtung bis zu ihrer Dehnungsgrenze zugbelastet (Gaziev). Oft wird für die Stahlstange ein Sicherheitskoeffizient benutzt, der als zulässige Höchstbelastung 2/3 des Dehnungsgrenzwertes vorsieht (Colombet - Bonazzi).

Der Höchstwert von C_b ist erreicht, wenn

$$\theta = \frac{\pi}{2} - \varphi$$

Bei der zweiten Methode werden folgende Annahmen gemacht:

$$\beta = \frac{\pi}{2} - \theta$$

$$R = \frac{Ne}{2}$$

Diese Methode wird vor allem im Falle eines sehr niedrigen Wertes von θ angewandt; es wird lediglich der Scherkraftwiderstand der Stange berücksichtigt, der Dehnungseffekt bleibt jedoch unberücksichtigt. Diese Methode ist zwar a priori pessimistisch, doch bei direkten Scherversuchen an schwach dehnbaren Diskontinuitäten, die mit senkrecht zur Diskontinuitätsfläche verlaufenden Stangen durchsetzt sind, wurden Werte gemessen, die gleich groß bzw. lediglich sehr geringfügig höher lagen (Azuar-Groupe Français). Die Stange wurde effektiv abgeschert (Abb. 15). Es muß ebenfalls angenommen werden, daß die Gesteinswände ohne Verfüllmaterial in Berührung sind. Haas hat in diesem Zusammenhang das Bild einer Schere verwendet, die zuviel Spiel hat und schlecht schneidet.

(Figure 15)

Abb. 15 - ABSCHERUNG EINER PASSIVEN STANGE BEI EINEM DIREKTEN SCHERVERSUCH

Bei der dritten Methode wird einfach das Prinzip der Maximalarbeit angewandt (Abb. 14). Dieses Verfahren wurde in der Bodenmechanik bei bestimmten Ankernägel-

Programmen eingesetzt (Schlosser).

Das Ende der Grenzbelastung R der Ankerstange liegt auf der Ellipse

$$\left(\frac{N}{Ne}\right)^2 + \left(\frac{S}{0,5\ Ne}\right)^2 = 1 \quad \text{(Trescasche Fließbedingung)}$$

Bei einer Bewegung in Richtung $\vec{u}$, die in einem Winkel δ zur Diskontinuitätsfläche verläuft, besagt das Prinzip der Maximalarbeit, daß für jeden zulässigen Wert $\vec{R}^*$ die folgende Beziehung gelten muß:

$$(\vec{R} - \vec{R}^*)\ \vec{u} > 0$$

Folglich ist die Extremität des Vektors R der Punkt der Ellipse, dessen Tangente senkrecht auf $\vec{u}$ steht.

Daraus schließt man

$$\frac{R}{Ne} = \left[\frac{1 + \frac{m^2}{16}}{1 + \frac{m^2}{4}}\right]^{\frac{1}{2}}$$

$$tg\,\beta = \frac{m}{4}$$

$$m = cotg\ (\theta + \delta)$$

In Abb. 16 wurden Nomogramme eingezeichnet, die R/Ne und β bei Winkelwerten δ von 0°, 5°, 10° und 20° und bei Werten von θ zwischen 0° und 60° liefern. Die Nomogramme zeigen deutlich, daß R/Ne bei einem sehr niedrigen Wert θ je nach Dehnungswert δ zwischen 0,5 und 0,8 liegt, was mit den Veruchsergebnissen übereinstimmt.

Wenn andererseits $\theta \geqq 45°$, ist der Winkel β geringer als 15° und R/Ne größer als 0,9.

(Figure 16)

Abb. 16 - Re/Ne UND β WERTE FÜR δ = 0°, 5°, 10°, 20° UND θ = 0° BIS 60°

Es ist aufschlußreich zu bestimmen, in welchem Winkel θ die Stangen ausgerichtet sein müssen, wenn der Höchstwert für C_b erreicht werden soll. Abb. 17 zeigt den Verlauf von C_b als Funktion von θ für drei Fälle.

Fall A $\varphi = 30°$ $\delta = 0$
Fall B $\varphi = 40°$ $\delta = 5°$
Fall C $\varphi = 50°$ $\delta = 10°$

In diesen drei Fällen wird der Höchstwert von C erreicht, wenn θ > 60°, θ = 60° bzw. wenn θ = 45°.

Je niedriger die Werte von δ und φ sind, desto größer ist der Winkel θ, der dem Höchstwert von C_b entspricht.

Es muß jedoch darauf hingewiesen werden, daß eine solche Optimierung nicht unbedingt auch eine wirtschaftliche Optimierung darstellt, bei der auch die

(Figure 17)

Abb. 17 - ÄNDERUNG VON Cb IN ABHÄNGIGKEIT VON θ FÜR DREI FÄLLE

Stangenlänge sowie die praktischen Einsatzbedingungen berücksichtigt werden müssen. Die Frage muß fallweise geprüft werden. In vielen Fällen kommt jedoch ein Winkelwert θ = 45° dem wirtschaftlich optimalen Wert ziemlich nahe.

Die Anwendung der obigen Formeln zur Bewertung des Beitrags eines passiven Ankers zur Scherfestigkeit einer Diskontinuität bedarf noch einiger Erläuterungen:

a) Es wird vorausgesetzt, daß die Grenzbelastung des Stahls und des betrachteten Gleitwinkels der Diskontinuität gleichzeitig aufgebracht werden können.

Nun haben wir jedoch gesehen, daß die Dehnungsgrenze des Stahls unabhängig von der Art der auf eine eingebettete Stange einwirkenden Spannung schon bei geringfügigen Bewegungen erreicht wird. Diese geringfügigen Bewegungen können im Falle stark verschuppter Diskontinuitäten ohne Verfüllmaterial ausreichen (Abb. 18a); bei schwach verschuppten oder aufgelockerten Diskontinuitäten (Abb. 18b) oder bei Diskontinuitäten, deren Gesteinswände mit einem nur schwach verformbaren Verfüllmaterial getrennt sind, trifft dies jedoch mit Sicherheit nicht zu.

b) Die Wahl des Dilatanzwinkels ist ebenfalls recht schwierig. Einserseits kann die Dilatanz bei geringfügigen Tangentialbewegungen je nach Beschaffenheit und Zustand der Diskontinuitäten erheblich variieren. Andererseits ist die Dehnung einer durch einen Anker verstärkten Diskontinuität örtlich geringer als unter natürlichen Bedingungen. Die Dehnung dn einer Diskontinuität hängt nämlich von den senkrecht auf die Diskontinuität aufgegebenen Spannungen ab.

Barton verwendet folgenden Ansatz:

$$d_n = 0{,}5\ K\ \log \frac{\tau_c}{\tau_n}$$

$$\text{wenn } 1 < \frac{\tau_c}{\tau_n} < 100$$

K - mit der Beschaffenheit der Diskontinuität und der Gesteinswändegeometrie verknüpfter Koeffizient (Wertbereich von K: 0 bis 20)

τ_c einfacher Kompressionswiderstand des Wandgesteins

τ_n lotrechte Spannung

Bei niedrigen Werten von $\tau_n \cdot \frac{\tau_c}{\tau_n} > 100$

$$dn = dn° = K$$

Das von Gaziev vorgeschlagene Gesetz hat eine andere Form:

$$d_n = d°_n \left[1 - \frac{\sigma_n}{\sigma_c}\right]^{4c}$$

(Figure 18)

a) Geschlossene, gut verschuppte Diskontinuität

b) Schwach verschuppte Diskontinuität

Abb. 18 - SCHERKURVEN FÜR GESTEINSDISKONTINUITÄTEN BEI KONSTANTER LOTRECHTER SPANNUNG
Scherspannung
Tangentialkomponente der Verschiebung
Normalkomponente der Verschiebung

In Anbetracht des Einflusses des Parameters δ auf die obigen Untersuchungen ist Vorsicht geboten, und es empfiehlt sich, für δ einen Wert anzusetzen, der niedriger ist als der für d_n mit Hilfe der üblichen Verfahren ermittelte Wert.

c) Schließlich wird bei den obigen Untersuchungen davon ausgegangen, daß die Stange richtig eingebettet ist. Wenn dies nicht der Fall ist, kommt es zu einer allmählichen Verformung der Stange, und es treten zwei plastische Knickpunkte auf, die sich auseinanderbewegen (Abb. 19). In diesem Fall ist der Beitrag der Stange zur Erhöhung der Scherfestigkeit erheblich reduziert (Azuar et al.).

(Figure 19)

Abb. 19 - ALLMÄHLICHE ABLÖSUNG EINER PASSIVEN STANGE IM VERLAUF EINES DIREKTEN SCHERVERSUCHES

2.4 Allgemeine Prinzipien für die Bewehrung einer Gesteinsmasse mittels passiver Stangen

Aus den obigen Untersuchungen ergeben sich mit aller Deutlichkeit die bei der Bewehrung einer Gesteinsmasse zu beachtenden Prinzipien.

a) Geringfügige Verschiebungen an den Diskontinuitäten genügen, um passive Anker örtlich bis zur Dehnungsgrenze zu belasten. Diese Bewegungen sind umso geringer, desto starrer das Gestein ist und desto dichter die Diskontinuitäten sind und folglich stärker zur Dilatanz neigen. In der Praxis ist es daher möglich, Verformungen an den Diskontinuitäten dadurch zu blockieren, daß die Gesteinsmasse Zug um Zug mit der Auffahrung verankert wird.

b) Durch Untersuchung der Verformungs-Kinematik der möglichen Bruchvorgänge und Beurteilung der Diskontinuitätseigenschaften kann die richtige Orientierung der Ankerstangen ermittelt und der Ankerausbau optimiert werden.
c) Es empfiehlt sich, der korrekten Einbettung der Ankerstangen beiderseits der Diskontinuitäten besondere Aufmerksamkeit zu schenken; dies läßt sich umso schwieriger durchführen, je offener die Diskontinuitäten sind und je größer die Gebirgsdruck-Entspannungseffekte sind. Ziehversuche erlauben es zwar, die Qualität des Verfüllmaterials und der Einbettungsmethode zu beurteilen, sind jedoch keine Garantie für die gesamte Verankerung.
d) In großen Abbauräumen, in denen Verformungen nicht vernachlässigt werden können, können die auf passive Anker wirkenden Spannungen die Dehnungsgrenze überschreiten und örtlich zu Ablösungen der Verankerung führen. Diese Lösung ist zwar für vorläufige Ausbauten durchaus akzeptabel, doch bei Dauerausbauten muß mit Rücksicht auf die Korrosionsgefahr wesentlich sorgfältiger vorgegangen werden. Während der Aushubarbeiten, nach der Inbetriebnahme und während der gesamten Standzeit der Anlage läßt sich die verankerte Gesteinsmasse am besten mit in Bohrlöchern eingebauten Extensometern überwachen. Nach Abschluß der Arbeiten kann die Notwendigkeit bestehen, die Anlage zwecks endgültiger Sicherung durch zusätzliche Anker zu verstärken. Bei großen Ausschachtungen mit Aushubhöhen von mehreren hundert Metern können die höchstgelegenen Teile praktisch nicht mehr mit vertretbaren Kosten verankert werden, es sei denn, daß der Zugang mit dem Fortschreiten des Aushubs erhalten geblieben ist.
e) Aus Gründen, die mit dem Verhalten der Anlage (homogenere Spannungsverteilung) und mit der Sicherheit der Anlage zusammenhängen (geringere Zerstörungsneigung), sollte bei gleichem Stahlgewicht einer größeren Anzahl von Stangen gegenüber einer geringeren Anzahl großkalibriger Stangen der Vorzug gegeben werden, auch wenn diese Lösung aufgrund höherer Bohrkosten insgesamt kostspieliger ist.

Aufgrund der Erfolge, die mit dem Einsatz von Ankerstangen unter Tage erzielt wurden, haben passive Ankerstangen auch Eingang in oberirdische Bauwerke gefunden, wo sie immer häufiger anzutreffen sind. 1974 haben P. Londe und D. Bonazzi auf dem Kongreß in Denver das Konzept des "bewehrten Gesteins" eingeführt. Seither wurden bei der Untersuchung der Wirkungsweise passiver Stangen im Gestein erhebliche Fortschritte erzielt. Die Wirkungsweise ist jedoch noch nicht völlig klar. Der Praktiker wünscht sich ein Verfahren, mit dessen Hilfe alle mechanischen Eigenschaften von bewehrtem Gestein im Zusammenhang mit der Verformbarkeit und den Bruchbedingungen bestimmt werden könnten und das beispielsweise auch in ein numerisches Modell eingebracht werden könnte.

Untersuchungen im Bereich der zusammengesetzten Materialien werden aktiv vorangetrieben. Meines Erachtens wird der Erforschung bewehrter Gesteinsmassen nicht genügend Aufmerksamkeit geschenkt, obwohl die wenigen Versuche mit Blockmodellen sehr ermutigend waren.

3 SCHLUSSBEMERKUNG

Zwischen vorgespannten Ankern und passiven Ankerstangen gibt es sowohl in bezug auf Konzeption und Funktionsweise als auch hinsichtlich der ihrem Einsatz zugrundeliegenden Sicherheits- und Haltbarkeitsüberlegungen grundlegende Unterschiede.

In diesem Vortrag habe ich selbstverständlich meine eigenen Theorien vertreten. Ich möchte jedoch nicht versäumen, allen meinen Kollegen in den französischen Forschungszentren und Ingenieurbüros, mit denen ich seit vielen Jahren auf diesem Gebiet zusammenarbeite, meinen Dank auszusprechen; von ihnen habe ich zahlreiche wertvolle Anregungen erhalten, die in meine Überlegungen eingeflossen sind.

LITERATURVERZEICHNIS (Siehe englischer Teil)

09:20 Kurzvorträge

Carrere, A., Nury, C., Pouyet, P., Frankreich
Der Beitrag eines nichtlinearen FE-Modells zur Bewertung einer geologisch komplizierten Baustelle für eine 200 m hohe Bogenstaumauer bei Lontan, China

Egger, P., Spang, K., Schweiz
Standsicherheitsuntersuchungen für eine mittels Felsbolzen verbesserte Talsperrengründung

Irfan, T.Y., Koirala, N., Tang, K.Y., Hong Kong
Progressiver Böschungsbruch in einem stark verwitterten Felsmassiv

Marchuk, A.N., Khrapkov, A.A., Zukerman, Y.N., Marchuk, M.A., U.S.S.R.
Das Verhalten der Fuge zwischen Beton und Felsgründung von hohen Talsperren mit einem am Fuss liegenden Kraftwerk

10:00 - 10:30 Pause

10:30

Martin, D.C., Kanada
Die Anwendung von Felsverankerung und künstlicher Abstützung im Tagebau

Nilsen, B., Norwegen
Knickung von hartem Fels - ein potentieller Bruchmechanismus in hohen Böschungen?

Rosengren, K.J., Friday, R.G., Parker, R.J., Australien
Vor dem Abbau installierte Kabelbolzen zur Stützung von Böschungen im Tagebau

Rowe, R.K., Armitage, H.H., Kanada
Methode für den Entwurf von Bohrpfählen in weichem Gestein (an 3 Fällen erprobt)

Savely, J.P., U.S.A.
Wahrscheinlichkeitsanalyse von intensiv zerstörten Felsmassen

Serra de Renobales, T., Spanien
Schichtenwölbung im Liegenden eines Kohletagebaus

Sharp, J.R., Bergeron, M., Ethier, R., U.K., Kanada
Aushub, Verbesserung und Messungen für eine 300 m hohe Felsböschung im Schiefer

Sueoka, T., Mauramatsu, M., Torii, Y., Shinoda, M., Tanida, M., Narita, S., Japan
Wirtschaftliche Konstruktion grosser Böschungen aus Schlammstein

Widmann, R., Promper, R., Österreich
Einfluss des Untergrundes auf die Spannungen von Bogenmauern nahe der Aufstandsfläche

12:00 - 13:30

Mittagspause und Postersitzung (Marquette)

13:30 - 15:00 Diskussion (Le Grand Salon)

Diskussionsleiter: K. Kovari (Schweiz)
Podiumsdiskussion: B.K. MacMahon (Australien)
R. Ribacchi (Italien)
R. Yoshinaka (Japan)

Felsgründungen und Böschungen - Thema II

Zusammenfassung des Diskussionsleiters

K. Kovari (Schweiz)

Bei Felsbauwerken wie Böschungen und Felsgründungen sind große Unterschiede hinsichtlich Ausdehnung und Gefährdungsgrad zu verzeichnen. Durchstiche für

Straßen und Eisenbahntrassen sind im allgemeinen nur bis zu mehreren Dutzend Metern hoch, da ansonsten ein Tunnel die weit rationellere Lösung darstellen würde. Große Staumauern können wohl Höhen bis zu 300 m erreichen, doch hängt der Gefährdungsgrad hier vom Fassungsvermögen der Staubecken ab. Da auch der Tagebaubetrieb in immer größerem Maßstab erfolgt, können die dafür erforderlichen Böschungen bis zu mehreren hundert Metern hoch sein, wobei jedoch der Gefährdungsgrad bei bergmännischen Böschungen von ganz anderen Faktoren abhängt als bei Staumauern. Schließlich können auch Erdrutsche - z.B. der Vajont-Erdrutsch - von überragender Bedeutung sein, besonders, wenn sie in unmittelbarer Nähe eines Wasserstaubeckens vorkommen. Diese Variationen der Größenordnung und des Risikogrades spielen bei der ingenieurtechnischen Lösung des jeweiligen Problems eine entscheidende Rolle, ebenso bei der Festlegung der Kriterien für die Errichtung derartiger Bauten. Es liegt auf der Hand, daß die Möglichkeit einer Einflußnahme auf das Verhalten dieser Felsbauten mit steigender Größe abnimmt; in manchen Fällen kann daher die einzige Lösung darin bestehen, daß das Projekt aufgegeben oder selbst minimale Gesteinsbeanspruchungen - z.B. durch Schwankungen des Stauspiegels - ausgeschaltet werden. Die Leistungsparameter obertägiger Bauten lassen sich durch quantitative Werte wie Sicherheitsfaktor, Bruchwahrscheinlichkeit, Spannungszustand im Beton und Wasserverlust zum Ausdruck bringen. Der letztgenannte Faktor kann auch die Standfestigkeit von Talsperren beeinflussen.

Baumaßnahmen bei Felsböschungen

Hier wäre zuerst der Böschungswinkel mit seinem großen Einfluß auf Wirtschaftlichkeit und Sicherheit der Baumaßnahme zu nennen. Während des Baus ist ein lokaler Bruch zu vermeiden, gleichzeitig muß jedoch der zufriedenstellende Fortschritt der Arbeiten sichergestellt werden. Bei hohen Böschungen ist die Beherrschung des Wassers auf der Geländeoberfläche und im Gestein die einzig mögliche Korrektivmaßnahme. Durch das Anlegen von Bohrlöchern und Entwässerungsstollen können sowohl der Grundwasserspiegel wie auch der Wasserdruck an potentiellen Rutschflächen gesenkt werden. Die Wirksamkeit solcher Maßnahmen läßt sich durch eine Beobachtung der Standrohrspiegelhöhe bestimmen. Bei der Felsverbesserung besteht die Tendenz, vollständig einbetonierten Ankern den Vorzug zu geben, da sie kostengünstiger sind und man von ihnen annimmt, daß sie korrosionsfester sind als aktive Anker. Der statische Effekt aktiver Anker läßt sich leicht erklären: Es wirken nur zwei Kräfte, deren Größe sich jederzeit kontrollieren läßt. Vollständig einbetonierte, d.h. passive Anker sind nur dann wirksam, wenn die Gesteinsverformung bereits stattgefunden hat. Kommt es nur an wenigen Klüften zur Deformation, dann ergeben sich an den betreffenden Stellen Spannungskonzentrationen in den passiven Ankern, während in aktiven Ankern keine nennenswerte Erhöhung der Spannungen zu verzeichnen wäre. Es bedarf daher weiterer Untersuchungen zum Verhalten passiver Anker. In vielen Fällen kann und soll einem Tunnel der Vorzug gegenüber einer Böschung gegeben werden.

Baumaßnahmen bei Staumauergründungen

Hierbei handelt es sich um das Abtragen von schwachem Gestein, dessen Vorhandensein ansonsten in steilen Tälern häufig die Standfestigkeit der Böschungen beeinträchtigen kann. Die Verfestigungsbetonierung soll die Steifigkeit und Festigkeit des Gesteins erhöhen, während es Ziel eines Dichtungsgürtels ist, die Sickerung im Gestein zu unterbinden. Eine der wohl umfangreichsten Betonierungsmaßnahmen überhaupt - insgesamt 11 km an Verpreßstollen und Bohrlöcher mit einer Gesamtlänge von 500 km - wurde kürzlich beim Bau des El Cajon-Damms in Honduras vorgenommen. In Karstgebieten ist die genaue Vorhersage der erforderlichen Betonierungsarbeiten äußerst schwierig, so daß in manchen Fällen die Prognose um 60 % überschritten wird. Dabei spielen nicht nur Standfestigkeitsprobleme der Böschungen eine Rolle, sondern in vielen Fällen muß die Betonierung selbst ausreichend gegen Abrutschung abgesichert sein. Gelegentlich können vollständig einbetonierte Anker hier Abhilfe schaffen. In anderen Fällen wieder wird ein Netzwerk von Strecken aufgefahren, die dann mit Beton vergossen werden, um so durch den Dübel-Effekt Scherwiderstand zu entwickeln. Dieser Dübel-Effekt wird manchmal eingesetzt, um die Steifigkeit des Gesteins zu erhöhen.

Entscheidungsfindung

Bei obertägigen Felsbauten ist eine derartige Vielfalt schwerwiegender Entscheidungen zu treffen, daß es sich lohnt, den Prozeß der Entscheidungsfindung näher zu betrachten. In vielen Fällen ist die Lösung so "augenfällig", daß man die Entscheidung auf rein empirischer Basis fällt und eine logische Prüfung der Situation für unnötig erachtet. Andererseits gibt es die "schwierigen" Situationen, bei denen die Erfahrung allein nicht zu genügen scheint und man das heranziehen muß, was wir Felsmechaniker unsere "Werkzeugkiste" nennen. Bevor wir uns näher mit dieser Werkzeugkiste befassen, soll erwähnt werden, daß eine Lösung häufig deshalb so "augenfällig" ist, weil die mit der Entscheidung Betrauten keine ausreichenden felsmechanischen Kenntnisse besitzen. Im übrigen sind die Begriffe "augenfällig" und "schwierig" eher subjektiv, und wie ausgiebig von der "Werkzeugkiste" Gebrauch gemacht wird, hängt sehr von der Erfahrung und den Kenntnissen der Projektingenieure ab. Die "Werkzeuge" können bei der Beurteilung der

- Eigenschaften des betreffenden Gesteins,
- der Grundwasserverhältnisse,
- des Sicherheitsfaktors und
- der Verformung der Bauwerke

behilflich sein.

Es ist bereits darauf hingewiesen worden, daß trotz geeigneter Werkzeuge ein gutes Urteilsvermögen nicht nur in allen Stadien der Planung, sondern auch bei der Auswertung der Ergebnisse vonnöten ist. Im Idealfall sollten alle Entscheidungen vor Baubeginn getroffen werden, denn dann können genaue Kostenvoranschläge erstellt, korrekte Auftragsunterlagen erarbeitet, passende Maschinen und Geräte gewählt sowie realistische Terminpläne aufgestellt werden. In vielen Fällen ist ein derartiges Vorgehen nicht möglich. Vielfach werden darum "baubegleitende" Planungsarbeiten befürwortet, ein Prinzip, das ohne Zweifel in vielen Tagebaubetrieben anzuwenden ist sowie dort, wo Eigentümer und Auftragnehmer derselben Organisation angehören. Wenden wir uns nun der felsmechanischen "Werkzeugkiste" zu und betrachten wir als Beispiel nur einen Aspekt des Materialverhaltens. Bei klüftungsabhängigen Standfestigkeitsproblemen in Böschungen oder Staumauern kann die Verformbarkeit von Diskontinuitäten eine entscheidende Rolle spielen. Bei geringer Verformbarkeit wird es zum Sprödbruch kommen, und selbst geringfügige Verschiebungen können zu einem bedeutenden Festigkeitsverlust führen und somit die Sicherheit erheblich beeinträchtigen. In solchen Fällen sind vorgespannte Anker passiven Ankern vorzuziehen, und in den Berechnungen sollte man eher Restfestigkeitswerte als Höchstwerte verwenden. Andererseits kann eine große Verformbarkeit und die sich daraus ergebende Fließdeformation mit darauffolgendem progressiven Bruch vollständig einbetonierte Anker erforderlich machen, so daß maximale Festigkeitswerte in die Berechnungen aufgenommen werden können. Ein in diesem Zusammenhang nur selten berücksichtigter Faktor ist natürlich das Materialverhalten in den Größenverhältnissen des zu errichtenden Bauwerks! Ein weiteres felsmechanisches Problem, auf das man fast immer stößt, ist die Auswirkung von Diskontinuitäten auf die Verformbarkeit eines Gesteins. Eine Möglichkeit, diese Frage experimentell zu beantworten, sind großmaßstäbliche Beaufschlagungsversuche (z.B. Vorspannversuche), bei denen das Formänderungsprofil entlang eines Bohrlochs in der Beaufschlagungsrichtung beobachtet wird. Die steilen Maxima in der Formänderungsverteilungskurve sind ein deutliches Zeichen dafür, daß sich aktive Klüfte und schwache Gesteinslagen mit dem Bohrloch schneiden. Betrachten wir nun die derzeit zur Analyse von Böschungen und Staumauergründungen eingesetzten Rechenmodelle. Alle statischen Überlegungen stützen sich auf Modelle; es gibt Modelle für das Materialverhalten unter statischen Bedingungen und bei verschiedenen Kraftbeaufschlagungen. Bei der Modellbildung wird die tatsächliche Situation auf die wesentlichen Elemente eines bestimmten Problems reduziert. Bei der Untersuchung von Grenzverhalten und Sicherheit herrscht das Starrkörper-Prinzip vor. Lineare elastische Annahmen werden in der Spannungs- und Deformationsanalyse angewendet. Nichtlineare Annahmen für das Gesteinsverhalten sind wohl bei bestimmten Rechenvorgängen zulässig, man erhält jedoch dann keine Aussagen über Grenzzustände (Einsturz).

In Wirklichkeit sind Felsbauten selten zweidimensional; in manchen Fällen sind jedoch zweidimensionale Modelle durchaus gerechtfertigt, z.B. beim Böschungsbruch entlang einer einfachen Bruchfläche oder bei der Spannungsanalyse einer Gewichtsstaumauer. Belastungen sind statischer oder dynamischer Natur; man ist sich jedoch auch heute noch nicht einig darüber, wann zur Simulierung eines Erdbebens eine Analyse der dynamischen Reaktion durchzuführen ist oder die Analyse einer gleichwertigen statischen Belastung ausreicht.

Analyseverfahren stützen sich hauptsächlich auf die gründliche Kenntnis der Mechanismen, die das strukturelle Verhalten von Böschungen und Felsgründungen beeinflussen. Bei der Auswahl der Rechenmodelle sowie bei der Auswertung der Ergebnisse sind einige praktische Regeln zu befolgen: Um auch den ungünstigsten Fall zu ermitteln, muß man häufig viele verschiedene Verhaltensmuster untersuchen.

Nächster Betrachtungspunkt sind Beobachtungen, die folgendes zum Ziel haben:

- Überwachung des Gleichgewichtszustands
- frühestmögliche Erfassung ungewöhnlicher Verhaltensmuster
- Klärung ungewöhnlicher Verhaltensmuster, um geeignete Gegenmaßnahmen treffen zu können
- Unterstützung bei der Sicherheitseinschätzung bereits vorhandener Bauten
  und
- Bruchprognose.

Oft werden mit solchen Beobachtungen mehrere Ziele gleichzeitig verfolgt. Zu den wichtigsten physikalischen Parametern zählen Verformungen (z.B. absolute und relative Verschiebungen, Formänderungen, Änderungen in der Böschungsneigung), die Standrohrspiegelhöhe und die von den Ankern ausgeübten Kräfte.

Verformungen werden meistens punktuell gemessen. Aus der räumlichen Verschiebung ausgewählter Punkte auf einer Böschungsfläche kann auf die Standfestigkeit der Böschung geschlossen werden. Weitaus ausführlichere Daten liefert die sogenannte "linienweise Beobachtung", bei der die Verteilung einer bestimmten Größe entlang einer Geraden oder einer Kurve gemessen wird. So kann man z.B. anhand der Verteilung der horizontalen Verschiebungen in einem Bohrloch in einer nicht standfesten Böschung die Gleitebene feststellen, die das unerwünschte Verhalten verursacht. Ähnlich zeigen Messungen der Formänderungsverteilung in einem Bohrloch in der Felsgründung einer Staumauer, wo bestehende Klüfte aktiv werden, d.h. sich mit den Schwankungen des Stauspiegels öffnen und schließen. Steile Maxima in der Formänderungskurve sind ein Zeichen dafür, daß sich solche aktive Klüfte mit dem Bohrloch schneiden. Das Wissen um das Vorhandensein solcher Klüfte ist für die Beurteilung des Sickerungsverhaltens sowie für die Sicherheitseinschätzung von Bedeutung.

Weiter dienen Böschungsbeobachtungen der Bruchprognose. In Böschungen mit großer Verformbarkeit und erheblichem Kriechen kommt es bevorzugt zum progressiven Bruch, wodurch sich in vielen Fällen nicht nur der Böschungsbruch, sondern auch die Wirksamkeit von Stützmaßnahmen vorhersagen läßt. Andererseits sind Verformungsbeobachtungen von geringem Nutzen, wenn heftige Regenfälle oder ein schweres Erdbeben den auslösenden Faktor darstellen. Die Analyse ist naturgemäß mit einer gewissen Unsicherheit hinsichtlich der Art der Diskontinuitäten, der Gesteinseigenschaften und der Grundwasserverhältnisse behaftet. Außerdem stehen selten genügend Daten zur Verfügung. Rechnerisch ermittelte Ergebnisse haben zwar in der Regel einen rein hypothetischen Charakter, doch lassen sich mit ihrer Hilfe Ursache und Wirkung erklären. Beobachtungsverfahren leiden wiederum darunter, daß nur vereinzelte bzw. überhaupt keine Meßwerte und außerdem vielfach noch keine geeigneten Meßverfahren zur Verfügung stehen. Empirische Beobachtungen liefern realitätsnahe Ergebnisse, machen die Realität jedoch nicht zwangsweise auch leichter verständlich. Ohne zusätzlichen Aufwand kann es geschehen, daß Beobachtungen für die angestrebte Klärung wertlos bleiben. Lassen sich jedoch Analyse und Beobachtung kombinieren, so trägt diese Kombination zum besseren Verständnis der Realität bei.

Um aus rechnerischen Analysen den größtmöglichen Nutzen zu ziehen, müssen die Modelle den Bedingungen des jeweiligen Problems gebührend Rechnung tragen; dabei ist die Zahl der Parameter so klein wie möglich zu halten, und wenigstens begriffsmäßig sollten diese Parameter meßbar sein. Ausgefeiltere Modelle führen zu komplexeren Auswertungsergebnissen, was die bauingenieurtechnischen Entscheidungen nicht unbedingt erleichtert. Zum Thema Beobachtungen wäre noch zu erwähnen, daß diese nur dann nutzbringend sind, wenn sie sich auf klare Begriffe stützen - mit anderen Worten ausgedrückt: "Keine Beobachtung ohne theoretische Untermauerung".

Zum Abschluß dieser kurzen Zusammenfassung soll auf die Mängel in unseren Ansätzen zur Problemlösung hingewiesen werden. Der Felsbau wird auch weiterhin Fortschritte in Theorie und Praxis zu verzeichnen haben. Ist es jedoch im Felsbau nicht so, daß der Felsbauingenieur trotz ungenügender Angaben und Informationen Entscheidungen treffen und große Verantwortung auf sich nehmen muß? Dieser Kongreß hat ohne Zweifel zur Ausräumung von Ungewißheiten in manchen Bereichen und zur stärkeren Untermauerung felsbaulicher Entscheidungen durch Tatsachen beigetragen.

Thema 3
Sprengen und Ausbruch

Analyse und Überwachung von Sprengungen; Bohrtechnik und Bohrlochkontrolle; Stückigkeit; Leistung von Tunnelvortriebsmaschinen; Bohrverfahren für Schächte mit großem Durchmesser

Vorsitzender: Tan Tjong Kie (China)

Schriftführer: W. Comeau (Kanada)

Sprecher: C.K. McKenzie (Australien)

Sprengen im Hartgestein: Diagnose und Modelle für Sprengschäden und Haufwerk

KURZFASSUNG: Es wurden Modelle für die Vorhersage von Felsreaktionen entwickelt, mit deren Hilfe alternative Sprengpläne für spezielle Anwendungen und Felsarten bewertet werden können. Für die Optimierung der Modelle sind jedoch noch weitere Beobachtungen und Messungen erforderlich, um auch die Detonationsfolge zu beherrschen und die wichtigsten Parameter der Felsreaktion zu identifizieren.

1. EINLEITUNG

In Australien wurde am Julius Kruttschnitt Mineral Research Centre (im folgenden als JKMRC bezeichnet) ein Modell für die Gesteinszertrümmerung durch Sprengen entwickelt, das auf den Ergebnissen von Messungen und Beobachtungen beruht, die in den vergangenen drei Jahren in über 200 Wochen Außenarbeit in mehr als 25 Gruben gesammelt wurden. Mit Hilfe des Modells konnten durch Sprengungen verursachte Felsschäden bestimmt werden; es wird außerdem für die Optimierung von Sprengplänen für spezielle Sprengarbeiten verwendet.

Das Modell beruht hauptsächlich auf der direkten Verwendung von Bruchdaten aus Bruchkartierungen und Stereobildern und hat bestehende Zerklüftungen als wichtigste Einflußgröße für die Zerkleinerung und die Schäden identifiziert. Ebenso wichtig ist vielleicht die Entwicklung von Beobachtungs- und Meßtechniken für die Gewinnung der für die Modellbildung erforderlichen Daten. Die Anwendung der Beobachtungstechniken hat einige der gängigsten Sprengfehler aufgezeigt, wie unzureichende Steuerung der Ladungszündung, Schwankungen der Sprengstoff- und Zündelementleistung, Fehlen genauer Verfahren für die Bewertung von Sprengplänen sowie unzulängliche Kenntnisse der Schadensmechanismen und der auf den Schadensumfang einwirkenden Einflußgrößen. In den meisten Anwendungsbereichen tritt die Modellbildung hinter grundlegenderen Problemen wie Ladeverfahren, Sprengpläne und Fertigungsspezifikationen zurück.

Es werden Ergebnisse einiger grundlegender Sprengbeobachtungen mitgeteilt, um die Anwendung von Beobachtungen auf die Optimierung der Sprengleistung aufzuzeigen. Die Beobachtung von Sprengungen hat zu erheblichen Produktivitätssteigerungen geführt, die die Vorhersagemöglichkeit derzeitiger Modelle übersteigt. Bergbaubetriebe berichten von Ersparnissen, die allein im Sprengstoff- und Bohrkostenbereich mehr als 2 Mio. Dollar betrugen; der spezifische Sprengstoffverbrauch war trotz Verbesserung der Stückigkeit, Schadensminderung und guter Beherrschung der Bodenerschütterungen um bis zu 50 % geringer.

2. MODELLBILDUNG FÜR DIE GESTEINSZERTRÜMMERUNG

Das vom JKMRC geschaffene Zerkleinerungs- und Schadensmodell beruht auf folgenden Beobachtungen:
1. Die Bruchbildung ist die wichtigste einzelne Einflußgröße auf die Gesteinszerkleinerung und die Schadenswirkung;
2. Der Zerkleinerungsgrad eines Gesteins hängt von der Korngröße und der Energie ab, der die Teilchen unterworfen wurden;
3. Während des Sprengvorgangs sind Brechwirkung und Rißbildung in erster Linie von der Brisanz abhängig, während die Gesteinsverschiebung von der Gasenergie bewirkt wird.

Für dieses Bruchmodell kann somit folgende Ausgangshypothese formuliert werden: Wenn Korngröße und Energiewerte bekannt sind, kann der Bruchgrad eines Teilchens vorausgesagt werden. Diese Hypothese beruht auf Beobachtungen des Bruchverhaltens von tausenden von Gesteinsteilchen bei Bruchversuchen im Labor. Das Hauptmerkmal dieses Modells ist vielleicht, daß es nicht auf theoretischen Ge- steins- bzw. Brucheigenschaften, sondern auf Messungen beruht und daß alle Vorhersagen des Modells durch Messungen überprüft werden können.

(Figure 1)

Abb. 1 Die Größenverteilung vor und nach dem Sprengen zeigt den Grad der Gesteinszerkleinerung

2.1 Zerklüftungsmessung

Der Zerklüftungsgrad wird durch Zeilenabtastung gemessen, wobei die Schnittpunkte aller natürlichen Bruchlinien mit einer Abtastlinie sowie Fallwinkel und Fallrichtung der Bruchlinie gemessen werden. Diese Bruchlinien werden dann in einem flächentreuen Stereobild festgehalten, und die In-situ-Blockgrößenverteilung für drei verschiedene Kluftabstandsverteilungen wird mittels einer Monte-Carlo-Simulierung errechnet. Der Fels wird somit als ein System unregelmäßig geformter Blöcke beschrieben, wobei die Größenverteilung der Blöcke durch die natürliche Grobbrüchigkeit des Gesteins gegeben ist.

Das Bruchmeßverfahren liefert auf direktem Wege die maximale Korngröße, die im Haufwerk beobachtet werden könnte, sowie den prozentualen Anteil der Partikel, die eine bestimmte Größe überschreiten. Wenn die gewünschte Bruchstück-Größenverteilung im Haufwerk bestimmt werden kann, ist ein Vergleich der beiden Verteilungen gemäß Abb. 1 möglich. Dieser Vergleich liefert einen direkten Hinweis auf den vom Sprengstoff zu fordernden Brechgrad, mit dessen Hilfe dann ein "Brechindex" bestimmt werden kann.

2.2 Verteilung der Stoßwellenenergie

Die Stoßwellenenergie wird aus der mit Hilfe von

..eophonen gemessenen Bodenerschütterung ermittelt.

Eine Reihe von Meßfühlern wird um die Sprengstelle herum angeordnet, und die Meßwerte werden mit Hilfe von Meßwertdruckern oder Analogregistriergeräten aufgezeichnet. Die Wellenform von Vollschwingungen wird benutzt, um die Energie in einem gegebenen Punkt zu berechnen, aufgrund der Felsdichte, der Ausbreitungsgeschwindigkeit der P-Welle und dem Zeitverlauf der Teilchengeschwindigkeit im betrachteten Punkt. Die Teilchengeschwindigkeit wird in einem relativ geringen Abstand vom Sprengpunkt (5 m bis 25 m) gemessen, damit es bei der Berechnung der Erschütterungsenergie zu keinen zu großen extrapolationsbedingten Fehlern kommt.

Anhand von Mehrfachaufzeichnungen von Teilchengeschwindigkeiten lassen sich die Erschütterungsausbreitungs-Gleichungen der Sprengung für eine gegebene Gestein/Sprengstoff-Kombination aufstellen. Die am JKMRC benutzten Gleichungen berücksichtigen sowohl Reibungsdämpfungs- als auch geometrische Dämpfungsmechanismen für Kugelwellen und erlauben detaillierte Extrapolationen der Form und Amplitude von Bodenerschütterungen in einem beliebigen Punkt. Durch Messungen und Berechnungen können somit die Stoßwellen- oder Erschütterungsenergie in einem beliebigen Punkt der Sprengung oder in der Umgebung der Sprengstelle sowie die Konturen der Bodenerschütterung im Bereich der Bohrlöcher bestimmt werden. Der Erschütterungsenergie werden die Ausbildung neuer Brüche und die Felszerkleinerung zugeschrieben.

2.3 Messung der Zerkleinerung

Programme für die Vorhersage der Zerkleinerungswirkung bei Sprengungen müssen durch Messungen verifiziert werden. Hierzu wird am JKMRC u.a. ein photographisches Verfahren angewandt, bei dem während der einzelnen Abtragsphasen zahlreiche Aufnahmen vom Haufwerk angefertigt werden; jedes einzelne auf den Aufnahmen erkennbare Partikel wird manuell digitalisiert, um Grundmaße wie Fläche, Umfang sowie projizierte maximale und minimale Sehnenlängen zu erhalten. Durch umfangreiche Kalibrierungen konnte eine Beziehung zwischen den digitalisierten Messungen und der Partikelgröße hergestellt werden, mit deren Hilfe die Größenverteilung errechnet werden kann. Mit diesem Verfahren lassen sich beobachtete Änderungen der Größenverteilung zwar genau beschreiben, das Verfahren ist jedoch sehr arbeitsaufwendig; es wird in der Tat bis zu einer Stunde benötigt, um mehrere hundert Partikel einer einzigen Aufnahme zu digitalisieren.

Die Messung der Zerkleinerung gewinnt im Hinblick auf die Anwendung moderner Sprengverfahren zunehmend an Bedeutung, bei denen es darum geht, spezielle Zerkleinerungsziele zu erreichen, damit über den Einsatz kostenwirksamer Bergbautechniken beim Abtransport des Haufwerks wie kontinuierliches Wegfüllen, mobile Vorbrecher und Stetigförderer entschieden werden kann. Es ist zwar möglich, die für den Einsatz dieser Verfahren erforderliche Stückigkeit zu definieren, es hat sich jedoch herausgestellt, daß es in der Praxis nicht möglich ist, die Stückigkeit so genau zu messen, daß Bergingenieure Sprengverfahren und Grobzerkleinerung weiter optimieren können. Dieses Problem eignet sich zweifelsohne für rechnergestützte Analysenverfahren, vorausgesetzt jedoch, daß die Probleme der Grenzdiskriminierung überwunden werden können.

2.4 Modellvorhersagen

Unter Verwendung aller oben beschriebenen Verfahren können in Verbindung mit Feldbeobachtungen und Messungen die Modellparameter bestimmt werden. Bei mehreren normalen Sprengungen wurden Erschütterungsverlauf, Stückigkeit des Haufwerks und Bruchverteilung vor dem Sprengen gemessen; die Ergebnisse wurden für die Bestimmung kritischer Parameter verwendet. Danach wurde der Sprengplan geändert, und das Modell wurde bei zwei verschiedenen Konfigurationen zur Vorhersage der Änderung der Stückigkeit des Haufwerks benutzt. Bei der ersten Konfiguration wurde der Bohrlochdurchmesser von 165 mm auf 114 mm verringert, und der spezifische Sprengstoffverbrauch wurde durch proportionale Verringerung der Vorgabe und des Bohrlochabstandes konstant gehalten. Im zweiten Versuch wurde der spezifische Sprengstoffverbrauch durch Erhöhung der Vorgabe und des Bohrlochabstands und Beibehaltung des Bohrloch-Standarddurchmessers von 165 mm um 20 % verringert. Abbildung 2 zeigt die Übereinstimmung zwischen der beobachteten und der unter Verwendung früherer Sprengmeßwerte vorhergesagten Stückigkeit.

(Figure 2)

Abbildung 2 Anwendung des Modells auf die Vorhersage der Zerkleinerungswirkung auf echte Sprengvorgänge - Vorhergesagte und gemessene Größenverteilung

2.5 Schadensmessung und Definition

Die mechanistische Modellbildung kann erwiesenermaßen genaue Vorhersagen der Zerkleinerungswirkung liefern und läßt sich ohne weiteres auch auf die Schadensvorhersage anwenden. Die Schadenswirkung kann demnach als Änderung der In-situ-Blockgrößenverteilung aufgrund benachbarter Sprengvorgänge definiert werden und mit dem weiter oben angesprochenen "Brechindex" in Beziehung gebracht werden, d.h. die auf den Zustand vor der Sprengung bezogene Veränderung der Blockgrößenverteilung nach der Sprengung. Ob die obige Definition ausreicht, hängt vom Anwendungsbereich der Schadensmessungen ab; daraus ergibt sich die Notwendigkeit einer expliziteren Definition des Schadensbegriffes. Die obige Definition genügt für den Vergleich der Wirkungen verschiedener Sprengpläne auf die Rißbildung im First oder auf den Mehrausbruch, liefert jedoch kaum Informationen über den entstandenen Standfestigkeitsverlust, der letztendlich zu Brüchen, Erzverunreinigungen und Ausbringungsverlusten führen kann.

3. GRENZEN DER MODELLBILDUNG

Bei Prognosemodellen müssen im Gegensatz zur retrospektiven Analyse Annahmen bezüglich des Sprengstoffverhaltens und der Maßhaltigkeit der Sprengplan-Spezifikationen (Durchmesser, Genauigkeit, Anordnung der Sprenglöcher usw.) gemacht werden. Wenn keinerlei Messungen vorgenommen werden, können weder die echte Sprengleistung noch die wirklichen Gründe für die beobachtete Sprengleistung ermittelt werden.

Die derzeit verwendeten Modelle für die Felszerkleinerung mittels Sprengungen gehen von der Annahme aus, daß geplante und effektive Sprengleistung identisch sind. Diese Annahme ist, wie sich zeigen wird, einer der entscheidenden Begrenzungsfaktoren der Modellbildung und eines der Haupthindernisse für die Entwicklung von Programmen zur Optimierung der Sprengpläne. Die auf dem Gebiet der Modellbildung und der rechnerischen Behandlung erzielten Fortschritte werden zunichte gemacht, wenn die Sprenginitiierung nicht immer genau gesteuert wird. Die einfache Beobachtung der Sprengleistung und die Fähigkeit, Funktionsstörungen bei der Initialzündung nachzuweisen, können zu einer erheblich verbesserten Kostenwirksamkeit der Sprengungen führen als die Verwendung von Prognosemodellen. Obgleich es mathematisch möglich ist, Bohrlochdurchmesser, Vorgabe,

Bohrlochabstand, Sprengleistung und Länge des Besatzes einzeln stark zu variieren, kann das in der Praxis jedoch leicht zu Resonanzzündungen, Desensibilisierung der Ladung bzw. zu Bohrlochabweichungen führen, die ohne genaue Beobachtung nicht festgestellt werden können. Zündintervalle können zwar theoretisch ebenfalls stark variiert werden, doch bei der Modellbildung werden die Auswirkungen eines falsch gewählten Zündintervalls auf die Zündfolgeumkehr, die Erschütterungsverstärkung oder die Versetzung benachbarter Ladungen nicht berücksichtigt.

4. BEOBACHTUNG DER SPRENGUNGEN

Mit Hilfe von in der Nähe der Sprengstelle aufgestellten Instrumenten kann der Detonationsverlauf der Sprengladungen genau aufgezeichnet werden. Abb. 3 zeigt den Verlauf einer Einzelsprengladung, die von einem in etwa 20 m Abstand von der Ladung fest in einem Bohrloch einzementierten Geophon aufgezeichnet wurde. Der genaue Zeitpunkt der Initialzündung, der Fehler bei der Initialzündung und die Schwingungsamplitude können in dieser Aufzeichnung leicht abgelesen werden. Die Genauigkeit der Initialzündung liefert Hinweise auf die Zeitfolgesteuerung, auf Zündversager und auf die Verzögerungsstreuung, während die Schwingungsamplitude Informationen über Sprengstoffleistung und Brechenergie liefert.

Abbildung 4 zeigt den Verlauf einer Sprengung mit Mehrfachverzögerung sowie die Ladungsverteilung. In diesem Beispiel wurden Ladungen mit Zwischenbesatz, d.h. mehrere Ladungen pro Bohrloch, verwendet; das Diagramm zeigt ferner die Soll- und Ist-Werte des Initiierungszeitpunktes jedes Zwischenbesatzes sowie die deutlich erkennbare Streuung der Initiierungszeitpunkte bei den beiden 300-ms-Ladungen. Trotz der Verzögerungsfehler liegt die Sprengleistung innerhalb der vorgesehenen Grenzen, und bei allen Ladungen war die Initialzündung von heftigen Schwingungsreaktionen begleitet.

(Figure 3)

Abb. 3 In etwa 20 m Abstand vom Bohrloch aufgezeichneter Schwingungsverlauf bei einer Einzelladung

(Figure 4)

Abb. 4 Schwingungsverlauf bei einer mehrfach verzögerten Sprengung

4.1 Funktionsstörungen bei der Initialzündung

Abbildung 5 zeigt drei der häufigsten Versagertypen bei Sprengungsarbeiten in Australien, Kanada, Chile und in den USA. Die Versagerquote wird stark von der Komplexität des Sprengplans, der Anzahl der verzögerten Ladungen, der Dauer des Sprengvorgangs und der geologischen Struktur beeinflußt. Probleme bei der Initiierung, mangelnde Zerkleinerungsleistung und erheblicher Mehrausbruch sind in erster Linie auf Überladungen zurückzuführen. Eine Verbesserung der Zerkleinerungsleistung kann am einfachsten durch Erhöhung des spezifischen Sprengstoffverbrauchs erreicht werden; das kann allerdings zu einer höheren Versagerquote und zu einem verstärkten Mehrausbruch führen, der je nach Zerklüftungsgrad des Gesteins zu grobstückiger Zerkleinerung tendiert.

Abbildung 6 zeigt die Häufigkeit von Schußversagern und anderen Zündproblemen: Von 20 Ladungen haben lediglich 9 planmäßig gezündet. Bei 20 % der Ladungen gab es Zündversager, in 35 % der Fälle fanden nach 5 und 8 ms Resonanzdetonationen statt, die zwei stark verdichtete Schwingungsimpulsbündel hervorgerufen haben.

Wenn die Ursache der Funktionsstörungen erkannt ist, können diese auch behoben werden; dadurch bekommt man die Steuerung der Initialzündung wieder voll in den Griff, und der Sprengstoffverbrauch wird in allen Fällen gesenkt. Abbildung 7 zeigt eine homogene Sprengung, die durch Modifikation des für die Aufzeichnung in Abbildung 6 zugrunde liegenden Sprengplans erzielt wurde. Schußversager konnten ausgeschaltet werden, die Anzahl kleiner Ladungen mit Zwischenbesatz wurde vermindert, der spezifische Sprengstoffverbrauch wurde um etwa 50 % verringert, und auch Schwingungsniveau, Schadensumfang und Mehraushub konnten reduziert werden.
Eine vierte weitverbreitete Funktionsstörung ist, besonders bei Verwendung von Sprengstoffen auf Emulsionsbasis, auf eine verminderte Sprengleistung zurückzuführen. Bei solchen Sprengstoffen wurden wiederholt Schwankungen im Vibrationsniveau, im Zerkleinerungsgrad und in der Haufwerkverschiebung festgestellt; auch Schußversager waren recht häufig.

Diese Fakten scheinen darauf hinzuweisen, daß die sprengtechnischen Kennwerte bei der Detonation nicht immer den Nenn- bzw. Maximalwerten entsprechen, und folglich lediglich eine erheblich verminderte Brechenergie zur Verfügung steht. Sprengstoffe auf Emulsionsbasis können nur dann kostenwirksam eingesetzt werden, wenn sie stets mit Nennleistung detonieren.

Angesichts des Ausmaßes der Initiierungs-Probleme, denen man bei Gewinnungs- sowie bei Aufschluß- und Tunnelsprengungen begegnet, scheinen Meßmethoden zur raschen Identifizierung der Probleme bei der Optimierung von Sprengplänen bessere Anwendungsmöglichkeiten zu haben als die Modellbildung. Durch Beobachtung der Sprengleistung konnte der Sprengstoffverbrauch im Schnitt um 25 %, in vielen Fällen sogar um etwa 50 % verringert werden, wobei gleichzeitig die Stückigkeit, die Erschütterungen und der Schadensumfang besser beeinflußt werden konnten.

5. BEWERTUNG DER SPRENGPLÄNE

Die Beobachtung der Sprengleistung hat mit aller Deutlichkeit gezeigt, wie wichtig eine straffe Kontrolle über Verteilung und Konzentration des Sprengstoffs ist. Auch bei einem geringen durchschnittlichen spezifischen Sprengstoffverbrauch kann es jedoch zu örtlichen Bruchbildungen kommen, die auf überladene Schüsse hinweisen. Mit Hilfe einfach zu entwickelnder Computer-Algorithmen können die mit der Planung der Schießarbeit befaßten Ingenieure die Bohrlöcher und Ladeschemas besser festlegen als mit den bisherigen Verfahren.

Abbildung 8(a) zeigt beispielsweise, wie die Sprengstoff-Konzentration in einem Schußabschnitt auf über den doppelten Durchschnittswert erhöht werden konnte. In diesem Falle wurde die Konzentration durch eine Vergrößerung des Bohrlochdurchmessers von 70 mm auf 115 mm verursacht, die vorgenommen wurde, um eine größere Bohrlochtiefe ohne Verlust an Bohrgenauigkeit zu erzielen. In diesem Schußabschnitt kommen Resonanzdetonationen häufiger vor, was wiederum Mehrausbruch und größere Schäden zur Folge hat. Abbildung 8(b) zeigt, wie diese Ladungskonzentration durch Verwendung von Sprengstoffen mit geringerer Sprengwirkung in großkalibrigen Bohrlöchern verringert werden kann.

Auch bei Fächerschießen unter Tage wurde durch Veränderung der Ladesäulen eine Verringerung des spezifischen Sprengstoffverbrauchs um über 10 % erzielt.

Solche einfachen Verfahren zur Konzeptionsbewertung bieten ferner Möglichkeiten zur Sprengplan-Optimierung, die den Anwendungsbereich gegenwärtiger Modelle überschreiten. Mit künftigen Modellen sollte es möglich sein, den Punkt zu definieren, bei dem es zu einer Resonanzdetonation bzw. zu einem Erlöschen der Detonation kommen kann.

(Figure 5)

Abb. 5 Typische Störungen bei der Initiierung von Detonationen (Beispiele aus der Praxis)

(Figure 6)

Abb. 6 Möglicher Umfang von Schußversagern (Beispiel aus der Praxis - Sprengung unter Tage)

(Figure 7)

Abb. 7 Homogene Initiierung von Ladungen ohne Schußversager (Beispiel aus der Praxis - Sprengung unter Tage)

(Figure 8(a))

Abb. 8(a) Durch Vergrößerung des Bohrlochdurchmessers verursachte Sprengstoffkonzentration während des Ladens

(Figure 8(b))

Abb. 8(b) Gleichmäßige Sprengstoffverteilung durch Anpassung der Sprengleistung in großkalibrigen Bohrlöchern

6. VERZÖGERUNGSSTREUUNG

Bei pyrotechnischen Zündmitteln werden Streuungen der Verzögerungswerte schon immer in Kauf genommen; dennoch werden sie bei der Ausarbeitung von Bohr- und Schießplänen und bei Modellbildungen nur selten berücksichtigt. Mehrere tausend nichtelektrische Verzögerungen wurden ausgewertet, um genaue Streumuster zu ermitteln, wobei die Verzögerungen jeweils elektronisch gesteuert und zum Teil auch mit Hilfe von Kurzzeitaufnahmen unterbaut wurden. Die Auswertung erfolgte mit Produkten aller australischer Lieferanten und mit fast allen Intervallzeiten, von 25 ms bis etwa 10 s.

Die Ergebnisse zeigen, daß die Streuwerte innerhalb eines einzelnen Loses sehr gering (etwa 3 bis 4 %), von Los zu Los jedoch wesentlich größer sind. Das bedeutet, daß bei Verwendung eines einzigen Loses und eines einzigen Verzögerungsintervalls die Streuung der Verzögerungswerte mit 3 oder 4 % als minimal betrachtet werden kann, während dieser Wert je nach Produkt erheblich höher sein kann (bis zu 13 %), wenn unterschiedliche Lose und Verzögerungsintervalle verwendet werden. Abbildung 9 zeigt die Streuung der Verzögerungswerte für zwei Lose mit Verzögerungsintervallen von 500 ms. Bei Mischung beider Lose ist der Gesamtstreuwert mindestens doppelt so hoch als der Streuwert jedes einzelnen Loses. Nach Auswertung von bis zu 40 verschiedenen Losen wurde festgestellt, daß die Streuung der Verzögerungswerte bei der Mischung von Losen je nach Produkt bis zu dreimal so hoch sein kann. Für den Zweck dieser Untersuchung wird für die Beschreibung der Verzögerungsintervall-Verteilung ein Wert von - 13 % angenommen, d.h. daß 95 % der Schüsse innerhalb einer Verzögerungsspanne von - 13 % des Nennwertes detonieren. Es wurde experimentell nachgewiesen, daß diese Definition der Streuung des Verzögerungswertes als realistisch betrachtet werden kann.

6.1. Einfluß der Verzögerungswert-Streuung auf den Bohr-und Schießplan

Wenn die Verzögerungswert-Streuung bekannt ist, kann auch die Wahrscheinlichkeit einer Zündfolgeumkehr beurteilt werden. Dieses Problem geht ebenfalls in die Bewertung des exakten Steuerbarkeit-Niveaus der Initiierung eines Sprengplanes ein.

Verzögerungen werden in der Sprengtechnik zu zweierlei Zwecken eingesetzt:
1. um sicherzustellen, daß Ladungen nicht gleichzeitig detonieren;
2. um die augenblickliche Initiierung von Ladungen zu unterstützen.

Durch Verwendung der für die Beschreibung der Verzögerungsstreuung gewonnenen Werte kann beurteilt werden, wie groß die Wahrscheinlichkeit ist, diese Ziele in einer gegebenen Situation zu erreichen. Abbildung 10 zeigt beispielsweise einen Aufschluß- oder Tunnelabschlag mit einer für einen Untertageschuß typischen Anordnung der Bohrlöcher und der Zündfolge. Bei der Untersuchung wird insbesonders untersucht, wie hoch die Wahrscheinlichkeit ist, daß eines der folgenden Ereignisse eintritt:
1. Die Kranzlöcher mit gleicher Zündnummer unterstützen sich gegenseitig dergestalt, daß ein gutes "Kommen" des Abschlags und glatte Stollenwände erreicht werden,
2. die Kranzlöcher detonieren erst dann, wenn die Löcher in der Streckenmitte bereits detoniert und freigemacht sind, wodurch die Beschädigung der Stollenwände und des Nebengesteins auf ein Minimum beschränkt bleiben.

Damit benachbarte Bohrlöcher sich gegenseitig so unterstützen können, daß eines gutes "Kommen" des Abschlags sichergestellt ist, müssen sie in einem kritischen Zeitabstand zünden, der von der Rißausbreitungsgeschwindigkeit des Gesteins abhängt. Wenn wir davon ausgehen, daß die Ausbreitungsgeschwindigkeit der P-Welle etwa 6000 m/s beträgt, die Rißausbreitungsgeschwindigkeit etwa ein Drittel der Geschwindigkeit der P-Welle beträgt, und daß der Abstand der Bohrlöcher nicht größer als 1 m ist, dann müssen benachbarte Löcher in nicht mehr als 0,5 ms Abstand zünden, damit sich die Ladungen gegenseitig unterstützen und der Abschlag gut kommt. Abbildung 11 zeigt bei einer angenommenen Verzögerungsstreuung von 13 % die Wahrscheinlichkeit einer gegenseitigen Unterstützung. Bei Kranzbohrlöchern mit hoher Zeitstufe ist die Wahrscheinlichkeit, daß benachbarte Bohrlöcher innerhalb von 0,5 ms zünden, praktisch gleich Null, und daß sie innerhalb von 5 ms zünden, geringer als 2 %. Deshalb wird es im Tunnelbau nie zu glatten Abschlägen kommen, wenn langperiodische Zünder mit hohen Zündzahlen eingesetzt werden.

Eine ähnliche Untersuchung glatter Abschläge an Streckenwänden bei Verwendung von Millisekundenzündern zeigt Abbildung 12, in der das Kranzloch ein zwischen den Reihen am Streckenkranz angeordnetes Loch ist. Um einen glatten Abschlag zu fördern, sollte dieses Loch im Idealfall gleichzeitig mit dem Kranzloch der dahinter liegenden Reihe zünden. Es hat sich gezeigt, daß die Wahrscheinlichkeit eines solchen Effekts bei Verwendung von 600-ms-Verzögerungen bei etwa 3 % liegt und bei 100-ms-Verzögerungen auf nicht mehr als 10 % ansteigt.

Es wurde ebenfalls untersucht, wie hoch die Wahrscheinlichkeit einer Zündfolgeumkehr bzw. einer Zündung der Kranzlöcher vor einer Zündung der benachbarten Bohrlöcher ist. Abbildung 13 zeigt die Wahrscheinlichkeitsfunktion für eine mögliche Überlappung der Verzögerungen, der eine Verzögerungsstreuung von etwa 13 % zugrunde gelegt wurde. Untersucht wurde die Wahrscheinlichkeit einer Überlappung sukzessiver Zündnummern. Es hat sich herausgestellt, daß die Wahrscheinlichkeit einer Überlappung bei Verzögerungsintervallen von ca. 6 s etwa 14 % beträgt. Es besteht somit eine relativ hohe Wahrscheinlichkeit, daß ein Kranzloch vor einem zur Streckenmitte hin gelegenen Bohrloch zündet, was zur Beschädigung des Nebengesteins, zu unregelmäßigen Strecken und zu erhöhten Grubenausbaukosten führen kann. Die Wahrscheinlichkeit eines Überlappens ist praktisch Null, wenn darauf geachtet wird, daß zwischen den Kranzlöchern und den Bohrlöchern in Streckenmitte mindestens zwei Zeitintervalle liegen.

Durch Verwendung von Ladungen mit unterschiedlichen Zeitverzögerungen ist jedoch nicht unbedingt sichergestellt, daß die Ladungen auch wirklich separat bzw. in der geplanten Reihenfolge zünden, und zwar besonders dann nicht, wenn die Standardintervallzeit durch Kombinieren von Zündstufen verkürzt wurde. Sprengmodelle müssen in erster Linie den Einfluß von Zündstufen auf die Stückigkeit und erst in zweiter Linie den Einfluß von Initiierungen außerhalb der Zündfolge auf Stückigkeit und Schaden beschreiben können.

6.2 Einfluß der Verzögerungsstreuung auf die Schwingungsdämpfung

Einer der Hauptgründe für die Verwendung von Zeitzündern in der modernen Sprengtechnik ist die Dämpfung und Minimierung von Schwingungen und von schwingungsbedingten Schäden. Zu diesem Zweck wird bei Langlochsprengungen jedes Bohrloch mit Mehrfachschüssen besetzt, wobei die Ladungen durch inertes Besatzmaterial voneinander getrennt und einzeln gezündet werden. Damit die erwartete Abschlagleistung pro Schuß auch effektiv erzielt werden kann, sind zahlreiche Zündstufen erforderlich. Da Verzögerungsreihen auf höchstens 40 Zündstufen beschränkt sind, werden in der Praxis die verfügbaren Serien durch Kombinieren von Zündstufen erweitert; dazu wird eine Zündstufe am Bohrlochmund angebracht, von der aus eine Zündschnur zu einer weiteren im Bohrlochtiefsten angeordneten Zündstufe führt. Dadurch wird zwar die Anzahl verfügbarer Zündstufen erhöht, die Intervallzeiten werden jedoch verkürzt, wodurch sich wiederum die Wahrscheinlichkeit eines Überlappens aufeinanderfolgender Zündstufen erhöht.

Alle derzeit zur Verfügung stehenden Schwingungsvorhersage-Gleichungen beziehen sich auf das maximale Ladungsgewicht pro Zündstufe, sagen jedoch wenig über das Minimalintervall der Verzögerungen aus. In der Praxis ist dieses Intervall ausschließlich von der Gesteinsart abhängig; bei Gesteinen mit hohem Elastizitätsmodul ist es relativ kurz und bei Gestein mit geringem Plastizitätsmodul wesentlich länger. Die Zeitspanne kann mit dem induzierten Schwingungsverhalten des Gesteins in Bezug gebracht werden. Das minimale Schwingungsintervall ist die Zeit zwischen Detonation und Rückkehr des Gesteins zu dem in Abbildung 3 dargestellten, nahezu schwingungsfreien Zustandes. Schwingungen, die durch die Detonation von Einzelladungen hervorgerufen werden, werden durch andere Ladungen nicht beeinflußt, und die Schwingungsniveaus werden gut beherrscht.

Die aufgrund der vorstehenden Überlegungen mögliche Vorhersage des Erschütterungsniveaus wird dann dazu benutzt, die Wahrscheinlichkeit einer Schwingungsverstärkung durch verschiedene Ladungen zu prüfen, und zwar, indem untersucht wird, wie groß die Wahrscheinlichkeit ist, daß Ladungen aufgrund von Verzögerungsstreuungen innerhalb eines gegebenen Zeitintervalls initiieren, der durch Beobachtung der Bodenerschütterungen mit Hilfe von Meßinstrumenten, mit denen der gesamte Schwingungsvorgang an jeder beliebigen Stelle aufgezeichnet werden kann, ermittelt wird. Mit dieser Methode konnten beispielsweise Schießpläne für Sprengungen mit zahlreichen gestreckten Zeitzünder-Ladungen vereinfacht werden. Probabilistische Untersuchungsverfahren wurden auf Schießpläne angewandt, um Vergleiche zwischen der Verwendung einer größeren Anzahl gestreckter Ladungen mit kurzen Verzögerungszeiten und der Verwendung einer kleinen Anzahl gestreckter Ladungen (mit jeweils größeren Lademengen) anzustellen. In vielen Fällen hat sich gezeigt, daß Ladungen mit höherer Ladungsmenge und größeren Verzögerungsintervallen geringere Bodenerschütterungen hervorrufen (bessere Schwingungsdämpfung) und den Ladevorgang vereinfachen und beschleunigen.

(Figure 9)

Abbildung 9 Streuung in der Intervallzeit bei verschiedenen Zünderlosen

(Figure 10)

Abbildung 10 Schießplan und Zündfolge für eine Aufschlußsprengung mit Bohrlochdurchmessern von 52 mm

(Figure 11)

Abbildung 11 Wahrscheinlichkeitsfunktion, die die Möglichkeit einer konstruktiven Unterstützung benachbarter Kranzlöcher bei einer Aufschlußsprengung beschreibt.

(Figure 12)

Abbildung 12 Wahrscheinlichkeitsfunktion, die die Möglichkeit einer konstruktiven Unterstützung benachbarter Kranzlöcher bei Sprengungen im versatzlosen Abbau beschreibt.

(Figure 13)

Abbildung 13 Wahrscheinlichkeitsfunktion, die die Möglichkeit der Überlappung sukzessiver Zündnummern beschreibt, die bei Kammersprengverfahren zu Initiierung der Wandlöcher außerhalb der Zündfolge führen können

7. SCHÄDEN DURCH GASPENETRATION

Beobachtungen von Sprengvorgängen haben gezeigt, daß sich hochgespannte Explosionsschwaden vom Bohrloch aus entlang von Rissen nach allen Richtungen hin ausbreiten. Die vordere Stoßfront hat eine schiebende Wirkung und fördert das Anschwellen des Haufwerks, während die hintere Stoßfront ein Anschwellen hinter dem Schuß bewirkt, was zu einem Reibungsverlust entlang der Bruchflächen und somit zu Stabilitätsverlusten führt. Die Gaseffekte hängen stark von der Klüftigkeit des Gesteins ab.

Mit Hilfe einfacher Versuche konnte der Verlauf der Gasströmungen hinter einem Schuß nachgewiesen und Aufschlüsse über die Ausbreitungsgeschwindigkeit und den zeitlichen Verlauf der Haufwerkbewegung in Abhängigkeit von den Kennwerten der Sprengung gewonnen werden. Bei diesen Versuchen wurden hinter den eigentlichen Bohrlöchern zusätzliche Bohrlöcher angesetzt. In diese Bohrlöcher wurden Druckgeber eingesetzt, und der Bohrlochmund wurde dann mit einem Betonstopfen verschlossen. In unmittelbarer Nähe des Bohrlochs, oder sogar im Bohrloch selbst, wurde ein Schwingungsmesser angebracht, mit dessen Hilfe der genaue Initiierungszeitpunkt der Ladung im benachbarten Bohrloch bestimmt wurde.

Abbildung 14 zeigt den Druckverlauf in einem Bohrloch in etwa 6 m Abstand von einer Reihe von Einbruch-Bohrlöchern. Trotz Schwingungskompensation sprach der Meßgeber auf die Detonation des nächstliegenden Bohrlochs leicht an; diesem schwachen Ausschlag folgte ein steiler Druckanstieg, dem sich ein allmählicher Druckabfall anschloß. Bei diesem Versuch hat sich die hochgespannte Druckwelle nicht entlang der natürlichen Bruchlinien, sondern entlang der Einbruch-Bohrlöcher, die sich bis hinter das letzte Bohrloch erstreckten, ausgebreitet.

In einem weiteren Versuch wurden Gasdruck und Erschütterungen hinter einer normalen Abraumsprengung in einem Kohletagebau gemessen. Abbildung 15 zeigt die dabei aufgezeichneten Schwingungsbilder. Die Schüsse wurden in fünf parallel zur Freifläche gebohrten Reihen abgetan; auf dem mit einem Radialmeßgeber aufgezeichneten Schwingungsbild vom Bohrloch BM 11 sind die von jeder Schußreihe hervorgerufenen Erschütterungen klar zu erkennen. Diese Ausschläge sind auch auf den beiden anderen Aufzeichnungen des Gasdruckverlaufs festzustellen, was darauf hinweist, daß die piezoelektrischen Druckgeber nicht einwandfrei schwingungskompensiert sind. Die Aufzeichnung des Gasdruckverlaufs von Bohrloch GM 7 zeigt ebenfalls sehr deutlich, daß sich dem Druckanstieg ein Druckabfall bis zu einem negativen Bohrlochdruck anschließt. Etwa 1 s nach Beginn des Einströmens des Gases in das Bohrloch hat sich der Umgebungsdruck wieder auf den ursprünglichen Wert eingependelt. Für die Gasausbreitungsgeschwindigkeit wurden bei zahlreichen Versuchen Werte von etwa 8 bis 12 m/s ermittelt.

Die negativen Druckwerte deuten auf eine Vergrößerung des Bohrlochvolumens hin, die wahrscheinlich auf eine Druckentlastungs-Relaxation zurückzuführen ist, die in dem Augenblick eintritt, in dem sich die Vorgabe vor dem letzten Bohrloch von der Strossenböschung löst. Messungen der Vorgabe-Bewegungszeiten, die auf dieser Annahme beruhen, zeigen, daß von der Detonation der Ladung bis zur ersten Vorgabe-Bewegung etwa 90 ms vergehen.

Gaspenetrationen wurden bis zu 8 m hinter dem Sprengpunkt gemessen und werden als eine der wichtigsten Ursachen für Schäden und Stabilitätseinbuße betrachtet. Die Gaspenetration läßt sich auf folgende Weise minimieren:

1. Minimierung der Druckbeaufschlagungsdauer (Maximierung der Vorgabebewegungsgeschwindigkeit);
2. Minimierung des maximalen Bohrlochdrucks;
3. Schaffung von Gasabzugsöffnungen, z.B. durch Vorzerklüftung.

Es wird angenommen, daß die Reduzierung der Dauer des Gasdrucks die geeignetste Form der Beherrschung der Gaspenetration ist. Vorgabe-Bewegungsgeschwindigkeiten können ohne Erhöhung der Spitzendrücke im Bohrloch durch Reduzierung des Vorgabevolumens erhöht werden. Damit die sprengbedingten Schäden auf ein Mindestmaß reduziert werden können, muß bei der Ausarbeitung des Sprengplans folgendes berücksichtigt werden:

1. Sprengen mit wenigen Bohrlochreihen, um eine unbehinderte Vorgabebewegung zu fördern;
2. Geringes Vorgabevolumen;
3. Vorzerkleinerung entlang kritischer Grenzen.

Mit Hilfe von Modellen sollte es letztendlich möglich sein, Schadenswirkung und Stückigkeit vorauszusagen. Für beide Faktoren ist der Zerklüftungsgrad des betrachteten Gesteins von entscheidender Bedeutung. Der Gasdruck spielt im Hinblick auf die endgültige Form des Haufwerks und auf den Umfang des gasdruckinduzierten Schadens eine entscheidende Rolle. Modelle müssen deshalb so beschaffen sein, daß mit ihrer Hilfe die bei einer Sprengung erzeugte Gasströmung sowie das Gasvolumen und der Gasdruck vorhergesagt werden können.

(Figure 14)

Abbildung 14 Verlauf des Gasdrucks in einem ungeladenen Bohrloch in 6 m Abstand von einer Reihe von Vorzerklüftungs-Bohrlöchern

(Figure 15)

Abbildung 15 Verlauf des Gasdrucks in einem ungeladenen Bohrloch hinter einem Gewinnungssprengen in stark zerklüftetem Gestein

8. SCHLUSSBEMERKUNG

Mathematische Modelle haben wesentlich zum besseren Verständnis einiger sprengungsinhärenter Mechanismen beigetragen. Im Gegensatz zu anderen Anwendungen mathematischer Modelle wurden Sprengmodelle ohne die für einen Beweis der Vorhersagen bzw. eine Bestätigung der wichtigsten Mechanismen erforderlichen Beobachtungs- und Meßverfahren entwickelt. Es wurde jeweils vorausgesetzt, daß Sprengstoffe und Zündmittel mit der geplanten bzw. angegebenen Genauigkeit funktionieren, und der Beurteilung von Faktoren, die die optimale Leistung negativ beeinflussen, wurde wenig Aufmerksamkeit geschenkt. Entscheidende Faktoren wie Stückigkeit und Schäden können in großem Maßstab nicht genau gemessen werden. Das Fehlen der elementarsten im Bergbau einsetzbaren Meßgeräte und -verfahren ist einer der Hauptgründe für die begrenzte Anwendung und Weiterentwicklung von Sprengmodellen. Für die Beobachtung der Leistung von Sprengladungen stehen jedoch die erforderlichen Instrumente zur Verfügung; diese bilden die Minimalausrüstung, wenn sichergestellt werden soll, daß die gewünschte bzw. vorhergesagte Sprengleistung auch wirklich den im Sprengplan vorgesehenen Werten entspricht.

09:20 Kurzvorträge

Bonapace, B., Österreich
Leistung und Grenzen des Fräsvortriebes beim Ausbruch eines 22 km langen Druckstollens

Brych, J., Ngoi Nsenga, Xiao Shan, Belgium, Zaïre, China
Die Zerstörbarkeit der Gesteine im Drehbohren

Farmer, I.W. Garrity, P., England
Vorhersage der Schnittleistung von Streckenvortriebs-maschinen aus Bruchzähigkeitsbetrachtungen

Ginn Huh, Kyung Won Lee, Han Uk Lim, Korea
Bestimmung der Sprengerschütterungen auf empirische Weise

10:00 - 10:30 Pause (Foyer)

Goto, Y., Kikuchi, A., Nishioka, T., Japan
Dynamisches Verhalten von Tunnelauskleidungen bei benachbarten Sprengungen

Grant, J.R., Spathis, A.T., Blair, D., Australien
Eine Untersuchung über den Einfluss von Sprengladungslänge auf Sprengschwingungen

Heraud, H., Rebeyrotte, A., Frankreich
Versuchvorspalten und Auswirkungsmessung im Granit des französischen Zentral- Massivs

Inazaki T., Takahashi, Y., Japan
Qualitätsbewertung von Gesteinsmassen durch seismische Tomographie

Maerz, N.H., Franklin, J.A., Rothenburg, L., Coursen, D.L., Kanada, U.S.A.
Haufwerkverteilung durch digital photographishe Analyse

Miranda, A.M., Mendes, F.M., Portugal
Gesteinsverwitterung als ein Bohrbarkeitsparameter

Mitani, S., Iwai, T., Isahai, H., Japan
Beziehungen zwischen den Gebirgsverhältnissen und der Ausführbarkeit der TBM

Singh, R.D., Virendra Singh, Khare, B.P., Indien
Optimierung von Sprengungen in einem geklüfteten mächtigen Kohleflöz

Wilson, W.H., Holloway, D.C., U.S.A.
Sprengversuche in instrumentierten Betonmodellen

12:00 - 13:30

Mittagspause und Postersitzung (Marquette)

13:30 - 15:00 Diskussion (Le Grand Salon)

Diskussionsleiter: P.A. Lindqvist (Schweden)
Podiumsdiskussion: R.F. Favreau (Kanada)
O.T. Blindheim (Norwegen)
C.D. da Gama (Brasilien)

Gesteinssprengen und mechanischer Vortrieb - III

Zusammenfassung des Diskussionsleiters

P. A. Lindqvist (Schweden)

1. EINLEITUNG

Der Themenbereich III des Sechsten Internationalen Kongresses der Felsmechanik befaßt sich mit dem Thema Sprengarbeiten und Ausbruch. In das Berichtwerk wurden insgesamt 30 Arbeiten zu den Themen Bohrarbeit, mechanischer Gesteinsaushub und Sprengen aufgenommen. In Tabelle 1 sind die in den Referaten behandelten Fachgebiete zusammengefaßt.

Tafel 1 - Behandelte Fachgebiete - Themenbereich III

Bohren	Vorgelegte Arbeiten
Bohren, allgemein	3
Hammer-/Schlagbohren	2
Drehbohren	1
Erdölbohrung	2
	8
Mechanischer Ausbruch	
Tunnelbohrmaschine	3
Streckenvortriebsmaschine	4
Blindschachtabteufung	1
Statischer Brecher	1
	9
Sprengen	
Gesteinscharakterisierung	1
Sprengen, allgemein	4
Tagebauliches Sprengen	3
Vorzerklüftung	1
Tunnelvortrieb, versch. Aspekte	4
	13

Wie man sieht, ist die Verteilung auf die drei Fachgebiete recht ausgervogen.

Aufschlußreicher ist vielleicht ein Blick auf die von den verschiedenen Referenten behandelten Probleme. In Tafel 2 wurden die zu den Themen Bohrarbeit und mechanischer Ausbruch eingereichten Arbeiten in vier Hauptgruppen unterteilt.

Tafel 2 - Themenwahl - Bohrarbeiten und mechanischer Ausbruch

	Anzahl der Vorträge
Vorhersage der Bohrleistung/Leistung/Durchführbarkeit	
Bernaola und Oyanguren	
Howarth und Rowlands	
Miranda und Mendes	
Farmer und Garrity	
Bonapace	
Mitani, Iwai und Isahai	6
Bohrerentwicklung/Bohrstangen/Ausrüstung	
Brysch, Nsenga und Shan	
Brighenti und Mesini	
Nishimatsu, Okubo und Jinno	
Xu, Tang und Zou	4
Bohrlochstabilität	
Kaiser und Maloney	1
Schneidmechanismen/Wasserstrahl	
Hood, Geier und Xu	
Iihoshi, Nakao, Torii und Ishii	
Ip und Fowell	
Kutter und Jütte	
Sekula, Krupa, Koci, Krepelka und Olos	
Fukuda, Kumasaka, Ohara und Ishijima	6
	17

In Tafel 3 wurden die von den Referenten behandelten Sprengprobleme zu 6 Themenkreisen zusammengefaßt. Bemerkenswert ist, daß keine Arbeiten über Computermodelle für Sprengvorgänge und empirische Zerkleinerungsmodelle vorgelegt wurden.

Tafel 3 - Themenwahl - Sprengen

Sprengtheorie	
Singh und Sastry	
Wilson und Holloway	2
Bohr- und Schießpläne	
R. D. Singh, V. Sing und Khare	1
Schwingungskriterien	
Fadeev, Glosman, Kartuzov und Safonov	
Huh, Lee und Lim	
Goto, Kikuchi und Nishioka	
Grant, Spathis und Blair	
Siskind	5
Vorzerklüftung	
Heraud und Rebeyrotte	1
Messung der Stückigkeit	
Maerz, Franklin, Rothenburg u. Coursen	1
Gesteinscharakterisierung/Kostenmodelle	
Inazaki und Takahashi	
Pöttler und John	
Mikura	3
	13

Zusammentassend kann angesichts der im Rahmen des Themenkreises III vorgelegten Referate festgestellt werden, daß auf diesem Gebiet in der ganzen Welt aktive und erfolgreiche Forschungs- und Entwicklungsarbeit geleistet wird. Erwähnt werden müssen einige wichtige Tendenzen, die sich seit dem letzten Felsmechanik-Kongreß ergeben haben. Dazu gehört der zunehmende Einsatz von Mittelhochdruck-Wasserstrahlverfahren zur Unterstützung von Streckenvortriebsmaschinen sowie die zunehmende Festlegung von Sprengschwingungskriterien, die auf örtlichen Annahmen beruhen, jedoch häufig von internationalen Erfahrungen beeinflußt sind. Zu den im Rahmen des Themenkreises III vorgelegten Arbeiten gehören auch einige wichtige Fallstudien, die einen Beitrag zur langfristigen Entwicklung von Spreng- und Aushubtechnologie leisten.

2 SCHLAGLICHTER

Angesichts des hohen Gesamtniveaus fällt es schwer, einzelne Arbeiten für eine Besprechung auszuwählen. Einige Referate verdienen es jedoch, besonders hervorgehoben zu werden.

Das Referat "Fragmentation studies in instrumented concrete models" (Fragmentationsuntersuchungen an instrumentierten Betonmodellen) von W.H. Wilson und D.C. Holloway beschreibt die Verwendung verschiedener Beobachtungstechniken, mit deren Hilfe neue und sichere Schlüsse gezogen werden können. Dieses Referat gehört zu den wichtigsten Arbeiten und leistet meines Erachtens einen wichtigen Beitrag zum Verständnis der Zerkleinerungs-Mechanismen.

In der von J. R. Grant et al. vorgelegten Arbeit "An investigation of the influence of charge length upon blasting vibrations" (Untersuchung des Einflusses der Ladungslänge auf die Bodenerschütterungen bei Sprengarbeiten) wird die Möglichkeit aufgezeigt, Detonations-Abrisse festzustellen, die derzeit zu den wichtigen Problemen im Untertagebau gehören.

Das Referat von N. H. Maerz et al. "Measurement of rock fragmentation" (Messung der Gesteinszerkleinerung) veranschaulicht den Einsatz von Computern für die innovative Lösung schwieriger bergbautechnischer Probleme.

In der von J. Byrch et al. vorgelegten Arbeit "Destructibility of rocks with rotation drilling bits" (Gesteinszerstörbarkeit bei Verwendung von Drehbohrern) wurden die wertvollen Ergebnisse mehrerer Diplom- und Doktorarbeiten zusammengefaßt.

3. AUSBLICK

Wenn von der Zukunft die Rede ist, müssen verschiedene Aspekte berücksichtigt werden, z.B. die betrachtete Zeitspanne. Für die Erörterung der Frage zukünftiger Forschungsarbeiten habe ich einen relativ kurzen Zeitraum gewählt. Nun läßt sich darüber streiten, ob sich Wissenschaftler überhaupt mit kurzfristigen Prognosen befassen sollten. Ich bin jedoch der Meinung, daß Entwicklungen bevorstehen, die die Wahl der Forschungsrichtungen entscheidend beeinflussen werden.

Bevor eine Forschungsarbeit in Angriff genommen wird, muß zunächst der Forschungsbereich abgesteckt werden. Anschließend muß das wissenschaftliche Problem festgelegt werden. Als letzter Schritt folgt die Auswahl der anzuwendenden Verfahren: Laborversuche, Feldversuche oder Computermodelle. Idealerweise sollten alle drei Verfahren zur Problemlösung eingesetzt werden, was jedoch gewiß nicht immer der Fall ist.

Im folgenden werden einige wichtige technische Entwicklungen und ihre Folgen aufgezeigt. Im Anschluß daran folgt eine Aufzählung noch zu lösender Forschungsaufgaben.

Künftige Entwicklung	Folgen
Neue Werkstoffe Mittelhochdruck-Wasserstrahl	Senkung der Bohrkosten durch Einsatz von Hochgeschwindigkeitsbohrern und Hochleistungsgeräten
Hochleistungsgeräte	Einsatz von Tunnelbohr- und Streckenvortriebsmaschinen für Aufschluß und Produktion in Hartgestein
	Hoher Verfügbarkeitsgrad der Geräte ist von entscheidender Bedeutung
Untertagebau und untertägige Ausbrucharbeiten in größerem Maßstab	Verwendung großerkalibriger Bohrlöcher (Ø 50 - 100 mm, Länge 10 - 50 m) auch in kleineren Betrieben
Zunehmender Einsatz von Automaten und Robotern	Größere Bohr- und Sprenggenauigkeit
Verstärkter Einsatz von Überwachungsverfahren und Bohrlochmessungen Datenverarbeitung	Erhebliche Zunahme bei der Erfassung, Auswertung, Speicherung und Aufbereitung von Geodaten
Verwendung von Präzisions-Sprengkapseln und von pumpfähigen Sprengstoffen mit variabler Sprengkraft	Erhebliche Zunahme der Gestaltungsmöglichkeiten für Bohr- und Sprengpläne

4. FORSCHUNGSZIELE

Unter Berücksichtigung künftiger Entwicklungstendenzen und der sich daraus ergebenden Folgen werden folgende Forschungsziele vorgeschlagen:

4.1 Bohren und mechanischer Aushub

Fallstudien mit Schwerpunkt auf neuen Erkenntnissen.

Entwicklung von Bohrklein-Abführungsgeräten.
Entwicklung von Vorhersagemodellen für Streckenvortrieb in Hartgestein.

Entwicklung von Schnellverfahren zur Messung von Bohrlochlängen und -abweichungen, Entwicklung von Modellen und Verfahren zur Reduzierung von Bohrlochabweichungen.

Entwicklung neuer Methoden zur Gewinnung von Geodaten:

Bohrbegleitende Messungen für Stoßbohrungen einschließlich Modellbildung.

Bohrbegleitende Entnahme von Bohrklein-Proben sowie chemische und physikalische Analyse einschließlich Modellbildung.

Stoßbohren mit anschließender geophysikalischer Bohrlochmessung einschließlich Modellbildung.

Entwicklung von Bohrmodellen und Modellen zur mechanischen Gesteinszertrümmerung.

Entwicklung leistungsfähiger integrierter Systeme zur Erfassung, Auswertung, Speicherung und Darstellung von Geodaten.

4.2 Sprengen

Datenerfassung, Weiterentwicklung empirischer Modelle sowie der Bohr- und Sprengtechnik unter besonderer Berücksichtigung aktueller Probleme und neuer Sprengstoffe.

Entwicklung von Schnellverfahren zur Beobachtung der Sprengleistung hinsichtlich Zeitfolge, Detonation der Gesamtladung, Erschütterungen und Gesteinszertrümmerung.

Weiterentwicklung von Vorzerkleinerungsverfahren und von Verfahren zur schonenden Sprengung bei Verwendung neuer Sprengstoffe.

Weiterentwicklung theoretischer Sprengmodelle unter Berücksichtigung von Gesteinsart, Gesteinsmasse und Sprengstoffverhalten.

Thema 4
Untertägige Hohlräume im überbeanspruchten Gebirge

Standortuntersuchungen und Charakterisierung im Fels; Entwurf; Felssicherung und Stützmaßnahmen; Gebirgsschläge und Seismik; Bruchvorgänge in untertägigen Hohlräumen, Gasausbrüche und andere katastrophale Ereignisse; Voraussagen, Überwachung und Rückrechnung

Vorsitzender: W. Bamford (Australien)

Schriftführer: J. Nantel (Kanada)

Sprecher: H. Wagner (S. Afrika)

Planung und Ausbau von Untertagehohlräumen im Fels mit hohen Druckspannungen

ZUSAMMENFASSUNG: Im Ingenieurbau und insbesondere im Bergbau ist das Auftreten von Bruchverformungen um unterirdische Hohlräume oft unvermeidbar. In diesen Fällen liegt das Hauptgewicht des Hohlraumentwurfs auf der Beherrschung der Bruchzone um den Hohlraum. Das Fehlen von Angaben über die Gebirgsfestigkeit und das Verhalten des Gebirges nach dem Bruch führt zur Anwendung semi-empirischer Entwurfsmethoden. In harten und spröden Gebirgsformationen bietet die Elastizitätstheorie einen guten Ausgangspunkt für den Hohlraumentwurf. Die Beherrschung der Bruchverformungen ist die Basis für die Wahl der Ausbaumethode, da sie nicht nur die Restfestigkeit des Gebirges bestimmt, sondern auch das Ausmass der Auflockerungszone beschränkt. Diese Aufgaben werden am besten mit einem integrierten Ausbausystem bestehend aus Gebirgsankern, Maschendraht und Spritzbeton erfüllt. Seismische Beanspruchungen von untertägigen Hohlräumen stellen im allgemeinen kein Problem dar. Die wesentliche Ausnahme sind Hohlräume in der Nähe des Herdes von seismischen Vorgängen.

EINLEITUNG

Der Mensch dringt beim Abbau mineralischer Rohstoffe, beim Bau von Infrastrukturen für den Transport und zur Deckung des Energiebedarfs unserer modernen Gesellschaft in immer grössere Tiefen der Erdkruste vor. So werden bereits Erkundungsbohrlöcher mit Erfolg bis zu Teufen von über 12 000 m abgesenkt, Tunnel durch Gebirgsstöcke mit einem Deckgebirge von über 3000 m vorgetrieben und Goldlagerstätten in Tiefen von über 3800 m unter der Oberfläche abgebaut. Der Felsausbruch ist bei diesen Teufen mit besonderen technischen Problemen und Gefahren verbunden.

Die meisten Daten über Felsausbrüche in grossen tiefen stehen wohl aus Südafrika zur Verfügung, wo jährlich über 100 Mill. Tonnen goldführender Gänge in einer Gesteinstiefe von durchschnittlich 1600 m abgebaut werden. Die tiefsten Feldstrecken befinden sich 3800 m unter der Oberfläche, und man ist gegenwärtig dabei, Schachtanlagen zum Abbau der Goldlagerstätten in Teufen von über 4000 m zu erstellen. In den südafrikanischen Goldbergwerken werden jährlich Schächte, Erzdurchgänge, Stollen und Förderstrecken mit einer Gesamtlänge von über 800 km aufgefahren, um den Abbau der geringmächtigen Goldlagerstätten in diesen Teufen zu erleichtern. Während die felsmechanischen Probleme bei geringmächtigen tafelförmigen Goldlagerstätten somit ganz spezifischer Art sind, gibt es in derartigen Teufen viele andere Gebirgsdruckprobleme, die für den Felsbau im allgemeinen von Interesse und Bedeutung sind.

In diesem Referat sollen ein Überblick über eine Reihe wichtiger Erkenntnisse hinsichtlich des Verhaltens von Hohlräumen bei hohen Gebirgsspannungen gegeben und daraus einige allgemeine Kriterien zum Entwurf und Ausbau derartiger Hohlräume abgeleitet werden. Viele der Aussagen über das Verhalten von Hohlräumen beziehen sich auf Felsausbrüche in relativ festem Quarzitgestein. Es werden zu Vergleichszwecken aber auch Aussagen über das Verhalten von Hohlräumen bei hohen Druckspannungen in anderen stark beanspruchten geologischen Formationen besprochen, um allgemeingültigere Schlussfolgerungen zu ermöglichen.

Da die meisten Erfahrungen über das Verhalten von Hohlräumen bei hohen Gebirgsspannungen aus tiefen Abbauen vorliegen, steht dieser Gesichtspunkt des Felsbaus in diesem Referat im Vordergrund. Dazu kommt noch, dass der Bergbau die Möglichkeit bietet, zahlreiche Aspekte des Verhaltens von Hohlräumen bei hohen Druckspannungen zu untersuchen, die aus verschiedenen Gründen bei unterirdischen Hohlräumen im Ingenieurbau nicht beobachtet werden können.

FELSBAULICHE ERWÄGUNGEN BEI HOHLRÄUMEN MIT HOHEN DRUCKSPANNUNGEN

In diesem Referat werden unter Felsbau in Gesteinsmassen mit hohen Druckspannungen jene Fälle verstanden, bei denen die durch den Bau eines unterirdischen Hohlraumes hervorgerufenen Spannungen die Gebirgsfestigkeit übersteigen. Es treten daher im Gebirge, das den Hohlraum umgibt, Bruchverformungen auf. In diesem Fall kommt dem Felsbau eine zweifache Aufgabe zu, nämlich zum einen die Form und Grösse des Hohlraumes derart zu bemessen, dass Bruchverformungen auf ein Mindestmass beschränkt bleiben, und zum anderen ein zweckmässiges, rationelles Ausbausystem zur Beherrschung der Bruchzone um Hohlräume zu wählen.

Daraus folgt, dass Verbrüche eine zwangsläufige Folgeerscheinung des Felsausbruchvorgangs bei hohen Gebirgsspannungen darstellen. Genaue Kenntnisse sowohl der Festigkeitseigenschaften des Gebirges als auch der Bruchvorgänge sind daher wesentliche Bestandteile des Entwurfsverfahrens. Gleichermassen wichtig ist auch die Kenntnis des Spannungszustandes innerhalb der Gesteinsmasse, in der der Hohlraum geschaffen wird. Beim Bergbau ist noch zu bedenken, dass nach dem Ausbruch infolge zusätzlicher Abbautätigkeiten Spannungsänderungen auftreten.

Ein wichtiger Gesichtspunkt der Planung und des Ausbaus von Hohlräumen bei hohen Druckspannungen ist die Bildung deutlicher Bruchzonen um den Hohlraum. Es zählt zu den Aufgaben des Felsbaus, die Ausbruchabfolge und die Geometrie derart festzulegen, dass man ausbaumässig die günstigste Bruchzone erhält. Demgegenüber ist die Anlage von Hohlräumen unter niedrigen bis mittleren Gebirgsspannungszuständen in hohem Masse von den vorgegebenen geologischen Diskontinuitäten abhängig, an denen unter dem Einfluss der Schwerkraft eine Gesteinsbewegung erfolgen kann.

Bei der Betrachtung des Verhaltens von Hohlräumen bei hohen Druckspannungen muss die Bildung neuer Bruchverformungen sowie die diese Bruchverformungen bestimmenden Faktoren berücksichtigt werden; man darf sich daher nicht nur auf eine Analyse vorgegebener geologischer Diskontinuitäten beschränken. Wenngleich die letzteren bei der Planung von untertägigen Hohlräumen mitberücksichtigt werden müssen, spielen sie doch bei hohen Gebirgsspannungen im allgemeinen eine geringere Rolle. Dennoch können derartige Diskontinuitäten bei hohen Gebirgsspannungen einen beträchtlichen Einfluss auf den Verbruchsvorgang haben. Dies gilt insbesondere auch dann, wenn die mechanischen Eigenschaften der verschiedenen Gesteinsschichten sehr voneinander abweichen. Entsprechende Beispiele sind aus tiefen Kohlengruben bekannt, wo sich die mechanischen Eigenschaften des Kohlenflözes erheblich von denen des Nebengesteins unterscheiden. Selbst relativ geringe Unterschiede können jedoch noch einen entscheidenden Einfluss auf die Bruchbildung um Hohlräume bei tiefliegenden Abbauen im Hartgestein ausüben.

GEBIRGSFESTIGKEIT UND BRUCHVERFORMUNGEN

Festigkeitseigenschaften

Man kann wohl mit Recht sagen, dass sich die bisherige Forschung weitaus mehr mit der Klärung der Festigkeitseigenschaften als mit den übrigen Aspekten des Gebirgsverhaltens beschäftigt hat. Dennoch bestehen nach wie vor Zweifel, ob verlässliche Prognosen über den Gesteinsverbruch um untertägige Hohlräume möglich sind.

Eine Reihe von Gründen lassen sich für diese Schwierigkeit anführen. Erstens werden die Festigkeitseigenschaften des Gebirges im allgemeinen aufgrund kleiner intakter Gesteinsproben bestimmt, die unter künstlichen Bedingungen auf Belastbarkeit geprüft werden. Die auf diese Weise ermittelten Ergebnisse lassen sich, wenn überhaupt, nur schwer für Planungszwecke extrapolieren, da das Vorhandensein von Klüften, Spalten und anderen Schwächeflächen die Festigkeit des Gebirges sehr beeinträchtigen kann. In den letzten Jahren sind erhebliche Bemühungen um eine angemessene Berücksichtigung dieser Einflüsse zu verzeichnen. Einen der erfolgversprechenderen Ansätze stellt das empirische Bruchkriterium nach Hoek und Brown (1982, S. 137) dar:

$$\sigma_1 \leq \sigma_3 + \sqrt{(m\sigma_c \sigma_3 + s\sigma_c^2)} \qquad (1)$$

Dabei sind σ_1 and σ_3 die beim Bruch wirksamen höchsten und niedrigsten Hauptspannungen, σ_c ist die einachsige Druckfestigkeit des intakten Gesteinsmaterials und 'm' und 's' sind Konstante, die von den Eigenschaften des Gesteins und dem Ausmass des Verbruchs vor Einwirkung der Spannungen σ_1 und σ_3 abhängen. Bei intaktem Gesteinsmaterial ist der Wert s = 1, während der Wert m je nach Gesteinsart zwischen 7 und 25 liegt.

Die zweite Schwierigkeit liegt in der Definition des Begriffes "Festigkeit", worauf bereits Salamon im Jahre 1974 hingewiesen hat. Die allgemein übliche Definition der Festigkeit beruht auf der Ungleichung

$$\sigma_1 \leq a\sigma_3^b + \sigma_c \qquad (2)$$

Dabei stellen der Parameter 'a' die Erhöhung der Gebirgsfestigkeit bei hydrostatischer Spannung σ_3 und der Exponent ein Mass der Linearität der Beziehung dar, wobei letzterer im allgemeinen einen Wert von fast 1 annimmt. In vielen Fällen kann man daher die Gleichung (2) wie folgt vereinfachen:

$$\sigma_1 \leq a\sigma_3 + \sigma_c \qquad (2a)$$

Aus dieser Ungleichung und übrigens auch aus dem von Hoek und Brown aufgestellten Bruchkriterium, d.h. Gleichung (1), geht somit hervor, dass beide zwar die obere Grenze der Gesteinsfestigkeit bei einer bestimmten hydrostatischen Spannung beschreiben, aber nach Erfüllung dieser Ungleichung keine Schlussfolgerungen über den Widerstand oder den Zustand des Gesteins zulassen.

Eine dritte Schwierigkeit ist darin zu suchen, dass die Gebirgseigenschaften bedeutenden Änderungen unterworfen sind, sobald diese Ungleichungen erfüllt sind. Um nun diese Änderungen zu erfassen und insbesondere die verringerte allseitige Einspannung aufgrund des gestörten Gebirges um einen unter hohen Druckspannungen stehenden Hohlraum zu berücksichtigen, haben Hoek und Brown (1982) ihr ursprüngliches Kriterium - Gleichung (1) - durch Einführung eines Kennwertes für das gestörte bzw. zerlegte Gebirge erweitert. Dabei verringern sich die Werte m and s für das zerlegte Gebirge je nach Ausmass der erfolgten Bruchverformung und Klüftung.

Nach Hoek und Brown kann man die Werte m und s für gestörtes und geklüftetes Gebirge entweder nach dem NGI- oder dem CSIR-Klassifizierungssystem schätzen. Abb. 1 zeigt die Beziehung zwischen dem CSIR-Gebirgskennwert und den Faktoren an, um die man die Werte m und s zur Berücksichtigung der Klüftung und Verwitterung des Gebirges verringern muss. Das Kriterium von Hoek und Brown für gestörte und geklüftete Gesteinsmassen wird bereits mit viel Erfolg zur Untersuchung von Hohlräumen in tiefliegenden Abbauen angewandt. Bei der Anwendung derartiger Korrekturen muss man jedoch mit grosser Sorgfalt vorgehen, da sie in hohem Masse von der Geometrie des hydrostatischen Spannungsfeldes und der Wirksamkeit des Ausbaus abhängen. Von Wagner und Schümann (1971) durchgeführte Untersuchungen haben beispielsweise ergeben, dass das tatsächliche Ausmass der Bruchzone oberhalb und unterhalb von stark beanspruchten Pfeilern im allgemeinen schmäler ist als die mit Hilfe der Gleichungen (1) oder (2) berechneten Werte. Dies liegt daran, dass diese Gleichungen das Auftreten örtlich begrenzter, durch die Ausdehnung des gestörten Gebirges hervorgerufener hydrostatischer Spannungen nicht berücksichtigen.

Viertens stellen die Abstände zwischen Klüften und andere tektonische Schwachstellen wie Schicht- und Trennflächen in bezug auf die kritische lineare Ausdehnung des Hohlraumes wichtige Faktoren dar, die das Ausmass der Bruchzone bestimmen. Dieser Gesichtspunkt ist von besonderer Bedeutung bei Ausbrüchen in Sedimentgesteinsformationen, z.B. Strecken in tiefen Zechen oder Untertage-Grossräumen in Quarziten oder dünnschichtigen Schiefer- und Sandsteinformationen (Jacobi, 1976; Piper, 1984).

Die meisten gegenwärtig benutzten Gebirgsklassifizierungssysteme haben grundsätzlich den Nachteil, dass sie den Einfluss der Hohlraumdimensionen nicht berücksichtigen (Piper, 1985).

Die von Hoek und Braun (1982) aufgestellte Beziehung zwischen der einachsigen Druckfestigkeit und dem Probendurchmesser lässt sich gut auf massive Felsgesteinsformationen anwenden.

$$\sigma_c = \sigma_{c50}(^{50}/d)^{0,18} \qquad (3)$$

Dabei sind σ_{c50} die einachsige Druckfestigkeit einer Gesteinsprobe von 50 mm Durchmesser und 'd' der Durchmesser der Probe.

Die in Gleichung (3) dargestellte Beziehung wurde bereits erfolgreich zur Ermittlung der voraussichtlichen Bruchverformung an Stollen von zwei bis drei Metern Durchmesser in massiven Quarziten angewandt (Abb. 2).

Fünftens kann das Vorhandensein weicher Gesteinsschichten an den Wandungen von Hohlräumen bei hohen Druckspannungen einen grossen Einfluss auf das Verhalten des Hohlraums und die Art der Bruchverformung ausüben (Jacobi, 1976; Kersten et al., 1983; Reuther, 1987). Weiche Schichten beeinflussen das Verhalten des Hohlraumes auf verschiedene Weise: 1. Ist die weiche Gesteinsschicht verhältnismässig dick, so erfolgt der grösste Teil der Gesteinsverformung innerhalb dieser Schicht. 2. Wegen des niedrigen Verformungsmoduls und der im allgemeinen hohen Poissonzahl bei weichen Gesteinsmaterialien entwickeln sich hohe Schubspannungen am Übergang von den harten zu den weichen Schichten, wenn sich die Belastungsverhältnisse ändern, wie dies bei Felsausbruchsarbeiten der Fall ist (Brummer, 1983; Abb. 3).

Da die Zugfestigkeit des Gesteins wesentlich geringer ist als seine Druckfestigkeit, treten in harten, zwischen weichen Lagen eingelagerten Gesteinsschichten Zugbrüche auf und Felswände gehen bei wesentlich geringeren Spannungszuständen zu Bruch, als dies nach den Druckversuchen an Gesteinsproben aus den harten Schichten zu erwarten wäre. Laubscher (1984) veröffentlichte ein Nomogramm, mit Hilfe dessen sich die Festigkeit der aus schwachen und festen Gesteinsschichten aufgebauten Felswände abschätzen lässt (Abb. 4).

Bruchverformungen

Abb. 5 zeigt die verschiedenen Gesteinsbrucharten, die an Hohlräumen bei hohen vertikalen Druckspannungen beobachtet werden. In massiven, spröden Gesteinen (Abb. 5a) entstehen Dehnungsbrüche in den Seitenwänden des Hohlraumes. Das Ausmass und die Dichte dieser Brüche hängt von der Grösse der Spannungen und der Sprödigkeit des Gesteins ab. Die Bruchverformung um Hohlräume in zerklüfteten Gesteinsmassen hängt von den Schubspannungen und gravitativen Kräften ab (Abb. 5b). Treten weiche Schichten an den Seitenwänden von Hohlräumen auf, so entstehen leicht Zugbrüche in den spröden, zwischen den weichen Lagen eingelagerten Gesteinsschichten. Diese Dehnungsbrüche treten gewöhnlich bei weit niedrigeren Spannungszuständen auf als beim Fehlen weicher Schichten. Ausserdem erstrecken sich diese Brüche viel weiter in die Seitenwände als bei massivem, spröden Gestein (Abb. 5c). Schichtflächen mit niedrigem Reibungskoeffizienten (Abb. 5d) begünstigen die Bildung von Keilbrüchen, die sich tief in die Seitenwände des Hohlraumes erstrecken können. Ausserdem werden dadurch hohe seitliche Spannungen in den umgebenden Schichten erzeugt. Je nach örtlichen Gegebenheiten sind Dehnungsbrüche oder ein Knicken und Falten der dünnschichtigen Gesteine zu beobachten (Jacobi, 1976; Reuther, 1987).

Abschliessend darf man wohl feststellen, dass trotz der vielen bereits durchgeführten Untersuchungen über die Festigkeitseigenschaften von Fels und Gebirge sowie über Bruchverformungen noch ein weites Betätigungsfeld vor uns liegt. Aus der Sicht des Planungsingenieurs ist die gegenwärtige Situation bei weitem nicht zufriedenstellend. Vielmehr bedarf es neuer Ansätze, um wirkliche Fortschritte zu erzielen. Mit der Entwicklung komplizierter numerischer Modelle und leistungsstarker Computer bietet sich die Rückanalyse von bereits bestehenden Felsausbrüchen als vielversprechende Lösung an. Salamon (1967) hat dafür bereits den entsprechenden Nachweis geliefert, indem er anhand einer gebirgsmechanischen Analyse von 125 intakten und zerstörten Kammer- und Pfeilerabbaubetrieben ein neues Kammer- und Pfeilerbauverfahren für südafrikanische Kohlengruben entwickelte. In diesem Zusammenhang sollen auch von Felsmechanikern in südafrikanischen Goldbergwerken durchgeführte Untersuchungen erwähnt werden, die darauf abzielten, eine Beziehung zwischen dem Verhalten der Stollenwände und der einachsigen Druckfestigkeit der Gesteinsformation sowie den auf diese Stollen wirkenden Spannungen festzustellen (Cook, 1972; Ortlepp et al., 1972; Wiseman, 1979; Hoek und Brown, 1982; Piper, 1984).

DER ENTWURF UNTERTÄGIGER HOHLRÄUME

Allgemeines

Die klassische Entwurfsmethode beruht beim Maschinen- und Ingenieurbau auf der Gewährleistung der Stabilität einer Maschine bzw. eines Bauwerks, indem die in jedem Bauteil der Maschine bzw. des Bauwerks auftretenden Spannungen jeweils geringer als die Festigkeit des betreffenden Bauteils sind. Je nachdem, wie wichtig die Maschine bzw. das Bauwerk ist und wie schwerwiegend die zu erwartenden Konsequenzen eines Versagens eingeschätzt werden, wird ein entsprechend hoher Sicherheitsfaktor festgelegt. Die Entwurfszielsetzung wird sodann durch die Wahl geeigneter Werkstoffe für die Maschine bzw. das Bauwerk und geeigneter Querschnitte für die beanspruchten Bauteile in der Maschine bzw. dem Bauwerk sowie durch Vermeidung ungünstiger Belastungsbedingungen verwirklicht. Demgegenüber wird im Felsbau eine derartige Handlungsfreiheit ganz wesentlich durch die spezifische Zweckbestimmung des untertägigen Hohlraums eingeschränkt, die vielfach den Standort des Hohlraumes und damit auch die gegebene geologische Formation vorschreibt. Darüber hinaus hängt das Spannungsfeld, in dem der Hohlraum angelegt werden muss, oftmals von den spezifischen Standortverhältnissen ab, so dass dem Planungsingenieur lediglich die Ausrichtung des Hohlraumes in bezug auf die geologischen Gegebenheiten und das Spannungsfeld freisteht. Es gibt jedoch zahlreiche Situationen, in denen nicht einmal diese Möglichkeit besteht. Im Extremfall kann der Ingenieur nur Hohlraumform, Ausbausystem und Ausbruchfolge wählen.

Stabile und instabile Bruchverformungen

Was nun den Entwurf und Ausbau von Hohlräumen bei hohen Druckspannungen betrifft, so wird unmittelbar deutlich, dass die herkömmlichen ingenieurtechnischen Massnahmen nicht angewandt werden können, weil nach der eingangs zitierten Definition die auf das Gestein um den Hohlraum wirkenden Spannungen die Festigkeit des Gesteinsmaterials übersteigen und ein Verbruch unvermeidbar ist. Das bedeutet aber noch nicht, dass damit die Standfestigkeit des Hohlraumes gefährdet ist. Es gibt vielmehr zahlreiche Fälle, in denen es im Gestein um untertägige Hohlräume teilweise zu Verbrüchen kommt und dennoch die Standfestigkeit dieser Hohlräume oft sogar ohne Abstützungen erhalten bleibt. Das Kriterium, demzufolge die auftretenden Spannungen stets geringer als die Gesteinsfestigkeit sein müssen, gewährleistet zwar die Standfestigkeit eines Bauwerks, ist aber nicht unbedingt erforderlich. Als Erklärung lässt sich anführen, dass die meisten Gesteine auch nach dem Verbruch immer noch eine gewisse Spannung aufzunehmen vermögen. Nach Jaeger und Cook (1979, S.80) befindet sich ein Körper in duktilem Zustand, so lange er eine Dauerverformung ohne Verlust seiner Belastungsfähigkeit aufnehmen kann. Ein Körper befindet sich in sprödem Zustand, wenn seine Belastungsfähigkeit mit zunehmender Verformung abnimmt.

Beim Auffahren von untertägigen Hohlräumen, insbesondere bei hohen Gebirgsspannungen, geht oft ein Teil des umgebenden Gesteins in den spröden Zustand über. Demzufolge gelangt man nur dann zu einem besseren Verständnis der bei untertägigen Gesteinsausbrüchen ablaufenden Vorgänge, wenn man das Verhalten des Gesteins nach dem Bruch - d.h. den durch den abfallenden Ast der Spannungs-Dehnungs-Kurve gekennzeichneten Bereich - mitberücksichtigt.

Abb. 6a zeigt eine idealisierte Belastungs/Deformations-Kurve für sprödes Gestein. Ebenfalls dargestellt ist die Kraft-Verschiebungs-Charakteristik des Belastungssystems, das durch die Steigung k der Belastungslinie definiert ist.

Abb. 6b zeigt die Steigungsänderung λ der Kraft-Verschiebungs-Kurve des Gesteins und die Steigung k des Belastungssystems. Nach Salamon (1974) befindet sich das System nur dann in stabilem Gleichgewicht, wenn

$$k + \lambda > 0 \qquad (4)$$

Da k definitionsgemäss positiv ist, zeigt der Ausdruck (4) an, dass unabhängig von der Steifheit des Belastungssystems ein stabiles Gleichgewicht herrscht, solange der Deformationswiderstand des Gesteins mit der Deformation zunimmt (Abschnitt OB). In dem den Zustand nach dem Bruch kennzeichnenden Abschnitt BC herrscht ein stabiles Gleichgewicht nur bis zu dem Punkt, von dem aus die Belastungslinie des Systems tangential zur Kraft-Verschiebungs-Kurve verläuft ($P = P_s$). Bei Erreichen dieses Punktes kommt es zu einem schlagartigen Verbruch im Gestein; jedoch tritt kein schlagartiger Verbruch auf, solange die Mindeststeigung der Kraft-Verschiebungs-Kurve für das Gestein (λ_{min}) die Beziehung $k + \lambda_{min} > 0$ erfüllt.

Daraus lässt sich ableiten, dass man sich beim Entwurf von Hohlräumen bei hohen Gebirgsspannungen nicht allein auf eine Analyse der Bruchzonen um die Hohlräume verlassen darf. Die Steifheit des umgebenden Gebirges muss ebenfalls berücksichtigt werden, da es von ihr und dem Gesteinsverhalten nach dem Bruch abhängt, ob ein schlagartiger Gesteinsbruch voraussichtlich stattfinden wird oder nicht.

Damit haben sich Jaeger und Cook (1979, S. 472-474) befasst, indem sie den idealisierten Fall eines kreisförmigen Hohlraumes, der unter einer hydrostatischen Spannung p stand, untersuchten. Dabei überstiegen die Spannungen, die in einem schmalen Gesteinsmantel einer Dicke t im Innern des Hohlraumes mit einem Radius R herrschten, die Festigkeit des Gesteins nur geringfügig. Nimmt man an, dass sich bei $r > R$ das Gestein um den Stollen linear und elastisch verhält, dann gilt folgendes für die Steifheit der Radialspannung, die durch das umgebende Gestein auf den Gesteinsmantel ausgeübt wird:

$$k_R = {}^{\sigma}R/\varepsilon_R = 2G \qquad (5)$$

wobei G der Steifheitsmodul des Gesteins ist. Daraus folgt

$$\left|\frac{t}{R}\,\frac{d\sigma}{d\varepsilon}\right| < |\,2G\,| \qquad (6)$$

als Kriterium für die Standfestigkeit eines zerstörten Gesteinsmantels in bezug auf die Spannungen, die das umgebende Gestein auf ihn ausübt. Da in den meisten Fällen $t/R \ll 1$, ist ein dünner Bruchgesteinsmantel im Vergleich zur Steifheit des umgebenden Gesteins "weich" und demzufolge standfest.

Obwohl es sich hierbei um ein sehr vereinfachtes Beispiel handelt, so veranschaulicht es doch, dass Gesteinsverbrüche um Stollen im geologisch ungestörten Gebirge voraussichtlich standfest sind. Diese Erkenntnis beruht auf zahlreichen untertägigen Beobachtungen, aus denen hervorgeht, dass es sich beim Verbruch um Stollen und Kavernen im allgemeinen um einen stabilen Vorgang handelt.

Es soll hier noch auf die Arbeiten von Deist (1965) und Ulgudur (1973) verwiesen werden, die im einzelnen die Problematik stabiler und instabiler Verbrüche um Stollen bei hohen Druckspannungen untersuchten.

Es ist darauf hinzuweisen, dass diese Beobachtungen nicht auf untertägige Hohlräume zutreffen, deren Geometrie wesentlich von der eines Stollens abweicht. Insbesondere kann es beim Bruch von Stützpfeilern in tafelförmigen Gesteinskörpern zu instabilen, katastrophenartigen örtlichen Verbrüchen kommen, bei denen beträchtliche Mengen seismischer Energie freigesetzt werden. Gleichermassen können sich auch Bruchverformungen beim Ausbruch von stark beanspruchten tafelförmigen Gesteinskörpern instabil verhalten, insbesondere, wenn sie im Bereich geologischer Störungen liegen (Heunis, 1980).

Der Faktor Energie

Anfang der sechziger Jahre wies Cook (1963) bereits darauf hin, dass beim Abbau bedeutende Energieänderungen auftreten. Insbesondere führte er aus, dass "sich die überschüssige potentielle Energie in Form eines Gebirgsschlags äussert". Seitdem werden verschiedene Energiegrössen und vor allem die pro Zeiteinheit freiwerdende Energie als Parameter bei der Entwurfsplanung von Abbauen in grosser Tiefe mitberücksichtigt.

Salamon untersuchte 1974 und 1984 einige der Hauptgesichtspunkte im Zusammenhang mit Energieänderungen, die infolge einer Hohlraumvergrösserung auftreten. Salamon weist darauf hin, dass die zunehmende Grösse und Anzahl der bergmännischen Ausbrüche Verschiebungen im umgebenden Gestein verursacht. Die durch diese Verschiebungen wirkenden äusseren sowie gebirgseigenen Kräfte verrichten ein gewisses Mass an Arbeit W. Diese Arbeit wird allgemein als "gravitative" oder "potentielle" Energieänderung bezeichnet. Darüber hinaus ist eine bestimmte Deformationsenergie im Gestein gespeichert, die durch den Abbauvorgang U_m freigesetzt wird. Die Summe ($W+U_m$) ist die Energiezufuhr, die zu berücksichtigen bzw. in irgendeiner Weise aufzuwenden ist.

Diese Energie wird zum Teil als Erhöhung der Deformationsenergie im umgebenden Gebirge U_c wirksam. Erfolgt ein Versatz bzw. ein Ausbau der Hohlräume, so wird bei der Verformung der Stützungen Arbeit W_s verrichtet. Geht man davon aus, dass das Gebirge ein elastisches Kontinuum darstellt, so wird beim Verbruch oder bei nichtelastischer Deformation keine Energie verbraucht. Die bei einem Abbauschritt nachweisbar aufgewandte Energie ist somit (U_c+W_s).

Salamon kam zu dem Schluss, dass diese einfachen Vorgänge nicht die gesamte Energiezufuhr berücksichtigen und dass eine bestimmte Energiemenge

$$W_r = (W+U_m) - (U_c+W_s) > 0 \qquad (7)$$

anderweitig abgegeben wird bzw. verloren geht. Die untere Grenze der freigesetzten Energie wird demnach durch die Beziehung

$$W_r \geq U_m > 0 \qquad (8)$$

dargestellt.

Beim Ausbruch von tafelförmigen Gesteinskörpern lässt sich die Energiemenge U_m und damit die untere Grenze von W_r aus den Komponenten durchschnittliche Spannung $\sigma_{k3}{}^{(p)}$ und Konvergenz $S_k{}^{(i)}$ vor und nach Abbau des Minerals aus der Fläche ΔA annähernd ermitteln, wobei $\sigma_{k3}{}^{(p)}$ (k = 1, 2, 3) die Komponenten des Spannungsvektors in A vor dem Abbau dieser Fläche und $s_k{}^{-(i)}$ die Komponenten des relativen Verschiebungsvektors nach Abbau desselben Bereiches sind.

$$\Delta U_m = -\int_{\Delta A} \tfrac{1}{2}\, S_k{}^{(i)}\; \sigma_{k3}{}^{(p)}\; \Delta A \qquad (9)$$

Diese Komponenten lassen sich sehr einfach entweder mit dem elektrischen Widerstandsnetz-Analogieverfahren oder mit einem der vielen, in den letzten Jahren entwickelten Digitalsimulatoren für den Abbau von tafelförmigen Gesteinskörpern berechnen (Ryder, 1986).

Praktische Erfahrungen in tiefliegenden Goldbergwerken haben gezeigt, dass das Verhältnis $\Delta U_m/\Delta A$

einen wichtigen Kennwert für die Prognose von Problemen darstellt, die sich beim Abbau in grosser Tiefe hinsichtlich einer ausreichenden Beherrschung der Schichten ergeben (Hodgson und Joughin, 1966; Heunis, 1980; Salamon und Wagner, 1979). Abb. 7 veranschaulicht die Beziehung zwischen $\Delta U_m/\Delta A$ und der Anzahl der Gebirgsschläge pro 1000 m^2 abgebauter Fläche.

Es gibt jedoch bisher nur wenige Untersuchungen über die Energieänderungen bei unterirdischen Felsausbrüchen in nicht tafelförmigen Gesteinskörpern. Salamon (1984) untersuchte die Energieänderungen, die bei der plötzlichen Vergrösserung des Radius eines kreisförmigen Stollens vom Wert a zum Wert c auftreten. Der folgende Ausdruck für die pro Zeiteinheit freigesetzte Energie W_r wurde für einen ausbaulosen Stollen ermittelt, auf den ein hydrostatischer Druck der Grösse p wirkte:

$$W_r = \frac{2(1 - \nu^2)p^2}{E} V_m \tag{10}$$

Dabei ist V_m das pro Längeneinheit des Stollens abzubauende Gesteinsvolumen:

$$V_m = \pi(c^2 - a^2) \tag{11}$$

Die potentielle kinetische Energie, die durch eine plötzliche Vergrösserung des Stollenradius freigesetzt wird, ergibt sich aus der Beziehung:

$$W_k = \frac{(1 + \nu)p^2}{E} \left(1 - \frac{a^2}{c^2}\right) V_m \tag{12}$$

Die Bedeutung dieses Ergebnisses liegt vor allem in der Erkenntnis, dass bei einer plötzlichen Vergrösserung eines Stollens eine beträchtliche Menge kinetischer Energie freigesetzt werden kann (Abb. 8).

Schliesslich sollen nachstehend noch die beim Abbau einer tafelförmigen Lagerstätte in einer Teufe von 3000 m freigesetzten Energiemengen mit denjenigen verglichen werden, die beim Stollenbau in derselben Teufe auftreten.

Die Obergrenze der aller Wahrscheinlichkeit nach bei umfangreichen Abbauvorgängen freigesetzten Energie ergibt sich aus dem Ausdruck:

$$\Delta U_m/\Delta A = S_m \sigma_v \tag{13}$$

Dabei ist S_m die Abbauhöhe und σ_v die Vertikalkomponente des primären Spannungstensors. Bei einer Abbauhöhe von S_m = 1 m und einer Vertikalspannung σ_v = 80 MPa ist $\Delta U_m/\Delta A$ = 80 MJ/m^2.

Demgegenüber liegt die Energie, die beim Bau eines Stollens mit einem Radius von 2 m in einem Spannungsfeld von 80 MPa pro Zeiteinheit freigesetzt wird, bei 0,2 MJ/m^3, d.h., sie ist mehr als zwei Grössenordnungen niedriger als die bei Abbauvorgängen freiwerdende Energie.

Aus diesem Beispiel wird deutlich, dass aller Wahrscheinlichkeit nach der Faktor Energie weniger beim Entwurf von tiefliegenden Stollen und Kavernen als bei der Abbauführung berücksichtigt werden muss. Es soll jedoch darauf hingewiesen werden, dass die beim Bau von tiefen Stollen pro Zeiteinheit freigesetzte Energie die gleiche Grössenordnung hat wie die Energie, die zum Bruch eines Kubikmeters Hartgestein bei Druckbelastung erforderlich ist.

Hohlraumform und auftretende Spannungen

Wie bereits erwähnt, gehört die Wahl der Hohlraumform zu den wichtigsten felsbaulichen Entscheidungen. Dies trifft besonders auf Hohlräume zu, die hohen Gebirgsspannungen ausgesetzt sind. Die Form des Hohlraumes und seine Raumstellung in bezug auf die Gebirgsspannungen bestimmen nicht nur die Grenzspannungen, sondern darüber hinaus auch die Bildung der Bruchzone und die erforderlichen Ausbaumassnahmen.

Bei niedrigen Spannungszuständen bestimmen normalerweise die geologischen und tektonischen Diskontinuitäten das Verhalten des Hohlraumes und damit die erforderlichen Ausbaumassnahmen. Unter derartigen Bedingungen wird normalerweise eine Hohlraumform gewählt, die sich zur Verhinderung von Gesteinsbewegungen an den Trennflächen und zur Erhaltung der natürlichen Festigkeit der Gesteinsformationen am günstigsten erweist. Beispielsweise ist es bei Schichtgesteinen besonders vorteilhaft, wenn deutliche Schicht- bzw. Trennflächen die Firste des Hohlraumes bilden oder wenn eine der festeren und tragfähigeren Schichten als natürliche Firste gewählt wird, selbst wenn dies spannungsmässig nicht die günstigste Geometrie ergibt.

Je grösser die Gebirgsspannungen im Vergleich zur Festigkeit des Gebirges im Umfeld des Hohlraumes sind, desto sorgfältiger muss auch die spannungsmässig günstigste Geometrie gewählt werden. Aus praktischen Erfahrungen beim Stollenbau in tiefliegenden südafrikanischen Bergwerken ergibt sich, dass die Seitenwände von Stollen mit quadratischem Querschnitt und einem Durchmesser von drei bis vier Metern nur geringe Ablöseerscheinungen aufweisen, wenn die Vertikalkomponente σ_v des Spannungstensors die in Laboruntersuchungen ermittelte, einachsige Druckfestigkeit σ_c des Gebirges von ca. 0,2 übersteigt. Starke Abblätterungserscheinungen treten an den Seitenwänden auf, wenn $\sigma_v/\sigma_c > 0,3$, und umfangreiche Ausbaumassnahmen werden dann erforderlich, wenn dieses Verhältnis ca. 0,4 übersteigt (Ortlepp et al., 1972).

Wenngleich sich das einfache, auf der Vertikalkomponente des Gebirgsspannungstensors beruhende Spannungskriterium im Goldbergbau gut bewährt hat, so hat es doch den Nachteil, dass es den Einfluss der übrigen Spannungskomponenten auf das Verhalten der unterirdischen Hohlräume vernachlässigt. Wiseman (1979) hat diese Spannungskomponenten dadurch berücksichtigt, dass er für die Seitenwände einen Spannungskonzentrationsfaktor SF aufstellte, den er wie folgt definierte:

$$SF = \frac{3\sigma_1 - \sigma_3}{\sigma_c} \tag{14}$$

Dabei sind σ_1 und σ_3 die höchsten bzw. niedrigsten, auf den Hohlraum wirkenden Hauptspannungen, und σ_c ist die einachsige, in Laboruntersuchungen bestimmte Druckfestigkeit des Gesteins. Wiseman stellte anhand einer detaillierten Untersuchung typischer Goldbergwerksstollen mit einer Gesamtlänge von mehr als 20 km fest, dass sich der Zustand nicht ausgebauter Stollen auffallend verschlechterte, sobald der Spannungskonzentrationsfaktor für die Seitenwände einen Wert von ca. 0,8 erreichte (Abb. 9).

Aus diesen Beispielen geht hervor, dass sich bei massiven Festgesteinen die auf die einachsige Festigkeit des Gesteins umgerechneten Grenzspannungen gut zur Aufdeckung eventueller Ausbruchsprobleme eignen. Es muss aber darauf hingewiesen werden, dass die Ermittlung eines angemessenen Näherungswertes für die einachsige Gebirgsdruckfestigkeit mit Schwierigkeiten verbunden ist. Obwohl diese Schwierigkeiten bereits erörtert wurden, soll an dieser Stelle doch nochmals darauf verwiesen werden.

Beim Entwurf muss daher zunächst eine Bestimmung der auf den Randbereich des geplanten Hohlraumes wirkenden Spannungen erfolgen. Voraussetzung dazu ist die Kenntnis der am Standort des geplanten Felsausbruches auftretenden Spannungen. Was den Bergbau anbetrifft, so kann man hier mit Modellen arbeiten, die den Einfluss der im allgemeinen sehr umfangreichen Abbauvorgänge auf die am vorgesehenen Standort des Hohlraumes herrschenden Spannungen darstellen. Es stehen dafür zahlreiche Boundary-Element- und Finite-Element-Modelle zur Verfügung. Beim Ingenieurbau können dagegen unter Umständen Spannungsmessungen in situ erforderlich werden. Wo dies nicht möglich ist, sollte man den Verlauf des Spannungsfeldes in der weiteren Umgebung bei der näherungsweisen Bestimmung der Spannungsverhältnisse am geplanten Ausbruchsort mitberücksichtigen. Als Faustregel gilt, dass die Vertikalkomponente des Spannungstensors der Deckgebirgsspannung entspricht, die sich aus der Mächtigkeit des Deckgebirges und der Dichte der Gesteinsformationen errechnet. Es besteht jedoch häufig viel Unklarheit über die Richtung und Grösse der Horizontalspannungen (Hoek und Brown, 1982).

Ein erster, oft ausreichender Näherungswert für die Grenzspannungen um einen unterirdischen Hohlraum lässt sich aus den folgenden Beziehungen ableiten (Brady und Brown, 1985, S. 195):

$$\sigma_A = p\left(1 - K + \sqrt{\frac{2W}{\rho_A}}\right) \tag{15}$$

$$\sigma_B = p\left(K - 1 + K\sqrt{\frac{2H}{\rho_B}}\right) \tag{16}$$

Hierin sind σ_A und σ_B die im Grenzbereich wirksamen Umfangsspannungen an der Seitenwand (A) und dem Scheitel (B) des Hohlraumes, ρ_A und ρ_B die Krümmungsradien um Punkte A und B, K das Verhältnis von Horizontal- zu Vertikalspannung (Abb. 10).

Massive Festgesteine

Bei massiven Festgesteinen begegnet man normalerweise keinen spannungsbedingten Ausbruchsproblemen, wenn die auf die Druckfestigkeit des Gebirges umgerechneten Randspannungen weit unter 1 liegen. Ergibt sich für die umgerechneten Grenzspannungen ein Wert nahe 1, dann wäre eine andere Hohlraumform angezeigt, um die Grenzspannungen zu verringern. In vielen Fällen sind jedoch die Gebirgsspannungen derart hoch, dass auch die günstigste Hohlraumgeometrie ein Zubruchgehen des Gesteins am Hohlraumrand nicht zu verhindern vermag. Dies tritt ein, wenn sich die maximale Hauptspannung σ_1 dem halben Wert der einachsigen Gesteinsdruckfestigkeit $\sigma_1 \cong 0,5\ \sigma_c$ nähert. In diesem Fall muss die Hohlraumform derart gewählt werden, dass das Ausmass der Bruchzone auf ein Mindestmass beschränkt bleibt und der Hohlraum leicht ausgebaut werden kann.

Man hat bei Stollen mit hohen Druckspannungen in massiven Quarziten bereits die Erfahrung gemacht, dass die günstigste Hohlraumgeometrie erheblich von der abweicht, die mit Hilfe einer Spannungsanalyse für elastische Körper ermittelt wurde. Als charakteristisches Merkmal hat sich dabei ergeben, dass bei Stollen mit gleichseitigem Querschnitt die Bruchverformung mit der Bildung von Dehnungsbrüchen in denjenigen Stollenwänden einsetzt, auf die die höchsten Druckspannungen einwirken. Je weiter sich die aus abgespaltenen Gesteinsplatten bestehende Bruchzone seitlich der Stollenwände erstreckt, desto mehr verringert sich im allgemeinen die Länge dieser Gesteinsplatten, so dass es schliesslich zu einem Gleichgewichtszustand kommt, bei dem keine weiteren Verbrüche mehr auftreten. Dieser Vorgang wurde im einzelnen von Fairhurst und Cook beschrieben (1966). Als Stollengeometrie ergibt sich eine gestreckte Ellipse, deren Hauptachse senkrecht zur Richtung der maximalen Druckspannung verläuft, somit genau das Gegenteil der günstigsten elastischen Form ist.

Die wohl umfassendste Analyse eines Stollens in einem extrem hohen Spannungsfeld wurde an einem Versuchsstollen durchgeführt, der im anstehenden Gestein vor einem Abbaustoss in einem tiefliegenden Goldbergwerk angelegt wurde, um die Seismizität in tiefen Bergwerken zu untersuchen (Ortlepp und Gay, 1984). Dieser Stollen wurde in einem Spannungsfeld vorgetrieben, das Werte zwischen ca. 50 MPa im abgebauten Grubenteil und 140 MPa im anstehenden Fels vor dem Abbauort aufwies. Durch die weiteren Abbauarbeiten erhöhten sich schliesslich die auf den Stollen wirkenden Spannungen auf ca. 230 MPa. Diese Spannungen entsprechen einer Teufe von mehr als 8000 m unter der Oberfläche. Der Stollen selbst lag in massiven Quarziten, deren einachsige Druckfestigkeit ca. 350 MPa betrug. Nach Umrechnung mit Hilfe der Gleichung (3) zur Berücksichtigung der Grösse des Hohlraumes erhielt man einen Wert von ca. 200 MPa für die einachsige Druckfestigkeit. Abb. 11 zeigt die Änderung der Gesteinsspannungen längs der Stollenachse zum Zeitpunkt des Stollenvortriebs sowie die gesprengten und endgültigen Stollenquerschnitte an verschiedenen Stellen des Stollens.

In Zonen hoher Druckspannungen konnte man an der Ortsbrust des Stollens etwa auf halber Höhe einen örtlich begrenzten Streifen intensiver Bruchverformungen beobachten. Diese Bruchzone verwandelte sich in eine spannungserhöhende "Kerbe", die sich beim Vortrieb des Stollens schnell von den Seitenwänden aus ins Gestein erstreckte und mit Knickvorgängen und dem Ablösen von kleinsten Gesteinsplättchen an der Kerbenspitze verbunden war. Dieser im wesentlichen nicht schlagartig verlaufende Bruchvorgang dauerte an, bis sich nach einiger Zeit ein charakteristisches, spitz zulaufendes ellipsenförmiges Profil herausbildete, das sich auf einer Breite von etwa 3-4,5 m stabilisierte. Als wichtige Beobachtung ergab sich dabei, dass die endgültige stabile Stollengeometrie mit einem Mindestmass an Ausbau erzielt wurde. In einem Stollenabschnitt, der gleich zu Anfang ausgebaut wurde, sollte der Ausbau im wesentlichen das Bruchgestein auf die Seitenwände des Stollens begrenzen. Als weiteres wichtiges Ergebnis wurden das Fehlen umfangreicher Verbrüche in der Firste und Sohle des Stollens und der ausserordentlich gute Firstenzustand festgestellt.

Diese Beobachtungen treffen auf Stollen in massiven, spröden Felsgesteinen zu, auf die hohe Vertikalspannungen wirken. Es muss darauf hingewiesen werden, dass ähnliche Bruchverformungen auch an Stollen beobachtet wurden, die sich in Gebieten hoher Horizontalspannungen befanden. In diesen Fällen erfolgte jedoch der Verbruch der Stollenwandungen in der Firste und Sohle des Stollens.

Allgemein lässt sich somit feststellen, dass sich Verbrüche bei Stollen mit hohen Druckspannungen in massiven, spröden Gesteinen weitgehend auf die Stollenwände beschränken, die parallel zur Richtung der maximalen Hauptdruckspannung verlaufen. Die bevorzugte Stollengeometrie bei Bruchverformungen an den Stollenwänden ist dabei eine Ellipse, deren Hauptachse senkrecht zur Richtung der maximalen Druckspannung verläuft. Das Längenverhältnis der Ellipsenachsen muss dabei in der Regel je nach Erhöhung der maximalen Druckspannung in bezug auf die Gesteinsfestigkeit vergrössert werden. Vom Gesichtspunkt der Bruchbeherrschung aus ist somit die günstigste Geometrie für Hohlräume in stark beanspruchten, spröden Gesteinen genau das Gegenteil der Geometrie, die sich aus einer Analyse der Spannungen ergibt, welche auf den Randbereich eines Hohlraumes in einem elastischen Körper wirken.

Ein erster Annäherungswert über das Ausmass der Bruchzone in den Seitenwänden von Stollen bei hohen Druckspannungen ergibt sich anhand einer Spannungsanalyse für elastische Körper sowie einer Bestimmung der Linien, bei denen die Spannungen um den Hohlraum eine der bekannten Bruchkriterien gemäss Gleichung (1) bzw. (2) erfüllen. Dieses Vorgehen hat jedoch den Nachteil, dass die Anderungen der Gesteinseigenschaften in den "Verbruchzonen" nicht berücksichtigt werden. Dennoch eignet sich dieses vereinfachte Verfahren erfahrungsgemäss gut zum Vergleich der jeweiligen Vorteile, die sich aus den verschiedenen Hohlraumformen bei hohen Gebirgsspannungen ergeben. Abb. 12 zeigt das Ausmass der Bruchzone ΔW_r in den Seitenwänden von vier Stollen mit unterschiedlichem Querschnitt in Abhängigkeit des Verhältnisses zwischen vertikaler Spannungstensorkomponente σ_v und einachsiger Gebirgsdruckfestigkeit. Den dargestellten Ergebnissen liegt ein Verhältnis von horizontaler zu vertikaler Spannung $\sigma_h/\sigma_v = 0,5$ zugrunde. Es wird dabei deutlich, dass bei niedrigen Spannungszuständen ($\sigma_v/\sigma_c < 0,5$) die Stollenquerschnitte in Form einer stehenden Ellipse und eines Quadrats günstiger sind. Sobald es aber zum Verbruch der Stollenwände kommt ($\Delta W_R/w > 0$), sind die Stollenquerschnitte in Form einer liegenden Ellipse oder eines Kreises günstiger, insbesondere wenn $\sigma_v/\sigma_c > 1$.

Eine Erklärung für das günstigere Verhalten von Stollen mit stark gekrümmten Seitenwänden ist aus Abb. 13 ersichtlich, in der die Verteilung der Hauptspannungen σ_1 und σ_3 entlang der Horizontalachse des Stollens dargestellt ist. Das wichtigste Merkmal dieser Spannungsverteilungen ist der flache Verlauf der Kurve σ_1/Δ_x für Stollen mit einem sehr grossen Seitenwand-Krümmungsradius, d.h., bei Stollenquerschnitten in Form einer stehenden Ellipse und eines Quadrats. Die Werte für die Mindesthauptspannung σ_3 dicht an den Seitenwänden weichen bei den vier Stollen ganz beträchtlich voneinander ab. Bei den Stollen mit dem kleinsten Krümmungsradius treten hohe Mindesthauptspannungen dicht an den Stollenwänden auf. Da die Gesteinsfestigkeit bei steigendem hydrostatischen Spannungszustand schnell ansteigt, bleiben Seitenwandbrüche um Hohlräume mit kleinen Krümmungsradien auf einen sehr schmalen Bereich beschränkt.

Ein weiterer Gesichtspunkt, der für Stollen mit kleinem Seitenwand-Krümmungsradius spricht, ist der Einfluss, den Spannungsgefälle auf die Gesteinsfestigkeit ausüben. Dieser Einfluss ist bereits aus Versuchen zur indirekten Messung der Zugfestigkeit von Gestein bekannt (Cook und Jaeger, 1979, S. 197-198), in denen festgestellt wurde, dass die Zugfestigkeit des Gesteins mit dem Spannungsgefälle zunimmt. Wagner und Schümann (1979) konnten ein ähnliches Ergebnis bei der Untersuchung der Gesteinsfestigkeit bei hohen Kontaktspannungen beobachten, nämlich, dass das Spannungsgefälle die Festigkeit von spröden Gesteinen stark beeinflusst. Aus diesen und anderen von Johnson (1985) veröffentlichten Beobachtungen geht hervor, dass der Einfluss von Spannungsgefällen beim Entwurf von Hohlräumen in grosser Tiefe nicht unberücksichtigt bleiben darf.

Geschichtete Festgesteine

Aus geschichteten Festgesteinen aufgebaute geologische Formationen sind häufig durch die Abfolge von harten und weichen Gesteinsschichten sowie das Vorhandensein von deutlichen Schwächeflächen gekennzeichnet. Da diese einen wesentlichen Einfluss auf das Verhalten von Hohlräumen ausüben können, muss man sie in der Entwurfs- und Ausbauphase berücksichtigen. Die mit Abstand umfangreichsten Kenntnisse über das Verhalten von Hohlräumen in geschichteten Festgesteinen liegen aus Kohlenbergwerken vor, so dass eine Untersuchung der in tiefen europäischen Kohlenbergwerken gesammelten Erfahrungen geboten scheint (Jacobi, 1976; Reuther, 1987).

Durch den Ausbruchsvorgang ändert sich die Spannungsverteilung im Gebirge, und das den Hohlraum umgebende Gestein verformt sich. Bei Hohlräumen in massiven, spröden Gesteinsformationen ging es vor allem um die Bildung von Bruchzonen infolge einer Überbeanspruchung des Gesteins. Demgegenüber kann man bei Schichtgesteinen feststellen, dass deutliche Schichtebenen, an denen Gesteinsbewegungen erfolgen können, das Hohlraumverhalten stark beeinflussen. Diese Problematik wurde bereits von vielen Autoren anhand von Modellversuchen im Labor mit entsprechenden Gesteinsproben untersucht (Everling, 1962; Buschmann 1964). In jüngerer Zeit hat Purrer (1984) numerische Modelle entwickelt, mit denen er den Einfluss von weichen Kohlenflözen und Flächen geringen Reibungswiderstandes auf das Hohlraumverhalten untersuchte.

Mit Hilfe einer einfachen Spannungsanalyse ermittelte Reuther (1987) Bereiche um eine typische Strecke in einem Steinkohlenbergwerk, in denen es zu reibungsabhängigen Gleitbewegungen an horizontalen Schichtflächen kommen kann (Abb. 14). Nach diesem Modell, dem ein hydrostatisches Spannungsfeld zugrunde liegt, treten Gleitbewegungen an den Schichtflächen im oberen Teil der Strecke auf und erstrecken sich bei niedrigem Reibungskoeffizienten μ unter Umständen bis tief in das Hangende bzw. den oberen Teil der Ulmen. Bei dickbankigen Felsgesteinen kann eine Gleitbewegung und Rissbildung über dem oberen Teil der Ulmen infolge einer Verschiebung der Hangendschicht auf der Scherfläche zum Verbruch führen. Bei dünnschichtiger Ausbildung liegt die Problematik im wesentlichen in der Festigkeit der Hangendschichten unter dem Einfluss des Überlagerungsdrucks und insbesondere im seitlichen Druck, der auf die Schichten wirkt und zu umfangreicher Knickung und Faltung der Schichten führen kann, wie es vielfach in tiefen Steinkohlenbergwerken beobachtet wird.

Die Axialspannung σ_a, bei der eine Gesteinsplatte einer Stärke t und einer Länge ℓ zu knicken beginnt, ergibt sich aus dem Ausdruck

$$\sigma_a = \frac{\pi^2 E}{12\zeta^2 (\ell/t)^2} \qquad (17)$$

Dabei sind E der Elastizitätsmodul des Gesteins, ℓ/t der Schlankheitsgrad der Gesteinplatte, und
ζ eine Konstante, die vom Endzustand der Platte abhängt und zwischen 0,5 (bei einer an beiden Enden eingespannten Platte) und 1 (bei einer Platte, deren beide Enden gelenkig gelagert sind) liegt.

Aus Gleichung (17) geht hervor, dass die Axialspannung, die eine Gesteinsplatte aufnehmen kann, ohne zu knicken, umgekehrt proportional zum Quadrat des Schlankheitsverhältnisses ist. Demzufolge knicken dünne Gesteinsplatten leichter als dicke Platten.

Von der Hohlraumplanung her betrachtet, ist eine mächtige Gesteinsschicht in der Firste des Hohlraumes erwünscht, um eine eventuelle Knickung der Hangendschichten so weit wie möglich zu verhindern. Wo dies nicht möglich ist, sollte man die Schaffung einer tragfähigen Firste durch umfangreichen Ankerausbau und Felsbewehrung in Erwägung ziehen.

Eine Möglichkeit der Reduzierung eventueller Knickungs- und Faltungsvorgänge in dünnschichtigen Gesteinen besteht darin, Entspannungsöffnungen in den oberen Ulmenbereich des Stollens zu bohren, um damit einen gewissen Schichtflächenschlupf ohne Überschiebung der Gesteinslagen in der Firste zu erreichen. Die teilweise entspannten Schichten werden daraufhin mit den darüberliegenden Schichten verankert und bilden somit ein tragfähiges Hangendes. Die Vorteile dieser Methode wurden bereits an Hand von Modellversuchen nachgewiesen und sollten genauer geprüft werden. Abb. 15 zeigt eine mögliche Ausgestaltung der Entspannungsbohrlöcher im Streckenbereich eines Steinkohlenbergwerks (Roest und Gramberg, 1981).

Obwohl Schäden an Hohlräumen in horizontal gelagerten Gesteinsschichten meistens in der Firste und Sohle der Hohlräume auftreten, so stellt doch die Höhe der Hohlraumwände einen kritischen Kennwert beim Entwurf dar, da sie weitgehend die seitliche Ausdehnung eines eventuellen Ulmenbruchs bestimmt. Dies wiederum beeinflusst die wirksamen Firstenspannweiten und den erforderlichen Firstenausbau. Von diesem Gesichtspunkt aus betrachtet ist die

Hohlraumhöhe in horizontal gelagerten Gesteinsschichten möglichst niedrig zu halten. Dieser Forderung kommt eine grössere Bedeutung zu, wenn eine Gleitbewegung längs der Schichtflächen unter geringem Reibungsdruck möglich ist. Liegen derartige Schichtflächen nicht vor, so ist die Forderung hinsichtlich der Hohlraumhöhe weniger kritisch.

Bei der bisherigen Erörterung ging es um den Hohlraumentwurf in horizontal gelagerten Gesteinsschichten mit hohen vertikalen Spannungen. Bei hohen horizontalen Spannungen stellt die Mächtigkeit der Gesteinsschichten im unmittelbaren Hangenden und Liegenden des Hohlraumes einen kritischen Faktor dar, da sie die Spannweiten bestimmt, bei der eine Knickung erfolgt. Im südafrikanischen Steinkohlenbergbau werden häufig Firstenbrüche mit einer Ausdehnung von über zehn Metern in sechs Meter breiten Strecken in horizontal gelagerten, dünnschichtigen Brandschiefern beobachtet, auf die relativ hohe horizontale Spannungen wirken. Diese Firstenbrüche endeten gewöhnlich an der ersten tragfähigen Sandsteinschicht.

AUSBAU VON HOHLRÄUMEN IM FELS MIT HOHEN DRUCKSPANNUNGEN

Bei der Besprechung der allgemeinen Entwurfsgesichtspunkte für Hohlräume in stark beanspruchten Gesteinsmassen wurde vor allem darauf hingewiesen, wie wichtig bei untertägigen Hohlräumen die Wahl der richtigen Geometrie ist. Aus Abb. 12 geht hervor, dass die Querschnittswahl dann besonders kritisch ist, wenn die maximale Druckspannung gleich der Druckfestigkeit des Gebirges ist oder sie übersteigt. Die Schwierigkeiten beim Ausbau in Gesteinsmassen mit hohen Druckspannungen hängen daher in hohem Masse von der Hohlraumgeometrie ab, da diese nicht nur die Form, sondern auch die Tiefe der den Hohlraum umgebenden Bruchzone bestimmt.

Die Hauptaufgabe des Ausbaus im Gebirge mit hohen Druckspannungen besteht darin, die Gebirgsfestigkeit so weit wie möglich zu mobilisieren und zu erhalten, damit das Gebirge selbsttragend wird. Dies lässt sich dadurch erreichen, dass man die nachträgliche Verformung des gestörten Gebirges um den Hohlraum begrenzt. So wird nicht nur die Restfestigkeit des gestörten Gebirges erhalten, sondern zusätzlich auch eine ungehinderte Ausdehnung der Bruchzone durch Sicherung des Gesteins ausserhalb dieser Zone verhindert. Am rationellsten lässt sich dies durch das Einbringen des Ausbausystems so bald wie möglich nach Abschluss der Ausbrucharbeiten erreichen.

Im Idealfall soll das Ausbausystem aktiver Art sein, d.h. es soll das Gestein innerhalb der Bruchzone begrenzen, ohne dass es zu einer Auflockerung der Bruchzone kommt, es soll Gesteinsverformungen um den Hohlraum durch schnellen Aufbau eines Ausbauwiderstandes verhindern und durch eine deutliche Fliessgrenze eine Überbeanspruchung des Ausbausystems entweder durch Vergrösserung der Bruchzone oder durch die gelegentlich in tiefen Bergwerken beobachtete dynamische Belastung verhindern.

Da sich eine Bruchzone um untertägige, unter hohen Druckspannungen stehende Hohlräume nicht unmittelbar bildet, sondern sich erst allmählich entwickelt, muss der anfängliche oder erste Ausbau ohne Verlust seiner Stützfähigkeit eine gewisse Gesteinsverformung aufnehmen können. Dieser Gesichtspunkt ist beim Ausbau von bergmännischen Hohlräumen besonders wichtig, da sich die Zahl, Grösse und Anordnung der Hohlräume zueinander in einem Abbau ständig ändern. Demzufolge ändern sich auch die auf einen bergmännischen Hohlraum wirkenden Gebirgsspannungen, solange der betreffende Hohlraum in Benutzung ist.

Bei unterirdischen Bauwerken im Rahmen des Ingenieurbaus, bei denen es nach Beendigung der Ausbruchsarbeiten normalerweise nicht mehr zu Gebirgsspannungsänderungen kommt, ist man dazu übergegangen, den endgültigen Ausbau bzw. die Hohlraumauskleidung erst dann einzubringen, wenn sich die Konvergenz der Hohlraumwände stabilisiert hat (Rabcewicz et al., 1972).

Wechselwirkung zwischen Gebirge und Ausbau

Die Ausbauplanung bereitet bei unterirdischen Hohlräumen deshalb Schwierigkeiten, weil der aus Ausbau und umgebendem Gestein bestehende Verband im allgemeinen eine statistisch unbestimmte Grösse darstellt, d.h., die auf den Ausbau wirkende Kraft wird durch die Verformung beider Verbandskomponenten bestimmt. Bisher gelang es jedoch nur in ein paar wenigen idealisierten Fällen, diese komplizierte Wechselwirkung rechnerisch zu lösen (Salamon, 1974; Brady und Brown, 1985).

Praktische Überlegungen

In der Praxis werden daher viele der Entscheidungen über den Ausbau von unterirdischen Hohlräumen aufgrund von Erfahrungswerten und Deformationsmessungen getroffen. Schwerwiegende Fehlentscheidungen hinsichtlich des erforderlichen Ausbaus von unterirdischen, hohen Druckspannungen ausgesetzten Hohlräumen lassen sich jedoch durch Beachtung einiger einfacher Grundsätze vermeiden.

Zunächst empfiehlt es sich, die Beurteilung mit einer Analyse der Spannungsverteilung um den Hohlraum zu beginnen. In vielen Fällen, insbesondere bei Hohlräumen in massiven, spröden Gesteinen, lässt sich diese Analyse mit Hilfe der Elastizitätstheorie durchführen. Ein Vergleich zwischen den sich ergebenden Spannungen und den Festigkeitseigenschaften des Gesteins zeigt an, ob infolge des Spannungszustandes ein Bruch um den Hohlraum zu erwarten ist und wie gross das voraussichtliche Ausmass des Bruchs sein wird. Bei dieser Analyse sollte auch geprüft werden, inwieweit das Ausmass der Bruchzone von Unsicherheitsfaktoren bei den Festigkeitseigenschaften des Gebirges abhängt (Abb. 16).

Zweitens sollte ein Vergleich mit anderen Felsausbrüchen angestellt werden, deren Verhalten unter ähnlichen Gesteinsverhältnissen bereits bekannt ist. Damit ein derartiger Vergleich auch sinnvoll ist, sollte er anhand einer Spannungs- und Bruchzonen-Analyse durchgeführt werden. Beim Vergleich der erforderlichen Ausbaumassnahmen für Hohlräume in der gleichen Gesteinsformation, aber mit unterschiedlichen Dimensionen, ist die Verringerung der Gebirgsfestigkeit bei zunehmender Hohlraumgrösse zu berücksichtigen (Piper, 1984). Dies geschieht beispielsweise durch entsprechende Anpassung der Kluftabstände im geomechanischen Klassifizierungssystem des CSIR an die jeweilige Höhe des Hohlraumes (Abb. 17).

Drittens empfiehlt es sich, den Einfluss von bekannten geolgischen Diskontinuitäten auf das Verhalten der Hohlraumwände entweder mit Hilfe von Spannungsanalysen oder mittels Block- und Keilmodellen zu bestimmen. Und letztens sollte man Modellvorstellungen über das mögliche Zubruchgehen der Hohlraumwände entwickeln und anhand dieser Modelle den Ausbauentwurf beurteilen.

Bewährte Ausbauverfahren

Praktische Erfahrungen aus den Goldbergwerken von Witwatersrand sowie mit tiefen Stollen haben gezeigt, dass ein integriertes, aus Felsanker, Maschendraht und Seilanker bestehendes Ausbausystem den meisten Ausbauerfordernissen bei Hohlräumen mit hohen Druckspannungen in massiven, spröden Gesteinen gerecht wird. Gelegentlich wird eine dünne Spritzbetonschicht als Ergänzung zu diesem Ausbausystem verwendet.

Für den Entwurf von Ausbausystemen unter derartigen Bedingungen hat sich die Anwendung folgender Grundsätze gut bewährt:

(i) Die Länge der Fels- und Seilanker sollte mindestens die halbe Breite des Hohlraumes beim

Firstenausbau bzw. ungefähr die halbe Hohlraumhöhe beim Ausbau der Seitenwände betragen. Diese Faustregel kann man entsprechend den Analysenergebnissen für die Bruchzonen um einen Hohlraum abwandeln, wenn daraus hervorgeht, dass eine der halben Hohlraumhöhe entsprechende Ankerlänge bei Stollen mit sehr hohen Druckspannungen nicht ausreicht.

(ii) Wenn möglich sollte der Höchstabstand zwischen den Ankern die halbe Ankerlänge nicht übersteigen. Bei sehr langen Fels- oder Seilankern empfiehlt es sich, zusätzlich kürzere Anker im Ausbauschema zu verwenden, um dadurch den Ankerabstand an der Hohlraumleibung zu verringern.

(iii) Beim Firstenausbau müssen die Festigkeit der Anker und der Ankerabstand eine Gesamttragkraft ergeben, die etwa doppelt so hoch ist wie das Eigengewicht der von den Ankern erfassten Schichten.

(iv) In geschichtetem oder bankigem Gestein sollten die Anker wenn möglich in einem Winkel von mindestens 45° zu den Schichten verlaufen.

(v) Der bei Hohlräumen mit hohen Druckspannungen verwendete flexible Ankerausbau sollte durch Maschendraht oder geschweisste Matten bzw. Drahtmatten ergänzt werden, um Felsbrüche zwischen den flexiblen Ankern zu verhindern. Wird ein dauerhafterer Ausbau benötigt, so empfiehlt sich eine dünne Spritzbetonschicht zur Verringerung von Verwitterungserscheinungen und Erhöhung der Schlusssteinwirkung.

Unter Verwendung der vorgenannten Ausbausysteme ist es bereits gelungen, Stollen mit einem Querschnitt von 4 m x 4 m in Spannungsfeldern von über 120 MPa zu sichern. Hepworth und Gay (1986) wiesen anhand von Untersuchungen nach, dass man die Konvergenz in Stollen mit hohen Druckspannungen durch Erhöhung des Ausbauwiderstandes im Stollen beträchtlich verringern kann. Der Ausbauwiderstand wird in kN/m^2 der Stollenwandung angegeben und hängt von der Tragkraft der Gesteinsanker und der Ausbaudichte ab. Zur Berücksichtigung der nach Einbringen der Stützungen auftretenden Spannungsänderungen haben die Autoren den Ausbauwiderstand in Abhängigkeit von den Spannungsänderungen angegeben. Abb. 18 zeigt, dass zwischen den umgerechneten Werten für den Ausbauwiderstand und der Konvergenz der Stollenwände ein deutlicher Zusammenhang besteht. Aus diesen Ergebnissen geht hervor, dass unterhalb eines kritischen Ausbauwiderstandswertes die Konvergenz in Bergwerksstollen rasch zunimmt, während eine Vergrösserung des Ausbauwiderstands oberhalb des kritischen Wertes nur geringfügige Verbesserungen mit sich bringt. Brady und Brown (1985) kamen zu ähnlichen Ergebnissen bei einer Untersuchung über die Wirkung des Ausbaudruckes auf die Deformation der Wände eines kreisförmigen Stollens, auf den ein hydrostatisches Spannungsfeld wirkte.

Wiseman (1979) untersuchte die Wirksamkeit verschiedener Stollenausbausysteme in tiefliegenden südafrikanischen Bergwerken. Die Ergebnisse seiner Untersuchungen sind in Abb. 19 zusammengestellt. Interessant ist dabei die ziemlich enttäuschende Leistung des starren Stahlausbaus sowie der flexiblen Gleitbögen im Vergleich zur Leistung der verschiedenen Arten von Gebirgsankern. Es ist daher nicht verwunderlich, dass starrer Stahlausbau und flexible Gleitbögen für tiefe Abbaue im Hartgestein praktisch keine Verwendung mehr finden.

Spritzbeton hat sich in Verbindung mit vollverklebten Ankern und Maschendraht als die beste Ausbautechnik bei hohen Spannungszuständen bewährt (Hepworth und Gay, 1986).

Bei Schichten mit geringer Festigkeit, wie man sie in den meisten europäischen Kohlengruben antrifft, werden vielfach Gleitbögen eingesetzt (Jacobi, 1976). Dieses Ausbausystem hat sich besonders dann bewährt, wenn die Strecken durch den Abbau der Kohlenflöze beim Strebbau hohen Druckspannungen ausgesetzt sind. Unter diesen Bedingungen werden häufig Streckenkonvergenzen von über 50 % beobachtet. Aber auch dann, wenn sich die Strecken ausserhalb des Einflusses der Widerlagerspannung eines Strebbaus befinden, kann die Konvergenz oft mehr als 30 % betragen. Dies liegt daran, dass in der Praxis ein enger Verbund zwischen Gleitbögen und umgebendem Gestein schwer herzustellen ist und die Bögen eine geringe Radialsteife aufweisen. Diese Faktoren begünstigen eine ziemlich ungehinderte Ausdehnung der die Strecken umgebenden Bruchzone (Abb. 20).

Erste Versuche mit Klebeankern in tiefen Kohlengruben in Deutschland waren soweit in technischer und wirtschaftlicher Hinsicht und vom Gesichtspunkt der Bergbausicherheit aus erfolgreich (Reuther, 1987). Diese Ausbauweise wird daher in Zukunft vermehrt in Kohlengruben angewendet und soll die Stahlbögen weitgehend ersetzen.

HOHLRAUMVERHALTEN BEI SEISMISCHEN BEANSPRUCHUNGEN

In tiefliegenden Bergwerken sind seismische Beanspruchungen von unterirdischen, unter hohen Druckspannungen stehenden Hohlräumen eine häufig zu beobachtende Erscheinung, die aber auch sonst vielfach von Bedeutung ist. Die Seismizität kann untertägige Hohlräume auf verschiedene Weise beeinträchtigen. Erstens wirken seismische Wellen bei ihrer Ausbreitung durch das Gestein auf den Hohlraum und verursachen eine dynamische Spannungszunahme. Zweitens können Bodenerschütterungen den Hohlraum verformen und die Hohlraumauskleidung belasten. Drittens können Bodenerschütterungen lockere Gesteinsteile ausbrechen und den Ausbau zusätzlichen Kräften aussetzen.

Dynamische Spannungsänderungen

Mow and Pao (1971) untersuchten die Wirkung gleichbleibender P-Wellen auf zylindrische Hohlräume, wenn die Wellenfront parallel zur Stollenachse verläuft. Die Autoren kamen zu dem Schluss, dass die durch harmonische Schwingungen bedingte höchste Spannungskonzentration die statischen Werte bei P-Wellen um 10-15 % und bei S-Wellen um 5 % übersteigt. Diese Höchstwerte treten bei relativ grossen Wellenlängen in einer Grössenordnung auf, die einem 10- bis 20-fachen Stollendurchmesser entspricht. Die infolge von P-Wellen auftretende maximale Normalspannung σ_P sowie die maximale Scherspannung τ_{xy} ergeben sich aus

$$\sigma_p = \frac{(1 - \nu)\, E}{(1 + \nu)\,(1 - 2\nu)} \cdot \frac{V_p}{c_p} \qquad (18)$$

sowie aus

$$\tau_{xy} = \frac{G\, V_p}{2\, c_p}$$

Hierin sind E und G der Elastizitätsmodul bzw. der Schermodul des Materials, ν ist die Poissonzahl, V_p die maximale Teilchengeschwindigkeit und c_p die Geschwindigkeit der P-Welle.

Die Spannungen σ_p und τ_{xy} geben die Anderung der Gebirgsspannungen an, die durch eine auftretende P-Welle verursacht werden. Wenn $E = 50$ GPa, $\nu = 0{,}2$, $V_p = 1\ ms^{-1}$ und $c_p = 5{,}10^3\ ms^{-1}$, so beträgt die Anderung der Druckspannung ca. 5 MPa, was im Vergleich zu den in grossen Teufen auftretenden Spannungen gering ist. Es ist darauf hinzuweisen, dass die Anderung der dynamischen Gebirgsspannung proportional zur maximalen Teilchengeschwindigkeit V_p ist. Nach McGarr et al. (1981) gilt für die maximale Teilchengeschwindigkeit V_p bei Erschütterungen in Bergwerken folgende Beziehung:

$$\log R\, V_p = 3{,}95 + 0{,}57\, M_L \qquad (19)$$

Dabei ist R die Entfernung vom Erschütterungsherd und M_L die Grösse des seismischen Vorganges.

Nach Gleichung (19) ist die maximale Teilchengeschwindigkeit umgekehrt proportional zur Entfernung des Hohlraumes vom Erschütterungsherd und in der Praxis gewöhnlich geringer als 1 m/sec. Von weit entfernt auftretenden seismischen Vorgängen verursachte dynamische Spannungsänderungen können daher praktisch vernachlässigt werden. Dynamische Spannungsänderungen sind nur dann von Bedeutung, wenn sich der Hohlraum in unmittelbarer Nähe starker Erschütterungen befindet. Für die maximale Stauchung ε_p infolge einer auf einen Hohlraum zueilenden P-Welle gilt:

$$\varepsilon_p = \frac{V_p}{c_p} \qquad (20)$$

Da V_p in der Grössenordnung von ms^{-1} und c_p in der Grössenordnung von km^{-1} liegt, betragen die von seismischen Vorgängen hervorgerufenen Stauchungen $\varepsilon_p < 10^{-3}$. Dies gilt jedoch nicht für Hohlräume, die sich in unmittelbarer Nähe eines grossen Erschütterungsherdes befinden.

Erforderliche Ausbaumassnahmen

Wagner (1984) hat sich mit den Ausbaukriterien bei seismischen Beanspruchungen von Stollen befasst. Bei seinen Modellvorstellungen über die erforderlichen Ausbaumassnahmen ging er davon aus, dass die die Stollen umgebenden Gesteinsplatten einer Masse m eine Beschleunigung erfahren, die dem Wert der maximalen Teilchengeschwindigkeit V_p entspricht, und dass der Ausbau imstande sein muss, die kinetische Energie W_k dieser Gesteinsplatten ohne Bruchgefahr zu absorbieren.

$$W_k = \frac{1}{2}\bar{m}\, V_p^2$$

Nimmt man für die flexiblen Anker eine Grenzbelastung Fy von 100 kN und eine Dehnung $\Delta\ell$ von 20 mm an, d.h., typische Werte für Gesteinsanker in Goldbergwerksstollen, so lässt sich die Dicke der Gesteinsplatten, die gegen seismische Beanpruchung gesichert werden können, aus Abb. 21 ermitteln. Aus dieser graphischen Darstellung ist erkennbar, dass die Möglichkeiten der Sicherung mächtiger Gesteinsplatten mittels herkömmlicher Gesteinsanker ziemlich begrenzt sind. Wagner (1984) kam zu dem Schluss, dass sich die Wirksamkeit der Ausbausysteme bei dynamischen Spannungen nur durch Dehnungserhöhung verbessern lässt. Dies erreicht man entweder durch die Verwendung glatter Verbundanker oder durch die Einplanung von Dehnungselementen im Ankerschema.

SCHLUSSBEMERKUNG

Praktische Erfahrungen bei tiefen Bergwerken haben gezeigt, dass es durchaus möglich ist, Hohlräume bei Spannungszuständen nahe der Druckfestigkeit des Gebirges aufzufahren und auszubauen. Aus der Sicht des Planungsingenieurs ist der wichtigste, wenngleich unbekannte Faktor die Gebirgsfestigkeit und ihre Abhängigkeit von der Hohlraumgrösse. Die einzige Möglichkeit wirklicher Fortschritte auf diesem Gebiet ist die Rückanalyse von Hohlräumen mit hohen Druckspannungen auf internationaler Basis. Für Hartgestein liegen schon umfangreiche diesbezügliche Daten vor, doch für andere Gesteinsarten ist diese Arbeit noch zu leisten.

Bei der Anlage von Hohlräumen in Gesteinsmassen mit hohen Druckspannungen muss das Schwergewicht auf der Beherrschung der diese Hohlräume umgebenden Bruchzone liegen, um dadurch die Sicherung der Bruchzonen zu erleichtern.

Ein frühzeitiger Ausbau sowie die Verwendung relativ steifer Ausbausysteme sind am günstigsten, da sie die Restfestigkeit des gestörten Gebirges durch Verringerung der Deformation nach dem Bruch erhalten und zudem noch den Aufbau eines dreiachsigen Spannungszustandes in der Nähe der Hohlraumwände ermöglichen.

Eine seismische Beanspruchung unterirdischer Hohlräume stellt im allgemeinen kein grosses Problem dar. Beim Bergbau besteht jedoch die Gefahr, dass Erschütterungen in unmittelbarer Nähe von Hohlräumen auftreten und starke dynamische Spannungen verursachen können, welche den Ausbau unter Umständen überbelasten. In diesem Falle ist die maximale Teilchengeschwindigkeit V_p der kritischste Kennwert, da sie sowohl die dynamischen Spannungsänderungen als auch die Wirksamkeit des Ausbaus bestimmt.

LISTE DER ABBILDUNGEN (Abbildungen sind im englischen Teil wiedergegeben)

Abbildung 15. Entspannungsbohrlöcher im Streckenbereich eines Steinkohlenbergwerks

Abbildung 16. Einfluss der Gesteinsfestigkeit auf die Bruchzone um einen quadratischen Stollen (nach Piper, 1984)

Abbildung 17. Einfluss der Hohlraumdimensionen und der Gebirgsspannung auf die Bruchtiefe um Hohlräume

Abbildung 18. Einfluss des Ausbauwiderstandes auf die Stollenkonvergenz (nach Hepworth und Gay, 1986)

Abbildung 19. Wirksamkeit verschiedener Ausbausysteme bei hohen Spannungszuständen (nach Wiseman, 1979)

Abbildung 20. Einfluss der Gebirgsspannung und des Ausbauwiderstandes auf den Radius der Auflockerungszone um einen Stollen im Steinkohlenbergbau (nach Jacobi, 1976)

Abbildung 21. Wirksamkeit des Ankerausbaus in einem Goldbergwerksstollen bei seismischen Beanspruchungen

LITERATURVERZEICHNIS (Siehe englischer Teil)

09:20 Kurzvorträge
Barla, G., Scavia, C., Antonellis, M., Guarascio, M., Italien
Charakterisierung von Gesteinsmassen mit Hilfe der geostatistischen Analyse in der Masua Grube

Borg, T., Holmstedt, A., Schweden
Eine felsmechanische Untersuchung des längsgehenden Durchbruchs in der Grube Kiirunavaara

Budavari, S., Croeser, R.W., Süd-Afrika
Auswirkungen vom Abbau tafelförmiger Erzkörper auf vertikale Schächte in tiefen Gruben

Ewy, R.T., Kemeny, J.M., Zheng Ziqiong, Cook, N.G.W., U.S.A.
Entwicklung und Analyse stabiler Hohlraumformen für hohe Druckspannungen im Fels

10:00 - 10:30 Pause (Foyer)

Gale, W.J., Nemcik, J.A., Upfold, R.W., Australien
Die Anwendung von Spannungskontrollmethoden im Entwurf von Kohlegruben in einem starken seitlichen Spannungsfeld

Guenot, A., Frankreich
Spannungs-und Bruchbedingungen um Erdölbohrungen

Ishijima, Y., Fujii, Y., Sato, K., Japan
Durch Tiefkohlenbergbau verursachte mikroseismische Bewegungen

Jager, A.J., Piper, P.S., Gay, N.C., Süd-Afrika
Felsmechanische Aspekte der Hinterfüllung in tiefen südafrikanischen Goldbergwerken

Kersten, R.W.O., Greeff, H.J., Süd-Afrika
Einfluss der geologischen Eigenschaften und Spannungen auf Schächte: ein Beispiel

Kusabuka, M., Kawamoto, T., Ohashi, T., Yoichi, H., Japan
Integrierte Stabilitätsbestimmung einer Untergrundkaverne

Kyung Won Lee, Ho Yeong Kim, Hi Keun Lee, Korea
Messung von Felsverformungen und die Bestimmung der Entspannungszone um einen Tunnel in einer Kohlegrube

Matsui, K., Ichinose, M., Uchino, K., Japan
Der Einfluss der Restfesten und Abbaukanten auf die Konvergenz von darunter liegenden Abbaustrecken

Petukhov, I.M., U.S.S.R.
Vorhersage von, und Kampf gegen Gebirgsschläge: neue Entwicklungen

12:00 - 13:30 Mittagspause und Postersitzung (Marquette)

Tanimoto, C., Hata, S., Fujiwara, T., Yoshioka, H., Michihiro, K., Japan
Die Beziehung zwischen Verformung und Auflastdruck durch Tunnelvortrieb in überspanntem Fels

Udd, J.E., Hedley, D.G.F., Kanada
Gebirgsschlagsforschung in Kanada -.1987

Van Sint Jan, M.L., Valenzuela, L., Morales, R., Chile. Biegsamer Ausbau für Untertagestrecken in einer Grube mit Blockbruchbau

Vervoort, A., Thimus, J.-F., Brych, J., Crombrugghe, O. de, Lousberg, E., Belgien
Die Bestätigung der Einflüsse auf den Zustand des Hangenden im Streb durch eine Finite-Element-Berechnung

Yuen, C.M.K., Boyd, J.M., Aston, T.R.C., Kanada
Verhalten des Tunnelausbaus während des Vortriebs mit einer Lovat Vortriebsmaschine im Donkin Bergwerk, Nova Scotia, Kanada

Zhu Weishen, Lin Shisheng, Zhu Jiaqiao, Dai Guanyi, Zhan Caizhao, China
Einige praktische Fälle der Rückanalyse von Verschiebungen in Untertagebauten mit Berücksichtigung von Zeit und Raum

14:30 - 15:00 Pause (Foyer)

15:00 - 16:30 Diskussion (Le Grand Salon)

DISKUSSIONSLEITER: B.N. Whittaker (England)
Podiumsdiskussion: V.M. Sharma (Indien)
P.K. Kaiser (Kanada)
V. Maury (Frankreich)

Diskussionsbeiträge sind nur im englischen Teil wiedergegeben.

Untertagehohlraume im überbelasteten Fels - Thema IV

Zusammenfassung des Diskussionsleiters

B. N. Whittaker (England)

KURZFASSUNG: Es wird ein Überblick gegeben über die wichtigsten Fragen und Probleme, die als gemeinsamer Ausgangspunkt für die Betrachtung und Erörterung des Themas "Untertägige Hohlräume in überbeanspruchtem Gebirge" gelten. In der Zusammenfassung, die sich im wesentlichen mit Vorschlägen für künftige Forschungsarbeiten befaßt, wird auch auf einzelne Kongreßreferate eingegangen.

EINLEITUNG

Unter "Überbeanspruchung" versteht man gewöhnlich eine Gebirgsspannung, die unter normalen Verhältnissen die natürliche Festigkeit des Gebirges übersteigt und zu gewissen Formen des Felsbruchs führen kann. Die Folgen einer Überbeanspruchung werden an spezifischen Beispielen dargestellt. So kann z.B. auf Strecken in tiefen Kohlenbergwerken eine Überbeanspruchung auf die geringe Festigkeit der umgebenden Gesteinsschichten zurückzuführen sein, was einen Felsbruch um den Hohlraum und meist zu einem gewissen Grad eine Querschnittsverminderung verursacht. Umgekehrt kann es bei überbeanspruchten Kohlenpfeilern aufgrund des Schlankheitsgrads der Pfeiler plötzlich zum Verbruch kommen, obwohl die meisten Kohlenpfeiler so entworfen sind, daß es bei Überbeanspruchung des Gebirges lediglich zu einer beschränkten Verformung kommt. Untertägige bergmännische Ausbrüche im Hartgestein sind verschieden grossen Überbeanspruchungen ausgesetzt - von lokalen Störungen im Hangenden oder in den Seitenwänden bis zur Schließung, wie sie bei Gebirgsschlägen auftreten kann. Wichtige Einflußfaktoren hierbei sind Gesteinseigenschaften (vor allem das Formänderungsverhalten), geologische Bedingungen und Spannungsfelder.

Bei Ausbruchsarbeiten ist man stets darauf bedacht, eine realistische Vorhersage des Umfangs der Schließung und der Standfestigkeit zu machen und Maßnahmen zur Bekämpfung der durch die Überbeanspruchung verursachten Probleme zu treffen. Für diese Zwecke stehen eine Reihe von Prognoseverfahren zur Verfügung, mit denen sich das vermutliche Spannungsfeld in der Umgebung des Ausbruch bestimmen läßt und mit denen neuerdings auch festgestellt werden kann, wie sich verschiedene Stützmaßnahmen auf die Festigkeit auswirken. Über die spezifischen Probleme bei tiefliegenden Ausbrüchen in überbeanspruchtem Gebirge, vor allem über die Möglichkeiten eines plötzlichen Felsbruchs oder sogar eines Gebirgsschlags herrscht noch beträchtliche Unklarheit. Besondere Aufmerksamkeit wurde der Verbesserung der Modelltechnik im Hinblick auf ihren Einsatz als Planungswerkzeug geschenkt. Modelle zur Beschreibung des Bruchgeschehens spielen dabei eine wichtige Rolle und werden ohne Zweifel in Zukunft breitere Anwendung finden, da Planungsingenieure über mögliche Bruchzonen besser informiert sein müssen. Modelle zur Beschreibung von Bruchvorgängen tragen zu einem besseren Verständnis der Auswirkungen einer Überbeanspruchung auf Planung und Sicherung von Ausbrüchen bei.

GRUNDLEGENDE FRAGEN UND PROBLEME

In Tabelle 1 sind die wichtigsten Fragen und Probleme zusammengefaßt. Geeignete felsbauliche Lösungsvorschläge können nur dann erbracht werden, wenn die im Gebirge auftretenden Spannungen und - was ebenso wichtig ist - die Reaktion des Gebirges auf die Ausbrüche bekannt sind.

Tabelle 1. Ungelöste Probleme und Fragen im Hinblick auf untertägige Hohlräume in überbeanspruchtem Gebirge

Grundlegende Probleme:

a) Vorhersage der durch Hohlräume im Gestein verursachten Spannungen

b) Anwendung der Spannungsanalyse auf die genaue Vorhersage der im Umfeld von Hohlräumen auftretenden Gebirgsbewegungen

Offene Fragen:

1. Wie wirkt sich der primäre Spannungszustand auf das Hohlraumverhalten aus?

2. Wie wirken sich die Schichtungsverhältnisse auf Spannungsumverteilung und Hohlraumverhalten aus?

3. In welchem Ausmaße kommt es bei bestimmten Spannungszuständen zum Verbruch in der Umgebung von Hohlräumen?

4. Welche plastischen Verformungseigenschaften besitzt gestörtes Gebirge und in welcher Weise beeinflussen sie die Planung und das Verhalten von Hohlräumen?

5. Welchen Einfluß hat die Zeit auf das Spannungsverhalten des Gesteins in bezug auf Deformation, Bruch und Gebirgsbewegung?

Die fünf in Tabelle 1 genannten, noch ungelösten Problembereiche werden im folgenden einzeln erörtert.

1. Es bedarf einer eingehenden Identifizierung geologischer Faktoren, z.B. größerer Schwächestellen und Diskontinuitäten, da das Spannungsfeld sonst falsch eingeschätzt werden kann. Von besonderer Dringlichkeit ist es, die Bedeutung verschiedener geologischer Störungen und Diskontinuitäten für allgemeine Spannungsfelder realistisch zu bewerten.

2. Die Lagerungsverhältnisse üben einen entscheidenden Einfluß auf das Bruchgeschehen und die resultierende Verformung in unterirdischen Hohlräumen aus. Es ist zu untersuchen, wie die Schichtenfolge die Fortpflanzung von Felsbrüchen verhindern und zu asymmetrischen Verformungseffekten führen kann.

3. Im allgemeinen kommt es im Laufe der Zeit in der Umgebung untertägiger Hohlräume zu Felsbrüchen; ein besseres Verständnis solcher Ereignisse ist daher erforderlich.

4. Überbeanspruchung bedeutet möglicher Bruch; Probleme ergeben sich daher häufig aus dem Verhalten des Gesteins nach dem Bruch.

5. Verwitterung und durch Druck-/Zugwechsel verursachte Alterung sind beim Fels zeitlich begrenzt. Es ist daher zu untersuchen, wie sich der Fels mit der Zeit verändert. Die Planung unterirdischer Hohlräume hat daher innerhalb zulässiger Grenzen zu erfolgen, die von Faktoren wie Form, Größe und Sicherung abhängen. Hieraus ergibt sich die Frage nach der Bedeutung dieser Faktoren für überbeanspruchtes Gestein.

ZUSAMMENFASSUNG DER WICHTIGSTEN VON DEN EINZELNEN REFERENTEN ANGESPROCHENEN FRAGEN DURCH DEN DISKUSSIONSLEITER

Die Referenten auf diesem Kongreß beschäftigten sich mit einer Vielfalt von Problemen im Zusammenhang mit dem Entwurf und der Festigkeit untertägiger Hohlräume, und die folgende Zusammenfassung konzentriert sich auf jene Punkte, die von besonderer Wichtigkeit für die Hauptthemen des Themenkreises IV sind. Bei den obengenannten Referenten handelt es sich um jene Autoren, deren Referate in Band 2 der Berichte des Sechsten Internationalen IGFM-Kongresses der Felsmechanik veröffentlicht werden. Der vollständige Text der Referate ist daher in Band 2 der Kongreßberichte zu finden. Die Themen der einzelnen Beiträge zu Themenkreis IV lassen sich in vier Hauptabschnitte einteilen, zu denen im folgenden einzeln Stellung genommen wird. Aus der großen Zahl ausgezeichneter Referate konnten in diesem Rahmen nur wenige zur Erläuterung der wichtigsten felsmechanischen Probleme im Zusammenhang mit Ausbrüchen in überbeanspruchtem Gebirge herangezogen werden.

1. Gebirgsfestigkeit sowie Methoden zum besseren Verständnis der Standfestigkeit untertägiger Ausbrüche

Bawden und Milne (Kanada) haben die Verwendung numerischer und empirischer Modelle zur Erhöhung der Standfestigkeit im Grubenbau überzeugend dargestellt. Ein weiterer wichtiger Beitrag zu diesem Thema sind numerische Modelle zum besseren Verständnis des Verhaltens von Kohlepfeilern (Duncan-Fama und Wardle/Australien).

Del Greco et al. (Italien) haben gezeigt, wie die Festigkeit des Firstes unterirdischer Hohlräume in bezug auf das Druckgewölbe bewertet werden kann.

Vervoort et al. (Belgien) haben bewiesen, daß die Methode der finiten Elemente erfolgreich zur Prognose von Firstbedingungen in bergmännischen Hohlräumen eingesetzt werden kann. Auch Cunha (Portugal) zeigt, daß Modellversuche ein wichtiges Instrument zur Einschätzung der Standsicherheit von Stollen darstellen.

Ewy et al. (USA) haben mit ihren Modellvorstellungen zur Bildung standfester Profile durch fortschreitenden Bruch einen wertvollen Beitrag zu diesem Thema geleistet.

Wichtige Erkenntnisse über die Rolle von Messungen bei der Bewertung von Konvergenz und Spannungsverhältnissen zur besseren Planung unterirdischer Ausbrüche verdanken wir Herget und Mackintosh (Kanada).

Hier soll der Bericht von Maury über die Standfestigkeit von Einzelstollen erwähnt werden, der besonders im Hinblick auf die Auswahl einer geeigneten Theorie für die Bestimmung der Standsicherheit von Bedeutung ist.

2. Ausbaumaßnahmen bei Ausbrüchen in überbeanspruchtem Gebirge

Seit vielen Jahren beschäftigen sich Bergingenieure mit dem Problem der Schachtsicherheitspfeiler und der dazugehörigen Abbaumethode. Ihm widmet sich der Beitrag von Budavari und Croeser (Südafrika), der zeigt, daß der Abbaubetrieb durch die Anwendung felsmechanischer Grundsätze leistungsfähiger gestaltet werden kann. Das Referat von McKinnon (ebenfalls Südafrika) ist eine Ergänzung zum obigen Beitrag.

Auch heute noch wird das felsbauliche Thema Stützmaßnahmen weltweit diskutiert. Tan (China) hat Interessantes über die Anwendung von NATM unter schwierigen Bedingungen zu berichten. Die Felsbewehrung ist auch weiterhin eine wichtige Aufgabe des Ausbaus, und wertvolle Beiträge zu dieser Frage stammen von Thompson (Australien), Aydan et al. (Japan) und Ballivy (Kanada).

Jager et al. (Südafrika) berichten, daß die Hinterfülltechnik ein äußerst wirksames Mittel darstellt, um die auf die Pfeiler wirkenden Spannungen zu reduzieren und so für größere Standfestigkeit und somit bergbauliche Sicherheit zu sorgen.

Weitere erwähnenswerte Beiträge zur Wechselbeziehung zwischen Gebirge und Stützbauten stammen von Van Sint Jan (Chile), Barroso und Lamas (Portugal), Fine (Frankreich), Korini (Albanien) und Yuen et al. (Kanada).

3. Verhalten, Einschätzung und Standfestigkeit von Ausbrüchen in überbeanspruchtem Gebirge

Die Untersuchungen von Barlow und Kaiser (Kanada) zeigen deutlich, wie wichtig die Auswertung von Konvergenzmessungen für die genauere Bewertung von Ausbruchs- und Verbauaspekten im Stollenbau ist. Wie Esterhuizen (Südafrika) berichtet, kann aus Spannungsfeldverlagerungen auf das Verhalten des Stollens geschlossen werden.

Das Stollenverhalten wird durch zeitabhängige Reaktions-Kennlinien gekennzeichnet. Wertvolle Erkenntnisse zu dieser Frage stammen von Gill und Ladanyi (Kanada), die zeigten, wie der Zeitfaktor bei der Prognose des Stollenverhaltens berücksichtigt werden kann. Auch Minh und Rousset (Frankreich) haben erfolgreiche Studien über den Einfluß des Zeitfaktors auf das Stollenverhalten angestellt. Laut Sharma et al. (Indien) gilt der Standfestigkeit von Stollen weiterhin große Aufmerksamkeit.

Die Bruchbildung im Grubenbau wurde von Ortlepp und Moore (Südafrika und Großbritannien) eingehend untersucht; ihnen verdanken wir wichtige Erkenntnisse über Geschwindigkeit, Art und allgemeine Eigenschaften der bei Brüchen stattfindenden Bewegungen. Lee und Kim (Korea) berichten über die Auswertung von Fließbereichdaten, aus der sie grundlegende Schlüsse hinsichtlich des Entspannungsgeschehens in der Umgebung von Stollen ziehen. Ishijima et al. (Japan) betonen die Bedeutung mikroseismischer Vorgänge als Hilfsmittel bei der Interpretation von Brucherscheinungen in der Umgebung von Ausbrüchen.

4. Gesteinscharakterisierung und -verhalten sowie Spannungsbeherrschung in unterirdischen Ausbrüchen

Die Gesteinscharakterisierung von Barla und Scavia (Italien) zeigt die Bedeutung geostatistischer Analysen für die Beurteilung des Einflusses von Gesteinsdiskontinuitäten auf das Verhalten des Gebirges in unterirdischen Hohlräumen.

Von besonderem Interesse an der Arbeit von Bandis et al. (Norwegen) ist ihre Interpretation des Spannungs- und Bruchverhaltens von Hohlräumen in schlechtem Gebirge. Ohne Zweifel rechtfertigen die von den Autoren erzielten Fortschritte weiterführende Untersuchungen bei anderen Gesteinsgefügen.

Gale et al. (Australien) haben bewiesen, wie wichtig eine Beherrschung der Spannungsverhältnisse im Gestein für den Entwurf von Untertagebetrieben, besonders für die Optimierung von Grubenzuschnitt und Abbauweise zur Reduzierung von Spannungseinflüssen ist.

Gebirgsschläge stellen weiterhin ein schwieriges felsbauliches Problem dar, und Udd und Hedley (Kanada) haben einen wichtigen Beitrag zum Verständnis dieser Erscheinung und zu ihrer Beherrschung geliefert. Von gleicher Wichtigkeit ist die von Suguwara et al. (Japan) geleistete Arbeit zur Erforschung plötzlicher Kohlenausbrüche und ihrer Beherrschung.

Kersten & Greef (Südafrika) weisen auf die Wechselbeziehung zwischen Gebirgsbeschaffenheit und Bergbau hin und zeigen einen Weg zum besseren Verständnis der Standfestigkeit beim Abteufen von Tiefschächten. Ihre Erklärung dieser Wechselbeziehung und ihre wichtigsten Beobachtungen sind auch für ähnliche Standfestigkeitsprobleme bei anderen untertägigen Ausbrüchen in überbeanspruchtem Gebirge von Bedeutung. Auch unser Hauptreferent, Horst Wagner (Südafrika), hat die Wichtigkeit dieses Aspekts betont und darauf hingewiesen, wie sich die Standfestigkeit in überbeanspruchtem Gebirge durch die Wahl der Ausbruchsform positiv beeinflußen läßt.

Abschließend komme ich nun wieder auf die Liste der noch offenen Fragen in Tabelle 1 zurück. Diesen fünf Fragenkomplexen sollten in zukünftigen Forschungsarbeiten gebührende Beachtung geschenkt werden.

ANHANG

DIE STANDFESTIGKEIT VON STOLLEN IN WEICHGESTEIN

Dieser Anhang wurde in den Bericht des Diskussionsleiters aufgenommen, um einen kurzen Überblick über die in einem felsmechanischen Forschungszentrum in Großbritannien zu diesem Thema geleisteten Arbeit zu geben.

Problemstellung

Wie alle anderen Teilgebiete der Mechanik beschäftigt sich auch die Felsmechanik mit dem dynamischen Verhalten von Stoffen und läßt sich daher in zwei grundlegende Bereiche unterteilen:

(i) den Unterschied in der Beschaffenheit eines Mediums zwischen zwei Punkten und die dadurch verursachte Ungleichheit dieses Mediums, die als "Treibpotential" bezeichnet wird,

und

(ii) die Bewegung des Mediums vom höheren zum niedrigeren Treibpotential, bis das Gleichgewicht wieder hergestellt ist.

In der Felsmechanik ist das Treibpotential der durch Ausbrüche im Gestein verursachte Spannungszustand, bei dem das Spannungsgleichgewicht durch das Nachrücken des Gesteins in den Hohlraum - mit oder ohne Ausbau - wieder hergestellt wird.

Die Prognose des Stollenverhaltens kann in zwei Problembereiche aufgegliedert werden:

a) Prognose der im Gestein durch den Hohlraum verursachten Spannungen und des Ortes potentieller Felsbrüche,

b) genaue, auf die Ergebnisse der Spannungsanalyse gestützte Prognose des Bewegungsablaufs in der Umgebung des Hohlraums.

Folgende Faktoren wirken sich auf die Standfestigkeit von Stollen im weichen Gestein von Kohlenlagern aus:

(i) der primäre Spannungszustand des unverritzten Gebirges,

(ii) die Festigkeit des Gebirges vor dem Bruch, wobei die Gebirgseigenschaften sich auf die Bildung der Bruchzone auswirken, und die Festigkeit des Gebirges in der Bruchzone, die als Verhalten nach dem Bruch bezeichnet wird,

(iii) die herrschenden Lagerungsverhältnisse und die relative Festigkeit der Streifenarten,

(iv) Form und Größe des Stollens sowie Art und Festigkeitskennwert des Stollenausbaus,

(v) die durch die Spannungsumlagerungen im Gebirge verursachte elastische Bewegung (vor dem Bruch) und plastische Bewegung (nach dem Bruch); die elastische Bewegung ist eine Folge des Poisson-Effekts und verhältnismäßig unbedeutend im Vergleich zur plastischen Bewegung des gestörten Gebirges, die von der auf die Spannungsumlagerung zurückzuführende Volumenzunahme verursacht wird (Whittaker und Frith, 1987).

Um die Modellbildung so realitätsnah wie möglich zu gestalten, sollten alle Faktoren in einer einzigen Entwurfmethode berücksichtigt werden. Dazu bedarf es jedoch noch umfangreicher Vorarbeiten hinsichtlich der Feststellung der für jeden einzelnen Faktor geltenden Gesetzmäßigkeiten. Da sich auf theoretischen Überlegungen aufbauende Gesetzmäßigkeiten sehr häufig als unzulänglich erwiesen haben, müssen auf Meßergebnissen beruhende empirische Gesetze aufgestellt werden. Das an der Universität von Nottingham entwickelte Prognosemodell ist daher stark empirisch ausgerichtet.

Die Genauigkeit von Planungsmodellen kann auch anhand von Konvergenz- und Fließbereichsmessungen in Stollen nachgeprüft werden. Da zur Gewinnung solcher Daten umfangreiche unterirdische Messungen erforderlich sind, werden zur Zeit von der Universität von Nottingham diesbezügliche Meßprogramme in mehreren Kohlenbergwerken in Großbritannien durchgeführt (Whittaker, 1981).

Gegenwärtiger Stand der Arbeiten an der Universität von Nottingham

Das Forschungsprojekt der Universität von Nottingham begann mit der Ausarbeitung von Methoden und Richtlinien für die Durchführung der Untersuchungsarbeiten an Kohlelagern nach dem Bruch, die dann zur Gewinnung von Daten für Vorhersagemodelle herangezogen werden. Mit Hilfe dieser Richtlinien wurde eine Reihe von Gesteinen unter besonderer Berücksichtigung der Festigkeit vor und nach dem Bruch und der räumlichen Formänderung untersucht. Aus den Ergebnisse wurden Gesetzmäßigkeiten für die Gebirgsfestigkeit und für die plastische Bewegung von gestörtem Gebirge (Whittaker et al., 1985) abgeleitet. Die Resultate dieser Arbeiten haben entscheidend zum Erfolg des Planungsmodells beigetragen, das nun etwas näher betrachtet werden soll.

Das derzeit verwendete Modell beruht auf den folgenden Annahmen:

(i) es gilt für Stollen mit Kreisquerschnitt von beliebiger Größe,

(ii) anfänglich werden hydrostatische Spannungszustände berücksichtigt,

(iii) der Stollenausbau übt unabhängig von der Gesteinsverformung einen konstanten Druck auf das Gestein aus,

(iv) den Lagerungsverhältnissen wird auf folgende Weise Rechnung getragen:

Speicherung der Daten in einem Computer;

Aufteilung der Gesteinsmasse um den Tunnel in kleine, voneinander unabhängige Elemente; jedem dieser Elemente wird die Gesteinseigenschaft des Gesteins zugewiesen, in dem sich das jeweilige Element befindet,

(v) die Gesteinsfestigkeit wird durch folgende Gleichung gegeben:

$$\sigma_1 = A\sigma_3^B + C$$

in der A, B und C nur für die jeweilige Gesteinsart und den jeweiligen Bruchzustand gelten; diese Gleichung wurde anhand der Daten aus den oben erwähnten Felsuntersuchungsprogrammen aufgestellt,

(vi) die räumliche Formänderung des gestörten Gebirges wird durch folgende Gleichung gegeben:

$$V_e = ae^{-\sigma_3/b} + c$$

in der a, b und c nur für die jeweilige Felsart gelten. Diese Gleichung liefert die maximale voraussichtliche räumliche Formänderung und stellt somit den schlimmstmöglichen Fall dar, was für den ersten Entwurf von Vorteil ist (Whittaker et al., 1984).

Die Spannungsanalyse erfolgt mit Hilfe der Differentialgleichung für die zylindrische Symmetrie bei Gleichgewichtsbedingungen. Bei dieser Gleichung handelt es sich um eine vereinfachte Spannungsgleichung für Stollen mit Kreisquerschnitt und hydrostatischen Spannungs-Ausgangsbedingungen. Daraus ergibt sich die Notwendigkeit für Annahme (i) und (ii). Annahme (iii) ergibt sich daraus, daß es im Hinblick auf eine leichtere spezifische Auflösung der sich ergebenden Spannungsgleichung einer Randbedingung bedarf. Da mit derartigen Gleichungen keine analytischen Lösungen möglich sind, ist für Annahme (iv) bis (vi) der Einsatz eines Mikrocomputers erforderlich. Das Verfahren ist daher eher numerischer als analytischer Art und wird als Methode der unabhängigen Elemente bezeichnet.

Ein allen derartigen prognostischen Methoden gemeinsames Problem ist die Gebirgsklassifikation. Es wird allgemein anerkannt, daß Laborwerte für die Festigkeit des ungestörten Gebirges um einige Größenordnungen höher liegen als in situ, ein Umstand, der auf die selektive Probenahme bei In-Situ-Versuchen zurückzuführen ist. Zur Angleichung der Laborwerte an die In-Situ-Werte sind wohl verschiedene Methoden vorgeschlagen worden, doch erfordern die meisten genaue geologische Informationen, die nicht immer zur Verfügung stehen. Es mußte daher ein anderer Weg eingeschlagen werden; ein solcher bietet sich tatsächlich auch in der Annahme, daß sich aus der Festigkeit des Gesteins nach dem Bruch auf seine ursprüngliche Festigkeit schließen läßt. Auch hier handelt es sich wieder um den schlimmstmöglichen Fall; ein Vergleich zwischen Meß- und Prognosedaten für die Stollenkonvergenz hat jedoch sehr befriedigende Korrelationen ergeben (siehe Abb. 1) (Kapusniak et al., 1984). Dadurch ist die Richtigkeit der obigen Annahme indirekt bewiesen, und diese wird jetzt in Nottingham routinemäßig als Planungsgrundlage im Stollenbau verwendet.

Das gegenwärtige Modell ist verhältnismäßig einfach und beruht auf fundierten, von Gesteinsmessungen abgeleiteten Gesetzmäßigkeiten für das Gesteinsverhalten; mit seiner Hilfe kann auch den Lagerungsverhältnissen des betreffenden Gebirges Rechnung getragen werden. Seine Gültigkeit ist durch einen Vergleich zwischen vorhergesagten und gemessenen Konvergenzwerten in einem Stollen mit Kreisquerschnitt nachgewiesen worden. Das Modell gilt jedoch nur für Stollen bei hydrostatischen Spannungen, wobei der Einfluß der Ausbauten durch einen einzigen Stützdruckwert dargestellt wird. Auf dieser Grundlage kann die Forschung nun weiter aufbauen.

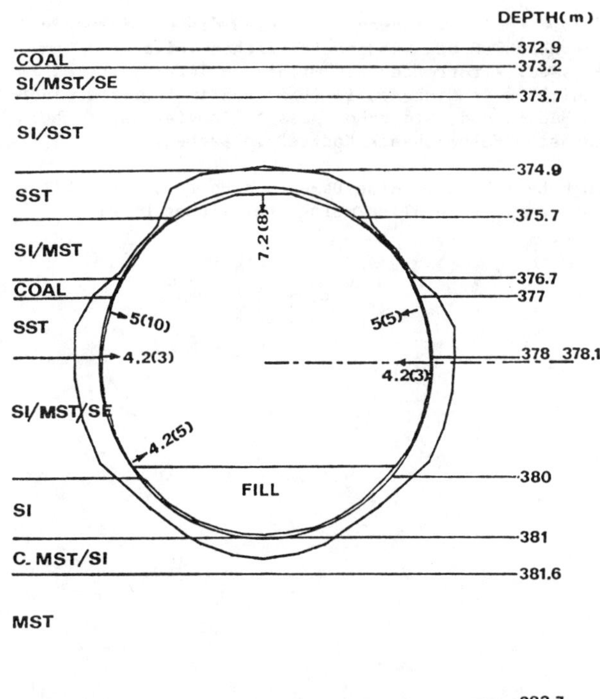

(ANMERKUNG: Radiale Bewegungswerte sind in cm angegeben.)

Abbildung 1. Berechnung der radialen Verformung und des Fließbereichs für einen Stollen mit Kreisquerschnitt in Schichtgestein

Ausblick

Der Streckenausbau ist ein wesentlicher Bestandteil der Felsmechanik und sollte in felsmechanische Betrachtungen voll einbezogen werden. Im Prinzip hat jede Ausbauart ein charakteristisches Spannungs-/Deformationsverhalten, das in die jeweiligen Modellvorstellungen eingehen muß, denn der betreffende Ausbau kann den Verformungskräften nur bis zu einer gewissen Grenze sicher widerstehen. Jenseits dieses Grenzwertes können die Deformationen die Standfestigkeit des Stollens beeinträchtigen, was eine andere Ausbauart erforderlich machen könnte. Zur Zeit sind Forschungsarbeiten im Gange mit dem Ziel, an Hand von maßstäblichen Modellen die jeweiligen Ausbaueigenschaften festzustellen und die gewonnenen Daten abzuspeichern und nach Bedarf abzurufen, um sie in die Planungsmodelle zu integrieren. Auf diese Weise können im Falle eines totalen oder partiellen Verlustes der Ausbaufestigkeit - was in tiefen Kohlelagern nach Felsbrüchen häufig der Fall ist - die Streckenverformungen genau vorausgesagt werden.

Eines der wichtigsten Anliegen ist es, den Einfluß verschiedener ursprünglicher Spannungen auf unterschiedliche Stollenformen festzustellen. Dazu bedarf es offensichtlich einer anderen Methode als der herkömmlichen Spannungsgleichungen. Ein Vorschlag geht zur Zeit dahin, die Methode der finiten Elemente in entsprechend abgewandelter Form auf die bereits aufgestellte Gesteinsfestigkeitsgleichung anzuwenden. Auf diese Weise könnten verläßliche Spannungsanalysen für verschiedene Stollenformen bei verschiedenen primären Spannungen durchgeführt werden. Der Anwendung der Methode der finiten Elemente sind jedoch Grenzen gesetzt, wenn nur elastische Verformungen berücksichtigt werden, da diese bei großen Fließbereichen unbedeutend sind. In diesen Fällen würde die Bewegungsanalyse auch weiterhin mit dem bereits bewiesenen Volumendehnungssatz durchgeführt werden, jedoch auf der Grundlage der verfeinerten spannungsanalytischen Methode. Eine solche Kombinationsmethode dürfte ein äußerst verläßliches Werkzeug für die Planung unterirdischer Hohlräume darstellen.

Von besonderer Wichtigkeit ist die Berücksichtigung verschiedener Stollenformen und Spannungszustände. Im britischen Kohlenbergbau sind Stollen mit Kreisquerschnitt noch ziemlich selten, und man betrachtet hydrostatische Spannungszustände nur als eine verallgemeinerte Regel. In letzter Zeit hat man festgestellt, daß anisotrope, auf geologische Verwerfungen und Gebirgsstörungen zurückzuführende Spannungszustände Stolleneinstürze verursacht haben. Weitere Untersuchungen zu diesem Thema sind daher von größter Wichtigkeit, damit an Hand verläßlicher Modelle wirksamere Ausbaumaßnahmen erstellt werden können.

Schlußwort

Die an der Universität Nottingham geleistete wissenschaftliche Arbeit hat mit der Erforschung des Gesteinsverhaltens und der Entwicklung von Berechnungsprogrammen eine realistische Grundlage für die Lösung felsmechanischer Probleme geschaffen. Das gegenwärtige Modell kann wegen der verwendeten Spannungsgleichungen vorerst nur beschränkt Anwendung finden, doch hat es schon zur Aufstellung wertvoller Korrelationen zwischen Konvergenzprognosen und -meßwerten beigetragen. Es ist noch sehr entwicklungsfähig, und Arbeiten in dieser Richtung dürften beträchtliche Fortschritte in der Stollenplanung erzielen.

LITERATURVERZEICHNIS (Siehe englischer Teil)

16:30 - 17:00 SCHLUSSVERANSTALTUNG
3. September 1987

Die Fachsitzungen waren am Donnerstag, dem 3. September 1987, zu Ende. Der Kongreßvorsitzende, Dr. G. Herget, bat den Ehrenvorsitzenden, Prof. Dr. B. Ladanyi, im Namen des kanadischen Organisationsausschusses den Vorsitzenden, Referenten, Diskussionsleitern und -teilnehmern für ihre vorzügliche Zusammenarbeit in den Fachsitzungen zu danken. Prof. Ladanyi dankte allen und überreichte den führenden Amtsträgern je eine Inuit- Plastik.

Anschließend hielt Prof. Dr. Wittke, Bundesrepublik Deutschland, sein Schlußreferat mit einer persönlichen Stellungnahme zu der seit der Gründung der Internationalen Gesellschaft für Felsmechanik vor 25 Jahren geleisteten Arbeit und den noch anstehenden Problemen.

Prof. E.T. Brown, Präsident der Internationalen Gesellschaft für Felsmechanik, beschloß den Kongreß mit folgenden Worten:

"Herr Vorsitzender, meine Damen und Herren!

Bei der Eröffnungsveranstaltung dankte ich im Namen der IGFM unseren kanadischen Kollegen für das herzliche Willkommen, das uns zuteil geworden war, und für die vorzügliche Vorbereitung des Kongresses. Nun, da wir unsere Beratungen abgeschlossen und miterlebt haben, wie ausgezeichnet alles organisiert war, möchte ich nochmals unseren herzlichsten Dank für alle ihre Mühewaltung zum Ausdruck bringen.

Wie Prof. Wittke, unser geschätzter Altpräsident, bereits sagte, hat sich unsere Disziplin in den 25 Jahren seit der Gründung der IGFM sehr rasch entwickelt, wenn auch noch immer sehr viel zu tun bleibt. Im Laufe des Kongresses hatte ich Gelegenheit, mich mit so manchen von der "alten Garde", zu unterhalten, d.h. mit Felsmechanikern in meinem Alter oder ein klein wenig älter, und wir konnten nicht umhin, über die Fähigkeiten und Leistungen vieler jüngerer Wissenschaftler zu staunen, die in unsere fachlichen Fußstapfen getreten sind, darunter einiger vielversprechender Kandidaten für die Manuel-Rocha-Medaille der IGFM. Die Felsmechanik wird immer mehr zu einem Gebiet, in dem die elektronische Datenverarbeitung eine wichtige Rolle spielt. Angesichts dieser Tatsache und in dem Wissen um unseren vorzüglichen Nachwuchs - besonders hier in Nordamerika - bin ich überzeugt, daß uns auf dem nächsten Kongreß in Aachen im Jahre 1991 weitere Überraschungen und Fortschritte erwarten. Ich darf Sie heute schon bitten, an diesem Kongreß teilzunehmen!

Wie wir sicher alle wissen, ergibt sich ein Kongreß wie der unsere nicht von selbst; es bedarf dazu unermüdlicher Planung und Organisation bis ins kleinste Detail - eine Aufgabe, deren sich der Organisationsausschuß und seine vielen Mitarbeiter bestens entledigt haben. Unser Dank gilt ganz besonders Prof. Branko Ladanyi, der bei den Vorbereitungen zu diesem Kongreß Pate gestanden und die Tätigkeit des Organisationsausschusses in einer Weise geleitet hat, die man vielleicht als gütig bezeichnen könnte - obwohl ich mir dessen nicht ganz sicher bin. Manche Mitglieder des Organisationsausschusses haben sich um die Vorbereitung der wissenschaftlich-technischen Seite des Kongresses besondere Verdienste erworben, vor allem Prof. Norbert Morgenstern, der ein Programm von ungewöhnlich hohem Niveau zusammengestellt hat. Ein besonderer Erfolg, so schien es mir bei vielen Gesprächen, sind die Workshops gewesen.

Dr. Bill Bawden ist der Organisator der ausgezeichneten Fachausstellung, eine der besten, die ich je auf einer felsmechanischen Veranstaltung gesehen habe. In diesem Zusammenhang zeichnet Will persönlich verantwortlich für den Dudelsackbläser, der den Minister bei der Eröffnung der Ausstellung am Montag begleitete.

Will organisierte auch die Fachsitzungen, wobei er den Vorsitzenden und Referenten Instruktionen zu geben und auf deren Einhaltung zu achten hatte. Die Wahl Will Bawdens für diese Aufgabe war genial, denn es hätte sie niemand besser lösen können.

Prof. Denis Gill trug als 2. Kongreßvorsitzender die Verantwortung für die sogenannten örtlichen Veranstaltungen, die meiner Meinung nach aber fast alles umfaßten, was vor sich ging. Denis zeigte sich allen Eventualitäten souverän gewachsen, und wir alle sagen ihm dafür herzlichen Dank.

Alle Fäden jedoch liefen in den Händen des Kongreßvorsitzenden, Dr. Gerhard Hergets, zusammen, dem alles zu gelingen scheint. Gerhards Arbeit und die des von ihm geleiteten Ausschusses begann vor einigen Jahren und ist noch lange nicht zu Ende. Wer von uns längere Zeit eng mit Gerhard zusammengearbeitet hat, wie z.B. der Generalsekretär und ich, weiß nur zu gut, wie viel er zu dem Erfolg dieses Kongresses beigetragen hat.

Wir danken euch - Branko, Gerhard und den anderen Mitgliedern des Organisationsausschusses - und beglückwünschen euch zu dieser gelungenen Veranstaltung.

Die IGFM ist wahrhaftig eine internationale Gesellschaft, hat sie doch drei offizielle und mehrere inoffizielle Arbeitssprachen, wobei zu den letzteren auch mein kurioses australisches Englisch zählt. Daher kommt es, daß die Verständigung bei den Fachsitzungen nur über die Simultandolmetschung erfolgen kann. Häufig wird diese von den Dolmetschern so brillant gemeisterte Aufgabe nicht gebührend gewürdigt; wir jedoch wollen uns kein solches Versäumnis zuschulden kommen lassen und danken somit den Damen und Herren in den Dolmetschkabinen für ihre hervorragende Leistung.

Und so kommen wir denn zum Schluß dieser Sitzung und zum Abschluß des wissenschaftlichen Teils des Kongresses. Wir treffen uns wieder zu IGFM-Symposien im Juni 1988 in Minneapolis (USA) und im September 1988 in Madrid und, wie schon gesagt, in vier Jahren beim nächsten Felsmechanik-Kongreß in Aachen.

So hebe ich nun, meine Damen und Herren, diese Sitzung auf. Ihnen allen meinen herzlichen Dank!

Jubiläumsbankett

Das Jubiläumsbankett fand anläßlich des 25. Jahrestages der Gründung der Internationalen Gesellschaft für Felsmechanik im Großen Salon sowie im Marquette-Saal und Jolliet-Saal des Hotels Queen Elizabeth statt. Die 575 Gäste begannen den Abend mit Cocktails und begaben sich hierauf in die Speisesäle. Nachdem die Ehrengäste an ihrem Tisch Platz genommen hatten, begrüßte der Kongreßvorsitzende, Dr. G. Herget, die Geladenen in drei Sprachen und wünschte ihnen eine angenehme Mahlzeit. Die Speisen (Menü unten) wurden unter der umsichtigen Leitung von J. Druda, dem Leiter der Catering-Abteilung des Hotels, zügig serviert.

Während Kaffee und Likör gereicht wurden, stellte der Vorsitzende, Dr. G. Herget, die Ehrengäste von rechts nach links vor:

P. Michaud, Montreal/Kanada, Geschäftsführender Leiter des Kanadischen Instituts für Bergbau und Metallurgie
N.F. Grossmann, Lissabon/Portugal, Generalsekretär der IGFM
Prof. Dr. Langer, Hannover/BRD, Altpräsident der Internationalen Vereinigung der Ingenieurgeologen
Prof. Bello M., Mexiko, Vizepräsident der IGFM (Nordamerika)
Dr. M.D. Everell, Ottawa/Kanada, Unterstaatssekretär im kanadischen Bundesministerium für Energiewirtschaft, Bergbau und Bodenschätze
Frau J. Udd, Ottawa/Kanada
Prof. Dr. Tan Tjong Kie, Beijing/China, Vizepräsident der IGFM (Asien)
Frau Gill, Montreal/Kanada
Prof. Dr. B. Ladanyi, Montreal/Kanada, Ehrenvorsitzender des Kongresses, Altpräsident der Kanadischen Vereinigung für Felsmechanik
Frau U. Herget, Ottawa/Kanada
Prof. Dr. E.T. Brown, London/England, Präsident der Internationalen Gesellschaft für Felsmechanik
Dr. G. Herget, Ottawa/Kanada, Kongreßvorsitzender
Frau N. Ladanyi, Montreal/Kanada
Prof. Dr. D. Gill, Montreal/Kanada, 2. Kongreßvorsitzender
Frau Tan Tjong Kie, Beijing/China
Dr. J. Udd, Präsident der Kanadischen Vereinigung für Felsmechanik
Frau A. Reid, London
Prof. Dr. B. Bamford, Melbourne/Australien, Vizepräsident der IGFM (Australien/Ozeanien)
Prof. Dr. F.H. Tinoco, Caracas/Venezuela, Vizepräsident der IGFM (Südamerika)
Prof. Dr. H. Wagner, Johannesburg/Südafrika, Vizepräsident der IGFM (Südafrika)
Dr. J.A. Franklin, Orangeville/Kanada, künftiger Präsident der IGFM
Dr. Bozozuk, Ottawa/Kanada, Präsident der Kanadischen Gesellschaft für Geotechnik

Hierauf dankte Dr. G. Herget allen Mitgliedern des Organisationsausschusses für ihren Einsatz und für ihre unermüdliche Arbeit zusätzlich zu ihren sonstigen Pflichten während der dreijährigen Vorbereitung des Kongresses. Jedem Mitglied des Ausschusses wurde Beifall gespendet, als sie einzeln zum Ehrentisch vortraten und in Anerkennung ihrer Verdienste ein kleines Geschenk erhielten:

Dr. B. Ladanyi, Ehrenvorsitzender/Mittelbeschaffung (Inuit-Skulptur)
Dr. D.E. Gill, 2. Kongreßvorsitzender/örtliche Veranstaltungen (Plakette)
Dr. N.R. Morgenstern, Technisch-wissenschaftliches Programm (Plakette)
Dr. J.A. Franklin, Öffentlichkeitsarbeit (Plakette)
T. Carmichael, Finanzen (Plakette)
Dr. W.F. Bawden, Ausstellungen (Plakette)
J. Gaydos, CIM, Leiter der Fachausstellung (Inuit-Skulptur)
Dr. A.T. Jakubick, Fachexkursionen (Plakette)
L. Geller, Übersetzungen (Plakette)
Frau J. Robertson, Programm für Begleitpersonen (Plakette)
Frau D. Grégoire, Anmeldung, CIM (Inuit-Skulptur)
Dr. J. Bourbonnais, Gastgeberfunktionen (Plakette)
J. Nantel, Mittelbeschaffung (Plakette)
Dr. S. Vongpaisal, Stellvertretender Herausgeber (Plakette)

Der Vorsitzende betonte, daß der so zum Ausdruck gebrachte Dank nicht nur den hier anwesenden Ausschußmitgliedern, sondern noch vielen anderen Mitarbeitern gebühre, die besonders in den letzten Monaten mit vollem Einsatz den Kongreß vorbereitet hatten.

Nach der Vorstellung der Mitglieder des Organisationsausschusses und den Dankesworten übergab der Vorsitzende das Wort an den scheidenden Präsidenten der Internationalen Gesellschaft für Felsmechanik, Prof. Dr. E.T. Brown:

"Danke, Gerhard. Guten Abend, meine Damen und Herren!

So gehen auch dieser Kongreß und unser gemeinsamer Aufenthalt in Montreal zu Ende, und ich stehe wieder vor Ihnen, doch zum letzten Mal als Präsident der IGFM. Der erfolgreiche Abschluß dieser wissenschaftlichen Veranstaltung ist zweifelsohne für uns alle eine große Genugtuung - für unsere kanadischen Kollegen, die auch erleichtert sind, daß alles gut gegangen ist, für mich, denn ich kann meine Geschäfte nun dem nächsten Präsidenten übergeben, und für Sie alle, da Sie in wenigen Minuten keine Reden - besonders meine - mehr anhören müssen.

Bevor ich nun tatsächlich zum Ende komme und mich niedersetze, möchte auch ich wie Gerhard den Mitgliedern des Organisationsausschusses im Namen der Nichtmontrealer danken. Besonders heute abend möchte ich jenen danken, die das gesellschaftliche und das Begleitpersonenprogramm zusammengestellt haben - vor allem Frau Joan Robertson und Dr. Jacques Bourbonnais. Wir haben Ihre Gastfreundschaft und die Sonderveranstaltungen genossen, die Sie für uns organisiert haben - Exkursionen, Empfänge, die Vorstellung von Les Ballets Jazz de Montreal am Montag abend, besonders den Pas de deux, den Gerhard Herget und Jacques Bourbonnais getanzt haben -, und wir genießen auch den heutigen Abend sehr. Vor allem die Wahl des Tagungsorts war eine glückliche, denn Montreal ist eine faszinierende Stadt, die jedem etwas bietet.

Bei der Schlußveranstaltung heute nachmittag dankte ich dem Ehrenvorsitzenden, Prof. Branko Ladanyi, und dem Kongreßvorsitzenden, Dr. Gerhard Herget, für ihre tragende Rolle bei der Ausrichtung dieses Kongresses. Im gegenwärtigen ungezwungeneren Rahmen möchte ich diesen Dank noch einmal im Namen aller aussprechen. Als ich heute abend zum Bankett hier eintraf, sah ich, wie Gerhard in Begleitung seiner Gattin wie immer still und unauffällig, aber auch in seiner gewohnten gründlichen Art und Weise hinter den Kulissen zum letzten Mal nach dem Rechten sah, damit das Bankett nur ja ein rechter Erfolg würde.

Schlußveranstaltung: Amtsantritt von J.A. Franklin als neuer Präsident der IGF, v.l. nach rechts - Hintergrund: A.A. Bello Maldonade (VP - Nordamerika), J.A. Franklin; Vordergrund: Frau U. Herget, E.T. Brown (Altpräsident)

Wie Sie wissen, feiern wir 1987 das silberne Gründungsjubiläum der IGFM, und das Bankett heute abend findet im Gedenken daran statt. Wir haben zu diesem feierlichen Anlaß alle Mitglieder der früheren Beiräte eingeladen; soweit sie kommen konnten, sitzen sie nun mit ihren Gattinnen an einem Ehrenplatz. Obwohl sie Leuchten unserer Wissenschaft sind, die einen wertvollen Beitrag zur Entwicklung und Fortdauer unserer Gesellschaft geleistet haben, muß ich doch zugeben, daß sie wohl keinem Schönheitswettbewerb standhalten würden. Trotzdem möchte ich sie bitten, einen Augenblick aufzustehen, so daß wir sie begrüßen und ihre Verdienste um die Felsmechanik und um die IGFM würdigen können. Meine Herren, ich bitte!

Seit 20 Jahren nimmt ein Sekretariat mit Sitz in Lissabon die Verwaltungsarbeiten der IGFM wahr. Unsere Gesellschaft kann sich besonders glücklich schätzen, eine Reihe ausgezeichneter Generalsekretäre gehabt zu haben, von denen zwei - Fernando Mello Mendes und Ricardo Oliviera - zu den eben vorgestellten Herren zählen. Seit vier Jahren ist Nuno Grossmann Generalsekretär. Wer von Ihnen in dieser Zeit an unseren Sitzungen teilgenommen hat, kennt seine gedrungene Gestalt, die immer dort erscheint, wo man sie gerade braucht, und die dafür sorgt, daß alle, ganz besonders aber ich, sich jederzeit korrekt und im besten Interesse der IGFM verhalten. Als Ihr Präsident weiß ich wie kein anderer, wie freudig, wie unermüdlich und wie treu er sich für die Belange der IGFM eingesetzt hat: Ich danke Dir, Nuno, in meinem eigenen Namen und im Namen der IGFM!

Noch jemandem im Sekretariat gebührt Anerkennung für die letzten 20 Jahre - Frau Maria de Lurdes Eusébio, der geschäftsführenden Sekretärin. Wenn es den Generalsekretären zu verdanken ist, daß Präsidenten und Beiratsmitglieder immer das Ihre taten, so hat Maria de Lurdes dafür bei den Generalsekretären gesorgt. Es zählt zu den angenehmsten Aufgaben eines Amtsträgers der IGFM, mit Maria de Lurdes zusammenarbeiten zu dürfen, denn ihr Takt und Charme, ihre Eleganz und Tüchtigkeit sind sprichwörtlich. Wir haben sie alle sehr ins Herz geschlossen und wüßten nicht, was wir ohne sie täten. Um ihren beispielhaften Einsatz für die Geschäfte der IGFM in den letzten zwei Jahrzehnten zu würdigen und als kleines Zeichen unserer Achtung und Wertschätzung möchte ich ihr im Namen der IGFM ein kleines Geschenk überreichen.

Schlußveranstaltung: v.l. nach rechts - Hintergrund: J.A. Franklin, W.M. Bamford (VP - Australasien); Vordergrund: G. Herget (Kongreßvorsitzender)

Die reibungslose Arbeit der IGFM hängt in großem Maße auch vom Beirat ab, zu dessen Mitgliedern auch je ein Vizepräsident für die sechs Regionen zählt. Ich bin den scheidenden Mitgliedern des Beirats persönlich zu Dank verpflichtet für die Unterstützung und den Rat, die mir von ihnen im Laufe meiner vierjährigen Amtsperiode zuteil geworden sind. Besonders danken möchte ich unserem Ersten Vizepräsidenten, Dr. Sten Bjurström (Schweden). Leider kann Sten heute nicht anwesend sein, da er aus dringenden beruflichen Gründen nach Stockholm zurückkehren mußte.

Schlußveranstaltung: v.l. nach rechts - Hintergrund: J.A. Franklin, H. Wagner (VP - Afrika); Vordergrund: G. Herget, Frau N. Ladanyi

Schlußveranstaltung: Les Sortileges

Manchen von Ihnen wird wohl aufgefallen sein, daß es während dieser Kongreßwoche in Montreal zu - sagen wir mal - Diskussionen über die Wahl des neuen Beirats gekommen ist, die heute abend vor dem Bankett vorgenommen wurde. Als letzte Amtshandlung darf ich Ihnen nun die neugewählten Mitglieder vorstellen, in deren Händen das Geschick der IGFM vom heutigen Tag bis zum 7. Kongreß in Aachen im September 1991 liegen wird:

Vizepräsident für Südamerika	- Dr. C. Dinis da Gama (Brasilien)
Vizepräsident für Nordamerika	- Dr. Jim Coulson (USA)
Vizepräsident für Europa	- Marc Panet (Frankreich)
Vizepräsident für Australien und Ozeanien	- Dr. Ian Johnston (Australien)
Vizepräsident für Asien	- S.L. Mokhashi (Indien) (nicht anwesend)
Vizepräsident für Afrika	- Dr. Oscar Steffen (Südafrika)

Zu guter Letzt möchte ich Ihnen meinen Nachfolger vorstellen - Dr. John Franklin aus Kanada. Ich möchte ihn zu seinem Wahlerfolg beglückwünschen und ihm die Geschäfte des Präsidenten übergeben."

Der künftige Präsident der Internationalen Gesellschaft für Felsmechanik, Dr. J.A. Franklin, wurde nun gebeten, einige Worte an die Kongreßteilnehmer zu richten.

Nachdem Dr. Franklin geendet hatte, erklärte der Kongreßvorsitzende, Dr. G. Herget, den offiziellen Teil des 6. IGFM-Kongresses als beendet. Er dankte den Teilnehmern für ihr Kommen und wies darauf hin, daß die Vorführung von "Les Sortilèges", der kanadischen Volkstanzgruppe, nun stattfinden würde. (Programm im englischen Teil)

Nach Beendigung der Vorstellung, die alle genossen hatten, luden die Tänzer die Gäste auf den Tanzboden ein - eine Einladung, der bald darauf zahlreiche Anwesende mit Freuden nachkamen. Damit fand der Sechste Internationale Kongreß der Felsmechanik seinen Abschluß; die meisten Teilnehmer verabschiedeten sich um ca. 23 Uhr.

Workshop
Durch Bergbaumaßnahmen verursachte Seismizität

R. Paul Young
Department of Geological Sciences
Queen's University, Kingston
Canada, K7L 3N6

Zum Gedenken an Fred Leighton

VORWORT

Der Gedanke an die Abhaltung eines Workshops zu diesem Thema stammt vom Unterausschuß Gebirgsschläge der Kanadischen Vereinigung für Felsmechanik und vom Organisationsausschuß des Sechsten Internationalen Kongresses der IGFM. Stan Bharti von der Firma Falconbridge und Will Bawden von der Firma Noranda traten an mich heran mit dem Vorschlag für einen eintägigen Workshop im Rahmen des vom 30. August bis 3. September 1987 in Montreal stattfindenden Sechsten Internationalen Kongresses der Felsmechanik. Der ins Auge gefaßte Termin schien passend, da seit dem Ersten Internationalen Symposium über Gebirgsschläge und Seismizität im Bergbau in Johannesburg/Südafrika fünf Jahre vergangen waren. Außerdem meinten wir, daß die Ergebnisse des Workshops auf dem 6. IGFM-Kongreß und auf dem 2. Internationalen Symposium über Gebirgsschläge und Seismizität im Bergbau im Juni 1988 an der University of Minnesota weiter diskutiert werden könnten.

Ich beschloß, den Workshop dem Gedenken an Fred Leighton zu widmen - dem 1986 verstorbenen Wissenschaftler und ehemaligen Leiter der Arbeitsgruppe Angewandte Mikroseismik im US Bureau of Mines am Denver Research Center in Colorado. Fred wurde in Colorado geboren, wo er auch die Schule und später die Colorado School of Mines besuchte. Mehr als 20 Jahre lang lieferte er wertvolle Beiträge zur Anwendung der AE/MS-Verfahren bei der Lösung von Problemen des Arbeitsschutzes und der technischen Sicherheit im Bergbau. Am bekanntesten ist wohl sein mehrbändiges, in Zusammenarbeit mit Dr. Reg Hardy von der Pennsylvania State University herausgegebenes Werk über die Anwendung von AE/MS-Verfahren in der Geotechnik. Ich sah Fred zum letzten Mal, als er mich im Sommer 1985 in meinem Laboratorium an der Queen's University anläßlich eines Treffens des Sachverständigenausschusses des Kanadischen Bundesinstituts für Naturwissenschaftliche und Technische Forschung besuchte. Wir unterhielten uns damals über meine geplanten Untersuchungen über bergbaulich verursachte seismische Erscheinungen, und ich verdanke ihm wertvolle Anregungen und Vorschläge.

Der Workshop zerfiel in drei Teile: Die Vormittagssitzung konzentrierte sich auf ausgewählte Themen zur bergbaulich verursachten Seismizität, gab einen Überblick über den gegenwärtigen Wissensstand auf diesem Gebiet und arbeitete Vorschläge für die Ausrichtung zukünftiger Forschungsarbeiten aus. Es referierten Dr. W. Blake über mikroseismische Instrumente, Dr. J. Niewiadomski über Verfahren zur Ortung des Erschütterungsherdes, Dr. A. McGarr über Verfahren zur Aufarbeitung und Auswertung seismischer Daten, und ich selbst über die Rolle der Geotomographie bei der Erforschung der durch Bergbaumaßnahmen verursachten Seismizität. Die Postersitzung zur Mittagszeit, der zweite Teil, bot den mit einschlägigen Arbeiten befaßten Bergbauunternehmen, Forschungsgremien und Hochschulen Gelegenheit, in Posterform über ihre Projekte, die dabei gewonnenen Resultate und deren Auswertung zu berichten. Die stattliche Zahl von 17 Postervorträgen wurde eingereicht. Ziel der Nachmittagssitzung, des dritten Teils des Workshops, war es, einen Überblick über ausgewählte nationale und internationale Forschungsprojekte auf dem Gebiet der durch den Bergbau verursachten Seismizität zu geben. Dr. H.R. Hardy referierte über den internationalen Anwendungsstand der AE/MS-Verfahren; dasselbe Thema wurde von Dr. D. Hedley für Kanada, von Dr. S. Spottiswoode für Südafrika und von Dr. B. Brady für die USA behandelt. Zum Abschluß wurden in einem von D. Ortlepp geleiteten Diskussionsforum aus Sachverständigen der Bergbauindustrie, nationaler Forschungsgremien und der Hochschulen wichtige Fragen und Probleme der bergbaulich verursachten Seismizität erörtert.

Das große Interesse an dem Workshop war nicht nur an und für sich erfreulich, sondern eröffnete auch vielversprechende Aussichten für die Zukunft. Über 125 Delegierte aus fünf Kontinenten nahmen am Workshop und am anschließenden Abendessen teil. Das Berichtswerk aus 25 Referaten mit insgesamt über 300 Seiten wurde beim Workshop verteilt und ist weiterhin von der Queen's University erhältlich. Sein Inhalt zeigt klar und deutlich, daß man die Bedeutung der bergbaulich verursachten Seismizität sowie die Notwendigkeit eines besseren Verständnisses dieser Erscheinungen erkannt hat. Da der Untertagebau in aller Welt in immer größere Teufen vordringt und die durch Bergbaumaßnahmen verursachte Seismizität immer häufiger wird, bedarf es angesichts der erhöhten Gebirgsschlaggefahr einer gründlicheren Erforschung der komplexen Wechselwirkungen zwischen der Seismizität einerseits und Grubenentwurf, Gesteinseigenschaften, lokalen und regionalen Druckspannungen sowie geotektonischen Verhältnissen andererseits. Professor S.J. Gibowicz, der in unseren Fachkreisen allgemein als der Meister unserer Disziplin schlechthin gilt, konnte an unserem Workshop leider nicht teilnehmen. Er ist zur Zeit mit der Herausgabe einer Sondernummer 1988-1989 des "Journal of Pure and Applied Geophysics" zum Thema "Seismizität im Grubenbau" befaßt und möchte das Berichtswerk des Workshops sowie mehrere überarbeitete Workshop-Referate in diese Sondernummer aufnehmen.

Zum Abschluß möchte ich allen jenen danken, die zum Erfolg des Workshops beigetragen haben - den Referenten, Dr. H. Brehaut für seine Rede beim Abendessen, den Sachverständigen des Diskussionsforums und allen Teilnehmern.

(Diskussionsbeiträge sind im englischen Teil aufgeführt.)

Posterkatalog

Der Technisch-Wissenschaftliche Ausschuß hatte um Beiträge zu den an jedem Kongreßtag stattfindenden Postersitzungen ersucht, die in einem weniger förmlichen, die Plenarsitzungen ergänzenden Rahmen Gelegenheit boten, die Autoren kennenzulernen und zum Einsendetermin noch nicht vorliegende Forschungsergebnisse nachträglich mitzuteilen. 58 Kongreßteilnehmer nahmen diese Gelegenheit wahr.

1. P.R. AGRAWAL (Indien): Felsverhalten der Verschiebungszone entlang des Wasserzuleitungssystems des Yamuna-Hydel-Projekts, Stufe II, Teil II

2. G. BALLIVY, B. BENMOKRANE, A. LAHOUD (Kanada): Ein integrales Verfahren zur Bemessung von Injektionsankern im Fels

3. K. BALTHASAR, M. HAUPT, CH. LEMPP, O. NATAU (West-Deutschland): Relaxationsverhalten von Steinsalz: Vergleiche von Laboruntersuchungen und In-Situ-Messungen

4. S. BANDIS, J. LINDMAN, N. BARTON (Norwegen): Dreidimensionales Druckspannungsfeld und Bruch um Hohlräume im überbelasteten Weichgestein

5. B. BAMFORD (Australien): Die Bohr- und Schneidfähigkeit von Gestein

6. W. BAWDEN, D. MILNE (Kanada): Eine geomechanische Entwurfsmethode für den unterirdischen Bergbau der Noranda Minerals Inc.

7. B. BJARNASON (Schweden): Nichtlineare und unstetige Spannungsveränderung mit der Tiefe in der oberen Kruste des baltischen Schildes

8. G. BORM (West-Deutschland): Bohrlochkonvergenz und Spannungsrelaxation im Steinsalzgebirge

9. T.C. CHAN, N.W. SCHEIER (Kanada): Simulation mittels Elementenmethode von Grundwasserbewegung und Wärme- und Radionuklidentransport in einem plutonischen Fels

10. F.H. CORNET, J. JOLIVET, J. MOSNIER (Frankreich): Identifizierung und hydraulische Kennzeichnung von Trennflächen in einem Bohrloch

11. EMMANEL DETOURNAY, L. VANDAMME, A. H-D. CHENG (USA): Fortpflanzung eines vertikalen hydraulischen Bruches in einer porös-elastischen Formation

12. C.H. DOWDING, K.M. O'CONNOR, M.B. SU (USA): Quantifizierung der Formänderungen im Fels mittels Zeitbereich-Reflexionsmessungen

13. M.E. DUNCAN FAMA, L.J. WARDLE (Australien): Eine numerische Analyse zur Standsicherheit von Zwischenpfeilern im Kohlenbergwerk

14. DYWIDAG SYSTEMS INTERNATIONAL (Kanada): Dywidag-Überbelastungsanzeiger und nachgiebiger Anker

15. J. ENEVER (Australien): Erfahrungsbericht über zehnjährige Messungen hydraulischer Bruchbelastungen in Australien

16. G.S. ESTERHUIZEN (Südafrika): Einschätzung der Beständigkeit bergmännischer Tunnels bei veränderlichen Gesteinsspannungen

17. E. FJAER, R.K. BRATLI, J.T. MALMO, O.J. LOEKBERG, R.M. HOLT (Norwegen): Optische Studien der Hohlraum-Deformation in weichem Sedimentgestein

18. R.J. FOWELL, C.K. IP (England): Der Mechanismus der Gewinnung von Gestein mit Spatenmeißel und mitwirkendem Wasserstrahl

19. S. GENTIER (Frankreich): Mechanisches und hydraulisches Verhalten einer Kluft unter Normaldruck

20. D.E. GILL, B. LADANYI (Kanada): Zeitabhängige Gebirgskennlinien für die Bemessung von Tunnelauskleidungen

21. GAO HANG, JING ZIGANG, SHEN GUANGHAN (Volksrepublik China): Untersuchungen in einer Kohlenzeche über gespanntem Grundwasser im nordchinesischen Kohlenrevier

22. T.F. HERBST (West-Deutschland): Erhöhte Sicherheit durch umfassende Überwachung der Stützwirkung in untertägigen Hohlräumen

23. François HEUZE, R.J. SCHAFFER, R.K. THORPE, A.R. INGRAFFEA, R.H. NILSON (USA): Simulierungen quasi-statischer und dynamischer flüssigkeitsgetriebener Brüche im geklüfteten Fels

24. R.M. HOLT, J. BERGEN, T.H. HANSSEN (Norwegen): Anisotrope Materialeigenschaften eines schlecht konsolidierten Sandsteins

25. D.R. HUGHSON, A.M. CRAWFORD (Kanada): Kaisereffektmessungen: Ein Verfahren zur akustischen Messung von Felsspannungen

26. OLDRICH HUNGR (Kanada): Die Stabilitätsanalyse für Böschungskeile mit Hilfe der Abschnittmethode

27. R.W. HUTSON, C.H. DOWDING (USA): Ein neues Verfahren zur Herstellung gleichartiger Klüfte in In-Situ-Gestein aller Härten

28. T. KYOYA, Y. ICHIKAWA, T. KAWAMOTO (Japan): Deformations- und Bruchprozesse von diskontinuierlichen Felsmassen und Schadensmechanik

29. YOZO KUDO (Japan): Physikalische Eigenschaften und Mikrostrukturen der Granitfelsen in Japan

30. KOKICHI KIKUCHI (Japan): Stochastische Schätzung und Modellkonstruktion der Kluftverteilung aufgrund statistischer Probenahmen

31. V. LABUC, W. BAWDEN, F. KITZIGER (Kanada): Entwicklung eines seismischen Überwachungssystems mit faseroptischer Signalübermittlung

32. LI XING CAI, SHI RU BIN, YU MU XING (China): Akustische Methoden für die Bestimmung der häufigsten Orientierung von Diskontinuitäten im Gelände

33. LI XING CAI (China): Null- oder negativer Effektivdruck und sein Einfluß auf physikalische Eigenschaften des Gesteins im Gelände

34. J.C.S. LONG, D. BILLAUX, K. HESTIR, J.P. CHILES (USA): Einige geostatistische Methoden für die Berücksichtigung von Raumstrukturen in Kluftnetzmodellen

35. PAUL H. LU (USA): Die Methode des Integrity Factor für den Entwurf von Pfeilern für den Strebbruchbau

36. J. MATHIS (Schweden): Statistische Kluftwerte: Ein Vergleich zwischen Tunnelkartierung und geräumigen Abbaukammern

37. KIKUO MATSUI, MASATOMO ICHINOSE, KENICHI UCHINO (Japan): Der Einfluß der Restfesten und Abbaukanten auf die Konvergenz von darunter liegenden Abbaustrecken

38. P. MACDONALD (Kanada): Seismische und numerische Simulierung bei INCO

39. D. MILNE (Kanada): Eine geomechanische Entwurfmethode für den Abbau bei Noranda Minerals Inc.

40. D.Z. MRAZ (Kanada): Vorgesehene Planung einer Hochdruck-Isolations-Struktur in Salzfelsen

41. H.-B. MÜHLHAUS (West-Deutschland): Stabilität von Untertagebauten in Gebirge großer Zugfestigkeit

42. T. NISHIOKA, A. NARIHIRO, O. KYOYA (Japan): Numerische Analyse einer auf weichem Fels errichteten Erdanker-Stützmauer

43. R.C.T. PAKANIS, H.D.S. MILLER, S. VONGPAISAL, T. MADIL (Kanada): Ein empirischer Ansatz zur Technik des Kammerbaus

44. DALE PREECE (USA): Messungen der Bohrlochkonvergenz und numerische Berechnungen für das Lager Big Hill, Texaco SPR

45. J.P. JOSIEN, J.P. PIGUET, R. REVALOR (Frankreich): Der Beitrag der Felsmechanik zum Verständnis von dynamischen Erscheinungen in Bergwerken

46. M. ROUSSIN, S. BURDICK (USA): Ein neues Werkzeug für den Abbau: Schlagreißentwicklung

47. S. SAKURAI, N. SHIMIZU (Japan): Stabilitätsanalyse von Böschungen mit Hilfe der Fuzzy-Set-Theorie

48. J.C. SHARP, C.H. LEMAY, B. NEVILLE (Kanada): Verhalten einer hohen Tagebauböschung in sprödem ultrabasischem Fels

49. P.L. SWANSON (USA): Widerstand zum Typ-I-Bruch in spröden Mehrkristallen: Ein Beitrag der Bruchflächenfestigkeit

50. M. THIERCELIN, J.C. ROEGIERS, T.J. BOONE, A.N. INGRAFFEA (USA): Untersuchung der die Ausbreitung von Bruchflächen bestimmenden Stoffparameter

51. SPE (Frankreich): Fels in großen Tiefen: Felsmechanik und Felsphysik in großen Tiefen

52. KOT F. UNRUG (USA): Die Auswirkungen von Sprengungen in einem Tagebau auf das Hangende einer Tiefbaugrube im selben Flöz

53. G. VENKATACHALAM, K. VINCENT PAUL (Indien): Die Bestimmung von Gesteinsklüften

54. B.S. VERMA (Kanada): Vorschläge zur unmittelbaren Bestimmung der Zugfestigkeit von tragfähigem Gestein

55. B.S. VERMA (Kanada): Die Bruchbildung im Umfeld von Drucklagen in benachbarten Mineralbildungszonen

56. T. VLADUT (Kanada): Das Retom LWD-Untersuchungssystem

57. D. WULLSCHLÄGER, O. NATAU (West-Deutschland): Das geankerte Gebirge als anisotroper Stoff - Materialverhalten und Dimensionierungsvorschlag für den Felshohlraumbau

58. TH. MUTSCHLER, B. FRÖHLICH (West-Deutschland): Analytische Berechnung des Festigkeitsverhaltens von Wechsellagerungen

4

Additional contributions / Contributions additionnelles / Ergänzende Beiträge

Influence de la vitesse de chargement sur la rupture du sel en extension

Effect of strain rate on the failure of rock salt in extension tests

Der Einfluß der Dehnungsgeschwindigkeit auf das Bruchverhalten von Hartsalz in Zugversuchen

J.BERGUES, Laboratoire de Mécanique des Solides, École Polytechnique, Palaiseau Cédex, France
J.P.CHARPENTIER, Laboratoire de Mécanique des Solides, École Polytechnique, Palaiseau Cédex, France

ABSTRACT: The authors studied experimentaly the effect of strain rate on the failure behaviour of rock salt in triaxial extension tests ($\sigma_1 < \sigma_2 = \sigma_3 = P$). The results are compared with some creep extension tests.

RÉSUMÉ: Les auteurs présentent des résultats expérimentaux relatifs à l'influence de la vitesse de chargement sur le comportement à la rupture du sel soumis à un chargement triaxial classique en extension ($\sigma_1 < \sigma_2 = \sigma_3 = P$). Les résultats sont comparés avec ceux obtenus au cours d'essais de fluage en extension.

ZUSAMMENFASSUNG: Die Verfasser haben den Einfluss des Dehnungsgeschwindigkeit auf das Bruchverhalten von Hartsalz in dreiaxiallen Zugversuchen untersucht ($\sigma_1 < \sigma_2 = \sigma_3 = P$). Die Ergebnisse werden mit dem Kriechverhalten in Zugversuchen verglichen.

I - INTRODUCTION

Dans le cadre des recherches entreprises au Laboratoire de Mécanique des Solides sur la stabilité des cavités souterraines, un large programme expérimental a été mis en oeuvre pour étudier le comportement mécanique du sel gemme.

Les essais sont réalisés sous contraintes triaxiales classique ($\sigma_2 = \sigma_3 = P$), de compression ($\sigma_1 > P$) ou d'extension ($\sigma_1 < P$). Un accent particulier est mis sur l'influence:

- de la température T,
- du temps t.

Le comportement différé est étudié à partir d'essais de fluage (σ = constant) et à partir d'essais d'écrouissage ($\Delta\ell/\ell$ = constant); ($\Delta\ell$ = variation de longueur de l'échantillon (ℓ)).

On présente ici une série de résultats expérimentaux relatifs à l'influence de la température (T) et de la vitesse de déformation sur la rupture du sel en extension.

II - RÉSULTATS EXPÉRIMENTAUX

II.1 - Généralités:

Le détail des moyens expérimentaux utilisés est présenté dans [1] (J. Bergues et Al., 1987) pour les essais d'écrouissage et dans [2] (J.P. Charpentier, 1983) pour les essais de fluage.

Les essais triaxiaux sont réalisés sur des éprouvettes cylindriques de 70 mm de diamètre et 180 mm de hauteur.

La mise en charge s'effectue en trois étapes:

1 - Mise en température progressive en trois heures environ;

2 - Mise en contrainte hydrostatique jusqu'à la pression de confinement de l'essai $\sigma_1 = \sigma 2 = \sigma_3 = P$;

3 - Mise en contrainte déviatorique avec $\sigma_1 < P$ = constante.

II.2 - Essais d'écrouissage:

Trois séries d'essais ont été réalisées avec des vitesses de mise en charge différentes:

- série n°1: $\dot{\Delta\ell}/\ell = 7{,}17 \times 10^{-8}\ s^{-1}$;
- série n°2: $\dot{\Delta\ell}/\ell = 7{,}17 \times 10^{-6}\ s^{-1}$;
- série n°3: $\dot{\Delta\ell}/\ell = 7{,}17 \times 10^{-5}\ s^{-1}$.

Pour tous les essais, la température est de 50°C.

Les résultats présentés sur la figure n°1 donnent dans le plan [σ_1, P] l'état de contrainte relatif à la rupture.

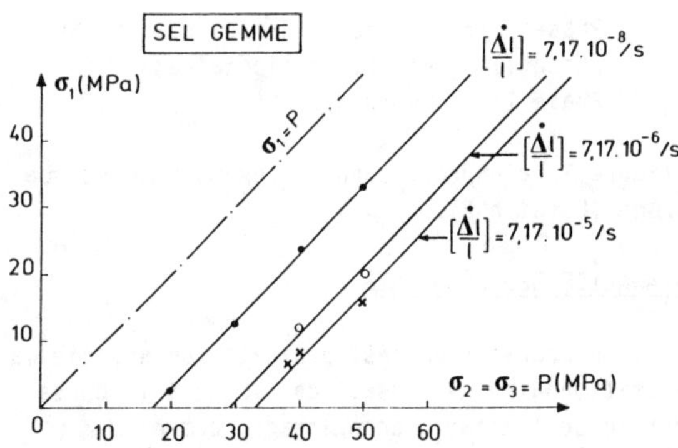

Fig.1 Ecrouissage. Influence de la vitesse de chargement sur la rupture

L'analyse de cette figure montre que pour chaque vitesse de chargement le critère de rupture peut être assimilé à un critère de Tresca, résultat déjà observé sur une autre série d'essais [1], (J. Bergues et Al., 1983).

Le tableau 1 donne la valeur de la cohésion C à la rupture, déterminée pour chaque série d'essais.

Sur ce tableau, on constate que la cohésion à la rupture augmente avec la vitesse de chargement.

Tableau 1

$\dot{\Delta\ell}$/ /sec.	2 C (MPa)
7,17 10^{-8}	17
7,17 10^{-6}	29
7,17 10^{-5}	33

II.3 - Essais de fluage:

Ces essais ont été réalisés pour des températures de 20°C et 100°C. Le tableau 2 donne le détail du programme expérimental.

Tableau 2.

Essai N°	T °C.	P (MPa)	σ_1 (MPa)
1	20	10	2,5
2	20	16	6
3	100	12,5	7,5
4	100	15	5
5	100	15	5

Mis à part l'essai n°1, qui a été arrêté pour des raisons techniques dans la phase de fluage stationnaire, tous les autres essais ont été menés jusqu'à la rupture avec les trois phases classiques de fluage:

- Phase I : Fluage transitoire;
- Phase II : Fluage stationnaire;
- Phase III: Rupture.

La figure 2, ci-après, donne un exemple de courbe de fluage (Essai n°2).

III - ANALYSE DES RÉSULTATS

Pour l'interprétation des résultats, on associe la phase stationnaire de l'essai de fluage à la phase de rupture de l'essai d'écrouissage, comme l'illustre le croquis de la figure 3.

Ainsi, pour chaque essai de fluage, on calcule:

- la vitesse du fluage stationnaire,
- la cohésion à la rupture correspondante (type Tresca), à partir de la figure 1.

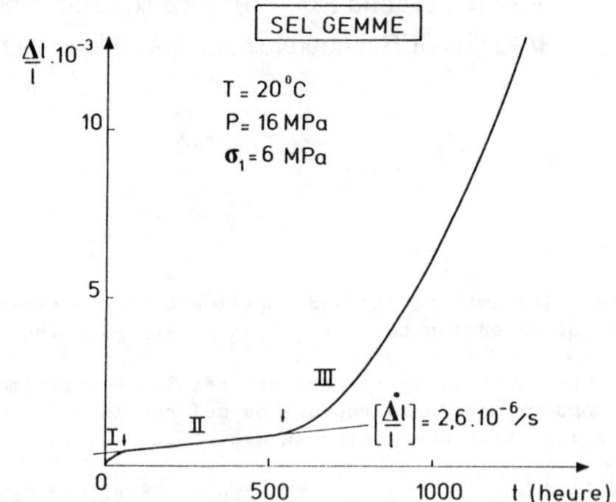

Fig.2. Fluage triaxial en extension

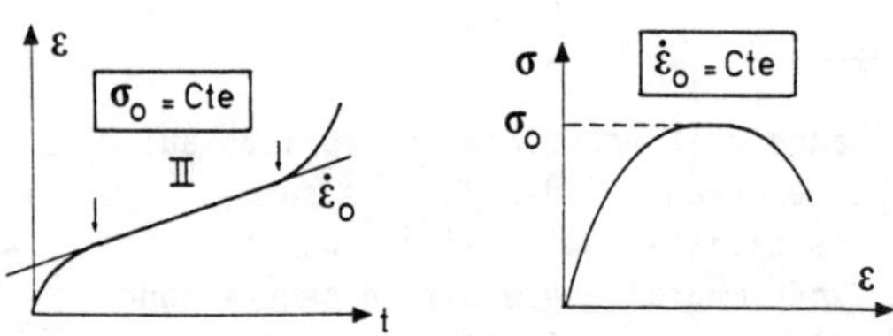

Fig.3_ Association écrouissage et fluage

Les valeurs sont consignées dans le tableau 3:

Tableau 3.

Essai N°	2 C (MPa)	$\dot{\Delta\ell}/\ell$ s^{-1}
1	7	1 x 10^{-10}
2	10	2,6 x 10^{-10}
3	5	3 x 10^{-10}
4	10	9 x 10^{-8}
5	10	2 x 10^{-8}

Sur la figure 4, sont reportés, dans le plan [2 C, $\dot{\Delta\ell}/\ell$], tous les résultats relatifs aux essais de fluage et d'écrouissage.

Ces résultats appellent les commentaires suivants:

- La cohésion à la rupture diminue avec la vitesse de chargement;
- La cohésion diminue lorsque la température augmente.

La cohésion tend à s'annuler pour des vitesses de mise en charge de l'ordre de 10^{-11} s^{-1}, quelle que soit la température.

Les auteurs proposent la forme empirique suivante pour déterminer la cohésion:

$$C = \log A + B\ (T)\ \log \frac{\dot{\Delta \ell}}{\ell} \quad .$$

Compte tenu du nombre relativement faible d'essais, il est encore prématuré de fixer les valeurs des deux constantes A et B.

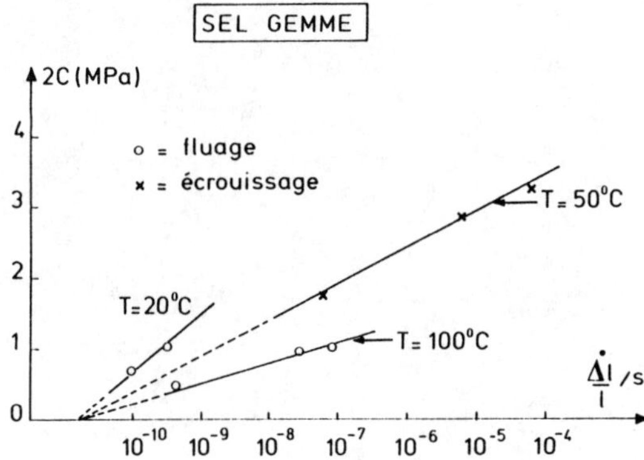

Fig.4 _Variation de la cohésion en fonction de la vitesse de chargement et de la température

Références:

[1] BERGUES, J., CHARPENTIER, J.P., DAO, M., "Étude expérimentale du comportement mécanique du sel en extension", 6ème Congrès du I.S.R.M. Montréal, 1987.

[2] CHARPENTIER, J.P., "Installation de fluage pour température élevée", 1er Colloque Annuel du Groupe Français de Rhéologie, Paris, 1983.

Deformation characteristics of rock mass in situ based on indirect tests

Caractéristiques de déformation du massif rocheux basées sur des essais indirects

Die Verformungseigenschaften des Gebirges, abgeleitet aus indirekten Versuchen

K.DROZD, Stavební geologie, Prague, Czechoslovakia

J.HUDEK, Projektový ústav dopravních a inženýrských staveb, Prague, Czechoslovakia

ABSTRACT: The knowledge of rock mass deformation characteristics is important to the designer of underground structures and of structures founded on rock mass. Different methods have been applied in order to determine the modulus of deformation D, modulus of elasticity E, and the Poisson's ratio ν, which are basic input data for all deformation calculations. The paper recommends a set of rock mass deformation characteristics based on the results from uniaxial compresive strength tests on rock samples, from the determination of the modulus ratio from the stress-strain curve according to D.U. Deere, and from the description of the fracture intercept of the rock mass in situ.

RÉSUMÉ: La connaissance des caractéristiques de déformation du massif rocheux est importante pour l'ingénieur, auteur de projets de structures souterraines et de constructions, fondées sur la roche. Plusieurs méthodes ont été employées pour déterminer le module de déformation D, le module d'élasticite E et le coefficient de Poisson ν, qui sont les données principales de tout calcul de déformation. Le rapport présenté recommande une série de caractéristiques de déformation du massif rocheux déterminées à partir d'essais en compression simple, sur échantillons de roche détermination du rapport de modules obtenu a partir de la courbe contrainte déformation selon D.U. Deere, et de la description de l'intersection des diaclases du massif rocheux en place.

ZUSAMMENFASSUNG: Die Kentniss der Verformungseigenschaften des Gebirges ist sowohl für den Entwurf von Untertagebauwerken als auch von Bauwerken, die auf Fels gegründet werden, wichtig. Für Bestimmung des Verformungsmoduls D, des Elastizitätsmoduls E und der Querdehnungszahl ν sind verschiedene Methoden angewendet worden. Der Bericht empfiehlt für untershiedene Gebirgsklassen einen Datensatz für dir Verformungseigenschaften des Gebirges, der auf der einachsigen Druckfestigkeit des Gesteins, auf dem Verhältnis E-Modul zu einachsiger Gesteinfestigkeit Q nach D.U. Deere und auf dem in situ beobachteten Trennflächenabstand basiert.

1.Input data

The paper recommends a set of rock mass deformation characteristics based on the results from simple tests and identification of the fracture intercept. The basic input data are:

a) the uniaxial compressive strength Q of the rock material, of which the rock mass is composed. The tested rock specimens are of regular shape and commonly used size.

b) the modulus ratio D/Q according to D.U. Deere. For this purpose the modulus of deformation D is determined from the tangent to the stress-strain curve for stress $\sigma_c = 0.5$ Q , obtained in the course of the uniaxial compressive strength test.

c) the fracture intercept of the rock mass in situ, according to the field observations.

The rock deformation characteristics are represented by the modulus of deformation D and Poisson's ratio ν .

2.The significance of the modulus of deformation

The modulus of deformation D for given rock mass in situ conveys both permanent and elastic deformations, that is it expresses the total deformation within a given range of stresses after the possible primary and secondary consolidation. To make it clear, the known Young's modulus of elasticity E, in contrast to the modulus of deformation D , is mostly for expressing only elastic deformation for rock specimens or for rock masses. Further, in a rock mass both the deformation of the rock material and the deformation caused by closure of joints are involved.

The recommended D values, represent the stress-strain relationship in the quasilineal part therefore they have meaning of the basic D values, which may be further modified according to the expected stress-strain relationship. For instance, if a certain rock mass hardening can be expected within the range of stress due to the closure of joints, that is in a case of a superlineal stress-strain curve, a known exponentional equation can be used thus giving higher moduli with the increasing stress level. On the other hand, a certain weakening not caused by water saturation alone can be expected when the stress deviator is increasing and the Mohr's circle of the principal stresses is approaching the Mohr's strenght envelope. That is, the sublineal stress-strain curve can be expected as well. Nevertheless in order to limit the application of the basic D values from table 2, they should not be applied without modifications for a stress level, which exceeds aproximately one half of the uniaxial compressive strength of the rock material of which the rock mass is composed. Certainly the reccommended D values cannot be used in a tension range of stresses or for rock materials which posses rheological behaviour (rock-salt), expansive behaviour (rock with anhydrite minerals or montmorillonite particles) or have a collapsing structure (chalk). In the meantime swelling rocks are excluded due to the shortage of deformation test data.

Another limitation is involved in the practical application of the recommended D values for a rock mass. The D values can be applied only for quasi-isotropic parts of the rock mass in situ, which are not affected by excessive compression of joints, as can be observed in rock masses affected by caving into underground openings, rockslides, blasting operations and so on.

Special attention should be given to fault zones which decrease the modulus of deformation. They may be evaluated as separate quasi-isotropic parts of the rock mass using classification procedures similar to those used for a sound rock mass. Because of the decreased strength of the rock material in the fault zone a lower strength and thus lower classification group may be applied. In using this system it is the responsibility of an engineering geologist to separate the total rock mass into appropriate quasi-isotropic parts. For example different strength of rock material due to weathering can be considered as dividing the rock mass into independent parts as well. We have used the term "quasi-isotropic" rather than the term "quasihomogeneous", because a quasihomogeneous rock mass can be ocasionally highly anisotropic. Nevertheless the orthotropic or completely anisotropic conditions can be considered in the selection of D values. Knowing in which direction we may expect the highest increase of stresses, in this direction we may put an orientated rock specimen to test, thus obtaining the basic Q and D of the rock material, which serve for further classification. As we suppose that the basic D values for a given rock mass will not be applied for high stress levels, the possible sliding on foliation or lamination planes, cannot be considered, as unfauvorable mutual orientation between the applied load vector and the weak planes may significantly decrease the uniaxial strength Q .

The recommended D values should reflect the quasilineal part of the stress-strain relationship, therefore a secant evaluation is better than a tangent one . Especially in the low range of loading the so called initial tangent moduli may be rather pessimistic in case of a superlineal stress-strain relationship. This of course is known from plate loading tests where additional deformations just in the contact area are involved in the testing procedures.

The D values presented in table 2, should be considered in such a way, that they express the virgin part of the stress-strain relation, that is the part wherein the stresses are increased. For example they can be used when a new structure transfers its load onto a rock subsoil and we want to calculate the settlement. It is known that in the unloading process, contrary to the loading process, the strains are lower. As they represent only the elastic part of the deformation, the given D value can be considered as the most unfavourable limit. Even if the relation between the modulus of deformation D and the modulus of elasticity E is not perfectly known, we may estimate, according to our experience, at least a 30% increase of E compared with D , thus making more realistic calculations around underground caverns for areas, where the stress decrease is expected, in contrast to areas, where stress increase is expected.

3.The significance of the Poisson's ratio

The Poisson's ratios ν are presented in table 3. The differences between the real ν values for a rock mass in situ and the recommended ν values can be rather high. At present we do not know perfectly the rules which influence the changing of ν values. Many authors have been trying to narrow the ranges of ν values, but until now we must still admit high discrepancies. Acccording to H. Link the ν values change with the stress level as well. However in the deformability calculations the Poisson's ratio plays no significant part and a small error can be accepted.

4.Basic classification system of the rock material

In table 1 the basic classification system is given, dividing the rock material into six groups in accordance with its uniaxial copressive strength Q. Rock types are given as examples, as well as a simplified method for rock classification according to the resistance to geological hammer blows or to scratching. The limits, as prescribed by the BGD (Basic Geological Description), have not been fully observed, as according to the engineering geologists the formal boundary between weak and hard rock is Q = 50 MPa and soils is Q = 1.5 MPa. Thus a slightly different division of strength limits has been applied in the classification.

In order to classify some rock types according to the strength of its rock material other simple tests are admitted in cases, where we know the relationship between the results. For instance the ratios between the point load index and the Q values are according to Z.T.Bieniawski 1 : 25, but practically in the range 1 : 15 to 1 : 30 depending on the core size. The ratios between the results from point load tests carried out on rock lumps of irregular shape and the Q values obtained from tests on specimens of regular shape give values in the range from 1 : 10 to 1 : 25. Sophisticated experts can give good Q classification even without any testing.

5.Fracture interecept

The classification of the fracture intercept has been observed according to the accepted BGD system. We may consider the decisive distance as a mean distance between joints of the joint system which has the highest joint density.

For the evaluation of the intercept of joints for a given rock mass in situ the dominant joint orientation had to be neglected. It was due to the necessary simplification of table 2 and 3 nevertheless a certain individual modification of the basic recommended D values can be observed. Higher D values can be used in case when main joints go parallel to the main stress vector. The same but in a moderate way can be reckoned for joints more or less inclined.

6.Verification of the recommended D a ν values.

In our classification we have not tried to verify directly the recommended D and ν values by presenting measured values from various loading tests as many authors have done. This was done deliberately as will be explained further.

Our recommendation is principally based on the results of hundreds of different loading tests carried out on different rock mass types either in Czechoslovakia as abroad within the last 60 years. The first loading tests used in our evaluation were those carried out in 1928 under the guidance of Quido Zaruba, the first president of the IAEG. Different rock mass types in the Barrandien Ordovician deposits in and around Prague were tested at that time. Later there were other loading tests on different dam sites,on sites of important factories,silos,bridges,tunnels, underground caverns and so on, so that many types of rock mass have been involved. Within the last 20 years additional large amounts of in situ tests have been conducted in connection with site investigations for the Prague Metro System. The tests have been carried out at normal temperature conditions. All these results have provided sophisticated data and experience to allow an evaluation of the deformability of the rock mass as it is introduced in a simplified form.

Some of our results have been verified from radial jack tests or from regressive deformability calculations where the deformations have been monitored.

A direct verification of the recommended data for a rock mass is difficult because we do not know with which values we should compare our recommended deformation characteristics. We can say that in all types of loading tests there are inherent some basic factors so that the deformation characteristics obtained from tests are more or less unlike of the real characteristics. The hierarchy of D values for a certain rockmass, from the highest ones to the lowest ones, obtained from different types of loading tests and testing procedures is known and has been many times documentated. The most sophisticated method, which the radial jack tests according to TIWAG procedures represent, is very expensive and cannot be applied everywhere. To make things more clear we must admit that for instance plate loading tests or tests with flat jacks fitted into crevices carried out on the

Table 1. Classification of rock specimens according to their uniaxial compressive strength Q (MPA) (according to M. Matula and M. Šamaliková)

group	strength interval Q (MPa)	strength	characteristics of hand specimens	rock types - examples
R1	>150	high to very high	corners can be broken off with difficulty using geologicall hammer	fresh: granitoides, diorites, gabbras, migmatites, granulites, amphibolites, andesites, basalts, silicites, orthogneisses, guartziferous paragneisses, crystaline limestones.
R2	50-150	high	can be broken into pieces with difficulty using geological hammer	fresh: limestones, dolomites, conglomerates, sandstones, greywackes, strong siltstones, paragneisses, mica schists, phyllites. Slightly weathered: R1.
R3	15- 50	moderate	can be broken easily into small pieces using geological hammer	fresh: clay shales, claystones, marlites, siltstones, volcanic tuffs, cataclastic rocks. Slightly weathered: R2. Moderately weathered: R1.
R4	5- 15	low	can be scratched with knife not with fingernail	fresh: chloritic and graphitic schists, mylonites (ultra) phyllonites, weakly cemented: sandstones, siltstones claystones. Slightly weathered: R3. Highly weathered: R1, R2.
R5	1.5- 5	very low	can be crumbled by hand	fresh: very weakly cemented: sandstones, siltstones, claystones, gouge material, tuffaceous rocks. Slightly weathered: R3. Completely weathered: R1, R2.
R6	0.5-1.5	extremely low	can be scratched easily with fingernail	Completely weathered: R3, R4, R5, eluvials.

Table 2. Recommended rock mass deformation characteristic - modulus of deformation D (MPa)

rock group	strength interval Q (MPa)	modulus ratio D/Q	fracturing intercept - distance between fractures (cm) > 200	200 - 60	60 - 20	20 - 6	6 - 2	< 2
R1	> 150	> 500	> 25 000	> 25 000	25 000	6 000	1 500	600
		200 - 500	> 25 000	> 25 000	13 000	3 500	850	400
		< 250	25 000	15 000	7 500	2 000	500	250
R2	50 - 150	> 500	25 000	20 000	8 500	2 500	600	300
		200 - 500	15 000	10 000	4 500	1 500	400	200
		< 200	7 500	5 000	2 500	800	250	130
R3	15 - 50	> 500	10 000	6 000	3 000	1 000	300	150
		200 - 500	4 500	3 200	1 500	600	200	100
		< 200	2 500	1 600	1 000	350	120	70
R4	5 - 15	> 500	3 000	2 000	1 000	400	150	80
		200 - 500	1 500	1 000	600	250	100	60
		< 200	750	550	300	140	60	40
R5	1.5 - 5	> 500	1 000	600	350	160	70	45
		200 - 500	500	330	200	100	40	30
		< 200	250	170	110	60	30	20
R6	0.5 - 1.5	> 500	300	200	130	70	35	25
		200 - 500	150	110	70	40	20	15
		< 200	75	60	40	25	15	10

unprotected walls of the testing galleries give either pessimistic values or optimistic values due to the fact that the rockmass in the wall is mostly in the limit state of stress conditions, not mentioning the influence of the excavation technology used for the testing gallery. On the other hand all borehole loading tests lack the possibility of testing a larger volume of rockmass and therefore the joints are only occasionaly involved in the deformability process. Many times the dilatometer tests results are influenced disadvantageously by radial craking of the rock mass in the tested borehole part and by the roughness of the borehole wall. In all loading tests the size effect is further involved as well as the testing procedures and methods of calculations.

The recommended D and ν values should be considered as guide data and give a certain direction for preliminary deformability calculations and do not mean that in situ loading tests, even if the tests are influenced by different inherent factors, should be abandoned.

The recommended evaluation of D and ν values has been applied in the Czechoslovak standard for settlement calculations for structures founded on rock masses.

Table 3. Recomended rock mass deformation characteristic - Poisson's ratio ν

rock group	strength interval Q (MPa)	modulus ratio D/Q	fracturing intercept distance between fractures(cm) >200	200-60	60-20	20-6	6-2	< 2
R1	> 150	> 500	0.10	0.10	0.10	0.10	0.10	0.10
		200-500	0.15	0.15	0.15	0.15	0.15	0.15
		< 200	0.20	0.20	0.20	0.20	0.20	0.20
R2	50-150	> 500	0.10	0.10	0.10	0.10	0.10	0.10
		200-500	0.15	0.15	0.15	0.15	0.15	0.15
		< 200	0.20	0.20	0.20	0.20	0.20	0.20
R3	15- 50	> 500	0.15	0.15	0.15	0.15	0.15	0.15
		200-500	0.20	0.20	0.20	0.20	0.20	0.20
		< 200	0.25	0.25	0.25	0.25	0.25	0.25
R4	5- 15	> 500	0.20	0.20	0.20	0.20	0.20	0.20
		200-500	0.25	0.25	0.25	0.25	0.25	0.25
		< 200	0.30	0.30	0.30	0.30	0.30	0.30
R5	1.5- 5	> 500	0.20	0.20	0.20	0.20	0.20	0.20
		200-500	0.25	0.25	0.25	0.25	0.25	0.25
		< 200	0.30	0.30	0.30	0.30	0.30	0.30
R6	0.5-1.5	> 500	0.25	0.25	0.25	0.30	0.30	0.30
		200-500	0.30	0.30	0.30	0.35	0.35	0.35
		< 200	0.35	0.35	0.35	0.40	0.40	0.40

REFERENCES

Bukovanský, M. 1970. Determination of elastic properties of rocks using various in situ and laboratory methods. Proc. 2 ISRM Congress, Beograd. Vol.1. Theme 2 - 9.

Doležalová, M. & Drozd, K. 1979. Site investigations and FEM Calculations for two underground caverns in Peru. Proc. 4. ISRM Congress, Montreux, Vol. 2. pp 105 - 112.

Drozd, K. 1978. Field loading tests on the Vrchlice dam site. Symp. on Rock Mechanics related to dam foundations. Rio de Janeiro, Theme II-21.

Dvořák, A. 1970. Seismic and static modulus of Rock Masses. Proc. 2 ISRM Congress, Beograd, Vol. 1, Theme 2 - 6.

Goodman, R.E. 1980. Introduction to rock mechanics. New York. John Wiley and Sons.

Kulhawy, F.H. Ingraffea A.R. 1978. Geomechanical Model for settlement of long dams on discontinuous rock masses. Symp. on Rock Mechanics related to dam foundations. Rio de Janeiro, Theme III-15.

Seeber, G. 1970. Ten years use of TIWAG radial jack. Proc. 2 ISRM Congress. Beograd,Vol. 1. Theme 2-22.

Záruba, Q., Bukovanský, M. 1966. Mechanical properties of Ordovician shales of Central Bohemia. Proc. 1. ISRM Congress, Lisbon. Vol. 1. pp 421 - 424.

Sliding of blocks in the top rock, a model for rockbursts of the regional type

Glissement des blocs au toit, un modèle des coups de terrain du type régional

Blockgleiten im Hangenden, ein Modell für Gebirgsschläge des regionalen Typs

W.MINKLEY, Institut für Bergbausicherheit, Leipzig, GDR

U.GROSS, Institut für Bergbausicherheit, Leipzig, GDR

ABSTRACT: In the presented model the reduction of the vertical supporting forces exerted by the pillars due to stress relaxation is considered to be the cause for a block sliding in the overlying rock. An unloading occuring in the extracted seam over an extended area can result in a sudden subsidence of the overlying rock and thus to the collapse of complete working areas when a critical size of the working area is attained.

RESUME: Dans le modèle presente, on considère la decroissance des forces d'appui verticales des piliers par suite d'une relaxation des contraintes comme la cause d'un glissement des blocs au toit. Au cas où les dimensions des panneaux arriveraient aux valeurs critiques, une decharge, se produisant sur une surface etendue par l'abattage de la couche, peut conduire à l'affaissement brutal du toit et en consequence à l'effondrement des panneaux entiers.

ZUSAMMENFASSUNG: Als Ursache für ein Blockgleiten im Hangenden wird im vorgestellten Modell die Reduzierung der vertikalen Stützkräfte der Pfeiler durch Spannungsrelaxation angesehen. Eine grossflächig eintretende Entlastung im abgebauten Floz kann bei Erreichen kritischer Baufeldgrössen zur plötzlichen Hangendabsenkung und damit zum Zusammenbruch ganzer Abbaufelder führen.

1. Problem

Rockbursts are in a close causal context to extraction when they occur as a consequence of an excessive strain of the supporting elements of the underground openings as for instance of the pillars or of the stress concentrations in the immediate vicinity of openings. There are a number of measures as well as starting points for combating and forecasting this kind of rockbursts which can be specified as mining rockbursts and can be assigned to the local type.

However, it is more difficult to prognosticate rockbursts of the regional type which affect both the underground opening and often the surface. These will be designated as tectonic rockbursts [1] or mining-tectonic rockbursts [2]. They are characterized by an unexpected collapse of complete working areas up to a size of km^2. Thereby considerable amounts of seismic energy can be released (on the order of magnitude up to 10^{12} J [1]). The failure-phenomena extend high into the top rock and penetrate sometimes up to the surface (figure 1). They indicate that these rockbursts develop as a result of an intensive interaction between the extraction-horizon and the hanging and the overlying rock respectively [3].

A failure in the overlying rock forms preferably on fault zones in the region above the boundary of the working area which represent particularly marked points of weakness inside the rock mass. There unfavourable stress redistributions take place induced by extraction. The steeply dipping faults will be unloaded in the horizontal direction above the boundaries of the working area and an additional shear stress will be imposed in the vertical direction (figure 3). Critical states and the loss of static friction on faults occur often not until water is present. This was found by investigations in the connection with the clarifaction of the collapse of fields in the Werra-Potash district [1].

2. Static condition of limiting equilibrium for sliding of a block of the overlying rock

The model conception for a failure of the overlying rock as a consequence of blocksliding starts from the following basic idea: The opening up of a seam in a large area represents a disturbance of the initial state of equilibrium inside the rock mass resulting in stress redistributions with a regional character. The overlying rock body takes part on the achievement of a new state of equilibrium depending on the restraining conditions and its deformation and strength properties. The pillars partially remove from loading in the interaction process between loading and supporting system by relaxation and creep processes. Consequently an areal unloading takes place in the vertical direction (Figure 2). The missing part of the supporting force will be redistributed onto the boundary and that regions in advance of the face by the supporting behaviour of the overlying rock. A difference between the weight-load of the overlying rock and the smaller vertical

Figure 1: Gaping cracks above ground surrounding the collapsed working area Wintershall-Heringen [4]

reaction forces produced by the pillars results in a shear loading in the overlying rock above the boundary of the working area (Figure 3).
The frictional resistance will be mobilized on the mechanical planes of weakness which are located there. It is assumed that always mechanical planes of weakness are present in the overlying rock above the boundary of the working area dipping vertically and limiting a block of the overlying rock of the size of the working area. As mechanical planes of weakness discontinuities will be adopted which penetrate the overlying rock completely and across which frictional and cohesive forces are effective. This is the most unfavourable case which provides a solution in the sense of a lower bound.
The weight of the rock mass-block, that has to be supported, will increase when the working area is extended until finally a critical size of the working area exists at which the state of limiting equilibrium for block-sliding in the overlying rock is attained. As a first approximation for block-sliding above a working area the condition for limiting equilibrium can be written as follows:

$$\left(\frac{A}{U}\right)_K = \frac{1}{2}\mu_K \cdot z_D\left(1-\frac{z_S^2}{z_D^2}\right)\cdot\left[\frac{(\lambda_D-\sigma_{S/K})\frac{\mu_S}{\mu_K}}{1-\frac{\overline{\sigma_{vD}}}{\sigma_{zD}}}-\frac{\nu}{1-\nu}\right] \tag{1}$$

in which is

$\left(\frac{A}{U}\right)_K$;	critical ratio of the size of the working area A to its circumference U
$\lambda_D = \frac{\frac{1}{2}(\sigma_x+\sigma_y)}{\sigma_z}$;	ratio between the average horizontal stress and the vertical stress (coefficient of lateral stress)
$\sigma_{S/K} = \frac{p}{\rho_G \cdot g \cdot z}$;	static and kinetic respectively joint water factor
$e_{vD} = 1-\frac{\overline{\sigma_{vD}}}{\sigma_{zD}}$; $1 > e_{vD} > 0$;	Unloading factor of the overlying rock (σ_{zD} = initial gravity-induced stress and $\overline{\sigma_{vD}}$ = average vertical stress above the working area due to extraction, see figure 2)
μ_S/μ_K ;	ratio of the static to the kinetic coefficient of friction
z_D ;	thickness of the overlying rock
$z_D - z_S$;	thickness of the supporting block of the overlying rock
ν ;	Poisson's ratio

3. Application of the model and conclusions

Back analyses of rock bursts that happened formerly were performed on the basis of the developed conceptions of the model. The known solutions for the infinite half-space subjected to a rectangular area load [5] were used to determine the extraction induced stress variations that had to be transfered from the seam horizon onto the overlying rock. A hydrostatic joint water pressure in the fault zone was assumed in all of the calculations. If partially higher values of the joint water pressure are adopted then the critical size of the working area can already be attained at an even smaller unloading in the working area.
Of fundamental importance is thereby the statement that the unloading in the working area, defined by the unloading-factor:

$$e_{Fl} = 1 - \frac{A_{pF}\cdot\overline{\sigma_{pF}}}{A_S\cdot\sigma_{zFl}} \tag{2}$$

where

$\overline{\sigma_{pF}}$	average vertical pillar stress in the working area
σ_{zFl}	lithostatical pressure in the seam horizon
A_{pF}	average value of the cross sectional area of the pillar
A_S	area of the system

can be influenced in an active way by the design of the pillars whereas the joint water pressure is primarily a passive influence factor in conjunction with mining and thus provides an additional source of danger.

Strong rainfalls preceded for instance in each case the rock bursts in the mines of the iron ore district in the Lorraine lying in shallow depths below surface [6]. A connection between the occurence of rockbursts and the build-up of a water pressure was also observed in the KOLAR GOLD FIELDS in India [7]. There is also the possibility to affect the hydrodynamic system in the rock mass negatively by means of technical actions. An additional endangering regarding the extraction but difficult to quantify arises for instance from the injection of fluids.
The approach to the static critical state in the overlying rock can be effected by the following processes:

- enlargement of the working area;
- increase of the unloading in the working area (unloading factors e_{Fl} and e_{VD} respectively in equation (1) increase);
- increase and fluctuation of the fluid pressure in the overlying rock (increase of the joint water pressure factor $\sigma_{S/K}$ in equation (1)).

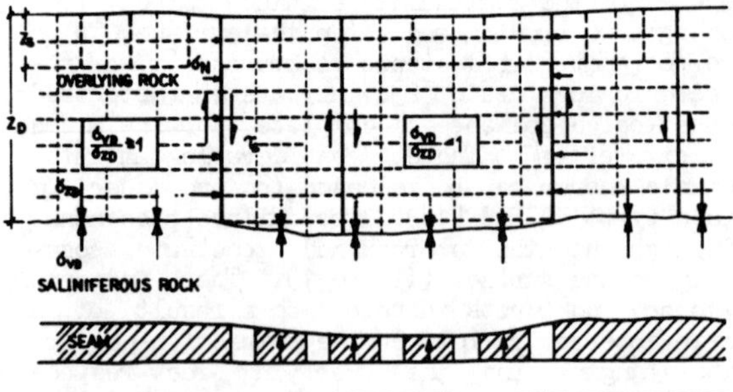

Figure 2: Model for the consideration of the equilibrium

As a rule, several of these factors will coincide. Unlike to rockbursts of the local type, which announce themselves by stress concentrations in the seam horizon, the development of rockburst of the regional type corresponding to the model conceptions can occur together with a too small loading of the extraction pillars for a long period. Not until a critical size of the working area is reached the load-carrying capacity of the top rock will be exhausted and the block of the overlying rock above the working area begins to slide down. The sliding block will undergo a sudden acceleration if thereby a loss of static friction along the mechanical planes of weakness bounding the block will occur [8]. The same process can be initiated according to the principle of effective stresses by pressure fluctuations in the hydrodynamic system so that a source of danger will arise for subcritical sizes of working areas. The sliding block of overlying rock will produce a sudden overload of the pillars in the working area if the rocks composing the pillar are prone to brittle fracture and a catastrophic collapse of the complete system will take place. Thereby considerable amounts of energy can be transfered which were stored before as strain energy or in the gravitational potential.

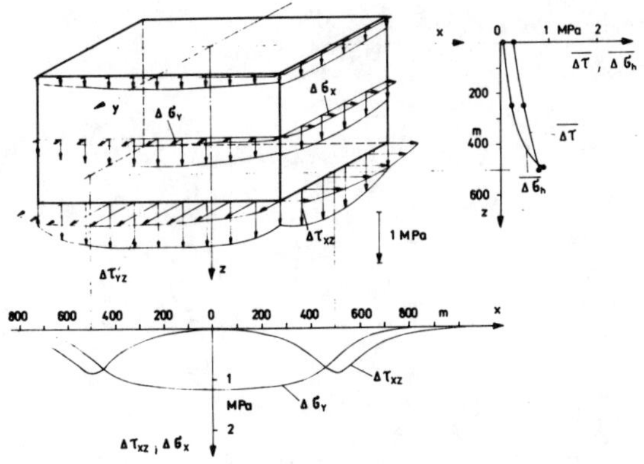

Figure 3: Changes in shear and horizontal stresses in the overlying rock above a working area of 500 x 1000 m for an unloading of 20 % in the working area

Extraction under more complicated geological and geomechanical conditions forces the designer to aim at a system of mining and planning which matches largely the definite conditions. The geomechanical realization of this objective requires the development and application of a new strategy in which the interaction of the behaviour of pillar and overlying rock can be taken into account as an active quantity controlling the extraction planning. The investigations carried out at the instance of the Supreme Mining Authority on the Council of Ministers of the GDR will be pursued in order to extensively penetrate the complex geomechanical problem and to recognize and control in time potential sources of danger for the safety in mines as well as to deduce also economic solutions for the exploitation of raw materials.

REFERENCES

[1] THOMA, K.; KNOLL, P.:
Neue Erkenntnisse bei der Beherrschung der Gebirgsschlaggefahr im Bergbau der DDR
Neue Bergbautechnik 10 (1980), 195 - 203

[2] LASAREWITSCH, L.M.; LASAREWITSCH, T.I.:
(Die Vorhersage bergbautektonischer Gebirgsschläge) in Russian
Bezopasnost' truda v promyslennosti, 3 (1987), 54 - 56

[3] KNOLL, P.; THOMA, K.; MINKLEY, W.; [illegible], U.:
Zur Wechselwirkung zwischen dem geomechanischen Verhalten des Deckgebirges und den Abbauarbeiten im Werra-Kalirevier
Neue Bergbautechnik 10 (1980), 209 - 214

[4] GIMM, W.; PFORR, H.:
Gebirgsschläge im Kalibergbau unter Berücksichtigung von Erfahrungen des Kohlen- und Erzbergbaus
Freiberger Forschungsheft A 173, Akademie-Verlag Berlin 1961

[5] VOCKE, W.:
Räumliche Probleme der linearen Elastizitat
VEB Fachbuchverlag Leipzig 1969

[6] TINCELIN, E.; SINOU, P.:
Der Einsturz des Hangenden bei Orterbau. Praktische Schlussfolgerungen und Versuch einer mathemathischen Erfassung der bei Gebirgsschlägen beobachteten Vorgänge
Internationaler Kongress für Gebirgsdruckforschung, Paris 16. - 20. Mai 1960, 605 - 623

[7] KRISHNA MURTHY; GUPTA, P.D.:
Rock mechanics studies on the problem of ground control and rockburst in the KOLAR GOLD FIELDS.
Rockburst: prediction and control, Published of THE INSTITUT OF MINING and METALLURGY, London 1983

[8] MINKLEY, W.; GROSS, U.:
Zum Haftreibungsverhalten von Felstrennflächen in Abhängigkeit von der Belastungsgeschichte
Felsbau 1988 (in preparation)

Rock mass deformability from back analyses of dam behaviour

Déformabilité des massifs rocheux dérivée de l'analyse des données de comportement des barrages

Die Felsverformung aufgrund der Rückanalyse von Staudammdeformationen

R.RIBACCHI, Department of Structural and Geotechnical Engineering, University of Rome, Italy

ABSTRACT: Indications on the deformation behaviour of rock masses at large scale can be obtained from back analyses of displacements data in dam monitoring. The paper discusses some problems which are met in the implementation of the models and in the analysis of the results and the general indications which can be drawn from various case histories. Finally, a number of identified rock moduli are compared with the design values, and a correlation between the moduli and the quality index RMR is presented.

RESUME: Sur la base des rétroanalyses des données de déplacement dans la surveillance des barrages on peut obtenir des indications sur le comportement de déformation des massifs rocheux à grande échelle. Cette mémoire prend en examen les problèmes que l'on rencontre dans l'implémentation des modèles et dans l'analyse des résultats, même que les indications générales que l'on peut tirer des différents cas. Enfin, on fait une comparaison entre les modules des rochers identifiés et les valeurs de projet, et on présente aussi une corrélation entre les modules et l'index de qualité RMR.

INHALTSANGABE: Die Daten über dem Verformbarkeitsverhalten der Bergmassive in grossem Umfang werden mit der Rückanalyse der Verschiebungsdaten durch Monitoring der Staumauern erhalten. Dieser Bericht verhandelt über einige Probleme, die man in der Verwirklichung der Modelle und in der Analyse der Ergebnisse triffts, und über die allgemeinen Angaben der verschiedenen vorhergehenden Fälle. Schiesslich vergleicht man einige Moduln von ermittelten Gesteine mit den Projekswerten und es wird eine Korrelation zwischen den Moduln und der Güteziffer RMR vorgebracht.

1 INTRODUCTION

The monitoring of dam foundations often includes a comparison of the observed displacement values with those predicted by deterministic structural models (FANELLI et al.,1979). Besides furnishing elements to check the safety conditions,a detailed examination of the results provides interesting information about the deformation behaviour of rock masses at a scale much larger than that involved in conventional in situ tests.

This report discusses some problems encountered in the implementation of structural models and in their numerical simulation; some general results which have been gathered from Italian case histories are also presented.

A well balanced monitoring scheme for gravity dams commonly used in recent projects is shown in Fig.1. The various components of displacements (rotation and horizontal displacement of the dam base, deformation of the structure) and the contribution of the various layers of rock can be easily identified.

Obviously in many old dams only a few of these instruments were installed at the time of construction; a (partial) integration is sometimes performed in the course of a reappraisal or rehabilitation of these old structures (BONALDI et al.,1982; BONALDI et al.,1985).

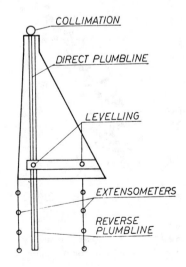

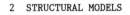

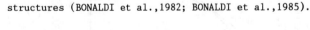

Fig. 1 - Monitoring the various displacement components in a gravity dam

2 STRUCTURAL MODELS

The analyses of displacement data are based on:

- statistical models;
- deterministic models;
- hybrid models.

Statistical methods find wide application in checking the safety of dams. However only a deterministic analysis gives full information on the influence of the various load components and (through back analysis) on the real mechanical parameters of the dam foundations, which can then be compared with those assumed in the design.

Also in the hybrid models, a deterministic analysis

of the effect of the loads related to impounding is performed, whereas the effect of temperature variations are evaluated by means of statistical models, usually because thermometric determinations within the structure are not available.

The deterministic analyses are now usually carried out by means of numerical (finite element) models because of their great flexibility; in the past, however, interesting results were also obtained, in the case of gravity dams, by the application of simplified analytical models (MARAZIO et al., 1966).

The results of a deterministic analysis can be conveniently represented by means of the "influence functions". The effect of the reservoir level, w, on the displacements (or rotations) u, measured by some monitoring instrument, can be written as:

$$u = \frac{1}{E_c} f_c \ (w) + \frac{1}{E_r} f_r \ (w)$$

The total effect can thus be considered as consisting of a contribution deriving from the deformation of the structure and of a contribution deriving from the deformation of the foundation, which are respectively proportional to the compliance of the concrete 1/Ec and to that of the rock, 1/Er (Fig.2). Obviously only the latter term is present for those measurement points which are entirely located within the foundation rock (reverse plumbline, extensometers).

The influence function related to the foundation rock can be conveniently broken down into one term which represents the effect of the water load on the upstream face of the dam, and another term which represents the effect of seepage forces within the rock mass (including uplift at the dam base). An example is shown in Fig.3.

From the results of many analyses carried out so far in various types of dams it was found that even strong variations of the ratio between concrete modulus and rock modulus have only a slight effect on the influence functions. This simplifies back analysis procedures for the identification of rock mass deformability; they can be performed by applying simple regression techniques to the observational data (Fig.4). It is thus also possible to evaluate the reliability of estimated values by means of their confidence intervals.

When independent back analyses are performed on the data of various monitoring points, consistent values

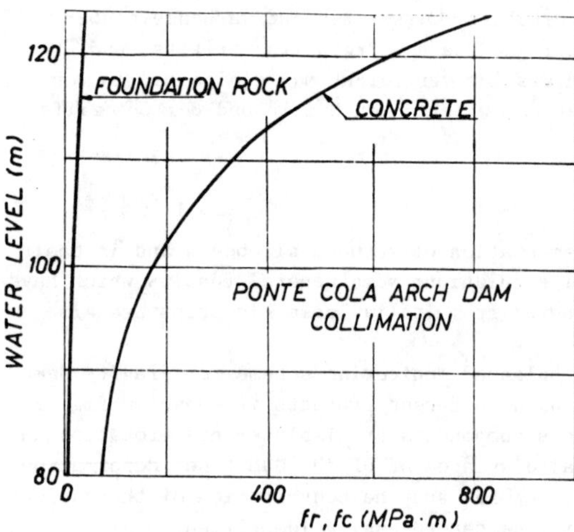

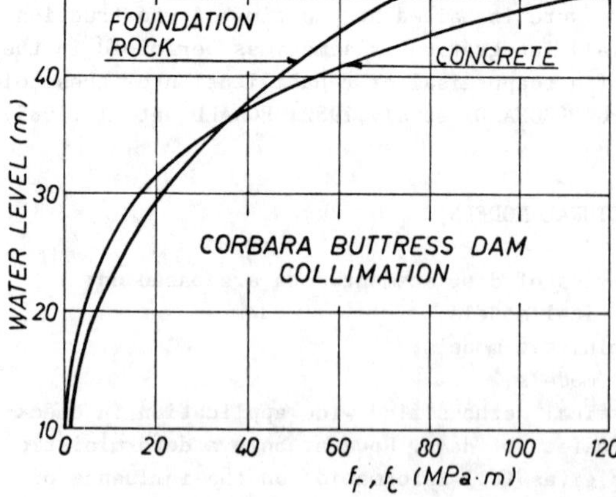

Fig. 2 - Influence functions for the collimation in the case of an arch dam and of a gravity buttress dam. The contribution of the foundation rock to the total displacements is markedly different for the two types of structure

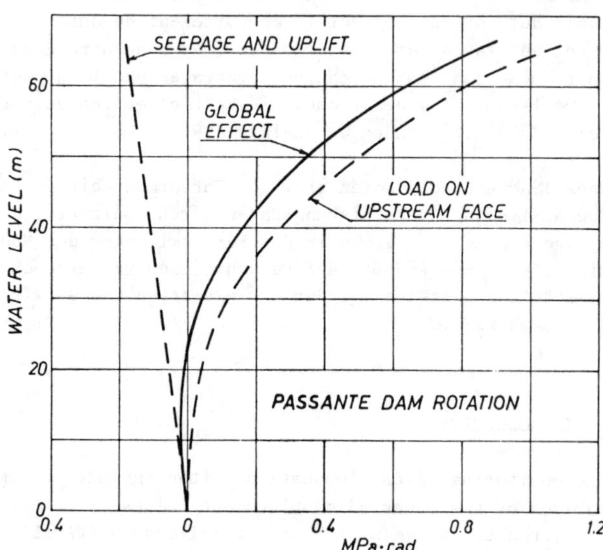

Fig. 3 - Influence functions for dam base rotation in the case of the Passante gravity dam (BONALDI et al., 1983)

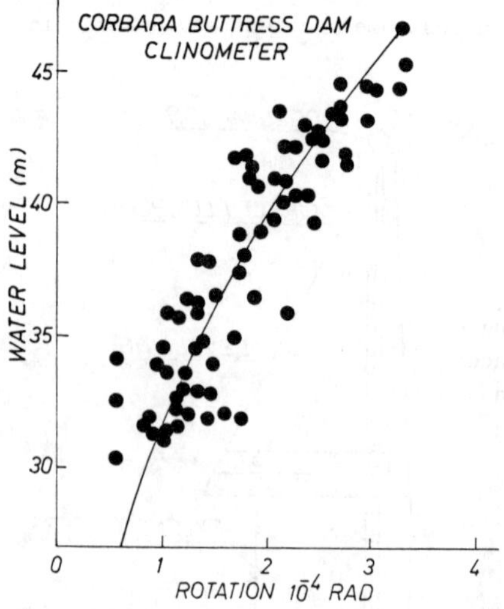

Fig. 4 - Fitting an influence function to observational data in order to identify the foundation moduli

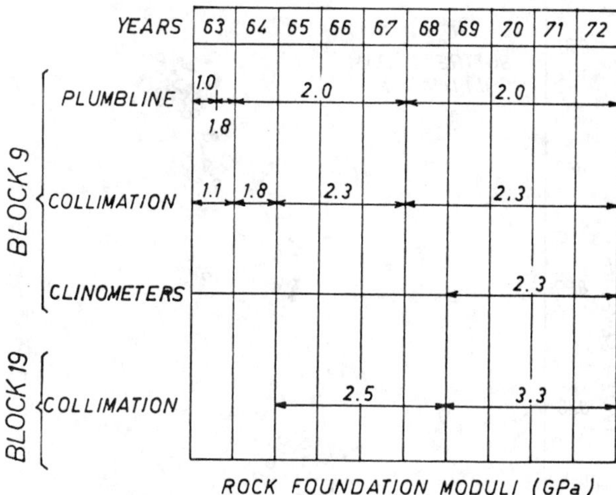

Fig. 5 - Consistent values of the rock foundation moduli obtained with different monitoring methods at the Corbara buttress dam (BONALDI et al., 1985)

for the deformability of the rock should be obtained, as is observed, for instance, in the case of Fig.5. Otherwise, the model is unsatisfactory, the most likely cause being a dishomogeneity of the rock which is differently felt by the various monitoring points (Fig.6).

In this case the application of more complex models accounting for rock dishomogeneity can be tried. At least in principle, if a model of the foundation rock which is characterized by n parameters is adopted, they could be identified when n different sets of measurements are available. However only models which are not very complex give reliable results, because the implementation of the model and the assumption of the loading conditions involve many uncertainties.

However, when direct measurements of the displacements in various points of the rock mass are available (for instance by means of extensometers) it is much easier to fully identify the rock moduli variations.

3 PROBLEMS IN THE IMPLEMENTATION OF THE MODEL

Careful attention must be paid to the boundary and loading conditions introduced in the model, because in some cases it is found that the adoption of alternative, but equally reasonable assumptions when the specific data are not available, can lead to widely differing results in the back analyses for the evaluation of the parameters of the rock mass.

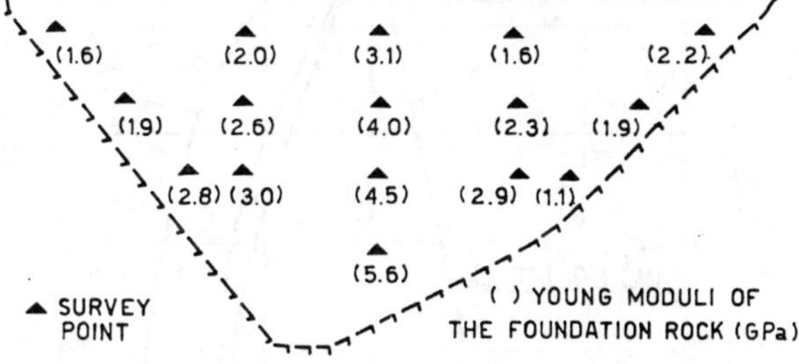

Fig. 6 - Values of the foundation moduli obtained at the Ponte Cola dam by independent back analyses of the displacements of various parts of the dam face (assuming rock homogeneity) (RIBACCHI & SELLERI, 1978)

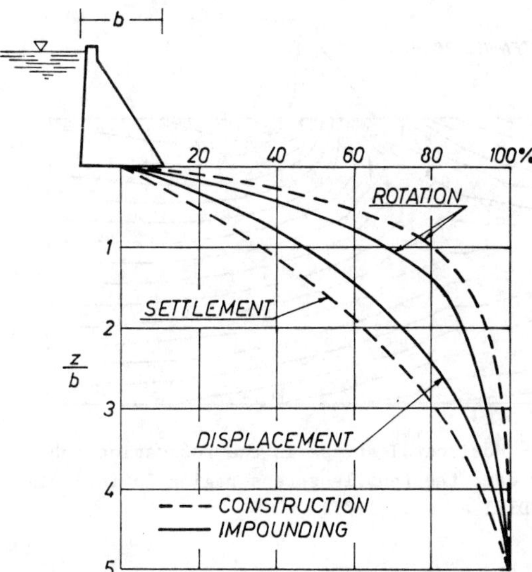

Fig. 7 - Rotation, settlement and horizontal displacement of a gravity dam during construction and impounding as a function of the thickness of the layer which is considered deformable; the values are scaled down with respect to those obtained for a ratio between the deformable layer and the width of dam base equal to 5.

The most important choices to be made can be summarized as follows:

- depth and extension of the mesh (mainly in plane models);
- existence and depth of a tensile zone at the heel;
- treatment of seepage effects;
- variation of moduli with depth.

3.1 Mesh boundaries

The mesh boundaries (especially in depth) are important when gravity dams are analyzed by means of simplified plane models. Their influence, however, depends also on the type of measurement which is being analyzed. In fact settlements during construction or horizontal displacements during impounding are sensitive to the depth of the mesh, whereas only the superficial layers give a significant contribution to the rotation of the dam base during construction and impounding (Fig.7).

In order to avoid these difficulties, a finite layer model can be adopted, in which both the thickness and the modulus of the layer are identified through back analysis. This procedure was adopted for instance in the analysis of the Passante dam (BONALDI et al., 1983) taking also into account (even if very roughly) the indications of the rock mass explorations which showed an improvement of rock quality with depth.

3.2 Tensile crack

Another factor to be considered in the model is the possible formation (at some critical water level) of a tensile crack or of a tensile strain zone near the upstream heel of the dam. Here, in fact, the stresses induced in the rock by the thrust of the dam alone are usually tensile (Fig.8); the formation and depth of the crack depends therefore on the preexisting state of stress within the rock mass, which is difficult to

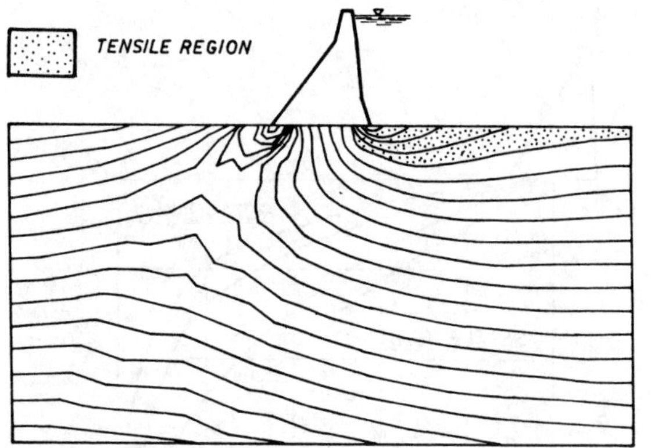

Fig. 8 - Horizontal stress in the foundation rock of a gravity dam; the tensile stress region is indicated (MANFREDINI et al., 1975)

be estimated at a given site. Its presence was assumed by some Authors in the numerical models of real dams (BOURBOUNNAIS and MORGENSTERN, 1974; SOUZA LIMA et al., 1985).

The formation of the tensile zone can be detected when the monitoring installation includes extensometers crossing this zone. Hydraulic monitoring (piezometric values and leakages from drain holes) can also be a sensitive indicator. In fact the permeability of rock masses is strongly influenced by the strains occurring either because of stress variation or because of "pseudoplastic" phenomena corresponding to an opening up of rock joints. As a consequence a markedly non-linear trend of the piezometric values or of the seepage flows as a function of the reservoir level, can be often observed (RIBACCHI and SELLERI, 1978; GOMEZ LAA and RODRIGUEZ, 1985). The formation of a tensile zone or a crack will be indicated by a sharp bend in these curves.

Various case histories of tensile zone formation and its influence on dam stability are discussed for instance by LUDESCHER (1985),WIDMAN and HEIGERTH (1976), TIZDEL (1970), LEUENBERGER (1976).

The influence of the tensile zone on dam behaviour can be large and lead to serious errors in back analyses when not properly taken into account.

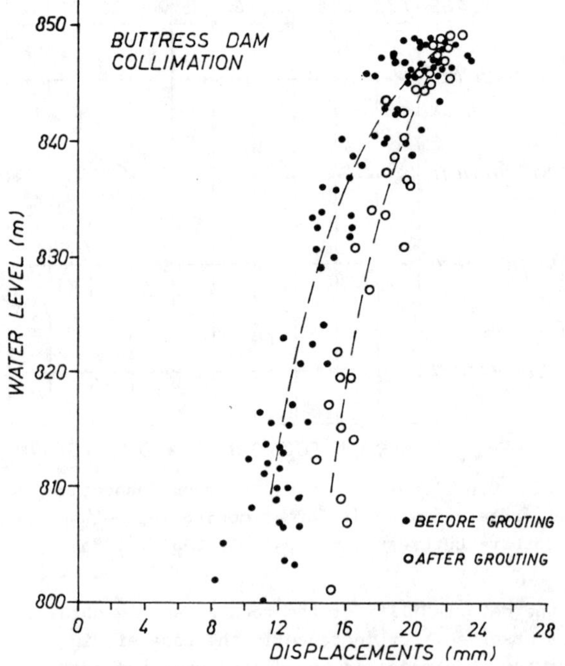

Fig. 9 - Collimation values (depurated of the thermal component) as a function of water level before and after grouting underneath the head of a buttress

A recent example is shown in Fig.9. Grouting operations (with cement and resins) were carried out below the heads of a buttress dam in Italy. This treatment was purposely effected at a high impounding level so that tensile strains prevailed in the zone to be grouted. After the treatment the maximum (downstream) displacement at high water levels did not markedly change, whereas there was a shift in the minimum values; this behaviour is clearly related to the filling of opened up joints in the zone below the upstream face. The apparent modulus of the rock mass (assumed as being homogeneous) almost doubled.

Similar situations were described by CANDIANI and GAVAZZI (1964) as a consequence of grouting operations performed at high reservoir levels at the Bissina dam.

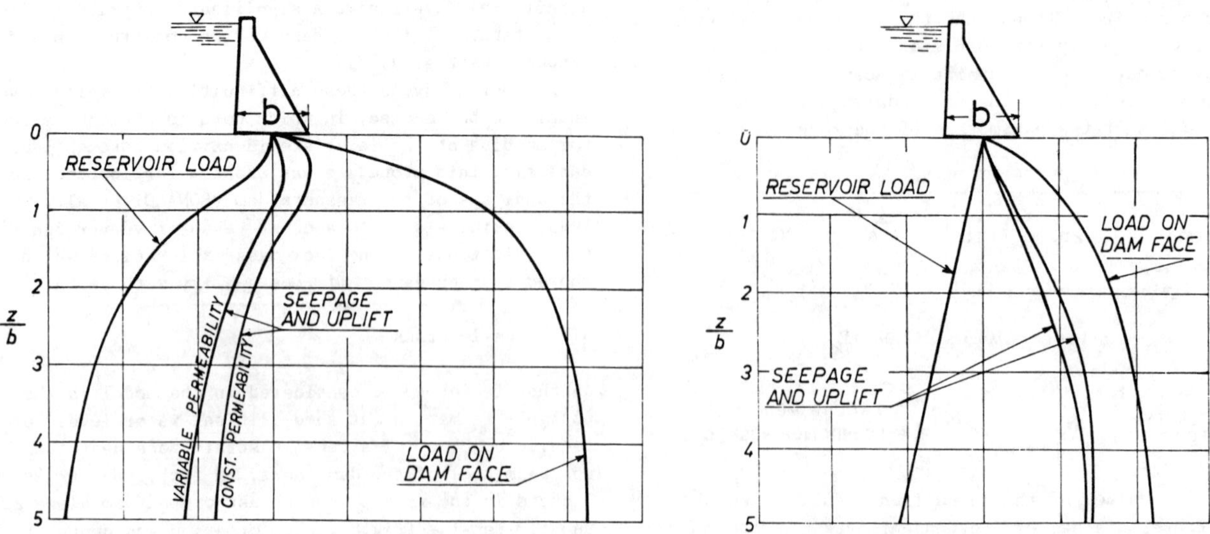

Fig. 10 - Dam base rotations (left) and horizontal displacements (right) at maximum impounding level as a function of the thickness of the deformable layer. The effect of the various load components is shown

3.3 Seepage conditions

Another possible difficulty in the implementation of the model is the correct representation of the seepage conditions within the foundation rock and below the reservoir. Sometimes the effect of impounding on the foundation rock has been represented as an external load at the bottom of the reservoir. In other studies it is simulated by introducing, within the rock, body forces corresponding to the seepage gradients.

Assuming a homogeneous foundation rock the mechanical response of these two models to water level variations is quite different (Fig.10). In both models, however, the global response of the structure at increasing water levels (including the effect of the load on the dam face) can be characterized initially by a slight upstream rotation, followed by a larger downstream rotation at higher water levels (Fig.3). This behaviour is observed in many Italian dams (MARAZIO et al., 1966).

We can state that the first model would be valid only if the water pressure variations at the reservoir bottom do not propagate into the rock mass at depth, a condition which depends both on rock mass permeability and on the frequency of the impounding cycle. It is likely however that in most cases an almost steady state seepage flow will be established and therefore, except in particular geotechnical situations (TIZDEL, 1970), the second model is the more realistic one. Piezometric monitoring within the rock mass (Fig.11) would make it possible to ascertain the real water pressure distribution, which is also influenced by the effectiveness of the grout curtain and of the drains.

Finally it must be observed that the effect of the seepage forces within the rock on the deformation of the dam is much influenced also by a systematic variation of rock deformability with depth, a situation which is frequently met in practice.

Fig.10 allows an immediate visualization of the influence of the various layers of rock below the dam. It is found that when the superficial rock layers are much more deformable than the rock mass in depth, the effect of seepage is a downstream rotation of the dam body, instead of an upstream one as in the case of homogeneous rock. This occurs because the effect of the uplift on the dam base is prevalent with respect to that of the body forces related to seepage, which is mostly felt in the deepest portions of the rock mass.

Therefore, in practice, the influence of water seepage on dam rotations can largely vary depending on the distribution of moduli with depth.

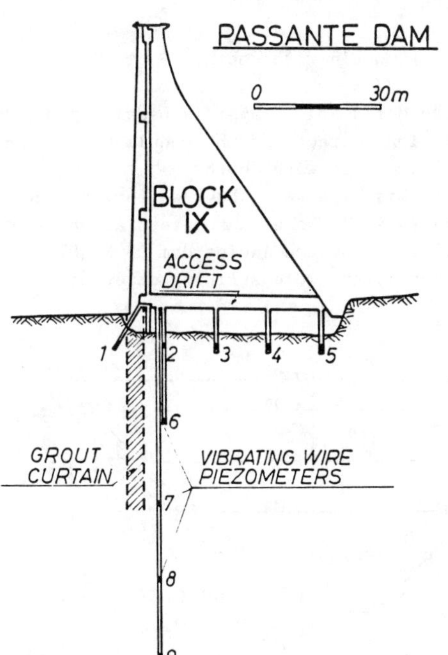

Fig. 11 - Piezometric monitoring in the foundation rock of the Passante dam (BONALDI et al., 1983)

4 TYPICAL CHARACTERISTICS OF ROCK MASSES

Some general characteristics of the deformation behaviour of the rock masses which have been inferred from various back analysis case histories will be now briefly commented on.

4.1 Rheologic and non linear behaviour

The first applications of load (weight of the dam or first impounding) is characterized by a much greater deformability with respect to that observed in the successive impounding cycles. The ratio between the total and the recovered displacements (that is between modulus of elasticity and modulus of deformation) varies greatly from one case to another; typically it falls in the range 1.2-2.2, as is shown in Fig.12.

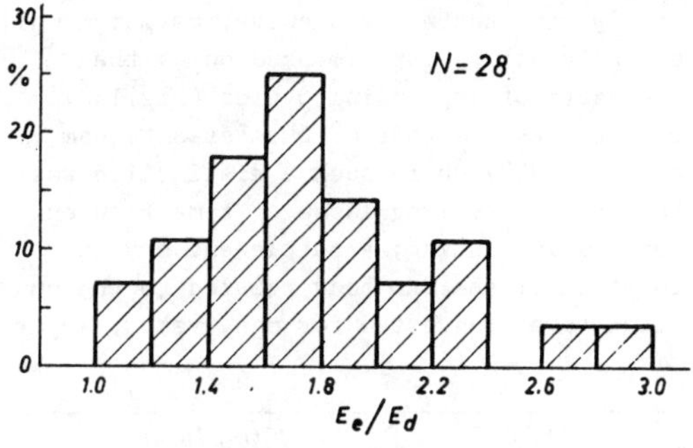

Fig. 12 - Histogram of the ratio between total and recovered deformation (that is between modulus of elasticity, Ee and modulus of deformation, Ed) for 28 case histories

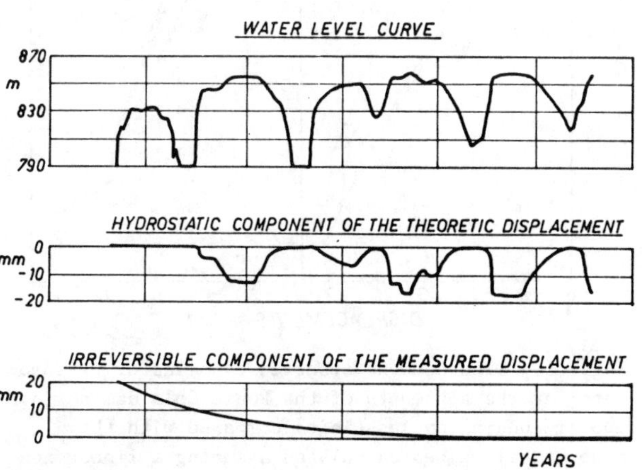

Fig. 13 - Determination of the irreversible component from the back analysis of the displacements at the Neves arch dam (FANELLI et al., 1979)

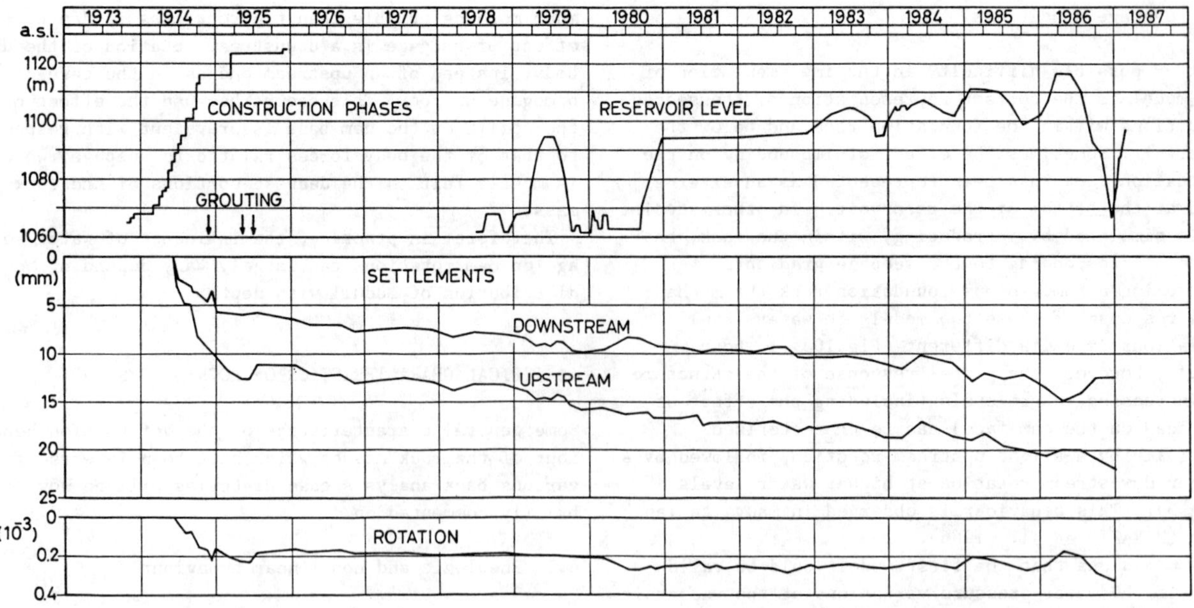

Fig. 14 - Settlements and rotation at the base of the Passante massive gravity dam during and after construction

Because of the variable loading conditions during the impounding cycles it is not always easy to distinguish the rheologic (time dependent) deformations from those corresponding to first loading conditions. Frequently the unrecovered displacements are empirically represented by a curve, asymptotically reaching a finite value, superimposed on the reversible effects of impounding cycles (Fig.13).

An interesting case is that of the Passante dam (BONALDI et al., 1983) where such a distinction was possible because of the long lapse of time between construction and impounding; a settlement and an upstream rotation of the dam body related to the construction load have progressed for many years, apparently at an almost constant rate, reaching a value which in almost twice the initial one (Fig.14).

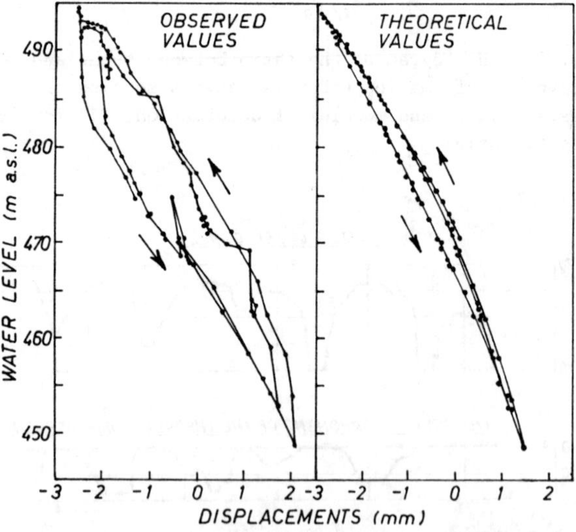

Fig. 15 - Displacement observed by means of extensometer at the abutments of the Ponte Cola dam during two impounding cycles (left) compared with the theoretical values calculated assuming a linear non hysteretic model (right). The small cycle present in theoretical model is due to the effect of thermal loading on the dam

After a number of impounding cycles, the displacements usually become fully recoverable, but a non linear behaviour of the rock mass is still indicated by a marked hysteresis area; as a consequence of this situation, the displacements are larger when the stress level decreases, than when it increases. This behaviour is more easily recognizable from extensometer readings, an example of which is shown in Fig.15 (Ponte Cola dam). Attenuation factors for the cyclic load were somewhat higher than those found in accurate plate loading cyclic tests (5-10%) in these same materials; the hysteretic behaviour is probably related partly to viscoelastic behaviour and partly to solid Coulomb friction.

4.2 Variation of moduli with depth

A loosened, more deformable, layer is usually present near the foundation contact and the moduli show a more or less gradual increase with depth.

Situations of this type were observed for instance in the back analysis of the extensometers at the Ponte Cola dam (Fig.16) and at the Ancipa dam (BONALDI et al., 1982). Probably the more striking observations

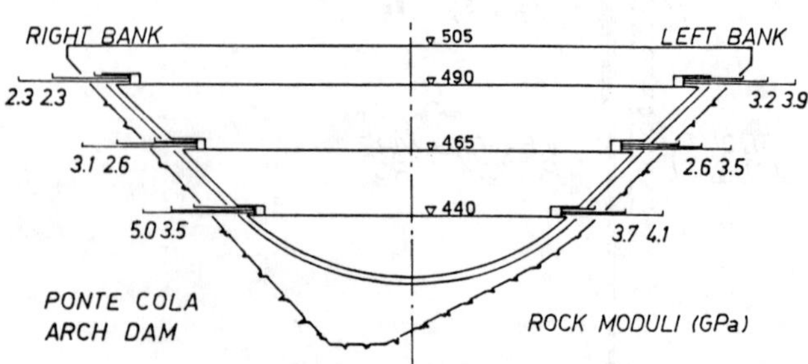

Fig. 16 - Moduli obtained from the back analysis of extensometers with different anchor depths in the abutments of Ponte Cola dam (BONALDI et al., 1982)

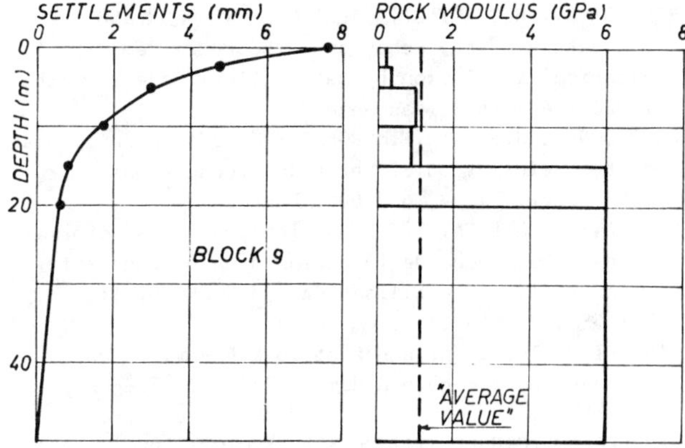

Fig. 17 - Extensometer readings during construction underneath a reinforcing buttress at the Corbara dam and corresponding back calculated moduli

were obtained at the Corbara buttress dam on the Tiber river north of Rome (BONALDI et al., 1985). Some doubts about the sliding stability of the dam and cracking of the concrete led to the construction of reinforcing buttresses downstream from the preexisting ones. Multipoint extensometers recorded the settlement caused by construction (Fig.17). The average values of the modulus for the new loading were almost the same as those found in the back analysis of the old structure at first impounding; however a detailed examination of the relative displacements of the various anchors shows that the foundation rock is really markedly dishomogeneous, with moduli increasing from 0.3 up to 7 GPa.

Three observations are to be made in this regard. Often the visual aspect of the rock in the loosened layer is not much different from that of the underlying rock. Besides, in various cases, the low modulus zone was not eliminated by the consolidation grouting which is usually performed in the rock mass near the contact with the foundation surface. Finally even where the original superficial layer is removed, a new loosened zone is easily formed during the excavation process (Fig.18).

The sensitivity of the rock deformability to even slight disturbance factors is also indicated by the observations at the Itaipu dam (PAES DE BARROS et al., 1985), where the excavation of the shear keys in a weak layer some meters below the foundation plane caused large settlements during the construction of the dam and a greater deformability during impounding. This occurred notwithstanding the limited excavated area with respect to the total foundation area.

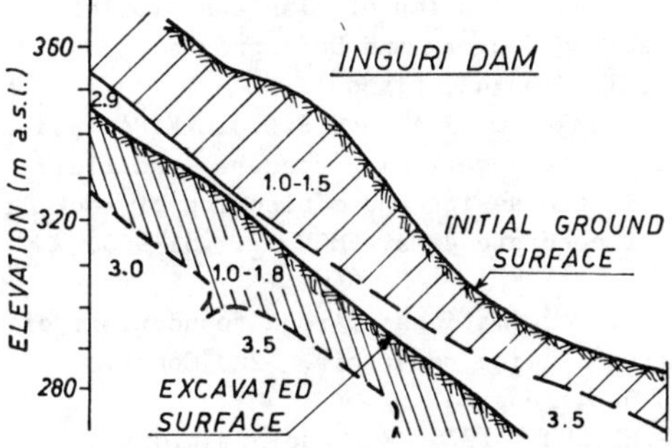

Fig. 18 - P-waves seismic velocities (km/s) underneath the initial ground surface and the subsequently excavated surface at the Inguri dam abutments (SAVICH et al., 1974)

4.3 Variation of moduli along the abutments

It is often found, especially in V-shaped valleys, that the deformability of the rock increases from the bottom of the valley towards the higher elevations. This phenomenon is probably related to a greater stress release in the upper parts of the valley and possibly to a longer exposure of the rock to weathering and loosening. This situation is typically exemplified by the extensometric observations at the Ponte Cola dam (Fig.16); similar situations were observed at the Ingouri dam (KALOUSTIAN, 1983) and at the Canelles dam (ALVAREZ, 1977).

4.4 Comparison between predicted and identified moduli

It is interesting to compare the moduli obtained from the back analyses with those determined by in situ tests or otherwise assumed in the design, usually on the basis of a subjective evaluation of the quality of the rock. Fig.19 shows various values taken either from Italian cases or from other published data. The agreement cannot be considered very good as about 1/3 of the identified moduli fall outside the range plus or minus 1.5 times the initially assumed values.

It is to be noted that the "identified" moduli are average values corresponding to the global response of the rock mass to the loads imposed by the structure after some years of service.

The data obtained from back analyses could find a wider utilization in future applications if it were possible to relate these numerical indications to the quality of the rock mass, expressed for instance by means of one of the well known quality indexes utilized in rock mechanics, such as for instance the RMR index which was introduced by Bieniaswkii (1979). Unfortunately the results of published back analyses are not always accompanied by a satisfactory description of the rock mass properties and by indications of the data provided by mechanical tests.

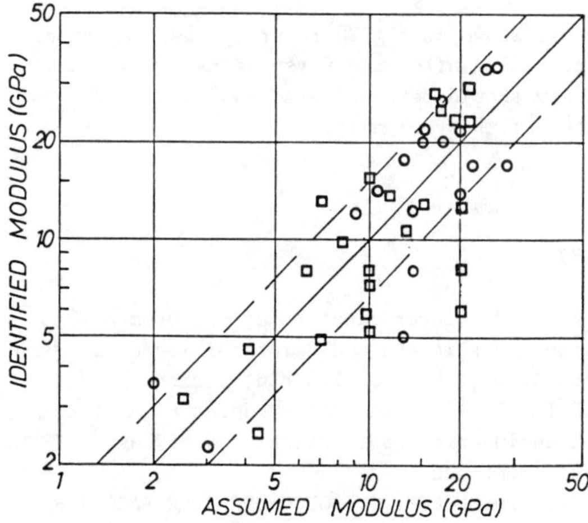

Fig. 19 - Relationship between moduli assumed in the design and back calculated moduli for various Italian sites (circles) and from other published data (squares). The dashed lines refer to a deviation of plus or minus 50%

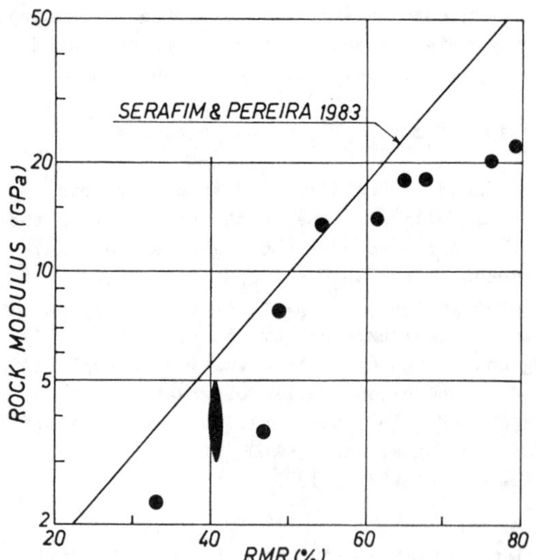

Fig. 20 - Correlation between rock moduli and quality index RMR for various Italian sites

Data for establishing such correlations are being gradually collected in the course of the investigations for the reappraisal of the conditions on many old dams. As an example, some preliminary indications based on the RMR values obtained in various Italian sites are shown in Fig.20. For a comparison with other quoted data it is to be noted that the RMR partial rating related to hydraulic conditions was not considered relevant in this setting, and was therefore ignored. The remaining rating values were scaled back again to 100%.

It can be noticed that on the whole the identified values of the moduli, although being on the low side, are in good agreement with the empirical relation suggested by SERAFIM and PEREIRA (1983). A better fit would be obtained by displacing the curve by some 5 RMR points to the right, that is by assuming a relationship between moduli and rock quality for the type:

$$\log E = (RMR - 15)/40$$

The moduli shown in fig.20 refer to the conditions of cyclic loading after the first years of service, that is they should be considered as "moduli of elasticity" of the rock mass.

REFERENCES

ALVAREZ A. 1977. Interpretation of measurements to determine the strength and deformability of an arch dam foundation, FMRM, 2: 825-836, Zurich

BIENIAWSKII Z.T. 1979. The geomechanics classification in rock engineering applications. 4th Congr. ISRM, 2: 41-48, Montreux

BONALDI P., GIUSEPPETTI G., GUCCIONE R, RIBACCHI R. & G.SELLERI 1982. Evaluation of rock foundation behaviour for two dams in operation. 14th ICOLD, 1: 227-242, Rio de Janeiro

BONALDI P., MANFREDINI G., MARTINETTI S., RIBACCHI R. & T.SILVESTRI 1983. Foundation rock behaviour of the Passante dam. 5th ISRM Congr., Melbourne

BONALDI R., RUGGERI G., VALLINO G. & G.FORZANO 1985. Evaluation of the behaviour of Corbara dam via numerical simulation by mathematical models. 15th ICOLD, 1497-1528, Lausanne

BOURBONNAIS J. & N.R.F.MORGENSTERN 1974. An analysis of the deformation of three dam foundations. 3rd ISRM Congr., II B, 685-690, Denver

CANDIANI G. & P.GAVAZZI 1964. Influence des déformations de la roche de fondation d'un barrage sur l'écran d'imperméabilisation. 8th ICOLD, 1: 571-584, Edinburgh

FANELLI M. 1983. Influence of rock behaviour on foundation of concrete dams. FMGM, 2: 839-863, Zurich

FANELLI M., GIUSEPPETTI O. & R.RICCIONI 1979. Experience gained during control of static behaviour of some italian dams. 13th ICOLD, 2: 663-683, New Delhi

GOMEZ LAA G. & J.A.RODRIGUEZ GONZALEZ 1985. In search of a deterministic hydraulic model of concrete dam foundation, 15th ICOLD, 1, 903-927, Lausanne

KALOUSTIAN E.S. 1983. Results of field studies of the Ingouri arch dam rock foundation with use of rock strain gauges, FMGM, 2: 901-906, Zurich

LUDESCHER H. 1985. A modern instrumentation for the surveillance of the stability of the Kolnbrein dam, 15th ICOLD, 797-812, 1, Lausanne

LEUENBERGER J. 1976. Les déformations des fondations des barrages-voûtes de grande portée et leur influence sur les infiltrations décrites dans l'exemple du barrage Santa Maria. 12th ICOLD, 727-747, Mexico

MANFREDINI G., MARTINETTI S. & R.RIBACCHI 1975. Mutual influence of water flow and state of stress in the analysis of dam foundations. Int. Symp. Numerical Analysis of Dams, 882-897, Swansea

MARAZIO A., CAPOZZA F. & P.PENTA 1966. Déformabilité de la roche de fondation dans le cas de quelques barrages italiens, 1st Congr. ISRM, 2: 603-615, Lisboa

PAES DE BARROS F., PANKOV R.H., & A.L.BARBI 1985. General behaviour of Itaipu dam foundations, 15th ICOLD, 1: 1153-1168, Lausanne

RIBACCHI R. & G.SELLERI 1978. The behaviour of Ponte Cola dam after ten years service. Int. Congr. "Rock Mechanics related to Dam Foundations", 1, Rio de Janeiro

SAVICH A.I., YASHCHENKO Z.G., KERESELIDZE S.B., A.A.GORBUNON & E.A.GRIGORIANTS 1974. Seismic survey of Ingouri dam pit - 3rd Congr. ISRM, IIB, 922-927, Denver

SERAFIM J.L. & J.P.PEREIRA 1983. Considerations of the geomechanical classification of Bieniawskii, Int. Symp. Engineering Geology and Underground Construction, 1: II33-42, Lisboa

SOUZA LIMA V.M., ABRAHAO R.A. & J.F.A.SILVEIRA 1983. Some aspects of the numerical procedures and field measurements in the design and monitoring of rock foundations of concrete dams, FMGM, 2: 955-966, Zurich

TIZDEL R.R. 1970. Deformations of rock foundations of high dams after filling reservoir. 2nd Congr. ISRM, 3: 563-566, Beograd

WIDMANN R. & G.HEIGERTH 1976. Rock deformations and seepage flows in the foundations of the Schlegeis arch dam. 12th ICOLD, 661-673, Mexico

ACKNOWLEDGEMENTS - This research was supported by a M.P.I. contribution

5

Annexes / Anhang

List of participants / Liste des participants / Teilnehmerliste

ARGENTINA

MARAGOTO, Carlos
307 No S1
Cipolletti 8324

MORETTO, Oreste
Luis S. Pena 250
1110 Buenos Aires

AUSTRALIA

BAMFORD, Bill
Civil Engineering Dept.
University of Melbourne
Parkville 3052

BRENNAN, Sean
3 Masters Place
Kambah, Act, 2902

DUNCAN FAMA, Mary E.
Division of Geomechanics
Private Bag 3
Indooroopilly
Queensland, 4068

ENDERSBEE, Lance
Monash Univ. Fac. of Eng.
Wellington Road
Clayton, V1C 3168

FRIDAY, Robin
11 Irving Ave.
Box Hill
Victoria, 3128

GRANT, John
c/o Dept. of Engineering
Melbourne University
Melbourne, Vict.

HELM, Donald
P.O. Box 54
MT. Waverley, Vic 3149

JOHNSTON, Ian W.
Dept. of Civil Engr.
Monash University
Clayton, Victoria
3168

LITTLE, Trevor Neil
3 Alderdice Place
Brisbane, 4068

McKENZIE, Cameron
J.K.M.R.C., Isles Rd,
Indooroopilly
Brisbane, 4068

AUSTRALIA

McMAHON, Barry
58 Bent Street
Lindfield NSW

NEMCIK, Jan Anton
P.O. Box 9
Corrimal, N.S.W. 2518

RAUERT, Neil
P.O. Box 444
Broken Hill, N.S.W., 2880

SIGGINS, Anthony
P.O. Box 54
Mount Waverley, Victoria

SIMMONS, John
James Cook Univ. of
North Queensland
Townsville QLD 481

STEVENSON, Harry
374A Warrigal Road
Asburton
Victoria 3147

WARDLE, Leigh J.
C.S.I.R.O.
P.O. Box 54
Mount Waverley
Victoria 3149

AUSTRIA

BONAPACE, Bruno
Tiroler Wasserkraftwerke
AG,
Landhausplatz 2
6010 Innsbruck

MAYR, Gotz
Rupertgasse 21
5020 Salzburg

PROMPER, Reinhard
Raphael-Donnerstr. 43
5026 Salzburg

RIENDESSL, Kurt
Rainerstrabe 29
5020 Salzburg

BELGIUM

BOLLE, Albert
Institut du Genie Civil
Quai Banning 6
B-4000 Liege

BELGIUM

BRYCH, Josef
30 rue du Bois
B-7000 Mons

CAUFRIEZ, Valhy
Rue Gretry 796
B-4020 Liege

NOMERANGE, Jules
Quai de Rome, 33-34,
B-4000 Liege

THIMUS, Jean-François
Dept. Genie Civil
1, Place du Levant
B-1348 Louvain-La-Neuve

VERVOORT, Andre
St-Corneliusstr. 23/3
B-3500 Hasselt

BRAZIL

DINIS DA GAMA, Carlos
P.O. Box 7141
Sao Paulo

DOBEREINER, Lorenz
Av Rio Branco 277-3
Enge-Rio, Rio de Janeiro

GONZALEZ, Francisco
Rua Martins Ferreira, 91
Rio de Janeiro

VARGAS, Euripedes Do
Amaral
RVA Marques S. Vincente
225, Rio de Janeiro

VARGENS, J. Rogerio
Av. Mandel Dias Da Silva
936-S/103 Pituba
Salvador-BA

CANADA

AMES, Douglas
260 Cedar St.
Sudbury, Ontario
P3B 3X2

ARCHAMBAULT, Guy
555 Boul. de l'Université
Chicoutimi, Quebec
G7H 2B1

CANADA

ARCHIBALD, J.F.
Dept. of Min. Eng.
Goodwin Hall
Queen's University
Kingston, Ontario
K7L 3N6

ARTEAU, Jean
505 W. de Maisonneuve
Montreal, Quebec
H3A 3C2

ASSIS, Andre
Dept. of Civil Engr.
Edmonton, Alberta
T6G 2G7

ASTON, Tim
219 George street
Sydney, Nova Scotia
B1P 1J3

AUBERTIN, Michel
C.P. 700
Rouyn, Quebec
J9X 5E4

BALLIVY, Gerard
Faculte des Sciences
Appliques
Sherbrooke, Quebec
J1K 2R1

BAWDEN, William
240 Hymus Blvd.
Pointe Claire, Quebec
H9R 1G5

BEAUCHAMP, Yves
#410, 224-4th Ave. South
Saskatoon, Saskatchewan
S7K 5M5

BELANGER, Roch
1023A Jorcan
St. Jean Chrysostome
Quebec, G6Z 2N3

BENMOKRANE, Brahim
Dept. de Genie-Civil
Universite de Sherbrooke
Sherbrooke, Quebec
J1K 2R1

BESNER, Jean
271 Cite des Jeunes
St-Clet, Quebec
J0P 1S0

CANADA

BHARTI, S.
907 Grandview
Sudbury, Ontario
P3A 2H3

BLAIS, Michel
2965 Cote Ste Catherine
App. 12
Montreal, Quebec
H3T 1C3

BLANCHET, Jean Yves
151 rue St Laurent
Robertsonville, Quebec
GON 1LO

BOOTHBY, Robert
22 Harnish Cr.
Willowdale, Ontario
M2M 2C1

BOSTOCK, Michael
#2-327 Johnson St.
Kingston, Ontario

BOUCHER, Bernard
200-4600 Cote Vertu
Montreal, Quebec
H4S 1C7

BOURBONNAIS, Jacques
665 Pine
St. Lambert, Quebec
J4P 2P4

CAJKA, Mary
1 Observatory Crescent
Ottawa, Ontario
K1A 0Y3

CARDIN, Jacques
1200 W. Boul. St. Martin
Laval, Quebec
H7S 2E4

CARMICHAEL, Thomas J.
KR 252, 800 Kipling Ave.
Toronto, Ontario
M7Z 5S4

CARVALHO, Jose
1951 Rathburn Rd. E. #68
Mississauga, Ontario

CHABOT, François
812 De La Colombiere Est
Quebec, Quebec
G1J 1E3

CHAUVIN, J.P.
Noranda Minerals Inc.
Matagami, Quebec

CHOQUET, Pierre
Dept. Min. & Met.
Quebec, Quebec
G1K 7P4

CLOSSET, Luc
555, Booth St. (BCC 10)
Ottawa, Ontario
K1A 0G1

COCHRANE, Lawrence B.
680 Camelot Drive
Sudbury, Ontario
P3B 3N1

CANADA

CODE, James A.
2045 Stanley St.
Montreal, Quebec
H3A 2V4

COMEAU, Wilfrid
7580 Boul Milan
Brossard, Quebec
J4Y 1H5

CORTHESY, Robert
24 Georgia Crescent
Pointe Claire, Quebec

CRAWFORD, Adrian
University of Toronto
Dept. of Civil Eng.
Toronto, Ontario
M5S 1A4

CRUDEN, David
University of Alberta
Edmonton, Alberta
T6G 2G7

CULLEN, Michael
General Delivery
Balmertown, Ontario
P0V 1C0

CURRAN,, John
Dept. of Civil Engr.
Toronto, Ontario
M5S 1A4

DAVIDGE, Glen
Falconbridge Limited
Sudbury Operations
Onaping, Ontario
P0M 2R0

DEBICKI, Ruth
100 Ramsay Lake Rd.
Sudbury, Ontario
P3E 5S9

DROLET, Andre
200 Dorchester Sud
4 Etage
Quebec, Quebec
G1K 5Z1

DUFOUR, Rene
Suite 1210
3400 de Maisonneuve W.
1 Place Alexis Nihon
Montreal, Quebec
H3Z 3B8

DUSSEAULT, Maurice
University of Waterloo
Waterloo, Ontario
N2L 3G1

ETHIER, Robert J-M
225 Cote
Asbestos, Quebec
J1T 1M3

FALLS, Stephen
Dept. of Geol. Sciences
Kingston, Ontario
K7L 3N6

CANADA

FALMAGNE, Veronique
Minnova Inc.
Division Lac Dufault
Rouyn-Noranda, Quebec
J9X 5B4

FARRELL, Lorraine
Queen's University
Kingston, Ontario
K7L 3N6

FINES, Alison
35 Park Side
Nepean, Ontario
K2G 3B6

FLEURY, Denis
5252 De Castille
Montreal Nord, Quebec
H1G 3E2

FORDHAM, Chris
Dept. of Earth Sci.
University of Waterloo
Waterloo, Ontario
N2L 3G1

FORIERO, Adolfo
4721 Beauvoir
St-Leonard, Quebec

FORTIN, Max
221 Domaine Du Roi
Chicoutimi, Quebec
G7H 6Y5

FRANKLIN, John
The Stream
R.R. 1,
Orangeville, Ontario
L9W 2Y8

GENDZWILL, Don
Dept. of Geological
Sciences
University of Saskatchewan
Saskatoon, Saskatchewan
S7N 0W0

GENDZWILL, Neil
105-902 Spadina Cres. E.
Saskatoon, Saskatchewan
S7K 3H5

GILL, Denis E.
C.P. 6079, Succ. A.
Montreal, Quebec
H3C 3A7

GOODALL, David
386 Rang 40,
Ormstown, Quebec
J0S 1K0

GRANT, Don
240 Hymus Blvd.
Pointe Claire, Quebec
H9R 1G5

GREGOIRE, Denise
Ste 1210
3400 de Maisonneuve W
1 Place Alexis Nihon
Montreal, Quebec
H3Z 3B8

CANADA

GRENIER, Richard A.
2700 Einstein
Ste Foy, Quebec

GU, Jianzhong
670 O'Connor St.
Ottawa, Ontario
K1S 3R8

GUAY, Chantal
9 Du Beau Site
Lauzon, Quebec

HADJIGEORGIOU, John
3575 University #509
Montreal, Quebec

HANSON, Douglas
P.O. Box 100
99 Spine Road
Elliot Lake, Ontario
P5A 2J6

HASSANI, Ferri
McGill University
Dept. Mining & Met. Eng.
Montreal, Quebec
H3A 2A7

HEDLEY, David G.F.
P.O. Box 100
Elliot Lake, Ontario
P5A 2J6

HEIDRICH, Henry
Domtar Chemicals
1136 Matheson Blvd
Mississauga, Ontario
L4W 2V4

HEINRICH, Werner
Pinawa, Ontario
R0E 1L0

HERBST, Thomas
65 Bowes Rd. Unit 8
Concord, Ontario
L4K 1H5

HERGET, Gerhard
555 Booth St. (BCC 10)
Ottawa, Ontario
K1A 0G1

HERNANDEZ, Pedro
630 Jacques Cartier N.
#24
Sherbrooke, Quebec
J1J 3A4

HERRON, Lorne
19000 Route
Trans-Canadienne
Baie D'Urfee, Quebec
H9X 3N8

HIVON, Elisabeth
10835-86 Ave. #205
Edmonton, Alberta
T6E 2N1

HOEK, Evert
Dept. of Civil Engr.
University of Toronto
Toronto, Ontario
M5S 1A4

CANADA

HOLLOWAY, Arthur
WNRE
Pinawa, Manitoba
ROE 1L0

HONG, Richard
P.O. Box 550
Kirkland Lake, Ontario
P2N 3J7

HRASTOVICH, William
240 Hymus Blvd.
Pointe Claire, Quebec
H9R 1G5

HUDYMA, Marty
6350 Stores Road
Vancouver, British
Columbia

HUGHSON, David
217 Rosedale Heights Dr.
Toronto, Ontario
M4T 1C7

HUNT, Gregory
1074 Webbwood Dr.
Sudbury, Ontario
P3C 3B7

HUTCHINSON, Douglas
#205, 10835-86 Ave.,
Edmonton, Alberta
T6E 2N1

HYUN-HA, Lee
4030 Mahogany Row
Mississauga, Ontario
L4W 2H6

ISAAC, Byron
110 Lancaster Cres.
St. Albert, Alberta
T8N 2N8

JAKUBICK, Alex T.
800 Kipling Ave
Toronto, Ontario
M8Z 5S4

JAURON, Richard L.
2 Place Quebec Suite 704
Quebec, Quebec
G1R 2B5

JONES, Donald S.
P.O. Box 1000
Halifax, Nova Scotia
B3J 2X4

JULIEN, Michel
4505 Dupuis Apt 6
Montreal, Quebec
H3T 1E7

JURKUS, Rimantas A.
855 Ste Catherine E
16 Etage
Montreal, Quebec
H2L 4P5

KAISER, Peter K.
Geom. Res. Ctr.
Laurentian Univ.
Sudbury, Ontario
P3E 2C6

CANADA

KAST, Gary
34720 Hamon Dr.
Abbotsford
British Columbia
V2S 1H5

KOCHHAR, Jagdish
P.O. Box 2500
Sydney, Nova Scotia
B1P 6K9

KRISTOFF, Brent A.
P.O. Box 3000
Bathurst, New Brunswick
E2A 4N5

KULLMANN, Dag
Dept. of Min. Met. & Pet.
Eng.
University of Alberta
Edmonton, Alberta
T6G 2G6

LABRIE, Denis
CANMET/MRL/CMTL
BCC, Building # 10
555 Booth Street
Ottawa, Ontario
K1A 0G1

LABUC, Vladimir
623 Main Rd., Box 20
Hudson, Quebec
J0P 1H0

LADANYI, Branko
C.P. 6079 Succ. A
Montreal, Quebec
H3C 3A7

LANDRIAULT, David A.
84 St. Brendan St.
Sudbury, Ontario
P3E 1K5

LANG, Peter Andrew
Pinawa, Manitoba
ROE 1L0

LAU, Joseph
555 Booth Street
Ottawa, Ontario
K1A 0G1

LAURIN, Richard
Les Mines d'Or Liena Ltee
C.P. 9000
Val D'Or, Quebec
J9P 6A5

LEBOEUF, Denis
Universite Sherbrooke
Dept. de Genie Civil
Sherbrooke, Quebec

LECOMTE, Paul
5655 De Marseille
Montreal, Quebec
H1N 1J4

LEE, Hyun-Ha
4030 Mahogany Row
Mississauga, Ontario
L4W 2H6

CANADA

LEFRANÇOIS, Gerald
C.P. 751-1216, 3E Ave.
Val D'Or, Quebec
J9P 4P8

LEITE, Maria Helena
5619, Av. Stirling
Montreal, Quebec
H3T 1R7

LEPAGE, Anick
700-9th St. S.W.
Apt. 1703
Calgary, Alberta
T2P 2B5

LEVAY, Jerry
855 Ste Catherine E
16 Etage
Montreal, Quebec
H2L 4P5

LEY, Gordon
3151 Wharton Way
Montreal, Quebec
H2L 4P5

MACDONALD, Paul
Mines Research
INCO Ltd.
Copper Cliff, Ontario
P0M 1N0

MACDONALD, Terry
P.O. Box 92
Bathurst, New Brunswick
E2A 3Z1

MACKINTOSH, A. Dave
Box 32
Vanscoy, Saskatchewan
S0L 3J0

MADSEN, Doug
Dept. of Min. Eng.
Goodwin Hall
Queen's University
Kingston, Ontario
K7L 3N6

MAERZ, Norbert
29 Wilton Pi.
Kitchener, Ontario
N2H 2S8

MAGASSOUBA, Filani Moudou
425 Boul. du College
Rouyn-Noranda, Quebec
J9X 5E5

MAGNI, Eric
1201 Wilson Ave.
Downsview, Ontario
M3M 1J8

MAKUCH, Tony
Campbell Red Lake Mines
Mine Road
Balmertown, Ontario
P0V 1C0

MALONEY, Sean
Dept. of Civil Eng.
Univ. of Alberta
Edmonton, Alberta
T6G 2G7

CANADA

MARCOUX, Louis
2700 Einstein
Ste-Foy, Quebec
L1P 3W8

McCREARY, Richard
Queen's University
Dept. of Geol. Sce.
Kingston, Ontario
K7L 3N6

McGAUGHEY, John
Dept. Geol. Sci.
Queen's University
Kingston, Ontario
K7L 3N6

McINTYRE, J. Grant
745 Montreal-Toronto Blvd.
Dorval, Quebec

MICHAUD, Pierre
Suite 1210
3400 de Maisonneuve W.
1 Place Alexis Nihon
Montreal, Quebec
H3Z 3B8

MILNE, Douglas
Centre de Recherche
Noranda
240 Blvd. Hymus
Pointe Claire, Quebec
H9R 1G5

MILNE, Lindsay
Suite 1210
3400 de Maisonneuve W.
1 Place Alexis Nihon
Montreal, Quebec
H3Z 3B8

MOMOH, Osman
240 Hymus Blvd
Pointe Claire, Quebec
H9R 1G5

MORGENSTERN, Norbert R.
Dept. of Civil Eng.
University of Alberta
Edmonton, Alberta
T6G 2G7

MORRISON, Douglas
1351 Bellevue Ave.
Sudbury, Ontario

MOSS, Allan
224 West-8th Avenue
Vancouver, B.C.
V5Y 1N5

MOTTAHED, Parviz
PCS Tower, Suite 500
122-1st Avenue South
Saskatoon, Saskatchewan
S7K 7G3

MRAZ, Denis
410 Jessop Avenue
Saskatoon, Saskatchewan
S7N 2S5

MUPPALANENI, S.N.
P.O. Box 1500
Elliot Lake, Ontario
P5A 2K1

CANADA

NANTEL, Jacques
240 Hymus Blvd.
Pointe Claire, Quebec
H9R 1G5

NEVILLE, Christopher J.
10015-83rd Ave. #7
Edmonton, Alberta
T6E 2C3

NIEWIADOMSKI, Janusz
P.O. Box 100
99 Spine Road
Elliot Lake, Ontario
P5A 2J6

O'HEARN, Brian
Cleland Bldg.
Onaping, Ontario
POM 2RO

OUELLET, Jacques
2890 Van Horne #2
Montreal, Quebec
H3S 1R1

OUELLET, Michel
4 Des Coquilles
Beauport, Quebec
G1E 4S8

PAKALNIS, Rimas
Univ. of B.C.
1441 McNair Dr.
Vancouver, B.C.
V7K 1X5

PATTON, Frank
507-E 3rd St.
North Vancouver, B.C.
V7L 1G4

PAYNE, Daniel A.
P.O. Box 1000
Halifax, Nova Scotia
B3J 2X4

PEAKER, Scott
41 Valecrest Drive
Islington, Ontario
M9A 4P5

PECK, Jonathan
12 Circle Rd.
Pointe Claire, Quebec

PELLET, Frederic
Universite de Sherbrooke
379 Vincent
Sherbrooke, Quebec
J1E 1W9

PFENDLER, Pascal
Universite de Sherbrooke
730 Des Ruisseaux
Sherbrooke, Quebec

PICIACCHIA, Luciano
1805 Muir St.
St. Laurent, Quebec
H4L 4T2

PINARD, Jean Paul
C.P. 546
Val D'Or, Quebec
J9P 4P5

CANADA

PRASAD, Anant
Edmonton, Alberta
T5J 3N3

PRASAD, Naresh
47-203 Range Road
Whitehorse, Yukon
Y1A 3A5

PUGSLEY, T.F.
Commerce Court West
Toronto, Ontario
M5L 1B4

QUESNEL, William
P.O. Box 670
Kirkland Lake, Ontario
P2N 3J9

ROCHON, Paul
P.O. Box 100,
99 Spine Road
Elliot Lake, Ontario
P5A 2J6

ROWE, R. Kerry
Univ. of Western Ontario
London, Ontario
N6A 3K7

SATHIAMOORTHY, Sinnathamby
31, St Dennis Dr.
Apt. 503
Mississauga, Ontario

SAVIC, Ivana
Ste 1210
3400 de Maisonneuve W.
1 Place Alexis Nihon
Montreal, Quebec
H3Z 3B8

SCOBLE, Malcolm
2480 University St.
Montreal, Quebec
H3A 2L4

SEKI, Frank
970 Burrard
Vancouver, B.C.
V6Z 1Y3

SEMADENI, Tom
P.O. Box 100
99 Spine Road
Elliot Lake, Ontario
P5A 2J6

SEYERS, William
970 Burrard St.
Vancouver, B.C.
V6Z 1Y3

SHEIKH, A.
P.O. Box 2600
Elliot Lake, Ontario
P5A 2K2

SINGH, S. Paul
School of Engineering
Sudbury, Ontario
P3E 2C6

SPROTT, David
250 Hymus Blvd.
Pointe Claire, Quebec
H9R 1G5

CANADA

STIMPSON, Brian
University of Manitoba
Winnipeg, Manitoba
R3T 2N2

SUGIHARA, Kozo
#12-30 Alexander
Pinawa, Manitoba
ROE 1LO

SUPERINA, Zdenko
855 Ste Catherine Est
16 Etage
Montreal, Quebec
H2L 4P5

TALEBI, Shahriar
Dept. of Geol. Science
Kingston, Ontario
K7L 3N6

TAN, Tjong-Kie
Xiao Yan Chen Triumf
4004 Westbrook Mall
Vancouver, B.C.
V6T 2A3

TERREAULT, Pierre H.
Malartic, Quebec
C.P. 1090
J0Y 1Z0

THOMPSON, Paul
Whiteshell Nuclear Res.
Establishment
Pinawa, Manitoba
ROE 1LO

TOWERS, Jeff
Dept. of Geol. Sciences
Kingston, Ontario
K7L 3N6

UDD, John E.
555 Booth Street
Ottawa, Ontario
K1A 0G1

UNEMBU, Diamba
950 Desblei
Sherbrooke, Quebec
J1E 3L4

URBANCIC, Ted
Rock Physics Lab
Dept. of Geo. Sci.
Kingston, Ontario
K7L 3N6

VANCE, Jim
c/o IMC, Canada
Estherhazy,
Saskatoon, Saskatchewan
S0A 0X0

VERMA, Bhagwat Sahai
43 Golf Links Road
Bedford, Nova Scotia
B4A 2J1

VLADUT, Thomas
300, 2421-37 Avenue
Calgary, Alberta
T2E 6Y7

CANADA

VONGPAISAL, Somchet
555 Booth St.
Ottawa, Ontario
K1A 0G1

WASERNA, E.
402 Isabey
Ville St. Laurent
Quebec
H4T 1V3

WETMILLER, Robert J.
1 Observatory Cr.
Ottawa, Ontario
K1A 0E8

WHITE, Owen L.
77 Grenville St.
Toronto, Ontario
M7A 1W4

WILES, Terry
16 Park St.
Copper Cliff, Ontario
POM 1NO

WOLOFSKY, Leir
RR1, Line 6
Niagara-On-The-Lake,
Ontario, LOS 1J0

YEO, T.K.
Dept. of Min. Eng.
Goodwin Hall
Queen's University
Kingston, Ontario
K7L 3N6

YOUNG, Paul
Dept. of Geol. Sci.
Kingston, Ontario
K7L 3N6

YU, Thian
P.O. Box 2002
Timmins, Ontario
P4N 7K1

YUEN, Clement
3151 Wharton Way
Mississauga, Ontario
L4X 2B6

CHILE

MULLER, Guillermo
Superintendencia De
Geologia
Div. Salvador, Santiago

VAN SINT JAN, Michel
Laboratorio de Ingenieria
Geo.
Casilla 6177
Santiago

CHINA

DING, Enbao
Inst. of Geol., Academia
Sinica
P.O. Box 634
Beijing

HUA, Anzeng
Xuzhou
Jiangsu

CHINA

HUANG, Rongzum
P.O. Box 920
Beijing

JIAYOU, Lu
P.O. Box 366
Beijing

SHEN, Guanghan
Shandong Inst. of Min. &
Tech. Taian
Shandong Province, P.R.

SUN, Guangzhong
Inst. of Geol. Academia
Sinica
P.O. Box 634
Beijing

SUN, Zonggi
9 Xizhang Hutong
Xizhimennei Dajie
Beijing

WANG, Yongjia
Northeast Univ. of
Technology Shenyang
Liaoning. P.R., 110006

WANG, Xianglin
CCMRI, Hepingli
Beijing

ZHOU, Weiyuan
Dept. of Hydraulic Eng.
Tsinghua University
Beijing

ZHU, Jingmin
Chongqing City

CZECHOSLOVAKIA

DROZD, Karel
Stavebni Geologie
Prague 1

FINLAND

ANTIKAINEN, Juha
Vuorimiehentie 2
02150 Espoo

COSMA, Calin
Kuparitie 12, 00440
00101 Helsinki 44

JAKOBSSON, Kai
P.O. Box 268
00101 Helsinki

JOHANSSON, Stig
Neste Oy
Keilaniemi
02150 Espoo

LAPPALAINEN, Pekka
86900 Pyhakumpu
Outdkumpy Oy

POLLA, Jukka
Betonimiehenkuja 1
02150 Espoo

FINLAND

SARKKA, Pekka
Laboratory of Min. Eng.
Vuorimehente 2 A
02150 Espoo

VAATAINEN, Anne
Helsinki Univ. of Tech.
Lab. Min. Engr.
Vuorimiehentie 2A
02150 Espoo

FRANCE

BEREST, Pierre
LMS
Palaiseau-Cedex
91128

BERGUES, Jean
LMS
Palaiseau-Cedex
91128

BOLLE, Gerard
13 Ave Morane Saulnier
Villa Cou blay
78140 Velizy

BORDES, Jean-Louis
2 rue Auguste Thomas
92600 Asnieres

BRENIAUX, Jean
11 Avenue D'Altkirch
68100 Mulhouse

CAMES-PINTAUX, Anne-Marie
14 rue de Chastillon
5100 Chalons/Marne

CHARLEZ, Philippe
Route De Versailles
Reuy Les Chevreuses
78470

COLIN, Pierre
Tour Aurore Cedex 5
Defense 2, Paris
92080

CORNET, François Henri
Laboratoire de Sismologie
I.P.G. Tour 14,
4 Place Jussiere
Paris-Cedex 75252

DESPAX, Damien
Tour Total
Paris

DUFFAUT, Pierre
130 rue De Rennes
Paris, 75006

FABRE, Denis
38402 St. Martin D'Mere
Cedex

FINE, Jacques
35 rue Saint Honore
77305 Fontainebleau

FOURMAINTRAUX, Dominique
Elf-Aquitaine
64018 Pau-Cedex

FRANCE

GENTIER, Sylvie
BRGM BP 6009,
Orleans-Cedex 45060

GOSSE, Marc
32 rue Du CLSO
Saint-Marcel
92330 Sceaux

GUENOT, Alain
Elf-Aquitaine
64018 Pau. Cedex

HABIB, Pierre
Ecole Polytechnique
Palaiseau 91128

HERAUD, Hubert
16 A Cours Sablon
Clermont-Ferrand
63000

HOMAND, Françoise
Rue Du Moyen Roubault
54500 Vandoeuvre

MAURY, Vincent
Elf-Aquitaine
64018 Pau-Cedex

LOUIS, Frederic
14-16 rue Miromesnil
75384 Paris
Cedex 08

PANET, Marc
L.C.P.C
58 Boulevard Lefevre
75732 Paris
Cedex 15

PIGUET, Jack Pierre
Lab. De Mecanique Des
Terrains
Ecole des Mines
54042 Nancy Cedex

REVALOR, Roger
Lab. De Mecanique Des
Terrains
Ecole des Mines
54042 Nancy Cedex

ROCHET, Louis
4 rue Clement Ader
69500 Bron

GERMANY, GDR

MINKLEY, Wolfgang
Friederikenstrasse 60,
DDR-7030 Leipzig

FED. REP. OF GERMANY

BORM, Guenter
Lehrstuhl für Felsmechanik
P.O. Box 6980
7500 Karlsruhe

DILLO, Michael
Mies-Van-Der-Rohe-Str. 1,
5100 Aachen

EFFENBERGER, Klaus
Ing. Geolog. Inst.
Dipl-Ing. S. Niedermeyer
8821 Westheim

FED. REP. OF GERMANY

FROEHLICH, Bernhard
Lehrstuhl für Felsmechanik
P.O. Box 6980
7500 Karlsruhe

HAUPT, Manfred
Lehrstuhl für Felsmechanik
7500 Karlsruhe

JOHN, Klaus W.
Im Haarmannsbusch 114 A
4630 Bochum 1

KIRSCHKE, Dieter
Hertzstr. 12
7505 Ettlingen

KLONNE, Heiner
Industriestr. 27,
4630 Bochum 7

KUTTER, Herbert K.
Arbeits Gruppe
Felsmechanik
Inst. Fuer Geol.
Ruhr-Universität
4630 Bochum

LANGER, Michael F.B.
Postfach 51 01 53
3000 Hannover 51

LEMPP, Christof
Lehrstuhl für Felsmechanik
P.O. Box 6980
7500 Karlsruhe

MOHN, Joachim
Ungererstrasse 19
8000 Munich

MUEHLHAUS, Hans-Bernd
Lehrstuhl für Felsmechanik
P.O. Box 6980
7500 Karlsruhe

MUTSCHLER, Thomas
Lehrstuhl für Felsmechanik
P.O. Box 6980
7500 Karlsruhe

NATAU, Otfried
Lehrstuhl für Felsmechanik
P.O. Box 6980
7500 Karlsruhe

PAHL, Arno G.
Postfach 51 01 53
3000 Hannover 51

PIMENTEL, Erich
Lehrstuhl für Felsmechanik
P.O. Box 6980
7500 Karlsruhe

PLISCHKE, Bertold
Homburger Str. 29
6238 Hofheim

RISSLER, Peter
Ruhrtalsperrenverein
Entwicklungsabeteilung
Kronprinzenstr. 37
4300 Essen 1

FED. REP. OF GERMANY

SCHUETZ, Hermann
Boden-Und Felsmechanik
Pauluskirchstrasse 7
5600 Wuppertal

SCHULTZ, Horst
Industriestr. 27
4630 Bochum 7

SPANG, Raymund M.
Mewer Ring 3
5810 Witten

WALLNER, Manfred
BGR
Postfach 510153
3000 Hannover 51

WENZ, Eberhard
Lehrstuhl für Felsmechanik
P.O. Box 6980
7500 Karlsruhe

WITTKE, Walter
Mies-Van-Der-Rohe-Str. 1
5100 Aachen

WULLSCHLAEGER, Dietrich
Lehrstuhl für Felsmechanik
P.O. Box 6980
7500 Karlsruhe

GREECE

BANDIS, Stavros
Faculty of Technology
Aristotle University
Thessaloniki

HONG KONG

KOIRALA, Naresh
6th Flr., Empire Ctr.
Mody Rd., Tsim Sha Tsui
East
Kowloon, Hong Kong

ICELAND

HARDARSON, Bjorn
Grensasvegur 9
108 Reykjavik

INDIA

CHINTAPALLI, Venkata
Ramana Murth
3-4-376/5, Lingampally
Hyderabad - 500 027

PHUKAN, Satyendra Nath
Neepco
Shillong - 793001
Meghalata

ITALY

BOTTO, Giuseppe
P.Ardigo 30
00143 Roma

BRIGHENTI, Giovanni
Viale Risorgimento, 2
40136 Bologna

ITALY

GIANI, Gian Paolo
24 Corso Duca Degli
Abruzzi
10129 Torino

GRASSO, Piergiorgio
Via Bobbio 12
10141 Torino

NARDOCCI, Alfredo
c/o Metroroma - Via Lima
51 Roma

NERI, Romolo
18 Sartena
00122 Roma

PAGLIARA, Paola
Via Val Pusteria 22
00141 Roma

RIBACCHI, Renato
Dipart. Ing. Strut. E
Geotecnica
Via Monte D'Oro
28-00186 Roma

ROSSI, Pier Paolo
Viale G. Cesare 29
24100 Bergamo

SCAVIA, Claudio
Politecnico Di Torino
Corso Duca Degli Abbruzzi
24
15129 Torino

VERDI, Fordinando
c/o Magia Travel-Via
Degli, Specche 3
00186 Roma

ZANINETTI, Attilio
Enel-Cris Via Ornato 90/14
20162 Milano

JAPAN

AOKI, Toshiro
Kumamoto 860

DENDA, Atsushi
3-4-7, Etchujima
Kohto-Ku, Tokyo

EGUCHI, Motokatsu
Tosetsu Doboku Consultant
Co.
1-6-15 Nishi-Shimbashi
Minato-Ku, Tokyo

FUHJII, Yoshiaki
Mizumoto, Muroran 050

FURUYA, Kazou
Dept. of Civil Engr.
Engr. Research Inst.
Sanda, Atsugi-Shi

GOTO, Yuji
2-13-9 Miyazaki
Miyamae-Ku, Kawasaki 213

HARITA, Kazvo
39 Moribiru, 2-4-5
Azabudai, Minatoku
Tokyo

JAPAN

IGURO, Mizuhito
Shimizu Construction Co.

IIBOSHI, Shigeru
Taisei Corp., 344-1,
Nase-Cho
Totsuka-Ku, Yokohama
Kanagawa

ISHII, Takashi
Mita 43, Mori Bldg. 13F
No. 13-16, Mita 3-Chome
Minato-Ku, Tokyo, 108

ISHIJIMA, Joyi
8 Chome Kitz 13J0
Kita-Ku, Sapporo-Shi
Hokkaido

ITO, Hiroshi
1646 Abiko
Abiko-City, Chiba

IWAMOTO, Takeshi
Instit of Weak Rock Engin.
2-13-12, Honkomagome
Bunkyoku, Tokyo 113

KANEKO, Katsuhiko
2-39-1 Kurokami,
Kumamoto-Shi
Kumamoto-Ken

KAWAMOTO, Toshikazu
Furo-Cho, Chikusa-Ku
Nagoya 464

KAZURAYAMA, Fumitharu
1-8-2 Marunouchi
Chiyoda-Ku, Tokyo

KIKUCHI, Kokichi
Serv. Co. Ltd, 1-4-6
Nichi-Shimbashi
Minato-Ku, Tokyo, 105

KITAGAWA, Takashi
1-20-10 Toranomon
Minato-Ku, Tokyo

KOBAYASHI, Akira
4-17-28, Yonohonmachi
Yono-Shi, Saitama

KOICHI, Sassa
Dept. of Min. Science &
Tech.
Kyoto University, Kyoto

KUDO, Yozo
Tokuyama-Shi
Tokuyama 745

KUSABUKA, Morito
17-23, Honmachi-Nishi
4-Chome, Yono, Saitama 338

KYOYA, Takashi
Dept. of Geotec Eng.
Chikusa, Nagoya 464

MATSUI, Kikuo
Hakozaki, Higashi-Ku
Fukuoka

JAPAN

MATSUKU, Nobuaki
39-Moribiru, 2-4-5
Azabudai
Minatoku, Tokyo

MICHIHIRO, Kazutoshi
Setsunan Univ.,
Neyagawa, Osaka 572

MIMAKI, Yoichi
2-19-1 Tobitakyu
Chofu-Shi, Tokyo

MITANI, Satoshi
17-1, Tsukudo-Cho
Shimjulcu-Ku, Tokyo

MITO, Yoshitada
Waseda Univ.
Tokyo 160

MIYAJIMA, Keiji
3-34-12 Tarumi-Cho
Suita-Shi, Osaka

NAKAGAWA, Kameichiro
1646 Abiko, Chiba 270-11

NISHIMATSU, Yuichi
Dept. of Mineral Dev.
Engr.
Tokyo 113

NISHIOKA, Toshimichi
Uchisaiwai 1-1-3
Tokyo

OBARA, Yuzo
2-39-1 Kurokami
Kumamoto, 860

ODA, Masanobu
Saitama 338

OHNISHI, Yuzo
Yoshida Hommachi
Sakyo-Ku, Kyoto 606

OHTOMO, Hideo
2-2-19, Daitakubo
Urawa, Saitama 336

SATO, Kazuhiko
Mizumoto, Muroran 050

SATO, Kuniaki
255, Okubo
Urawa-Shi, Saitama 338

SAITO, Toshiaki
6-1-110 Kawaradamachi
Matsugasaki, Sakyo-Ku
Kyoto

SAKURAI, Shunsuke
Dept. of Civil Engr.
Kobe 657

SASAKI, Sigeru
39-Moribiru, 2-4-5-
Azabuda
Minataku, Tokyo

SHIMIZU, Norikazu
Dept. Civil Eng.
Kobe University
Kobe 657

JAPAN

SUEOKA, Toru
344-1, Nase-Cho, Tosuka-Ku
Yokohama
Nase, YoKohama 245

SUGAWARA, Katsuhiko
Kurokami 2-39-1
Kumamoto 860

TANAKA, Soichi
Tokyo Office, Oyo Corp.
3-2-1, Ohtsuka
Bunkyo-Ku, Tokyo 112

TANAKA, Tatsukichi
Geotec Inst, Oyo Corp.
2-19 Daitakubo, Urawa
Saitama 336

TANIMOTO, Chikaosa
Kyoto Univ. Dept of Civil
Eng.
Sakyo, Kyoto 606

UEDA, Takao
2-5-14 Minamisuna Koto-Ku
Tokyo 136

WATANABE, Kunio
255, Shimo -Obkubo
Urawa-Shi, Saitawa 338

YOSHINAKA, Ryunoshin
255 Shimo - Obkubo
Urawa-Shi, Saitawa 338

KOREA

LEE, Kyung-Won
219-5 Garibong-Dong
Guro-Gu, Seoul 150-5

LIM, Han-Uk
51-502 Banpo Apt
Banpodong, Kagnamku, Seoul

MEXICO

BELLO, Arturo
Rio Becerra 27, 5th Piso
03810 Mexico, D.F.

ESPINOSA-GRAHAM, Leopoldo
Chimalpa 25, P. Genova 7
14390 Mexico, D.F.

NETHERLANDS

GRAMBERG, John
Mijnbouwstraat 120
2628 RX, Delft

ROEST, Johannese
Mijnbouwstr 120
2628 RX Delft

SPIERS, Christopher J.
Budapestlaan 4,
3508 TA Utrecht

NORWAY

BARTON, Nick
P.O. Box 40 Taasen
0801 Oslo 8

NORWAY

BLINDHEIM, Olav T.
P.O. Box 4342
N7002 Trondheim

BY, Tore Lasse
P.O. Box 40 Taasen
0801 Oslo 8

CHRYSSANTHAKIS, Panayiotis
Postboks 40 Taasen
0801 Oslo 8

FREDERIKSEN, Ulf
P.O. Box 9884 Ila
0132 Oslo

HANSSEN, Tor Harald
Sintef-Div. of Rock &
Min. Eng.
7034 Trondheim

LUNDE, Halvor
Liakroken 34,
5090 Nyborg

NIELSEN, Bjorn
Dept. of Geology
7034 Trondheim-Nth

VIK, Gunnar
P.O. Box 40
Taasen, 0801
Oslo 8

PORTUGAL

BARROSO, Manuel
Av. Do Brasil
Lisboa-Codex 1799

CHARRUA-GRACA, Jose
Gabriel LNEC
101 Av. Brasil
Lisboa-Codex 1799

COSTA PEREIRA, Alberto
Rua Carolina M.
Vasconcelos 10
Lisboa 781610-1500

CRUZ, Antonio Augusto
Apartado 5058-1702
Lisboa Codex

DOURADO EUSEBIO, Maria De
Lurdes
c/o LNEC, Av. Brasil 101
Lisboa-Codex 1799

ESTEVES, Joaquim Moura
c/o LNEC, Av. Brasil 101
Lisboa Codex 1799

GROSSMANN, Nuno Feodor
c/o LNEC, Av. Brasil 101
Lisboa Codex 1799

LOUREIRO-PINTO, Jose
c/o LNEC, Av. Brasil 101
Lisboa Codex 1799

MARTINS, Julio B.
Universidade Do Minto
Enga., Braga Codex 4719

PORTUGAL

MELLO-MENDES, Fernando
Rua Cidade De Cabina,
22-30
Lisboa 1800

OLIVEIRA, Ricardo
LNEC Av. Brasil
Lisboa Codex 1799

PERES-RODRIGUES, Fernando
Lab. Nacional de Eng.
Civil, 101 Av. Brasil
Lisboa Codex 1799

PINTO DA CUNHA, Antonio
101 Av. Brasil
Lisboa Codex 1799

SOUTH AFRICA

BUDAVARI, Sandor
Dept. of Min. Eng.
Univ. of The Witwatersrand
P.O. Wits 2050
Johannesburg

CUNNINGHAM, Claude
P.O. Box 1122
Johannesburg 2000

ESTERHUIZEN, Gabriel S.
Dept. Mining Engineering
Pretoria 0002

GREFF, Hennie
4 Joseph Street
Flamwood, Ventersdorp 2570

KLOKOW, Johannes
Private Bag X2011
Carletonville 2500

McKINNON, Stephen D.
P.O. Box 91230
Auckland Park 2006

MENDECKI, Aleksander
Rock Mechanics Dept.
P.O. Box 2083
Welkom

MOORE, Brian
P.O. Box 1167
Johannesburg

MORE O'FERRALL, Roger
P.O. Box 641
Stilfontein 2550

O'FERRALL, R.C. More
P.O. Box 641
Stilfontain 2550

ORTLEPP, W. David
Rock Mechanics Dept.
Box 61587
Marshalltown 2107

RIEMER, Kevin
West Driefontein Goldmine
Private Bag X2011
Carletonville 2500

SELDON, Sean
West Driefontein Goldmine
Private Bag X2011
Carletonville

SOUTH AFRICA

SPANN, Hans Peter
P.O. Box 1210
Johannesburg 2000

SPENCER, David
Private Bag 82279
Rustenburg, 0300 Transvaal

SPOTTISWOODE, Dr. S.M.
P.O. Box 91230
Auckland Park 2006

STACEY, Dick
P.O. Box 8856
Johannesburg 2000

STEFFEN, Oskar
P.O. Box 8856
Johannesburg

VAN BERS, Marcus
6 Janahof Pretorius St.

WAGNER, Horst
P.O. Box 91230
Auckland Park 2006

WIGGETT, Colin
Private Bag X555
Vandyksdrif 2245

SPAIN

LAIN HUERTA, Ricardo
Ribera Del Mansanares
Madrid

FUENTE RAMIREZ, Santiago
De La
Avda De Burgos No. 11
28036 Madrid

MONTE, Jose-Luis
Menorca 49
Madrid

ROMANA, Manuel
Camino De Vera S/N
46701 Valencia

SERRA, Ma Teresa
Santa Teresa, 5-11 B
33005 Oviedo

URIEL ROMERO, Santiago
Alfonso X11, 3
28016 Madrid

SWEDEN

ALMEN, Karl-Erik
Box 5864
S-10248 Stockholm

BERGMAN, Magnus
Ringvagen 45
13300 Saltsjobaden

BJARNASON, Bjarni
Div. of Rock Mechanics
University of Lulea
951 87 Lulea

BJURSTROM, Sten
Box 5864
102 48 Stockholm

SWEDEN

BORG, Torgny
58249 Linkoping

FINKEL, Menachem
Research Foundation -
Befo, Box 5501
114 85 Stockholm

FRANZEN, Tomas
Research Foundation -
Befo, Box 5501
114 85 Stockholm

HAKANSSON, Ulf
Dept. of Soil & Rock
Mechanics
100 44 Stockholm 70

HASSLER, Lars
Dept. of Soil & Rock
Mechanics
100 44 Stockholm 70

HOLMSTEDT, Anders
LKAB
98186 Kiruna

LINDQVIST, Per Arne
Befo, Laboratoriegrand 11,
95164 Lulea

MAKI, Kenneth
FACK
104 84 Stockholm

MAKINEN, Ilpo
98186 Kiruna

MARTNA, Juri
Vallingby
16287 Stockholm

MATHIS, James
Div. of Rock Mechanics
Univ. of Lulea
951 87 Lulea

OUCHTERLONY, Finn
P.O. Box 32058
12611 Stockholm

STEPHANSSON, Ove
Lulea Univ. of Techn.
95187 Lulea

SWITZERLAND

BUCHER, Felix
Inst. of Foundation
Engineering and Soil
Mechanics
Hoggerberg 8093
Zurich

DESCOEUDRES, François
EPFL-ISRF
1015 Lausanne

EGGER, Peter
Labor. de Mecanique des
Roches, EPFL
1015 Lausanne

FRITZ, Peter
ETH-Hoggerberg
8093 Zurich

SWITZERLAND

GYSEL, Martin
Parkstrasse 27
825045 Baden

KAPP, Hans E.
Hohenweg 362
9000 St. Gall

KOVARI, Kalman
ETH-Hoggerberg 1
8093 Zurich

LAUGHTON, Christopher
Groupe GC-Div. Lep
Cern Meyrin
1211 Geneva 23

LOMBARDI, Giovanni
3 Via Ciseri
6601 Locarno

NATEROP, Daniel
Solexperts Ltd.,
Ifangstrasse 12
8603 Schwergenbach

SPANG, Konrad
Lab. de Mecanique des
roches, EPFL-DGC,
Ecublens
1015 Lausanne

STEINER, Walter
Balzari & Schudel AG
Kramburgstrasse 14, 3000
Bern 16

THUT, Arno
Solexperts Ltd.,
Ifangstrasse 12
8603 Schwergenbach

THAILAND

SAIHOM, Nibondh
59/135 Muangthongnives
3 Jaengwatana Road
Nonthaburi-11120

TURKEY

PASAMEHMETOGLU, A. Gunhan
Mining Eng. Dept.
Middle East Tech.
University, Ankara

U.K.

BROWN, Edwin
Dept. of Mineral Resources
Eng.
Imperial College
London SW7 2BP

DAW, Graham
Cementation House, Denham
Way, Maple Cross
Rickmansworth, Herts
WD3 2SW

DYKE, Christopher
Mineral Resources Eng.
Dept.
Imperial College
London SW7 2AZ

U.K.

FOWELL, Robert
Dept. Mining Eng.
Upon Tyne NE1 7RU

HACKETT, Peter
Pool, Redruth
Cornwall TR15 3SE

HUDSON, John
7 the Quadrangle
Welwyn Garden City
Hertj AL8 6SG

SANTARELLI, Frederic
Joseph
Min. Res. Eng. Dept.
Imperial College
London SW7 2BP

SHARP, John C.
Coin Varin, St Peter
Jersey, Channel Islands

WHITTAKER, Barry N.
Dept. of Mining Engineer
University of Nottingham
Nottingham NG9 3DV

U.S.A.

BANKS, Don
P.O. Box 361
Vicksburg, MS 39180-0631

BELLWALD, Philippe
Dept. of Civil Engineering
Rm 1 - 383,
Cambridge, MA 02139

BEZAT, Frederick
MTS Systems Corp
Box 24012, Minneapolis
MN 55424

BIENIAWSKI, Z.T.
110 Mineral Sciences Bldg.
University Park, PA 16802

BILLAUX, Daniel
Lawrence Berkeley Lab.
Bldg. 50F
Berkeley, CA 94720

BINNALL, Eugene P.
Lawrence Berkely Lab.
Bldg. 50B, RM 4235
Berkeley, CA 94720

BLAKE, Wilson
P.O. Box 928
Hay Lake, ID 83835

BLANKENSHIP, Douglas
P.O. Box 725
Rapid City, SD 57709

BOONE, Thomas
Rand Hall, Rm 120
Cornell University
Ithaca, NY 14850

BRADY, Barry
1313 5th St. S.E.
Minneapolis, MN 55414

U.S.A.

BRADY, Brian T.
Denver Federal Center Bldg
20 Mail Stop 3087
Denver, CO 80225

BURDICK, J. Scott
118 Edge Water Dr.
N. Pekin, IL 61554

BYRNE, R. John
4104-148th Ave. N.E.
Redmond, WA 98025

CHOI, Dai
4000 Brownsville Rd.
Library, PA 15129

CORDING, Edward
2221 Newmark Ce Lab.
208N Romine St.
Urbana, IL 61801

COULSON, James
Tennessee Valley Authority
400 W. Summit Hill Dr.
W3C 127, Knoxville, TN 37

COURSEN, D. Linn
1901 Dorcas Lane
Wilmington, DE 19806

CRAMER, Lynne
2101 Constitution Ave.
N.Washington, DC 20418

DANEK, Edward R.
508 Pleasant St. #3
Worcester, MA 01609

DELA CRUZ, Rodolfo
1509 University Ave.
Madison, WI 53706

DERSHOWITZ, William
4104-148th Ave. N.E.
Redmond, WA 98052

DESCOUR, Jozef
Earth Mechanics Institute
Colorado School of Mines
Golden, CO 80401

DIXON, Jay
Spokane Research Center
E. 315 Montgomery Avenue
Spokane, WA 99207-2291

DOE, Thomas W.
4804-148th Aven. N.E.
Redmond, WA

EINSTEIN, Herbert
Room 1 - 330 MIT
Cambridge, MA 02139

ELSWORTH, Derek
119 Mineral Sciences Bldg.
University Park, PA 16802

EWY, Russell
318 Hearst Mining Bldg.
Univ. of California
Berkeley, CA 94720

FAIRHURST, Charles
Dept. of Civil & Mineral
500 Pillsbury Drive S.E.
Minneapolis, MN 55455-022

U.S.A.

FARMER, Ian
P.O. Box 44302
Tucson, AZ 85733

FERRIGAN, P. Michael
9800 S. Cass Avenue
Argonne, IL 60439

FOWLER, Matthew
1625 Van Ness Ave.
San Francisco, CA 94109

GRAINGER, Gerald
Southern Electric Int.
P.O. Box 2625, Bin B242
Birmingham, AL 35202

GREGORY, Christine
Westinghouse Hanford Co.
P.O. Box 1970
Richard, WA 99352

HAMBLEY, Douglas
9700 S. Cass Ave.
Argonne, IL 60439

HARDY JR., H. Reginald
117 Mineral Sciences Bldg.
University Park, PA 16802

HEDSTROM, Jay
597 Pleasant St. #3
Worcester, MA 01609

HEUZE, François E.
L-200 LLNL
Livermore, CA 94550

HOPKINS, Deborah
90/3147 Lawrence Berkeley,
Berkeley,
CA 94720

HUTSON, Robert
1001 Cedar Lane
Northbrook, IL 60062

IANNACCHIONE, Anthony
P.O. Box 18070
Pittsburgh, PA 15236

IMAMURA, Satoshi
Lawrence Berkeley Lab.
50E Bldg.
Berkeley, CA 94720

JENKINS, Michael F.
Spokane Research Center
E. 315 Montgomery Avenue
Spokane, WA 99207-2291

KAFRAKIS, Mario
Univ. of Wyoming
P.O. Box 3314
Univ. Stat
Laramie, WY 82071

KARZULOVIC, Antonio
161 Wilson St. #69
Albany, CA 94710

KEMENY, John
Lawrence Berkeley Lab.
Bldg. 50E
Berkeley, CA 94720

U.S.A.

KIM, Kunsoo
P.O. Box 1970, PBB/1100
Richland, WA 93352

LA POINTE, Paul
11 Eastlane Place
Plano, TX 75074

LEE, Fitzhugh
Denver Federal Ctr. M.S.
Denver, CO 80225

LINDER, Ernest
439, 31th Street
Chicago, IL 60616

LONG, Jane
Lawrence Berkeley Lab.
Bldg. 50E
Berkeley, CA 94720

LU, Paul
Bldg. 20, Denver Federal
Denver, CO 80225

MAHTAB, Ashraf
918 Mudd
Columbia University
New York, NY 10027

McCAIN, Richard
P.O. Box 800
Richland, WA 99352

McFARLAND, Russell
8220 Russell Road
Alexandria, VA 22309

MUFF, Oliver
3004 Hillegass Ave.
Berkeley, CA 94705

MUNSON, Darrell E.
P.O. Box 5800
Albuquerque, NM 87112

MYER, Larry
Lawrence Berkeley Lab.
Bldg. 50E
Berkeley, CA 94720

NELSON, Jeffrey
17625 Goose Creek Rd
Olney, MD 20832

NELSON, Priscilla
1308 Harriet Ct.
Austin, TX 78756

NOLTE-PYRAK, Laura
Lawrence Berkeley Lab.
Bldg. 50E
Berkeley, CA 94720

NOLTING, Rick
1625 Van Ness Ave
San Francisco, CA 94109

OHYA, Satoru
7334 N. Gessner Rd.
Houston, TX 77040

OLSON, Roy
1800 G St. N.W.
Washington, DC 20550

U.S.A.

OSNES, John
P.O. Box 725
Rapid City, SD 57709

OUYANG, Zhihua
3720 Plaza Dr.
State College
PA 16801

PANEK, Louis
Mining Eng. Dept.
Houghton, MI 49931

PAULSSON, Bjorn
P.O. Box 446
La Habra, CA 90631-0446

PENTZ, David
15509-152nd Ave. Ne.
Woodenville, WA

PIGGOTT, Andrew
1311 Plaza Dr.
State College
PA 16801

PINCUS, Howard
P.O. Box 27630
San Diego, CA 92128

PRATT, Howard
10210 Campus Point Dr.
San Diego, CA 92121

PREECE, Dale S.
P.O. Box 5800
Albuquerque, NM 87185

ROSS-BROWN, Dermot
245-2929 Kenny Rd.
Columbus, OH 43221

SAGE, Joseph Douglas
Dept. Civil Eng.
100 Institute Rd
Worchester, MA 01609

SALAMON, Miklos
Mining Engineering Dept.
Golden, CO 80401

SCHATZ, John F.
10210 Campus Pt. Dr.
M/S:San Diego, CA 92121

SENSENY, Paul
P.O. Box 725
Rapid City, SD 57709

SHAFFER, Ronald J.
L-200, LLNL
Livermore, CA 94550

SHARP, Robert
P.O. Box 503
Los Alamos, NM 87544

STRANSKY, Terry
P.O. Box 27168
Cincinnati, OH 45227

STYLER, Neil
14 Milbren Ct.
Verona, PA 15147

SUNDARAM, Panchanatham
N.45, Fremont St.
San Francisco, CA 94119

U.S.A.

SWANSON, Peter L.
Denver Federal Ctr.
Bldg.Mail Stop 3087
Denver, CO 80225

THOMPSON, T. William
270-1626 Cole Blvd.
Golden, CO 80401

THORPE, Richard K.
P.O. Box 808, L-200
Livermore, CA 94550

TINUCCI, John
160 Spear St., 31250
San Francisco, CA 94105

TSUKASA, Matsumoto
345 Allerton Ave.
San Francisco, CA

UBBES, William
439, E.-31th Street
Chicago, IL 60616

UNRUG, K.
Dept. of Mining Eng.
Univ. of Kentucky
KY 40506-0046

VANDAMME, Luc
P.O. Box 2710
Tulsa, OK 74101

ROEGIERS, Jean-Claude
P.O. Box 2710
Tulsa, OK 74105

VAN EECKHOUT, Ed
Los Alamos, NM 87545

VAN SAMBEEK, Leo
P.O. Box 725
Rapid City, SD 57709

VERSLUIS, W. Scott
7000 S. Adams Street
Willowbrook, IL 60521

VOSS, Charles
Suite 800, 2030 M. St. NW
Washington, DC 20036

WAHI, Krishan
1709 Moon St. N.E.
Albuquerque, NM 87112

WAWERSIK, Wolfgang
Geomechanics Division
Sandia National Labs.
Albuquerque, NM 87185

WAWRZYNEK, Paul
120 Rand Hall
Cornell Univ.
Ithaca, NY 14853

WHYATT, Jeff
Spokane Research Center
E. 315 Montgomery Avenue
Spokane, WA 99207-2291

WILSON, William
Mechanical Engr. Dept.
College Park, MD 20742

U.S.A.

WONG, Ivan G.
250-1390 Market St.
San Francisco, CA 94102

WONG, I.H.
Two World Trade Ctr.
New York, NY 10048

XIANG, Jiannan
3720 Plaza Dr.
State College
PA 16801

U.S.A.

YOW, Jesse
P.O. Box 808
Livermore, CA 94550

ZHENG, Zigiong
Lawrence Berkeley Lab.
Bldg. 50E
Berkeley, CA 94720

U.S.S.R.

KAZIKAEV, Djak
46 Kostivkov Str.
Belgorod 308012

U.S.S.R.

MGALOBELOV, Juri
Volokolmskoe Str 2
Moscow 125812

MOROZOV, Alexander
Volokolmskoe Str 2
Moscow 125812

VENEZUELA

TINOCO, Fernando
Apartado 66252
Caracas 1061-A

YUGOSLAVIA

HUDEC, Mladen
Pierotijeva 6
4100 Zagreb

JASAREVIC, Ibrahim
Zagreb-Gradevinski
Institut
4100 Zagreb

VUJEC, Slavko
University of Zagreb
Mining, Geotechnics &
Petroleum
Pierotijeva 6
4100 Zagreb

Complete subject index / Index des matières complet / Komplettes Themen Verzeichnis

D

Complete author index / Index des auteurs complet / Komplettes Autoren Verzeichnis

Complete contents / Contenu complet / Kompletter Inhalt
Volume 1 / Tome 1 / Band 1

1 Fluid flow and waste isolation in rock masses
Ecoulement de fluides et enfouissement de déchets dans les massifs rocheux
Flüssigkeitsbewegung und Abfallisolierung im Fels

2 Rock foundations and slopes
Fondations et talus rocheux
Felsgründungen und Böschungen

3 Rock blasting and excavation
Sautage et excavation
Sprengen und Ausbruch

Volume 2 / Tome 2 / Band 2

4 Underground openings in overstressed rock
Souterrains en massifs rocheux sous grandes contraintes
Untertägige Hohlräume im überbeanspruchten Gebirge

Contents / Contenu / Inhalt
Volume 3 / Tome 3 / Band 3

1 English section

2 La partie française

3 Deutscher Teil

4 Additional contributions / Contributions additionelles / Ergänzende Beiträge

5 Annexes / Anhang